ENGINEERING DRAWING

(FOR DEGREE, DIPLOMA AND AMIE COURSES)

K. VENUGOPAL

PUBLISHING FOR ONE WORLD

NEW AGE INTERNATIONAL (P) LIMITED, PUBLISHERS

New Delhi • Bangalore • Chennai • Cochin • Guwahati • Hyderabad
Jalandhar • Kolkata • Lucknow • Mumbai • Ranchi

Visit us at www.newagepublishers.com

Published by New Age International (P) Ltd., Publishers
First Edition: 2010

Branches:

- 37/10, 8th Cross (Near Hanuman Temple), Azad Nagar, Chamrajpet, **Bangalore**-560 018. Tel.: (080) 26756823, Telefax: 26756820, E-mail: bangalore@newagepublishers.com
- 26, Damodaran Street, T. Nagar, **Chennai**-600 017. Tel.: (044) 24353401, Telefax: 24351463 E-mail: chennai@newagepublishers.com
- CE-39/1016, Carrier Station Road, Ernakulam South, **Cochin**-682 016. Tel.: (0484) 2377004, Telefax: 4051303, E-mail: cochin@newagepublishers.com
- Hemsen Complex, Mohd. Shah Road, Paltan Bazar, Near Starline Hotel, **Guwahati**-781 008. Tel.: (0361) 2513881, Telefax: 2543669, E-mail: guwahati@newagepublishers.com
- No. 105, 1st Floor, Madhiray Kaveri Tower, 3-2-19, Azam Jahi Road, Nimboliadda, **Hyderabad**-500 027. Tel.: (040) 24652456, Telefax: 24652457, E-mail: hyderabad@newagepublishers.com
- RDB Chambers (Formerly Lotus Cinema) 106A, Ist Floor, S.N. Banerjee Road, **Kolkata**-700 014. Tel.: (033) 22273773, Telefax: 22275247, E-mail: kolkata@newagepublishers.com
- 16-A, Jopling Road, **Lucknow**-226 001. Tel.: (0522) 2209578, 4045297, Telefax: 2204098, E-mail: lucknow@newagepublishers.com
- 142C, Victor House, Ground Floor, N.M. Joshi Marg, Lower Parel, **Mumbai**-400 013. Tel.: (022) 24927869 Telefax: 24915415, E-mail: mumbai@newagepublishers.com
- 22, Golden House, Daryaganj, **New Delhi**-110 002. Tel.: (011) 23262370, 23262368, Telefax: 43551305, E-mail: sales@newagepublishers.com

ISBN: 978-81-224-3140-7

Rs. 275.00

C-10-06-4966

Printed in India at Print 'o' Pack, Delhi.
Typeset by Author.

PUBLISHING FOR ONE WORLD

NEW AGE INTERNATIONAL (P) LIMITED, PUBLISHERS
4835/24, Ansari Road, Daryaganj, New Delhi - 110002
Visit us at **www.newagepublishers.com**

PREFACE

This textbook **ENGINEERING DRAWING** is intended to be used as a classroom textbook by the beginners of the Engineering Curriculum, both of Degree and Diploma level, including AMIE course. The drawings have been updated as per the latest specifications prescribed by the Bureau of Indian Standards.

Numerous solved problems from various leading Universities with the unique features of detailed constructional STEP-BY-STEP procedures and SIDE-BY-SIDE simplified drawings are presented in this book such that the book is a self-learning tool.

Another notable and interesting feature of this book is that numerous practical engineering components, *viz.,* bolt and nut, L-angle, funnel, table, stool, overhead tank, gland, stage, cavity block, drilling jig, cast iron and steel castings, hopper, chimney, duct, chute, V-block, bearing cap, open bearing, dovetail, fork, rocker arm, lock plate, Tri-square, jig plate, flange, elbow, shaft bracket, machine clamp, tailstock, angle plate, electromagnet, I-beam, rod support, tray, etc., are dealt, instead of only the conventional prisms, pyramids, cylinders and cones in various chapters - thus the beginners are exposed to numerous engineering components in the First Year itself.

The ultimate judges of any book are the students who use it and I am quite confident that this book will dispel the common illusion that the subject Engineering Drawing is a tough one.

I am immensely thankful to my Publishers Shri. R.K. Gupta, B.Sc. (Engg.), Chairman and Shri Saumya Gupta, B.E., M.S. (US), Managing Director, M/s. New Age International (P) Ltd., New Delhi, for their most co-operative style, meticulous attitude and persistent effort for bringing out the book in the present form in a short period with a nice getup.

I solicit comments and suggestions for improving the book.

K. VENUGOPAL

2010

ENGINEERING DRAWING
Made Easy

HIGHLIGHTS

- This Text Book is in **simple English,** hence **easy-to-understand** and **easy-to-follow**.
- **It is self-learning,** because it is **self-explanatory**.
- It is **user-friendly** with **step-by-step** procedures.
- "A neat diagram with clarity is worth pages of explanation." Sincere efforts have been made to simplify the drawings that they will be more clear and striking and give clear conception of the subject.
- It is **students' delight,** because **step-by-step** procedure is presented with simplified, easy-to-visualize **side-by-side** drawing–both contained in the same/adjoining page itself.
- It is **readers' delight**, because both procedure and drawing of any problem are contained in the same/adjoining page itself. Thereby, the reader need not strain at all in 'turning over to the next page' for the continuation of the procedure/drawing of any problem.
- This textbook is **unique** in the sense that each Solved Problem is graded and also followed by a **similar** Exercise Problem, for the students to practise confidently and grasp the fundamental principles much easier.
- The Book is **exhaustive**, as **Additional Problems** are also added.
- It is an **excellent Guide** for an average student, because it highlights the important **Points to Remember**, **Note**, **Rule**, **Hint**, etc.
- It is **examination-oriented!** Yes, it is illustrated with 800 solved University Problems with illustrations.

☞ Is imagination of the shape of objects difficult for you?

Is visualizing the shape of objects from pictorial views difficult for you?

Does visualizing the shape of objects from orthographic views seem to be almost impossible for you?

Are you poor in communicating your ideas in the form of freehand sketching?

👍 This book will certainly dispel the common illusion among the students that the subject Engineering Graphics is a tough one.

Yes! ENGINEERING DRAWING is indeed Made Easy!!

* * *

Perhaps, too much is spoken about the book and it may be superfluous too!

Because, ***"The taste of pudding lies in eating!"***

CONTENTS

UNIT – I

SYLLABUS

UNIT – I

Introduction to Engineering Graphics – Construction of Ellipse, Parabola and Hyperbola – Cylindrical Curves

1

IMPORTANCE OF ENGINEERING GRAPHICS

"Not knowing English language is not a sin! But not willing to know English is definitely a sin!!"

1. ENGINEERING GRAPHICS

1.1 ENGINEERING

"Scientists study the world as it is !
Engineers create the world that never has been !!"

Engineer is derived from the Greek word *Ingenious* !
Ingenious means *intelligent / imaginative / innovative / creative* !!

Engineering means *Applied Science, i.e., Technology*; Technology is the technical aspects and concepts of man-made machines, structures, sky-rise buildings, electrical and high-tech electronic machines, robotics, aeronautics, instruments, etc.

1.2 GRAPHICS

Language is the medium of communication. *Graphics* mean *drawings*. *Graphic Language* is the mode of communication by means of sketches. Communication of ideas through sketches is probably the oldest form of communication among humans.

Prehistoric man used a stick as equipment to scratch out a message in the sand. Early Egyptians used pictures as a form of communication with the equipment of paint and brush; Whether the tools be a simple pencil and sketch pad or a computer controlled plotting device – the basic need to draw pictures is still present.

1.3 ENGINEERING GRAPHICS

Engineering Drawing: The representation of an object containing details like shape, size, etc., is called *Engineering Drawing.*

Engineering Graphics: Engineering Drawing is a *graphical* medium of technical expression. Hence it is also known as *Engineering Graphics*. It is a study that requires special equipment or tools (either manual drafting tools or CAD software) to form the images.

2. IMPORTANCE OF ENGINEERING GRAPHICS

This is an era of globalization and emerging of Multi-National Companies (MNCs). Technological development of a country is the deciding factor for its export potential. Engineering Graphics prepared in a country is adopted in any other country. Hence, ENGINEERING GRAPHICS IS CALLED THE UNIVERSAL GRAPHIC LANGUAGE OF ENGINEERS, spoken, read and written in its own way.

2.1 CLASSIFICATION

(i) **Plane Geometry :** It is the art of representation of geometrical objects having Two Dimensions (2D), *i.e.,* length and breadth on a Two Dimensional (2D) drawing sheet.

(ii) **Solid Geometry :** It is the representation of geometrical objects of Three Dimensions (3D), *i.e.,* length, breadth and height on a Two Dimensional (2D) drawing sheet.

We make use of objects in our day-to-day life such as bolts and nuts, bearings, utensils, furniture, pumps, cranes, etc., (Mechanical Engineering / Production Engineering), scooter engines (Automobile Engineering), buildings, bridges, etc., (Civil Engineering), motor, generator, etc., (Electrical Engineering), TV, computer (Electronics Engineering), dish antenna (Communication Engineering), watches, speedometer, etc., (Instrumentation Engineering), sky-rise buildings, stadium, fly-over, etc., (Architecture), etc.

All objects of production and construction have to be conceived, designed and drawn by the engineers. *Yes!* Engineers are certainly the *Builders of the Nation.*

2.2 VISUALIZATION CONCEPT

Engineering Graphics is the graphic language, from which an engineer can *visualize* the object and then effectively communicate his/ her ideas and designs to others in the form of pictorial drawings. Reading, understanding and interpreting an engineering drawing requires a lot of skill of imagination and the ability of visualization of the actual object.

Engineering graphics offers students an insight into the methods of attacking engineering problems. Its lessons teach the principles of accuracy, exactness and positiveness with regard to the information necessary for the production of an engineering component. Finally, it develops the engineering imagination and visualization that are so essential to the creation of successful design.

2.3 COMMUNICATION SKILL

See Fig. 1. The engineer, to whatever branch he / she may belong to, must have perfect knowledge and *communication skill* in drawing. As the bad language is unpleasant to read and communicate, so also a wrong drawing will worry a trained eye. It is doubtful if any engineer can fulfill his / her functions completely and efficiently without the ability either to read or to prepare, *i.e.*, to communicate others in the form of drawings.

FIG. 1 MANUAL DRAFTING

Yes ! Many good technical ideas of yours may get rejected because of poor communication skill in the graphic language.

2.4 DOCUMENTATION

Documentation is the process of executing the drafting (manually or computer-aided) of engineering components accurately and legibly using imagination, visualization and communication skills and presentation of neat drawings.

2.5 ENGINEERING GRAPHICS – APPLICATIONS

Engineering Graphics is a ***core subject*** for all the branches of engineering. It is used in different engineering disciplines to design and draw:

(i) Machine components, Transmission systems, CNC machine tools, Mechanics of solids, Robotics, etc., in *Mechanical and Production Engineering.*

(ii) Structures, Plan of buildings, Multi-storeyed complex, Bridges, Stadium, etc., Town planning, Mapping, Contour plotting in *Civil and Architectural Engineering.*

(iii) Circuit layouts, Electrical-wiring diagrams, Panel design, Control schematics, etc., in *Electrical Engineering.*

(iv) Measuring instruments, Sensors, etc., in *Instrumentation Engineering.*

(v) Schematic diagrams of Printed circuits, Microprocessors, Integrated circuits, etc., in *Electronics including Computer Science Engineering.*

(vi) Communication network, Satellite transmitting pictures, TV telecasting, etc., in *Communication Engineering.*

(vii) Internal Combustion Engines, Kinematics, Steering, Hydraulic circuits, etc., in *Automobile Engineering.*

(viii) Spacecraft, Jet propulsion, Flight simulator, Lofting, etc., in *Aeronautical Engineering.*

(ix) Fashion design of garments in Textile Technology

(x) Metallurgical Engineering, Mining Engineering, Marine Engineering, etc.

3. MANUAL DRAFTING *Vs* COMPUTER AIDED DRAFTING

FIG. 2 COMPUTER AIDED DRAFTING

Can Computer Aided Drafting (CADr) replace the Manual Drafting Tools for the learning process?

See Fig. 2. Compared to Manual Drafting, Computer Aided Drafting is faster as well as accurate. No doubt, much time can be saved by using the computer to make even complicated engineering drawings. However, remember, the *learning process* is accomplished through the traditional manual drafting tools only. The computer cannot replace the drafting board and equipment as a learning tool.

Yes! Computer Aided Drafting is not a substitute for Manual Drafting. It is only a tool that can be used to supplement traditional drafting tools. The underlying basic concepts of construction of 2D and 3D drawings remain the same regardless of the tools chosen to create the graphics.

It is ideal to first master the fundamentals of engineering drawings. Having grasped the fundamentals and acquired sufficient practice on the drawing board, then try your hands on your user- friendly computer. The computer would then come handy in doing the task.

Parallel Example

A somewhat parallel example is the usefulness of calculators. Calculators are used to solve math problems, but knowledge of math is still required by the user to solve problems. What is eliminated is the tedious task of performing longhand mathematical manipulations!

*

2

DRAWING INSTRUMENTS AND THEIR USES

"Plan your work and Work your plan!"

1. LIST OF DRAWING INSTRUMENTS

Drawing Instruments (*Drafting Tools*) are used to prepare neat and scaled drawings. Accuracy of engineering drawings depends largely on the quality of the drawing instruments.

LIST OF DRAWING INSTRUMENTS FOR MANUAL DRAFTING :

1. **Drawing Board**
2. **Drawing Sheets**
3. **Standard Set-Scales**
4. **Mini-Drafter**
5. **Instrument Box**
6. **Set-Squares (45° & 30° - 60°)**
7. **Protractor-cum-Procircle**
8. **Drawing Pencils (HB, H and 2H)**
9. **Sharpener and Emery Paper**
10. **Adhesive Tape (*Cello-Tape*) / Clips**
11. **Pencil Eraser**
12. **French Curves**

Important Note : ***Check your drawing instruments while buying.***

The accuracy of the drawing mainly depends on the quality of the instruments. Unsuitable or inaccurate instruments often lead to inaccurate results, consequently dampen the enthusiasm of young students during drawing. On the other hand, a set of instruments of superior quality shall prove an asset and pleasure to work with. When procuring the instruments, they should be carefully scrutinized for accuracy, structural features and finish.

1.1 DRAWING BOARD (IS 1444 : 1989)

Drawing board is made up of strips of well-seasoned soft wood about 25 mm thick. It is provided at the back by two battens to prevent warping.

For use in Engineering Colleges, drawing board is of standard size 650 mm x 470 mm as recommended by IS 1444 : 1989.

1.2 DRAWING SHEETS (IS 10711 : 2001)

The drawing sheet should be tough and strong. When eraser is used on it, its fibres should not disintegrate.

Sizes of Drawing Sheets: Standard sizes of *trimmed* drawing sheets recommended by **IS 10711 : 2001** (Technical Documentation – Sizes of drawing sheets) are given in the Table below : (Area of A0 size is 1 sq. m. of length to width ratio $\sqrt{2} : 1$. Successive smaller sizes are obtained by halving previous size sheets, with the above constant length to width ratio.)

Designation	Dimension, mm Trimmed size
A0	841 x 1189
A1	594 x 841
A2 *	420 x 594
A3	297 x 420
A4	210 x 297

* ***A2 size drawing sheet is preferred for class use.***

1.3 STANDARD SET-SCALES

BIS recommends the following set-scales (plastic / cardboard), eight in number with designations M1, M2 and so on, with the scales graduated on each edge. *See Table below*:

	M1	M2	M3	M4	M5	M6	M7	M8
Scale on one edge	1:1	1:2.5	1:10	1:50	1:200	1:300	1:400	1:1000
Scale on other edge	1:2	1:5	1:20	1:100	1:500	1:600	1:800	1:2000

Important Note : ***Do not use the scales as a straight edge for drawing straight lines.***

1.4 MINI- DRAFTER : *(Fig. 1)*

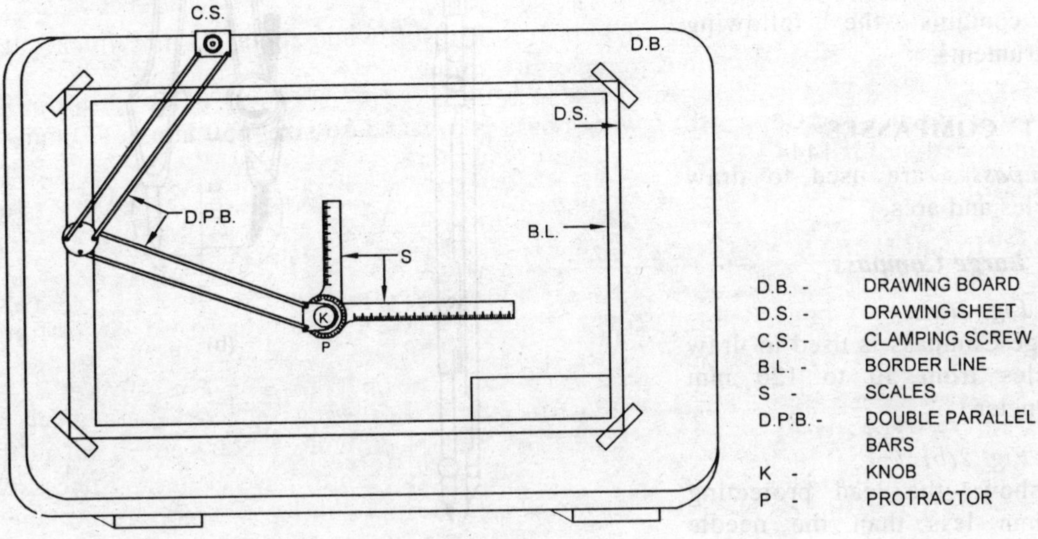

FIG.1 MINI - DRAFTER

A Drafting Machine is used by professional draftsman to prepare drawings. It combines the functions of T-Square, Set-Squares, Scales, Clinograph and Protractor. A miniature version of the drafting machine known as Mini-Drafter is used for drafting by students.

Mini-Drafter : *See Fig. 1.* One end of the mini-drafter is clamped by means of a Clamping Screw (C.S.) to the longer edge of the Drawing Board (D.B.). At its other end, an adjustable Knob (K) having Protractor (P) markings is fitted. Two Scales (S) of transparent celluloid, set at right angles to each other are attached to the knob.

How to use your Mini-Drafter?

1. *Clamping the Mini-Drafter:* Set the reference mark on the protractor head to read O^o and tighten the knob. Then, clamp the drafter to the top edge of the drawing board using the clamping screw, such that the horizontal scale is in-line with the horizontal edge of the Drawing Sheet (D.S.).
2. Now, the scales along with the protractor head as a unit can be moved easily to any location on the drawing sheet without altering the position of the two scales. The Double Paraliel-Bar (D.P.B.) mechanism, pivoted as shown, keeps the scales in every position to remain parallel to their previous position.
3. To draw horizontal and vertical lines, the reference mark on the protractor head has to be set to O^o. To draw inclined lines, the scales should be fixed with the reference mark, indexing the required angle. Do all the positioning by one hand and draw the lines by the other.

In short, the mini-drafter is used for different drafting operations, viz., to draw horizontal, vertical and inclined lines, parallel and perpendicular lines and also for measuring lines and angles.

1.5 INSTRUMENT BOX

It contains the following instruments:

1.5.1 COMPASSES

Compasses are used to draw circles and arcs.

(i) Large Compass

See Fig. 2(a).
Large Compass is used to draw circles from 30 to 120 mm diameter.

See Fig. 2(b).
It shows the lead projecting 1 mm less than the needle point. Also, keep the inclined cut of the lead turned outside.

See Fig. 2(c).
For drawing circles of 120 to 300 mm diameter, both needle point leg and pencil lead should be bent at the knuckle joints so as to be perpendicular to the paper.

(ii) Lengthening Bar

See Fig. 2(d).
For drawing larger circles of more than 300 mm diameter, *Lengthening Bar* attachment is used.
For this, the lower part of the pencil leg is detached and lengthening bar is inserted in its place. The detached part is then fitted at the end of the lengthening bar.

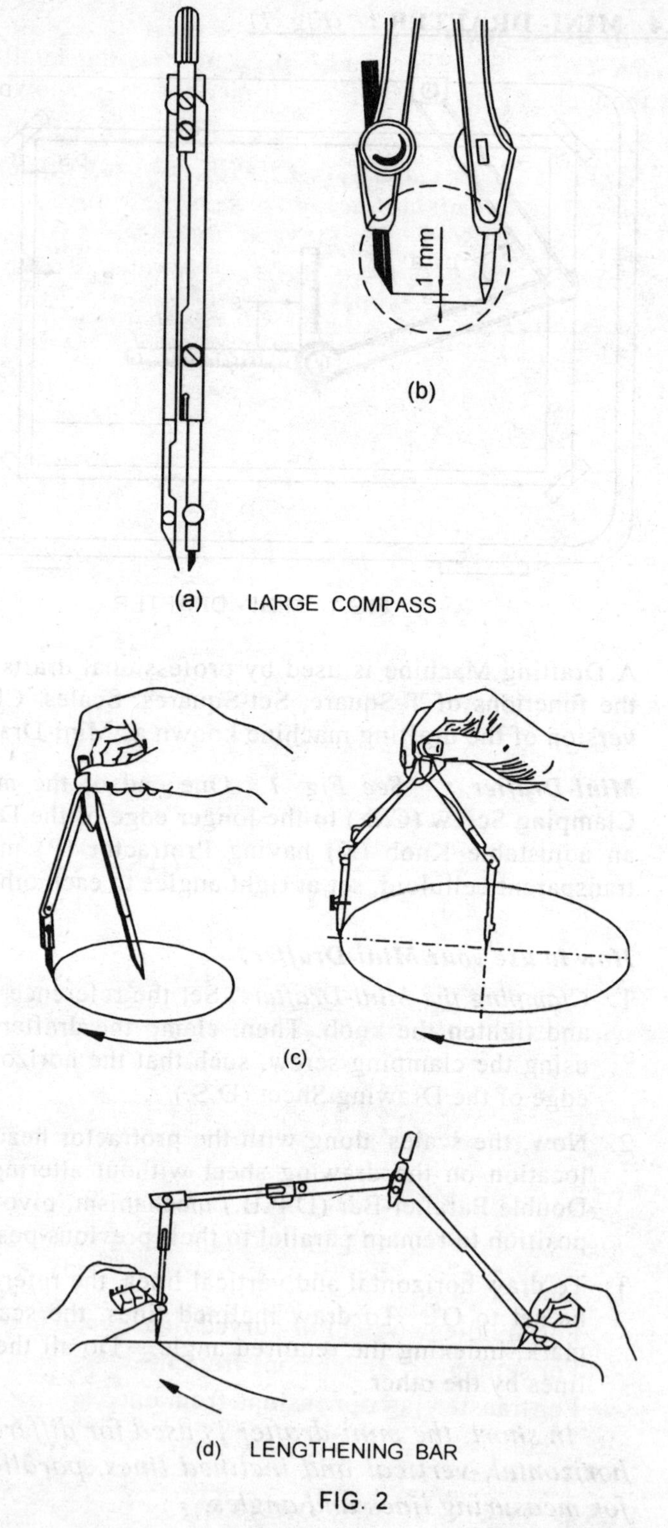

FIG. 2

(iii) Bow Compass

See Fig. 3. Bow Compass is used for drawing small circles upto 30 mm diameter. In this, the inclined cut of the lead is turned inside as shown.

NOTE : *The curves drawn with compass should be of the* ***same thickness*** *as that of straight lines in a drawing. It is difficult to exert the same amount of pressure on the lead in the compass as on a pencil. It is therefore advisable to use a softer lead about one grade lower in the compass than the pencil used for drawing straight lines.*

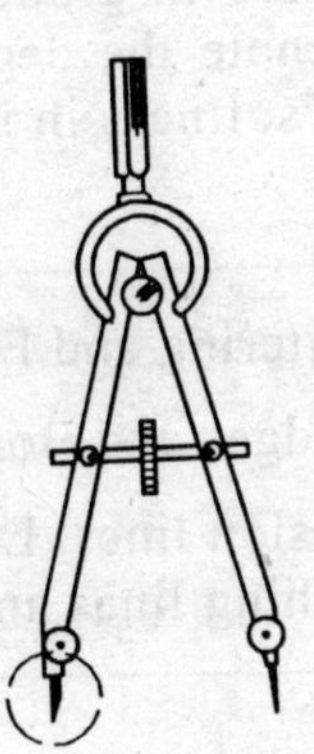

BOW COMPASS

FIG. 3

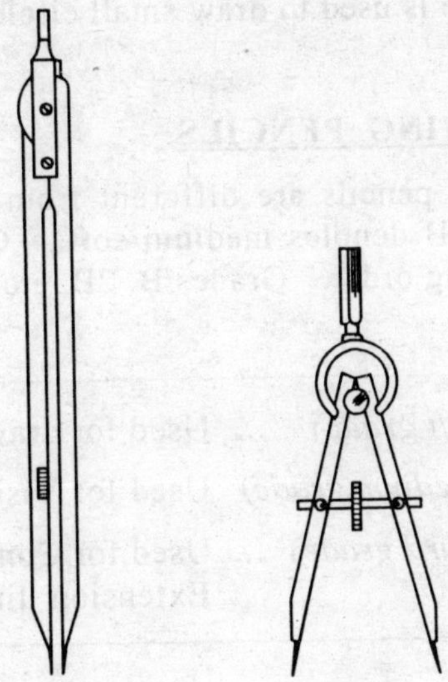

(a) LARGE DIVIDER

(b) BOW DIVIDER

FIG. 4

1.5.2 DIVIDERS

Fig. 4(a) shows ***Large Divider.*** *Fig. 4(b) shows* ***Bow Divider.***

Dividers are used :

- For dividing straight or curved lines into a number of equal parts,
- For setting-off dimensions from the set-scale to the drawing and
- For transferring dimensions from one part of the drawing to another part.

1.5.3 LEAD CASE

1.5.4 SMALL SCREW DRIVER

1.6 SET-SQUARES (45° and 60°-30° triangles)

Set-Squares are made of transparent celluloid or plastic. Those made of transparent celluloid are preferred as the lines underneath can be seen easily. Edge of the set-squares may be bevelled or levelled.

Set-squares may have *French Curves* cut in them.

1.7 PROTRACTOR - CUM - PROCIRCLE

Protractor is divided into 0.5° division. It is used to mark or measure accurately such angles that cannot be precisely drawn or measured with the mini-drafter.

Procircle is used to draw small circles of different diameters without using the compass.

1.8 DRAWING PENCILS

Drawing pencils are different from ordinary pencils and are in grades HB, H, 2H, etc. Grade HB denotes medium soft. Grades H, 2H, etc., denote the degree of hardness in increasing order. Grades B, 2B, etc., denote the degree of softness in increasing order.

HB - (*Soft grade*) ...	Used for drawing Border Lines, Lettering and Freehand sketching
H - (*Medium grade*)	Used for Visible outlines, Visible edges and Boundary lines
2H - (*Hard grade*) ...	Used for Construction lines, Dimension lines, Leader lines, Extension lines, Centre lines, Hatching lines and Hidden lines

NOTE : *Never use ordinary cheap quality pencils on your drawing sheet.*

Keep your pencil sharp: A dull pencil produces fuzzy lines. Only a sharp pencil produces clean-cut black and sharp lines that sparkle with clarity.

(a) CONICAL POINT

(b) CHISEL EDGE

FIG. 5

See Figs. 5(a) and (b). Lead is sharpened to *conical point or chisel edge* by rubbing on an emery paper. *Conical point* is used for lettering and freehand sketching. *Chisel edged pencil* is used for drawing lines.

***Micro - tip pencils / Clutch pencils** with 0.5mm thick leads of grades HB, H and 2H are preferred than wooden pencils, as they need no sharpening.*

1.9 SHARPENER AND EMERY PAPER

Drawing pencils are sharpened by the pencil sharpener. Emery paper is used to obtain conical or chisel edge of the lead.

1.10 ADHESIVE TAPE (*CELLO-TAPE*) OR CLIPS

A thin transparent adhesive tape known as ***cello tape*** is used to fix the drawing sheet on the drawing board. Best quality clips can also be used.

1.11 PENCIL ERASER

Pencil Eraser (*It is not 'rubber'! Call it 'eraser'!*) is used to rub-off unwanted lines, etc. Quality eraser called *Eraser Soft* should be used.

1.12 FRENCH CURVES (*Irregular Curves*)

See Fig. 6.

French Curves are also known as Irregular Curves. These are available in different sizes and shapes. French Curves are also inscribed in the setsquares.

French curves are used to draw non-circular curves, which cannot be drawn with a compass.

FIG. 6

2. HOW TO BEGIN YOUR DRAWING ?

1. Clean the drawing board and all your drawing instruments using handkerchief.
2. Fix your drawing sheet over a **padding sheet** on the drawing board.
3. Fix the mini-drafter on the drawing board on its left top corner and set it to *zero angle* position so that the longer scale is parallel to the bottom edge of the drawing board.
4. Check that the scales and the protractor head of the mini-drafter move to any position on the drawing sheet.
5. Draw the **Border Lines** and the **Title Block** on the drawing sheet using HB pencil.
6. Print the **Problem Number** on the **left top side** of the figure. Enclose the number in a small circle of about 10 mm diameter. Then commence your drawing work.
7. Captions like the Name of the Object, Views, etc., should be written below the drawing.
8. Spacing of drawings between two problems / views is to be planned before the commencement of the drawing.

3. POINTS TO REMEMBER

3.1 DOs

- ✓ *Always* clean hands are the first necessity; (*"Neatness is a personal habit"*).
- ✓ *Always* begin the work after wiping-off the drawing board.
- ✓ *Always* use *appropriate* drawing pencils / leads. (The grade of the pencil / lead is decided by the amount of graphite mixed with clay.)
- ✓ *Always* maintain the sharpness of the pencil lead by periodically sharpening it. (Micro-tip pencils are recommended for fast drafting.)
- ✓ *Always* keep a padding sheet under the drawing sheet in use.
- ✓ *Always* keep your handkerchief at hand to clean any dirt on the drawing sheet.
- ✓ *Always* use the divider for transferring measurements from the set-scale to the drawing for getting dimensional accuracy.
- ✓ *Always* remember that you are NOT writing, but you are printing the text on your drawing sheet.
- ✓ *Always* remember: *"Small things make perfection, but perfection is no small thing."*
- ✓ *Always* remember : *"Time is money in industries."* You should not only draw your drawings accurately, but also swiftly to save time.

3.2 DON'Ts

- × *Never* use cheap quality pencil / lead and eraser on the drawing sheet.
- × *Never* use set-scale as a ruler for drawing lines or for taking measurements directly.
- × *Never* sharpen the pencil over the drawing board or drawing sheet.
- × *Never* forget to clean every time, the sliding instruments (mini-drafter, set-squares and protractor) placed on the drawing sheet. Sliding of the instruments on the drawing sheet should be reduced to a minimum.
- × *Never* keep the instruments, *viz.*, compass, divider, etc., outside the instrument box, while not in use. Some students have the habit of carrying only a compass and a divider in an *ordinary tin geometry box* to the classroom, thereby spoiling the precision of the instruments.

  Therefore, *always keep the instruments in the respective velvet-seats inside your instrument box.*

*

3

BIS CONVENTIONS AND SPECIFICATIONS

"Standards connect the world !"

1. BIS CONVENTIONS AND SPECIFICATIONS

The representation of any matter by some sign or mark on the drawing is known as *Convention* or *Code*. These conventions are specified by Bureau of Indian Standards (BIS) and are called *BIS Specifications*.

Engineering Graphics is the *Universal Graphic Language of Engineers*, spoken, read and written in its own way. Every language has its own rules of grammar. Engineering graphics also has its grammar in the types of lines and lettering, its idioms in the conventional practices, its punctuations in the types of dimensioning, its abbreviations in the symbols and its descriptions in the theory of projections and constructions.

1.2 I.S.O.

In the 21st century, growing international engineering and technological developments, technical collaboration between mutual countries, exchange and export of technologies and globalization leading to the establishment of various Multi-National Companies (MNCs) have necessitated the inter-nationalization of engineering drawing standards. INTERNATIONAL STANDARDS ORGANIZATION (I.S.O.), Geneva, Switzerland, have formulated (1970) international standards for engineering drawing, for which our country has also given the approval and accordingly modified the Indian Standards IS 696 : 1972.

1.3 B.I.S.

Indian Standards Institution (I.S.I.) is renamed as Bureau of Indian Standards (B.I.S.) in 1983. Engineering Drawings Committee (E.D.C.) of B.I.S. has adopted the I.S.O. standards in *toto* for the various topics given in the following Table in engineering drawing:-

Sl. No.	BIS Codes	Topics
1.	IS 10711 – 2001	Technical products Documentation – Size and Layout of Drawing Sheets
2.	IS 9609 (Parts 0 & 1) – 2001	Technical product Documentation – Lettering
3.	IS 11669 – 1986	Dimensioning on Technical Drawings
4.	IS 15021 (Parts 1 – 4) – 2001	Technical Drawings – Projection Methods

2. SIZES, LAYOUT AND FOLDING OF DRAWING SHEETS

2.1 SIZES OF DRAWING SHEETS

The standard sizes of drawing sheets recommended by IS 10711 : 2001 are designated as A0, A1, A2, A3 and A4. Surface area of A0 size is 1 square meter.

See Fig. 1. Successive sizes are obtained by halving along the length. The areas of two successive sizes are in the ratio 1 : 2. The width to length ratio for all sizes of the drawing sheets are maintained as 1 : √2.

Designation	Dimension, mm Trimmed size
A0	841 x 1189
A1	594 x 841
A2 *	420 x 594
A3	297 x 420
A4	210 x 297

* A2 size drawing sheet of size 420 mm x 594 mm is preferred for class use.

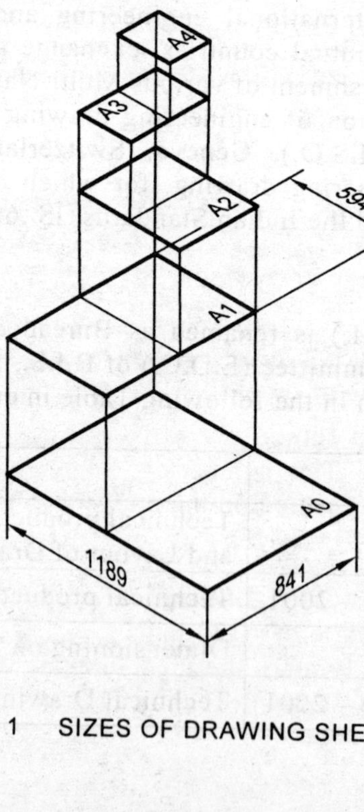

FIG. 1 SIZES OF DRAWING SHEETS

2.2 LAYOUT OF DRAWING SHEETS

2.2.1 LAYOUT OF TRIMMED DRAWING SHEET : (IS 10711 : 2001)

IS 10711 : 2001 recommends the Layout of a Drawing Sheet, as the standard form of arrangement of particulars on it, *viz.*, Borders and Frame, Grid References and Title Block.

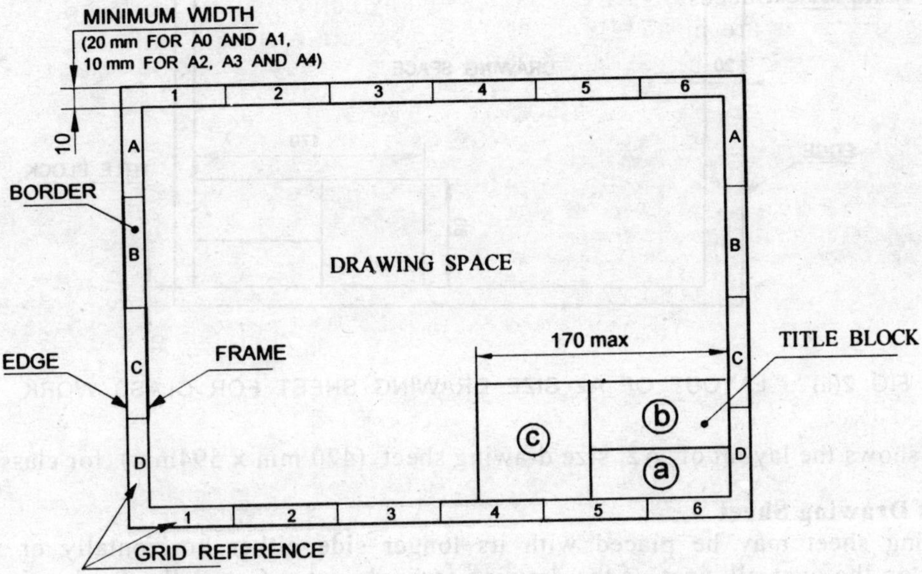

FIG. 2(i) LAYOUT OF TRIMMED DRAWING SHEET

Borders and Frame : *See Fig. 2(i)*
Borders enclosed by the edges of the trimmed sheet and the Frame limit the drawing space. Borders have a minimum width of 20 mm for A0 and A1, and 10 mm for A2, A3 and A4.

Grid References
Grid Reference System permits easy location on the drawing of details, modifications, etc. Number of divisions should be divisible by two and be chosen in relation to the complexity of the drawing. Length of any side of the grid rectangles shall vary from 25 mm to 75 mm.

The rectangles of the grid should be referred by capital letters along one edge and numerals along the other edge. The numbering direction may start at the sheet corner opposite to the title block and be repeated on the opposite sides.

Title Block
Title Block is provided at the bottom right-hand corner of the drawing sheet. It should be visible when folded. It is a rectangle of 170 mm x 95 mm *or* 170 mm x 80 mm size. It is divided into boxes a, b and c to mark the identification details of production drawings in industries, *i.e.*, (a) Registration or Identification number of the drawing, (b) Title of the drawing and (c) Name of the legal owner of the drawing.

2.2.2 LAYOUT OF A2 SIZE DRAWING SHEET FOR CLASS WORK : *[Fig. 2(ii)]*

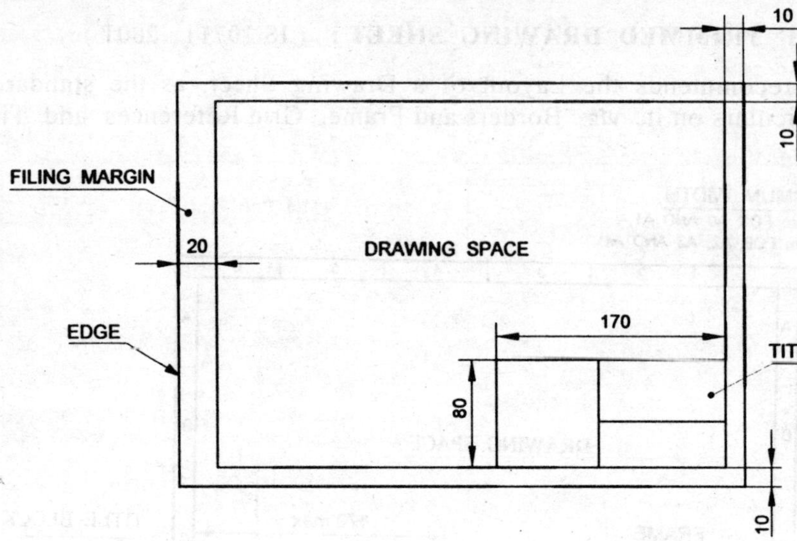

FIG. 2(ii) LAYOUT OF A2 SIZE DRAWING SHEET FOR CLASS WORK

Fig. 2(ii) shows the layout of A2 size drawing sheet (420 mm x 594mm) for class work.

Placing of Drawing Sheet
The drawing sheet may be placed with its longer side either horizontally or vertically depending on the over all sizes of the drawing to be drawn. Generally the drawing sheet is placed with its longer side horizontally for class work.

Borders
Border is the space left all around in between the trimmed edges of the drawing sheet and the frame. The border has a minimum width of 10 mm for A2 size drawing sheet.

Filing Margin
Filing Margin is provided for taking perforations for the purpose of filing the drawing sheets. It shall have a minimum width of 20 mm *with the border included* therein as shown.

Spacing of Drawings
Before commencing your drawing work, some idea about the spacing of drawings is to be formed. When only one figure is to be drawn on a sheet it should be drawn in the center of the drawing space. For more than one figure, the drawing space should be planned and divided into suitable blocks.

Title Block
Title Block of 170 mm x 80 mm size is preferred for the A2 size drawing sheet for class work. It is divided into boxes to insert all the necessary details about you and your drawing.

Scale of a Drawing
Scale of a Drawing is the ratio of the linear dimension of an object represented in the drawing to its real linear dimension. It is indicated in the title block as 1 : 1 for Full Size, or 1 : X for Reduced Scales or X : 1 for Enlarged Scales.

2.2.3 TITLE BLOCK FOR CLASS WORK : *(Fig. 3)*

Title Block is a rectangular block. It should contain the following particulars as shown in *Fig. 3*:

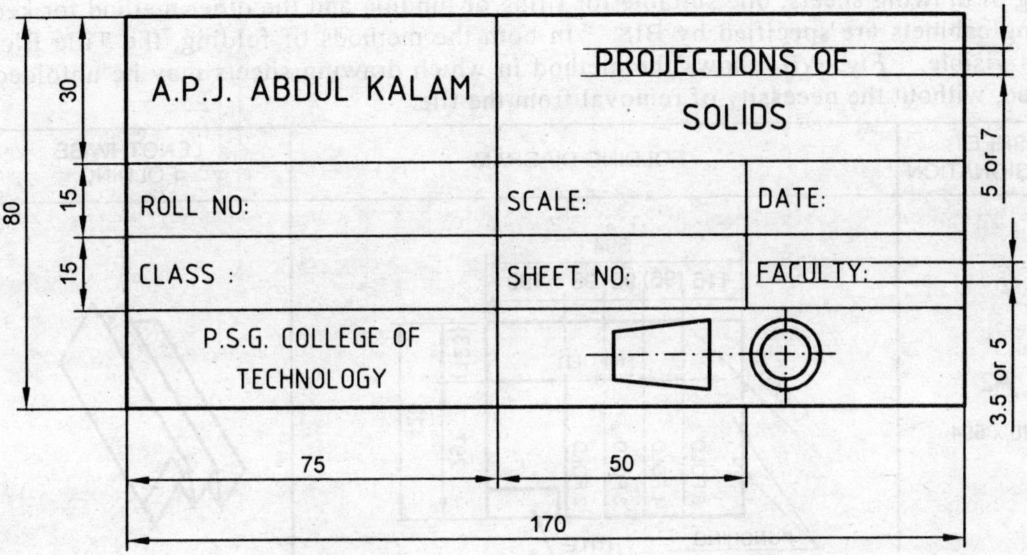

FIG. 3 SUGGESTED TITLE BLOCK FOR CLASS WORK

See Fig. 3. Left Hand Side is the identification zone with following details :

Your Name, Roll No., Class and Branch

Name of your Institution

Right Hand Side is the information zone with following details :

Title of the Drawing,

Scale, Sheet No.*, Date of submission and Signature of Faculty;

Symbol indicating First Angle Projection.

**Sheet No:* If more than one drawing sheet is used for a particular TITLE OF THE DRAWING,

Sheet No. is n / p, where n is the sequential sheet number and p is the total number of sheets, *i.e.,* 1/2, 1/3, etc.

If only one sheet is used, Sheet No. is 1/1.

2.3 FOLDING OF DRAWING SHEETS

Fold the drawing sheet after completing the drawings and file it neatly for submission of drawing assignments to your Faculty and for future revisions.

IS : 11664 – 1999 specifies the method of folding drawing sheets. Two methods of folding of drawing sheets, one suitable for filing or binding and the other method for keeping in filing cabinets are specified by BIS. In both the methods of folding, the Title Block is always visible. *Fig. 4(i)* shows the method in which drawing sheets may be unfolded and refolded, without the necessity of removal from the file.

SHEET DESIGNATION	FOLDING DIAGRAM	LENGTHWISE FOLDING
A2 420 x 594	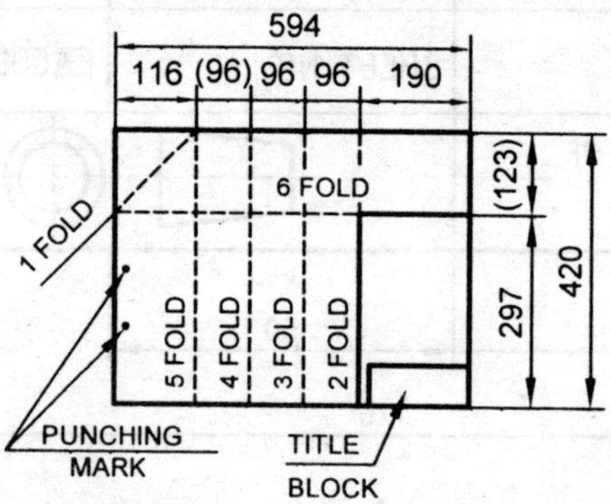	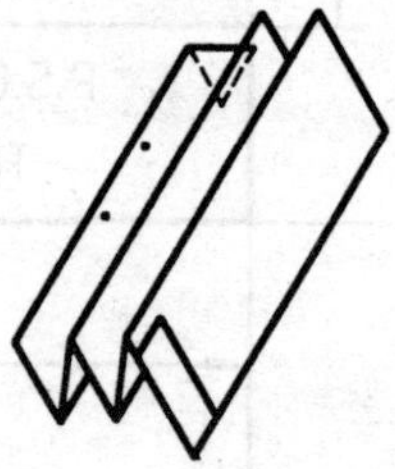

FIG. 4 (i) FOLDING OF DRAWING SHEET FOR FILING OR BINDING

Fig. 4(ii) shows folding of drawing sheets for storing in filing cabinet in industries.

A2 420 x 594	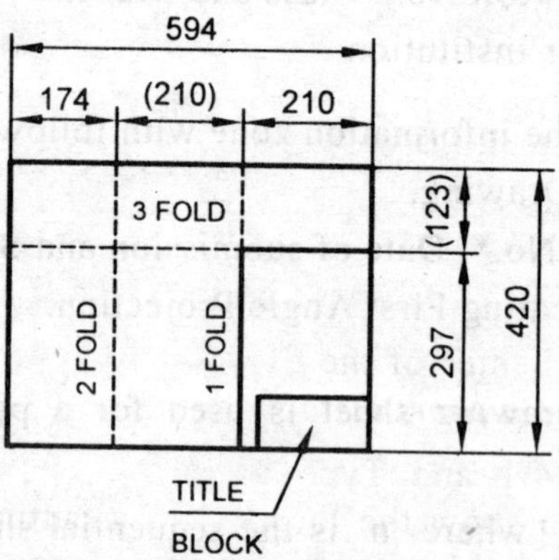	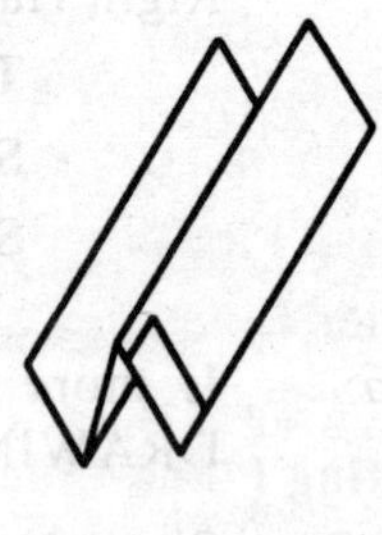

FIG. 4 (ii) FOLDING OF DRAWING SHEET FOR STORING IN FILING CABINET

3. LINES [IS 10714 (Part 20) : 2001 and SP 46 : 2003]

"Small things make perfection! But perfection is no small thing!!"

Just as in your English text book, you are taught to use the correct words for making correct sentences. Similarly, in Engineering Graphics, the details of various objects are drawn by different types of lines. Each line has a definite meaning and sense to convey.

IS 10714 (Part 20) : 2001 (General principles of presentation on technical drawings) and ISO 128-20 : 1996 specify the following types of lines and their applications :

- **Visible Outlines, Visible Edges :** Type 01.2 (CONTINUOUS WIDE LINES)
  The lines drawn to represent the visible outlines / visible edges / surface boundary lines of objects should be outstanding in appearance.
- **Dimension Lines :** Type 01.1 (CONTINUOUS NARROW LINES)
  Dimension Lines are drawn to mark dimensions.
- **Extension Lines :** Type 01.1 (CONTINUOUS NARROW LINES)
  These are extending slightly beyond the respective dimension lines.
- **Construction Lines :** Type 01.1 (CONTINUOUS NARROW LINES)
  Construction Lines are drawn for constructing drawings and should not be erased after completion of the drawing.
- **Hatching / Section Lines :** Type 01.1 (CONTINUOUS NARROW LINES)
  Hatching Lines are drawn for the sectioned portion of an object. These are drawn inclined at an angle of 45° to the axis or to the main outline of the section.
- **Guide Lines :** Type 01.1 (CONTINUOUS NARROW LINES)
  Guide Lines are drawn for lettering and should not be erased after lettering.
- **Break Lines :** Type 01.1 (CONTINUOUS NARROW FREEHAND LINES)
  Wavy continuous narrow line drawn freehand is used to represent break of an object.
- **Break Lines :** Type 01.1 (CONTINUOUS NARROW LINES WITH ZIGZAGS)
  Straight continuous narrow line with zigzags is used to represent break of an object.
- **Dashed Narrow Lines :** Type 02.1 (DASHED NARROW LINES)
  Hidden edges / Hidden outlines of objects are shown by dashed lines of short dashes of equal lengths of about 3 mm, spaced at equal distances of about 1 mm. *The points of intersection of these lines with the outlines / another hidden line should be clearly shown.*
- **Center Lines :** Type 04.1 (LONG-DASHED DOTTED NARROW LINES)
  Center Lines are drawn at the center of the drawings symmetrical about an axis or both the axes. *These are extended by a short distance beyond the outline of the drawing.*
- **Cutting Plane Lines :** Type 04.1 and Type 04.2
  Cutting Plane Line is drawn to show the location of a cutting plane. It is long-dashed dotted *narrow* line, made *wide* at the ends, bends and change of direction. The direction of viewing is shown by means of arrows resting on the cutting plane line.
- **Border Lines**
  Border Lines are continuous wide lines of minimum thickness of 0.7 mm.

3.1 TYPES OF LINES AND THEIR APPLICATIONS [IS 10714 (Part 20) : 2001]

No.	LINE DESCRIPTION AND REPRESENTATION	APPLICATIONS
01.1	CONTINUOUS NARROW LINE	Dimension lines, Extension lines
		Leader lines, Reference lines
		Short center lines
		Projection lines
		Hatching
		Construction lines, Guide lines
		Outlines of revolved sections
		Imaginary lines of intersection
01.1	CONTINUOUS NARROW FREEHAND LINE	Preferably manually represented termination of partial or interrupted views, cuts and sections, if the limit is not a line of symmetry or a center line[a].
01.1	CONTINUOUS NARROW LINE WITH ZIGZAGS	Preferably mechanically represented termination of partial or interrupted views, cuts and sections, if the limit is not a line of symmetry or a center line[a].
01.2	CONTINUOUS WIDE LINE	Visible edges, Visible outlines
		Main representations in Diagrams, Maps, Flow charts
02.1	DASHED NARROW LINE	Hidden edges
		Hidden outlines
04.1	LONG-DASHED DOTTED NARROW LINE	Center lines / Axes, Lines of symmetry
		Cutting planes (Line 04.2 at ends and changes of direction)
04.2	LONG-DASHED DOTTED WIDE LINE	Cutting planes at the ends and changes of direction Outlines of visible parts situated in front of cutting plane
05.1	LONG-DASHED DOUBLED-DOTTED NARROW LINE	Outlines of adjacent parts
		Parts situated in front of a cutting plane

[a] *It is recommended to use only one type of line on one drawing.*

Fig. 5 shows types of lines.

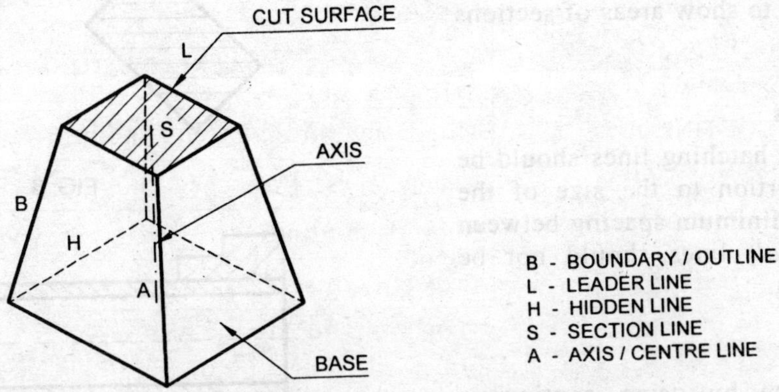

FIG. 5 TYPES OF LINES

3.2 LINE WIDTHS (IS 10714 : 2001)

Line Width means line thickness.

Choose Line Widths according to the size of the drawing from the following range: **0.13, 0.18, 0.25, 0.35, 0.5, 0.7** and **1 mm**.

BIS recommends *two* Line Widths on a drawing. Ratio between the *thin* and *thick* lines on a drawing shall not be less than **1 : 2**.

Note that the ratio of successive line widths is $1 : \sqrt{2}$.

3.3 DRAWING PENCILS

HB - (*Soft grade*)	...	Used for drawing Border Lines, Lettering and Freehand sketching
H - (*Medium grade*)		Used for Visible outlines, Visible edges and Boundary lines
2H - (*Hard grade*)	...	Used for Construction lines, Dimension lines, Leader lines, Extension lines, Center lines, Hatching lines and Hidden lines

3.4 PRECEDENCE OF LINES

1. When a Visible Line coincides with a Hidden Line or Center Line, draw the Visible Line. Also, extend the Center Line beyond the outlines of the view.
2. When a Hidden Line coincides with a Center Line, draw the Hidden Line.
3. When a Visible Line coincides with a Cutting Plane, draw the Visible Line.
4. When a Center Line coincides with a Cutting Plane, draw the Center Line and show the Cutting Plane line outside the outlines of the view at the ends of the Center Line by thick dashes.

3.5 HATCHING OF SECTIONS

Hatching is used to show areas of sections in a drawing.

Spacing of Lines

Spacing between hatching lines should be chosen in proportion to the size of the hatched areas. Minimum spacing between hatching (*parallel*) lines should not be less than 0.7 mm.

See Fig. 6.
Hatching is done by using continuous narrow lines (Type 01.1) at an angle of 45° to the principle outlines or line of symmetry of the sections as shown.

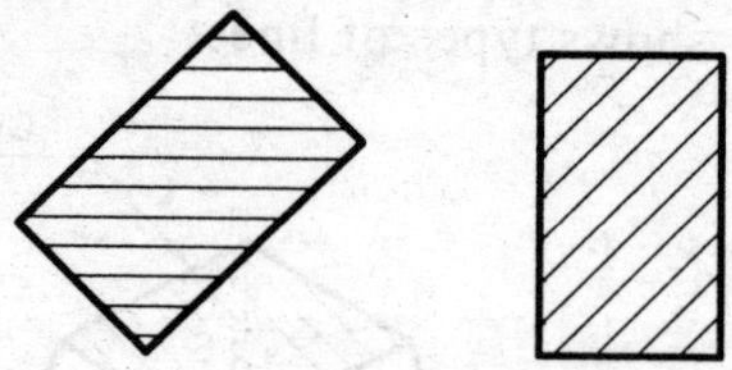

FIG. 6

See Figs. 7 and 8.
Separate areas of a section of the same component shall be hatched in an identical manner. The hatching of adjacent components shall be carried out with different directions or spacings as shown.

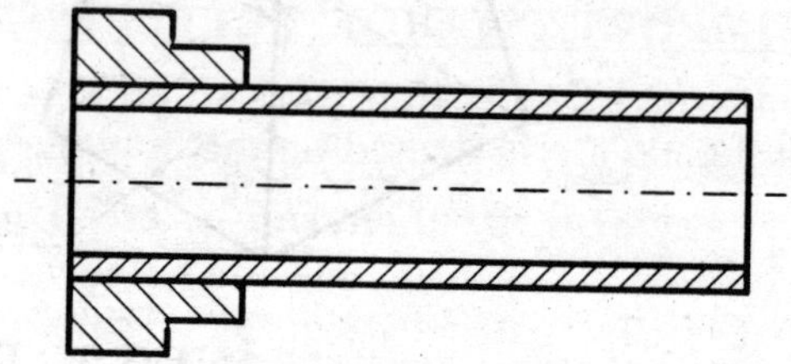

FIG. 7

See Fig. 8.
In the case of *large areas*, the hatching may be limited to a zone following the contour of the hatched area as shown.

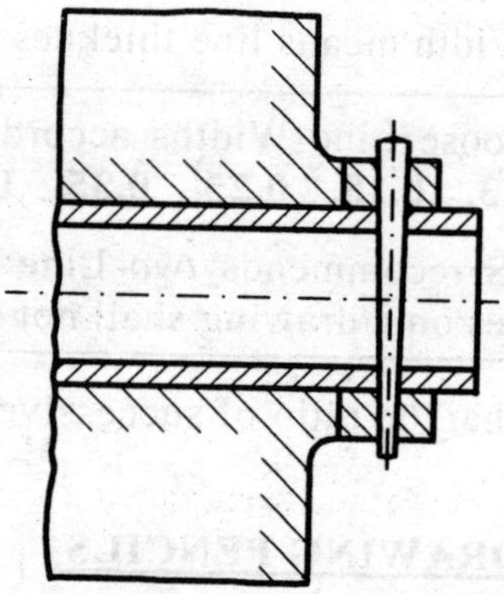

FIG. 8

See Fig. 9.
Hatching shall be interrupted when it is not possible to place inscriptions outside the hatched area as shown.

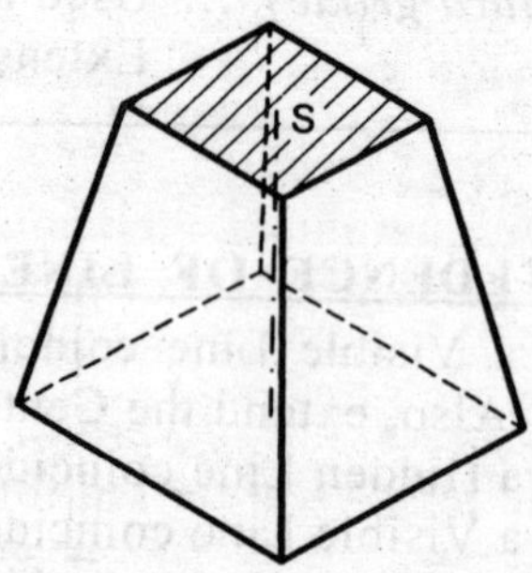

FIG. 9

4. LETTERING [IS 9609 (Part 0) : 2001 and SP 46 : 2003]

"Practice makes you perfect!"

4.1 LETTERING

Lettering is defined as writing of titles, sub-titles, dimensions, etc., on a drawing.

4.2 IMPORTANCE OF LETTERING

To undertake production work of an engineering component as per the drawing, the size and other details are indicated on the drawing. This is done in the form of notes and dimensions.

Main Features of Lettering are Legibility, Uniformity and Rapidity of execution. Use of drawing instruments for lettering consumes more time. Also, time consuming ornamental letters shall not be used. Plain letters facilitate rapid reading. Lettering, therefore, should be done freehand with speed. *Yes*, **Lettering is freehand drawing and not writing.**

Remember ! Practice makes you perfect!! Practice accompanied by continuous efforts would certainly improve the lettering skill and style. Poor lettering will mar the appearance of an otherwise acceptable drawing.

4.3 BIS AND ISO CONVENTIONS

IS 9609 (Part 0) : 2001 and SP 46 : 2003 (Lettering for technical drawings) specify lettering in technical product documentation. This BIS standard is based on ISO 3098 – 0 : 1997.

4.3 SINGLE STROKE LETTERS

The word single-stroke should not be taken to mean that the lettering should be made in one stroke without lifting the pencil. It means that the thickness of the letter should be uniform such as is obtained in one stroke of the pencil.

4.4 TYPES OF SINGLE STROKE LETTERS

1. Lettering Type A : (i) Vertical and (ii) Sloped (at 75° to the horizontal)
2. **Lettering Type B* : (i) Vertical**$^{+}$ and (ii) Sloped (at 75° to the horizontal)

* ***Type B preferred :***
In Type A, height of the capital letter is divided into 14 equal parts, while in Type B, height of the capital letter is divided into 10 equal parts. Type B is preferred for easy and fast execution, because of the division of height into 10 equal parts.

\+ ***Vertical Letters preferred :***
Vertical letters are preferred for easy and fast execution, instead of sloped letters.

NOTE : *Lettering in drawing should be in CAPITALS (i.e., Upper-case letters). Lower-case (small) letters are used for abbreviations like mm, cm, etc.*

4.5 SIZE OF LETTERS

- Size of Letters is measured by the height **h** of the CAPITAL letters as well as numerals.
- Standard heights for CAPITAL letters & numerals recommended by BIS are given below :

1.8, 2.5, 3.5, 5, 7, 10, 14 and **20 mm**.

Recommended Size (height h) of Letters / Numerals		
MAIN TITLE	:	**5 mm** *or* **7 mm** *or* **10 mm**
SUB-TITLES	:	**3.5 mm** *or* **5 mm**
Dimensions, Notes, etc.,	:	**2.5 mm, 3.5 mm** *or* **5 mm**

Ratio between successive heights is **1 : √2**. The multiple of **√2** in the range of heights for lettering is derived from the standardized progression of dimensions for drawing sheet sizes. Size of the letters may be selected based upon the size of the drawing.

4.6 GUIDE LINES

- In order to obtain correct and uniform height of letters and numerals.
- Guide Lines are drawn, using 2H pencil with light pressure.
- HB grade conical end pencil is used for Lettering.

4.7 PROCEDURE FOR LETTERING

1. Draw thin horizontal guide lines at a distance **h** apart.
2. *Lettering Technique :* Horizontal lines of the letters are drawn from left to right. Vertical, Inclined and Curved lines are drawn from top to bottom.
3. After lettering has been completed, don't erase the guide lines.

4.8 DIMENSIONING OF TYPE B LETTERS

BIS denotes the characteristics of lettering as:

h (height of capital letters),
c_1 (height of lower-case letters),
c_2 (tail of lower-case letters),
c_3 (stem of lower-case letters),
a (spacing between characters),
b_1 & b_2 (spacing between baselines),
e (spacing between words) and
d (line width) as shown in *Figs. 10, 11(a)* & *(b)* and in the **Table** below.

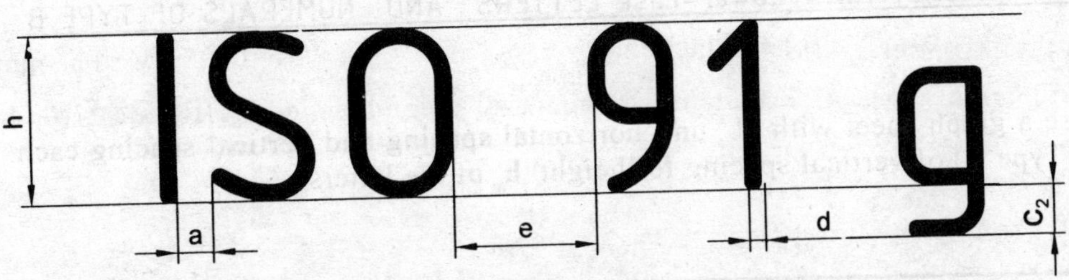

FIG. 10

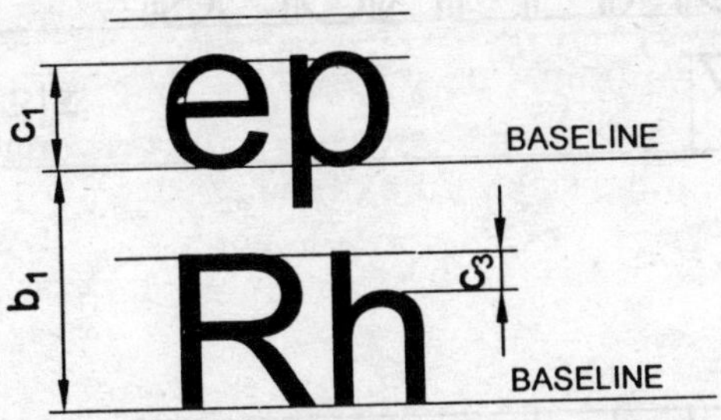

FIG. 11 (a)

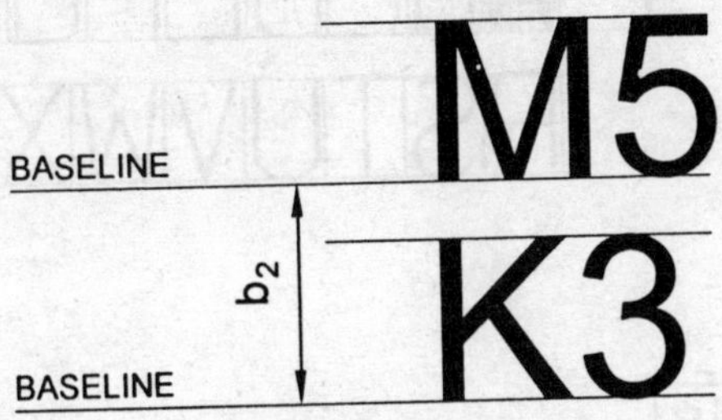

FIG. 11 (b)

TABLE

Characteristics of Lettering		**Multiple of *h***	**Dimensions (mm)**				
Lettering height	h	(10/10) h	2.5	3.5	5	7	10
Height of lower-case letters	c_1	(7/10) h	1.75	2.5	3.5	5	7
Tail of lower-case letters	c_2	(3/10) h	0.75	1.05	1.5	2.1	3
Stem of lower-case letters	c_3	(3/10) h	0.75	1.05	1.5	2.1	3
Spacing between characters	a	(2/10) h	0.5	0.7	1	1.4	2
Minimum spacing between baselines[1)]	b_1	(15/10) h	3.75	5.25	7.5	10.5	15
Minimum spacing between baselines[2)]	b_2	(13/10) h	3.25	4.55	6.5	9.1	13
Spacing between words	e	(6/10) h	1.5	2.1	3	4.2	6
Line width	d	(1/10) h	0.25	0.35	0.5	0.7	1

1) Lettering style: Upper-case & lower-case letters *2) Lettering style: Upper-case letters only*

4.9 VERTICAL CAPITAL & Lower-case LETTERS AND NUMERALS OF TYPE B

See Fig. 12.

Use a graph sheet with 1 unit horizontal spacing and vertical spacing each. Mark 10 units (Type B) of vertical spacing for height **h** of the letters.

FIG. 12

VERTICAL CAPITAL & Lower-Case LETTERS AND NUMERALS OF TYPE B

4.10 POINTS TO REMEMBER

CAPITAL Letters

- Ratio of height to width for most of the CAPITAL letters is approximately = **10 : 6**
- However, for **M** and **W**, the ratio ≈ **10 : 8**. For **I** the ratio = **10 : 2**

Lower-case Letters

- Height of lower-case letters *with* stem / tail (b, d, f, g, h, j, k, l, p, q, t, y) = c_2 = c_3 = **h**
- Ratio of height to width for lower-case letters *with* stem or tail ≈ **10 : 5**
- Height of lower-case letters *without* stem or tail **c_1** is approximately = **(7/10) h**
- Ratio of height to width for most lower-case letters *without* stem or tail ≈ **7 : 5**
- However, for **m** and **w**, the ratio = **7 : 7**. For **i** and **l**, the ratio = **10 : 2**

Numerals

- For numerals **0** to **9**, the ratio of height to width ≈ **10 : 5**. For **1**, ratio = **10 : 2**

Spacing

- Spacing between characters = **a** = **(2/10) h**
- Spacing between words = **e** = **(6/10) h**

4.11 EXAMPLE IN LETTERING PRACTICE

Write freehand the following, using single stroke vertical CAPITAL letters of 5 mm (h) size:

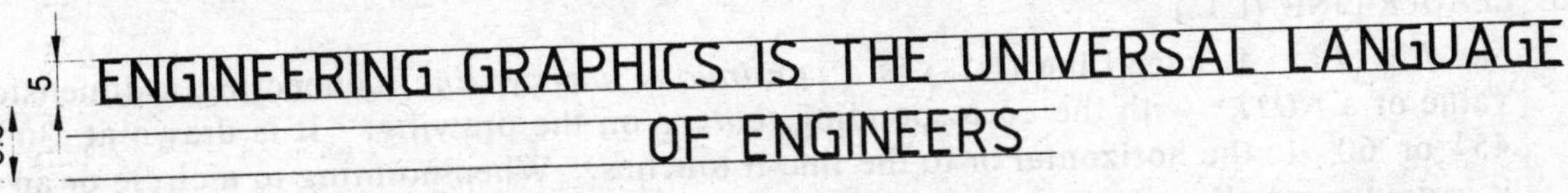

Minimum spacing between baselines (Upper-case letters only) = **b_2** = **(13 / 10) h = 6.5**

EXERCISES

Write freehand the following, using single stroke vertical (CAPITAL and lower-case) letters:

1. Alphabets (Upper-case & Lower-case) and Numerals 0 to 9 (h = 3.5, 5 and 7 mm)
2. USE CAPITAL LETTERS, UNLESS OTHERWISE SPECIFIED. (h = 3.5 and 5 mm)
3. SMALL THINGS MAKE PERFECTION, BUT PERFECTION IS NO SMALL THING. (h = 5 & 7 mm)
4. STANDARDS CONNECT THE WORLD. (h = 5 and 7 mm)

5. DIMENSIONING [IS 11669 : 1986 and SP 46 : 2003]

Dimensioning : To manufacture an engineering component, we need its shape and sizes like length, breadth, height, diameter, angle, etc. Indicating the sizes of the *features* (*See Fig. 13.* Circle is a feature.) of the object and other details essential for its construction and function on a drawing by the use of lines, numerals, symbols, notes, etc., is called *Dimensioning.*

IS 11669 : 1986 and SP 46 : 2003 specify the General Principles of Dimensioning on Technical Drawings.

5.1 ELEMENTS OF DIMENSIONING

1. DIMENSION LINE (D.L.)

See Fig. 13. Dimension Line (D.L.) is a *continuous narrow line,* drawn parallel to the edge or surface whose measurement should be shown. It should be placed *outside* the views. Sometimes they may be placed *inside* the views, if the drawing becomes very clear by doing so.

Dimension line is placed at least 6 to 10 mm away from the outline of the drawing. It is terminated by arrowheads at its ends.

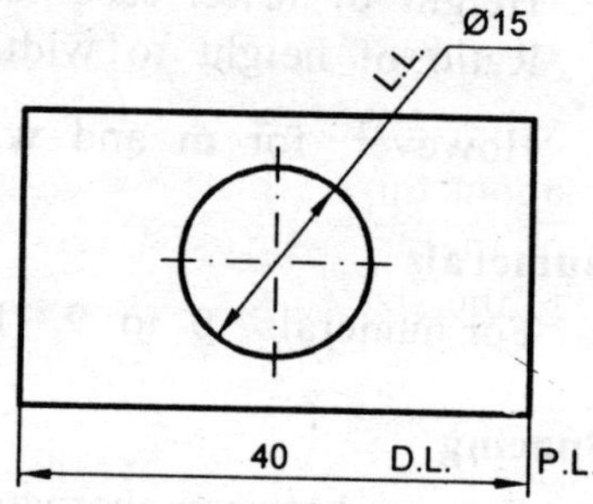

FIG. 13

2. PROJECTION LINE (P.L.) *or* EXTENSION LINE

See Fig. 13. Projection Line (P.L.) or Extension Line is a *continuous narrow line* drawn perpendicular to the outline to be dimensioned and without leaving a gap from the outline. It is drawn extending slightly beyond the dimension line by about 2 mm.

3. LEADER LINE (L.L.)

See Fig. 13. Leader Line (L.L.) is a *continuous narrow line*, connecting a dimensional value or a ***NOTE**** with the corresponding *feature* on the drawing. It is drawn at 30° or 45° or 60° to the horizontal or to the line it touches. When pointing to a circle or an arc, it is drawn radially.

If the leader line ends on the outline of a drawing, it shall have an arrowhead [*Fig. 14 (i)*].
If it ends within the outline of the drawing, it shall have a dot at the end [*Fig. 14 (ii)*].
If it ends on the dimension line, it terminates without arrowhead or dot [*Fig. 14 (iii)*].

* ***NOTE*** gives information relating to a feature, *e.g.,* Ø 15, 2 HOLES, etc.

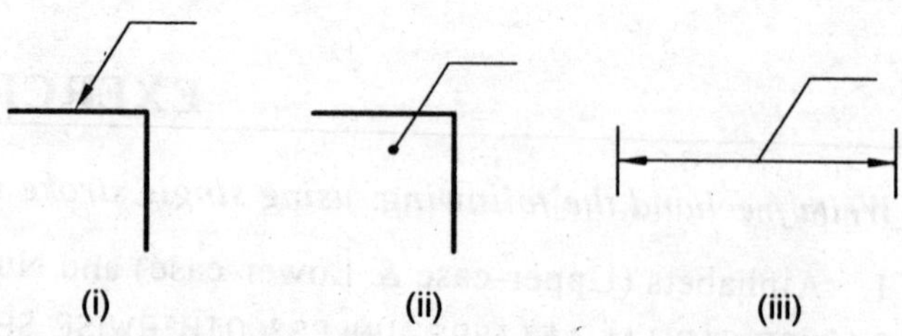

FIG. 14

4. ARROWHEADS

See Fig. 15. An arrowhead is placed at each end of a dimension line. The arrowhead may be open, closed or closed and filled in. **Open type arrowhead is preferable for fast execution.**

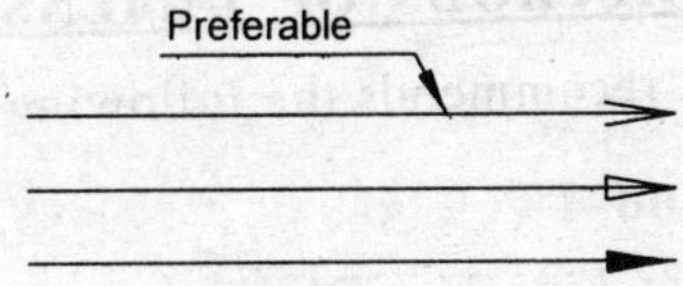

FIG. 15

In a drawing, only one style of arrowhead termination should be used.

Shape of the arrowhead

See Fig. 16. Length **l** of an arrowhead is about three times the width **w**. The size of the arrowheads should be proportionate to the size of the drawing.

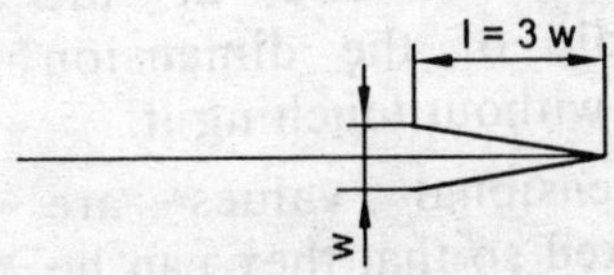

FIG. 16

See Fig. 17. Place the dimensional value parallel to and at the middle of dimension line. When the length of the dimension line is not sufficient, place arrowheads outside (*e.g.,* dimensional value 7) *or* place the dimensional value (3) above the extended portion of the dimension line beyond the arrowhead.

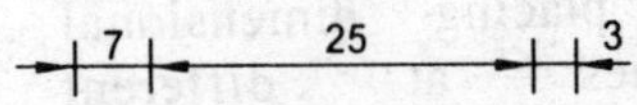

FIG. 17

Oblique stroke

See Fig. 18. Instead of arrowheads, the ends of the dimension lines may be marked by *oblique stroke* drawn as a short line inclined at an angle of 45°.

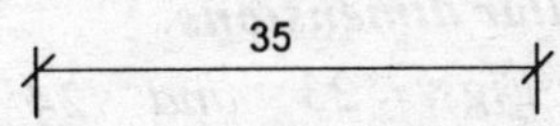

FIG. 18

See Fig. 19. Where space is too small for an arrowhead, the oblique stroke or a dot may be substituted as shown.

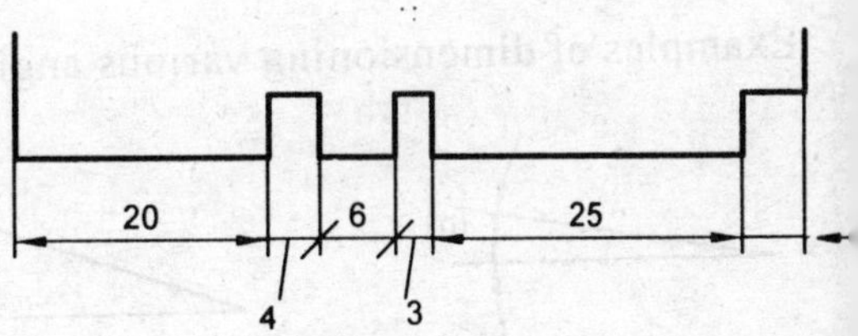

FIG. 19

Origin indication: *See Fig. 20.* Origin indication is drawn as a small circle of about 3 mm diameter.

FIG. 20

5.2 METHODS OF DIMENSIONING

B.I.S. recommends the following ***two*** methods of placing dimensional values on a drawing:

Method I

- *See Fig. 21.* Dimensional value is placed above the dimension line without breaking it. It is placed parallel to and at the middle of the dimension line without touching it.

  Dimensional values are marked so that they can be read *either from the bottom or from the right hand side of the drawing.*

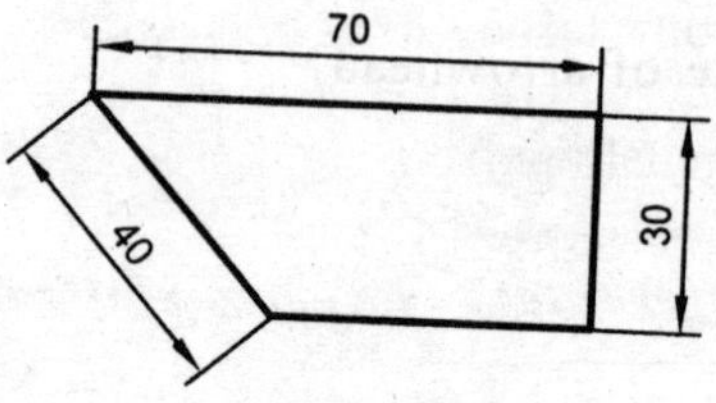

FIG. 21

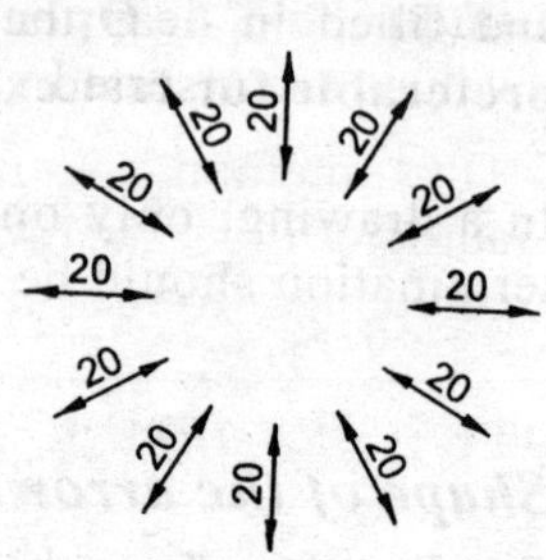

FIG. 22

- *Fig. 22* shows the method of placing dimensional values at different positions on the dimension lines.

- ***Angular dimensions***

  See Figs. 23 and 24. Angular dimensions may be oriented either as in *Fig. 23* or *Fig. 24.*

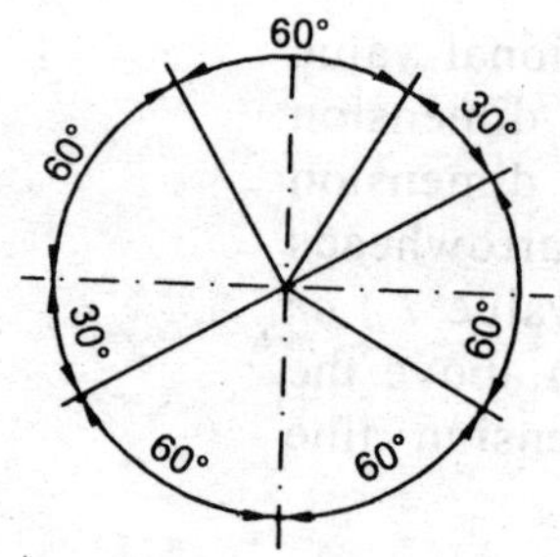

FIG. 23

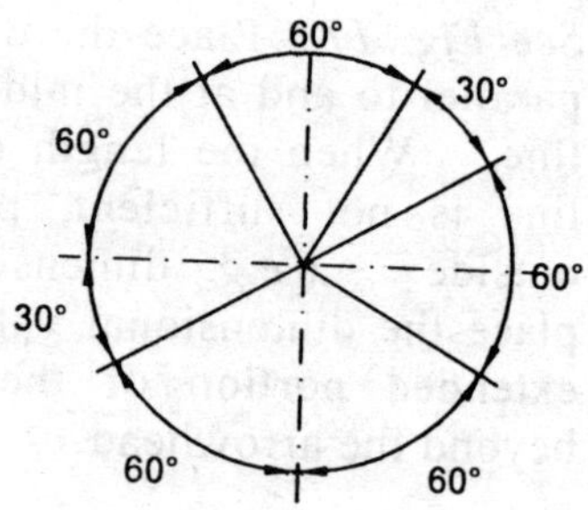

FIG. 24

Examples of dimensioning various angles are shown in *Fig. 25.*

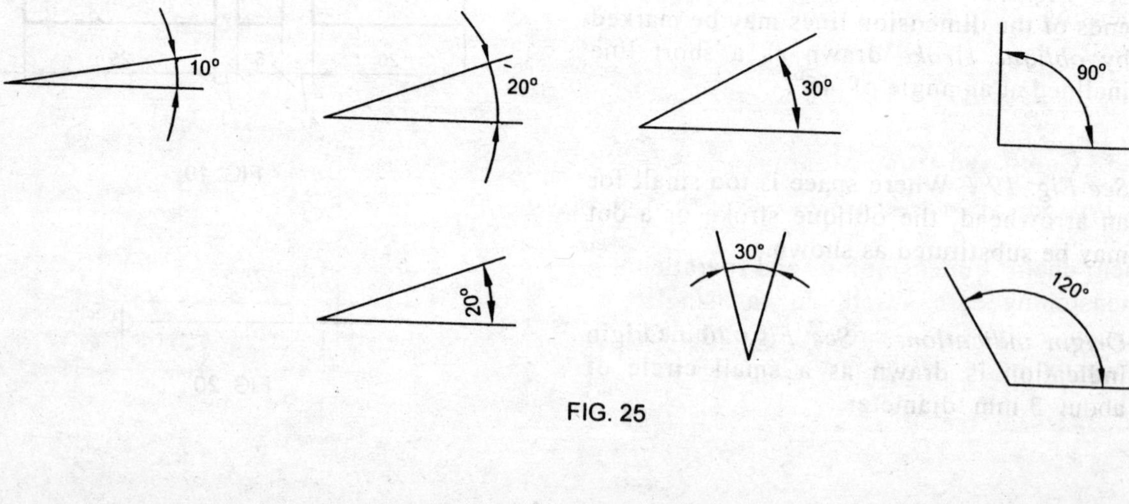

FIG. 25

Method II

- *See Fig. 26.* Dimensional values are indicated so that they may be read from the bottom of the drawing. Non-horizontal dimension lines are interrupted near the middle and the dimensional values are inserted.

  This method is used on large drawings of automobiles, aircrafts, etc., where it may not be convenient to read the dimensions from right side.

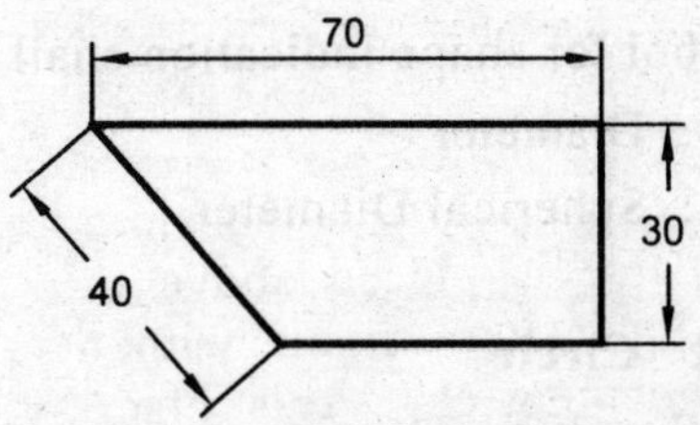

FIG. 26

NOTE : ***Method I is preferable.*** *Only one method should be used on a drawing.*

5.3 ARRANGEMENT OF DIMENSIONS

5.3.1 Chain Dimensioning

See Fig. 27.

When successive dimensions are arranged in a straight line, the method is known as *Chain Dimensioning.* Note that the chain of dimensions shall be arranged in a *continuous* straight line as shown.

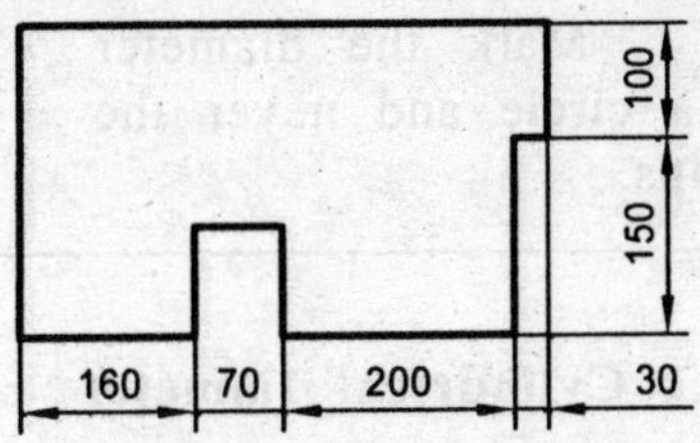

FIG. 27

5.3.2 Parallel Dimensioning

See Fig. 28.

Parallel Dimensioning is the placement of a number of single dimension lines parallel to one another from a common origin. It is used where a number of dimensions have a common origin.

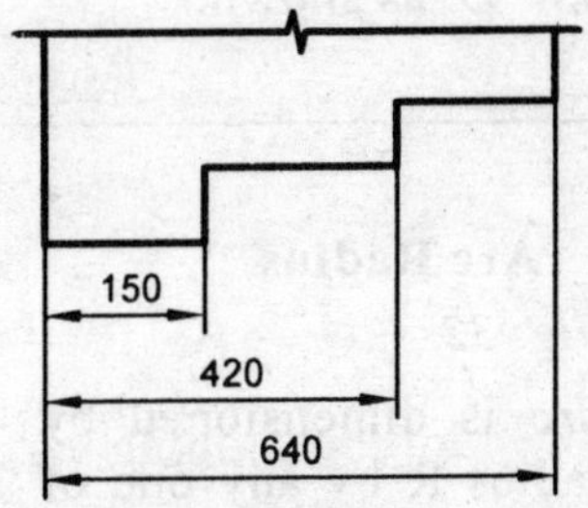

FIG. 28

5.3.3 Combined Dimensioning

See Fig. 29.

When chain dimensioning and parallel dimensioning are used in a single drawing, it is called Combined Dimensioning.

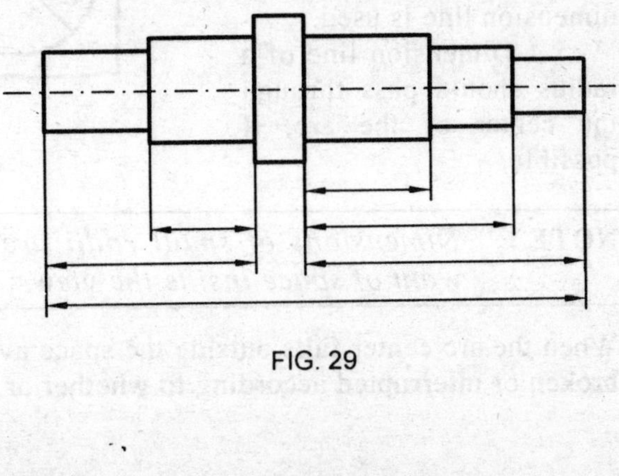
FIG. 29

5.4 SYMBOLS FOR SHAPE INDICATION

Symbol for shape indication shall precede the dimension value. Symbols used are:

Ø : Diameter
SØ : Spherical Diameter
R : Radius
SR : Spherical Radius
□ : Square

5.4.1 Circle

See Fig. 30.

A *circle* is dimensioned using the symbol Ø, by any one of the ways shown. Ø precedes the dimension value.

Mark the diameter of a circle and never the radius.

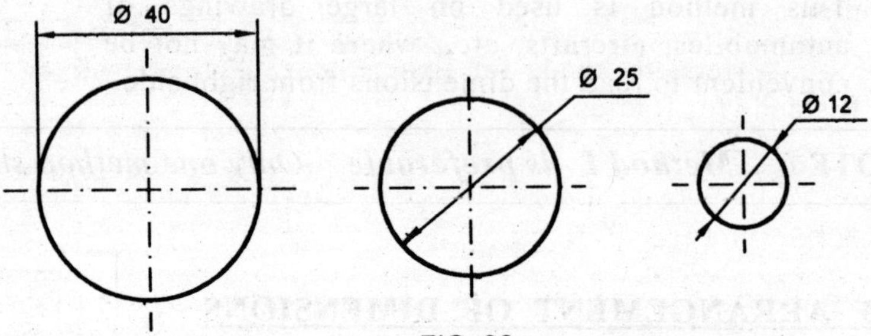

FIG. 30

5.4.2 Cylindrical Diameter

See Fig. 31.

Cylindrical Diameter uses symbol Ø as shown.

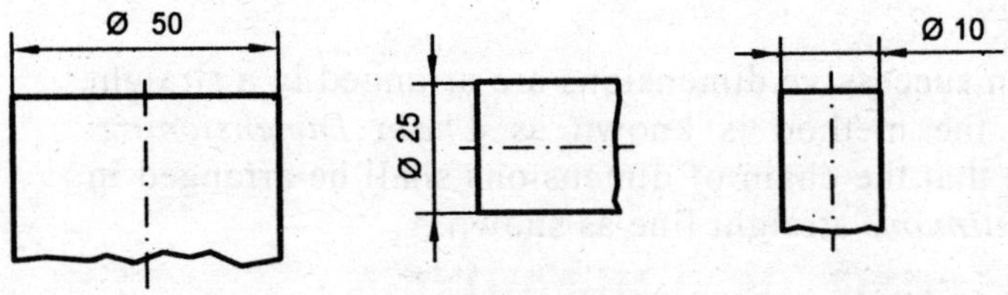

FIG. 31

5.4.3 Arc Radius

See Fig. 32.

An *arc* is dimensioned by its Radius R by any one of the ways shown. Only one arrowhead with its point on the arc end of the dimension line is used.

Dimension line of a radius should pass through the center of the arc, if possible.

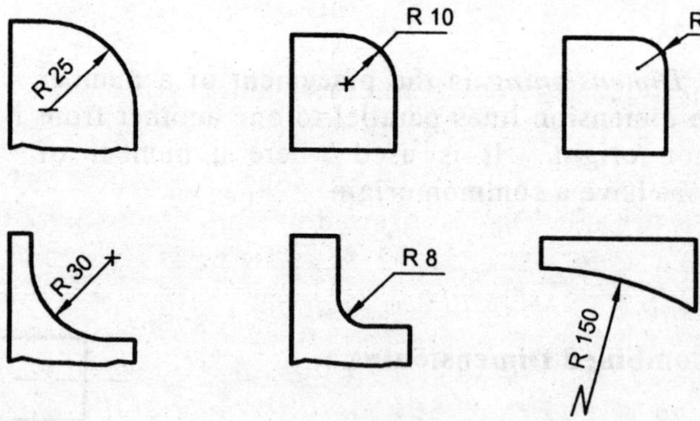

FIG. 32

NOTE : *Dimensions of small radii are shown preferably outside the view, due to want of space inside the view.*

When the arc center falls outside the space available, the dimension line of the radius shall be broken or interrupted according to whether or not it is necessary to locate the center.

5.4.4 Sphere

See Fig. 33.

Use symbol SØ to represent Spherical Diameter and SR for Spherical Radius. The symbols SØ and SR should precede the dimensional value.

Here, S stands for Sphere and R stands for Radius.

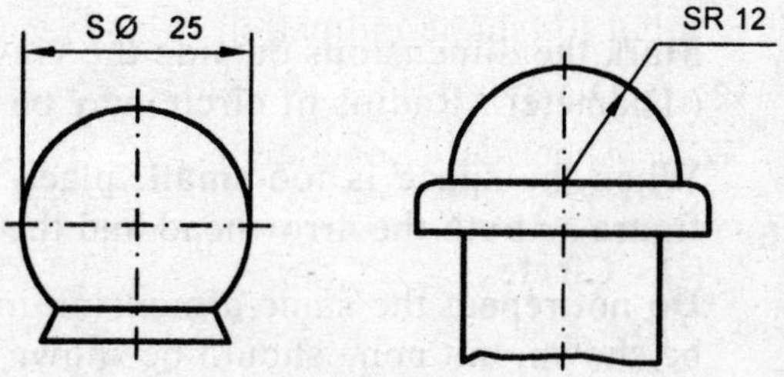

FIG. 33

5.4.5 Square

See Fig. 34.

Symbol □ is used to represent a square and it should precede the dimensional value. Square ends may be indicated by diagonals drawn as *thin lines*.

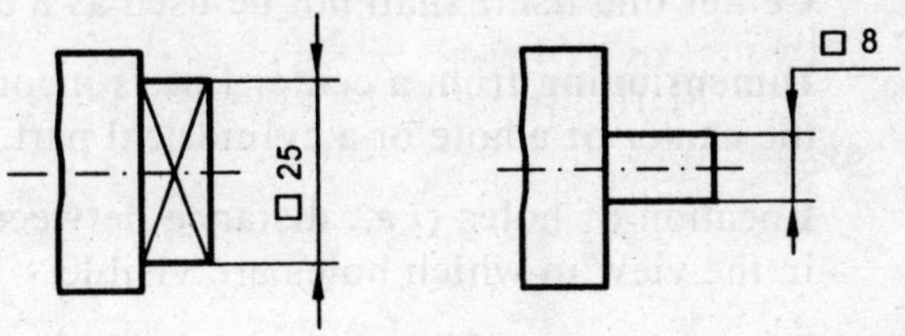

FIG. 34

5.5 UNIT OF DIMENSIONING

All dimensions in one particular drawing should be expressed in *one unit* only. B.I.S. recommended unit is *millimetre*. Abbreviation **mm** for millimetre need not be shown on each dimension of the drawing.

Example: A dimension **25** means **25 mm**. A foot note stating 'ALL DIMENSIONS IN mm' is written in a prominent place near the Title Block.

When it is not convenient to give dimensions in millimetres and another unit, say, centimetres is used, only the dimensional values are written. And a foot note stating 'ALL DIMENSIONS IN cm' is inserted in a prominent place near the Title Block.

NOTE : i. *Height of the dimensional values should be 3.5 mm or 5 mm.*

ii. *The decimal point in a dimension should be quite distinct and written in line with the bottom line of the dimension value.*
Example: ***0.5 and not 0˙5***

iii. *When dimension is < 1, a zero must always precede the decimal point.*
Example: ***0.5 and not .5***

iv. *Abbreviation for millimeters is **not mms**, but **mm only**. (s denotes seconds). In other words, s is not added for the plural of any units.*

5.6 RULES FOR DIMENSIONING

1. **M**ark the dimensions outside the view.
(Diameter / Radius of circle may be shown inside.)

2. **W**hen the space is too small, place either the arrowhead outside with angle inside the figure *or* both the arrowhead and the angle outside the figure.

3. **D**o not repeat the same dimension in different views. Note that every dimension must be shown, but none should be shown more than once.

4. **A** circle shall be dimensioned by its diameter symbol ∅ (and an arc by its radius symbol R). Convention ∅ or R shall be placed before the dimensional value.

5. **C**enter line itself shall not be used as a dimension line with arrowheads at its ends.

6. **D**imensioning from a center line is incorrect, except when the center line passes through the center of a hole or a cylindrical part.

7. **L**ocation of holes (*i.e.*, distance between the centers of the holes) shall be dimensioned in the view in which holes are visible.

8. **D**imensions shall be given to visible lines and not to hidden (invisible) lines

9. **I**f dimensioning inside a hatched portion of a drawing is unavoidable, the hatching lines should not cut the dimensional text.

10. *Notes* should always be written horizontally.

11. *Leader line* shall be inclined at an angle of 30°, 45° or 60° to the horizontal.

12. *Overall Dimensioning:* Overall dimension shall be placed outside the intermediate dimensions. i.e., smaller dimensions shall be placed nearer the view and the larger further away so that extension lines do not cross dimension lines. (Extension lines may cross each other or the outlines of the drawing, such as the case of leader lines.)
When an overall dimension is shown, one of the intermediate dimensions should not be marked.

13. **A**lways mark the overall length, width and height of an object.

14. **D**imension lines should not cross any other line of the drawing. Mutual crossing of dimension lines should be avoided.

15. **S**cale of the drawing, other than 1 : 1, should be written below the drawing or in the space provided inside the title block. At the same time, whatever may be the scale of the drawing, the ***actual given dimensions*** should be written on the dimension lines of the drawing.

*

4

ENGINEERING SCALES

1. STANDARD SET-SCALES

Set of Standard Scales M1 to M8 available for measuring is already discussed in *Chapter 3.*

2. SPECIAL SCALES

When you require to make measurements other than those given by the set of standard scales M1 to M8, *Special Unusual Scales* have to be constructed.

This chapter deals with the construction of such special unusual scales, hereafter called as simply *Construction of Scales*.

2.1 CONSTRUCTION OF SCALES

To construct a scale on the drawing sheet, the following data are required :

(i) Representative Fraction (R.F.) of the scale
(ii) Measuring Units of the scale
(iii) Maximum measurement that the scale has to read

2.2 REPRESENTATIVE FRACTION (R.F.)

Ratio of the drawing size of an object to its actual size is called ***Representative Fraction***.

$$\text{i.e., R.F.} = \frac{\text{Drawing size of an object}}{\text{Its actual size}} \quad \text{(in same units)}$$

For ***Reducing Scale***, the drawings will have R.F. values of less than unity. For example, if 1 centimetre on a drawing represents 1 metre length of an object,

$$\text{then R.F.} = \frac{1 \text{ cm}}{1 \times 100 \text{ cm}} = \frac{1}{100} < 1$$

For drawings using ***Enlarging Scale***, R.F. values will be > 1. For example, when 1 mm length of an object is shown by a line of length of 1 cm on the drawing,

$$\text{then R.F.} = \frac{1 \times 10 \text{ mm}}{1 \text{ mm}} = \frac{10}{1} = 10 > 1$$

2.3 METRIC MEASUREMENTS

10 decimetres	(dm)	=	1 metre	(m)
10 metres	(m)	=	1 decametre	(dam)
10 decametres	(dam)	=	1 hectometre	(hm)
10 hectometre	(hm)	=	1 kilometre	(km)

3. TYPES OF SCALES

1. PLAIN SCALES
2. DIAGONAL SCALES
3. VERNIER SCALES
4. COMPARATIVE SCALES

4. PLAIN SCALES

A *Plain Scale* is simply a line divided into a number of equal parts, the first of which is further sub-divided into small parts. It represents either two units or a unit and its fraction such as km and hm, etc.

5. DIAGONAL SCALES

Plain Scales are used to read lengths in two units or to read to the accuracy correct to first decimal. Diagonal Scales are used either to measure very minute distances such as 0.1 mm, etc., or to measure in three units such as dm, cm and mm.

6. VERNIER SCALES

Vernier scale is a modified form of a diagonal scale. It is a short scale used when a diagonal scale is inconvenient to use due to lack of space. It consists of two parts, i.e., **Main Scale** and a **Vernier Scale**. The main scale is a *Plain Scale.* Vernier scale slides on the side of the main scale and both of them are used together to measure small divisions upto three divisions as in the case of diagonal scale.

Uses

Vernier scales are used in mechanical measuring instruments like screw gauge, calipers, etc., and also used in surveying.

*

5

GEOMETRICAL CONSTRUCTIONS

This chapter deals with some of the important 2D geometrical constructions which will be frequently used in engineering drawings.

1. TO DIVIDE A LINE INTO NUMBER OF EQUAL PARTS

Problem 1: *(Fig. 1)* *Divide a given line of 100 mm length into six equal parts.*

1. Draw a line AB of 100 mm length.
2. Draw a *thin* line AC of any length inclined at some convenient acute angle, say, 30° or 45° to AB.
3. From A and along AC, step-off 6 equal divisions of any convenient length using divider. Join B and 6.
4. Draw lines parallel to B6 through 1, 2, 3, 4 and 5 to cut AB at 1′, 2′, 3′, 4′ and 5′. The points 1′, 2′, etc., are division points dividing AB into six equal parts.

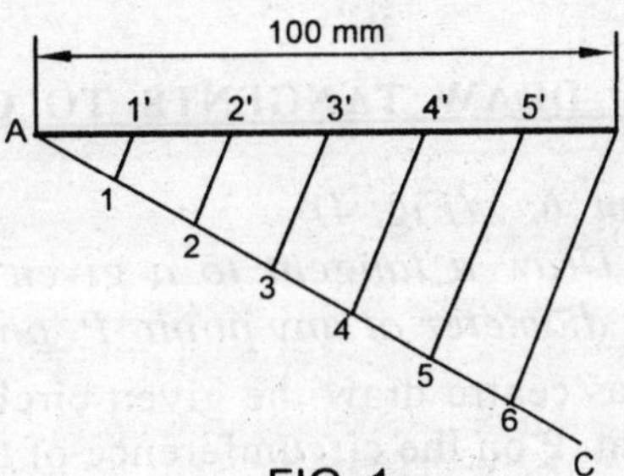

FIG. 1

Problem 2: *(Exercise)* *Divide a line of 108 mm length into seven equal parts.*

2. TANGENT ARCS TO LINES AND TANGENTS TO CIRCLES

2.1 TO DRAW ARCS TANGENTIAL TO LINES

Problem 3: *(Fig. 2)* *Draw an arc of radius 25 mm to touch 2 given lines at 90° to each other.*

1. Draw AB and AC at right angles. A as centre and given radius R, draw arcs to cut AB at M and AC at N.
2. M and N as centers and radius R, draw arcs to intersect each other at O.
3. O as centre and the same radius R, draw the tangent arc NM to touch AC and AB.

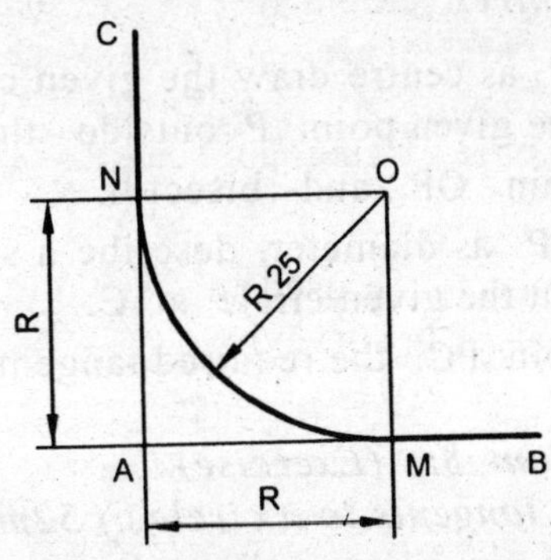

FIG. 2

Problem 4: *(Fig. 3)*

Draw an arc of a given radius 20 mm touching two given straight lines which make an acute angle 59° between them.

1. Draw the given acute angle CAB.
2. Draw a line PQ parallel to and at a distance = R (given radius) from AB.
3. Draw a line DE parallel to and at a distance of R from AC meeting PQ at O.
4. O as centre and R as radius, draw the tangent arc PD to touch AB and AC.

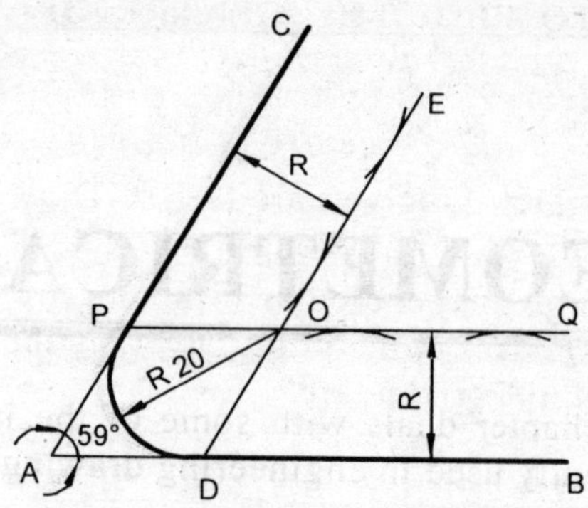

FIG. 3

Problem 5: *(Exercise)*

Draw an arc of radius 22 mm touching two given straight lines which make an obtuse angle of 135° between them.

HINT: Follow the *same* procedure given in Problem 4 for *obtuse angle* too.

2.2 TO DRAW TANGENTS TO CIRCLES

Problem 6: *(Fig. 4)*

Draw a tangent to a given circle of 50 mm diameter at any point P on it.

1. O as centre draw the given circle. Mark point P on the circumference of the circle.
2. Join OP. Through P draw line RS perpendicular to PO. RS is the tangent at P.

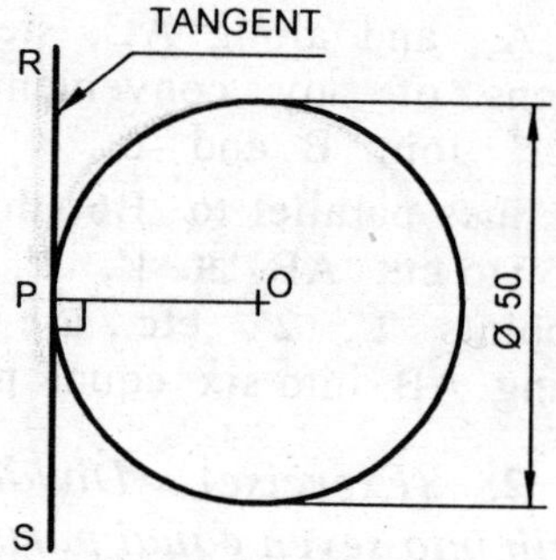

FIG. 4

Problem 7: *(Fig. 5)*

Draw a tangent to a circle of 48 mm diameter from a point P 58 mm away from its centre.

1. O as centre draw the given circle. Mark the given point P outside the circle.
2. Join OP and bisect it.
3. OP as diameter, describe a semi-circle to cut the given circle at C.
4. Join PC, the required tangent.

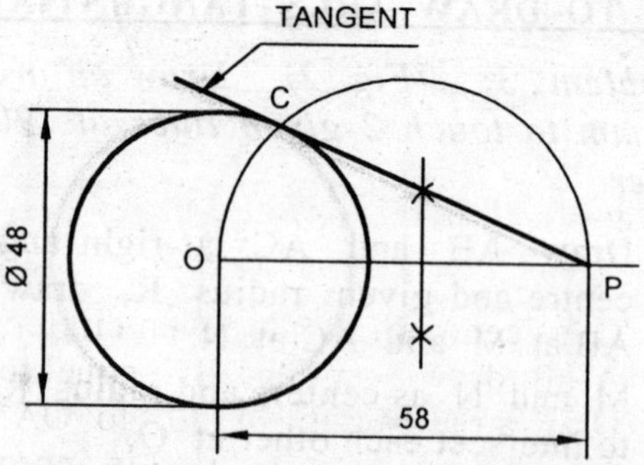

FIG. 5

Problem 8: *(Exercise)*

Draw tangents to a circle of 52mm diameter

i. at any point on it, and

ii. at a point 52 mm away from its centre.

2.3 TO FIND THE CENTRE OF AN ARC

Problem 9: *(Fig. 6)*
Find the centre of a given arc.

1. Draw the given arc AB. Draw two chords PQ and RS of any length on AB.
2. Draw perpendicular bisectors of PQ and RS to intersect at the point O. O is the required centre of the given arc.

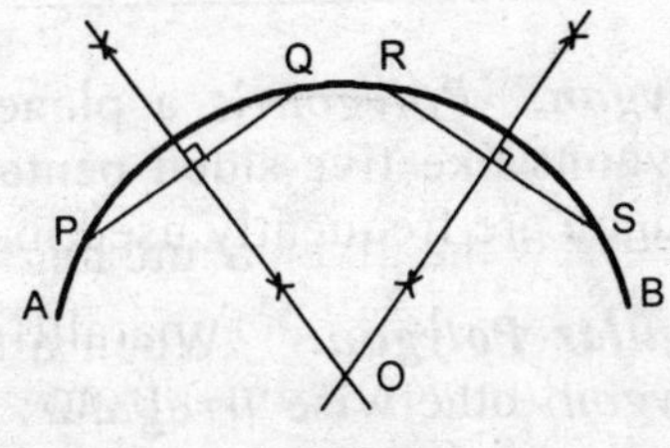

FIG. 6

2.4 TO DRAW AN ARC PASSING THROUGH THREE GIVEN POINTS

Problem 10: *(Fig. 7) Draw an arc passing through three given points not in a straight line.*

1. Mark the given points A, B and C.
2. Join AB and BC. Draw perpendicular bisectors of AB and BC to meet at O.
3. O as center and radius as OA / OB / OC, draw the arc, passing through the three points.

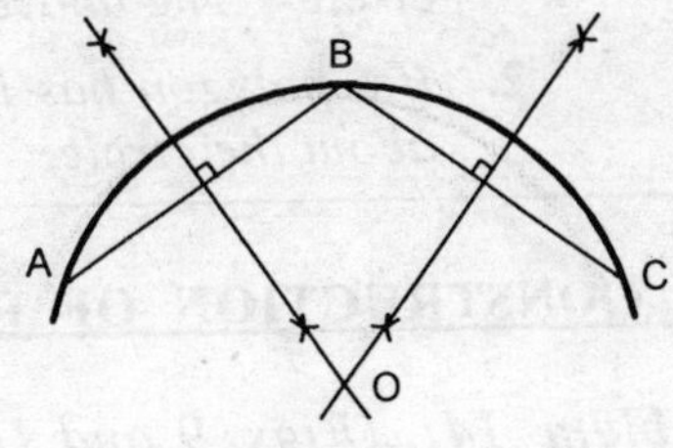

FIG. 7

NOTE : *Above construction is known as* ***"Describing a Circle about a Δ ABC".***

Problem 11: *(Exercise) Describe a circle about a triangle of sides 60, 70 and 75 mm.*

3. LENGTH OF THE CIRCUMFERENCE OF A CIRCLE

Problem 12: *(Fig. 8) Find the length of circumference of a circle of 40 mm diameter.*

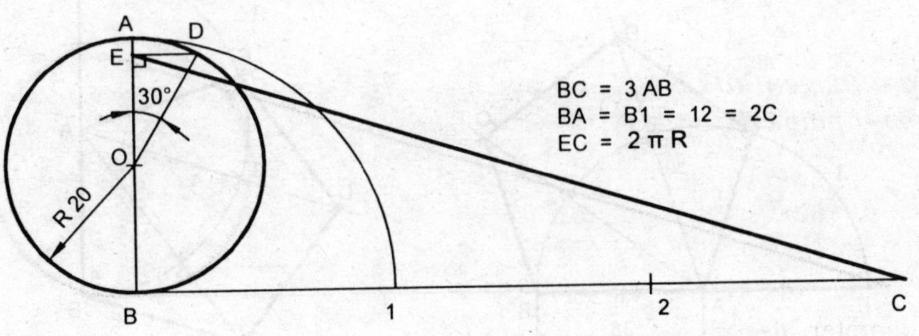

FIG. 8

1. O as centre draw the given circle. Draw a vertical diameter AB through O.
2. At B draw a tangent BC equal to 3 times AB as shown in *Fig. 8*.
3. Draw a radius OD at 30° to OA. From D, draw DE perpendicular to AB.
4. Join E and C. Now EC is approximately = the circumference of the given circle.

Problem 13: *(Exercise) Find the circumference of a circle of 52 mm diameter.*

4. POLYGONS

Polygon: *Polygon* is a plane figure bounded by more than four straight lines. Regular polygons like five-sided pentagon, six-sided hexagon, seven-sided heptagon and eight-sided octagon are frequently used in engineering drawings.

Regular Polygon: When all the sides of a polygon are equal, it is known as a *regular polygon*, otherwise *irregular*. **Unless specified as irregular, *polygon* means *regular polygon* only.**

NOTE: 1. *If a polygon has all its vertices on a circle, it is said to be* ***inscribed*** *in the circle. The above circle is said to be* ***circumscribed*** *about the polygon.*

2. *If a polygon has its sides tangential to a circle, it is said to be* ***circumscribed*** *about the circle. The circle is said to be* ***inscribed*** *in the polygon.*

4.1 CONSTRUCTION OF REGULAR POLYGONS

Problem 14: *(Figs. 9 and 10) Construct a regular **polygon** of any number of sides given the length of its sides 35 mm.*

Example : Consider the construction of a *regular* ***pentagon.***

Method I – Arc Method

Fig. 9(a) shows the pentagon when one side AB is ***horizontal***, and *Fig. 9(b)* shows the pentagon when one side AB is ***vertical.***

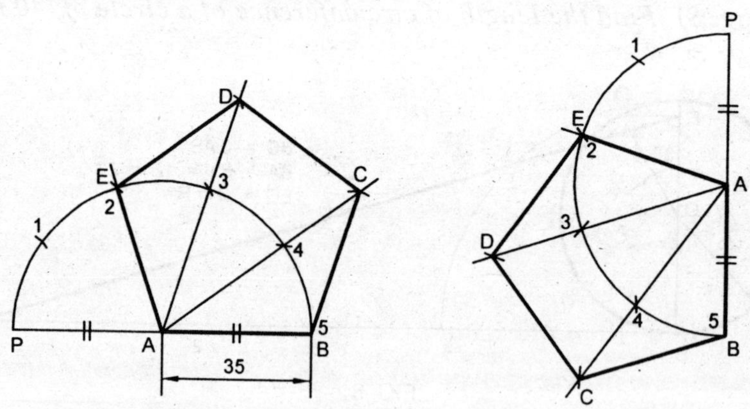

(a) AB - HORIZONTAL (b) AB - VERTICAL

FIG. 9 PENTAGON - METHOD I
ARC METHOD

1. Draw AB = 35 mm. Extend BA and mark P such that AP = AB. With A as centre and AB as radius draw a semi-circle as shown in *Figs. 9(a) and (b).*

2. Divide it into 5 equal parts (equal to the number of sides of the pentagon) and name the points as 1, 2, 3, 4 and 5 starting from P.
3. Join A2. Now A2 = AB = AE. Join A3 and A4 and extend them.
4. AB as radius and B as centre, draw an arc to cut the extension of A4 at C.
5. E as centre and same radius draw an arc to cut the extension of A3 at D.
6. Join BC, CD and DE. ABCDE is the required pentagon.

NOTE: *Always, the line joining the second division point (i.e., 2) and the point A gives the second side of the pentagon (polygon).*

Method II – Inscribe Circle Method

1. *See Fig. 10.* Draw AB = 35 mm. Extend AB and mark P such that AB = BP.
2. B as centre and BA as radius draw a semi-circle.
3. Divide it into 5 equal parts (equal to the number of sides of the pentagon) and name the points as 1, 2,....,5 starting from P. Join B2. Now AB = B2 = BC.
4. Draw the perpendicular bisectors of AB and BC to intersect at O.
5. O as centre and OA = OB = OC as radius draw a circle.

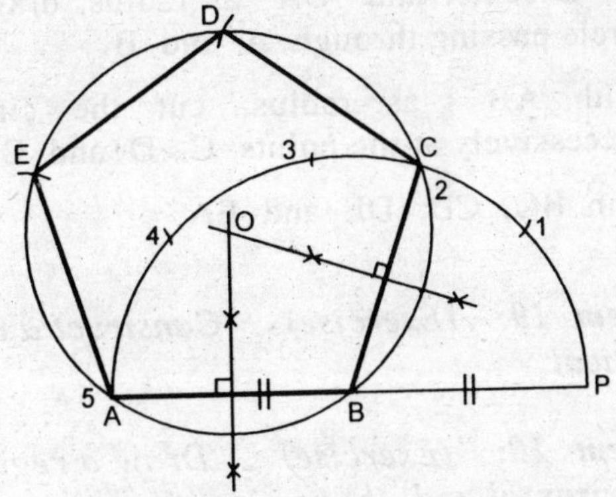

FIG. 10 PENTAGON - METHOD II
INSCRIBE CIRCLE METHOD

6. A and C as centres and AB as radius, draw arcs to cut the circle at E and D respectively.
7. Join CD, DE and EA. ABCDE is the required pentagon.

NOTE : 1. *In Method II, for any polygon, irrespective of the number of sides, always join B to the second division (i.e., 2).*

2. *Above two methods may be used for the construction of* ***any polygon.***

Problem 15: *(Exercise) Construct a regular hexagon of 34 mm side with one side (i) horizontal and (ii) vertical.*

Problem 16: *(Exercise) Construct a regular heptagon of 32 mm side.*

Problem 17: *(Exercise) Construct a regular octagon of 30 mm side.*

4.2 CONSTRUCTION OF REGULAR PENTAGON

Problem 18: *(Fig. 11) Construct a regular pentagon given the length of side 35 mm.*

1. Draw AB = 35 mm.
2. Draw an isosceles triangle OAB with AB as base and base angles of 54°. [For the pentagon, angle subtended at the center of the isosceles triangle = 360°/2 = 72°. Hence, the base angle = (180° – 72°) / 2 = 54°.]
3. O as center and OA as radius, draw a circle passing through A and B.
4. With AB as radius, cut the circle successively at the points C, D and E.
5. Join BC, CD, DE and EA.

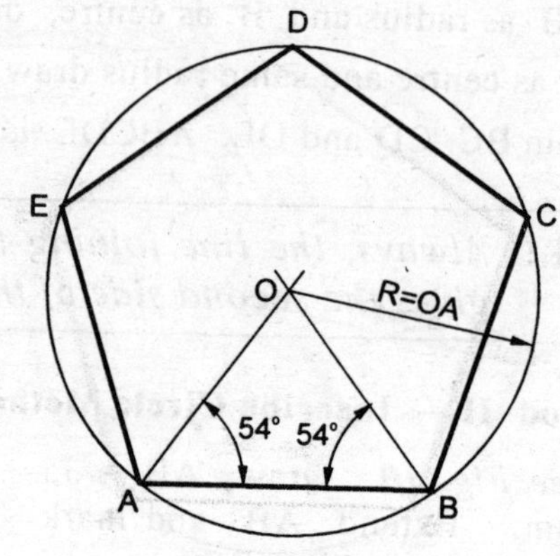

FIG. 11

Problem 19: *(Exercise) Construct a regular pentagon of side 32 mm when one side is vertical.*

Problem 20: *(Exercise) Draw a regular pentagon of side 36 mm when one side is (a) horizontal and (b) vertical.*

4.3 CONSTRUCTION OF REGULAR HEXAGON

Problem 21: *[Figs. 12(a) & (b)] Construct a regular hexagon of side 32 mm when one side is (a)* ***horizontal*** *and (b)* ***vertical.***

1. Let AB be the given side. With A and B as centers and AB as radius, draw two arcs to meet at O.
2. O as centre and AB as radius describe a circle.
3. With the same radius and A and B as centers, draw arcs to cut the circle at F and C respectively.
4. With the same radius and C and F as centers, draw arcs to cut the circle at D and E respectively.
5. Join BC, CD, DE, EF and FA. ABCDEF is the required hexagon.

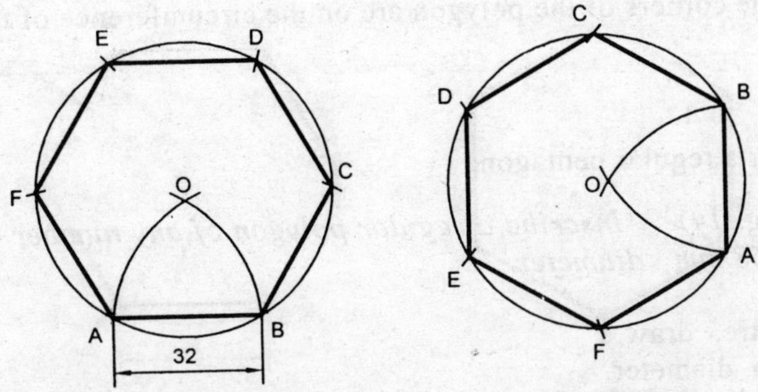

FIG. 12

Problem 22: *(Exercise) Construct a regular hexagon of side 34 mm when one side is (a) horizontal and (b) vertical.*

4.4 CONSTRUCTION OF REGULAR OCTAGON

Problem 23: *(Fig. 13) Draw an* ***octagon*** *given the length of side 25 mm.*

1. Draw AB = 25 mm. Draw perpendiculars through A and B.
2. **Bisect the external angle at A by the line AH as follows:**
   A as centre and any length as radius, draw an arc cutting A1 at 1 and A2 at 2.
   1 and 2 as centres and the same or any length as radius, draw arcs to cut each other at 3. Join A3 and extend it to H.
3. Bisect the external angle at B by the line BC such that AH = BC = AB.
4. Draw HG & CD = AB and parallel to the perpendiculars drawn through A and B.
5. G and D as centres and AB as radius cut the perpendiculars drawn through A and B at F and E respectively.
6. Join DE, EF and FG. ABCDEFGH is the required octagon.

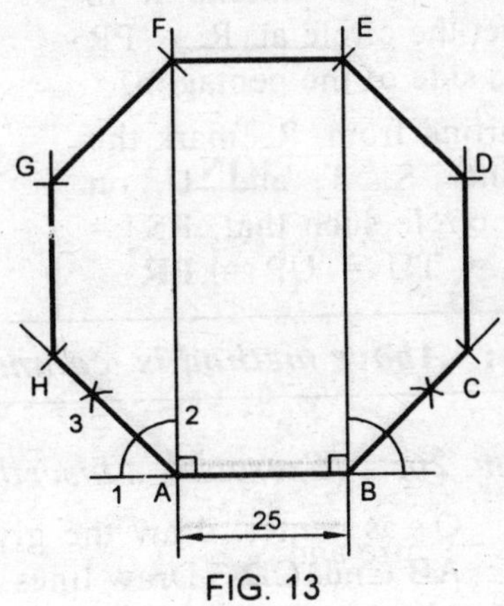

FIG. 13

Problem 24: *(Exercise) Construct an octagon of side of length 32 mm.*

5. INSCRIPTION OF POLYGONS

Inscribing a regular polygon *in* a given circle is the construction of the largest polygon inside that circle so that the corners of the polygon are on the circumference of the circle.

5.1 METHOD I

Example: Consider a regular pentagon.

Problem 25: *(Fig. 14)* *Inscribe a regular polygon of any number of sides, say 5, in a given circle of 68 mm diameter.*

1. With O as centre, draw a circle of 68 mm diameter.
2. Draw the horizontal diameter PQ and divide it into 5 equal parts (same number of parts as the number of sides) and number them as 1, 2, 3 and 4.
3. With P and Q as centers and PQ as radius, draw arcs cutting each other at M.
4. Join M2. Extend it to meet the circle at R. PR = a side of the pentagon.
5. Starting from R, mark the points S, T and U on the circle such that RS = ST = TU = UP = **PR**.

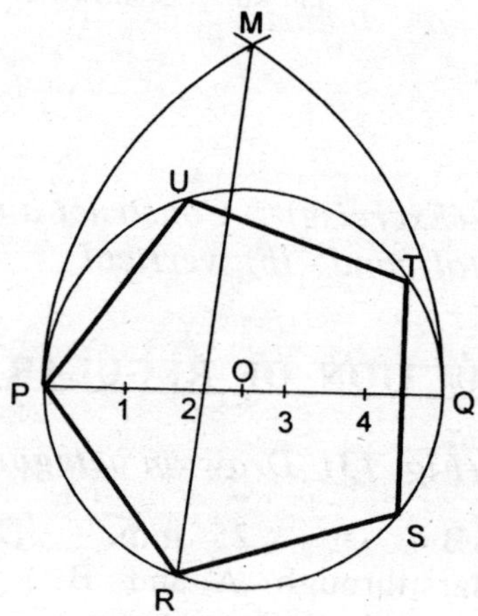

FIG. 14

NOTE : ***Above method is common for the inscription of all polygons.***

Problem 26: *(Exercise)* *Inscribe a square in a circle of 60 mm diameter.*

HINT : O as center, draw the given circle. Draw the vertical and horizontal diameters AB and CD. Draw lines AC, CB, BD and DA to obtain the square.

Problem 27: *(Exercise)* *Inscribe a regular hexagon in a circle of 72 mm diameter.*

Problem 28: *(Exercise)* *Inscribe a regular heptagon in a circle of 76 mm diameter.*

Problem 29: *(Exercise)* *Inscribe a regular octagon in a circle of 78 mm diameter.*

5.2 METHOD II

Problem 30: *(Exercise) Inscribe a regular polygon of sides, say 5, in a circle of 64 mm diameter.*

1. Draw the given circle of 64 mm diameter.
2. Draw radial lines from the center such that the angle between any two of them is equal to 360° / 5 = 72°. (Number of sides is 5)
3. Mark the points of intersection of the radii with the circumference of the circle.
4. Join the above points in proper sequence. The required pentagon is obtained.

NOTE : ***The above two methods are common for all polygons.***

Problem 31: *(Exercise) Inscribe a (a) hexagon, (b) heptagon and (c) octagon in circle of 68 mm diameter.*

6. INSCRIPTION OF PENTAGON, HEXAGON, HEPTAGON AND OCTAGON

6.1 INSCRIPTION OF PENTAGON

Problem 32: *(Fig. 15) Inscribe a regular pentagon in a given circle of 70 mm diameter.*

1. O as center, draw a circle of 70 mm diameter.
2. Draw diameters DP and TS perpendicular to each other.
3. Bisect TO at the point M.
4. M as center and MD as radius, draw an arc to cut the diameter TS at N.
5. D as center and DN as radius, draw an arc to cut the circle at the points C and E.
6. C and E as centers and DN as radius, draw arcs to cut the circle at the points A and B respectively.
7. Join AB, BC, CD, DE and EA, thus completing the required pentagon.

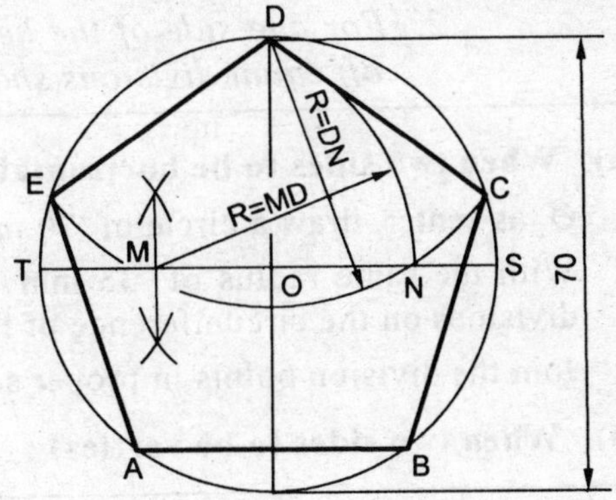

FIG. 15

Problem 33: *(Exercise) Inscribe a regular pentagon inside the given circle of diameter 80 mm.*

6.2 INSCRIPTION OF HEXAGON

Case I : Given the Diameter of the Circle

Problem 34: *(Fig. 16) Inscribe a regular hexagon in a given circle of 70 mm diameter, when two sides of the hexagon are (a) horizontal and (b) vertical.*

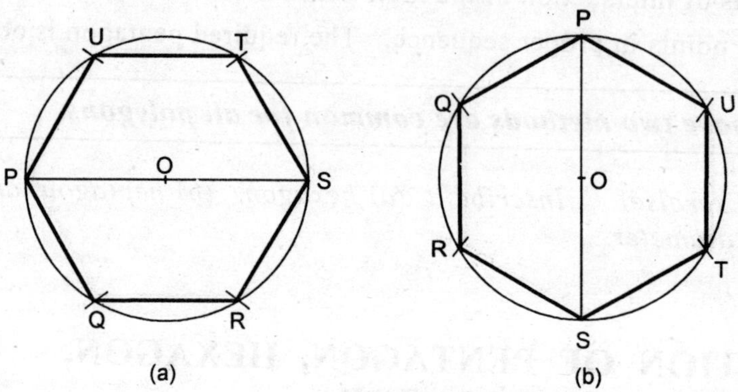

FIG. 16

NOTE : 1. *For one side of the hexagon to be **horizontal**, the starting point for marking equal divisions should be on an end of the **horizontal** diameter.*
2. *For one side of the hexagon to be **vertical**, the starting point for stepping-off equal divisions should be on an end of the **vertical** diameter.*

(a) When two sides to be horizontal : [Fig. 16 (a)]

1. O as center, draw a circle of 70 mm diameter. Mark the horizontal diameter PS.
2. With the same radius of 35 mm and starting from point P on the circle, set-off six divisions on the circumference of the circle using the divider.
3. Join the division points in proper sequence to obtain the hexagon P Q R S T U.

(b) When two sides to be vertical : [Fig. 16 (b)]

NOTE : *Above method may also be used for the construction of hexagon, given length of a side. In this case, draw the circle of radius = the given length of side.*

Problem 35: *(Exercise) Inscribe a regular hexagon in a circle of 34 mm radius, when two sides of the hexagon are required to be (i) horizontal and (ii) vertical.*

HINT: (i) First, draw the diameter of the given circle *horizontally*.
(ii) First, draw the diameter of the given circle *vertically*.

Case II : Given the Distance Across Corners of the Hexagon

Problem 36: *(Fig. 17) Construct a regular hexagon given the distance across corners is 60 mm.*

1. O as center, draw a circle of diameter 60 mm equal to the given distance across corners of the hexagon.
2. Draw the horizontal diameter AD.
3. With centers A and D and radius equal to 30 mm (= radius of the circle), draw arcs to cut the circumference of the circle at the points B, F and C, E respectively.
4. Join AB, BC, CD, DE, EF and FA to obtain the required hexagon.

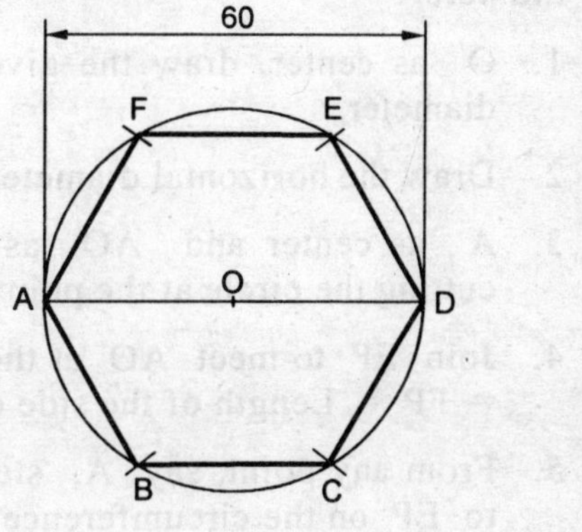

FIG. 17

Problem 37: *(Exercise) Construct a regular hexagon given the distance across corners as 64 mm.*

Case III : Given the Distance Across Flats of the Hexagon

Problem 38: *(Fig. 18) Construct a regular hexagon given the distance across flats as 60 mm.*

1. O as center, draw a circle of diameter equal to the given distance across flats of the hexagon 60 mm.
2. Through O, draw a vertical diameter VV.
3. Draw the lines XX and YY at an angle of 30° to the horizontal diameter.
4. Mark the points of intersection 1 & 4, 6 & 3, 5 & 2 of the circle and the lines VV, XX and YY respectively.
5. Draw tangents to the circle at the points 1, 2, 3, 4, etc., and obtain the required hexagon as shown.

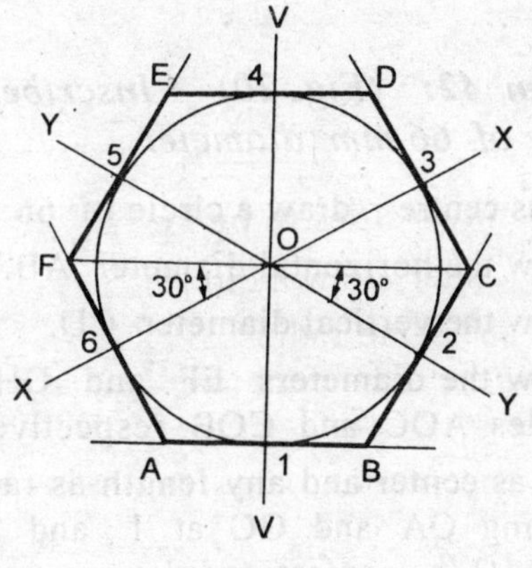

FIG. 18

NOTE : In this case, the hexagon is ***described*** about a circle whose diameter is equal to the given distance across flats of the hexagon.

Problem 39: *(Exercise) Construct a regular hexagon given the distance across flats as 64 mm.*

6.3 INSCRIPTION OF HEPTAGON

Problem 40: *(Fig. 19) Inscribe a regular heptagon inside the given circle of 70 mm diameter.*

1. O as center, draw the given circle of 70 mm diameter.
2. Draw the horizontal diameter AB.
3. A as center and AO as radius, draw an arc cutting the circle at the points E and F.
4. Join EF to meet AO at the point P. Now EP = FP = Length of the side of the heptagon.
5. From any point, say A, step-off divisions equal to EP on the circumference of the circle.
6. Join the division points A1, 12, 23, 34, 45, 56 and 6A. Thus obtain the heptagon.

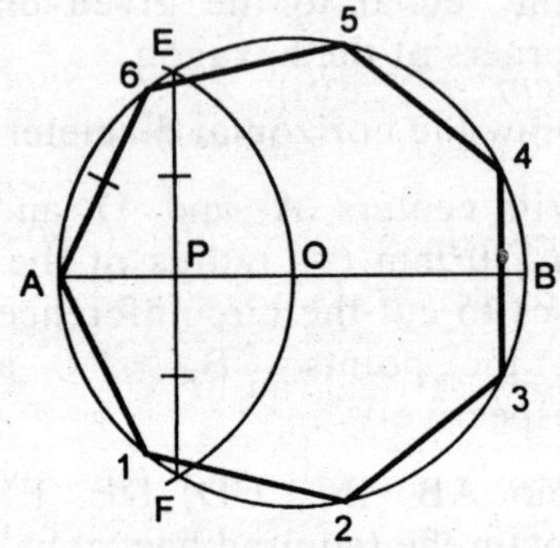

FIG. 19

Problem 41: *(Exercise) Inscribe a regular heptagon inside the given circle of diameter 74 mm.*

6.4 INSCRIPTION OF OCTAGON

Problem 42: *(Fig. 20) Inscribe an octagon in a circle of 66 mm diameter.*

1. O as centre , draw a circle of 66 mm diameter.
2. Draw the horizontal diameter AB.
3. Draw the vertical diameter CD.
4. Draw the diameters EF and GH bisecting the angles AOC and COB respectively as shown.

   [O as center and any length as radius, draw arcs cutting OA and OC at 1 and 2 respectively. 1 and 2 as centers and the same or any length as radius, draw arcs to cut at 3. O3 bisects the angle AOC.]
5. Draw AE, EC, CH, HB, BF, FD, DG and GA.
6. Thus complete the regular octagon as shown.

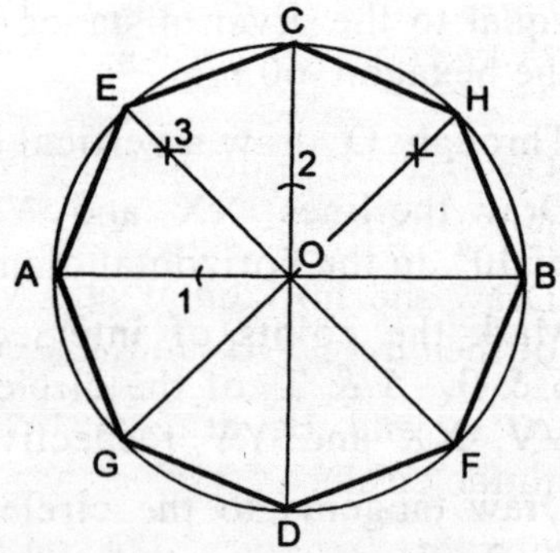

FIG. 20

Problem 43: *(Exercise) Inscribe a regular octagon in a circle of 68 mm diameter.*

7. SUPERSCRIPTION / DESCRIPTION / CIRCUMSCRIPTION OF POLYGONS

Superscribing / Describing / Circumscribing a regular polygon about a circle is the construction of the polygon, keeping the sides tangential to the given circle.

Problem 44: *(Fig. 21) Superscribe / Describe / Circumscribe an equilateral triangle about a circle of 50 mm diameter.*

1. Draw the given circle of 50 mm diameter with center O.
2. Draw vertical radius OA.
3. Draw radii OB and OC such that angles AOB and AOC are each equal to 120°.
4. At A, B and C, draw tangents to the circle to obtain the required equilateral triangle PQR.

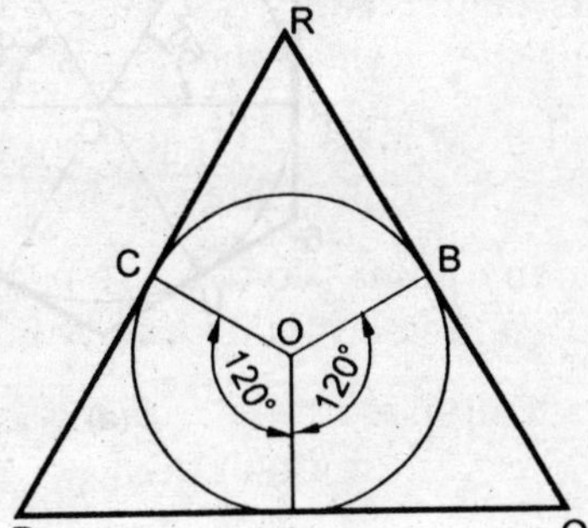

FIG. 21

Problem 45: *(Exercise) Superscribe / Describe / Circumscribe an equilateral triangle about a circle of 44 mm diameter.*

Problem 46: *(Fig. 22) Superscribe / Describe / Circumscribe a square about a given circle of diameter 58 mm.*

1. O as center, draw the given circle of 58 mm diameter.
2. Draw the horizontal and vertical diameters of the circle AB and CD respectively.
3. At A and B, draw vertical lines tangential to the circle.
4. Similarly, at C and D draw horizontal lines tangential to the circle.
5. The intersecting points obtained are P, Q, R and S. PQRS is the required square.

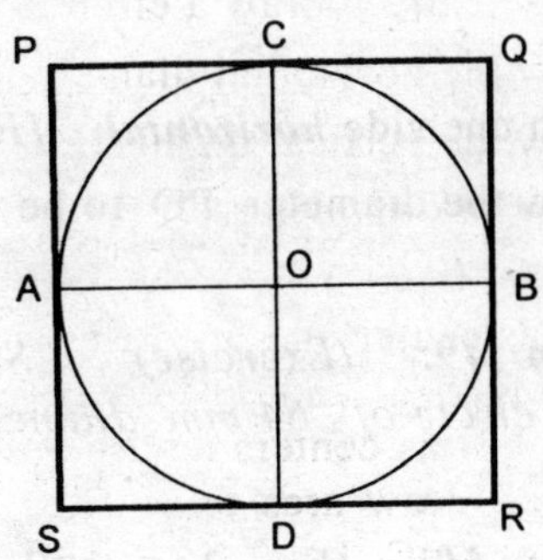

FIG. 22

Problem 47: *(Exercise) Superscribe / Describe / Circumscribe a square about a given circle of diameter 56 mm.*

Problem 48: *(Fig. 23) Superscribe / Describe / Circumscribe a hexagon about a circle of 60 mm diameter,* with one side (a). *vertical and (b) horizontal.*

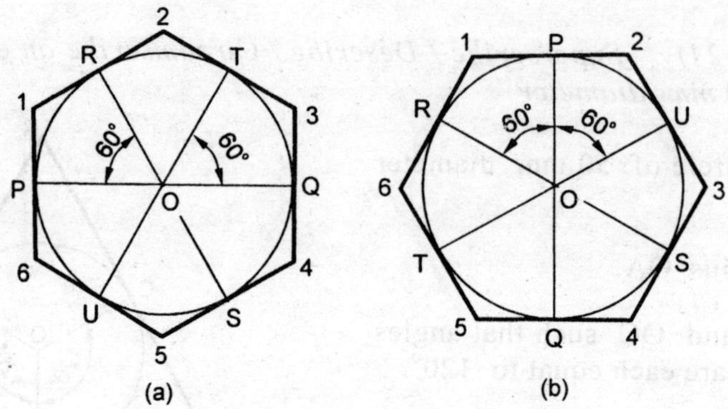

FIG. 23

a) With one side ***vertical***: *[Fig. 23(a)]*

1. Draw the given circle with centre O.
2. Draw the *horizontal* diameter PQ and diameters RS and TU making 60° (= 360°/6) with PQ at the centre O as shown.
3. Draw tangents to the circle at all the 6 points P, Q, R, S, T and U intersecting at 1, 2, ..., 6.
4. 123456 is the required hexagon.

b) With one side ***horizontal***: *[Fig. 23(b)]*

Draw the diameter PQ to be vertical and complete the construction as above.

Problem 49: *(Exercise) Superscribe / Describe / Circumscribe a regular hexagon about a circle of 64 mm diameter.*

Problem 50: *(Fig. 24) Superscribe / Describe / Circumscribe a regular pentagon about a circle of diameter equal to 70 mm.*

1. Draw a circle of 70 mm diameter.
2. Divide the circumference of the circle into a number of equal parts equal to the sides of the required polygon, say 5.

3. Name the division points as 1, 2, 3, etc.
4. Join O1, O2, etc., and extend them.
5. Through the points 1, 2, etc., draw perpendiculars to O1, O2, etc., intersecting at the points Q, R, S, T and P respectively. PQRST is the required pentagon.

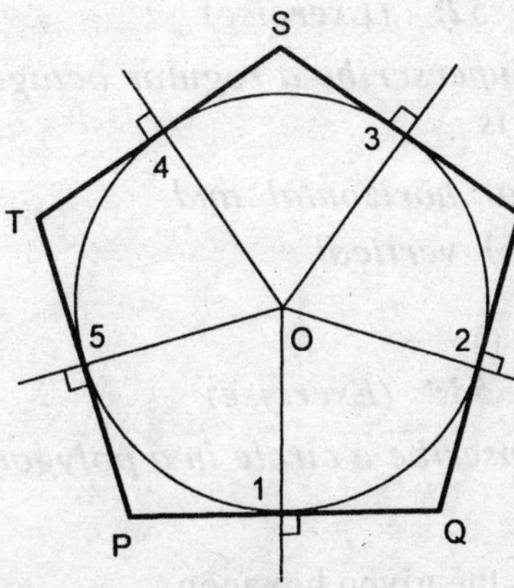

FIG. 24

NOTE : *Above method is common for the construction of all polygons.*

Problem 51: *(Fig. 25) Superscribe a regular octagon about a given circle of diameter equal to 70 mm.*

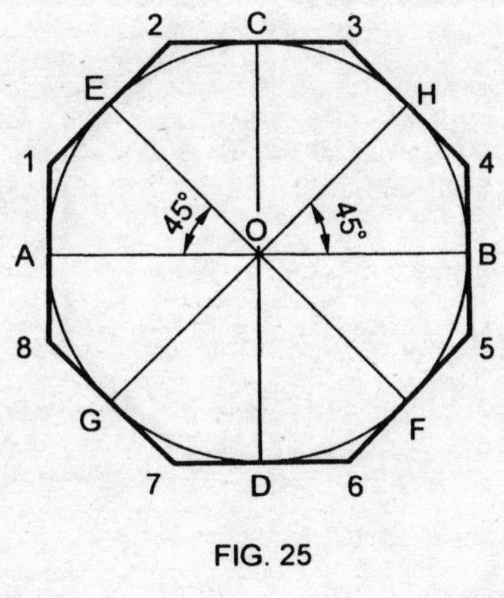

FIG. 25

1. Draw the given circle with centre O and diameter equal to 70 mm.
2. Draw the horizontal diameter AB and the vertical diameter CD.
3. Draw the diameters EF and GH inclined at an angle of 45° to AB or CD.
4. Draw 8 tangents to the circle at the ends of the diameters A, B, C, ..., H to intersect one another at the points 1, 2, 3, ..., 8.
5. Complete the required octagon 1 2 3 ... 8.

Problem 52: *(Exercise)*

Superscribe a regular octagon about a circle of diameter equal to 72 mm, when one side is :

(a) horizontal and

(b) vertical.

Problem 53: *(Exercise)*

Inscribe a circle in a polygon of any number of sides, say 6, of length 28 mm.

HINT

1. Draw the given hexagon.
2. Bisect any two angles by drawing lines intersecting each other at the point O.
3. From O, draw a perpendicular to any one side of the hexagon cutting it at the point R.
4. O as center and OR as radius, draw the required circle.

Problem 54: *(Exercise)*

Inscribe a circle in a polygon of any number of sides, say 6, of length 28 mm.

*

6

CONIC SECTIONS

1. ENGINEERING CURVES

In engineering practice, we come across various types of non-circular curves. These non-circular curved shapes are used in finding the path of space vehicles (aircrafts, missiles, etc.), machine components (elliptical gears, etc.), civil structures (bridges, arches, etc.), electrical equipment (reflectors, etc.), electronic transmitters and receivers (radar antenna, etc.), solar concentrators, etc. These non-circular curves cannot be drawn by compass alone.

Sufficient number of points are first located and then a smooth curve passing through them are drawn freehand or by using French Curve.

Examples : **Conic Sections, Cycloids, Involutes, Spirals, Helices, etc.**

2. CONIC SECTIONS

See Fig. 1. A *cone* is a surface generated by the rotation of a straight line whose one end is in contact with a fixed point while the other end is in contact with a closed curve, not lying in the plane of the curve. A *right circular cone* is a cone having its axis perpendicular to its base.

Apex or Vertex is the top point of the cone.

Axis is imaginary line joining apex & centre of base.

Generator is the straight line which is generating the surface of the cone.

Base of the cone is the closed curve.

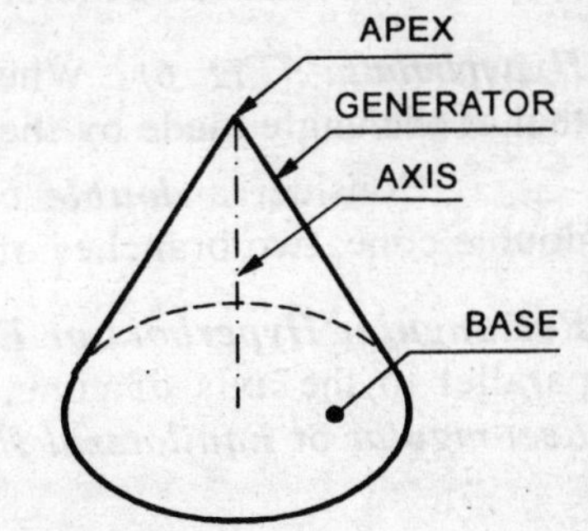

FIG. 1

2.1 CONIC SECTIONS : Definition I

A right circular cone can be cut by section (cutting) planes in different positions relative to the axis of the cone to obtain cut sections of different shapes, namely, Circle, Ellipse, Parabola, Hyperbola and Rectangular Hyperbola. These cut sections seen on the cone are called Conic Sections or simply Conics.

In *Fig. 2* the cutting plane is shown in different positions, *viz.*, AA, BB, CC, DD and EE.

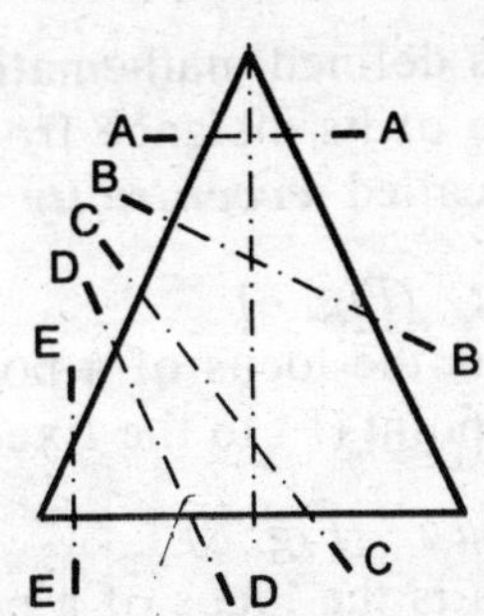

FIG. 2

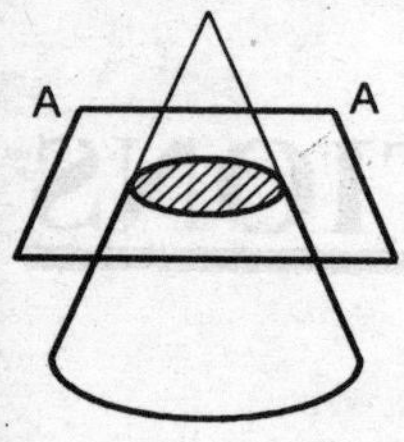

FIG. 3

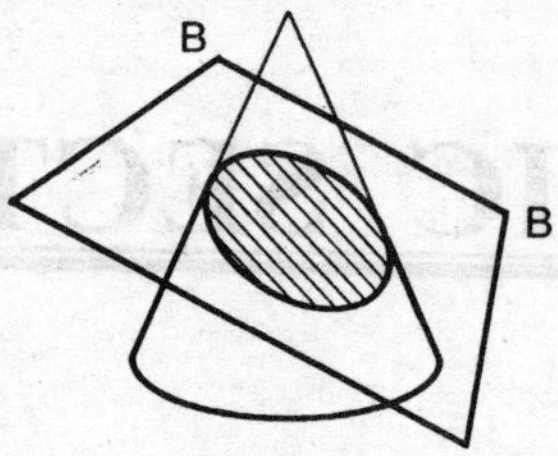

FIG. 4

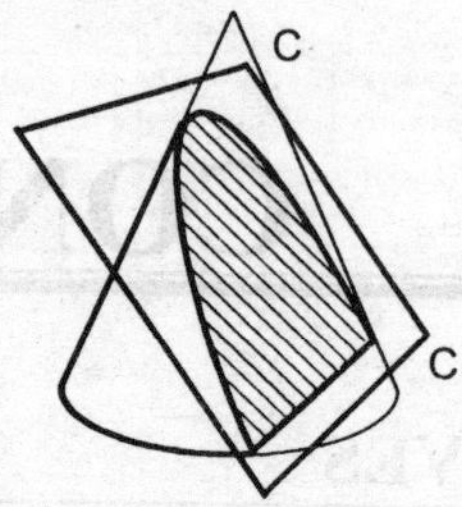

FIG. 5

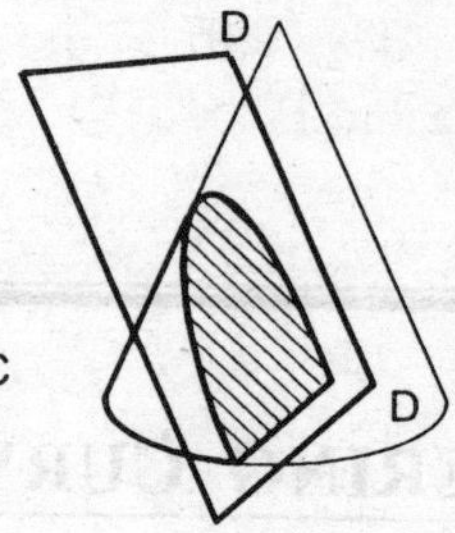

FIG. 6

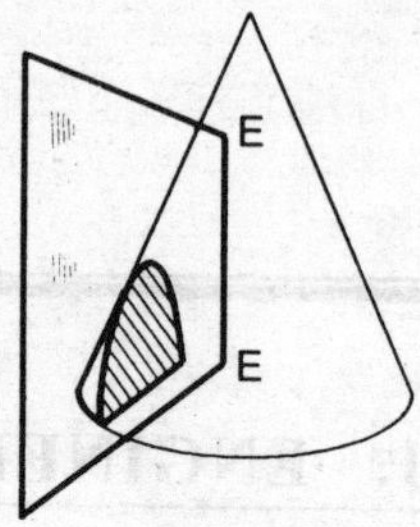

FIG. 7

Circle : *(Fig. 3)* When the cutting plane AA is perpendicular to the axis and cuts all the generators, the section obtained is a *Circle*.

Ellipse : *(Fig. 4)* When the cutting plane BB is inclined to the axis of the cone and cuts all the generators on one side of the apex, the section obtained is an *Ellipse*. Only when the cutting plane cuts all the generators of the cone, the elliptical section formed is a closed one.

Parabola : *(Fig. 5)* When the cutting plane CC is inclined to the axis of the cone and parallel to one of the generators, the section obtained is a *Parabola*.

Hyperbola : *(Fig. 6)* When the cutting plane DD makes a smaller angle with the axis than that of the angle made by the generator of the cone, the section obtained is a *Hyperbola*.

Consider a ***double cone***. If the cutting plane lies on one side of the axis of the double cone, two branches of hyperbola are obtained.

Rectangular Hyperbola or Equilateral Hyperbola : *(Fig. 7)* When the cutting plane EE is parallel to the axis of cone, but does not pass through the apex, the section obtained is a *Rectangular* or *Equilateral Hyperbola*.

2.2 CONIC SECTIONS DEFINED AS LOCI : Definition II

Conic is defined mathematically as the locus of a point moving in a plane in such a way that the ratio of its distance from a fixed point to a fixed straight line is always a constant. This ratio is called ***eccentricity*** (e).

Ellipse : *(Fig. 8)*
Ellipse is the locus of a point moving in a plane in such a way that the ratio of its distance from a point (F) to the fixed straight line (DD) is a constant. It is always < 1.

Parabola : *(Fig. 8)*
Parabola is the locus of a point moving in a plane in such a way that the ratio of its distance from a fixed point (F) to the fixed straight line (DD) is a constant and is always equal to 1.

Hyperbola : *(Fig. 8)* It is the locus of a point moving in a plane in such a way that the ratio of its distance from a fixed point (F) to the fixed straight line (DD) is constant and is > 1.

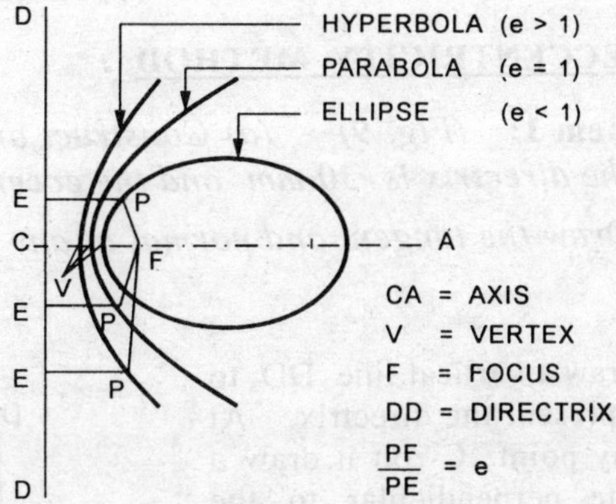

FIG. 8

Focus (F)
The fixed point is called the Focus.

Directrix (DD)
The fixed line is called the Directrix.

Axis (CA)
Axis is the line passing through focus and perpendicular to the directrix.

Vertex (V)
Vertex is a point at which the conic cuts its axis.

Eccentricity : $e = \dfrac{\text{Distance of the moving point from the focus}}{\text{Distance of the moving point from the directrix}}$

NOTE : If the eccentricity (e) is < 1, the curve obtained is called *Ellipse*.
If the eccentricity (e) is = 1, the curve obtained is called *Parabola*.
If the eccentricity (e) is > 1, the curve obtained is called *Hyperbola*.

Engineering Applications

Ellipse : Elliptical shape is used in the construction of arches, bridges, dams, elliptical gears of textile machines and printing presses, ends of cylindrical tanks, tops or bottoms of utensils, man-holes, glands, stuffing boxes, flanges of pipes, monuments, etc.

Parabola is used for suspension bridges, reflectors for parallel beams such as head lights of automobiles, solar concentrators, including in machine tool structures, etc. Path of a thrown object or missile and path of a jet of water issuing from vertical orifice are parabolic shape.

Hyperbola is used in the design of cooling towers, hydraulic channels, electronic transmitters and receivers like radar antenna, etc. ***Rectangular Hyperbola*** is used to represent the Boyle's Law expansion curve or the Theoretical Indicator Diagram of an Engine.

2.3 METHODS OF CONSTRUCTION

- Eccentricity Method .... (*Ellipse / Parabola / Hyperbola*)
- Intersecting Arcs or Arc of Circles or Foci Method / Concentric Circles Method / Parallelogram Method using Conjugate Diametres / Four Centres Method (*Ellipse*)
- Rectangle or Oblong Method .... (*Ellipse / Parabola*)
- Parallelogram Method / Tangent Method .... (*Parabola*)
- Foci & Transverse Axis Method / Ordinate, Abscissa & Transverse Axis Method (*Hyperbola*)
- Asymptotes and a Point on the Curve Method ....(*Rectangular Hyperbola*)

3. ELLIPSE

3.1 ECCENTRICITY METHOD

Problem 1: *(Fig. 9) (a) Construct an ellipse when the distance between the focus and the directrix is 30 mm and the eccentricity is 3/4.*

(b) Draw the tangent and normal at any point P on the curve using directrix.

(a)

1. Draw a vertical line DD to represent the directrix. At any point C on it draw a line perpendicular to the directrix to represent the axis CC′.
2. Distance between the focus and the directrix is 30 mm. Hence mark F_1, the focus such that $CF_1 = 30$ mm.
3. e = 3/4. So construct a right-angled Δ CXY such that

$$\frac{XY}{CX} = \frac{3 \text{ units}}{4 \text{ units}}$$ (X is any point on the axis).

4. From F_1 draw a 45° line to intersect line CY at S.
5. From S erect vertical to intersect CF_1 at V_1, the vertex. Now $SV_1 = F_1V_1$.
6. Δ CXY is similar to Δ CV_1 S.

$$\therefore \quad \frac{SV_1}{CV_1} = \frac{XY}{CX} = \frac{F_1V_1}{CV_1} = 3/4$$

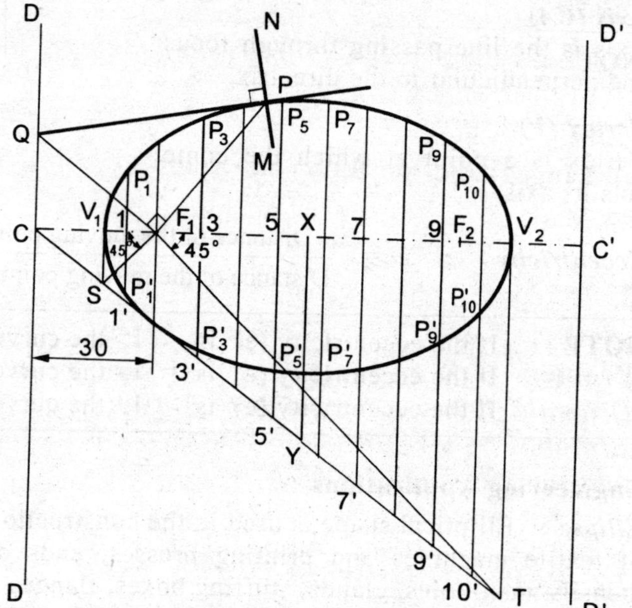

F_1 & F_2 - FOCI

V_1 & V_2 - VERTICES

DD & D'D' - DIRECTRICES

FIG. 9 ELLIPSE - ECCENTRICITY METHOD

7. From F_1 draw another 45° line to intersect the extension of CY at T. From T erect vertical to intersect the axis at V_2, another vertex. V_1V_2 = Major Axis.
8. Along the major axis, mark points 1, 2, ..., 10 at approximately equal intervals. Through these points erect verticals to intersect CY (produced if necessary) at 1′, 2′, ..., 10′.
9. With 11′ as radius and F_1 as centre draw two arcs on either side of the axis to intersect the vertical line drawn through 1 at P_1 and P'_1.

10. With 22′ as radius and F_1 as centre draw two arcs on either side of the axis to intersect the vertical line drawn through 2 at P_2 and P_2' .

11. Repeat the above and obtain P_3 and P_3',......, P_{10} and P_{10}' corresponding to 2, 3,..., 10 respectively and draw a *smooth ellipse* passing through V_1, P_1, ..., P_{10}, V_2, P_{10}' ,......, P_1', V_1.

12. *To mark another Focus* F_2 : Mark F_2 on the axis such that $V_2 F_2 = V_1 F_2$.

13. *To mark another Directrix* D′D′: Mark C′ along the axis such that $C' V_2 = CV_1$. Through C′ draw a vertical line D′D′.

NOTE : *First draw thin freehand curve followed by smooth thick curve using French Curve.*

---

b) **Tangent and Normal at any point P on the ellipse using Directrix :**

1. Mark the given point P and join PF_1. At F_1 draw a line perpendicular to PF_1 to cut DD at Q. Join QP and extend it. QP is the tangent at P.
2. Through P, draw a line NM perpendicular to QP. NM is the normal at P.

Problem 2: *(Exercise) A fixed point F is 7.5 cm from a fixed straight line. Draw the locus of a point P moving in such a way that its distance from the fixed straight line is 2/3 times its distance from F. Name the curve. Draw normal and tangent at a point 6 cm from F.*

3.2 RELATIONSHIP BETWEEN MAJOR AXIS, MINOR AXIS AND FOCI

Ellipse may also be defined as the locus of a point which moves such that the sum of its distances from two fixed points is a constant and is equal to the length of the major axis.

See Fig. 10. F_1 and F_2 are foci. The line AB which passes through F_1 and F_2 is the Major Axis. CD which bisects the major axis at 90° is called the Minor Axis. Now, $CF_1 = 1/2$ AB.

Using this, if major axis and minor axis are given, the two fixed points F_1 and F_2 can be located.

Note : $CF_1 = CF_2 = DF_1 = DF_2 = AB/2$

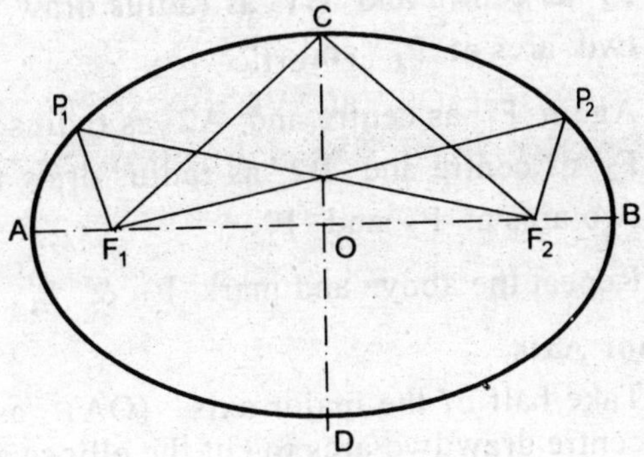

AB - MAJOR AXIS
CD - MINOR AXIS
F_1 & F_2 - FOCI

RESULT : $CF_1 = CF_2 = DF_1 = DF_2 = \frac{1}{2}$ AB

FIG. 10

3.3 INTERSECTING ARCS or ARCS OF CIRCLES or FOCI METHOD

Problem 3: *(Fig. 11) Construct an ellipse when the major axis is 120 mm and the distance between the foci is 108 mm. Determine the length of the minor axis. (UQ)*

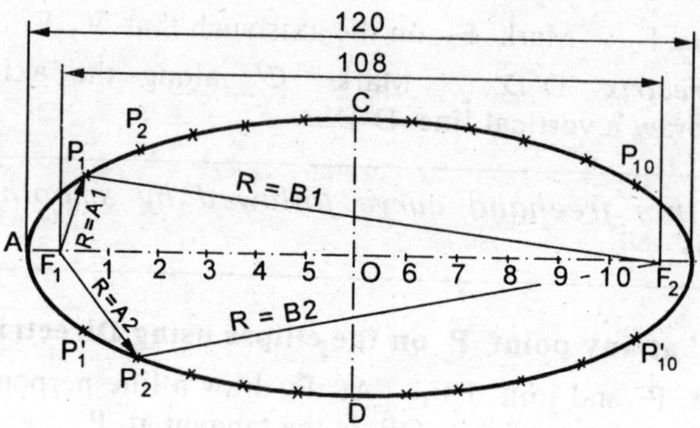

RESULT : MINOR AXIS CD = 50 mm

FIG. 11 INTERSECTING ARC METHOD

1. Draw the major axis AB = 120 mm and bisect it at O.
2. Mark the foci F_1 and F_2 on AB such that O F_1 = O F_2 = 108/2 = 54 mm.
3. Mark any number of points 1, 2, ..., (*say upto*) 10 at approximately equal intervals on AB in between F_1 and F_2.
4. F_1 as centre and A1 as radius draw two arcs on either side of AB.
5. F_2 as centre and B1 as radius draw two arcs on either side of AB to cut the previous two arcs at P_1 and P'_1.
6. Again F_1 as centre and A2 as radius draw two arcs on either side of AB.
7. F_2 as centre and B2 as radius draw two arcs on either side of AB to cut the previous two arcs at P_2 and P'_2.
8. Repeat the above and mark P_3 & P'_3, ..., P_{10} & P'_{10} and draw a *smooth ellipse*.

Minor Axis

9. Take half of the major axis (OA) as radius and any one of the foci (F_1 or F_2) as centre draw two arcs to cut the ellipse at C and D.
10. Join CD which is the **Minor Axis = 50 mm.**
11. **Check :** CD should pass through O. **[$CF_1 = CF_2$ = ½ AB]**

Problem 4: *(Exercise) The major axis of an ellipse is 100 mm and the minor axis is 55 mm. Find the foci and construct the ellipse by Intersecting Arcs Method.*

3.4 CONCENTRIC CIRCLES METHOD

Problem 5: *(Fig. 12)* *A flowerbed in a botanical garden is elliptical in shape. Major and minor axes are 9 m and 5.5 m respectively. Draw the profile of the flowerbed to a scale of 1:100.* *(UQ)*

1. Draw the major axis AB = 90 mm.
2. Bisect it at O.
3. Through O draw a vertical line CD = 55 mm. the minor axis.
4. O as centre and major and minor axes as diameters draw two concentric circles.
5. Divide the circles into 12 equal parts.
6. Draw the radial lines to intersect the major and minor axes circles at 1, 2,...,12 and 1′, 2′, ..., 12′ respectively.
7. From 1 draw a vertical line parallel to CD. From 1′ draw a horizontal line parallel to AB. Both intersect at P_1.

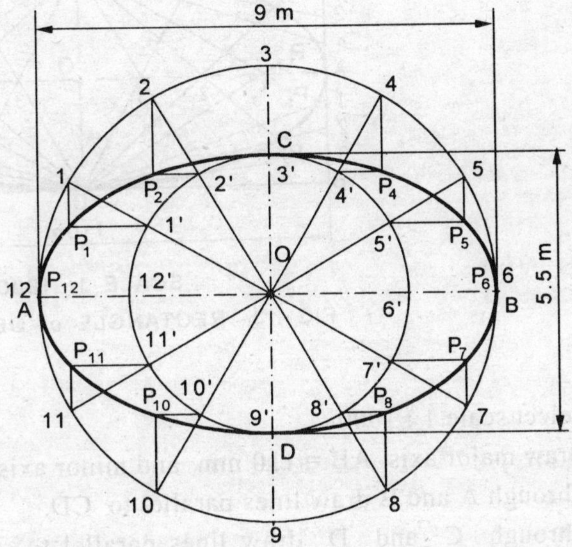

SCALE 1 : 100

FIG. 12 CONCENTRIC CIRCLES METHOD

8. Repeat the above and obtain the points P_2, ..., P_{12} corresponding to 2 and 2′, ... , 12 and 12′ respectively.
9. Draw a *smooth ellipse* through P_1, P_2, ..., P_{12}, P_1.

Problem 6: *(Exercise)* *Construct an ellipse when its major axis is equal to 100 mm and minor axis is equal to 65 mm. Find its foci, directrix and eccentricity.* (UQ)

HINT :
1. Draw the ellipse by Concentric Circles Method / Rectangle Method.
2. Mark F_1 and F_2.
3. At any point P on it, draw a tangent using F_1 and F_2.
4. At F_1 draw a perpendicular to PF_1 to cut the tangent at Q.
5. Draw a vertical line through Q to obtain the directrix.
6. Measure $e = V_1F_1 / V_1C = 2/3$.

3.5 RECTANGLE METHOD or OBLONG METHOD

Problem 7: *(Fig. 13) A plot of ground is in the shape of a rectangle 110 m x 50 m. Inscribe an elliptical lawn in it. Take a suitable scale.* (UQ)

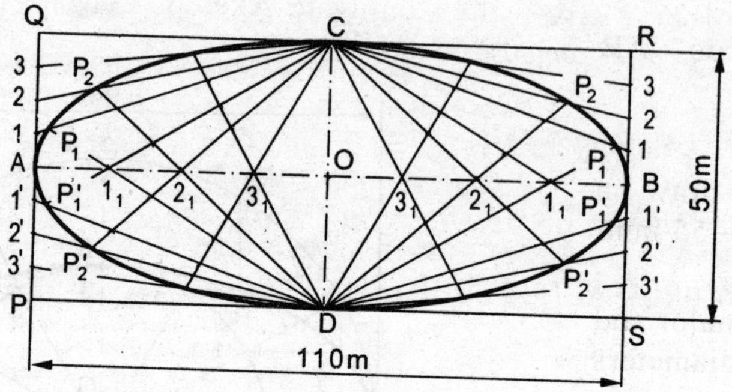

SCALE 1 : 1000

FIG. 13 RECTANGLE or OBLONG METHOD

1. Select scale 1 : 1000.
2. Draw major axis AB = 110 mm and minor axis CD = 50 mm. Both bisect at O.
3. Through A and B draw lines parallel to CD.
4. Through C and D draw lines parallel to AB and construct the rectangle PQRS. Now PS = AB and SR = CD.
5. Divide AQ and AP into any number of equal parts (*say 4*) and name the points as 1, 2, 3 and 1′, 2′, 3′ respectively starting from A on AQ and AP.
6. Divide AO into same (4) number of equal parts, and name the points as 1_1, 2_1, 3_1 starting from A on AO.
7. Join 1, 2, 3 with C.
8. Join $D1_1$ and extend it to intersect C1 at P_1.
9. Similarly extend $D2_1$ and $D3_1$ to intersect C2 and C3 at P_2 and P_3 respectively.
10. Join 1′, 2′, 3′ with D.
11. Join $C1_1$ and extend it to intersect D1′ at P_1'.
12. Similarly extend $C2_1$ & $C3_1$ to intersect D2′ & D3′ at P_2' & P_3' respectively.
13. Draw a smooth curve through C, P_3, P_2, P_1, A, P_1', P_2', P_3', D and obtain *one half* (*left-half*) of the ellipse.
14. Repeat the above and *draw the right-half of ellipse, symmetrical to the left-half.*

Problem 8: *(Exercise) The major and minor axes of an elliptical fish pond are 10 m and 6 m respectively. Draw one-half of the ellipse by Oblong Method and the other half by Concentric Circles Method.* (UQ)

3.6 PARALLELOGRAM METHOD

What are Conjugate Axes?

Conjugate Axes are two lines intersecting each other at the centre of an ellipse when they are parallel to the tangents drawn at their extremities. An ellipse may have an unlimited number of pairs of *Conjugate Axes.* Given the Conjugate Axes (Conjugate Diameters), the ellipse may be constructed by *Parallelogram Method.*

Problem 9: *(Fig. 14)*

a) *Construct an ellipse when a pair of conjugate axes are equal to 110 mm and 50 mm respectively. The angle between the conjugate axes is 70°.*

b) *Determine the lengths of the major and minor axes.* (UQ)

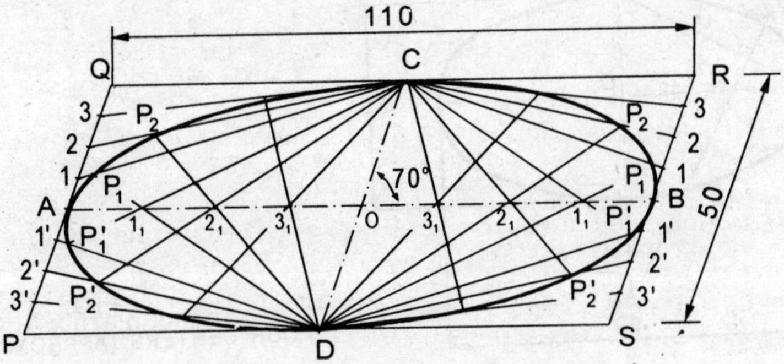

AB AND CD : CONJUGATE AXES

FIG.14 PARALLELOGRAM METHOD

a)

1. Draw a conjugate diameter AB = 110 mm and bisect it at O.
2. Angle between the conjugate diameters is 70°. So draw another conjugate diameter CD through O such that ∠ COB = 70°.
3. Through A and B draw lines parallel to CD.
4. Through C and D draw lines parallel to AB and construct a parallelogram PQRS.
5. Repeat the procedure given in *Step Nos. 5 to 14 in Problem 7 (Rectangle Method)* and complete the construction of the ellipse inside the parallelogram PQRS.

***b)* To find the Major and Minor axes of the above ellipse drawn with Conjugate Axes**

1. O as centre and a distance more than OC (or OD) as radius, draw a circle to cut the ellipse at 1, 2, 3 and 4. Join 1-2 and 2-3.
2. Through O draw $A_1 B_1$ and C_1D_1 parallel to 1-2 and 2-3 respectively to represent the major and minor axes respectively.
3. Now $A_1 B_1$ and C_1D_1 intersect at right angles at O. Measure the major axis = **$A_1 B_1$ = 112 mm** (> 110 mm) and minor axis = **C_1D_1 = 47 mm** (< 50 mm).

Problem 10: *(Exercise)* *A plot of land in a Five Star Hotel is in the shape of a parallelogram of 20 m x 12 m. Angle between the sides is 55o. Inscribe an elliptical swimming pool in it.*

(UQ)

3.7 FOUR CENTERS (approximate) METHOD (GIVEN MAJOR AND MINOR AXES)

Problem 11: *(Fig. 15) Construct an ellipse when the major and minor axes are 90 mm and 50 mm respectively.*

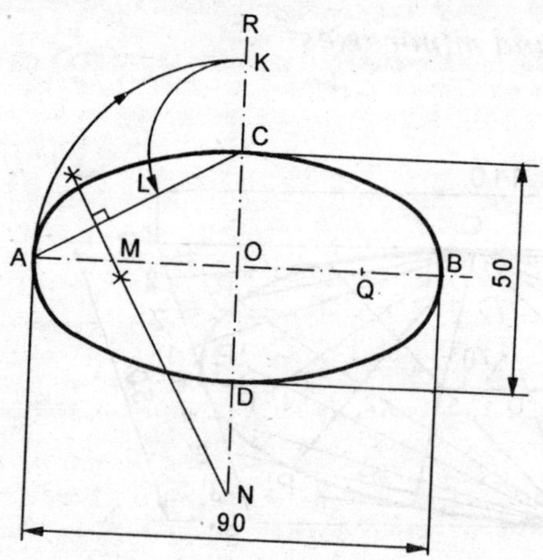

FIG. 15
FOUR CENTERS (APPROXIMATE) METHOD

1. Draw major axis AB = 90 mm and minor axis CD = 50 mm.
2. Join AC.
3. With O as centre and OA as radius draw an arc to intersect the extension of CD at K.
4. With C as centre and CK as radius draw an arc to cut AC at L.
5. Draw a perpendicular bisector to AL.
6. Extend it to cut AB at M and the extension of CD at N.
7. Locate Q along AB such that QB = AM.
8. Locate R along the extension of CD such that CR = DN. Now M, N, Q and R are the *four centres* to construct the ellipse.
9. With N and R as centres and NC as radius draw two large arcs.
10. M and Q as centres and AM as radius draw two arcs. Obtain the *ellipse*.

Problem 12: *(Exercise)* *Construct an ellipse when the major and minor axes are 100 mm and 55 mm respectively.*

4. PARABOLA

4.1 ECCENTRICITY METHOD

Problem 13: *(Fig. 16) Construct a parabola when the distance between focus and the directrix is 40 mm. Draw tangent and normal at any point P on your curve.*

1. Draw the directrix DD. At any point C on it draw a line CA ⊥r to DD to represent the axis.
2. Distance between the focus and the directrix is 40 mm. So mark F the focus, such that CF = 40 mm.
3. For parabola, **e** is 1; so construct the right-angled Δ CXY such that

$$\frac{XY}{CX} = 1 \quad (\text{X is any point on axis})$$

4. From F draw a 45° line to intersect CY at S.
5. From S erect vertical to cut CF at vertex V. Now SV = FV.
6. From similar Δs CXY and CVS,

$$\frac{SV}{CV} = \frac{XY}{CX} = \frac{FV}{CV} = 1.$$

7. Along the axis CA mark the points 1, 2,.., 5, etc., at approximately equal intervals.
8. Through 1, 2,.., 5 erect verticals to intersect the line CY (produced if necessary) at 1′, 2′, ..., 5′, etc., respectively.

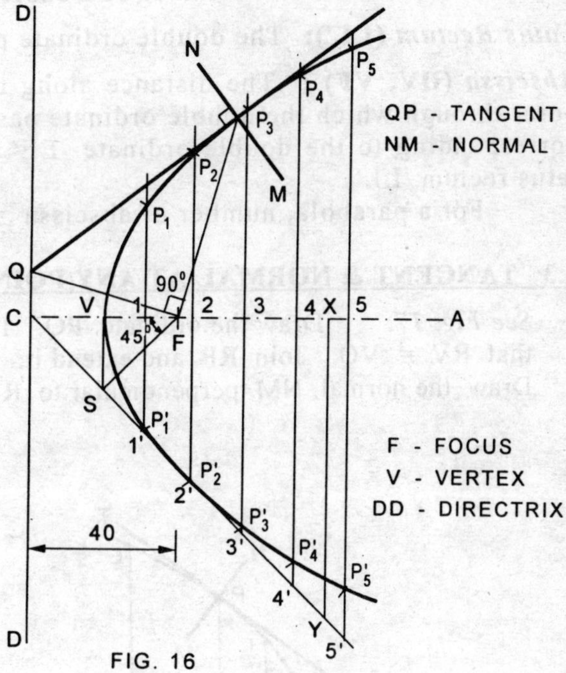

FIG. 16
PARABOLA - ECCENTRICITY METHOD

9. With 11′ as radius and F as centre, draw two arcs on either side of the axis to intersect the vertical line drawn through 1 at P_1 and P_1'.
10. Repeat the above and obtain P_2 & P_2', P_3 & P_3', ..., P_5 & P_5' corresponding to points 2, 3, 4, 5 respectively. Draw a smooth *parabola* through P_5, ..., P_1, V, P_1', etc.

NOTE : *Refer* ***Problem 1*** *to draw the tangent and normal at any point on the parabola.*

Problem 14: *(Exercise) A fixed point F is 7.5 cm from a fixed straight line. Draw the locus of a point P moving in such a way that its distance from the fixed straight line is equal to its distance from F. Name the curve. Draw normal and tangent at a point 6 cm from F.*

4.2 PARABOLA – TERMINOLOGY

See Fig. 17. Parabola is an open unlimited curve. CA is the Axis. V is the Vertex.

Ordinate **(ED)** : Any line ED drawn from E on the parabola such that it is perpendicular to the axis CA is called *Ordinate.*

Double Ordinate **(EE′)** : The ordinate ED is extended to cut the parabola at E′. EDE′ is *Double Ordinate,* which is twice the length of the ordinate, *i.e.*, EE′ = 2ED. A parabola has unlimited number of Double Ordinates.

Latus Rectum **(LL′):** The double ordinate passing through the focus is *Latus Rectum.*

Abscissa **(DV, VF)** : The distance along the axis (CA) between vertex (V) and a point through which the double ordinate passes is called *Abscissa.* DV is the abscissa corresponding to the double ordinate EE′ and VF is the abscissa corresponding to latus rectum LL′.

For a parabola, number of abscissa = number of double ordinates.

4.3 TANGENT & NORMAL AT ANY POINT P WHEN F & DD ARE NOT KNOWN

1. *See Fig. 17.* Draw the ordinate PQ. Find the abscissa VQ. Mark R on CA such that RV = VQ. Join RP and extend it. TR is the tangent.
2. Draw the normal NM perpendicular to RP at P.

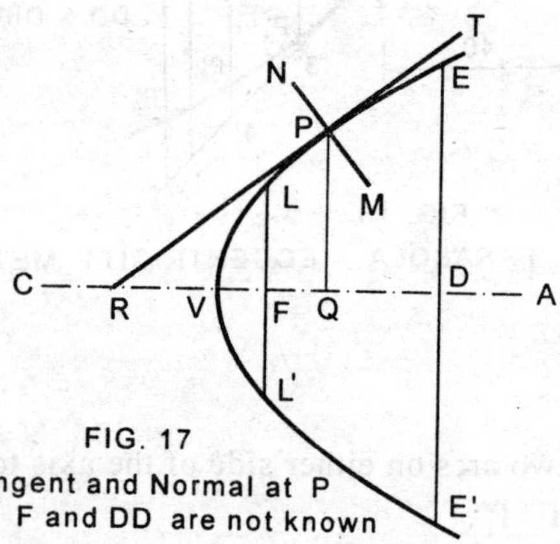

FIG. 17
Tangent and Normal at P
when F and DD are not known

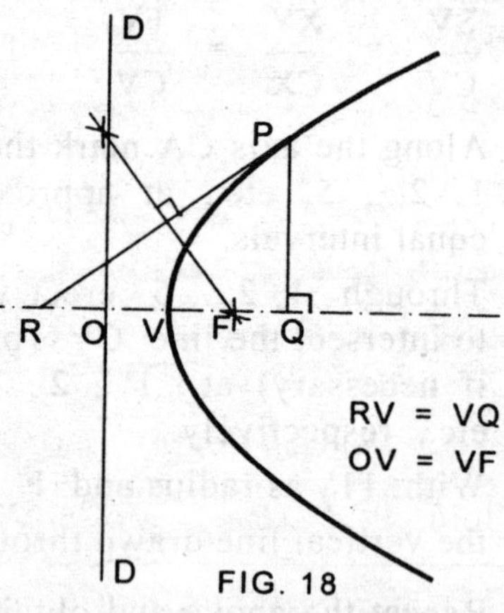

FIG. 18

4.4 TO FIND THE FOCUS & THE DIRECTRIX OF A PARABOLA GIVEN ITS AXIS

1. *See Fig. 18.* Mark any point P on the parabola. Draw a perpendicular PQ to the given axis as shown. Mark a point R on the axis such that RV = VQ.
2. *Focus :* Join RP. Draw a $\perp^r$ bisector of RP cutting the axis at F. F is the Focus.
3. *Directrix :* Mark O on the axis such that OV = VF. Through O draw the *Directrix* DD perpendicular to the axis.

4.5 RECTANGLE METHOD GIVEN THE BASE (DOUBLE ORDINATE) AND AXIS

Problem 15: *(Fig. 19) A cricket ball thrown from the ground level reaches the Wicket Keeper's gloves. Maximum height reached by the ball is 5 m. The ball travels a horizontal distance of 11 m from the point of projection. Trace the path of the ball.* (UQ)

1. The ball travels a horizontal distance of 11 m. Take a scale of 1: 100.
2. Draw PS = 11 cm to represent the double ordinate. Bisect PS at O.
3. The ball reaches a maximum height of 5 m. Hence from O erect vertical and mark the vertex V such that OV = 5 cm. Now OV is the abscissa.
4. Draw rectangle PQRS such that PS is double ordinate and PQ = RS = VO (abscissa).
5. Divide PQ and RS into 8 equal parts as 1, 2, ..., 8 and 1′, 2′, ..., 8′ respectively, starting from P on PQ and S on SR.

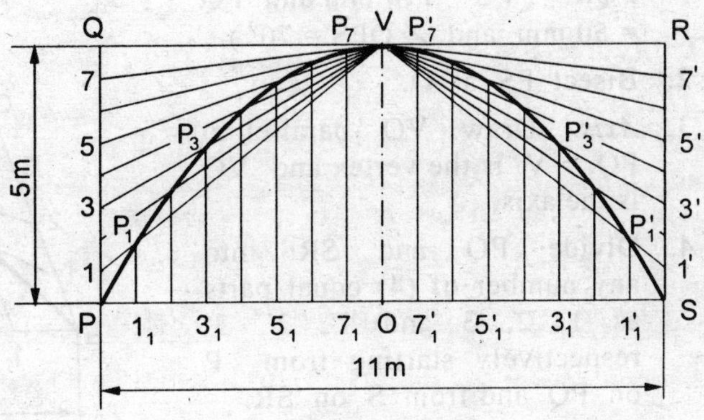

FIG. 19 RECTANGLE METHOD

6. Join 1, 2, ..., 8 and 1′, 2′, ..., 8′ with V.
7. Divide PO and OS into 8 equal parts as 1_1, 2_1,..., 8_1 and $1'_1$, $2'_1$, ..., $8'_1$ respectively, starting from P on PO and from S on SO.
8. From 1_1 erect vertical to meet the line V1 at P_1.
9. Similarly from 2_1, ..., 8_1 erect verticals to meet the lines V2,..., V8 at P_2 ,..., P_8 respectively.
10. Also erect verticals from $1'_1$, $2'_1$, ..., $8'_1$ to meet the lines V1′ , V2′ ,..., V8′ at P'_1, P'_2, ..., P'_8 respectively.
11. Join P, P_1, P_2, ..., P_7, V, P'_7 , ..., P'_1 and S to represent the path of the ball *(parabola)*.

NOTE : *Path of a projectile when thrown / falling due to gravitational force is Parabola.*

Problem 16: *(Exercise) A rocket shot up in the air, reaches a maximum height of 800 m and falls on the ground at a distance of 1800 m from the point of projection. Trace the path of the rocket in space. What is the angle of the projectile?* (UQ)

HINT : Draw the parabola by *Rectangle Method*. Angle of the projectile is obtained from the direction of motion of the rocket with reference to the horizontal. Measure the angle of projectile by drawing a tangent at the point of projection. It is equal to **58°**.

4.6 PARALLELOGRAM METHOD

Problem 17: *(Fig. 20) Construct a parabolic arch within a parallelogram of sides 110 mm x 50 mm. One of the included angle between the sides is 70°. Locate the axis, focus and direcrtix of the parabola.* (UQ)

1. Construct the parallelogram PQRS (PS = 110 mm and PQ = 50 mm and ∠ QPS = 70°).
2. Bisect PS at O.
3. *Axis:* Draw VO parallel to PQ. V is the vertex and VO is the axis.
4. Divide PQ and SR into any number of (4) equal parts as 1, 2, 3 and 1′, 2′, 3′ respectively starting from P on PQ and from S on SR.
5. Join V1, V2 and V3.
6. Also join V1′, V2′ and V3′.
7. Divide PO and OS into 4 equal parts as 1_1, 2_1, 3_1 and $1'_1$, $2'_1$, $3'_1$ respectively starting from P on PO and from S on SO.
8. From 1_1 draw a line parallel to PQ to meet the line V1 at P_1.
9. Similarly obtain P_2 and P_3.
10. Also from $1'_1$, $2'_1$, $3'_1$ draw lines parallel to RS to meet the lines V1′, V2′ and V3′ at P'_1, P'_2 and P'_3 respectively and draw a *smooth parabola*.

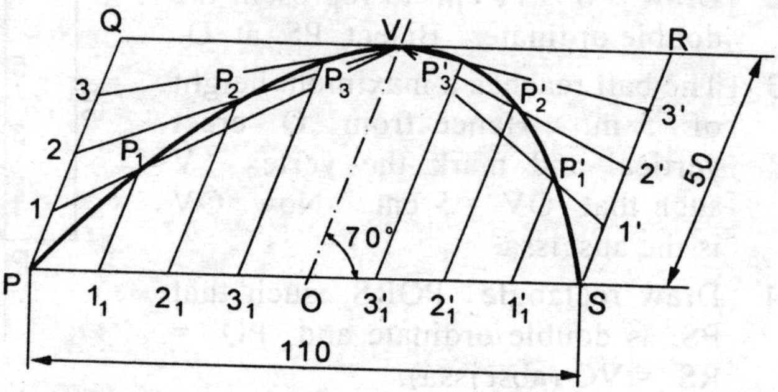

FIG. 20 PARABOLA
PARALLELOGRAM METHOD

NOTE : *Refer Fig. 18 to find the focus and directrix of the parabola.*

Problem 18: *(Exercise) The headlight reflector of a four-wheeler has a maximum rim diameter of 115 mm and a maximum depth of 90 mm. Draw the shape of the reflector. Draw a tangent and normal at any point on the curve.* (UQ)

HINT : Shape of the reflector is parabolic.
Maximum rim diameter of the four-wheeler = Base of the parabola.
Maximum depth of the reflector = Axis of the parabola.
Use *Parallelogram Method* or *Rectangle Method*.

4.7 TANGENT METHOD

Problem 19: *(Fig. 21) A fountain jet discharges water from ground level at an inclination of 50° to the ground. The jet travels a horizontal distance of 9 m from the point of discharge and falls on the ground. Trace the path of the jet.* (UQ)

1. Select a scale 1 : 100.
2. Draw PQ = 9 cm.
3. Jet discharges water at an angle of 50° to the ground. Hence, at P and Q draw 50° lines to intersect at R. PQR is an isosceles triangle.
4. Bisect PQ at O.
5. At O, erect vertical to pass through R.
6. Bisect OR at V, the vertex.
7. Divide PR into, *say*, 8 equal parts as 1, 2, ..., 7 starting from P on PR.
8. Divide RQ into the *same* number of (8) equal parts as 1′, 2′, ..., 7′ starting from R on RQ.
9. Join 11′ and also 77′. Both will meet the vertical OR at a point.
10. Join 22′ and also 66′. Both will meet the vertical OR at *another* point.
11. Join 33′ and also 55′. Both will meet the vertical OR at a *third* point.
12. Join 44′ to meet OR at V.
13. Draw a *smooth parabola* through P, V, Q such that the curve is tangential to the lines 11′, 22′, ..., 77′.

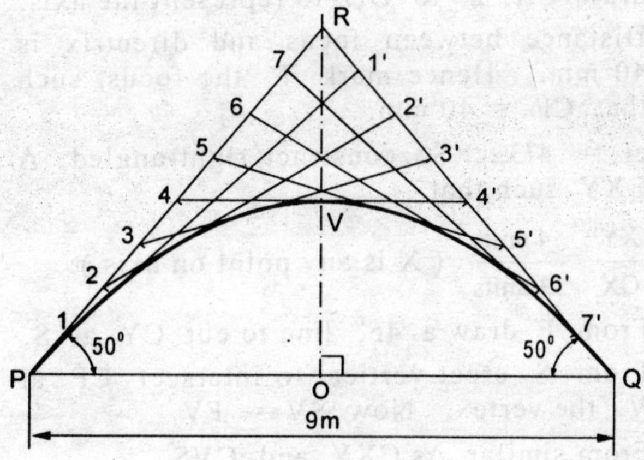

SCALE : 1:100

FIG. 21

PARABOLA - TANGENT METHOD

Problem 20: *(Exercise)*

(a) An aircraft is shot at an angle of 45° to the horizontal from the ground level. It is hit at a maximum height of 255 m of the trajectory. Trace the path of the shot in space using Tangent Method or Rectangle Method. Name the curve.

(b) Draw a tangent and normal at any point on the curve. (UQ)

5. HYPERBOLA

5.1 ECCENTRICITY METHOD

Problem 21: *(Fig. 22) Construct a hyperbola when the distance between the focus and directrix is 40 mm. The eccentricity is 4/3. Draw a tangent and normal at any point on the hyperbola.* (UQ)

1. Draw directrix DD. At any point C on it draw CA ⊥r to DD to represent the axis.
2. Distance between focus and directrix is 40 mm. Hence mark F the focus, such that CF = 40 mm.
3. e = 4/3. So construct right-angled Δ CXY such that

$$\frac{XY}{CX} = \frac{4 \text{ units}}{3 \text{ units}} \quad (\text{X is any point on axis})$$

4. From F draw a 45° line to cut CY at S.
5. From S erect vertical to intersect CF at V, the vertex. Now SV = FV.
6. From similar Δs CXY and CVS,

$$\frac{SV}{CV} = \frac{YX}{CX} = \frac{FV}{CV} = \frac{4}{3}$$

7. Along CA mark 1, 2, 3, 4, 5 at approximately equal intervals.
8. Through 1, 2, 3, 4, 5 erect verticals to intersect CY (produced if necessary) at 1′, 2′, ..., 5′ respectively.
9. With 11′ as radius and F as centre draw two arcs on either side of the axis to intersect the vertical line drawn through 1 at P_1 and P_1'.
10. Repeat the above and obtain P_2 and P_2', ..., P_5 and P_5' corresponding to 2, 3, 4 and 5 respectively.
11. Draw a *smooth hyperbola* through P_5, P_4, ..., P_1, V, P_1', ..., P_5'.

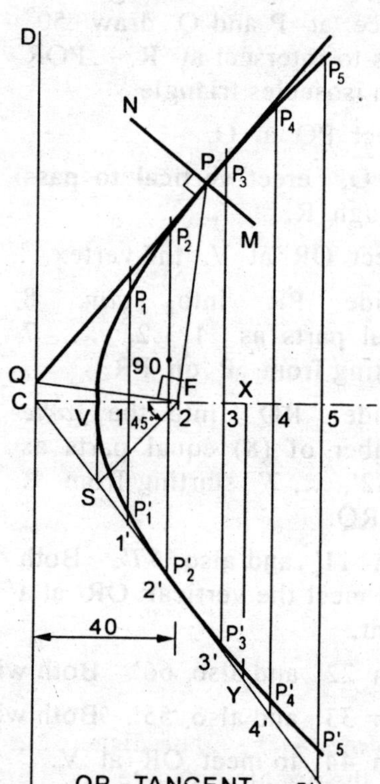

FIG. 22 HYPERBOLA
ECCENTRICITY METHOD

Problem 22: *(Exercise) A fixed point F is 7.5 cm from a fixed straight line. Draw the locus of a point P moving in such a way that its distance from the fixed straight line is 3/2 times its distance from F. Name the curve. Draw normal and tangent at a point 6 cm from F.*

5.2 HYPERBOLA – TERMINOLOGY

Hyperbola has two branches. Each branch has the same eccentricity and therefore this curve is also known as Central Conic. *A hyperbola has two foci and two directrices.*

Hyperbola is also defined as the locus of a point P which moves such that the difference of its distances from two fixed points called Foci (F_1 and F_2) is always a constant and is equal to the distance between the vertices of the *two branches of Hyperbola*, *i.e.*, equal to the length of the transverse axis (AB). *i.e.*, $PF_2 \sim PF_1 = AB$.

Refer Fig. 23.

Vertices : Points A and B where the hyperbola cuts the axis are the Vertices. When point A is considered instead of P on the curve, $F_2A - F_1A = AB$ (since $F_1A = F_2B$).

Transverse Axis or Major Axis (AB)
It is a horizontal axis and is equal to the distance between the two vertices of a pair of hyperbolas. Thus by definition, $AB = PF_2 \sim PF_1$.

Centre (O)
Centre O is midpoint of transverse axis AB.

Foci (F_1 and F_2) : Foci F_1 and F_2 lie on the extension of transverse axis.

Ordinate (PQ)
It is a vertical line drawn from P perpendicular to the axis.

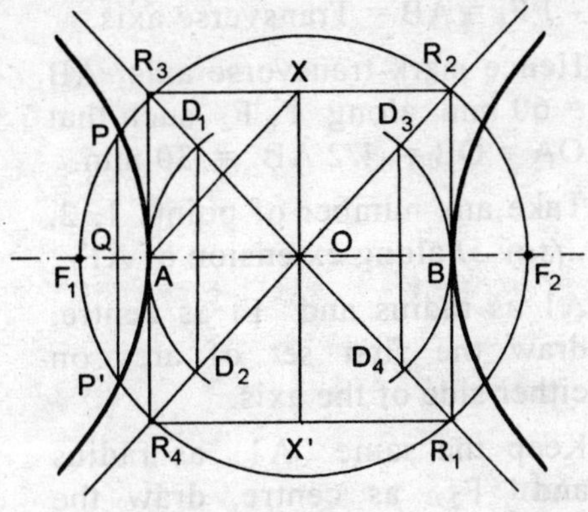

FIG. 23
HYPERBOLA - TERMINOLOGY

Double Ordinate (PP')
Ordinate PQ is extended to cut hyperbola at P'. Hence Double Ordinate = PP' = 2PQ

Abscissa (AQ) : The distance along the axis between the vertex A and a point Q through which the double ordinate PP' passes is called *Abscissa.*

Conjugate Axis (XX') : It is a line perpendicular to the transverse axis and passes through the centre of the transverse axis. When XX' is the conjugate axis, $AX = AX' = OF_1$. Conjugate Axes are situated such that each is parallel to the tangents at extremities of other.

Asymptotes (R_1R_3 and R_2R_4)
Draw a circle with O as centre and radius OF_1. Draw perpendiculars to AB at A and B cutting the circle at R_3, R_4, R_1 and R_2. The diagonals R_1R_3 and R_2R_4 are the *Asymptotes.* When they are extended on either side, they never touch the curve, but approach nearer and nearer to the curves and will be tangential to the curves at infinity.

Directrices (D_1D_2 and D_3D_4) : O as centre and OA or OB as radius, draw arcs to cut the asymptotes at D_1, D_2, D_3, D_4. D_1D_2 and D_3D_4 are the *Directrices.*

5.3 FOCI AND TRANSVERSE AXIS METHOD

Problem 23: *(Fig. 24) The distance between two fixed points is 90 mm. A point P moves such that the difference of its distances from two fixed points always remains constant and is equal to 60 mm. Draw the loci of P. Draw the tangent and normal at any point on the hyperbola.* (UQ)

1. *See Fig. 24.* Draw $F_1 F_2$ = 90 mm and bisect it at O.
2. A point P moves such that the difference of its distances from F_1 and F_2 = 60 mm. But $PF_2 - PF_1$ = AB = Transverse axis.
3. Hence mark transverse axis AB = 60 mm along $F_1 F_2$ such that OA = OB = 1/2 AB = 30 mm.
4. Take any number of points 1, 2, ..*(say 5)* along extension of AF_1.
5. A1 as radius and F_1 as centre, draw the first set of arcs on either side of the axis.
6. Keep the same A1 as radius and F_2 as centre, draw the second set of arcs on either side.

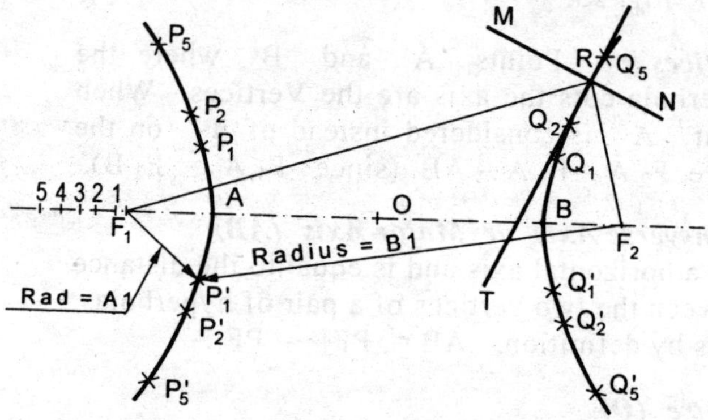

F_1F_2 = 90mm, $PF_2 \sim PF_1$ = AB = 60mm

FIG. 24 A PAIR OF HYPERBOLAS

7. B1 as radius and F_2 as center, cut the first set of arcs at P_1 and P_1'.
8. Keep the same B1 as radius and F_1 as center, cut the second set of arcs at Q_1 and Q_1'.
9. Repeat the above and obtain the points P_2, P_2', P_3, P_3', Q_2, Q_2', Q_3, Q_3', etc., corresponding to 2, 3, etc.
10. Join P_5, ..., P_1, A, P_1', ..., P_5' and Q_5, ..., Q_1, B, Q_1', ..., Q_5' and obtain a *pair of hyperbolas* as shown.

Tangent and Normal at any point R on the Hyperbola

11. Mark R on the hyperbola. Join RF_1 and RF_2.
12. Bisect $\angle F_1 RF_2$. The bisector TR is the tangent at R.
13. Draw the normal MN perpendicular to TR through R.

Problem 24: *(Exercise) Draw a pair of hyperbolas when the distance between the foci is 70 mm and the length of the transverse axis is 45 mm. Locate the asymptotes and directrices: Find the eccentricity of the curve.* (UQ)

HINT : Length of transverse axis = Distance between vertices = 45 mm; e = 1.56.

5.4 ORDINATE, ABSCISSA AND TRANSVERSE AXIS METHOD

Problem 25: *(Fig. 25) Draw a hyperbola when its double ordinate is 90 mm, abscissa is 35 mm and half the transverse axis is 45 mm.* (UQ)

1. Draw OBQ such that OB = half the transverse axis = 45 mm and BQ = abscissa = 35 mm.
2. Through Q erect vertical such that PP′ = double ordinate = 90 mm = 2PQ.
3. Construct the rectangle PP′ R′ R.
4. Divide PR and P′ R′ into any number of equal parts *(say 4)* as 1, 2, 3 and 1′, 2′, 3′ starting from P on PR and P′ on P′ R′ respectively.
5. Join B1, B2, B3, B1′, B2′ and B3′.
6. Divide the ordinates PQ and QP′ into the *same* number of equal parts as 1_1, 2_1, 3_1 and $1_1'$, $2_1'$, $3_1'$ starting from P on PQ and P′ on P′ Q respectively.
7. Join $O1_1$ to meet B1 at P_1.
8. Join $O2_1$ and $O3_1$ to meet B2 and B3 at P_2 and P_3 respectively.
9. Similarly join $O1_1'$, $O2_1'$ and $O3_1'$ to meet B1′, B2′ and B3′ at P_1', P_2' and P_3' respectively.
10. Join P, P_1, P_2, P_3, B, P_3', P_2', P_1' and P′ by a *smooth hyperbola*.

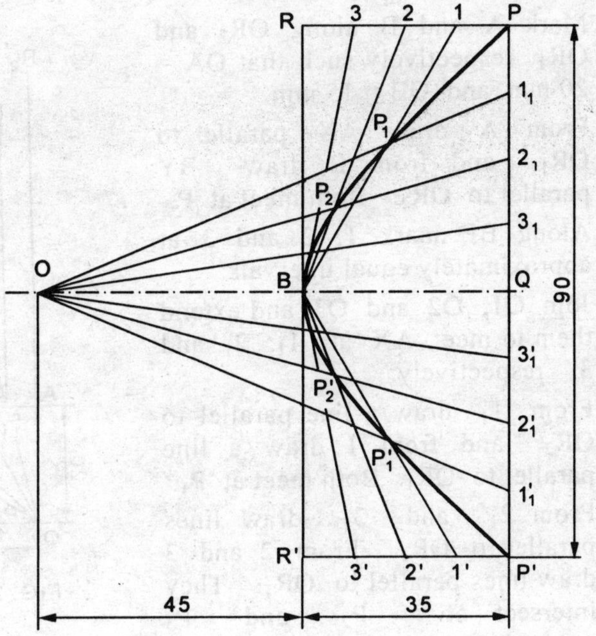

FIG. 25 HYPERBOLA

Problem 26: *(Exercise) Construct a branch of hyperbola whose double ordinate is 150 mm, abscissa is 60 mm and half the transverse axis is 90 mm.* (UQ)

Problem 27: *(Exercise) Construct a hyperbola given half the major axis equal to 40 mm, single ordinate 60 mm and abscissa 48 mm.*

HINT : Major axis means Transverse axis

5.5 RECTANGULAR HYPERBOLA or EQUILATERAL HYPERBOLA

When angle between the asymptotes is **90°**, the curve is called *Rectangular* or *Equilateral Hyperbola* of **e = √2.** It is the curve traced by a point moving such that the product of its distances from two fixed lines (asymptotes) at right angles to each other is a constant.

Problem 28: *(Fig. 26) Construct a rectangular hyperbola when a point P on it is at a distance of 15 mm and 20 mm respectively from the two asymptotes.* (UQ)

1. Draw OR_1 and OR_2 such that $\angle R_1OR_2$ is 90°.
2. Mark A and B along OR_2 and OR_1 respectively such that OA = 20 mm and OB = 15 mm.
3. From A draw AX parallel to OR_1 and from B draw BY parallel to OR_2. Both meet at P.
4. Along BP mark 1, 2 and 3 at approximately equal intervals.
5. Join O1, O2 and O3 and extend them to meet AX at 1_1, 2_1 and 3_1 respectively.
6. From 1_1 draw a line parallel to OR_2 and from 1 draw a line parallel to OR_1. Both meet at P_1.
7. From 2_1 and 3_1 draw lines parallel to OR_1. From 2 and 3 draw lines parallel to OR_1. They intersect at P_2 and P_3 respectively.
8. Then along PA mark points 4_1 and 5_1 approximately at equal intervals.

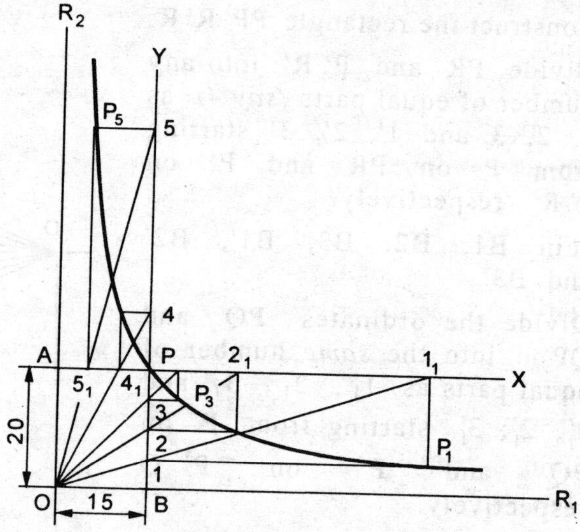

FIG. 26 RECTANGULAR or EQUILATERAL HYPERBOLA

9. Join $O4_1$ and $O5_1$ and extend them to meet BY at 4 and 5 respectively.
10. From 4_1 and 5_1 draw lines parallel to OR_2 and from 4 and 5 draw lines parallel to OR_1 to intersect at P_4 and P_5 respectively.
11. Join P_1, P_2, P_3, P, P_4, P_5 by a *smooth Rectangular Hyperbola.*

NOTE : *Extend the curve on either side such that it does not touch the asymptotes.*

Problem 29: *(Exercise) The asymptotes of a hyperbola make 80° with each other. A point P on the curve is at a distance of 30 mm from the horizontal asymptote and 45 mm from the inclined asymptote when measured horizontally. Plot the hyperbola. Draw tangent and normal at any point R on the curve.*

HINT : Procedure for this problem is exactly similar to that of Rectangular Hyperbola.

Problem 30: *(Fig. 27) An ideal gas expands according to Boyle's Law pv = a constant. Pressure p is 0.3 kg per sq. cm. and volume v is 0.06 cu. cm. Select a suitable scale and plot the expansion curve.* (UQ)

1. Product of p and v is a constant. Hence the curve traced by satisfying this condition is a *hyperbola*. Mark p and v on two mutually perpendicular axes OX and OY.
2. Select the Scale as follows:

   X axis: 1 mm = 0.01 kg / cm^2

   Y axis: 1 mm = 0.001 cm^3
3. Draw asymptotes OX and OY.
4. Locate the point P such that OA = 0.3 kg / sq. cm. and OB = 0.06 cu. cm.
5. Along AP mark 1, 2 and 3 at approximately equal distance.
6. Join O1 and extend it to meet $B1_1$ at 1_1.
7. From 1_1 draw a line parallel to OY. From 1 draw a line parallel to OX. Both intersect at P_1.
8. Similarly obtain the points P_2, P_3, P, P_4 and P_5 as explained in the Problem 37.
9. Join P_1, P_2, P_3, P, P_4 and P_5 in sequence by a *smooth Rectangular Hyperbola*.

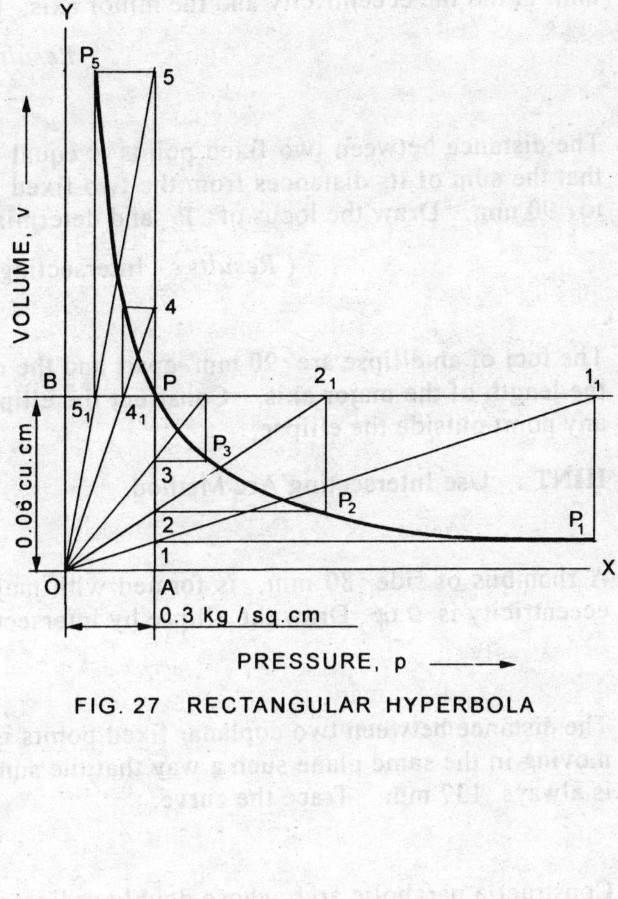

FIG. 27 RECTANGULAR HYPERBOLA

Problem 31: *(Exercise) For a perfect gas the relation between pressure P and volume V in an isothermal expansion is given by the law PV = constant. Draw the curve of the isothermal expansion of an enclosed volume of the gas if 0.056 m^3 of gas correspond to a pressure of 0.355 kg / cm^2. Name the curve.* (UQ)

HINT : Scale : X axis: 1 mm = 0.01 kg/cm^2

Y axis: 1 mm = 0.001 m^3

6. ADDITIONAL PROBLEMS

32. The directrices of an ellipse are 200 mm apart. Distance between the vertices is 150 mm. Find the eccentricity and the minor axis. Use Concentric Circles Method. (UQ)

(*Results* : e = 3/4 ; Minor Axis = 94 mm)

33. The distance between two fixed points is equal to 75 mm. A point P moves such that the sum of its distances from the two fixed points is always a constant and is equal to 90 mm. Draw the locus of P and determine the minor axis. (UQ)

(*Results* : Intersecting Arcs Method; Minor axis = 50 mm)

34. The foci of an ellipse are 90 mm apart and the minor axis is 72 mm long. Determine the length of the major axis. Construct the ellipse. Draw a tangent to the ellipse from any point outside the ellipse. (UQ)

HINT : Use Intersecting Arc Method.

35. A rhombus of side 80 mm, is formed with major and minor axes as diagonals. The eccentricity is 0.6. Draw the ellipse by intersecting arcs method.

36. The distance between two coplanar fixed points is 90 mm. Trace the path of a point P moving in the same plane such a way that the sum of the distances from the fixed points is always 137 mm. Trace the curve.

37. Construct a parabolic arch whose double ordinate is 110 mm and abscissa is 85 mm. Locate the focus and directrix of the parabola using Tangent Method.

38. The span and rise of a parabola are 1800 mm and 1320 mm respectively. Construct the parabola. Draw a tangent and normal to the curve at a point on it, located 45 mm from the vertex on the right side. (UQ)

HINT : Use Rectangle Method

39. The distance between a fixed point and fixed line is 54 mm. Trace the path of a point moving in the same plane such that its distance between the fixed point and the fixed line is always equal. If the point moves upto 72 mm from the fixed point, plot the curve. (UQ)

HINT : Use Eccentricity Method

40. A roof arch is in the shape of a parabola. Span of the arch is 40 m and the rise is 8 m. Plot the profile. (UQ)

HINT : Span = Base and Rise = Height of the parabola

41. A jet of water is issuing through an orifice of 25 mm diameter fitted to a vertical side of a tank. The centre of the orifice is 0.75 m above the ground and the centre of the jet touches the ground at a distance of 1.25 m from the vertical orifice. Draw the locus of the centre of the jet, which is just issuing from the orifice till it reaches the ground. Name the curve. (UQ)

(Ans : Parabola *)*

42. Construct a parabola with the double ordinate as 108 mm and abscissa as 81 mm. Draw a tangent and normal to the curve at any point from the base.

43. A boy throws a cricket ball from the top of a building 4 m high crosses the top of a palm tree 9 m high and falls on the ground. The distance between the building and the tree is 3 m. Plot the path of the projectile. (UQ)

HINT : Use Rectangle Method

44. A Δ PQR has sides PQ = 80 mm, QR = 65 mm and RP = 80 mm. Construct a parabola passing through the points P, Q and R such that side QR is horizontal. (UQ)

HINT : Draw QR horizontal of 65 mm length and PQ = RP = 80 mm length. Δ PQR is isosceles. Points P, Q and R lie on the parabola of Base QR. Construct the parabola by Rectangle Method.

45. A highway suspension bridge of parabolic shape has a span of 25 m and a dip of 5 m at the center. Make out a profile of the bridge.

46. Two points P and Q are located on a plane of paper and are 90 mm apart. A point X moves on the sheet of paper such that the difference of its distances from P and Q always remains 45 mm. Trace the locus of X and name the curve. Mark the asymptotes and directrices. (UQ)

(*Ans:* Hyperbola)

47. Given that the distance between foci is equal to 90 mm and the transverse axis is equal to 50 mm. Construct one branch of the hyperbola and mark its asymptotes and directrix. (UQ)

HINT : Use Fixed Points and Difference of the Distances Method.

48. The two asymptotes of a hyperbola are at an angle of 100°. A point is 30 mm and 65 mm from each asymptote as measured parallel to them and is a point on the curve. Construct the two branches of the curve and determine the following:

i. Length of the transverse axis
ii. Eccentricity
iii. Directrices

Also, draw a tangent and normal to the curve at a point on it 50 mm from the directrix.

49. The asymptotes of a rectangular hyperbola are equally inclined to a horizontal line. A point X on the curve is 50 mm and 30 mm away from the two asymptotes. Draw the hyperbola passing through the point X. Also locate the focus and directrix of the curve.

*

7

CYCLOIDS AND INVOLUTES

1. SPECIAL ENGINEERING CURVES

The Table below gives the classifications of the family of Special Types of Engineering Curves:

SPECIAL ENGINEERING CURVES

- Cycloidal Curves
  - Cycloids, Epicycloids, Hypocycloids
  - Trochoids, Epitrochoids, Hypotrochoids
- Involutes
- Spirals
  - Archimedian
  - Logarithmic
- Helices
  1. Cylindrical
  2. Conical

2. CYCLOIDAL CURVES

Cycloidal Curve : A *Cycloidal Curve* is traced by a fixed point on the circumference of a circle which rolls without slipping along a fixed straight line or a circle.

Rolling Circle or Generating Circle: The circle which rolls along the line is called *Rolling Circle,* also known as *Generating Circle.*

Directing Line / Base Line or Directing Circle / Base Circle: The fixed line *or* circle is called *Directing Line / Base Line or Directing Circle / Base Circle.*

3. INVOLUTES

Involute

Involute is a single-curved line traced out by an end of a string when unwound itself from a straight line or a circle or a polygon, the string being kept taut.

Uses

Involute of a circle is used as the profile of gear teeth wheel. Casings of centrifugal pumps and Cams are of involute shape.

8

CYLINDRICAL CURVES

1. HELICAL CURVES

Helix is defined as a curve generated by a point moving around and along the surface of a right circular ***cylinder*** or ***cone*** with uniform angular velocity about the axis and with a uniform linear velocity in the direction of the axis.

Engineering Applications
Helix is widely used in the design of helical springs, worm gears, screw threads of V and square types, spiral stairways, etc. Helical springs are used in instruments, machines, etc., for various purposes. Helical grooves of square shape is used on cylindrical cams, screw conveyors, etc.

Types of Helical Curves

1. Helices
   a. Cylindrical Helix, simply known as Helix
   b. Conical Helix
2. Helical Springs
   a. Cylindrical Springs
   b. Conical Springs
3. Screw Threads
   a. V-Threads
   b. Square Threads

1.1 CYLINDRICAL CURVES

Curves obtained around the surface of a right circular cylinder is known as *Cylindrical Curves*.

Types of Cylindrical Curves

1. Cylindrical Helix, simply known as Helix
2. Helical Cylindrical Springs
   a. Helical Spring of Square Section
   b. Helical Spring of Circular Section
3. Screw Threads
   a. V-Threads (Single Start and Double Starts)
   b. Square Threads (Single Start and Double Starts)

2. CYLINDRICAL HELIX (HELIX)

See Fig. 1(a). If the point moves around and along the surface of a cylinder, the curve obtained is called *Cylindrical Helix* or simply *Helix*.

Lead of the Helix : Lead of the helix is the axial distance moved by the generating point on the surface of a cylinder in one revolution.

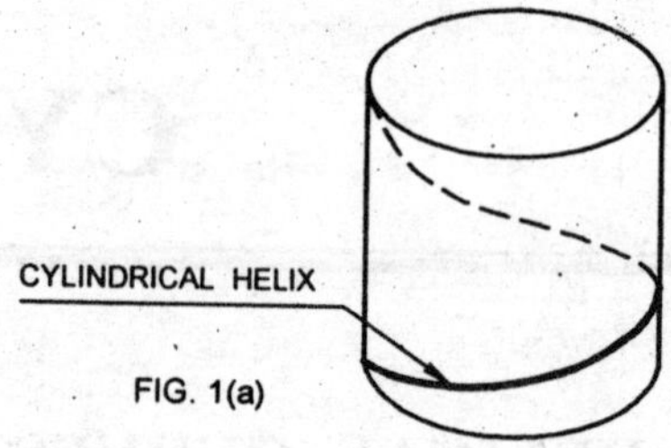

FIG. 1(a)

Pitch : For a single start helix, this axial movement is called *Pitch, i.e.*, lead and pitch are equal. For a double start helix, pitch is half the lead. Thus for a multi-start helix, pitch = Lead / Number of starts.

A helix may be either right-handed or left-handed as shown in Fig. 1(b).

Right-handed Helix : In a right-handed helix, the point which generates the helix, moves clockwise and towards the observer when cylinder is viewed axially from the top.

Left-handed Helix : In a left-handed helix the point moves anti-clockwise and away from the observer when the cylinder is viewed axially from the top.

Convolution : It is the helix corresponding to one revolution of a moving point.

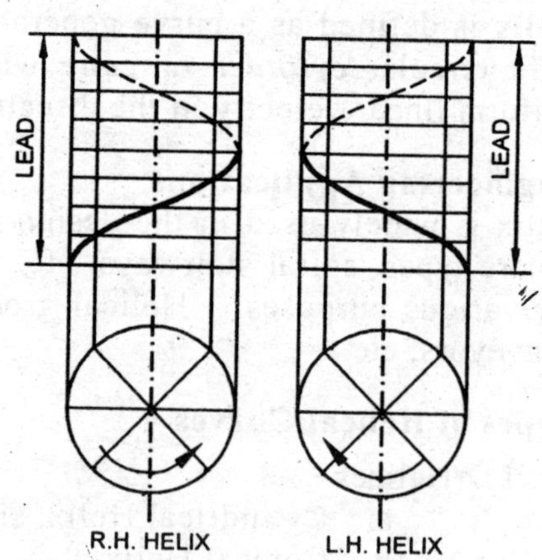

FIG. 1(b) CYLINDRICAL HELICES

Helix Angle (θ) : *See Fig. 1(c).* When the cylindrical surface upon which a helix is generated, is stretched out into a plane, it will have the shape of a right-angled triangular lamina ABC. Length of the helix becomes the hypotenuse of the triangle, height of the triangle BC equal to the lead (pitch) of the helix and the base of the triangle AB equal to the circumference of the cylinder of diameter D.

Angle θ made between the length of the helix and the base of the triangle is called *Helix Angle*.

tan θ = Pitch / Circumference of the circle

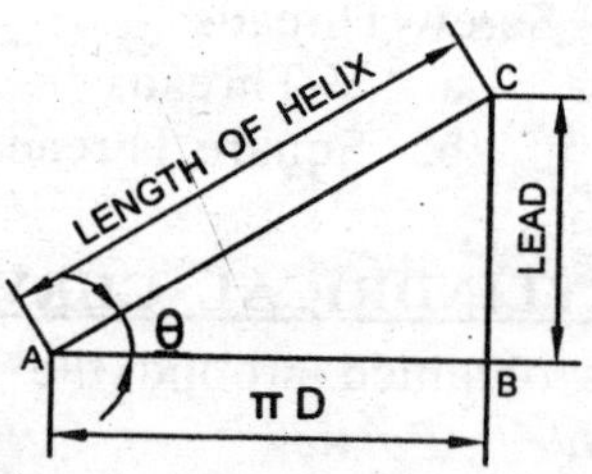

FIG. 1(c)

Problem 1: *(Fig. 2) An ant is moving on a vertical cylinder in a circumferential direction at a constant angular speed in the clock-wise direction as seen axially from the top and with simultaneous uniform rate of advance in the axial direction.*

If the diameter of the cylinder is 50 mm and the ant moves 75 mm in the axial direction during one turn, draw the path traced by the ant. (UQ)

1. Assume the cylinder stands vertically.
2. Draw a circle of 50 mm diameter (top view of the cylinder).
3. Divide the circle into 12 equal parts as 1, 2, 3, . . . , 12, in CW direction as the ant moves CW when viewed from top.
4. Draw a rectangle of sides equal to 50 mm (diameter) and 75 mm (cylinder height = pitch). This is the front view of the cylinder.
5. Divide the pitch also into same (12) equal parts as 1′, 2′, 3′, ..., 12′ and draw horizontal lines through each point.
6. Draw vertical projectors from 1, 2, . . . , 12 to intersect the corresponding horizontal projectors at P_1', P_2', ..., P_{12}'.
7. Passing through these points, draw a *smooth helical curve.*

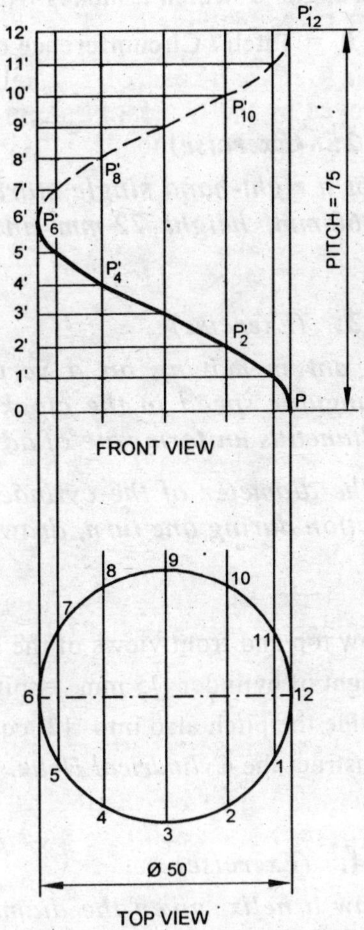

FIG. 2
L.H. CYLINDRICAL HELIX

Invisible Portion of the Curve: *Portion of the curve from* P_6' *to* P_{12}' *is on rear side of cylinder / cone and will not be visible.*

NOTE : *To draw helix for two revolutions, the cylinder height shall be taken as equal to twice the pitch. Then divide the cylinder height into 24 equal parts.*

To draw the development of the surface of the cylinder

Helix is drawn as a straight line. It is the hypotenuse of a right-angled triangle having the base equal to the circumference of the circle and the vertical side equal to the pitch of the helix. The angle θ which it makes with the base, is known as the Helix Angle.

$$\tan \theta = \text{Pitch / Circumference of the circle}$$

Problem 2: *(Exercise)*

Plot a right-hand single start helix for one convolution on a cylinder of base diameter 60 mm, height 72 mm and pitch 72 mm.

Problem 3: *(Exercise)*

An ant is moving on a vertical cylinder in a circumferential direction at constant angular speed in the clockwise direction as seen axially from the top and with simultaneous uniform rate of advance in the axial direction.

If the diameter of the cylinder is 50 mm and the ant moves 75 mm in the axial direction during one turn, draw the path of the ant.

HINT

1. Draw top and front views of the cylinder and divide it into 12 equal parts.
2. Height of cylinder 75 mm = pitch of helix = distance moved by the ant in 1 turn.
3. Divide the pitch also into 12 equal parts.
4. Construct the *Cylindrical Helix.*

Problem 4: *(Exercise)*

Draw a helix, given the diameter of the cylinder 54 mm and pitch 36 mm. Obtain the helix angle.

Problem 5: *(Exercise)*

A right circular vertical cylinder of 60 mm diameter and height 90 mm rotates uniformly. An insect crawls at a uniform speed on the surface of the cylinder along a straight line parallel to the axis from the bottom to the top of the cylinder. It travels a height of 90 mm when the cylinder rotates once.

Show the path and name it.

2.1 CONICAL HELIX

Cases in design occur where a helical type curve has to be formed on the outside surface of a shape other than a cylinder. The method of construction, is similar to cylindrical helix, the difference being that each point is constructed on a different radius. A *conical helix* is generated as the point moves on the surface of a cone such that it moves around the axis and simultaneously moves towards the vertex.

Problem 6: *(Fig. 3) Draw a helix of one convolution around a cone, diameter of base 50 mm, height 80 mm and pitch 65 mm.*

1. Draw the top view of the cone as a circle of diameter 50 mm.
2. Divide the circle into 12 equal parts as 1, 2, 3, . . . , 12.
3. Join 1, 2, etc., with center O.
4. Draw projectors from these points and draw a Δ of base 50 mm and height 80 mm. O′ is the apex.
5. Mark 1′, 2′, ..., 12′ on the above base of the cone in the front view.
6. Mark A on axis 65 mm from the base to represent pitch of helix.
7. Draw a horizontal through A to cut the generator (O′12′) at A′.
8. Divide pitch distance 3′A (65 mm) into 12 equal parts as 1″, 2″, etc.
9. Let P be the starting point. When it moves around through 30°, it should have moved up through one division to a point P_1' on the generator O′1′ obtained by drawing a horizontal line through 1″.
10. Project P_1' downwards to cut top view of the generator O1 at P_1.
11. Similarly obtain all other points and draw smooth curves through them in both views.

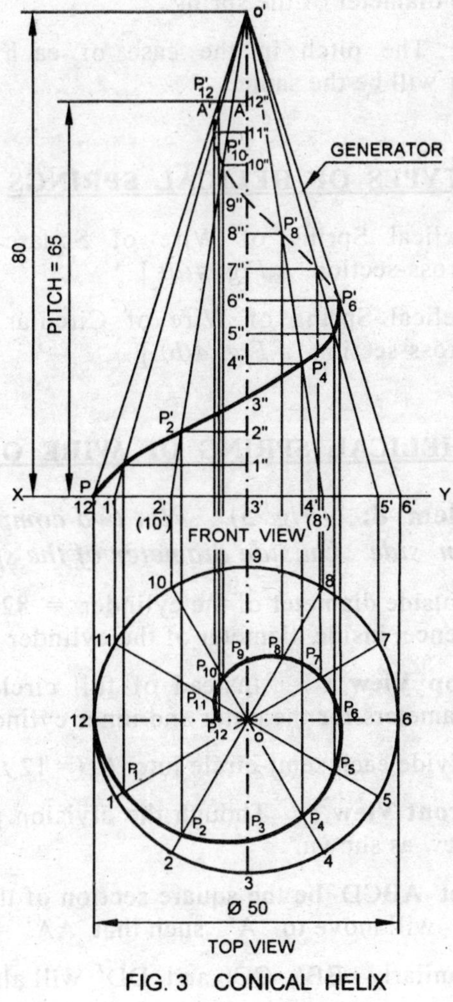

FIG. 3 CONICAL HELIX

Problem 7: *(Exercise) Draw a helix of one convolution around a cone of diameter 50 mm. Height and pitch of the helix are 60 mm. Also show the curve in the top view and in the development.*

3. HELICAL SPRINGS

In a helical spring having a wire of square cross-section, the outer two corners of the section may be assumed to be moving around the axis, on the surface of a cylinder having a diameter equal to the outside diameter of the spring.

The inner two corners of the section will move on the surface of a cylinder having a diameter, equal to the inside diameter of the spring.

The pitch in the case of each corner will be the same.

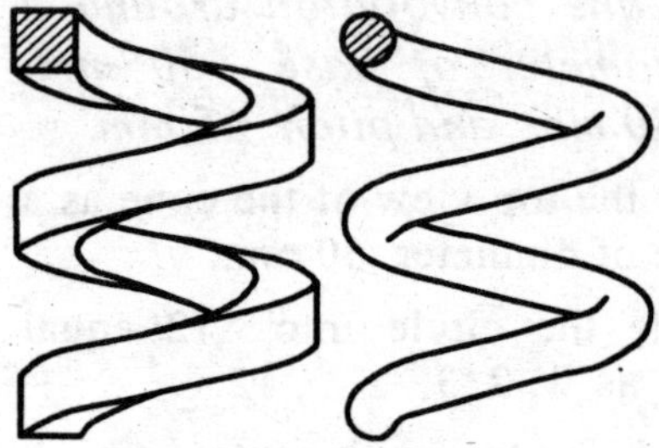

(a) (b)

FIG. 4

3.1 TYPES OF HELICAL SPRINGS

1. Helical Spring of Wire of Square Cross-section [*Fig. 4(a)*]
2. Helical Spring of Wire of Circular Cross-section [*Fig. 4(b)*]

3.2 HELICAL SPRING OF WIRE OF SQUARE CROSS-SECTION

Problem 8: *(Fig. 5) Plot two complete turns of a wire spring of square section of 15 mm side. Outside diameter of the spring is 82 mm and pitch is 45 mm.* (UQ)

1. Outside diameter of the cylinder = 82 mm;
   Hence, inside diameter of the cylinder = 82 – (2 x 15) = 52 mm.
2. **Top View :** Instead of full circles, draw semi-circles of 82 mm and 52 mm diameters for the outer and inner cylinders of the spring in the top view.
3. Divide each semi-circle into 6 (= 12 / 2) equal parts as 1, 2, . . . , 6.
4. **Front View :** Though the division points, draw the vertical projectors on to the front view as shown.
5. Let ABCD be the square section of the wire of the spring. In one revolution, the point A will move to A′ such that AA′ = pitch = 45 mm.
6. Similarly BB′, CC′ and DD′ will also be equal to 45 mm.
7. Divide CC′ into 12 equal parts as 1′, 2′, 3′, etc.
8. Plot the helices traced out by the points A and C on the outer surface of the cylinder and by the points B and D on the inner surface. Outer helices will be parallel to each other.
9. Mark the visible and invisible parts of the helices as shown.

NOTE

1. *Helical spring of square cross-section has four helices. The two outer corners of the square section generate helices of diameter equal to the outer diameter of the spring.*

   The two inner corners of the square section generate helices of diameter equal to the inner diameter of the spring.

   However, the pitch of all the helices will be the same.

2. *Outside diameter of the spring is the sum of inside diameter and twice the thickness of the square section of the spring.*

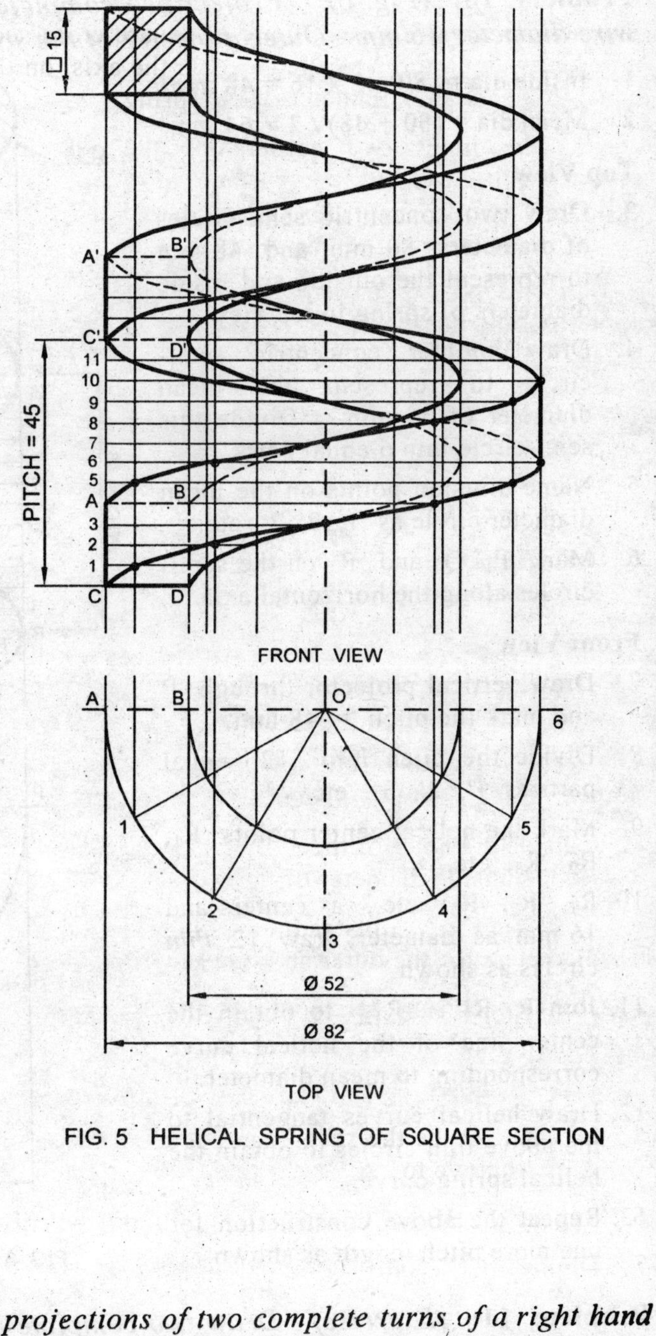

FIG. 5 HELICAL SPRING OF SQUARE SECTION

Problem 9: *(Exercise)* *Draw the projections of two complete turns of a right hand helical spring of wire of square cross section of 20 mm side. The outside diameter of the spring is 100 mm and its pitch is 60 mm.*

3.3 HELICAL SPRING OF WIRE OF CIRCULAR CROSS-SECTION

Problem 10: *(Fig. 6) Project two complete turns of an open coil helical spring of wire diameter 16 mm. Outer diameter of the spring is 80 mm and pitch 48 mm.* (UQ)

1. Inside dia = 80 – 2 x 16 = 48 mm.
2. Mean dia = (80 + 48) / 2 = 64 mm.

Top View

3. Draw two concentric semi-circles of diameters 80 mm and 48 mm to represent the outside and inside diameters of spring in top view.
4. Draw another concentric semi-circle to represent the mean diameter of 64 mm. Divide this semi-circle into 6 equal parts.
5. Name division points on the mean diameter circle as 1, 2, 3, etc.
6. Mark P, Q and R on the semi-circles along the horizontal axis.

Front View

7. Draw vertical projector through P and mark the pitch = 48 mm.
8. Divide the pitch into 12 equal parts as 1′, 2′, 3′, etc.
9. Mark the helical center points R_1, R_2, R_3, etc.
10. R_1, R_2, R_3, etc., as centers and 16 mm as diameter, draw 12 *thin* circles as shown.
11. Join R_1 R_2 ... R_{12} to obtain the center line of the helical curve corresponding to mean diameter.
12. Draw helical curves tangential to the above thin circles to obtain the helical spring curve.
13. Repeat the above construction for one more pitch length as shown.

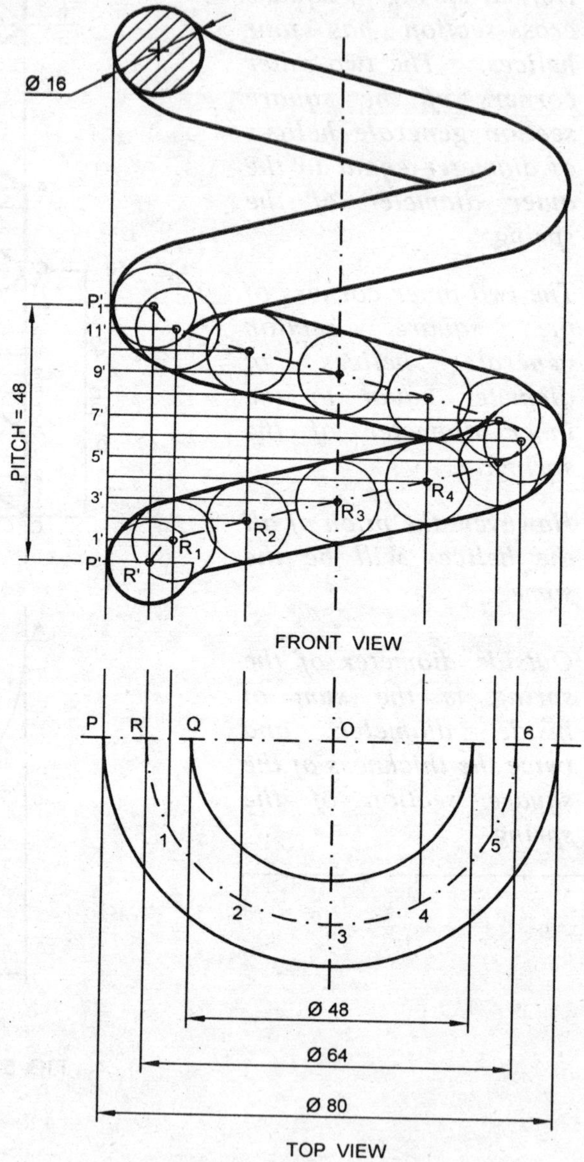

FIG. 6 HELICAL SPRING OF CIRCULAR SECTION

Problem 11: *(Exercise) Draw two complete turns of a helical spring of wire of 20 mm diameter. Spring index is 6 and the pitch is 50 mm.*

HINT : Spring Index = Mean diameter of the coil / Diameter of wire.
Therefore, Mean diameter of coil = 6 x 20 = 120 mm.

4. SCREW THREADS

Scıew threads are drawn on the principle of the helix. Threaded fasteners are temporary fasteners. The reason is that parts connected by threaded fasteners can be readily separated as and when required. Common examples of threaded fasteners are Bolts, Nuts, Screws, etc.

4.1 TYPES OF FORMS OF SCREW THREADS

1. V-Threads [*Fig. 7(a)*]
2. Square Threads [*Fig. 7(b)*]

Definitions

Pitch

Pitch of a screw thread is defined as the distance from a point on a thread to the corresponding point on the adjacent thread. It is measured parallel to the axis.

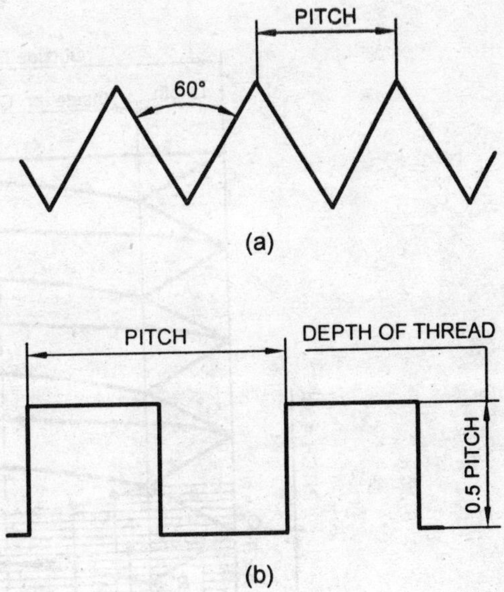

FIG. 7

Lead

Lead of the screw is defined as the actual distance advanced by a nut for its one complete revolution over a screw rod.

External Thread

It is a continuous helical ridge on the external surface of a cylinder. Threads on bolts, studs, screws, etc., are external threads.

Internal Thread

It is a thread on the internal surface of a cylinder. The thread on the surface of a hole of a nut is an internal thread.

External threads of a bolt or a stud engage with the corresponding internal threads of a nut.

Single Start and Double Start Threads

For a Single Start Thread, pitch is equal to lead, that is, pitch of the screw is equal to the pitch of the helix. Unless stated otherwise, screws are always assumed to be single started.

For a Double Start Thread, lead is equal to twice the pitch.

4.2 V-THREADS

Triangular Thread: *V-Thread* is also known as *Triangular Thread.*
Thread Angle: It is the angle contained by the V profile of the thread.
Types of V-Threads: (i) Single Start V or Triangular Thread and
(ii) Double Start V or Triangular Thread

4.2.1 SINGLE START V OR TRIANGULAR THREAD

Problem 12: *(Fig. 8) Draw four turns of single start sharp V-Thread of BIS profile, with outside diameter 80 mm, pitch 16 mm and the angle of the thread 60°.* (UQ)

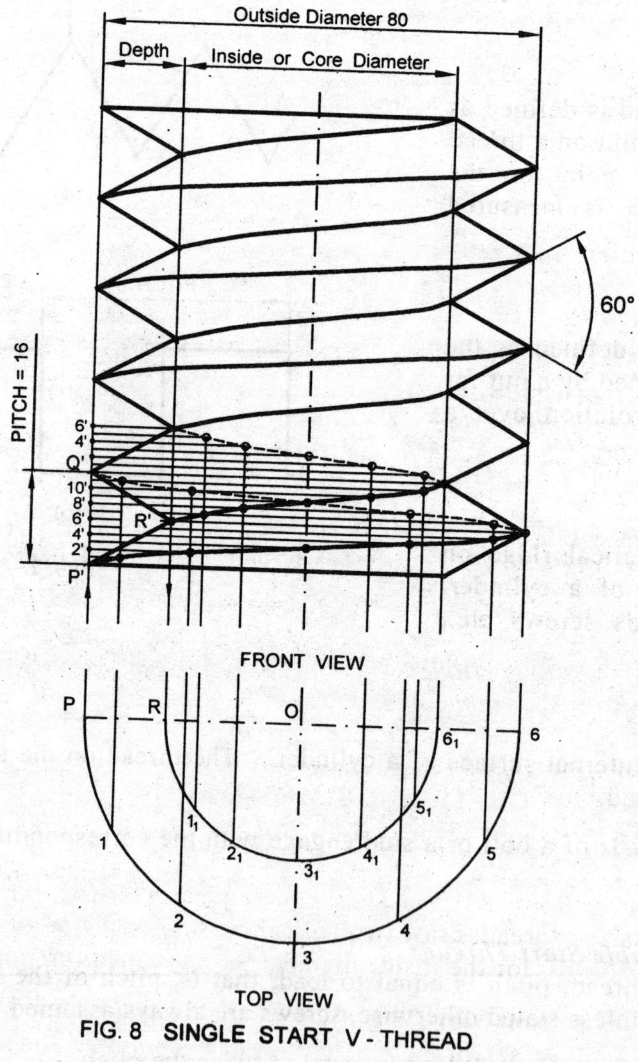

FIG. 8 SINGLE START V - THREAD

1. Instead of full circle, draw a semicircle of 80 mm diameter, representing the outside diameter of the V-thread in the top view.
2. To find the *Inside Diameter* (also known as *Core Diameter*) of the V-thread graphically in the front view:
   - Mark P′ Q′ = pitch 16 mm.
   - Construct equilateral Δ P′ Q′ R′ with P′ Q′ as the base and angle P′ R′ Q′ = 60°.

Note : Altitude of the Δ P′ Q′ R′ = Depth of the V-Thread.

   - Inside (Core) Diameter = Outside Diameter – (2 x Depth of V-Thread)
3. Now draw the vertical projector through R′ to cut horizontal axis of the top view at R.
4. O as center and OR as radius, draw a semicircle to represent the inside diameter in the top view.
5. Divide the two semicircles drawn in the top view into 6 (= 12 / 2) equal parts.
6. Name the division points on the two semicircles as 1, 2, 3, etc., and 1_1, 2_1, 3_1, etc.
7. In the front view, divide the pitch into 12 equal parts as 1′, 2′, 3′, etc.
8. Draw the helical curve traced by the points P′ and Q′ for the outside diameter.
9. Draw the helical curve traced by the point R′ for the inside diameter as shown.
10. Following the rules of visibility, mark the invisible portions of the curves by *thin* dashes and visible portions by *thick* continuous lines.

Problem 13: *(Exercise)*

Construct five complete turns of a single start V-Thread of ISO profile of 60° thread angle and pitch 20 mm. The outside diameter of the threaded rod is 90 mm.

Problem 14: *(Exercise)*

Project two complete turns of a single start triangular thread of outside diameter 125 mm, pitch 25 mm and angle 60°.

4.2.2 DOUBLE START V OR TRIANGULAR THREAD

In Double Start V-Thread, Pitch of the Helix = Number of starts x Pitch of the thread
= 2 x Pitch of the thread

For double start V-thread, draw two helical curves for the points P′ and Q′ for the outside diameter and for R′ for the inside diameter as explained for single start V-thread.

Then mark in the front view, the Pitch of the helix (= Lead) = 2 x Pitch of the thread. Construct another set of helical curves and complete the double start V-thread profile.

4.3 SQUARE THREADS

Types of Square Threads

1. Single Start Square Thread
2. Double Start Square Thread

4.3.1 SINGLE START SQUARE THREAD

Problem 15: *(Fig. 9) Draw a single start square thread of two turns. Outside diameter is 90 mm and pitch is 34 mm.* (UQ)

NOTE : *In a square thread, depth or thickness of thread is always equal to half the pitch of the thread.*

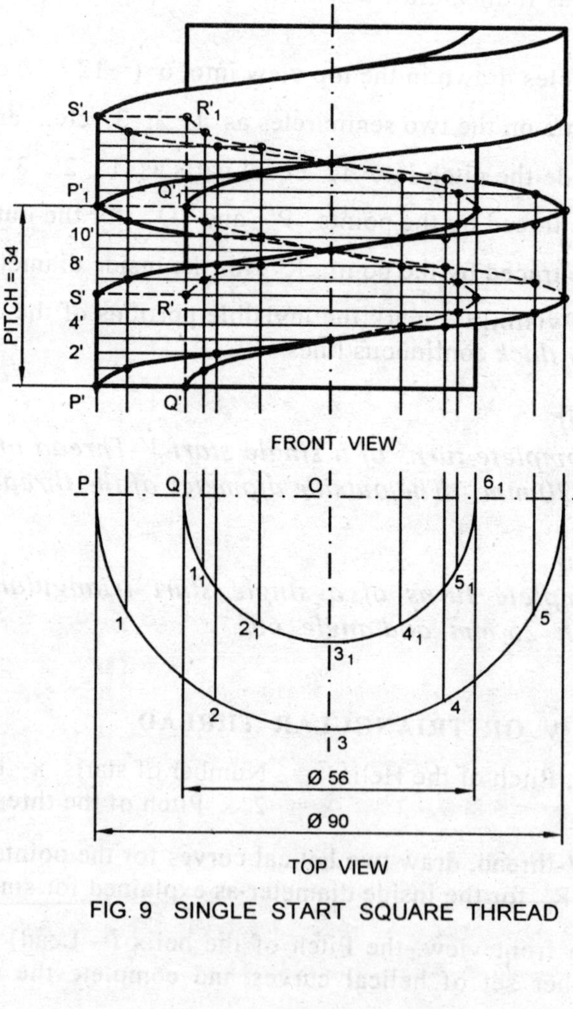

FIG. 9 SINGLE START SQUARE THREAD

1. Depth or thickness of the square thread = 0.5 x pitch = 0.5 x 34 = 17 mm.
2. Hence, the section of thread is a square of 17 mm side.

NOTE : 1. *Procedure to draw the single start square thread is the same as the helical spring of square cross section.*
2. *However, the square threaded screw differs from the helical spring in having a* ***solid*** *cylinder inside.*
3. *This inner solid cylinder completely hides the rear portions of the helical curves.*

---

3. Single start square thread is drawn by plotting four helical curves starting from each of the corners of the square section P′, Q′, R′ and S′ as follows:
4. To draw the helical curves starting from the outside two corners of the square section, assume that the two outer corners of the section move around the curved surface of an imaginary cylinder of diameter = Outside diameter 90 mm of the thread.
5. Similarly to draw the helical curves starting from the inside two corners of the square section, assume that these two inner corners move around the curved surface of an imaginary cylinder of diameter = Inside diameter of the thread.
6. Inside (Core) diameter of the thread = Outside diameter – Pitch = 90 – 34 = 56 mm.

Top View

7. Draw the outside semicircle of diameter 90 mm. Divide it into 6 (= 12/2) equal parts as 1, 2, 3, etc.
8. Draw the inside semicircle of diameter 56 mm. Divide it into 6 (= 12/2) equal parts as 1_1, 2_1, 3_1, etc.

Front View

9. Draw the vertical projectors from the top view to obtain the front view.
10. Mark the square section of the thread in the front view as P′ Q′ R′ S′ as shown.
11. Mark P' P_1' = pitch = 34 mm.
12. Divide the pitch P' P_1' into 12 equal parts. Name the division points as 1′, 2′, 3′, etc.
13. Draw the helical curves traced by the points P′ and S′ for the outside diameter.
14. Draw the helical curves traces by the points R′ and Q′ for the inside diameter.
15. Mark the visible and invisible portions of the curves as shown.
16. Draw the second turn of the square thread starting from P_1' , Q_1', R_1' and S_1'.

Problem 16: *(Exercise) Project two complete turns of a single start square thread having outside diameter 120 mm and pitch 30 mm.*

4.3.2 DOUBLE START SQUARE THREAD

In a Double Start Square Threaded Screw, two threads of the same size run parallel to each other. The lead, *i.e.*, the actual distance advanced by the thread per revolution is twice the lead of the single start square threaded screw.

However, the pitch of the square thread remains the same in both the types.

In double start square threaded screw, pitch of the helix = Lead
= 2 x pitch of the screw.

*

9

FREEHAND SKETCHING TECHNIQUES

1. FREEHAND SKETCHING

Freehand sketching is a means of rapidly communicating / recording your technical ideas to your colleagues in your organization. It means making sketches freehand without the use of drawing instruments. It is not drawn to scale, but should be in *good proportion* by eye judgment. It does not mean that it is a crude or inaccurate drawing without care for the rules of projection.

In fact, in a freehand sketch, all the rules of the graphic language (i.e., B.I.S. Conventions) are as rigidly observed as in a scale-drawing. It should contain all the necessary details such as dimensions, etc.

Yes, a fair knowledge of orthographic projections, sections of solids, isometric projection, etc., is a pre-requisite for a good freehand sketching.

1.1 IMPORTANCE OF FREEHAND SKETCHING

1. ***Sketches are more effective than words.*** Proportioning is as important as scale for a drawing.
2. Often engineers are required to draw rapidly freehand sketches of either orthographic views or pictorial views to convey their technical ideas pertaining to a new design of a new product to the fellow engineers, designers, draftsmen, supervisors and technicians.
3. The freehand sketches are later converted into actual scale drawings for production work. Freehand sketch is like *a manuscript in literature prior to printing*.
4. Design engineer prepares pictorial sketches to visualize the shape of any object. Technical ideas and alternative design approaches can be easily shared and discussed between colleagues with the help of freehand sketches.
5. Often, when changes are proposed in a design, old sketches are to be improved and new sketches are prepared freehand. Also, when repairs are to be made in a machine, structure or system, you may have to sketch and explain your ideas to your technical staff.
6. Sales and service engineers use pictorial sketches when writing field reports on client's needs and suggestions.

2. PROPORTIONING IN SKETCHING

2.1 THREE 'P's

The three 'P's essential for a good freehand sketching are: ***Proportion, Practice*** and ***Proficiency***.

NOTE: *To acquire skill in freehand sketching, practice sketching of horizontal, vertical and inclined lines, circles, ellipses, etc., approximately proportional.*

2.2 *THINKING WITH A PENCIL . . .*

Thinking with a pencil is an exclusive mannerism of engineering personnel while engaged in designing an engineering component or sketching it freehand.

Freehand sketching should be done as easily as writing so that you focus on the design aspects by *thinking with a pencil* and not on the techniques of sketching. A young person preparing for an engineering career should understand that he / she must be also to prepare detailed sketches to present his / her ideas to colleagues in an understandable manner.

2.3 *PRACTICE MAKES YOU PERFECT . . .*

A proficiency in freehand sketching is an untold asset in the profession. Students may find how poor their first attempts are in sketching. The proverb ***Practice makes you perfect*** applies very strongly in this case.

Some students can develop a keen eye for proportions in sketching with a little amount of practice. Others, of course, may have poor skill in proportioning a sketch. ***Practice, Practice*** and ***Practice...***

To sketch rapidly, learn to judge lengths, recognize proportions and compare dimensions by *eye judgment*. ***Until you are able to do this, you may practice using rectangular grid paper or follow the following graphical methods:***

*

UNIT – II

SYLLABUS

UNIT – II

Orthographic Projections of Points, Lines and Planes – Axis Inclined to One Plane and Inclined to Both the Planes

10

ORTHOGRAPHIC PROJECTIONS

1. SOLID GEOMETRY

Plane Geometry
In the preceding Chapters 2 to 7, the attention of the student has been dawn to Plane Geometry, where the constructions of the geometrical figures having only two dimensions were discussed.

Solid Geometry

By the term ***Solid Geometry*** or more correctly ***Descriptive Geometry*** is meant, Geometry as used by the engineers for the graphic representation of solids of three dimensions on plane surfaces of two dimensions.

The subject now is developed further to an object or solid having three dimensions, viz., length, breadth and height. But the drawing sheet is a two dimensional surface. The problem is to represent a three dimensional object or solid on a two dimensional drawing sheet, unless shown pictorially. (In the pictorial drawings the depth of the object is visualized.)

Solid Geometry deals with the representation of

(1) ***Points,***
(2) ***Lines,***
(3) ***Plane Surfaces*** and
(4) Solids

in the form of projections on your drawing sheet.

1.1 3-D OBJECTS

We, the engineers design, develop and manufacture mechanical / electrical / electronics machines or structures or buildings. For this purpose, each and every information about the shapes and sizes of the various components (obviously, 3-D objects) of the machines has to be furnished in detail. The graphic language is used to record as well as to communicate these particulars to the manufacturing sector.

The task of recording the shapes and the sizes of 3-D objects on your 2-D drawing sheets is done by the method of projection.

1.2 TECHNICAL DRAWING

An object has Three Dimensions (3D), namely,

- **Length,**
- **Width** and
- **Height / Thickness**

Engineers record the shapes and the sizes of Three-Dimensional objects on a Two-Dimensional (Length and Width) drawing sheet. This is referred to as *Engineering Graphics*, also known as *Technical Drawing.*

Engineers conceive new ideas, design or modify the existing design, draw the drawings of the objects and finally develop the engineering components like machine elements, structures, electrical and electronic components, instruments, etc., and direct the manufacture of them. Engineering drawings of any object, drawn in true sizes and shapes, serve as the original source of information and will be required at each and every stage of the manufacture of the object.

1.3 PICTORIAL DRAWINGS

It will not be possible to show a 3-D object in its three-dimensional true form on your 2-D drawing sheet. However, solids may be drawn on a drawing paper pictorially in its 3-D form, but not in its true shape and size.

Remember ! Pictorial drawings do *not* show the true shape of the solid. Also, it will be rather difficult to understand the *interior features* of the object.

Therefore, pictorial drawings, as such, cannot be used effectively in engineering field for the manufacturing technology, for the simple reason that the drawings of engineering components have to be necessarily in their true shapes and sizes.

2. THEORY OF PROJECTION

2.1 PROJECTION

The concept of *Projection* invented by the French scientist Monge in the year 1800 has made it possible to prepare engineering drawings of engineering components in their true shapes and sizes.

The word ***projection*** is of Latin origin and means to *throw forward.* Thus, *Projection* is defined as an image of an object thrown forward on to a plane by means of straight lines or visual rays. The figure formed by joining various points thus obtained on the plane, is the image of the object and is termed as ***Projection of the object.***

Method of Projection
The task of recording shapes and sizes of 3-D objects on the 2-D drawing sheet is done by the *Method of Projection.*

2.2 THEORY OF PROJECTION

Observer, Object and Visual Rays
In the study of Optics in Physics, you are already familiar with the terms, namely

(i) ***Observer,***
(ii) ***Object*** and
(iii) ***Visual Rays***, *that is*, an object is seen when the light rays called *Visual Rays* coming from the object strike the observer's eye. The size of the image formed in the retina depends on the distance of the observer from the object.

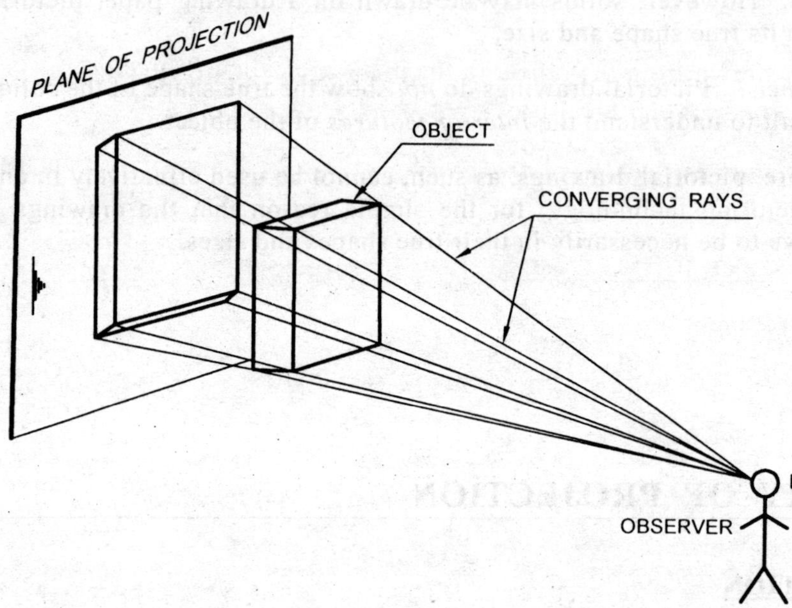

FIG. 1 EYE - OBJECT - PLANE CONCEPT

Imaginary Transparent Plane

See Fig. 1. Suppose, now an *Imaginary Transparent Plane* is introduced such that the object is in between the observer and the plane. The image obtained on the plane is as shown. This is called ***Perspective View***. The arrangement is termed as EYE – OBJECT – PLANE concept.

Here, straight lines (*rays*) are drawn from various points on the contour of the object to meet the transparent plane, thus the object is said to be *projected* on that plane from the eye of the observer.

2.3 TERMINOLOGY

View

The figure formed on the plane by the projection of all the faces of an object is called a *View*. The size of the view depends on the distance from the observer to the plane and the distance from the plane to the object.

Projection
The figure or view formed by joining, in correct sequence, the points at which the rays meet the plane is called the *Projection of the object*. (It is obvious that the outlines of the shadow on the plane show the projection of the object).

Projectors
The lines or rays drawn from the object to the plane are called *Projectors*.

Plane of Projection
The transparent plane on which the projections are drawn is known as Plane of Projection.

2.4 METHODS OF PROJECTION

1. Pictorial Projections
   - Perspective Projection
   - Isometric Projection
   - Oblique Projection
2. Orthographic Projection

2.5 PICTORIAL PROJECTIONS

The projections in which the description of the object is completely understood in one view is known as *Pictorial Projection*. The pictorial projections have the advantage of conveying an immediate impression of the general shape and details of the object, but not its true dimensions or sizes.

Perspective Projection
In Perspective Projection, the projecting lines or visual rays converge at a point as shown in *Fig. 1*. Hence it is termed as *Convergent Projection*. The picture is formed on the picture plane by the piercing points of the projecting lines from the eye to the object.

Perspective Projections are *not* used by engineers for manufacturing and construction, because the perspective view does not reveal the exact size and shape. Perspectives may be used in marketing where a natural view of a product is desirable.

Isometric Projection
Isometric Projection gives the true shape of the object, but not the true size. It shows three sides of the object in one projection.

Oblique Projection
In this, the projectors are not perpendicular to the plane of projection.

2.6 ORTHOGRAPHIC PROJECTIONS

In orthographic projection, an object is represented by two or three views on the mutually perpendicular projection planes.

3. PRINCIPLES OF ORTHOGRAPHIC PROJECTION

What is meant by Orthographic Projection ?

ORTHO means ***Right-angle;***
GRAPHIC means ***Drawing***;
ORTHO-GRAPHIC means ***Right-angled Drawing***.

When the projectors are perpendicular to the plane on which the projection is obtained and are also parallel to each other, it is known as ***Orthographic Projection***. It is the projection system that engineers use for all types of manufacturing and construction drawings.

To obtain the view of an object to its *same size*, the visual rays should be made parallel to each other and perpendicular to the plane of projection. This will be possible only when the observer is at ***infinity***.

NOTE : 1. *Orthographic Projection is the standard drawing form of the industrial world. The form is unreal in that we do not see an object as it is drawn orthographically.*
2. *Pictorial drawing, however, has photographic realism.*

3.1 VERTICAL PLANE (V.P.)

See Fig. 2(a). Imagine that the observer looks at the object (*a rectangular block*) from an ***infinite*** distance. The rays will now be *parallel* to each other and perpendicular to both the front surface of the object and the plane. (When the observer is at a *finite* distance from the object, the rays converge to the eye as in the case of Perspective Projection.)

The observer looks at the block from the front as shown. The front surface **F** of the block of Length L and Height H, is seen in its true shape and size. Extend the rays, called *projectors* further to meet a ***transparent Vertical Plane*** (*denoted as* ***V.P.***) located behind the object. Join the points at which the projectors meet the plane, in correct sequence, to obtain the orthographic projection.

Orthographic View

Since the visual rays, called projectors are perpendicular, *i.e.*, orthogonal to the plane of projection, the projected view is called *Orthographic View*.

3.2 FRONT VIEW

See Fig. 2(a). The orthographic view is obtained by projecting the object when the observer is viewing it from the front. Hence it is called the *View from the front* or **Front View (F_V)** of the object. Both F and F_V are of exactly the same shape and size.

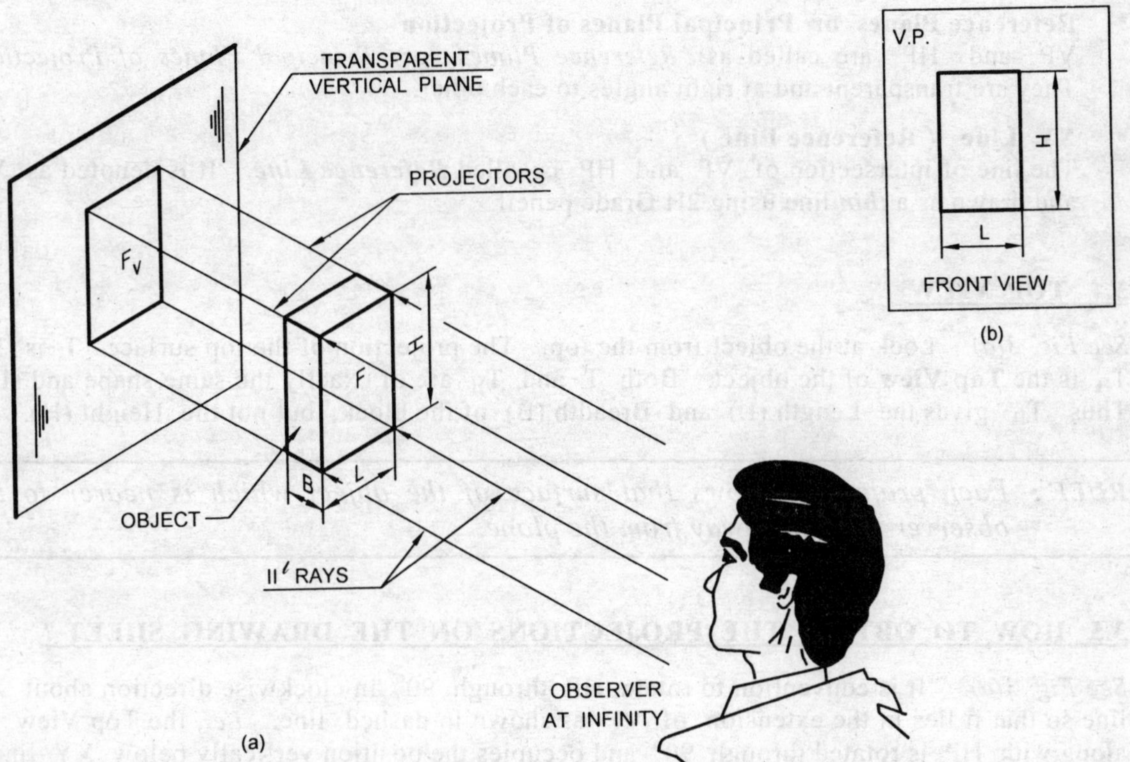

FIG. 2
ORTHOGRAPHIC PROJECTION - FRONT VIEW

See Fig. 2(b). Front View is shown separately with its Length (L) and Height (H).

Front View shows only two dimensions of the object, *viz.,* Length (L) and Height (H). It does not show the Breadth (B). Thus Front View alone cannot describe the shape and sizes of the object completely, as it shows only two out of three dimensions, *viz.,* L and H. It does not show the Breadth (B) of the object, since it lies along the direction of sight while looking from the front.

3.3 HORIZONTAL PLANE (H.P.)

In order to show the Breadth (B) of the object, one more view is necessary. Generally, the View from above or View from top, called Top View is projected to show the breadth of the object.

See Fig. 3(a). Let us assume another plane, called Horizontal Plane (HP) hinged perpendicular to VP such that the object is in front of VP and above HP

* **Reference Planes or Principal Planes of Projection**
  VP and HP are called as *Reference Planes* or *Principal Planes of Projection.* They are transparent and at right angles to each other.

* **XY Line (Reference Line)**
  The line of intersection of VP and HP is called ***Reference Line.*** It is denoted as XY and drawn as a *thin* line using 2H Grade pencil.

3.4 TOP VIEW

See Fig. 3(a). Look at the object from the top. The projection of the top surface T is T_H. T_H is the **Top View** of the object. Both T and T_H are of exactly the same shape and size. Thus T_H gives the Length (L) and Breadth (B) of the block, but not the Height (H).

RULE : *Each projection shows that surface of the object which is nearer to the observer and far away from the plane.*

3.5 HOW TO OBTAIN THE PROJECTIONS ON THE DRAWING SHEET ?

See Fig. 3(a). It is convention to rotate HP through 90° in clockwise direction about XY line so that it lies in the extension of VP as shown in dashed line. *i.e.*, the Top View T_H along with HP is rotated through 90° and occupies the position vertically below XY line.

See Fig. 3(b). Now the two projections F_V and T_H are drawn on your two dimensional drawing sheet as shown.

Thus all the details regarding the shape and size, *viz.*, Length (L), Height (H) and Breadth (B) of a simple object may be represented by means of the orthographic projections, *viz.*, Front View and Top View.

NOTE:

1. ***Front View:*** *Projection on the VP is known as Front View.*
   Front view shows Length (L) and Height (H) of the object.
2. ***Top View:*** *Projection on the HP is known as Top View.*
   Top View shows Length (L) and Breadth (B) of the object.
3. ***Front and Top Views are drawn as thick lines (H Grade Pencil).***

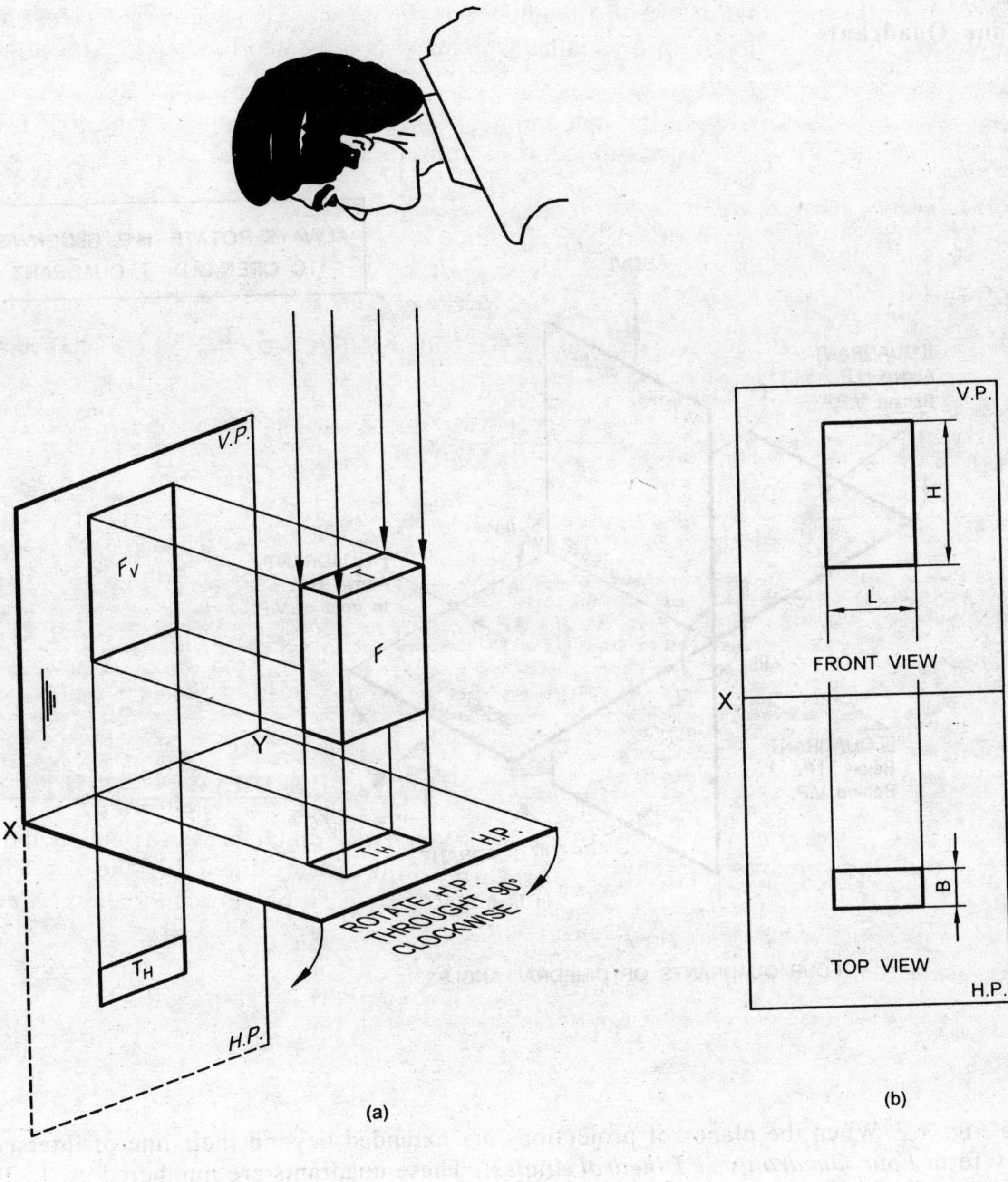

FIG. 3

FRONT VIEW AND TOP VIEW

4. FIRST ANGLE AND THIRD ANGLE PROJECTIONS

Four Quadrants

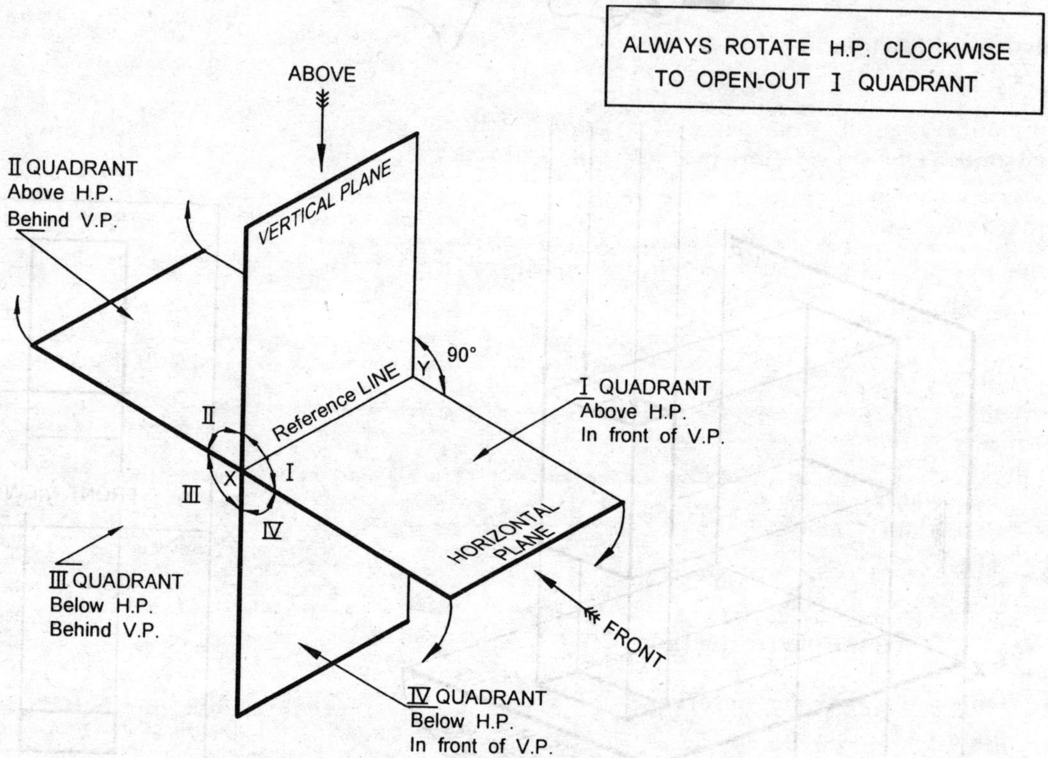

FIG. 4
FOUR QUADRANTS OR DIHEDRAL ANGLES

See Fig. 4. When the planes of projections are extended beyond their line of intersection, they form *Four Quadrants* or *Dihedral Angles*. These quadrants are numbered as I, II, III and IV in counter-clockwise direction as shown.

Position of the Observer

The object may be situated in any one of the four quadrants. The position of the object relative to the reference planes is described as *in front of VP* and *above HP* in First Quadrant and so on.

The observer will always be on the right side of the four quadrants and in front of VP to obtain the Front View and above HP to obtain the Top View, irrespective of the position of the object in any of the four quadrants.

First Angle Projection

NOTE : *Fig. 3 shows* ***FIRST ANGLE PROJECTION,***
- object lies in front of VP and above HP, i.e., in ***FIRST QUADRANT.***

Third Angle Projection

See Fig. 4. In Third Angle Projection, the object is situated in the ***Third Quadrant***. The planes of projection, namely, HP and VP lie in between the object and the observer. The front view comes below the XY line and the top view above it.

Points to Remember

1. In orthographic projection, the frontal and horizontal projection planes are assumed to extend infinitely in space, four dihedral angles (90°) are formed and are designated as the First, Second, Third and Fourth Quadrants. Orthographic projection system is used by engineers for manufacturing and construction drawings.
2. Perspective projection is not used by engineers for manufacturing and construction drawings, because the perspective view does not reveal the exact size and shape. Perspective views are used in marketing where a natural view of a product is desirable.

4.1 FIRST ANGLE PROJECTION *Vs* THIRD ANGLE PROJECTION

First Angle Projection	Third Angle Projection
1. Object lies between observer and the planes of projection.	1. Planes of projection lie between the object and observer.
2. Object is situated on or above the HP and in front of VP.	2. Object is situated on or above the ground, *i.e.*, below HP and behind VP.
3. Front view comes above top view.	3. Top View comes above front view.
4. Symbol: 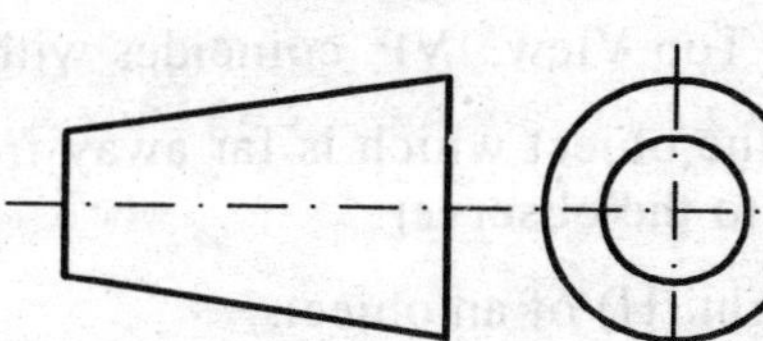	4. Symbol: 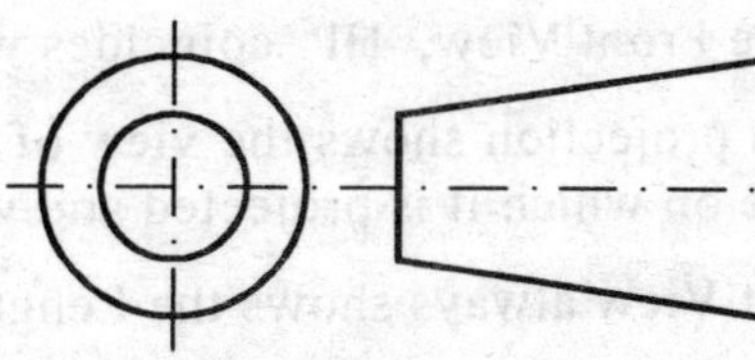

5. BIS CONVENTIONS [IS 15021 (Parts 1 - 4) – 2001]

B.I.S. has recommended the use of First Angle Projection in Engineering Drawing in line with the specifications of I.S.O., being adopted by all the countries of the world, since 1983. BIS Conventions are specified by IS 15021 (Parts 1 - 4) – 2001 Technical Drawings – Projection Methods.

- ***Projectors***

  In pictorial views, draw the *projectors* from the points in space to HP or VP in *thin* lines. Lines from the projections to the XY line are also called *projectors*, though they are the projections of the projectors.

- Projectors as well as the XY line are drawn as *thin* lines.

- HP and VP may also be written on either side of the XY line in orthographic views.

5.1 CONVENTIONS OF FIRST ANGLE PROJECTION

When the object is situated in First Quadrant, *i.e.*, in front of VP and above HP, the projections obtained on these principle planes is called **First Angle Projection**.

Figs. 2 and 3 explain the ***concept*** of First Angle Projection as follows:-

1. Object lies in between the observer and HP and VP, *i.e.*, in front of VP and above HP.
2. Object is then projected on VP to obtain front view and on HP to obtain top view.
3. First and Third Quadrants are always opened out while rotating the HP.
4. For simplicity sake, view from front is called Front View and view from above Top View.
5. When the views are drawn in their relative position, Top View comes below Front View.
6. Front View is drawn above the XY line and Top View is drawn below the XY line. (Here, above XY line represents the VP and below XY line represents the HP).
7. In the Front View, HP coincides with XY and in Top View, VP coincides with XY.
8. Each projection shows the view of that surface of the object which is far away from the plane on which it is projected and which is nearest to the observer.
9. Front View always shows the Length (L) and Height (H) of an object.
10. Top View always shows the Length (L) and Breadth (B) or Width (W) of it.

6. LAYOUT OF VIEWS

For laying the views of a 3-D object on your 2-D drawing sheet, the reference planes are rotated or revolved away from the object. Two orthographic views, *viz*., front view and top view, may be necessary to represent a symmetrical object. However, complex objects require three views, namely, front view, top view and a side view.

6.1 PROFILE PLANES AND SIDE VIEWS

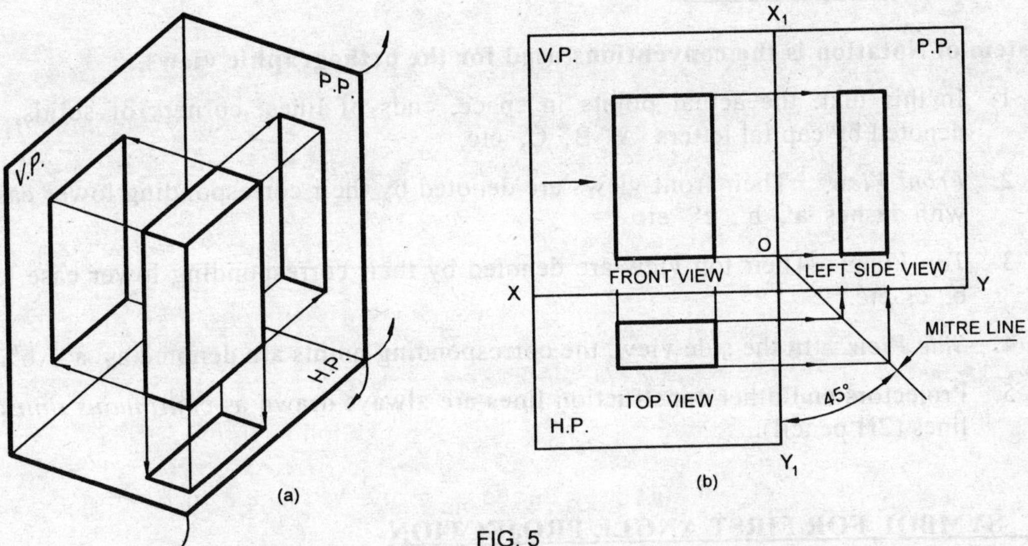

FIG. 5

FRONT VIEW, TOP VIEW AND LEFT SIDE VIEW

Profile Plane (PP): Sometimes some features of an object lie on its left or right side. Top and Front Views will not show these features in their true shape. In such cases, you are required to project the object either on a Left or Right Side Plane, known as *Profile Plane (PP)* and the view projected on it is known as *Profile View* or *Side View*.

Right Side PP and Left Side PP : In the First Angle Projection, the object lies in between the observer and the plane. Therefore, *Left Side View* will be projected on the Right Side Profile Plane and drawn on the right side of the front view. *Right Side View* is projected on the Left Side Profile Plane and drawn on the left side of the front view.

Left Side View: *See Fig. 5(a).* It shows HP, VP and a Right Profile Plane (RPP) drawn perpendicular to both VP and HP. RPP is used to project the Left Side View as shown. *See Fig. 5(b).* Draw Front & Top Views. Look from left. Project Left Side View on RPP.

Rotate the RPP in the direction of the arrow shown in *Fig. 5(a)*, so as to make it to coincide with the VP [*Fig. 5(b)*]. Look the object from the left to obtain the Left Side View and draw it on the right side of the Front View.

How to draw Left Side View using 45° Mitre Line ?

1. Draw X_1Y_1 perpendicular to XY to meet at O. Through O draw a 45° (Mitre) line.
2. Draw horizontal projectors from all corners of the Top View to intersect the 45° line.
3. Through these points of intersection, draw vertical projectors as shown in *Fig. (b).*
4. Also, draw horizontal projectors from all the corners of the Front View to intersect the previously drawn vertical projectors.
5. Join the corresponding points and complete the Left Side View as shown in *Fig. (b).*

6.2 SYSTEM OF NOTATION

System of Notation is the conventions used for the orthographic views.

1. In this text, the actual points in space, ends of lines, corners of solids, etc., are denoted by capital letters A, B, C, etc.
2. *Front View:* Their front views are denoted by their corresponding lower case letters with dashes a′, b′, c′, etc.
3. *Top View:* Their top view are denoted by their corresponding lower case letters a, b, c, etc.
4. *Side View:* In the side view, the corresponding points are denoted as a″, b″, c″, etc.
5. Projectors and other construction lines are always drawn as *continuous thin (narrow)* lines (2H pencil).

6.3 SYMBOL FOR FIRST ANGLE PROJECTION

BIS recommends the Front View and Left Side View of the frustum of a cone placed with its axis horizontal as the Symbol of First Angle Projection.

This symbol shall be indicated in the space provided for the purpose in the Title Block of the drawing sheet. *See Fig. 6.*

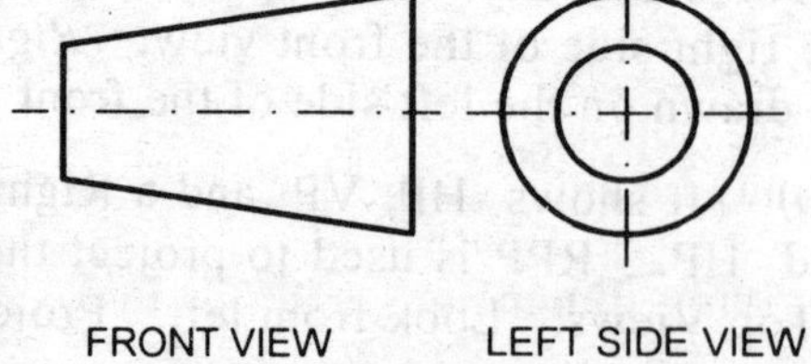

FIG. 6

6.4 LAYOUT OF VIEWS

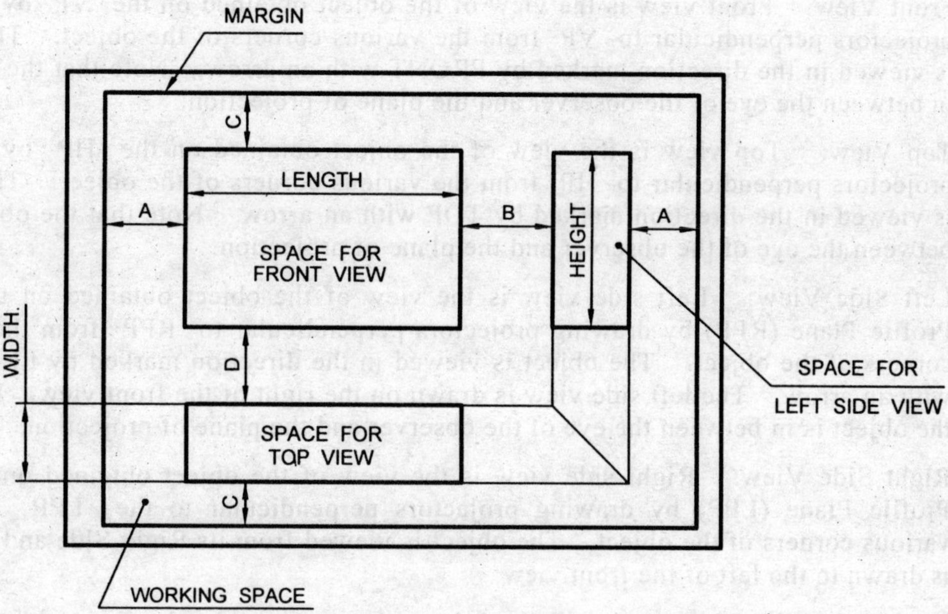

FIG. 7 LAYOUT OF VIEWS

Fig. 7 shows the layout of a three view drawing. Spacing for the three views, *viz.*, front view, top view and left side view are shown as rectangular blocks.

Spaces A shown in the left and right sides of the drawing sheet may be kept approximately equal. Space B may be equal to or slightly greater than either of the spaces A.

Spaces C shown in the top and bottom of the drawing sheet may be kept approximately equal. Space D may be equal to or slightly greater than the space C.

Further adjustments, for giving dimensions and any notes on the particular drawing, may be made accordingly.

6.5 POINTS TO REMEMBER

1. Object is situated in the first quadrant with its main faces parallel to the three principal planes of projections. Therefore, these parallel faces will have the true shape and size in their respective views.

2. Front View: Front view is the view of the object obtained on the VP by drawing projectors perpendicular to VP from the various corners of the object. The object is viewed in the direction marked by FRONT with an arrow. Note that the object is in between the eye of the observer and the plane of projection.

3. Top View: Top view is the view of the object obtained on the HP by drawing projectors perpendicular to HP from the various corners of the object. The object is viewed in the direction marked by TOP with an arrow. Note that the object is in between the eye of the observer and the plane of projection.

4. Left Side View: Left side view is the view of the object obtained on the Right Profile Plane (RPP) by drawing projectors perpendicular to RPP from the various corners of the object. The object is viewed in the direction marked by LEFT SIDE with an arrow. The left side view is drawn on the right of the front view. Note that the object is in between the eye of the observer and the plane of projection.

5. Right Side View: Right side view is the view of the object obtained on the Left Profile Plane (LPP) by drawing projectors perpendicular to the LPP from the various corners of the object. The object is viewed from its Right Side and the view is drawn to the left of the front view.

6. Transfer the depth dimensions from the top view to the side view, and vice versa, by means of the 45° mitre line.

7. In practice, the depth dimensions from the top view to the side view, or vice versa, are transferred by using the divider, which is faster and more accurate than the 45° line method.

*

11

REPRESENTATION OF 3-D OBJECTS

1. REPRESENTATION OF 3-D OBJECTS

Engineers use Orthographic Views, viz., Front, Top and Left or Right Side Views to represent the shape and size of 3D objects on a 2D drawing sheet.

1.1 ONE VIEW DRAWINGS

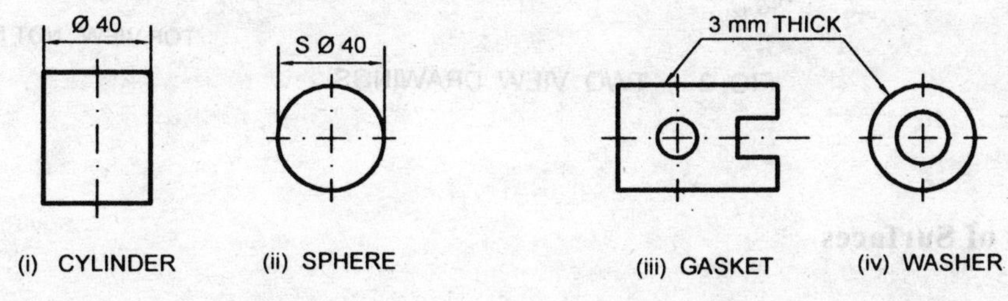

FIG. 1

See Fig. 1. A single view is sufficient for simple objects, if a *qualifying note* is added.

For example, a cylinder is drawn as a rectangle and the symbol ϕ for diameter is added to the dimension *[Fig. (i)]*.

A sphere looks as a circle in all the views. Hence a circle with a note Sϕ denotes a sphere *[Fig. (ii)]*.

Flat objects like washers, shims, gasket, etc., are represented with thickness mentioned *[Figs. (iii) and (iv)]*.

1.2 TWO VIEW DRAWINGS

Two views can explain many objects *symmetrical* about their axes, provided they are correctly chosen. Careful analysis of the object will enable you to determine whether a top or a side view is necessary after the front view is selected.

When two views of an object are insufficient to explain all the details of the object, then the third view becomes necessary.

Problem 1: *(Fig. 2) Represent the Steel Block in orthographic views.*

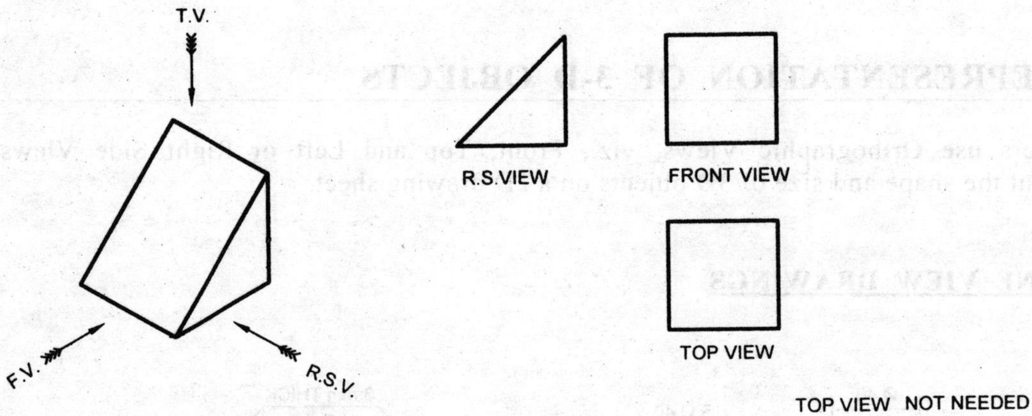

FIG. 2 TWO VIEW DRAWINGS

Identification of Surfaces

Problem 2: *(Fig. 3) Draw the orthographic views of the Wedge Piece shown in Fig. 3(i):*

1. *See Fig. 3(a).* Identify or recognize the visible surfaces of the object in its pictorial view. Name them using numerals or alphabets.
2. The surface 1 is parallel to VP and perpendicular to HP. Its true shape is visible in the front view.
3. But it is seen as *lines* in the front and top views. *See Fig. 3(b).* These lines may be indicated by the corresponding numerals enclosed in circles.
4. Surface 2 is seen as *rectangles* in the front view and top view. But it is seen as an *arc* in the right side view.

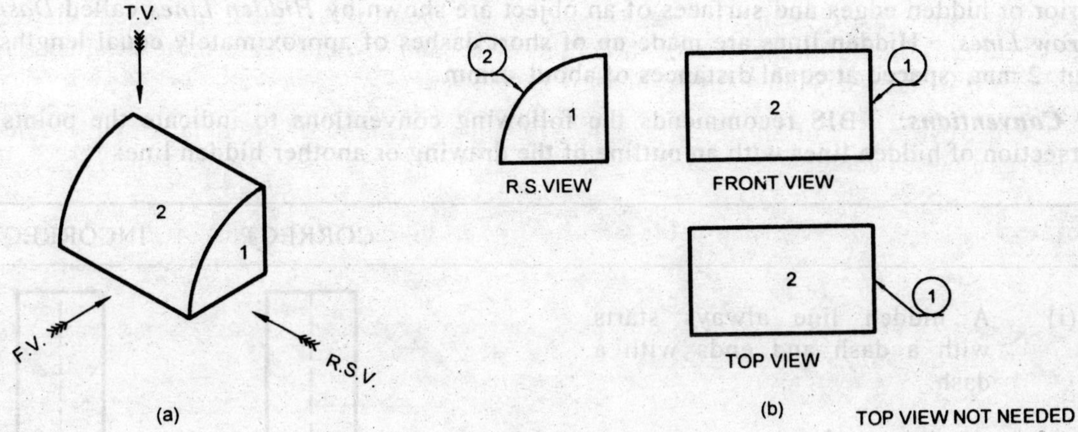

FIG. 3 TWO VIEW DRAWINGS

1.3 THREE VIEW DRAWINGS

Irregular or unsymmetrical objects require three views for complete shape description.

See Fig. 4.

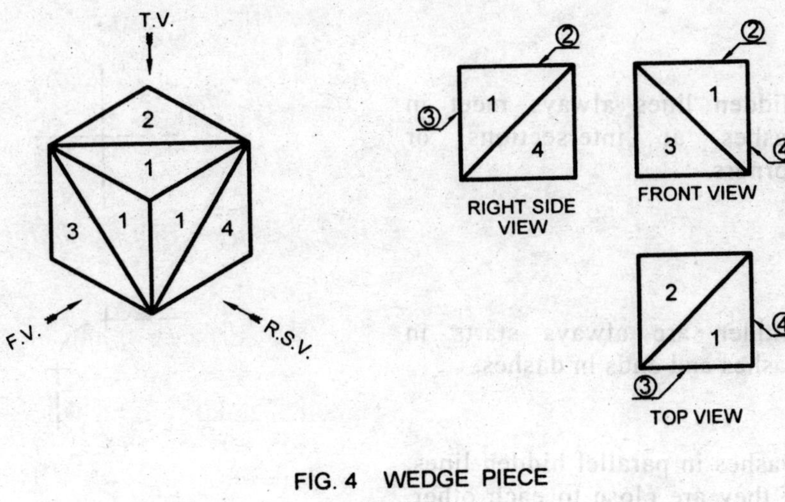

FIG. 4 WEDGE PIECE

2. HIDDEN LINES

Interior or hidden edges and surfaces of an object are shown by *Hidden Lines* called *Dashed Narrow Lines*. Hidden lines are made up of short dashes of approximately equal lengths of about 2 mm, spaced at equal distances of about 1 mm.

BIS Conventions: BIS recommends the following conventions to indicate the points of intersection of hidden lines with an outline of the drawing or another hidden line:

	CORRECT	INCORRECT
(i) A hidden line always starts with a dash and ends with a dash.	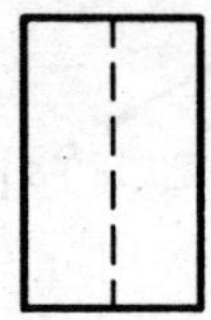	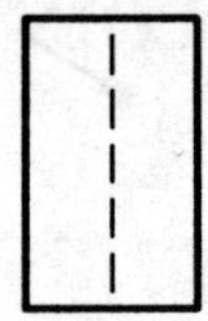
(ii) A hidden line, when drawn in continuation of a visible line, always starts with a space.	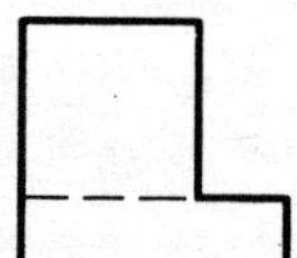	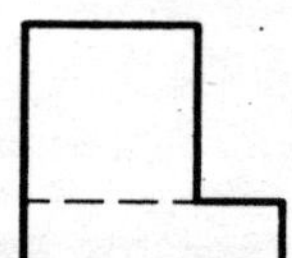
(iii) Hidden line jumps the visible line.	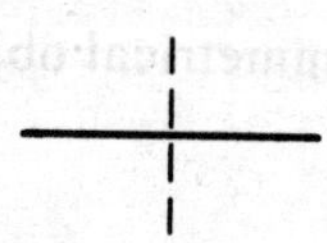	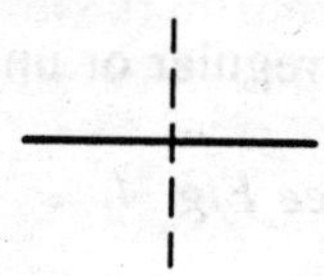
(iv) Hidden lines always meet in dashes at intersections or corners.		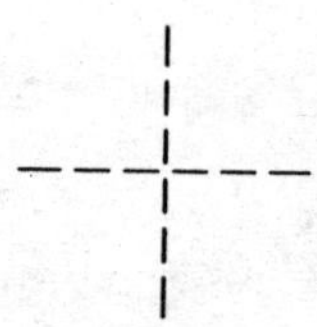
(v) Hidden arc always starts in dashes and ends in dashes.	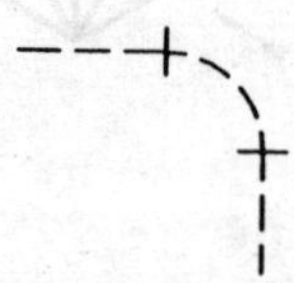	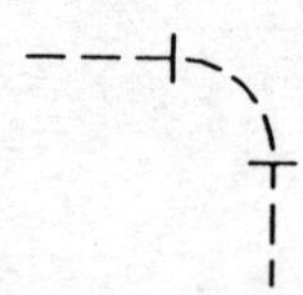
(vi) Dashes in parallel hidden lines, if they are close to each other, shall be drawn staggered.	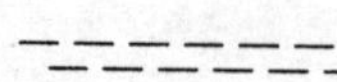	

*

12

PROJECTIONS OF POINTS

1. POINTS

Solid – Plane Surface – Line – Point

A *solid* is formed by three dimensions measured in three mutually perpendicular directions. If one of the dimensions is made zero, the object is converted into a 2-D *plane surface*. If one more dimension is also reduced to zero, the plane is converted into a *line*. Finally, if the remaining dimension is also removed, the line is of zero length and becomes a *point*. Therefore, a point in 3-D geometry is the smallest dimensionless size of a solid.

1.1 POINTS IN SPACE

A point may lie in space, in any one of the four quadrants formed by two reference planes HP and VP. A *point* is represented by a *dot* in the drawings.

1.2 LOCATIONS OF A POINT

When a point lies in the **First Quadrant,** it will be **above HP** and **in front of VP.**

When the point lies in the **Second Quadrant,** it will be **above HP** and **behind VP.**

When the point lies in the **Third Quadrant,** it will be **below HP** and **behind VP.**

When the point lies in the **Fourth Quadrant,** it will be **in front of VP** and **below HP.**

After knowing the distances of a point from HP and VP, projections on HP and VP are found by extending the projectors perpendicular to both the reference planes. Projection on HP is called **Top View** and projection on VP is called **Front View.**

NOTE :

1. *Students are advised to make their own paper / cardboard / perspex model of HP and VP. The model will facilitate developing a good concept of the relative positions of the points lying in any of the four quadrants.*
2. *Since the Projections of Points, Lines and Planes are the basic Chapters for the forthcoming topics on Solids (viz., Projections of Solids, Sections, Development of Surfaces, etc.), the students should follow these basic Chapters carefully to draw the orthographic projections.*
3. *In this Chapter, a series of pictorial views are illustrated to visualize easily the different positions occupied by a point with respect to HP and VP.*

2. BIS CONVENTIONS

BIS recommends the following conventional representations in orthographic projections, similar to that of a language. These conventions are obeyed internationally and a variation will be treated as a spelling or grammar mistake in the graphics language:

1. The actual *points* are denoted by capital letters A, B, C, etc., in the pictorial view.
2. In the *Front View*, the points are denoted by the corresponding lower-case letters with dash a′, b′, c′, etc. In the *Top View*, the points are denoted by the corresponding lower-case letters a, b, c, etc. In the *Side View*, the points are denoted by the corresponding lower-case letters with double dashes a″, b″, c″, etc.
3. The intersection line of HP and VP is shown as the reference line XY
4. Projectors are to be necessarily shown in the orthographic views in *thin lines* (2H pencil).

3. POINT IN FIRST QUADRANT

Problem 1: *(Fig. 1) Point A is 30 mm above HP and 45 mm in front of VP. Draw its Front View and Top View.*

1. *See Fig. (i).* Look at the Pictorial View. Point **A** lies in the first quadrant.
2. **To obtain the Front View a′, look at the Pictorial View from the front**

   Point **A** is 30 mm above HP. **Aa′** is the projector perpendicular to the VP. Hence **a′** is the Front View of the point **A** and it is 30 mm above the XY line.
3. **To obtain the Top View a , look at the Pictorial View from the top**

   Point **A** is 45 mm in front of VP. **Aa** is the projector perpendicular to the HP. Hence **a** is the Top View of the point **A** and it is 45 mm in front of XY.
4. **To convert projections a′ and a from Pictorial View into Orthographic Projections**
   - Rotate HP about XY line through 90° in the CW direction as shown in *Fig. (i).*
   - After rotation, the first quadrant is opened out and the HP occupies the position (dashed lines) vertically below the VP.
   - Also, the point **a** on HP will trace a quadrant of a circle with **o** as center and **oa** as radius.
   - Now, **a** occupies the position just below **o**.
   - *See Fig .(ii).* Line joining **a′** and **a**, called *projector*, is perpendicular to XY.

How to draw the orthographic projections on the drawing sheet : *Fig. (iii).*

5. **Front View :** *See Fig. (iii).* Draw the XY line and draw a projector at any point on it. Mark the Front View **a′** 30 mm above XY on the projector.
6. **Top View :** On the *same* projector, mark the Top View **a** 45 mm below XY.

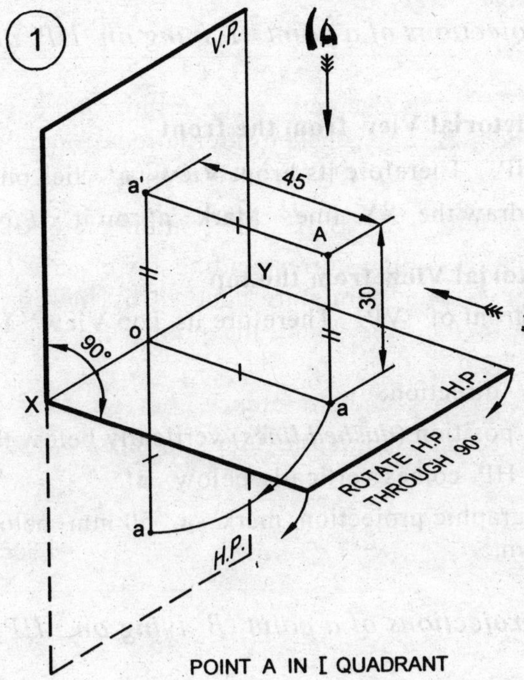

POINT A IN I QUADRANT

(i)

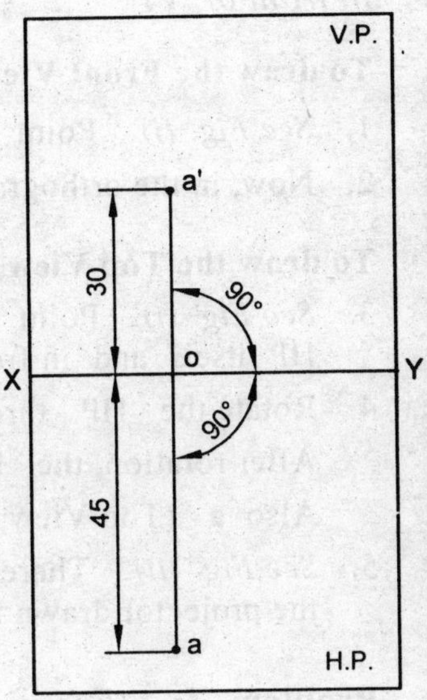

H.P. AFTER ROTATION

(ii)

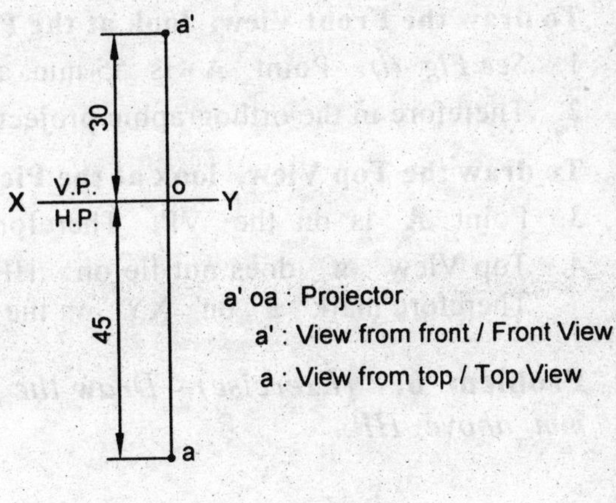

(iii) ORTHOGRAPHIC PROJECTION

FIG. 1

RULE : 1. *XY represents HP in Front View and VP in Top View. So, while drawing Front and Top Views on your drawing sheet, the squares or rectangles for individual imaginary planes HP and VP are unnecessary and are to be discarded.*

2. *Draw only the orthographic views shown in Fig. (iii) and not Figs. (i) and (ii).*

Problem 2: *(Exercise) A point M is 35 mm above HP and 40 mm in front of VP. Draw its projections.* *(Projections mean Front and Top Views.)*

Problem 3: *(Fig. 2) Draw the projections of a point A lying on HP and 50 mm in front of VP.*

To draw the Front View, look at the Pictorial View from the front

1. *See Fig. (i).* Point **A** is lying on HP. Therefore its Front View **a′** lies on XY line.
2. Now, in the orthographic projection, draw the XY line. Mark **a′** on it. *Fig. (ii).*

To draw the Top View, look at the Pictorial View from the top

3. *See Fig. (i).* Point **A** is 50 mm in front of VP. Therefore its Top View **a** lies on HP itself and in front of XY.
4. Rotate the HP through 90° in CW direction.

   After rotation, the HP occupies the position (*dashed lines*) vertically below the VP.

   Also **a** (Top View of the point) on HP comes vertically below **a′**.
5. *See Fig. (ii).* Therefore, in the orthographic projection, mark **a** 50 mm below XY on the projector drawn from **a′** as shown.

Problem 4: *(Exercise) Draw the projections of a point B lying on HP and 55 mm in front of V.P.*

Problem 5: *(Fig. 3) Draw the projections of a point A lying on VP and 55 mm above HP*

To draw the Front View, look at the Pictorial View from the front

1. *See Fig. (i).* Point **A** is 55 mm above HP. Hence **a′** is 55 mm above XY.
2. Therefore in the orthographic projection *(Fig. ii)*, mark **a′** 55 mm above XY.

To draw the Top View, look at the Pictorial View from the top

3. Point **A** is on the VP. Therefore, its Top View lies on XY itself.
4. Top View **a** does not lie on HP. Thus, in this case, the HP need *not* be rotated. Therefore mark **a** on XY on the projector drawn from **a′**.

Problem 6: *(Exercise) Draw the projections of a point Q lying on VP and 58 mm above HP.*

Problem 7: *(Fig. 4) Draw the projections of a point F which lies in both the HP and the V.P.*

See Fig. 4 for Solution.

NOTE: 1. *Distance of Front View from XY = Distance of the point from HP.*
2. *Distance of Top View from XY = Distance of the point from VP.*

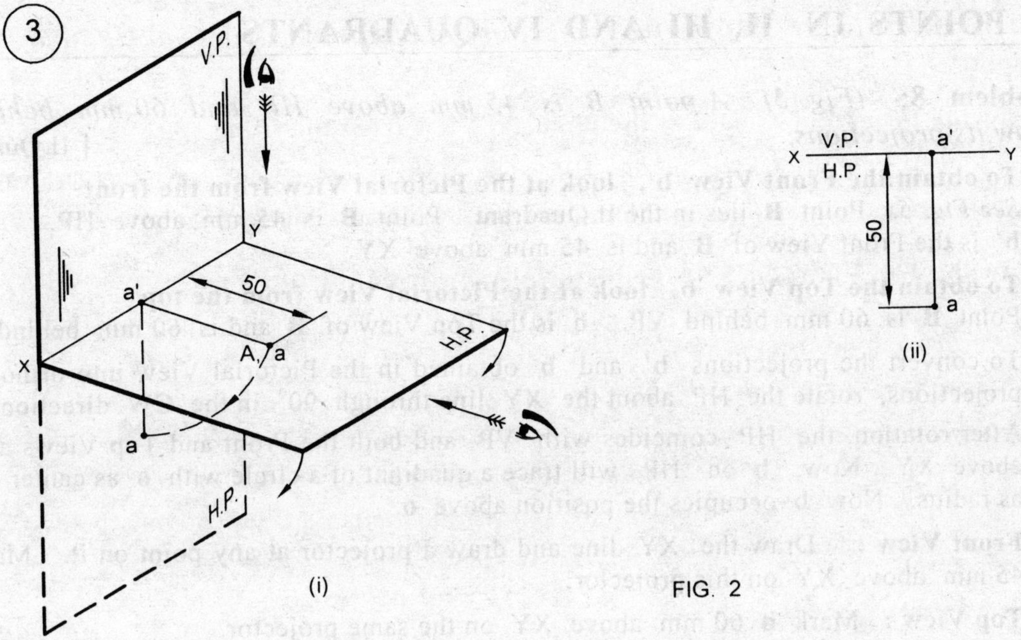

FIG. 2

FIG. 3

FIG. 4

4. POINTS IN II, III AND IV QUADRANTS

Problem 8: *(Fig. 5) A point B is 45 mm above HP and 60 mm behind VP. Draw its projections.*

[II Quadrant]

1. **To obtain the Front View b′, look at the Pictorial View from the front**
   See Fig. 5. Point **B** lies in the II Quadrant. Point **B** is 45 mm above HP. **b′** is the Front View of **B** and is 45 mm above XY.
2. **To obtain the Top View b, look at the Pictorial View from the top**
   Point **B** is 60 mm behind VP. **b** is the Top View of **B** and is 60 mm behind XY.
3. To convert the projections **b′** and **b** obtained in the Pictorial View into orthographic projections, rotate the HP about the XY line through 90° in the CW direction.
4. After rotation, the HP coincides with VP and both the Front and Top Views are seen above XY. Now, **b** on HP will trace a quadrant of a circle with **o** as center and **ob** as radius. Now **b** occupies the position above **o**.
5. **Front View :** Draw the XY line and draw a projector at any point on it. Mark **b′** 45 mm above XY on this projector.
6. **Top View :** Mark **b** 60 mm above XY on the same projector.

Problem 9: *(Exercise) A point S is 35 mm above HP and 55 mm behind VP. Draw its projections.*

Problem 10: *(Fig. 6) A point C is 35 mm below HP and 25 mm behind VP. Draw its projections.*

[III Quadrant]

Front View : *See Fig. 6.* **C** is 35 mm below HP. Hence **c′** is 35 mm below XY.
Draw XY and draw a projector at any point on it.
Mark **c′** 35 mm below XY on the projector.

Top View : **C** is 25 mm behind VP. Therefore **c** is 25 mm behind XY.
Now in the orthographic projections mark **c** 25 mm above XY on the above projector.

Problem 11: *(Exercise) A point D is 35 mm below HP and 35 mm behind VP. Draw its projections.*

Problem 12: *(Fig. 7) A point D is 45 mm below HP and 60 mm in front of VP. Draw its projections.*

[IV Quadrant]

Front View : *See Fig. 7.* **D** is 45 mm below HP. Hence **d′** is 45 mm below XY.
Draw XY line and draw a projector perpendicular to it.
Mark **d′** 45 mm below XY on the projector.

Top View : **D** is 60 mm in front of VP. Therefore **d** is 60 mm in front of XY.
Now mark **d** 60 mm below XY.

Problem 13: *(Exercise) A point M is 60 mm below HP and 45 mm in front of VP. Draw its projections.*

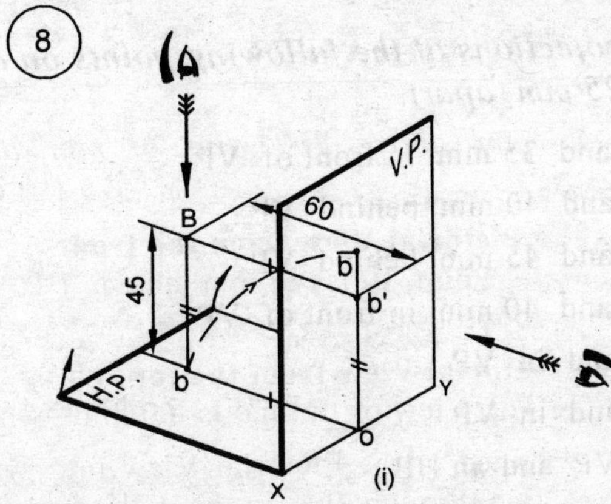

(i)

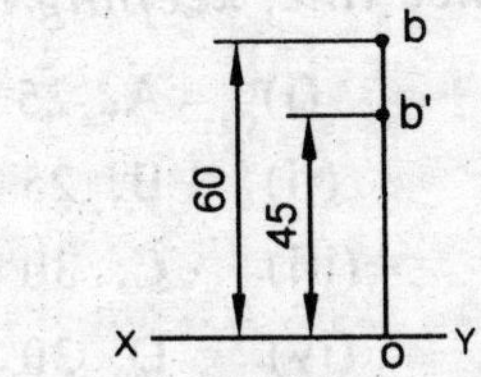

(ii)

FIG. 5
II QUADRANT

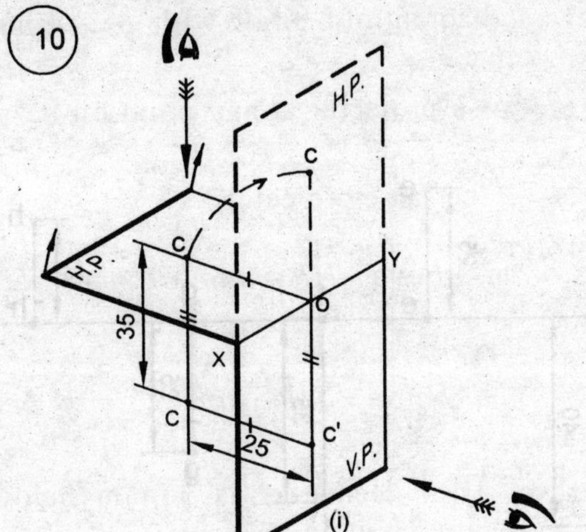

(i)

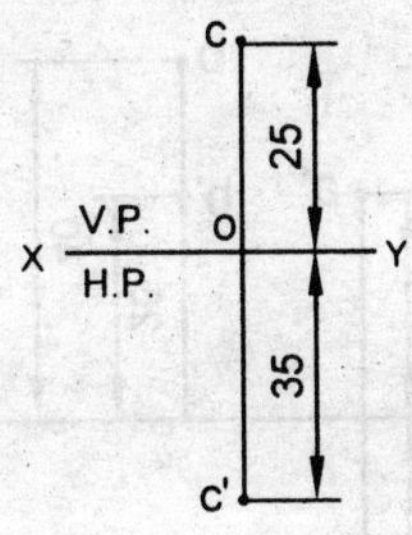

(ii)

FIG. 6
III QUADRANT

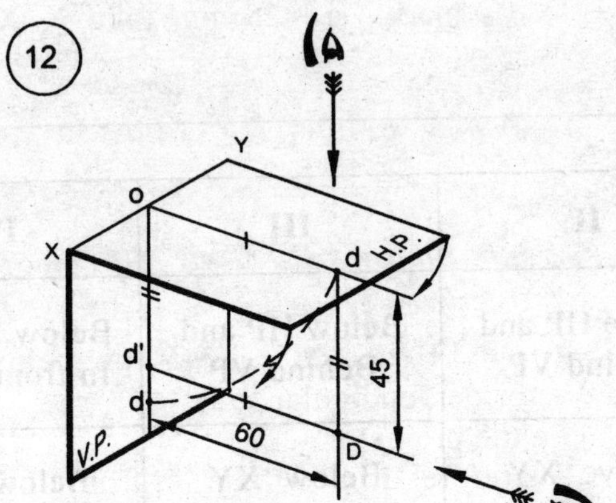

(i)

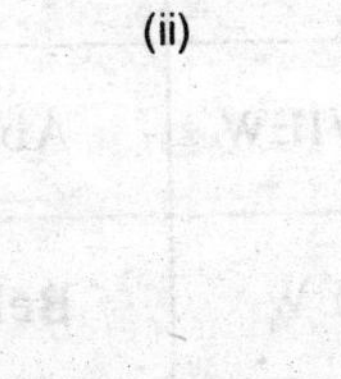

(ii)

FIG. 7
IV QUADRANT

Problem 14: *(Fig. 8) Mark the projections of the following points on a common reference line, keeping the projectors 25 mm apart.*

(i) A, 25 mm above HP and 35 mm in front of VP

(ii) B, 25 mm above HP and 40 mm behind VP

(iii) C, 30 mm below HP and 45 mm behind VP

(iv) D, 30 mm below HP and 40 mm in front of VP

(v) E, 25 mm above HP and in VP

(vi) F, 35 mm below HP and in VP

(vii) G, 25 mm in front of VP and in HP

(viii) H, 20 mm behind VP and in HP

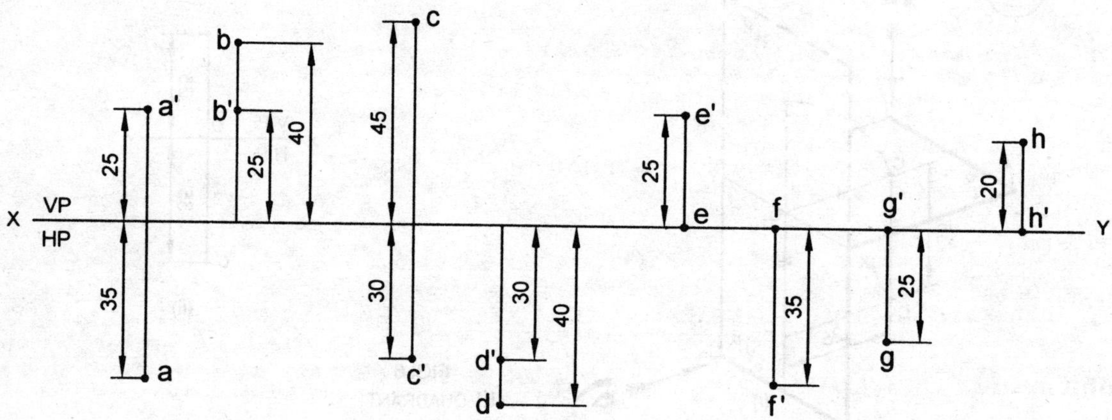

FIG. 8 SOLUTION FOR PROBLEM 14

5. SUMMARY

QUADRANT	I	II	III	IV
POSITION OF THE POINT	Above HP and In front of VP	Above HP and Behind VP	Below HP and Behind VP	Below HP and In front of VP
FRONT VIEW	Above XY	Above XY	Below XY	Below XY
TOP VIEW	Below XY	Above XY	Above XY	Below XY

6. ADDITIONAL PROBLEMS

Problem 15: *(Exercise) Draw the projections of the following points on the same reference line, keeping the projectors 30 mm apart:*

(i) A, 30 mm above HP and 30 mm in front of VP
(ii) B, 40 mm above HP and 30 mm behind VP
(iii) C, 45 mm below HP and 30 mm behind VP
(iv) D, 40 mm below HP and 30 mm in front of VP
(v) E, 40 mm above HP and in VP
(vi) F, 45 mm below HP and in VP
(vii) G, 40 mm in front of VP and in HP
(viii) H, 45 mm behind VP and in HP
(ix) I, on both HP and VP.

Problem 16: *(Exercise) Mention the relative positions of the projections of the following points with respect to the reference line XY :*

(i) P in the first quadrant
(ii) Q in the third quadrant
(iii) R in the second quadrant
(iv) S in the fourth quadrant.

Problem 17: *(Exercise) Name the quadrants in which the following points are located :*

(i) P -- Front view above XY and Top view below XY
(ii) Q – Front and Top views are above XY
(iii) R – Front and Top views are below XY
(iv) S – Front view below XY and Top view above XY

Problem 18: *(Exercise)* *The projections of a point coincide and both fall below XY. In which quadrant is the point situated?* [Ans: IV QUADRANT]

Problem 19: *(Exercise)*

Two points P and Q are in the HP. The point P is 35 mm in front of the VP, while Q is behind the VP. The distance between their projectors is 75 mm and the line joining their top views makes an angle of 45° with XY.

Find the distance of the point Q from the VP.

*

13

PROJECTIONS OF STRAIGHT LINES

1. STRAIGHT LINES

A *straight line* is the shortest distance between two given points. Thus, the two ends of a straight line are points. Projection of a Straight Line can therefore be drawn by joining the respective projections of its *end points*.

NOTE: ***Straight Lines are also called simply as Lines.***

The shape of an object is formed by different surfaces. These surfaces are formed by moving or rotating lines in different ways. Thus, line forms the basic geometrical entity using which any engineering component is made. Therefore, clear understanding of Projections of Lines is a necessary requirement for an engineer to understand the projections of 3-D objects. Also, true lengths of edges of any object can be determined from a study of Projection of Lines.

BIS Conventions for Line Thickness

1. XY line, Projector lines, Locus line, Construction line: *Narrow Thin Lines.*
2. Front View, Top View, True Length line: *Thick Lines.*

1.1 POSITIONS OF A LINE WITH RESPECT TO HP AND VP

A line may occupy an infinite number of positions in space with respect to HP and VP. The various positions may be classified into the following *seven* types:

1. **Parallel to Both the Planes HP and VP**
2. **Contained in One Plane and Parallel to the Other**
3. **Contained in Both the Planes HP and VP**
4. **Perpendicular to One Plane and Parallel to the Other**
5. **Parallel to One Plane and Inclined to the Other**
6. **Contained by One Plane and Inclined to the Other**
7. **Inclined to Both the Planes HP and VP**

2. LINE PARALLEL TO BOTH HP AND VP

Fig. 1(i) shows the Pictorial View of a line CD in the First Quadrant.

1. Look from front to get Front View. As CD is parallel to VP, c′ d′ = True Length of CD.
2. Look from top to get the Top View. As CD is parallel to VP, cd = True Length of CD.
3. The position of the line CD and its projections on HP and VP are shown in *Fig. 1(i)*.
4. Rotate HP (along with the projection on it) through 90° in CW direction as shown.
5. *See Fig. (ii)*. Projection of CD on VP is Front View and projection on HP is Top View.

Problem 1: *(Fig. 1) A line CD 30 mm long is parallel to both the planes. The line is 40 mm above HP and 25 mm in front of VP. Draw its projections.*

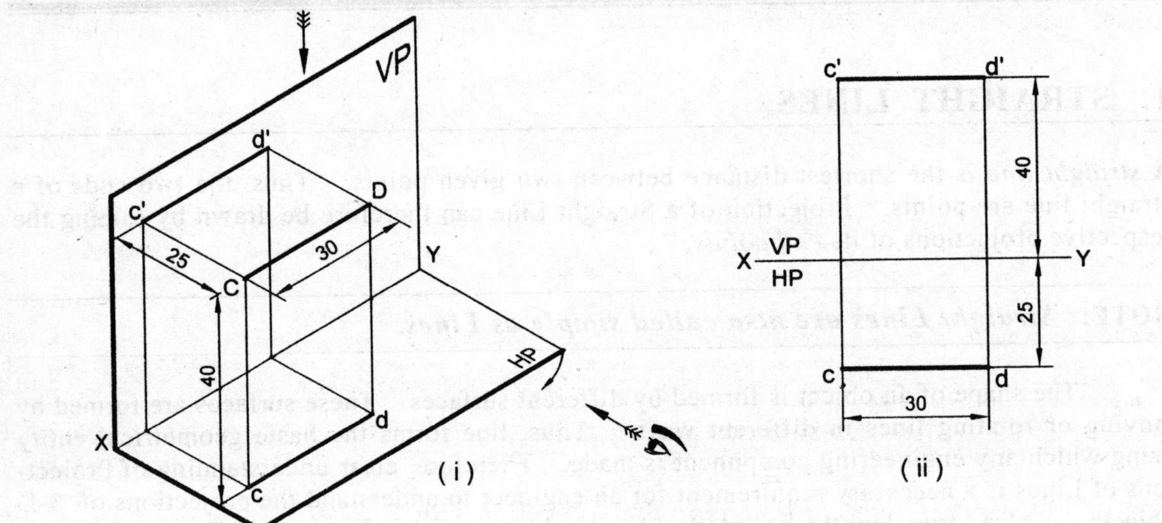

(i) (ii)

FIG. 1

1. *See Fig. (ii)*. Draw reference line XY and draw a projector at any point on it (*thin lines*).

Front View

2. *See Fig. (i)*. Line CD is 40 mm above HP, *i.e.*, point C is 40 mm above HP. Therefore mark c′ at a distance of 40 mm above XY in *Fig. (ii)*.
3. Line CD is parallel to HP. Front View will be in True Length and parallel to XY. *See Fig. (ii)*. Draw c′ d′ parallel to XY such that c′ d′ = True Length CD = 30 mm.

Top View

4. *See Fig. (i)*. Line CD is also parallel to VP and 25 mm in front of VP. Hence on the projector from c′, mark c at a distance of 25 mm below XY as in *Fig. (ii)*.
5. *See Figs. (i) & (ii)*. Top View is also in True Length and parallel to the XY line. Hence, draw cd parallel to XY such that cd = True Length = CD = 30 mm.
6. Mark the dimensions as shown in *Fig. (ii)*.

Problem 2: *(Exercise) A line AB 50 mm long is parallel to both the planes. The line is 40 mm in front of VP and 25 mm above HP. Draw its projections.*

3. LINE CONTAINED IN ONE PLANE & PARALLEL TO OTHER

3.1 LINE CONTAINED IN HP AND PARALLEL TO VP

Problem 3: *(Fig. 2) A line RS 60 mm long lies in HP and 45 mm in front of VP. Draw its projections.*

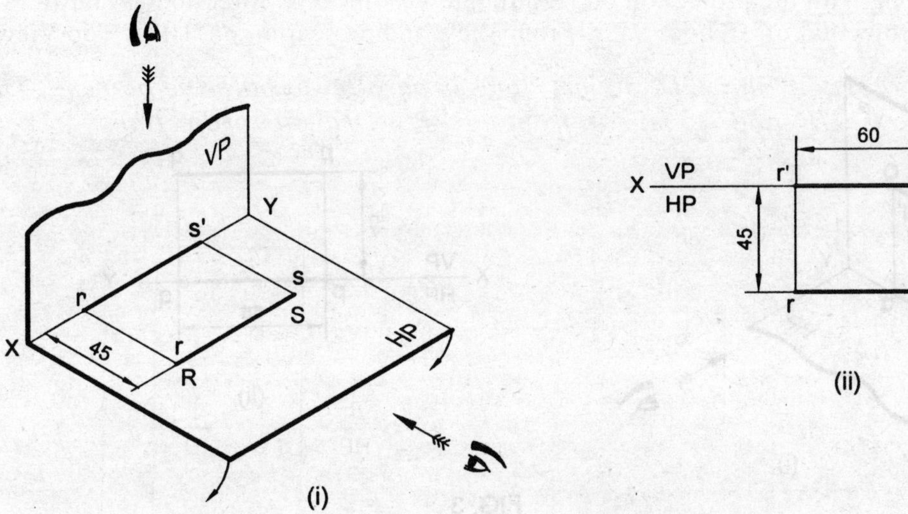

FIG. 2

1. *Fig. (i)* shows the Pictorial View of line RS lying in HP and 45 mm in front of VP.

NOTE: ***From the given data, decide the view to be drawn first.***
In this Problem, since the line is lying in HP, True Length of the line is seen in the Front View. Therefore draw the Front View first.

---

2. Draw the XY line as shown in *Fig. (ii)*.

To draw the Front View r′ s′, look at the Pictorial View from the front

3. Line RS lies on HP. Hence mark the Front View r′ s′ (= RS = True Length = 60 mm) on XY itself.

To draw the Top View rs, look at the Pictorial View from the top

4. Line RS is 45 mm in front of VP. Rotate the HP through 90° in CW direction. Hence mark the Top View rs (= RS = True Length = 60 mm) at a distance of 45 mm below XY on the projectors drawn from the Front View.
5. Mark the dimensions as shown in *Fig. (ii)*.

Problem 4: *(Exercise) A line MN 55 mm long lies in HP and 48 mm in front of VP. Draw its projections.*

3.2 LINE CONTAINED IN VP AND PARALLEL TO HP

Problem 5: *(Fig. 3) A line PQ 55 mm long is lying in VP and 45 mm above HP. Draw its projections.*

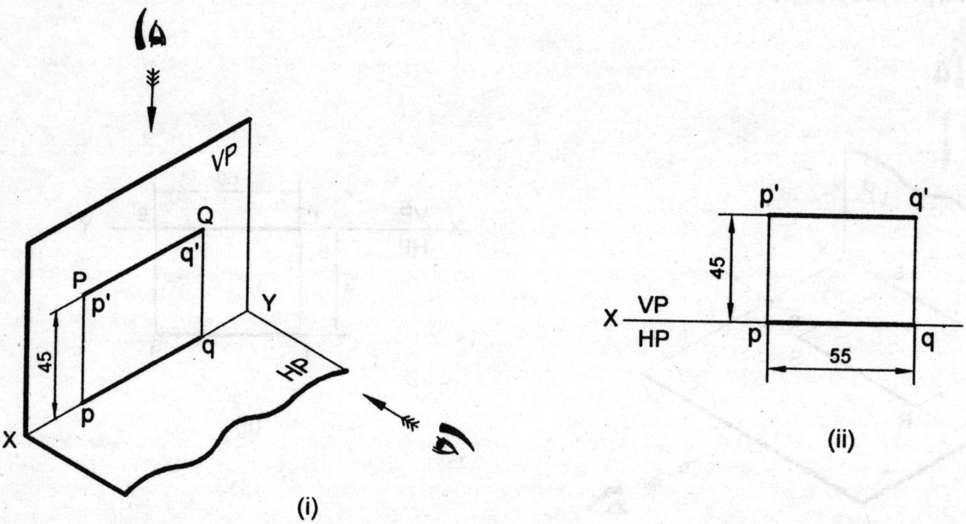

FIG. 3

1. *Fig. (i)* shows the Pictorial View of line PQ lying in VP and 45 mm above VP.

NOTE: ***From the given data, decide the view to be drawn first.***
In this Problem, since the line is parallel to HP, the True Length of the line is seen in the Top View. Therefore draw the Top View first.

---

2. Draw the XY line as shown in *Fig. (ii)*.

To draw the Top View pq , look at the Pictorial View from the top

3. Line PQ is parallel to HP. Hence mark the Top View pq (= PQ = True Length = 55 mm) on XY itself.

To draw the Front View p′q′ , look at the Pictorial View from the front

4. Line PQ is lying on the VP and 45 mm above the HP. Hence mark the Front View p′q′ (= PQ = True Length = 55 mm) at a distance of 45 mm above XY on the projectors drawn from the Top View.
5. Mark the dimensions as shown in *Fig. (ii)*.

Problem 6: *(Exercise) A line AB 50 mm long is lying in VP and 50 mm above HP. Draw its projections.*

4. LINE CONTAINED IN BOTH HP AND VP

Problem 7: *(Fig. 4)* *A line AB 55 mm long is lying on both HP and VP. Draw its projections.*

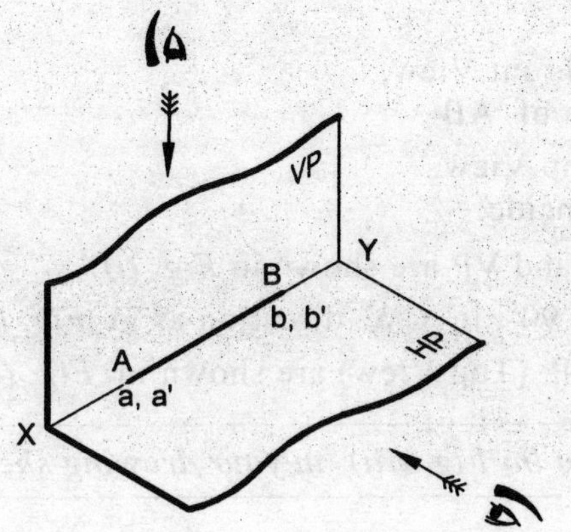

(i)

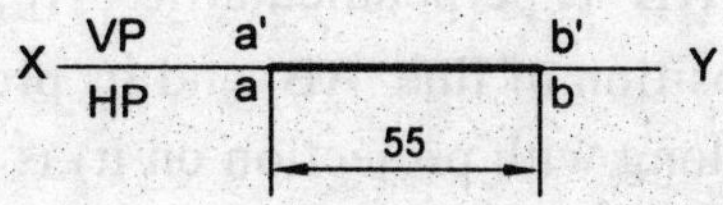

(ii)

FIG. 4

1. *See Fig. (i).* Line AB is contained by both HP and VP. Thus, AB is lying on XY.
2. *See Fig. (ii).* Front View a' b' is lying on XY and a' b' = AB = 55 mm.
3. *See Fig. (ii).* Top View ab is also lying on XY and ab = AB = 55 mm.
4. Both the Front View a' b' and Top View ab coincide with each other and with the line AB, which is also in XY.

Problem 8: *(Exercise)* *A line CD 58 mm long is lying on both HP and VP. Draw its projections.*

NOTE : ***True Length of a Line :*** *When a line in space is projected on HP or VP, the projected length of the view of the line will be either equal to or less than the actual length of the line, depending on the position of the line with reference to the planes.*

When the line is held parallel to HP or VP , its view projected on that plane will be equal to the actual length of the line called True Length. (For the line to be parallel to the HP or VP, it should be placed such that both the ends of the line are equidistant from the plane of projection.)

When the line is held inclined to HP or VP, its view projected on that plane will be less than the True Length of the line.

---

5. LINE PERPENDICULAR TO ONE PLANE AND PARALLEL TO THE OTHER

5.1 LINE PERPENDICULAR TO HP AND PARALLEL TO VP

Fig. 5(i) shows the Pictorial View of a line AB.

1. Look at the pictorial view from the front to get the Front View.
   Since AB is parallel to VP, a′ b′ = True Length of AB.
2. Look at the pictorial view from the top to get the Top View.
   Since AB is perpendicular to HP, a and b coincide.
3. The position of line AB and its projections on HP and VP are shown in *Fig. (i)*.
4. HP (along with projection on it) is rotated through 90° in CW direction as in *Fig. (ii)*.
5. Projections of AB on VP (Front View) and on HP (Top View) are shown in *Fig. (iii)*.

NOTE: *Draw only the orthographic projections shown in Fig. (iii) in your drawing sheet.*

Problem 9: *(Fig. 5) A line AB 25 mm long is parallel to VP and perpendicular to HP. Point A is 35 mm above HP and 20 mm in front of VP. Point B is 10 mm above HP. Draw the projections of the line AB.*

NOTE: ***From the given data, decide the view to be drawn first.***
In this Problem, since the line is parallel to VP, the True Length of the line is seen in Front View. Hence draw the Front View first.

---

1. *See Fig. (ii).* Draw the reference line XY and draw a projector at any point o on it.

To draw the Front View a′ b′, look at the Pictorial View from the front : *Fig. (i)*

2. B is 10 mm above HP. Hence mark b′ at a distance of 10 mm above XY.
3. A is 35 mm above HP. So mark a′ 25 (= 35-10) mm above b′ on the *same* projector.

To draw the Top View a b, look at the Pictorial View from the top : *Fig. (i)*

4. A is 20 mm in front of VP. Hence mark a 20 mm below XY.
5. When AB is seen from top, A is visible and B is invisible. Top Views of A and B coincide. The Top View of the invisible point B is represented within bracket as (b).

RULE : *Always mark the invisible points within brackets.*

---

6. Mark the dimensions as shown in *Fig. (iii)*.

Problem 10: *(Fig. 6) A line CD 25 mm long is parallel to VP and perpendicular to HP. Point C is 45 mm above HP and 15 mm in front of VP. Draw its projections.*
Refer Fig. 6 for Solution.

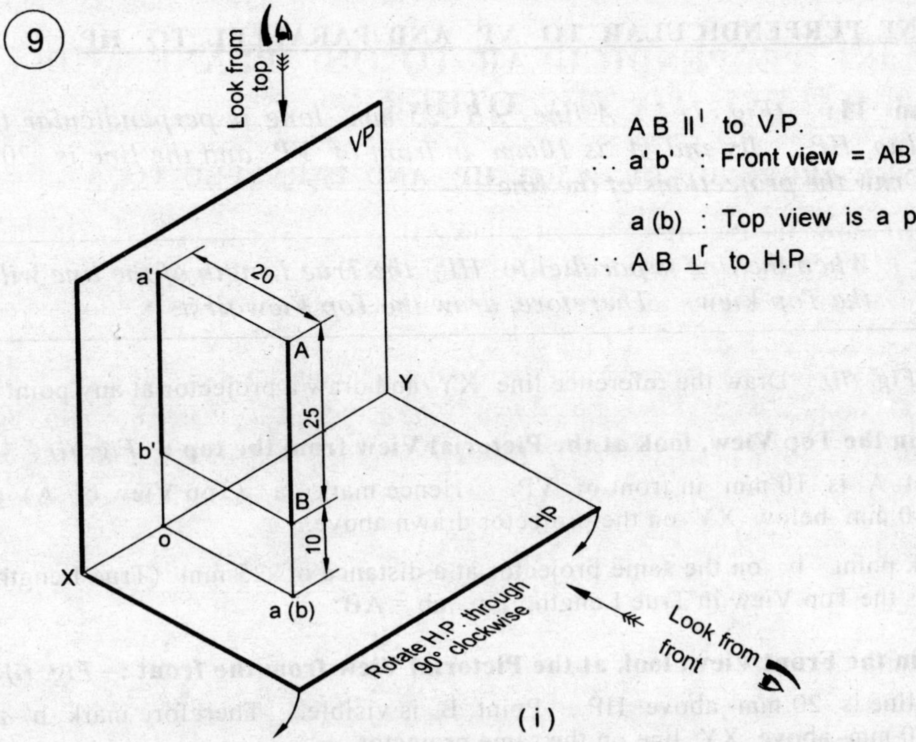

(i)

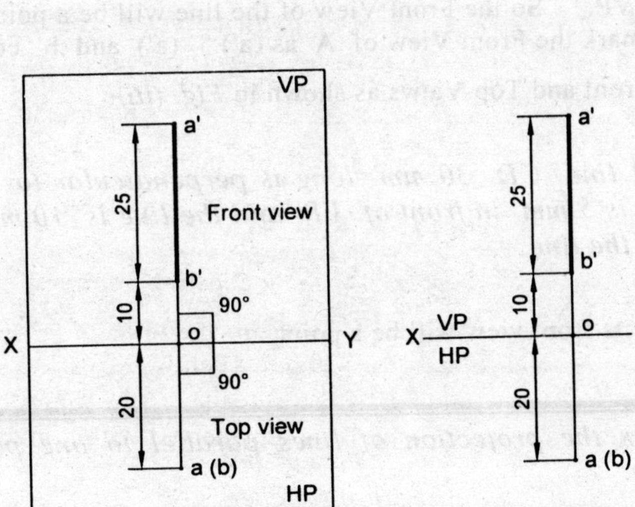

(ii) After Rotation

(iii)

FIG. 5

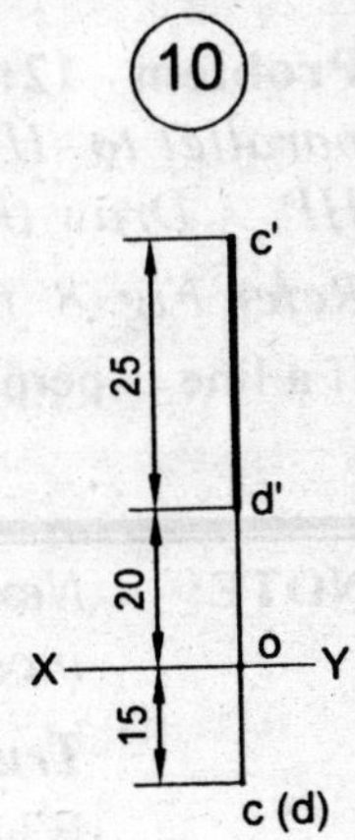

FIG. 6

5.2 LINE PERPENDICULAR TO VP AND PARALLEL TO HP

Problem 11: *(Fig. 7) A line AB 25 mm long is perpendicular to VP and parallel to HP. Its end A is 10mm in front of VP and the line is 20 mm above HP. Draw the projections of the line.*

NOTE : *When the line is parallel to HP, the True Length of the line will be seen in the Top View. Therefore, draw the Top View first.*

---

1. *See Fig. (ii).* Draw the reference line XY and draw a projector at any point o on it.

To obtain the Top View, look at the Pictorial View from the top : *Fig. (i)*

2. Point A is 10 mm in front of VP. Hence mark a (Top View of A) at a distance of 10 mm below XY on the projector drawn above.
3. Mark point b on the same projector at a distance of 25 mm (True Length) below a. ab is the Top View in True Length, *i.e.*, ab = AB.

To obtain the Front View, look at the Pictorial View from the front : *Fig. (i)*

4. The line is 20 mm above HP. Point B is visible. Therefore mark b′ at a distance of 20 mm above XY line on the same projector.
5. The line is perpendicular to VP. So the Front View of the line will be a point. Point A is hidden by B. Hence mark the Front View of A as (a′). (a′) and b′ coincide.
6. Mark the dimensions in the Front and Top Views as shown in *Fig. (iii).*

Problem 12: *(Fig. 8) A line CD 30 mm long is perpendicular to VP and parallel to HP. Its end C is 5 mm in front of VP and the line is 10 mm above HP. Draw the projections of the line.*

Refer Fig. 8 for Solution.

If a line is perpendicular to VP, its front view will be a point.

NOTE: *Next, let us discuss the projection of lines parallel to one plane and inclined to the other.*

True Inclination of a Line

When a line is inclined to HP or VP, then the view of the line projected on the other plane of projection will be inclined to XY line at the True Angle, called True Inclination. This is true only if the line is placed parallel to the plane of projection on which it is projected. Also, the view of the line will be in True Length.

---

(11)

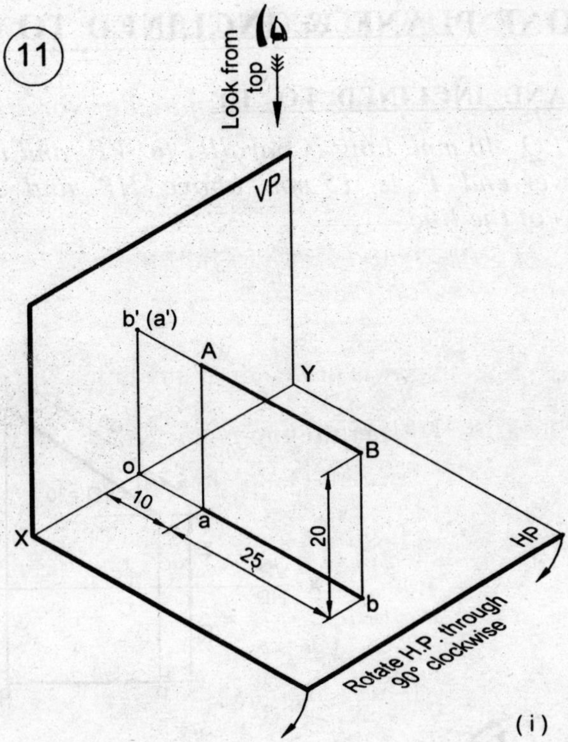

A B $\perp^r$ to V.P.

$\therefore$ b' (a') is a point

A B || to H.P.

$\therefore$ a b = A B

& a b $\perp^r$ to X Y

(i)

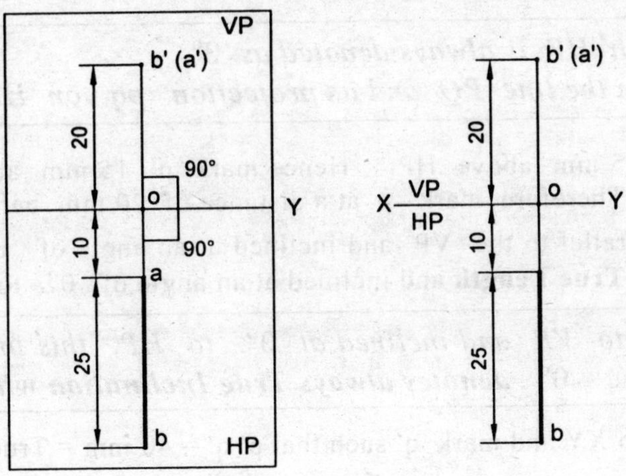

(ii) (iii)

FIG. 7

(12)

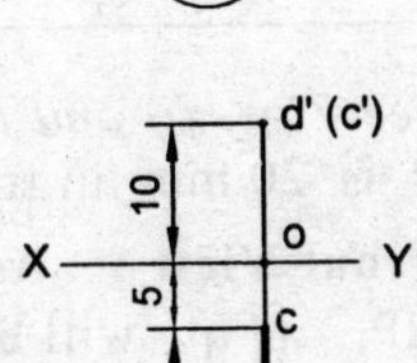

FIG. 8

6. LINE PARALLEL TO ONE PLANE & INCLINED TO OTHER

6.1 LINE PARALLEL TO VP AND INCLINED TO HP

Problem 13: *(Fig. 9)* *A line PQ 40 mm long is parallel to VP and inclined at an angle of 30° to HP. The lower end P is 15 mm above HP and 20 mm in front of VP. Draw the projections of the line.* (UQ)

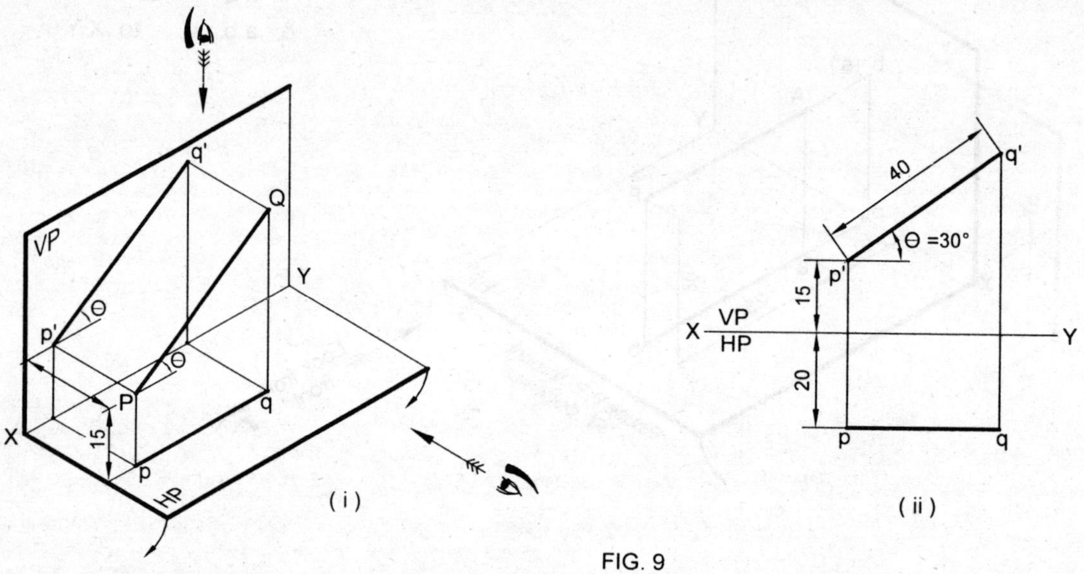

FIG. 9

NOTE : *Inclination of a line with HP is always denoted as θ°.*
θ° is the angle between the line PQ and its projection pq on HP.

---

1. *See Figs. (i) and (ii).* P is 15 mm above HP. Hence mark p′ 15 mm above XY. P is 20 mm in front of VP. Therefore, mark p at a distance of 20 mm below XY.
2. **Front View :** As PQ is parallel to the VP and inclined at an angle of θ° to the HP, p′ q′ will be equal to its **True Length** and inclined at an angle of θ° to XY.

RULE: *When a line is parallel to VP and inclined at θ° to HP, this inclination θ° is seen in Front View.* ***θ° denotes always True Inclination with HP.***

So from p′ draw a line at 30° to XY and mark q′ such that p′ q′ = 40 mm = True Length.

3. **Top View :** When the line is inclined to HP, its projection on HP, *i.e.*, the Top View will be in ***reduced*** length. From q′ draw a projector to intersect the horizontal line drawn from p at q. pq is the Top View of PQ.

Problem 14: *(Exercise)* *A line MN 55 mm long is parallel to VP and inclined at 30° to HP. End M is 25 mm above HP and 20 mm in front of VP. Draw the projections.*

6.2 LINE PARALLEL TO HP AND INCLINED TO VP

Problem 15: *(Fig. 10) Draw the projections of a line EF 40 mm long parallel to HP and inclined at 35° to VP. E is 20 mm above HP and 15 mm in front of VP.*

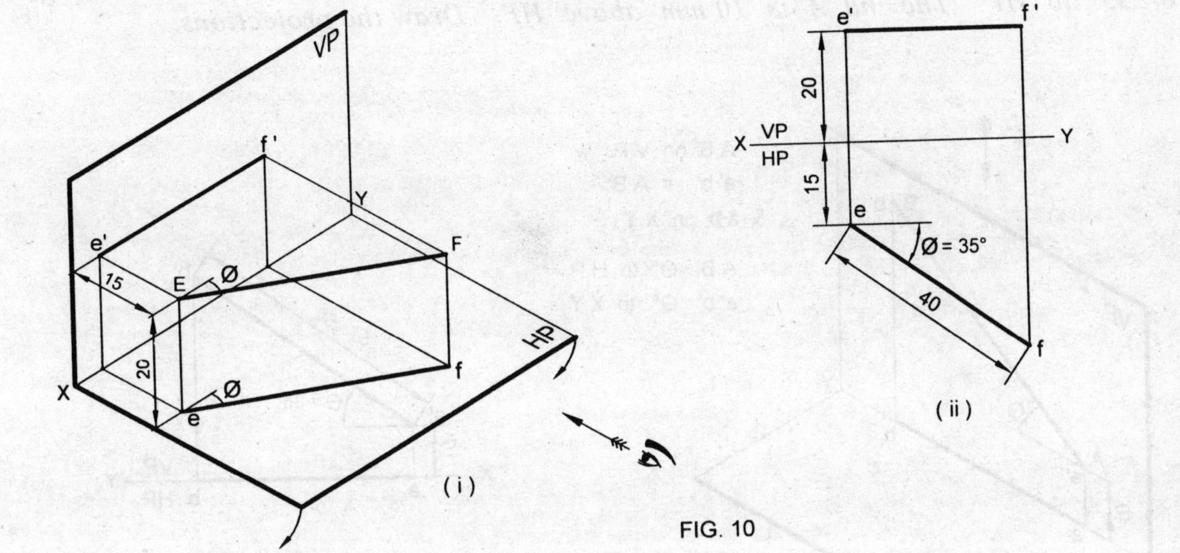

(i) (ii)

FIG. 10

NOTE : *Inclination of a line with VP is always denoted as ϕ.*
ϕ is the angle between the line EF and its projection e′ f′ on VP.

1. *See Figs. (i) & (ii).* E is 20 mm above HP. Therefore mark e′ 20 mm above XY. E is 15 mm in front of VP. Therefore mark e 15 mm below XY.
2. **Top View :** As EF is parallel to HP and inclined at an angle of ϕ to VP, ef will be equal to the **True Length** EF and inclined at an angle of ϕ to XY.

RULE : *When a line is parallel to HP and inclined at ϕ to VP, this inclination ϕ will be seen in the Top View.* ***ϕ denotes always True Inclination with VP.***

Hence from e draw a line 35° to XY and mark f such that ef = 40 mm = True Length.

3. **Front View :** When the line is inclined to VP its projection on VP, *i.e.*, Front View will be in ***reduced*** length. From f draw a projector to intersect the horizontal line drawn from e′ at f′. e′ f′ is the Front View of EF.

Problem 16: *(Exercise) A line MN measuring 60 mm has one of its extremities 25 mm in front of VP and 15 mm above HP. It is parallel to HP and inclined at an angle of 45° to VP. Draw its projections.*

7. LINE CONTAINED BY ONE PLANE & INCLINED TO OTHER

7.1 LINE CONTAINED BY VP AND INCLINED TO HP

Problem 17: *(Fig. 11) A line AB 50 mm long is in VP and inclined at an angle of 35° to HP. The end A is 10 mm above HP. Draw the projections.*

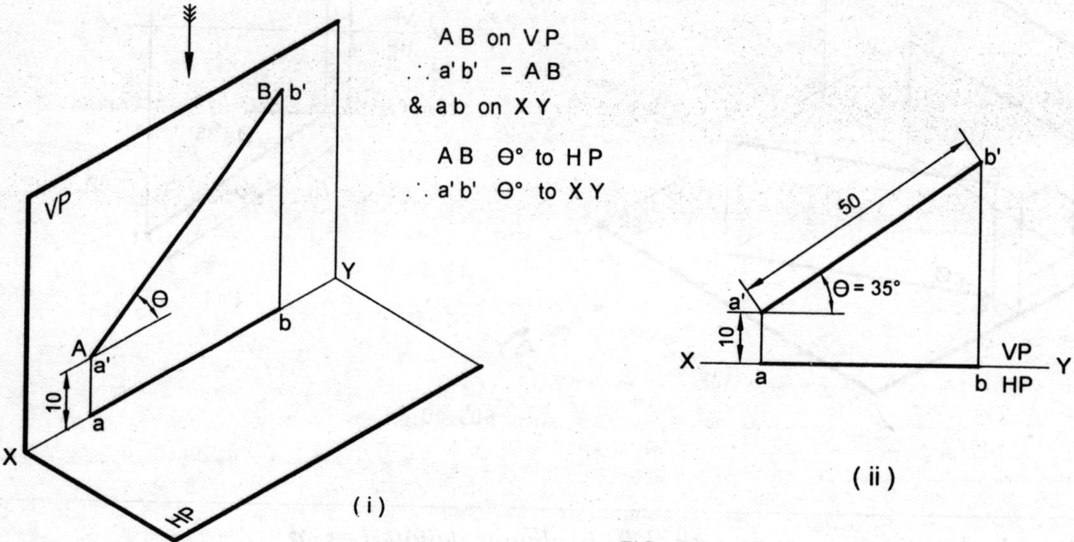

FIG. 11

1. *See Figs. (i) & (ii).* A is 10 mm above HP. Hence mark a′ 10 mm above XY.
2. Line AB is in VP. Therefore mark a on XY itself.
3. As the line AB is in VP and inclined at 35° to HP, a′ b′ = True Length = AB and a′ b′ will be inclined at $\theta = 35°$ to XY.

   Therefore from a′ draw a line at an angle of 35° to XY and mark b′ such that a′ b′ = 50 mm = **True Length**.
4. Since the line is inclined to HP, the Top View will be in ***reduced length***. Since the line is in VP, Top View will lie on XY itself.

   Therefore from b′ draw a projector to intersect the XY line at b. ab is the Top View of AB lying on XY.

Problem 18: *(Exercise) A line MN 68 mm long is lying on VP and inclined at an angle of 36° to HP. The end M is 12 mm above HP. Draw the projections of the line.*

7.2 LINE CONTAINED BY HP AND INCLINED TO VP

Problem 19: *(Fig. 12) A line RS measuring 52 mm is in HP and inclined at an angle of 45° to VP. The end R is 10 mm in front of VP. Draw the projections.* (UQ)

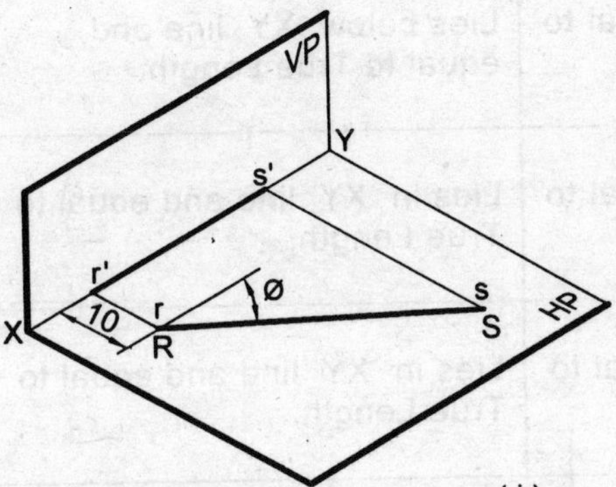

(i)

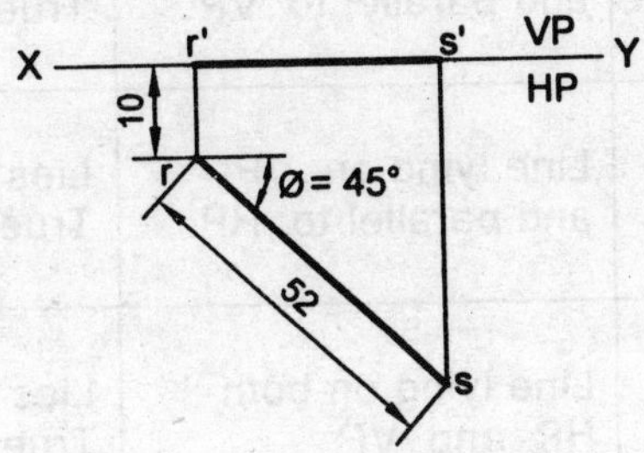

(ii)

FIG. 12

RS on HP	∴ rs = RS & r′ s′ on XY
RS ϕ^{o} to VP	∴ rs ϕ^{o} to XY

1. *See Figs. (i) and (ii).* Mark the Top View r at a distance of 10 mm below XY.
2. Mark the Front View r′ on XY itself.
3. As the line RS is in HP and inclined at an angle of 45° to VP, rs will be equal to the True Length RS (= 52 mm) and inclined at an angle of ϕ = 45° to XY.
4. As the line is in HP, its Front View r′ s′ will lie on XY.

Problem 20: *(Exercise) A line AB 66 mm long is lying on HP and inclined at an angle of 40° to VP. The end A is 18 mm in front of VP. Draw the projections.*

8. POINTS TO REMEMBER

S. No.	Position of Line	Front View	Top View
1.	Line parallel to both HP and VP.	Parallel to XY and having True Length of the line.	Parallel to XY and having True Length of the line
2.	Line lying on HP and parallel to VP.	Lies in XY line and equal to True Length.	Lies below XY line and equal to True Length.
3.	Line lying on VP and parallel to HP.	Lies above XY and equal to True Length.	Lies in XY line and equal to True Length.
4.	Line lying on both HP and VP.	Lies in XY line and equal to True Length.	Lies in XY line and equal to True Length.
5.	Line parallel to VP and perpendicular to HP	Perpendicular to XY and equal to True Length.	A point.
6.	Line parallel to HP and perpendicular to VP	A point.	Perpendicular to XY and equal to True Length.
7.	Line parallel to VP and inclined at θ^o to HP.	Inclined at θ^o to XY line and is equal to the True Length.	Parallel to XY line and shorter than the True Length.
8.	Line parallel to HP and inclined at Φ^o to VP	Parallel to XY line and shorter than the True Length.	Incline at Φ^o to XY line and is equal to the True Length.
9.	Line lying in VP and inclined at θ^o to HP.	Inclined at θ^o to XY line and is equal to the True Length.	Lies in XY line and shorter than the True Length.
10.	Line lying in HP and inclined at Φ^o to VP.	Lies in XY line and shorter than the True Length.	Inclined at Φ^o to XY line and is equal to the True Length.

9. ADDITIONAL PROBLEMS – I

21. *(Fig. 13)* The length of the top view of a line MN parallel to VP and inclined at 45° to HP is 50 mm. Point M is 12 mm above HP and 25 mm in front of VP. Draw the projections of the line. Find its true length.

MN is parallel to VP. Hence its Top View will be parallel to XY and Front View will give True Length and True Inclination with HP.

See Fig. 13. Mark m 25 mm below XY and m′ 12 mm above XY. Draw mn parallel to XY and 50 mm long. Draw a projector through n. From m′ draw a line at 45° with XY to meet the projector through n at n′. Now m′n′ is the Front View giving the *True Length* of the line. m′n′ = MN = 71 mm

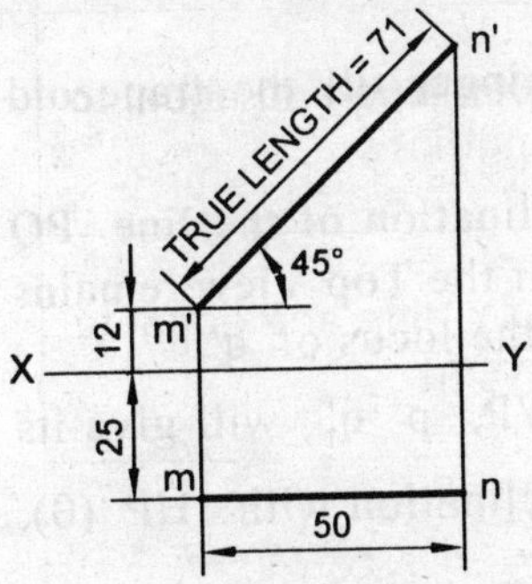

FIG. 13

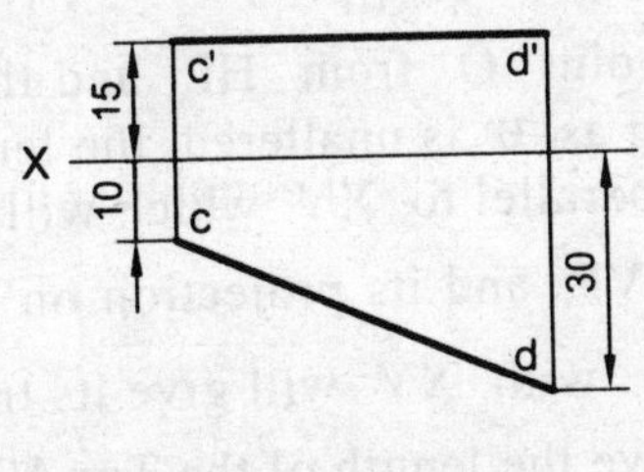

FIG. 14

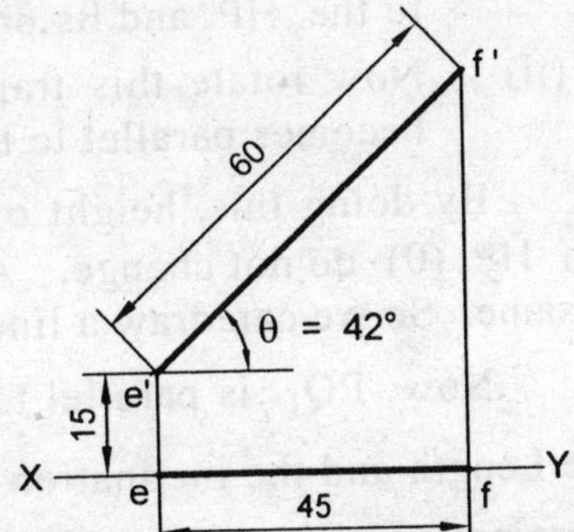

FIG. 15

22. *(Fig. 14)* Draw the projections of a line CD 50 mm long, parallel to HP and inclined to VP. The end C is 10 mm in front of VP and D is 30 mm in front of VP. The line is 15 mm above HP.

23. *(Fig. 15)* A line EF 60 mm long is in VP and inclined to HP. The top view measures 45 mm. The end E is 15 mm above HP. Draw the projections of the line. Find its inclination with HP. *(Result : $\theta = 42^o$)* (UQ)

24. A line GH 45 mm long is in HP and inclined to VP. The end G is 15 mm in front of VP. The length of the front view is 35 mm. Draw the projections of the line. Determine its inclination with VP. *(Result : $\phi = 38^o$)*

25. A line AB 60 mm long is parallel to HP. The point P is 20 mm above HP and 35 mm in front of VP. The length of the front view is 50 mm. Determine its true inclination with VP. *(Result : $\phi = 35^o$)* (UQ)

26. A line CD is parallel to VP and inclined at 40° to HP. C is in HP and 25 mm in front of VP. Top view is 50 mm long. Find its true length. *(Result : CD = 65 mm)*

27. An electric bulb B and its switch S fixed on a wall are 2.25 m apart. The distance between B and S measured parallel to the floor is 1.8 m. If S is 0.75 m above the floor, what is the height of B ? Also determine the inclination of the line joining B and S with the floor. Select a suitable scale.

10. LINE INCLINED TO BOTH HP AND VP

When a line is inclined to both HP and VP, the projections are *shorter* than the True Length and the inclination angles with XY are *greater* than the true inclinations of the line.

See Fig. 16. PQ is inclined to both the planes in first quadrant. It makes an angle of θ with the HP and ϕ with the VP. Point P is above the HP and in front of the VP.

Projections of this line can be obtained by rotating it into two simple positions, *viz.,*

a) Parallel to the VP and b) Parallel to the HP

10.1 TO FIND LENGTH OF TOP VIEW pq_1 AND LOCUS OF Q IN FRONT VIEW

(i) *See Fig. 16.* Imagine a trapezoid made up of the line PQ, its projectors P_p, Q_q to the HP and its projection pq on the HP.

(ii) Now rotate this trapezoid PQ qp with Pp as the hinge, till the trapezoid becomes parallel to the VP as $PQ_1 q_1p$.

By doing this, height of the point Q from HP and the inclination of the line PQ with HP (θ) do not change. As long as θ is unaltered, the length of the Top View remains the same. So we can draw a line cd parallel to XY which will give the locus of q'.

Now PQ_1 is parallel to the VP and its projection on the VP, $p'\ q_1'$ will give its True Length and the inclination $p'\ q_1'$ with XY will give its true inclination with HP (θ). Its projection pq_1 on the HP will give the length of the Top View.

10.2 TO FIND FRONT VIEW LENGTH $p'q_2'$ AND LOCUS OF Q IN TOP VIEW

See Fig. 17. Imagine the trapezoid made up of the line PQ, its projectors Pp', Qp' to the VP and its projection p' q' on the VP.

Now rotate this trapezoid PQ q' p' with Pp' as the hinge, till the trapezoid becomes parallel to the HP as $PQ_2\ q_2'\ p'$.

By doing this, the distance of Q from VP and the inclination of the line PQ with VP (ϕ) do not change. As long as ϕ is unaltered, the length of the Front View remains the same. Now we can draw a line gh parallel to XY which will give the locus of q.

Now PQ_2 is parallel to the HP and its projection on the HP. pq_2 will give its True Length and inclination pq_2 with XY will give its true inclination (ϕ) with the VP. Its projection on the VP, $p'\ q_2'$ will give the length of the Front View.

NOTE : i) *From Fig.16, it can be seen that if the line PQ makes an angle θ to HP, for various values of ϕ the Front View of Q will always lie on cd and the length of the Top View pq_1 remains unaltered.*

ii) *Similarly from Fig. 17, it can be seen that if the line PQ makes ϕ to VP, for various values of θ the Top View of Q will always lie on gh and the length of the Front View $p'q_2'$ remains unaltered.*

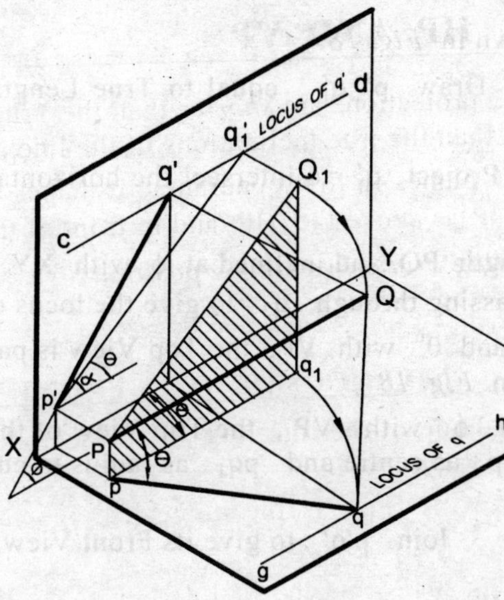

FIG. 16

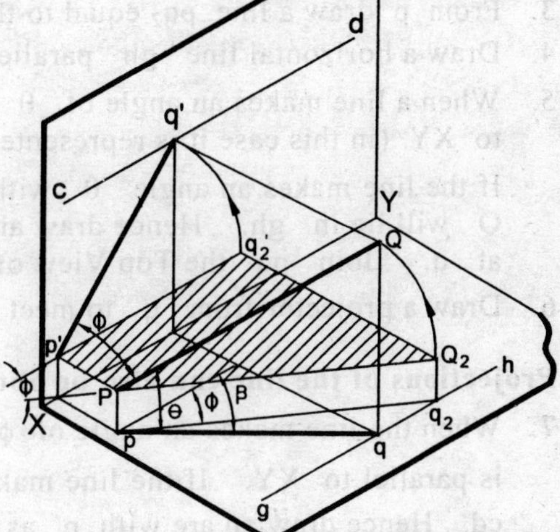

FIG. 17

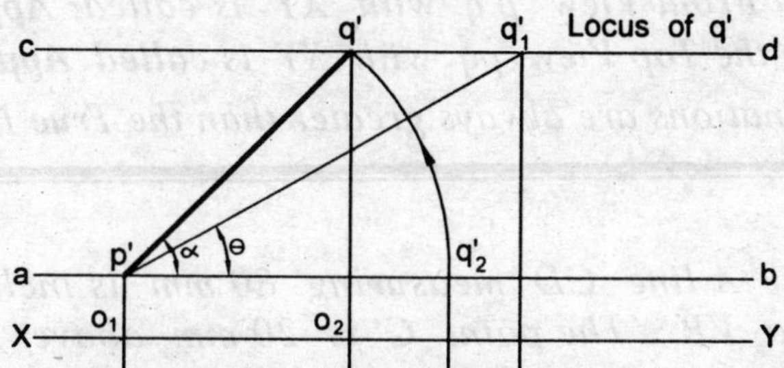

FIG. 18

Now the above two cases can be integrated as shown in *Fig. 18.*

1. Mark p′ above XY and p below XY. Draw $p' q_1'$ equal to True Length and inclined at θ to XY.
2. From q_1' draw cd to give the locus of q′. Project q_1' to intersect the horizontal line drawn from p at q_1.
3. From p draw a line pq_2 equal to the True Length PQ and inclined at ϕ with XY.
4. Draw a horizontal line gh parallel to XY passing through q_2 to give the locus of q.
5. When a line makes an angle of θ with HP and 0° with VP, the Top View is parallel to XY (in this case it is represented by pq_1 in *Fig. 18*).

   If the line makes an angle θ with HP and ϕ with VP , the Top View of the end Q will lie in gh. Hence draw an arc with p as centre and pq_1 as radius to cut gh at q. Join pq the Top View of PQ.
6. Draw a projector from q to meet cd at q′. Join p′q′ to give its Front View.

Projections of the line can also be obtained as follows :

7. When the line makes an angle of ϕ with VP and 0° with HP the Front View ($p'q_2'$) is parallel to XY. If the line makes ϕ with VP and θ with the HP, q′ will lie on cd. Hence draw an arc with p′ as centre and $p' q_2'$ as radius to meet cd at q q′.
8. From q′ draw a projector to intersect gh at q. Join pq which is Top View of the line.

NOTE : *1. Angle made by Front View p′q′ with XY is called* ***Apparent angle α.***
Angle made by the Top View pq with XY is called ***Apparent angle β.***
2. Apparent Inclinations are always greater than the True Inclinations.

Problem 28: *(Fig. 19) A line CD measuring 80 mm is inclined at an angle of 30° to HP and 45° to VP. The point C is 20 mm above HP and 30 mm in front of VP. Draw the projections of the straight line.* (UQ)

1. *See Fig. 19.* C is 20 mm above HP. Therefore mark c′ 20 mm above XY.
2. C is 30 mm in front of VP. Therefore mark c 30 mm below XY.

To obtain the locus of d′ and Top View length :

3. To start with, assume the line to be inclined to only one plane (say, HP), keeping θ constant and to be made parallel to the other (VP). Now we can obtain the locus of d′. From c′ draw 30° line to XY and mark d_1' such that $c' d_1'$ = **True Length** = 80 mm.
4. Draw a horizontal line through d_1' which will be the locus of D in Front View.
5. Draw a horizontal line from c.
6. From d_1' draw a projector to intersect the horizontal drawn from c at d_1. Now cd_1 is the length of the Top View.

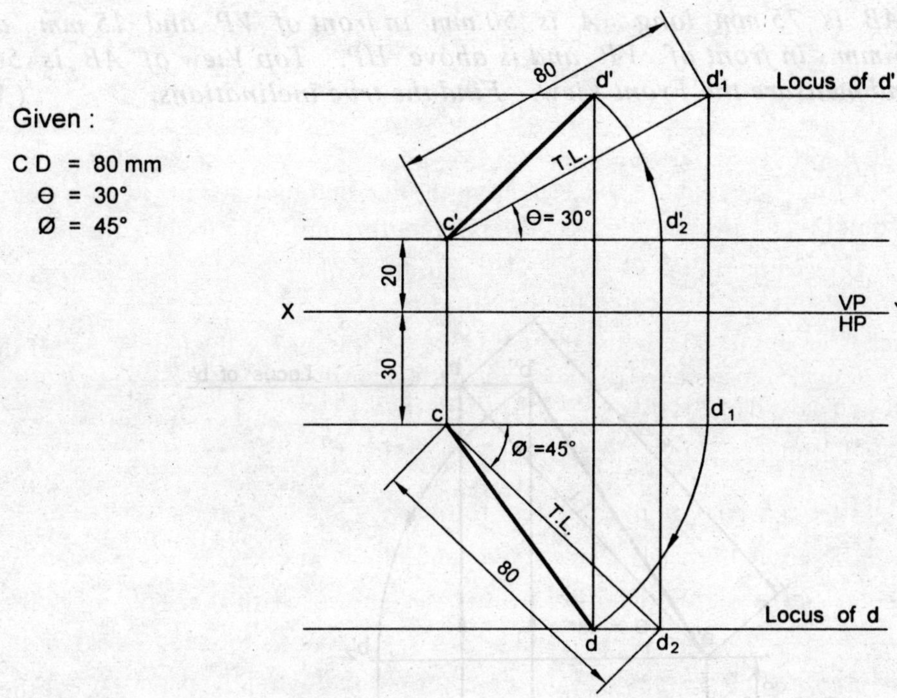

FIG . 19

To obtain the locus of d and Front View length

7. Next assume the line to be parallel to HP and inclined at ϕ to VP to obtain locus of d. From c draw a line 45° to XY and mark d_2 such that cd_2 = **True Length** = 80 mm.
8. Draw a horizontal line through d_2 which will be the locus of D in Top View.
9. From d_2 draw a projector to intersect the horizontal line drawn from c′ at d'_2. $c' d'_2$ is the length of the Front View.

Top View

10. With c as centre and cd_1 as radius draw an arc to intersect the locus of d at d. Now cd is the Top View.

Front View

11. With c′ as centre and $c' d'_2$ as radius draw an arc to intersect the locus of d′. Now c′ d′ is the Front View.

Check : *Now d′ and d will be on the same projector.*

Problem 29: *(Exercise) A line PQ 75 mm long has its end P in both HP and VP. It is inclined at an angle of 30° to HP and 45° to VP. Draw projections of the line.*

Problem 30: *(Fig. 20)*

A line AB is 75 mm long. A is 50 mm in front of VP and 15 mm above HP. B is 15 mm in front of VP and is above HP. Top View of AB is 50 mm long. Draw and measure the Front View. Find the true inclinations. (UQ)

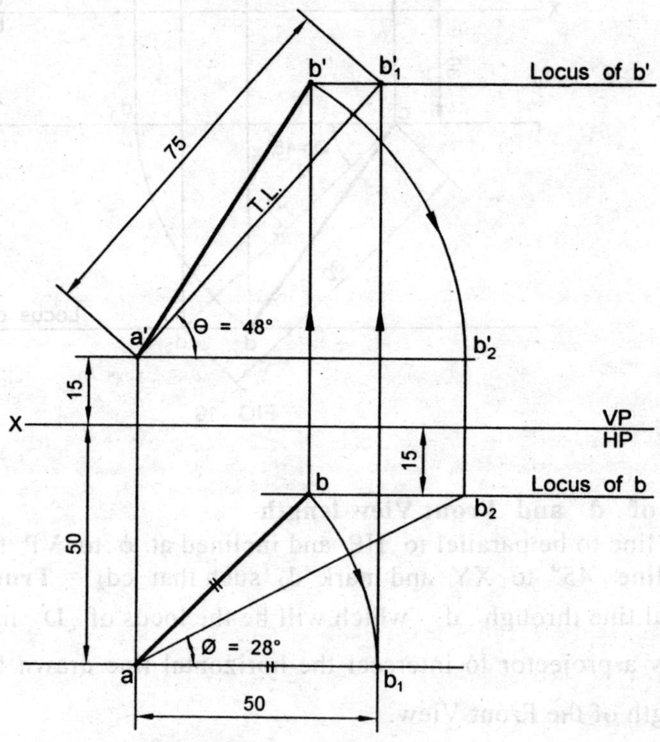

RESULTS :

a' b' = 66 mm

Θ = 48°, Ø = 28°

FIG. 20

1. A is 15 mm above HP and 50 mm in front of VP. Therefore mark a' 15 mm above XY and a 50 mm below XY.

To draw Top View, given Top View length

2. B is 15 mm in front of VP. Therefore draw a horizontal line 15 mm below XY to represent the locus of b.

3. Top View of the line is 50 mm. Hence with a as center and 50 mm as radius, draw an arc to intersect locus of b at b. Now ab is the Top View of the given line.

To find True Inclination with HP (θ)

NOTE : *Now the Top View is made parallel to XY. i.e., the line is made parallel to VP. A is fixed, θ is fixed. Therefore the Front View corresponding to this Top View* ab_1 *will give the True Length and True Inclination θ.*

---

4. a as center and ab as radius draw an arc to meet the horizontal drawn through a at b_1.

5. True Length is given as 75 mm. Therefore with a′ as center and 75 mm as radius, draw an arc to intersect the projector drawn from b_1 at b_1'.

6. Join a′ b_1'. Measure **θ = 48°**.

Front View

7. Draw a horizontal line through b_1' to represent the locus of b′.

8. From b draw a projector to intersect the locus of b′ at b′.

9. Join a′ b′ which is the Front View of the given line. **a′ b′ = 66 mm.**

To find True Inclination with VP (φ)

10. *See Fig. 20.* a as center and 75 mm as radius cut a point b_2 on the locus of b.

11. Join a b_2.

12. Measure **φ = 28°**.

Problem 31: *(Exercise)*

A line measuring 80 mm long has one of its ends 60 mm above HP and 20 mm in front of VP. The other end is 15 mm above HP and in front of VP. The front view of the line is 60 mm long. Draw the top view.

Problem 32: *(Fig. 21)*

Draw the projections of a straight line AB of 100 mm long when one of its ends is touching the VP and the other end touching HP. The angles of inclination with HP and VP are 40° and 50° respectively. (UQ)

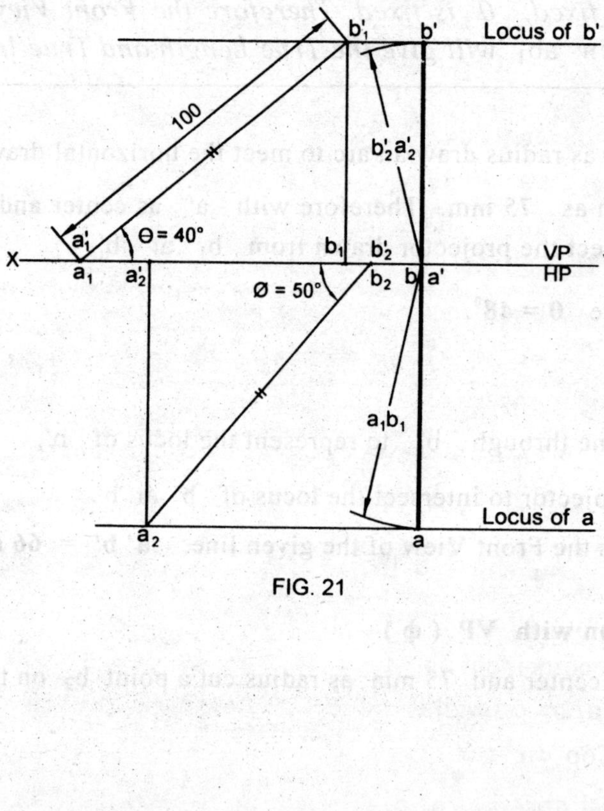

FIG. 21

To obtain Top View length and locus of b′

- Let A be the end which is touching HP and B be the end touching VP. Therefore a′ and b will be on XY.
- With reference to B, rotate the line to be parallel to VP, keeping θ constant. Now end A will be on XY. Front View will give the True Length and θ.
- We can obtain the locus of b′ and Top View length. Top View will coincide with XY.

1. Now mark any point a_1' on XY to represent the Front View of A in the above position.
2. Draw $a_1' b_1'$ = 100 mm at an angle of 40° to XY.
3. From b_1' draw a horizontal to represent the locus of b′.
4. a_1b_1 : corresponding top view length.

To obtain Front View length & locus of a

- Again consider the given position of the line.
- Now with reference to A and keeping ϕ constant, rotate AB until it is in HP *i.e.*, B is in XY. Top View will give True Length and ϕ. Front View coincides with XY.
- Now obtain the locus of a and the Front View length.

5. Take another point b_2 on XY.
6. Draw a line from b_2 at an angle of 50° (ϕ) to XY.
7. Mark $a_2 b_2$ = 100 mm = True Length.
8. Draw a horizontal through a_2 to represent the locus of a.
9. Draw and measure the Front View $a_2' b_2'$ on XY corresponding to $a_2 b_2$.

To obtain the final projections

10. Mark some other point a′ on XY.
11. With a′ as center and $a_2' b_2'$ as radius draw an arc to cut the locus of b′ at b′.
12. From the configuration, a′ b′ = $b_1 b_2'$.
13. Therefore draw a′ b′ (Front View) perpendicular to XY.
14. Note that b will coincide with a′.
15. b as center and $a_1 b_1$ as radius draw an arc to cut the locus of a at a.
16. Now ab is the Top View.
17. From the configuration, ab = $a_2' a_2$. Hence ab is perpendicular to XY.

Problem 33: *(Exercise)*

Draw the projections of a line JK 70 mm long and touching both HP and VP. It is inclined at 40° to HP and 35° to VP.

Problem 34: *(Fig. 22)*

*The **mid-point** of a straight line AB is 60 mm above HP and 50 mm in front of VP. The line measures 80 mm long and inclined at an angle of 30° to HP and 45° to VP. Draw its projections.* (UQ)

Given :

AB = 80 mm

θ = 30°

Ø = 45°

Mid - point M 60mm above HP & 50mm in front of VP.

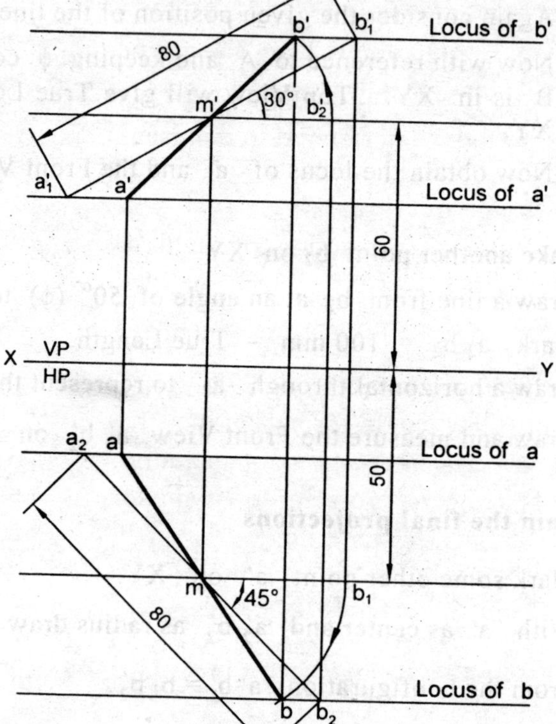

FIG. 22

1. Mark m′ 60 mm above XY and m 50 mm below XY.

To obtain the loci of a′ and b′

2. To start with, assume the line to be inclined to only one plane, say, HP and made parallel to VP with reference to mid-point M.
3. Thus obtain the loci of b′ and a′. The whole line is now 50 mm in front of VP.
4. Top View corresponding to this will be parallel to XY and passes through m.
5. Through m′ draw a line 30° (θ) to XY.

6. Mark $a_1' b_1'$ = 80 mm = **True Length** on this line such that $a_1' m' = m' b_1'$ = 40 mm.
7. From a_1' draw a horizontal to represent the locus of a′.
8. From b_1' draw a horizontal to represent the locus of b′.

To obtain the loci of a and b

9. Next assume that the line is inclined to VP only and made parallel to HP with reference to M. We can thus obtain the loci of b and a.
10. Through m draw a line, 45° (φ) to XY.
11. Mark $a_2 b_2$ = 80 mm = **True Length** on this line such that $a_2m = mb_2$ = 40 mm.
12. From a_2 draw a horizontal to represent the locus of a.
13. From b_2 draw a horizontal to represent the locus of b.

To obtain the projections

– Consider the projection of right half of the line and remaining half is exactly symmetrical.

14. Now consider the Front View $m' b_1'$ = 40 mm inclined at 30° to XY.
15. From b_1' draw a projector to cut the horizontal drawn through m at b_1. mb_1 is the Top View length for half of the straight line.
16. With m as center and mb_1 as radius draw an arc to cut the locus of b at b.
17. From b draw a projector to cut the locus of b′ at b′.
18. Join bm and extend this till it touches the locus of a at a. (amb is the Top View of AB).
19. Join b′ m′. Extend this till it touches locus of a′ at a′. (a′ m′ b′ is the Front View of AB).

CHECK : *Now a' and a will be on the same projector.*

Problem 35: *(Exercise) The mid-point of a line 80 mm long is 25 mm above HP and 30 mm in front of VP. The line is inclined at 30° to HP and 40° to VP. Draw its projections.*

Problem 36: *(Fig. 23)*
*A line AB 120 mm long is inclined at 45° to the HP and 30° to the VP. Its **mid-point** C is in VP and 20 mm above HP. The end A is in third quadrant and B is in first quadrant. Draw the projections of the line.* (UQ)

To obtain the loci of a′ & b′ and Top View length

1. Projections of mid-point C in the VP : Mark c′ 20 mm above XY and c on XY.
2. Keeping θ constant and with reference to C rotate AB to lie on VP. Through c′ draw a line at 45° to XY. Mark $a_1' c' b_1'$ = 120 mm = T.L. such that $a_1' c' = c' b_1'$ = 60 mm.

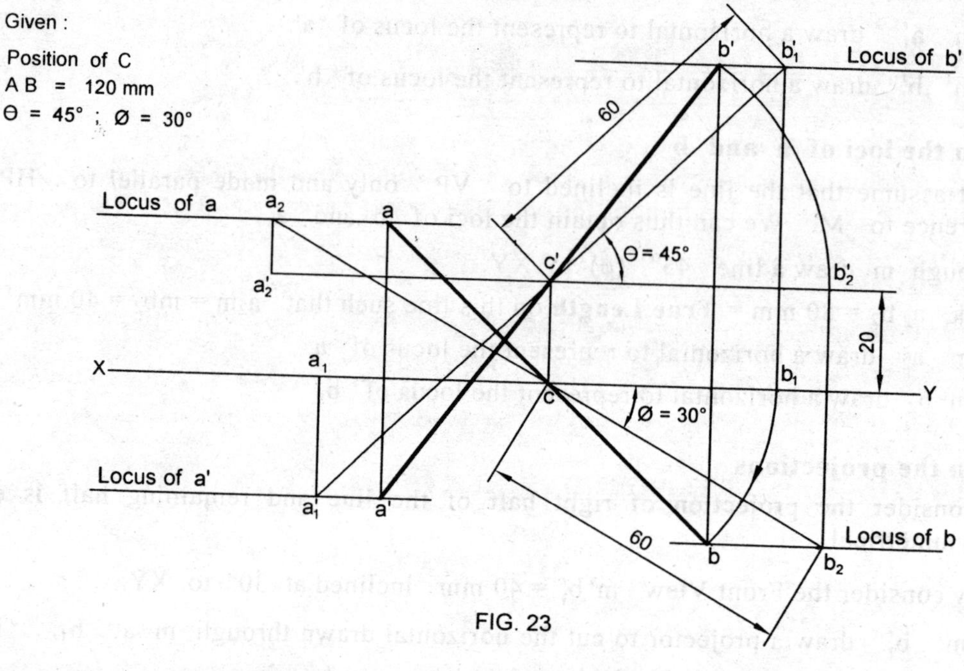

FIG. 23

3. Through b_1' & a_1' draw horizontal lines to represent the loci of b' & a' respectively.
4. From b_1' draw a projector to intersect XY line at b_1. Now cb_1 is the Top View length for half of the straight line.

To obtain the loci of a and b

5. Keeping ϕ constant and with reference to C rotate the line parallel to HP. Now the line is 20 mm above HP.
6. Through c draw a line at 30° to XY and mark a_2cb_2 = 120 mm = True Length, such that a_2c = cb_2 = 60 mm.
7. Draw a horizontal from a_2 to represent the locus of a.
8. Draw a horizontal from b_2 to represent the locus of b.
9. **Front View :** With c as center and cb_1 as radius draw an arc to intersect the locus of b at b. Draw a projector from b to intersect the locus of b' at b'. Join $b'c'$ and extend this to intersect the locus of a' at a'. Now $a'c'b'$ is the Front View.
10. **Top View:** Join bc & extend this to intersect locus of a at a. Now acb is Top View.

CHECK : *a and a' will be on the same projector.*

Problem 37: *(Exercise) A line MN is 70 mm long. Its mid point is 30 mm above HP and 25 mm in front of VP. If the line is inclined at 45° to HP and 35° to VP, draw its projections.*

11. TRUE LENGTH AND TRUE INCLINATIONS

11.1 GIVEN THE PROJECTIONS OF A LINE, TO FIND TRUE LENGTH AND TRUE INCLINATIONS WITH HP & VP

When a line is parallel to a plane, its projection on that plane will give its True Length and will make an angle with XY equal to True Inclination with other plane. After making the line parallel to a plane, its True Length is obtained by any one of the following *two* methods:

11.2 METHOD – I : ROTATING LINE METHOD

When the orthographic view of a line is parallel to XY, its other view will have True Length and True Inclination. In Rotating Line Method, each view is made parallel to XY line and the other view is projected from it. This method is just the reverse of the method used for drawing the projections. (*Refer* : Previous Article on ***Lines Inclined to Both the Planes.***)

11.3 METHOD – II : TRAPEZOIDAL PLANE METHOD

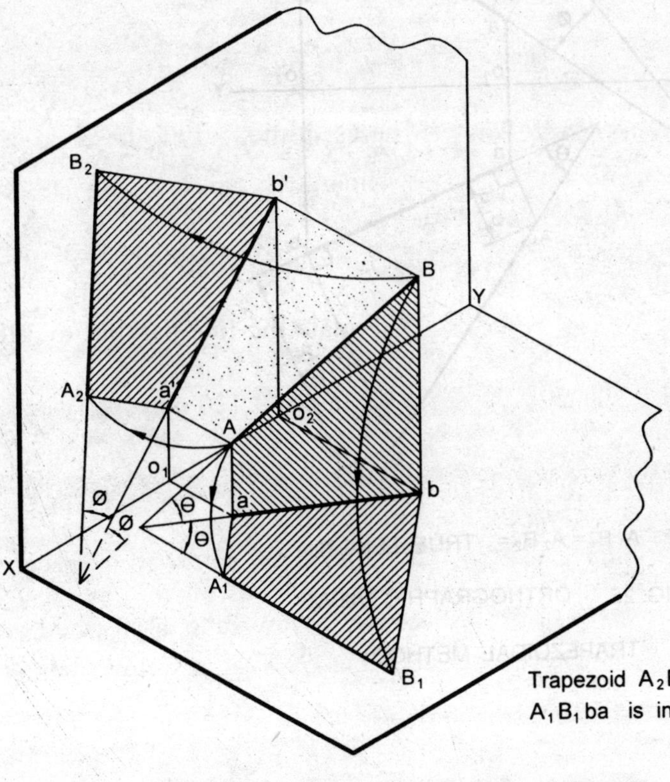

Trapezoid $A_2B_2b'a'$ is in VP
A_1B_1ba is in HP

FIG. 24 PICTORIAL VIEW - TRAPEZOIDAL METHOD

Rotate the line about its projections till it lies in HP or VP.

1. *See Fig. 24.* ab is the Top View and a′ b′ is the Front View.
2. In the trapezoid AB b′ a′, a′ A and b′ B are equal to ao_1 and bo_2 respectively and also perpendicular to a′ b′.

NOTE : *ao_1 and bo_2 are the distances of a and b from XY.*

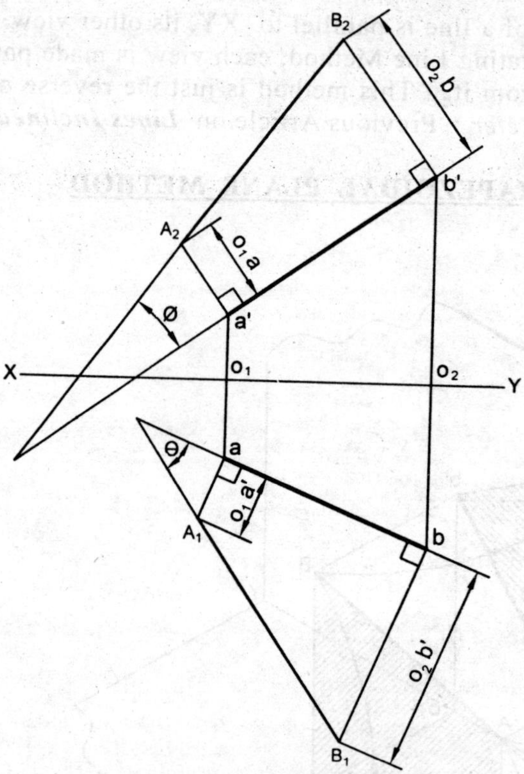

$A_1 B_1 = A_2 B_2 =$ TRUE LENGTH

FIG. 25 ORTHOGRAPHIC VIEWS

TRAPEZOIDAL METHOD

3. Also the angle between AB and a′ b′ is ϕ.

4. Assume that the trapezoid is rotated about a′ b′ until it lies in the VP.

5. *See Fig. 25.* This trapezoid is obtained by drawing perpendiculars to a′ b′, *i.e.*, $a'A_2 = ao_1$ and $b'B_2 = bo_2$ and then joining $A_2 B_2$.

6. Now A_2B_2 is the True Length of AB and its inclination with a′ b′ is equal to ϕ.

7. Similarly the trapezoid ab BA may be assumed to be rotated about ab as axis, until it lies in HP.

8. In *Fig. 25* this trapezoid is obtained by erecting perpendiculars to ab, *i.e.*, $aA_1 = a'o_1$ and $bB_1 = b'o_2$ and then joining A_1B_1.

9. A_1B_1 = True Length of AB.

10. Inclination of A_1B_1 with ab = θ.

RULE : *When one end of the line is in First Quadrant and the other end is in Fourth Quadrant, 'erect' the perpendiculars in the same direction from Front View and erect the perpendiculars in the opposite direction from Top View.*

11.4 POINTS TO REMEMBER

1. Given the front and top views of a line, its true length and true inclinations may also be found by *Auxiliary Plane Method.* (Refer Chapter 15)

2. The top and front views of a straight line can be drawn from the following data:

   i. Distance between the end projectors.

   ii. Distance of one or both the end points from the HP and the VP.

   iii. Apparent inclinations of the line.

3. Trapezoidal Plane Method is also termed as Rotating Trapezoidal Plane Method.

4. When a line is parallel to a reference plane, its projection on that plane will show its true length and the true inclination with the other plane.

   The line may be made parallel to a reference plane, and its true length obtained by any one of the following three methods:

   Rotating Line Method, Trapezoidal Method and Auxiliary Plane Method.

5. When a view of a line is parallel to XY, its other view will be in true length and at true inclination.

11.5 ROTATING LINE METHOD

Problem 38: *(Fig. 26)*

The distance between the projectors of two points A and B is 70 mm. Point A is 10 mm above HP and 15 mm in front of VP. Point B is 50 mm above HP and 40 mm in front of VP.

Find the shortest distance between A and B. Measure the true inclinations of the line AB with VP and HP. (UQ)

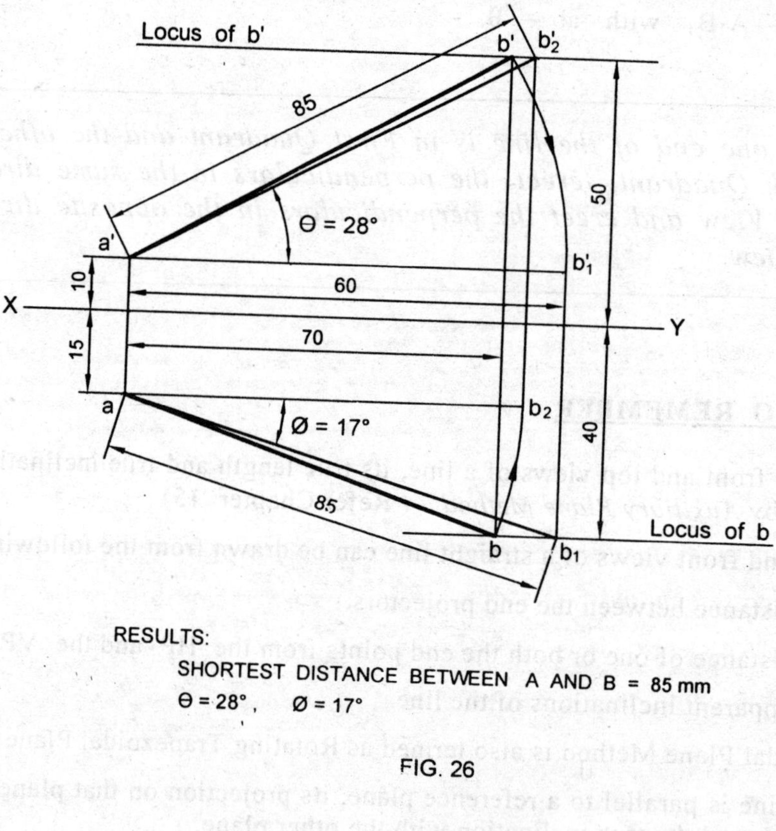

RESULTS:

SHORTEST DISTANCE BETWEEN A AND B = 85 mm

Θ = 28°, Ø = 17°

FIG. 26

Front and Top Views

1. Mark a′ 10 mm above XY and a 15 mm below XY.
2. Draw another projector 70 mm from the projector of A.
3. Mark a′ 50 mm above XY and b 40 mm below XY on the second projector.
4. Join a′ b′ and ab.

To obtain the Shortest Distance between A and B (= True Length of AB)

5. Through b′ draw the locus of b′.

NOTE : *Make the top view parallel to XY and draw the corresponding front view to get the shortest distance.*

---

6. With a as center and ab as radius draw an arc to intersect the horizontal line drawn from a at b_2.
7. From b_2 draw a projector to intersect the locus of b′ at b'_2.
8. Join a′ b'_2.
9. Now a′ b'_2 = AB = **Shortest distance between A and B = 85 mm.**
10. Measure $\theta = 28°$ and $\Phi = 17°$.

CHECK

11. Draw the locus of b.
12. With a′ as centre and a′ b′ as radius draw an arc to intersect the horizontal line drawn from a′ at b'_1.
13. From b'_1 draw a projector to meet the locus of b at b_1.
14. Join ab_1. Now ab_1 = AB = a′ b'_2.

Problem 39: *(Exercise)*

The distance between the projectors of two ends of a straight line is 60 mm. One end is 15 mm above HP and 50 mm in front of VP. The other end is 60 mm above HP and 10 mm in front of VP.

Draw the projections and find the true length of the line.

11.6 TRAPEZOIDAL PLANE METHOD

Problem 40: *(Fig. 27)* *Solve Problem No.38 by Trapezoidal Plane Method.*

Repeat **Step Nos. 1 to 4** of **Problem 38.**

Shortest Distance

5. At a′ draw a perpendicular to a′ b′ and mark a′A = 15 mm = distance of a from XY.
6. Similarly draw a perpendicular to a′ b′ at b′.
7. Mark b′B = 40 mm = distance of b from XY.
8. Join AB. Now AB = 85 mm.

CHECK

9. At a draw a perpendicular to ab and mark aA = 10 mm = distance of a′ from XY.
10. At b draw a perpendicular to ab.
11. Mark b B = 50 mm = distance of b′ from XY
12. Join AB. **Measure AB (= 85 mm).**

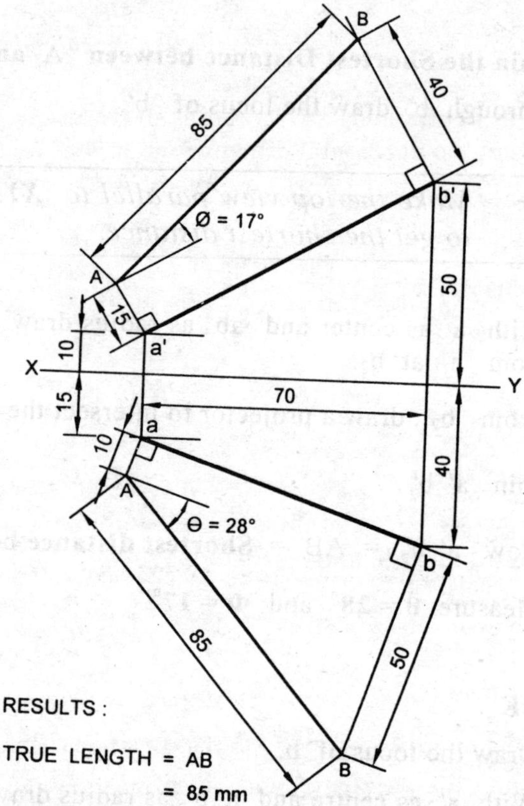

RESULTS :

TRUE LENGTH = AB = 85 mm

Θ = 28° , Ø = 17°

FIG. 27

NOTE : *The results obtained in Problems 38 and 40 are the same.*

Problem 41: *(Exercise)* *The distance between the projectors of two ends of a straight line is 40 mm. The lower end is 15 mm above HP and 10 mm in front of VP. The upper end is 40 mm above HP and 40 mm in front of VP. Find the true length and the true inclinations of the line by :*

(i) Rotating Line Method and (ii) Trapezoidal Plane Method.
(iii) Compare your results. (UQ)

11.7 GIVEN FRONT AND TOP VIEWS, TO FIND TRUE LENGTH AND TRUE INCLINATIONS

Make the top view parallel to XY and draw the corresponding front view to obtain the true length and θ.

Refer Fig. 28.

1. In this *Figure*, top and front views are given.
2. a as centre and ab as radius draw an arc to cut the horizontal drawn from a at b_1.
3. Draw the locus of b′.
4. From b_1 draw a projector to intersect the locus of b′ at b'_1.
5. Join $a'b'_1$. Now $a'b'_1$ = True Length AB.
6. Measure angle θ between $a'b'_1$ and XY.
7. Make the front view parallel to XY.
8. Draw the corresponding top view to obtain the true length and ϕ.

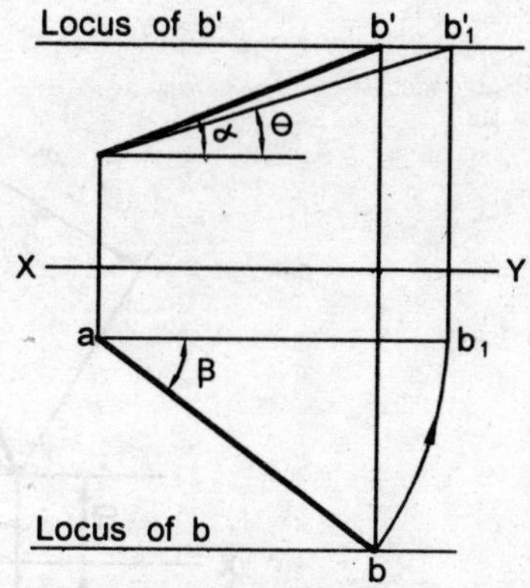

FIG. 28

Refer Fig. 29.

1. The projections of the line AB are given.
2. With a′ as centre and a′ b′ as radius draw an arc to intersect the horizontal line drawn from a′ at b'_2.
3. Draw the locus of b.
4. From b'_2 draw a projector to intersect the locus of b at b_2.
5. Join ab_2. Now ab_2 = AB = True Length.
6. Measure the angle ϕ between the line ab_2 and XY which is the True Inclination of the line AB with VP.

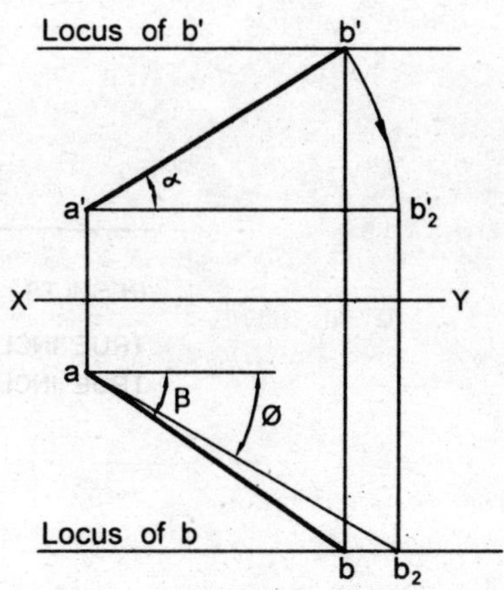

FIG. 29

Problem 42: *(Fig. 30)*

A line LM 70 mm long, has its end L 10 mm above HP and 15 mm in front of VP. Its Top and Front Views measure 60 mm and 40 mm respectively. Draw projections of the line. Find its inclinations with HP and VP. (UQ)

Given : LM = 70 mm; **l′** and **l** positions; lm = 60 mm; l′ m′ = 40 mm.

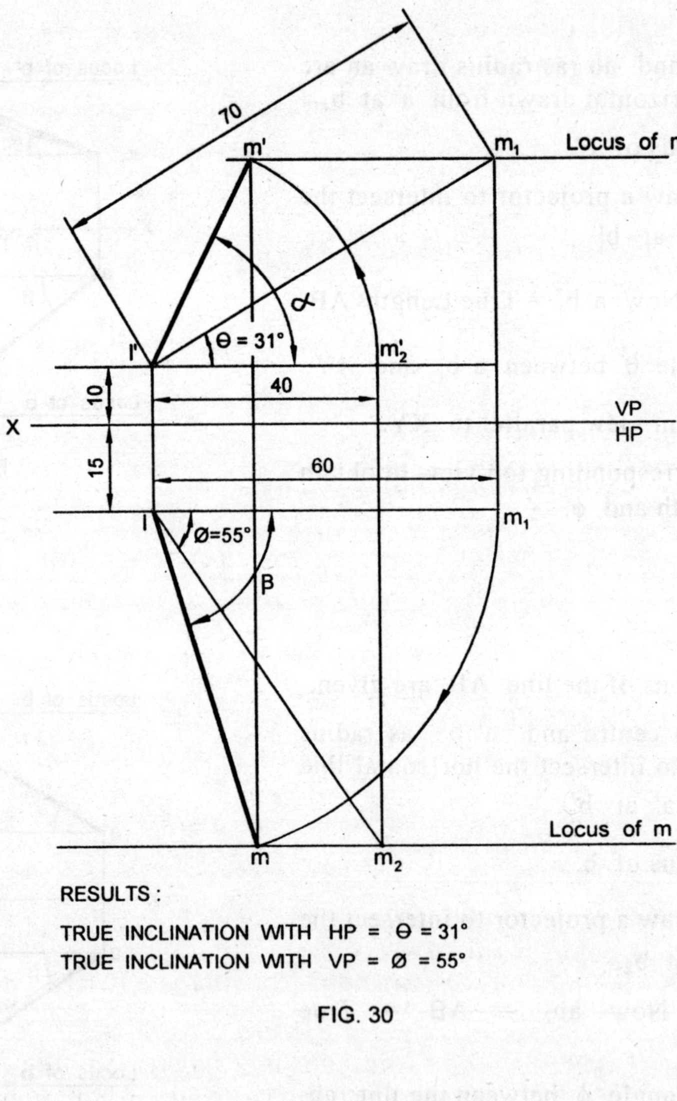

RESULTS :

TRUE INCLINATION WITH HP = Θ = 31°

TRUE INCLINATION WITH VP = Ø = 55°

FIG. 30

1. Mark l′ 10 mm above XY and **l** 15 mm below XY.

To find θ and l′ m′

2. From l draw a line parallel to XY.
3. Mark lm_1 = 60 mm = Top view length.

NOTE : *Front view corresponding to this top view will give the true length and θ.*

---

4. From m_1 draw a projector.
5. l′ as center and true length 70 mm as radius draw an arc to cut above projector at m'_1.
6. Join $l'm'_1$. Now measure θ = **31°**.
7. Draw the locus of m′.
8. l′ as centre and front view 40 mm as radius draw an arc to cut the locus of m′ at m′.
9. Join l′ m′. This is the front view of LM.

To find ϕ and lm

10. Draw $l'm'_2$ = 40 mm = Front view length and parallel to XY.

NOTE : *Top view corresponding to this front view will give the true length and ϕ.*

---

11. From m'_2 draw a projector.
12. l as center and true length 70 mm as radius, draw an arc to cut the above projector at m_2.
13. Join lm_2. Measure ϕ = **55°**.
14. Through m_2 draw the locus of m.
15. With l as center and 60 mm (top view length) as radius draw an arc to cut the locus of m at m.
16. Join lm, the top view.

NOTE : 1. *Angle made by the front view with XY is called* ***Apparent Angle*** *α.*
Angle made by the top view with XY is called ***Apparent Angle*** *β.*
2. *True inclinations are always less than the apparent inclinations.*

Problem 43: *(Exercise) A line AB measuring 85 mm has its end A 25 mm above HP and 20 mm in front of VP. The front and top views of the line measure 70 mm and 55 mm respectively. Draw the projections of the line and determine its true inclinations.* (UQ)

(*Results :* θ = 49° and ϕ = 35°)

Problem 44: *(Fig. 31)* ***(Crossing Quadrant Problem)***

One end of the line UV is in the first quadrant, 20 mm above HP and 30 mm in front of VP. The other end is 40 mm behind VP and 50 mm below HP. The distance between the projectors of the line is 60 mm.

Draw the projections of the line and find its true length. (UQ)

Given : U - I Quadrant ; V - III Quadrant

Solution by Rotating Line Method and Trapezoid Method for comparison :

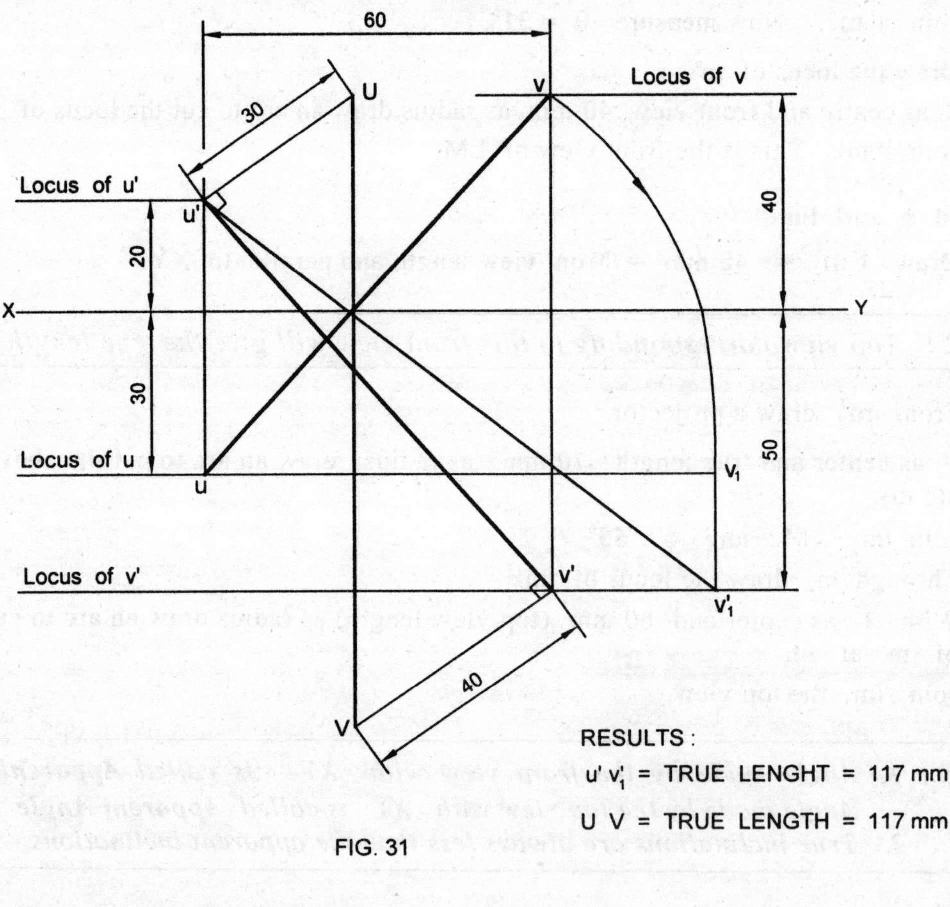

FIG. 31

To draw Front and Top Views

1. U is 20 mm above HP and 30 mm in front of VP. Therefore mark u′ 20 mm above XY and u 30 mm below XY.
2. Draw another projector which is 60 mm from u′ u.

3. V is 40 mm behind VP and 50 mm below HP. Therefore mark v 40 mm above XY and v′ 50 mm below XY.
4. Join u′ v′ **(Front View)** and uv **(Top View)**.

To find the True Length

5. Draw the loci of v, u′, u and v′.
6. Rotate the top view uv to be parallel to XY and obtain uv_1.
7. The corresponding front view $u'v_1'$ gives the true length and is obtained by drawing the projector from v_1 to cut the locus of v′ at v_1'.
8. Join $u'v_1'$. Now $u'v_1'$ = **True Length = 117 mm.**

To check the True Length

9. At u′ draw a perpendicular to $u'v_1'$ and mark U such that u′U = 30 mm.
10. At v′ draw a perpendicular to $u'v_1'$ in opposite direction.
11. Mark V such that v′ V = 40 mm.
12. Join UV. Now UV = **True Length = 117 mm.**

NOTE : *UV is obtained by Trapezoid Method and Rotating Line Method and the results are compared.*

Problem 45: *(Exercise) The end A of a line AB is in the HP and 15 mm in front of the VP. The end B is 50 mm behind the VP and 40 mm below the HP. The distance between the end projectors is 50 mm. Draw the projections of AB and determine its true length and true inclinations with the two planes.*

11.8 POINTS TO REMEMBER

Front and Top Views of a line may be drawn in different combinations as follows:

- (i) Distance of one or both end points of the line from HP and VP,
- (ii) Length of Front or Top View,
- (iii) Distance between end projectors and
- (iv) Apparent inclinations of line.

Position of Line	Front View	Top View
Line inclined to both HP and VP.	Shorter than True Length, Inclined to XY and Angle is not True Inclination.	Shorter than True Length, Inclined to XY and Angle is not True Inclination.

12. TRACES OF A LINE

Trace : When a line is inclined to a reference plane, it will meet that plane, produced if necessary. The point at which the line or line produced meets the plane is called its *Trace*.

Horizontal Trace (H.T.) : The point of intersection of the line or line produced with HP is called ***Horizontal Trace*** (H.T.)

Vertical Trace (V.T.) : The point of intersection of the line or line produced with VP is called ***Vertical Trace*** (V.T.).

12.1 TO FIND H.T. AND V.T. OF A LINE FOR ITS VARIOUS POSITIONS WITH RESPECT TO HP AND VP

1. Line Parallel to One Plane and Perpendicular to the Other
2. Line Parallel to One Plane and Inclined to the Other
3. Line Parallel to Both HP and VP
4. Line Inclined to Both HP and VP

12.2 LINE PARALLEL TO ONE PLANE AND PERPENDICULAR TO OTHER

1. LINE PARALLEL TO HP AND PERPENDICULAR TO VP

1. *See Fig. 32.* Produce the line PQ to meet the VP at V.T.
2. Mark V.T. in the front view to coincide with p' (q').
3. Line when produced will not meet HP, since it is parallel to HP. Hence it has no H.T.

Problem 46: *(Fig. 32) A line PQ 25 mm long is parallel to HP and perpendicular to VP. The end Q is 10 mm in front of VP and the line is 20 mm above HP. Draw the projections of the line and find its traces.*

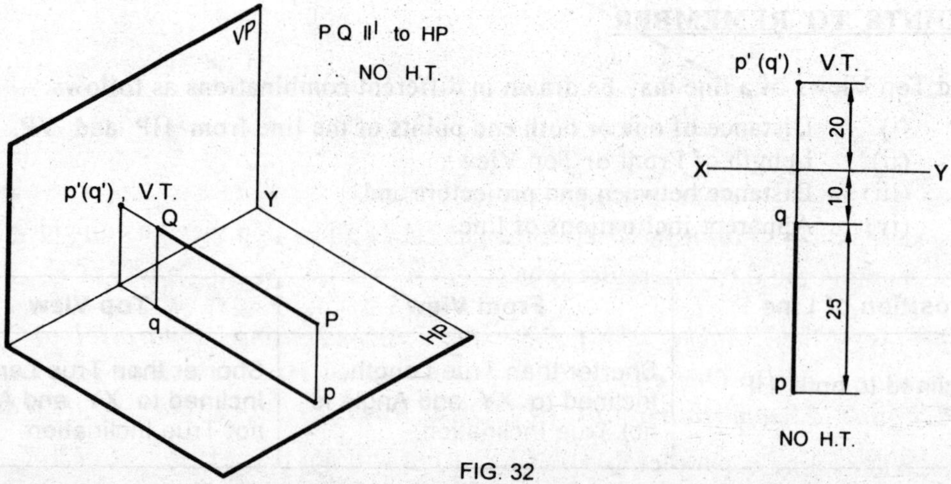

FIG. 32

1. *See Fig. 32.* Draw the front view p′ (q′) and top view pq.
2. PQ is perpendicular to VP. Hence mark V.T. in the front view to coincide with p′(q′).
3. PQ is parallel to HP. Therefore it has no H.T.

Problem 47: *(Exercise) A line CD 50 mm long is parallel to HP and perpendicular to VP. The end C is 20 mm in front of VP and the line is 40 mm above HP. Draw the projections of the line and find its traces.*

2. LINE PARALLEL TO VP AND PERPENDICULAR TO HP

1. *See Fig. 33.* Produce CD to meet HP at H.T.
2. Mark H.T. in top view to coincide with c(d).
3. Line when produced will not meet VP, since it is parallel to VP. Hence it has no V.T.

Problem 48: *(Fig. 33) A line CD 25 mm long is parallel to VP and perpendicular to HP. End C is 35 mm above HP and 20 mm in front of VP. End D is 10 mm above HP. Draw the projections of the line CD and find its traces.*

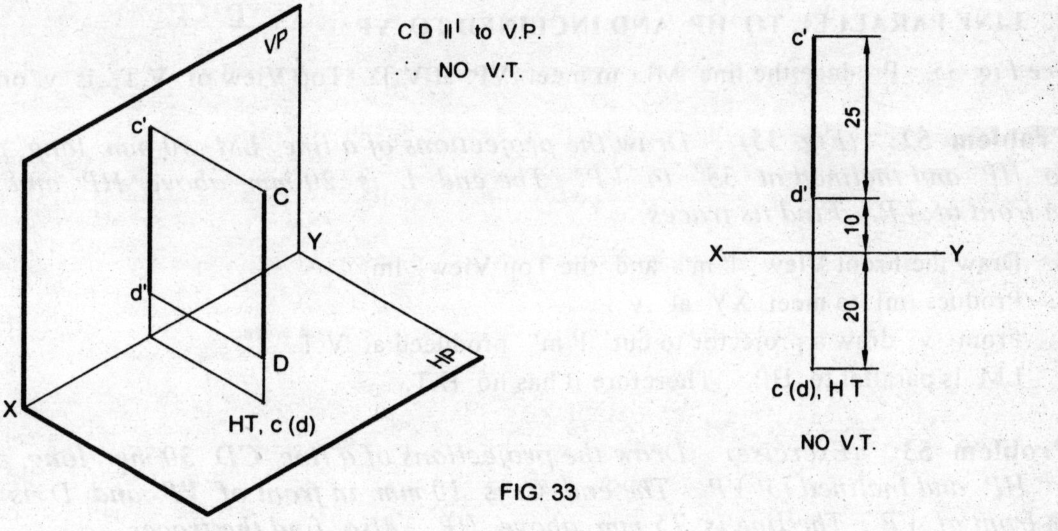

FIG. 33

1. Draw the front view c′ d′ and top view c(d).
2. CD is perpendicular to H.P. Therefore mark H.T. in the top view to coincide with c(d).
3. CD is parallel to V.P. Therefore it has no V.T.

Problem 49: *(Exercise) A line PQ 40 mm long is parallel to VP and perpendicular to HP. One end Q is 15 mm above HP. Another end P is 55 mm above HP and 25 mm in front of VP. Draw the projections and find its traces.*

RULE : When a line is parallel to HP (or VP) it will not have H.T. (or V.T.).

12.3 LINE PARALLEL TO ONE PLANE AND INCLINED TO THE OTHER

1. LINE PARALLEL TO VP AND INCLINED TO HP

See Fig. 34. Produce line BA to meet HP at H.T. Front view of H.T. is h′ on XY.

Problem 50: *(Fig. 34) A line AB 40 mm long is parallel to VP and inclined at 35° to HP. The end A is 15 mm above HP and 20 mm in front of VP. Draw the projections of the line and find its traces.*

1. Draw the front view a′ b′ and top view ab.
2. Produce b′ a′ to meet XY at h′.
3. From h′ draw a projector to intersect ba produced at H.T.
4. AB is parallel to VP. Therefore it has no V.T.

Problem 51: *(Exercise) A line MN 50 mm long is parallel to VP and inclined at 45° to HP. The end M is 20 mm above HP and 15 mm in front of VP. Draw the projections of the line and find its traces.*

2. LINE PARALLEL TO HP AND INCLINED TO VP

See Fig. 35. Produce the line ML to meet VP at V.T. Top View of V.T. is v on XY.

Problem 52: *(Fig. 35) Draw the projections of a line LM 40 mm long, parallel to HP and inclined at 35° to VP. The end L is 20 mm above HP and 15 mm in front of VP. Find its traces.*

1. Draw the Front View l′ m′ and the Top View lm.
2. Produce ml to meet XY at v.
3. From v draw a projector to cut l′ m′ produced at V.T.
4. LM is parallel to HP. Therefore it has no H.T.

Problem 53: *(Exercise) Draw the projections of a line CD 30 mm long, parallel to HP and inclined to VP. The end C is 10 mm in front of VP and D is 20 mm in front of VP. The line is 15 mm above HP. Also, find the traces.*

12.4 LINE PARALLEL TO BOTH HP AND VP

Problem 54: *(Fig. 36) A line RS 40 mm long is parallel to both planes. It is 20 mm above HP and 15 mm in front of VP. Draw projections and find its traces.*

1. Draw the front view r′ s′ and top view rs.
2. The line RS is parallel to both the planes. If the line is extended, it will not meet both HP and VP, since it is parallel to both the planes. Therefore it has no H.T. and V.T.

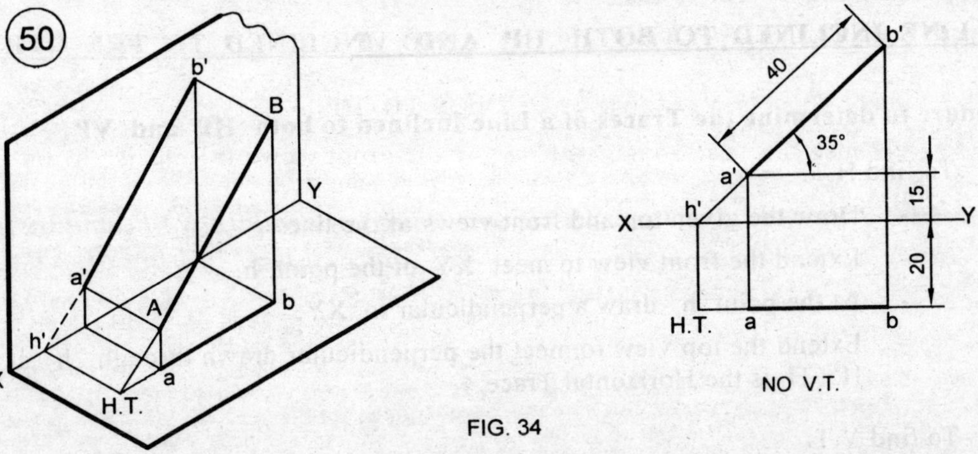

FIG. 34

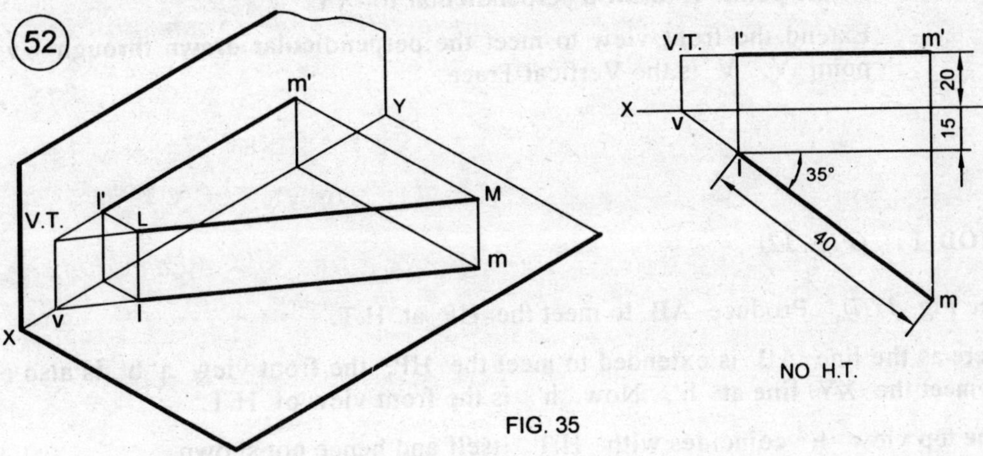

FIG. 35

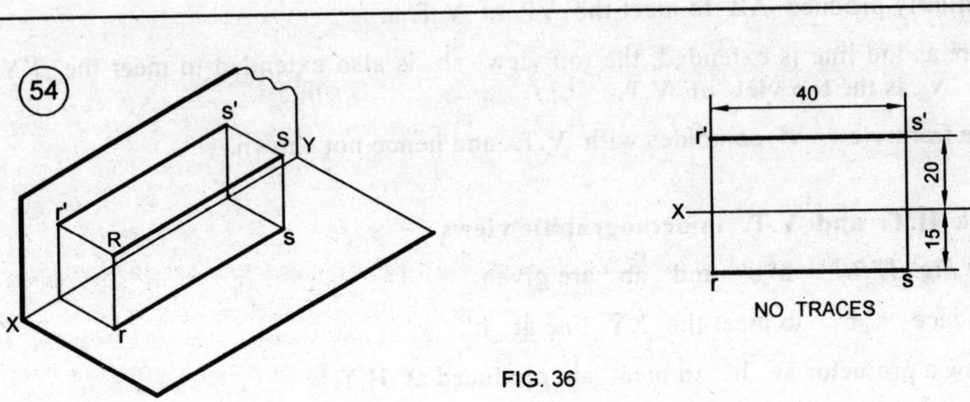

FIG. 36

12.5 LINE INCLINED TO BOTH HP AND VP

Procedure to determine the Traces of a Line Inclined to both HP and VP

- To find H.T.
  - Draw the given top and front views of the line
  - Extend the front view to meet XY at the point h.
  - At the point h, draw a perpendicular to XY.
  - Extend the top view to meet the perpendicular drawn through h at the point H. H is the Horizontal Trace.
- To find V.T.
  - Extend the top view to meet XY at the point v.
  - At the point v, draw a perpendicular to XY.
  - Extend the front view to meet the perpendicular drawn through v at the point V. V is the Vertical Trace.

METHOD I : ***(Fig. 37)***

1. *See Fig. 37(i).* Produce AB to meet the HP at H.T.
2. Here as the line AB is extended to meet the HP, the front view a′ b′ is also extended to meet the XY line at h′. Now h′ is the front view of H.T.
3. The top view h coincides with H.T. itself and hence not shown.
4. Similarly produce AB to meet the VP at V.T.
5. Here as the line is extended, the top view ab is also extended to meet the XY line at v. v is the top view of V.T.
6. The front view v′ coincides with V.T. and hence not shown.

To mark H.T. and V.T. in orthographic views

7. *See Fig. 37(ii).* a′ b′ and ab are given.
8. Produce a′ b′ to meet the XY line at h′.
9. Draw a projector at h′ to meet ab produced at H.T.
10. Similarly produce ab to meet the XY line at v.
11. At v draw a projector to meet a′ b′ produced at V.T.

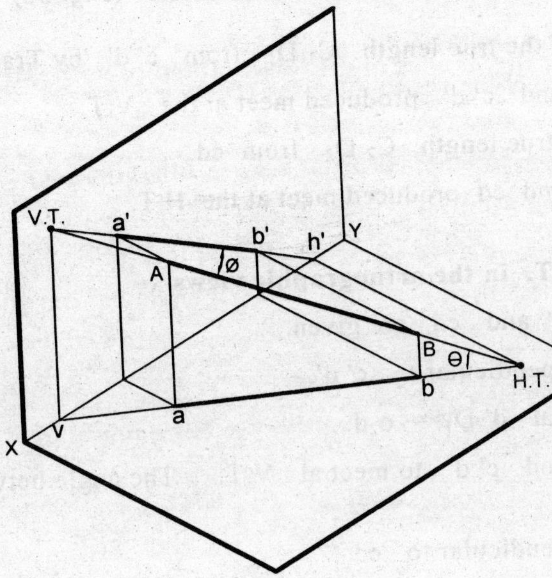

(i)

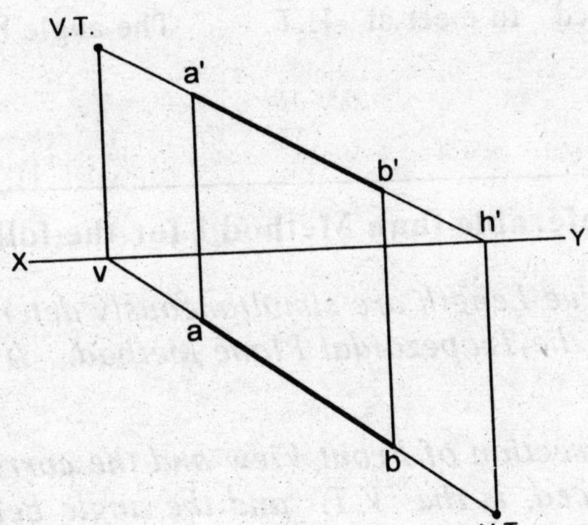

(ii)

FIG. 37

METHOD II TRAPEZOIDAL PLANE METHOD : *(Fig. 38)*

1. *See Fig. (i).* Find the true length $C_1 D_1$ from $c' d'$ by Trapezoidal Plane Method.
2. $C_1 D_1$ produced and $c' d'$ produced meet at the V.T.
3. Similarly find the true length $C_2 D_2$ from cd.
4. $C_2 D_2$ produced and cd produced meet at the H.T.

To mark H.T. and V.T. in the orthographic views

5. *See Fig. (ii).* $c' d'$ and cd are given.
6. At d' draw a perpendicular to $c' d'$.
7. Mark D_1 such that $d' D_1 = o_2 d$.
8. Produce $C_1 D_1$ and $c' d'$ to meet at V.T. The angle between $C_1 D_1$ and $c' d'$ is ϕ.
9. At c draw a perpendicular to cd.
10. Mark C_2 such that $c C_2 = o_1 c'$.
11. At d draw a perpendicular to cd.
12. Mark D_2 such that $d D_2 = o_2 d'$.
13. Produce $C_2 D_2$ and cd to meet at H.T. The angle between cd and $C_2 D_2$ is θ.

NOTE : Method II is preferable than Method I for the following reasons :

1. *θ, ϕ and True Length are simultaneously determined along with H.T. and V.T. by the Trapezoidal Plane Method. It is not so in the case of Method I.*
2. *Point of intersection of Front View and the corresponding True Length line, if produced, is the V.T. and the angle between these gives True Inclination ϕ.*
3. *Point of intersection of Top View and the corresponding True Length line, if produced, is the H.T. and the angle between these gives True Inclination θ.*
4. *When the projections of the line are perpendicular to XY i.e., when the sum of its inclinations with HP and VP is 90°, it is possible to find the traces by Method II only (Refer Problem in the next page).*

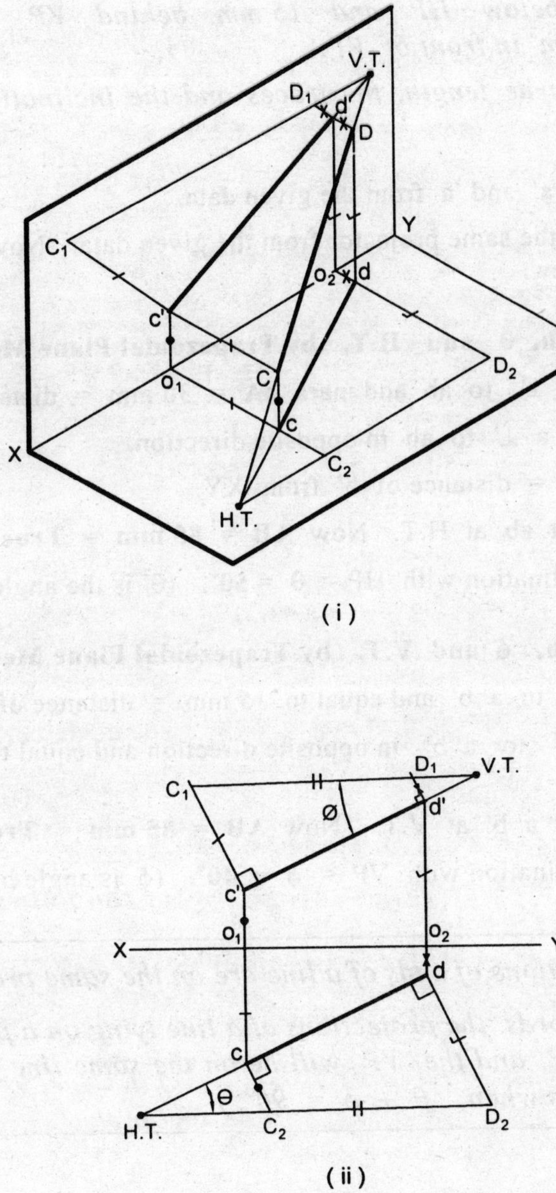
V.T.
D_1
d'
D
Y
C_1
o_2
d
c'
C
o_1
D_2
c
X
C_2
H.T.
(i)
D_1
V.T.
C_1
Ø
d'
c'
o_1
o_2
X
Y
d
c
Θ
H.T.
C_2
D_2
(ii)

FIG. 38

Problem 55: *(Fig. 39)*

The projections of the ends of a line AB are on the same projector. The end A is 30 mm below HP and 15 mm behind VP. The end B is 35 mm above HP and 40 mm in front of VP.

Determine its true length, the traces and the inclinations with the reference planes.

(UQ)

1. *See Fig. 39.* Mark a′ and a from the given data.
2. Mark b′ and b on the same projector from the given data. Now ab is the top view and a′ b′ is the front view.

To find the True Length, θ and H.T. by Trapezoidal Plane Method

3. From a draw a line ⊥r to ab and mark aA = 30 mm = distance of a′ from XY.
4. Similarly at b draw a ⊥r to ab in opposite direction.
5. Mark bB = 35 mm = distance of b′ from XY.
6. Join AB to intersect ab at H.T. Now **AB = 85 mm = True Length.**
7. Measure the true inclination with HP = θ = **50°**. (θ is the angle between ab and AB)

To find the True Length, ϕ and V.T. by Trapezoidal Plane Method

8. At a′ draw a′A ⊥r to a′ b′ and equal to 15 mm = distance of a from XY.
9. At b′ draw b′ B ⊥r to a′ b′ in opposite direction and equal to 40 mm = distance of b from XY.
10. Join AB to intersect a′ b′ at V.T. Now **AB = 85 mm = True Length.**
11. Measure the true inclination with VP = ϕ = **40°**. (ϕ is angle between a′ b′ and AB).

RULE : *If the projections of ends of a line are on the same projector, $\theta + \phi = 90^o$.*

(In other words, the projections of a line lying on a plane perpendicular to both the HP and the VP will be on the same line perpendicular to XY. This happens when $\theta + \phi = 90^o$.)

Problem 56: *(Exercise)*

The projections of the ends of a line EF are on the same projector. The end E is 10 mm above HP and 15 mm in front of VP. The end F is 35 mm above HP and 45 mm in front of VP. Draw the projections and find the true length, true inclinations and traces.

(UQ)

(Results: True Length. = 39 mm, $\theta = 40^o$ and $\phi = 50^o$)

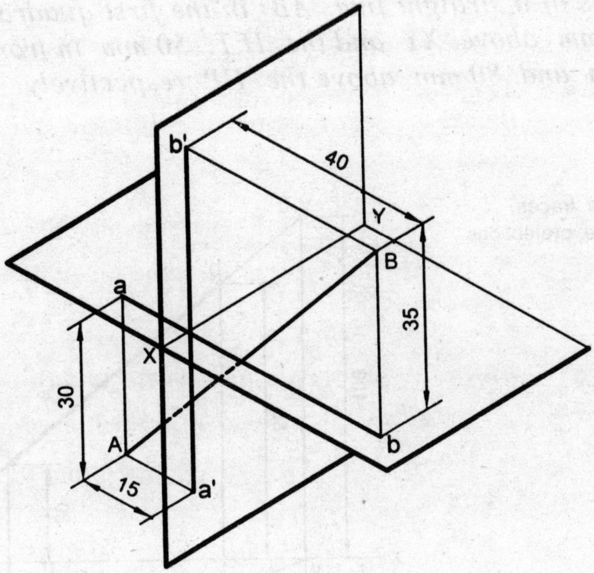

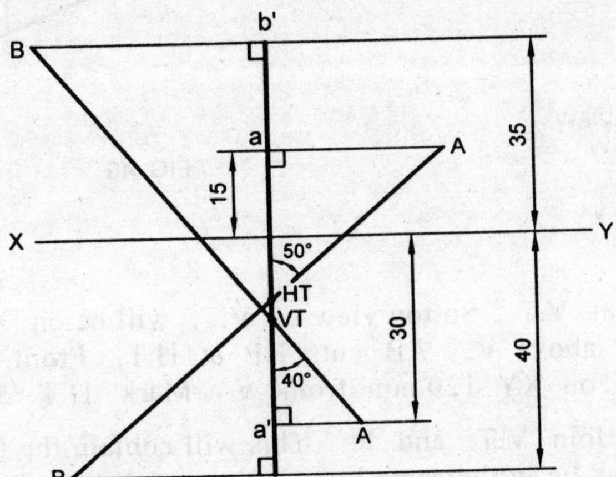

FIG . 39

RESULTS :

TRUE LENGTH = A B = 85 mm

θ = 50°, ϕ = 40°

Problem 57: *(Fig. 40)* *The projections on the XY line of the horizontal and vertical traces of a straight line AB in the first quadrant are 120 mm apart. The V.T. is 100 mm above XY and the H.T. 50 mm in front of XY. The points A and B are 30 mm and 80 mm above the HP respectively. Draw the projections.* (UQ)

Given the traces, to find the projections

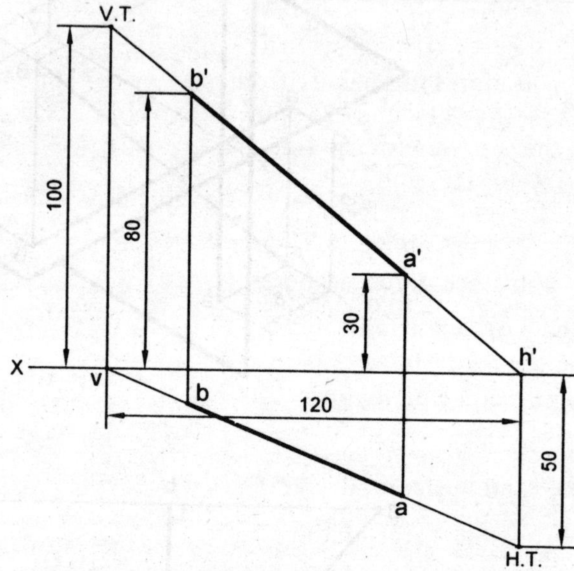

RESULTS :

a' b' : FRONT VIEW
a b : TOP VIEW

FIG. 40

1. AB cuts VP at V.T. So top view of V.T. will be on XY. Mark v on XY. Mark V.T. 100 mm above v. AB cuts HP at H.T. Front view of H.T. will be on XY. Hence mark h′ on XY 120 mm from v. Mark H.T. 50 mm from h′ below XY.
2. **Front view :** Join V.T. and h′. This will contain the front view a′ b′. V.T., a′ b′ and h′ always lie in the same line. Now mark b′ 80 mm above XY and a′ 30 mm above XY on this line.
3. **Top view :** H.T., ab and v will always lie in the *same* line. Therefore join v and H.T. This line will contain ab. Draw the projectors from a′ and b′ to cut the above line at a and b respectively.

Problem 58: *(Exercise)* *A line AB measuring 12 cm long has its V.T. 8 cm above XY line and the H.T. 10 cm in front of XY line. The projections of the H.T. and V.T. on the XY line are 11 cm apart. If the point A is 1 cm above HP, draw the projections of AB and find its true inclinations.*

(Results : $\theta = 29^{o}$ and $\phi = 36.5^{o}$)

13. ADDITIONAL PRACTICAL PROBLEMS – II

59. A line AB 100 mm long has its front view a′ b′ inclined at an angle of 45° to XY. The point A is in VP and 25 mm above HP. The length of the front view is 60 mm. Draw the top view of the line and measure its length. Also, find the inclinations of the line AB to HP and VP. (UQ)

1. Mark a′ 25 mm above XY and a on XY.
2. α and a′ b′ are given. Therefore draw a′ b′ = 60 mm at 45° to XY.

To obtain the Top View and θ

3. Draw the locus of b′.
4. a′ as centre and true length 100 mm as radius draw an arc to cut the locus of b′ at b_1'.
5. Join $a'b_1'$ and measure **θ = 25.5°**. From b_1' draw a projector to meet XY at b_1.

   Now a b_1 is the corresponding top view of $a'b_1'$.
6. With a as centre and a b_1 as radius draw an arc to intersect the projector from b′ at b. Join ab and measure **ab = 90 mm**.

To find ϕ

7. Draw the locus of b.
8. a as centre and true length = 100 mm as radius draw an arc to cut the locus of b at b_2.
9. Join ab_2 and measure **ϕ = 53°**.

60. One end R of a straight line RS is 35 mm above HP and 25 mm in front of VP. The other end S is 20 mm below HP and 55 mm behind VP. The distance between the projectors is 50 mm. Determine the true length, inclinations of the line RS and its traces. (UQ)

Results : True Length = 110 mm, True Inclinations with HP = 30° & with VP = 45°

61. A line CD 80 mm long has its end C 55 mm in front of VP and 15 mm above HP. The line is inclined at 50° to HP and 40° to VP. Draw the projections of the line and find its traces.

HINT : *Since the sum of the inclinations (θ + ϕ) of the line with the HP and VP is 90°, the projections of the line lie in the same projector.*

62. *(Fig. 41)* An ornamental light O is placed 10 m above the floor and in the center of an auditorium 40 m x 50 m x 35 m high. Determine graphically its distance from one of the corners between the roof and two adjacent walls. (UQ)

1. *See Fig. 41.* The projections of the line joining the light O and a corner between the roof and the adjacent walls are drawn first.
2. Then, by Trapezoidal Plane Method, the distance between the light and that corner is determined.

63. *(Fig. 42)* A fan is hanging in the center of a room 4 m x 4.5 m x 4 m high. The center of the fan is 0.6 m below the ceiling. Determine graphically the shortest distance of the fan from one of the corners of the floor. (UQ)

1. *See Fig. 42.* Take a Scale : 1 m = 0.5 m.
2. Draw top and front views of the room as shown.
3. Locate the positions of the fan in the top view and front view as a and a′.
4. Locate the positions of corner of the floor as b and b′.
5. Join ab and a′ b′.
6. With a as centre, rotate ab to b_1 such that a b_1 is parallel to XY.
7. Project b_1 upwards to meet the locus of the corner b′ of the floor at b_1' in the front view.
8. Join a′ b_1' which gives the distance of the fan from the corner of the floor and is equal to **4.5 m**.

64. An electric bulb is hung vertically from the center of a terraced room at a height of 0.5 m below the ceiling. The room measures 5 m long, 4 m wide and 3 m high. Determine graphically the true distance between the bulb and any one of the corners below. (UQ)

(Result: 4.1 m)

65. An Auditorium of a college is of 48 m length, 42 m width and 36 m height. Determine graphically the distance between a top corner and the bottom corner diagonally opposite to it. (UQ)

(Result: 73 m)

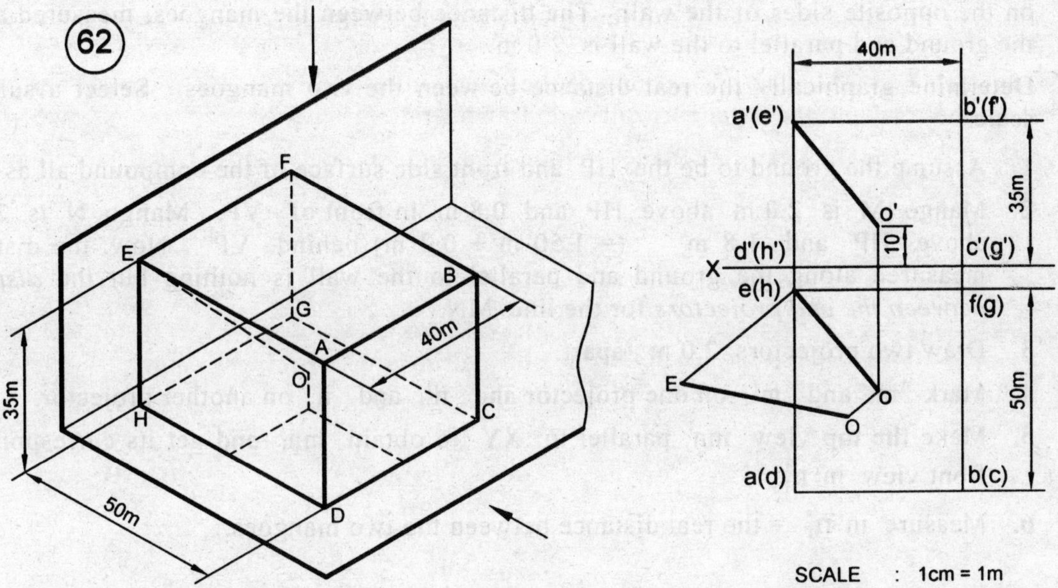

SCALE : 1cm = 1m

RESULT : OE = 40.5m

FIG. 41

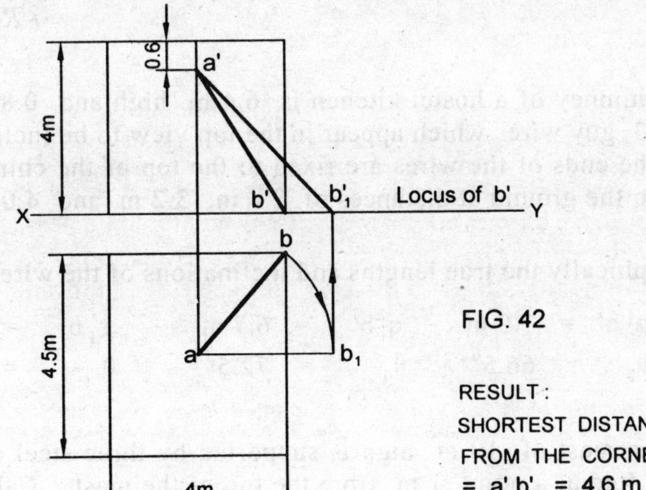

FIG. 42

RESULT :

SHORTEST DISTANCE OF THE FAN FROM THE CORNER OF THE FLOOR $= a' b'_1 = 4.6$ m

66. *(Fig. 43)* Two mangoes on a tree are 2.0 m and 2.5 m respectively above the ground and 0.8 m and 1.50 m from a 0.3 m thick compound wall of a bungalow, but on the opposite sides of the wall. The distance between the mangoes, measured along the ground and parallel to the wall is 2.0 m.

Determine graphically the real distance between the two mangoes. Select a suitable scale. (UQ)

1. Assume the ground to be the HP and front side surface of the compound all as VP.
2. Mango M is 2.0 m above HP and 0.8 m in front of VP. Mango N is 2.5 m above HP and 1.8 m (= 1.50 m + 0.3 m) behind VP. Now, the distance measured along the ground and parallel to the wall is nothing but the *distance between the end projectors* for the line MN.
3. Draw two projectors 2.0 m apart.
4. Mark m and m′ on one projector and n′ and n on another projector.
5. Make the top view mn parallel to XY to obtain mn_1 and get its corresponding front view $m'n_1'$.
6. Measure $m'n_1'$ = the real distance between the two mangoes.

67. Two apples A and B on an apple tree are 3 m and 2 m respectively above the ground. Apple A is 0.8 m and 1.2 m from the adjacent compound walls P and Q respectively while the corresponding distances of the apple B are 4 m and 8 m.

Draw the projections of the apples if the two walls are at right angles to each other. Find the true distance between the apples. (UQ)

(Result : AB = 7.6 m)

68. *(Fig. 44)* A chimney of a hostel kitchen is 6.4 m high and 0.8 m in diameter. It is supported by 3 guy wires which appear in the top view to be inclined at equal angles to each other. The ends of the wires are fixed to the top of the chimney. The other ends are anchored to the ground at distances of 2.4 m, 3.2 m and 4.0 m from the center of the chimney.

Determine graphically the true lengths and inclinations of the wires with ground. (UQ)

Results :	$p'a'$ = 7.0 m	$q_1'b'$ = 6.7 m	$r_1'b'$ = 7.3 m
	θ_p = 66.5°	θ_q = 72.5°	θ_r = 60.5°

69. A TV antenna mast of 10 m high is supported by three steel cables R, S and T. The cables are tied at a point 1 m from the top of the mast. Cable R is anchored on the ground, S on a roof corner of a 4 m high building and T on a compound wall of 2 m height. In the top view, all the cables are of the same length of 6 m and make 120° between them.

Find graphically the true lengths and inclinations of the cables.

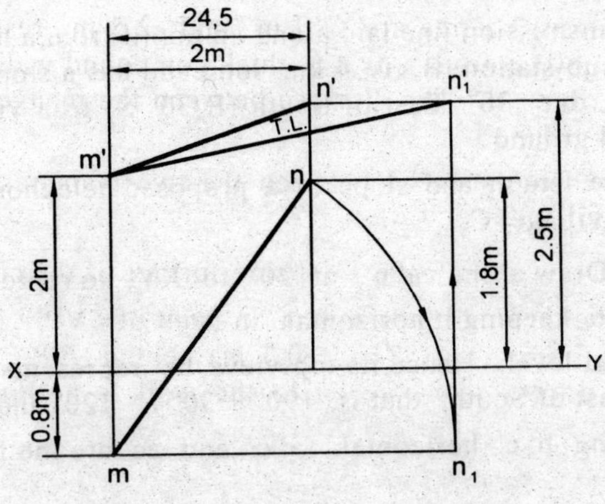

FIG. 43

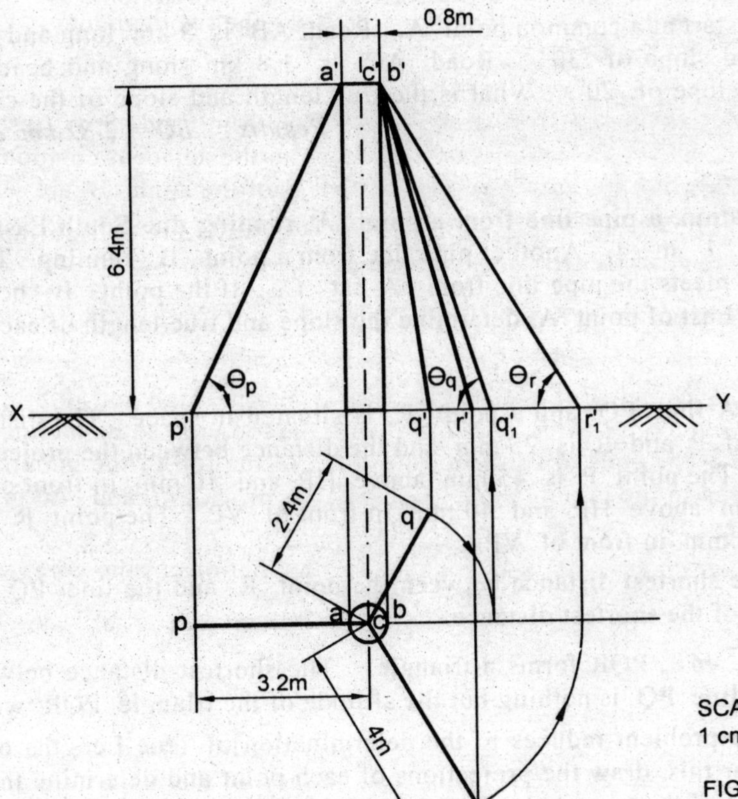

SCALE :
1 cm = 1.0 m

FIG. 44

70. *(Fig. 45)* An electric transmission line laid along an uphill from a hydroelectric power station A due East to a sub-station B, is 4 km long and has a slope of 20°. Another line from the sub-station, due 30° East of South (E 30° S) to a village C is 3 km long, but is laid on a level ground.

Determine graphically the length and slope of a proposed telephone line joining the power station A and the village C. (UQ)

1. Mark any point a'. Draw a line a' b' at 20° to XY, to represent AB.
2. Project its top view ab keeping it horizontal.
3. Line BC is on ground level. Hence its top view bc represents 3 km. Draw bc in a direction 30° East of South, that is, (90° + 30°) = 120° inclined to ab.
4. Project c to c' making b' c' horizontal. a' c' and ac are the front and top views of the line AC.
5. Determine the true length a' c_1' and the angle θ which are respectively the length and slope of the telephone line AC.

71. Two roads start at a common point A. Road AB is 3 km long and bears 50° SE on a downward slope of 30°. Road AC is 1.8 km long and bears 65° NE on a downward slope of 20°. What is the true length and slope of the connecting road B and C ? *(Results : BC = 2.82 km and Slope = 18°)*

72. In a hill station, a pipe line from a point A, running due South East has a downward gradient of 1 in 4. Another pipe line from a point B running 20° East of South (S 20° E) meets the pipe line from A at C. If the point B be in level with and 16 km due East of point A, determine the slope and true length of each pipe line.

73. *(Fig. 46)* A line PQ and a point R is situated in space. The distance between the projectors of P and R is 25 mm and the distance between the projectors of P and Q is 70 mm. The point P is 45 mm above HP and 10 mm in front of VP. The point Q is 20 mm above HP and 40 mm in front of VP. The point R is 15 mm above HP and 50 mm in front of VP.

Find out the shortest distance between the point R and the line PQ. Also, draw the projections of the shortest distance. (UQ)

1. *See Fig. 46.* PQR forms a triangle. The shortest distance between the point R and the line PQ is nothing but the altitude of the triangle PQR with PQ as base.
2. Now the problem reduces to the determination of True Lengths of PQ, QR and RP. For this, draw the projections of each point and determine the True Length of each side of the triangle by *Trapezoidal Plane Method.*
3. Construct the triangle in its actual size separately and find the shortest distance between R and PQ.
4. *Result:* Shortest distance between R and line PQ = 32 mm

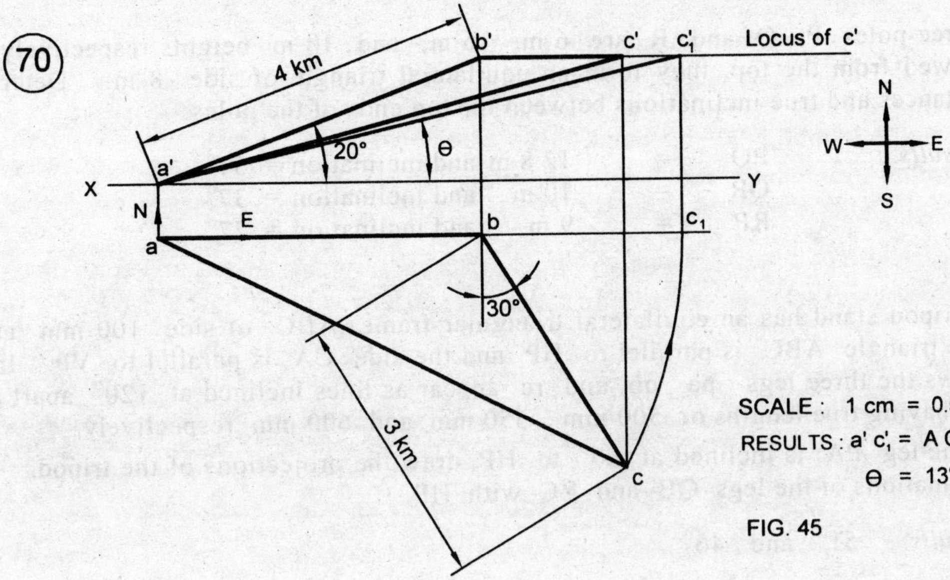

SCALE : 1 cm = 0.5 km

RESULTS : a' c'1 = A C = 6 km

Θ = 13°

FIG. 45

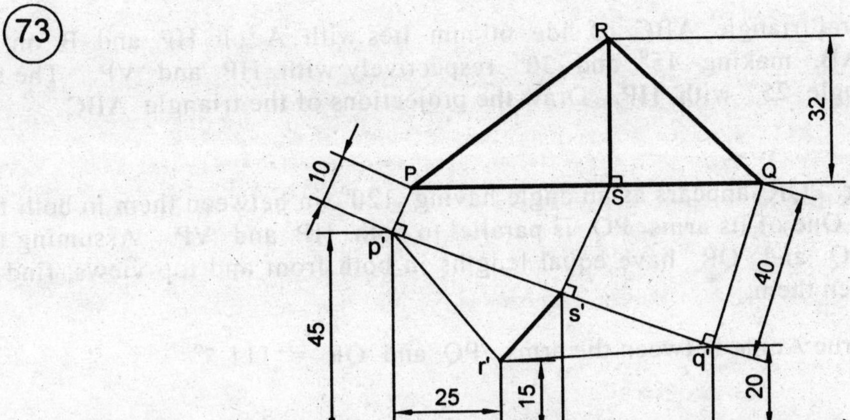

FIG. 46

THE SHORTEST DISTANCE BETWEEN R and LINE PQ = RS = 32 mm

74. Three poles P, Q and R are 6 m, 16 m, and 10 m heights respectively. When viewed from the top, they form an equilateral triangle of side 8 m. Determine the distances and true inclinations between the top ends of the poles. (UQ)

Results:

PQ	=	12.8 m	and inclination = 51°
QR	=	10 m	and inclination = 37°
RP	=	9 m	and inclination = 27°

75. A tripod stand has an equilateral triangular frame ABC of side 100 mm at its top. The triangle ABC is parallel to HP and the side CA is parallel to VP. In the top views the three legs pa, qb and rc appear as lines inclined at 120° apart and they are having true lengths of 500 mm , 550 mm and 600 mm respectively.

If the leg PA is inclined at 60° to HP, draw the projections of the tripod. Find the inclinations of the legs QB and RC with HP. (UQ)

Results: 51° and 46°

76. An equilateral triangle ABC of side 60 mm lies with A on HP and B on VP and with side AB, making 45° and 30° respectively with HP and VP. The side CA makes an angle 25° with HP. Draw the projections of the triangle ABC.

77. A bent lever PQR appears as an angle having 120° in between them in both front and top views. One of its arms PQ is parallel to both HP and VP. Assuming that both the arms PQ and QR have equal lengths in both front and top views, find the true angle between them. (UQ)

Result : True Angle between the arms PQ and QR = 111.5°

78. A flag post F is 10 m high. It is supported by three guy wires. The top ends of the wires are tied to the flag post and their lower ends are tied to the top of three pegs A, B and C fixed to the ground. The pegs A, B and C are 0.5 m, 0.75 m and 1 m high respectively. In the top view, three guy wires appear to be at 120° to each other and the lengths of their top views are oa = 3 m, ob = 4 m and oc = 5 m.

Draw the projections of the guy wires and the pegs and determine the true lengths of the guy wires and the true inclinations they make with the ground.

*

14

PROJECTIONS ON AUXILIARY PLANES

1. AUXILIARY PLANES

Principal Planes

It has already been discussed that the two views of an object, *namely*, the Front View and the Top View can be projected on **two Principal Planes**, *viz.*, **HP** and **VP**.

Auxiliary Planes

Sometimes, either of these two views may not show the different edges and faces of an object in their true sizes, when these edges and faces are *not* parallel to either HP or VP, that is, when these edges and faces are *inclined* to HP and VP. In order to show such edges and faces in their true sizes, it becomes necessary to introduce additional Planes of Projection (*in addition to HP and VP*) in such positions which will show them in true sizes.

It is therefore obvious that if an edge or a face is to be shown in true size, it should be parallel to the plane of projection. Thus the additional planes of projection are set up so as to be parallel to the edges or faces which should be shown in true sizes. These additional planes of projection which are set up to show the true sizes are known as ***Auxiliary Planes***.

Profile Planes

Students should not be confused between the Profile Planes and the Auxiliary Planes. *Profile Planes* are always perpendicular to the HP and VP and are used to project the Left and Right Side Views.

But *Auxiliary Planes* are introduced such that they are parallel to the edge or face (*inclined to HP / VP*) which is to be shown in true size and perpendicular to either HP or VP.

1.1 AUXILIARY VIEWS

Additional views called **Auxiliary Views,** are obtained by projecting the object on other planes known as **Auxiliary Planes,** *which are inclined to one of the Principal Planes and perpendicular to the other*.

1.2 TYPES OF AUXILIARY PLANES

(i) **Auxiliary Vertical Plane (A.V.P.)**

If the Auxiliary Plane selected is perpendicular to the HP and inclined to the VP, the view of the object projected on Auxiliary Plane is called as ***Auxiliary Front View;*** And the Auxiliary Plane is called ***Auxiliary Vertical Plane*** (A.V.P.).

(ii) **Auxiliary Inclined Plane (A.I.P.)**

If the Auxiliary Plane selected is perpendicular to the VP and inclined to the HP, the view of the object projected on Auxiliary Plane is called as the ***Auxiliary Top View;*** And the Auxiliary Plane is called ***Auxiliary Inclined Plane*** (A.I.P.).

NOTE: P.P. is also considered as an A.P. which is perpendicular to both HP and VP.

New Reference Lines (X_1Y_1 and X_2Y_2)

Line of intersection of HP with A.V.P. or VP with A.I.P. is called *New Reference Line* X_1Y_1. Line of intersection of A.I.P. and A.V.P. is called *Another Reference Line* X_2Y_2.

2. PROJECTIONS OF POINTS ON AUXILIARY PLANES

2.1 PROJECTION OF A POINT P ON A.V.P.

1. *See Fig. 1.* Consider the given point P and the two principal planes HP and VP. p′ is the projection of P on VP and p is the projection of P on HP.
2. Consider HP and A.V.P. separately. They intersect each other at right angles at the new reference line X_1Y_1.
3. *Refer Fig. (i).* Look at the point P such that the eye ray is perpendicular to the A.V.P. Now the projection of P on A.V.P. is p_1' and is called the ***Auxiliary Front View*** of point P.
4. ***Note that the orthographic views of the Auxiliary Projections are obtained by rotating the Auxiliary Plane about the HP or VP to which it is perpendicular.***

   Refer Fig. (ii). Since A.V.P. is perpendicular to HP, rotate the A.V.P. about X_1Y_1 to coincide with the HP. Now the angle between XY and X_1Y_1 is ϕ^o.

NOTE: *Distance of* p_1' *from* X_1Y_1 = *Distance of* p′ *from* XY, *i.e.,* $p_1' o_1 = p'o$.

Problem 1: *[Fig. 1(iii)]*

A point P is 40 mm above HP and 30 mm in front of VP. Draw an Auxiliary Front View of P on an A.V.P. inclined at 30° to the VP.

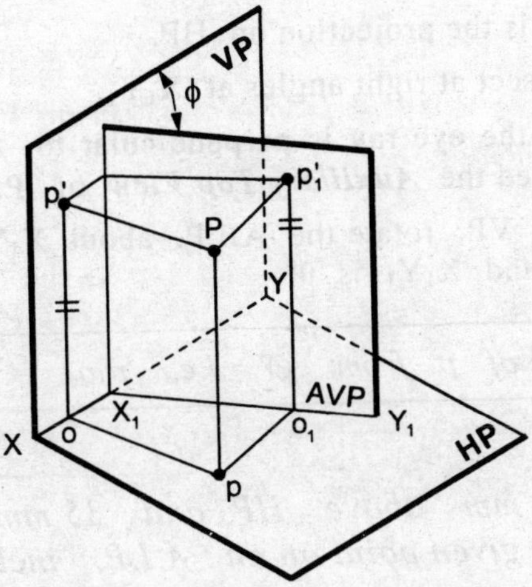

(i)

$o_1 p_1' = op'$

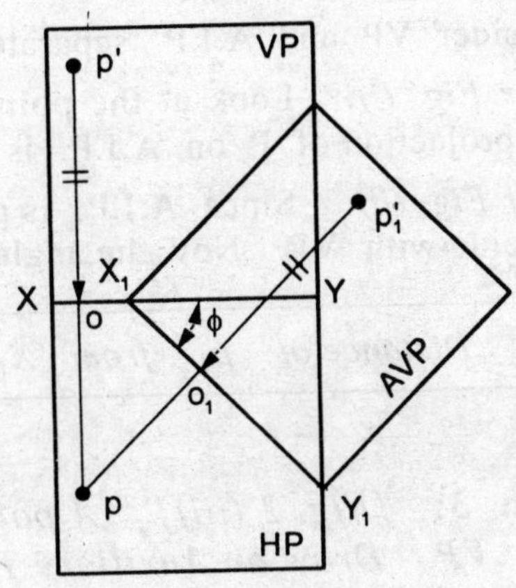

(ii)

p_1' - Auxiliary front view of P

1. Draw XY line.
2. Draw a projector through o on it.
3. Mark p' 40 mm above XY and p 30 mm below XY.
4. Draw a new reference line X_1Y_1 at *any convenient point* on XY at 30° (φ) to it.
5. From p draw a projector perpendicular to X_1Y_1 to intersect it at o_1.
6. Mark p_1' on this projector such that $p_1' o_1 = p' o = 40$ mm. Now p_1' is the required Auxiliary Front View.

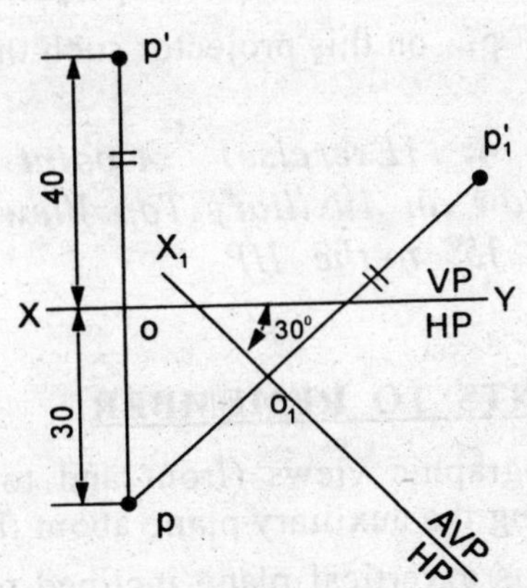

(iii)

FIG. 1

Problem 2: *(Exercise)*

A point Q is 25 mm above HP and 35 mm in front of VP. Draw an Auxiliary Front View of Q on an A.V.P. inclined at 45° to VP.

2.2 PROJECTION OF A POINT P ON A.I.P.

1. *See Fig. 2.* p′ is the projection on VP and p is the projection on HP.
2. Consider VP and A.I.P. separately. They intersect at right angles at X_1Y_1.
3. *Refer Fig. (i).* Look at the point P such that the eye ray is perpendicular to A.I.P. The projection of P on A.I.P. is p_1 and is called the ***Auxiliary Top View*** of P.
4. *Refer Fig. (ii).* Since A.I.P. is perpendicular to VP, rotate the A.I.P. about X_1Y_1 to coincide with VP. Now the angle between XY and X_1Y_1 is θ^o.

NOTE : *Distance of p_1 from X_1Y_1 = Distance of p from XY, i.e., p_1o_1 = po.*

Problem 3: *[Fig. 2 (iii)] A point P is 25 mm above HP and 35 mm in front of VP. Draw an Auxiliary Top View of the given point on an A.I.P. inclined at an angle of 25° to the HP.*

1. *See Fig. 2(iii).* Draw XY and a projector perpendicular to it at o.
2. Mark p′ 25 mm above XY and p 35 mm below XY on it.
3. Draw a new reference line X_1Y_1 at any convenient point on XY at 25° (θ) to it.
4. From p′ draw a projector perpendicular to X_1Y_1 to intersect it at o_1.
5. Mark p_1 on this projector such that p_1o_1 = po. p_1 is the Auxiliary Top View.

Problem 4: *(Exercise) A point R is 35 mm above HP and 25 mm in front of VP. Draw an Auxiliary Top View of the given point on an A.I.P. inclined at an angle of 35° to the HP.*

2.3 POINTS TO REMEMBER

1. Orthographic views (front and top views) of the auxiliary projections are drawn by rotating the auxiliary plane about that principle plane to which it is perpendicular.
2. AVP is a vertical plane inclined to VP. The view obtained on it is called Auxiliary Front View. Auxiliary Front View of a point lies on a line drawn through the top view, perpendicular to the new reference line $X_1 Y_1$ and at a distance from it, equal to the distance of the first front view from XY.
3. AIP is inclined to HP. The view obtained on it is called Auxiliary Top View. Auxiliary Top View of a point lies on a line drawn through the front view, perpendicular to $X_1 Y_1$ and at a distance from it, equal to the distance of the first top view from XY.
4. The distances of all the front views of the same point, (projected from the same top view), from their respective reference lines are equal.
5. The distances of all the top views of the same point, (projected from the same front view), from their respective reference lines are equal.

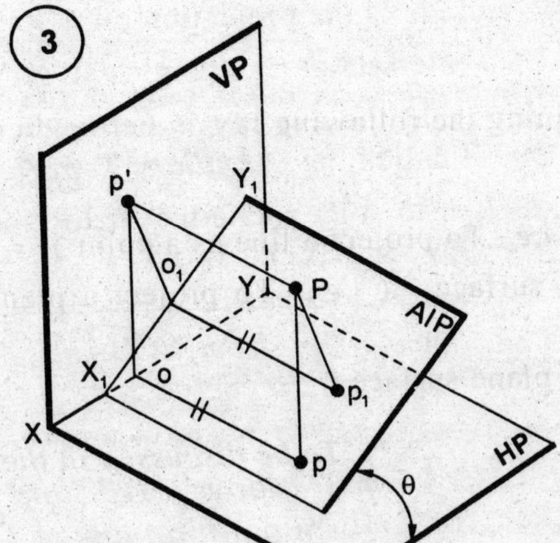

(i)

$o_1p_1 = op$

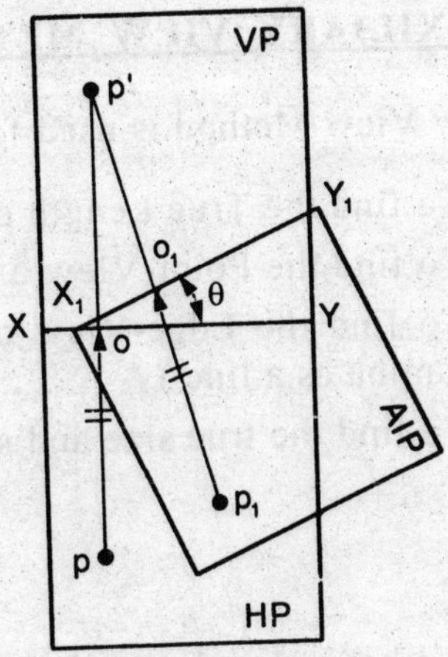

(ii)

p_1 - Auxiliary top view of P

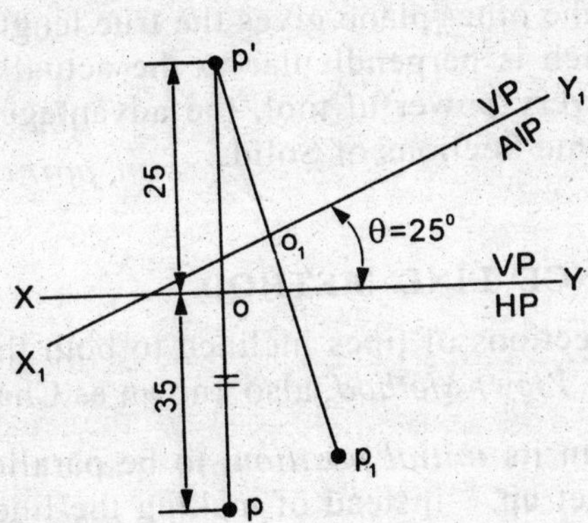

(iii)

FIG. 2

3. PROJECTIONS OF LINES ON AUXILIARY PLANES

3.1 AUXILIARY VIEW METHOD

Auxiliary View Method is used for determining the following :

1. To find the True Length of a line
2. To find the Point-View of a line (*i.e.*, To project a line as a point)
3. To find the Edge-View of a plane surface (*i.e.*, To project a plane surface or a lamina as a line) *
4. To find the true size and shape of a plane surface *

* *To be discussed in the next Chapter.*

We have already studied that when a straight line is parallel to either HP or VP, its projection on that plane gives the true length. Its other projection is parallel to the reference line XY. When a line is inclined to both HP and VP, none of the projections show its true length or is parallel to the reference line XY. Therefore, in order to determine the true length of a line, an auxiliary plane is introduced parallel to the line and its projection on that plane gives the true length of the line.

If a line is perpendicular to either HP or VP, its projection on that plane will be a point. Its projection on the other plane gives the true length. In other words, the projection of a line on a plane, which is perpendicular to the actual line, will be a point. Thus, the auxiliary projection is a very powerful tool, the advantages of which can be appreciated in the Projections of Solids and Sections of Solids.

CHANGE OF REFERENCE LINE METHOD

The problems on the projections of lines inclined to both the principle planes may be solved by the *Auxiliary Plane (or View) Method*, also known as Change of Reference Line Method.

Assume the line, in its *initial position*, to be parallel to both HP and VP. Then, two auxiliary planes are set up. Instead of making the line inclined to one of the reference planes, assume an auxiliary plane inclined to the line. Therefore, one auxiliary plane is introduced perpendicular to VP and inclined at θ^{o} to HP. This is called A.I.P. In other words, draw a new reference line and project the view on it.

Another auxiliary plane will be then introduced perpendicular to HP and inclined at ϕ^{o} to VP. This is known as A.V.P.

Auxiliary Projections of Straight Lines, Planes and Solids at given inclinations to one or both the Principal Planes (HP / VP), may be similarly obtained by the use of Auxiliary Planes. This is known as *Change of Reference Line Method.*

3.2 LINE INCLINED TO HP AND PARALLEL TO VP

Problem 5: *(Fig. 3) A line AB of 60 mm length, has its end A 20 mm above HP and 25 mm in front of VP. The line is inclined at an angle of 45° to HP and parallel to VP. Draw its projections by auxiliary plane method.*

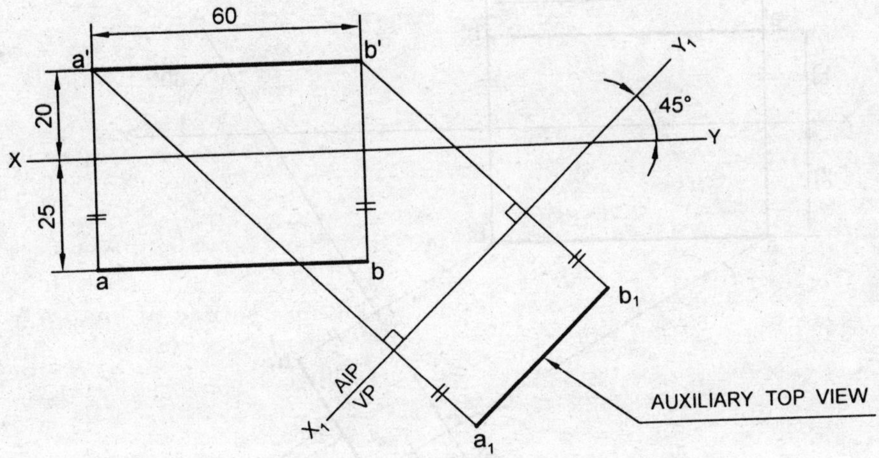

FIG. 3

1. Assume that the line is kept parallel to both HP and VP.
2. *See Fig. 3.* Draw XY line.
3. Mark the front view a′ 20 mm above XY and top view a 25 mm below XY.
4. Draw the front and top views of the line AB in true length and parallel to XY.
5. Draw the auxiliary reference line X_1Y_1 inclined at 45° to XY to represent an AIP.
6. From the front view of the ends a′ and b′, draw projectors perpendicular to X_1Y_1.
7. Mark a_1 on the projector such that the distance of a from XY = 25 mm = distance of a_1 from X_1Y_1.
8. Similarly, mark b_1 such that the distance of b from XY = 25 mm = distance of b_1 from X_1Y_1.
9. Join a_1 and b_1 to get the required Auxiliary Top View corresponding to the front view a′ b′.

Problem 6: *(Exercise) A straight line AB of 65 mm length is inclined at 30° to HP. The end A of the line is 25 mm above HP and 20 mm in front of VP. Draw the projections using auxiliary projection method.*

3.3 LINE INCLINED TO VP AND PARALLEL TO HP

Problem 7: *(Fig. 4) A line 60 mm long has its one end 20 mm above HP and 25 mm in front of VP. It is inclined at 60° to VP and parallel to HP. Draw its projections by auxiliary projection method.*

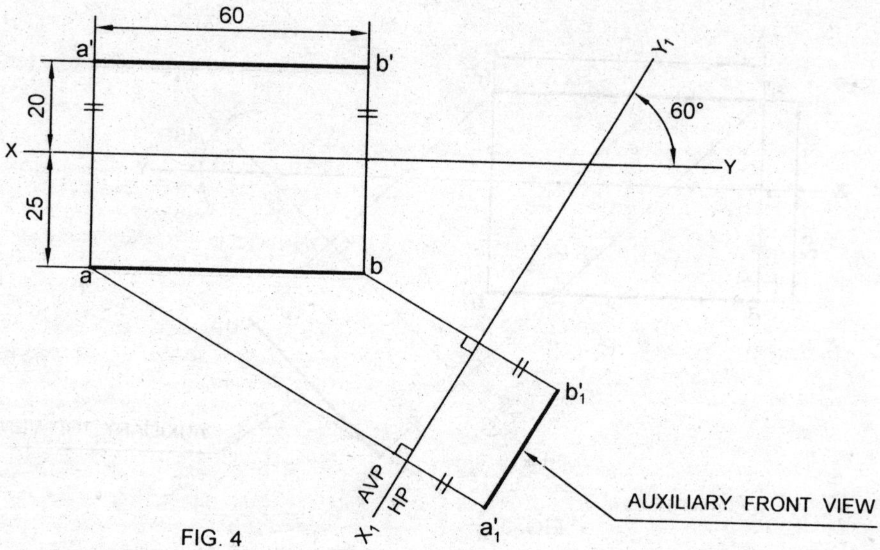

FIG. 4

1. Assume that the line AB is kept parallel to both HP and VP.
2. *See Fig. 4.* Draw the XY line. Draw the front view a′ of the end A 20 mm above XY and the top view a 25 mm below XY.
3. Draw the front and top views of the line having true length 60 mm and parallel to XY.
4. Draw the auxiliary reference line X_1Y_1 inclined at 60° to XY to represent an AVP.
5. From the top view of the ends a and b, draw projectors perpendicular to X_1Y_1.
6. Mark the distance a′ to XY (= 20 mm) from X_1Y_1 to get a'_1.
7. Similarly, mark the distance b′ to XY (= 20 mm) from X_1Y_1 to get b'_1.
8. Join a'_1 and b'_1 which is the required Auxiliary Front View corresponding to the top view ab.

Problem 8: *(Exercise) A line PQ of 60 mm length is inclined at 45° to VP. The end P is 20 mm above HP and 15 mm in front of VP. Draw the projections by auxiliary plane method.*

3.4 LINE INCLINED TO BOTH HP AND VP

Problem 9: *(Fig. 5) A line PQ 48 mm long is inclined at an angle of 30° to HP. Its top view makes an angle 60° with XY. End P is 20 mm above HP and 15 mm in front of VP. Draw the projections of the line by auxiliary plane method.* (UQ)

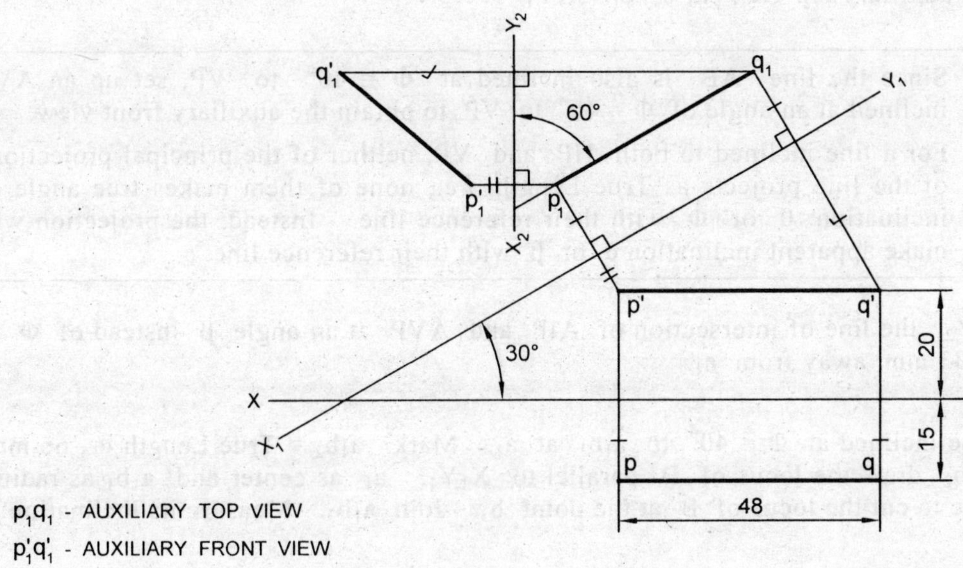

FIG. 5

1. *See Fig. 5.* Assume the line PQ to be parallel to both HP and VP. Draw the front view p′ q′ and top view p q of the line PQ.
2. Draw a new reference line X_1Y_1 corresponding to an A.I.P. making 30° with XY.
3. Project the Auxiliary Top View p_1q_1 on X_1Y_1.
4. Draw a second reference line X_2Y_2 representing an A.V.P. making 60° with p_1q_1.
5. Project the Auxiliary Front View $p'_1\, q'_1$ on X_2Y_2.
6. p_1q_1 and $p'_1\, q'_1$ are the Auxiliary Top View and Auxiliary Front View respectively.

Problem 10: *(Exercise) A line AB 62 mm long has its end A in HP and 20 mm in front of VP. If the line is inclined at 30° to VP and the front view makes 45° to the reference line, draw its projections by auxiliary plane method.*

Problem 11: *(Exercise) A line AB, 66 mm long, has its end A 22 mm above HP and 33 mm in front of VP. It is inclined at 30° to HP and 40° to VP. Draw the projections of the line by Auxiliary Plane Method.*

1. Assume the line AB to be parallel to both HP and VP.
2. Draw the corresponding front and top views.
3. Set up an AIP at $\theta = 30^\circ$, the angle made by the line AB with HP.
4. Draw X_1Y_1 inclined at $\theta = 30^\circ$ to XY and 22 mm away from a′.
5. Project the auxiliary top view a_1 b_1 on X_1Y_1.

NOTE:

1. Since the line AB is also inclined at $\Phi' = 40^\circ$ to VP, set up an AVP inclined at an angle of $\Phi = 40^\circ$ to VP to obtain the auxiliary front view.
2. For a line inclined to both HP and VP, neither of the principal projections of the line projects as True Length, *i.e.*, none of them makes true angle of inclination θ or Φ with their reference line. Instead, the projection will make apparent inclination α or β with their reference line.

---

6. Draw X_2Y_2, the line of intersection of AIP and AVP at an angle β instead of Φ to a_1b_1 and 33 mm away from a_1.
7. **To find β**

   Draw a line inclined at $\Phi = 40^\circ$ to a_1b_1 at a_1. Mark a_1b_2 = True Length = 66 mm. Through b_2 draw the locus of B parallel to X_1Y_1. a_1 as center and a_1b_1 as radius, draw an arc to cut the locus of B at the point b_3. Join a_1b_3. Measure its inclination β with a_1b_1.
8. Draw X_2Y_2 inclined at an angle of β to a_1b_1.
9. Draw the projectors through a_1 and b_1 perpendicular to X_2Y_2.
10. Mark a_1' at a distance equal to distance of a′ from X_1Y_1.
11. Similarly, mark b_1' at a distance equal to distance of b′ from XY.
12. Join a_1' b_1', the required auxiliary front view.

NOTE: Method of determination of Apparent Angle β is already discussed in the previous Chapter 13. Therefore, the above Problem is given as an Exercise, of course, with procedure and without the figure.

Problem 12: *(Exercise) A line MN of 55 mm length, has its end M 18 mm above HP and 22 mm in front of VP. The line is inclined at an angle of 30° to HP and 45° to VP. Draw the projections of the line by auxiliary plane method.*

3.5 TRUE LENGTH AND TRUE INCLINATION

Problem 13: *(Fig. 6) The end A of a line AB is 100 mm above HP and 150 mm in front of VP. The end B is 225 mm above HP and 225 mm in front of VP. The distance between the end projectors is 300 mm. Draw the projections of the line. Find its true length and true inclinations by auxiliary plane method.* (UQ)

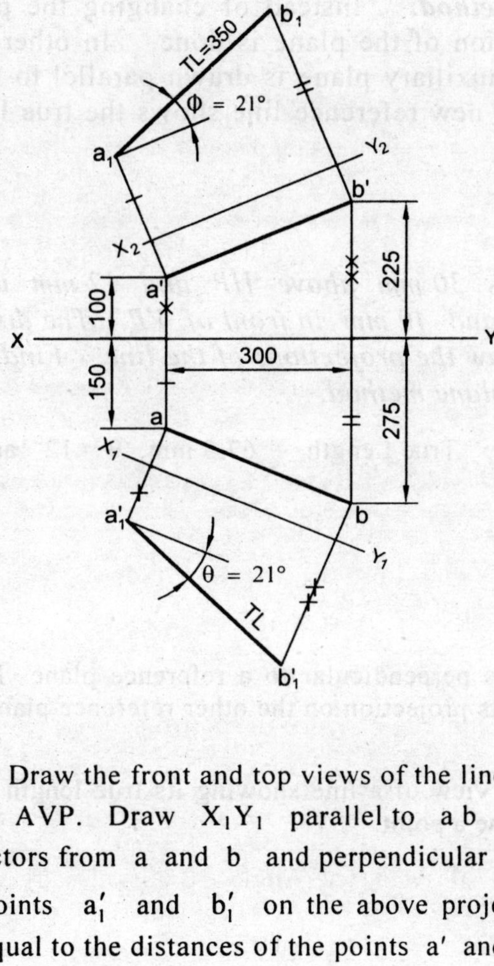

RESULTS :

TRUE LENGTH = 350 mm

θ = 21°, Φ = 21°

FIG. 6

1. *See Fig. 6.* Draw the front and top views of the line AB from the given data.
2. Consider an AVP. Draw X_1Y_1 parallel to a b at a convenient position.
3. Draw projectors from a and b and perpendicular to X_1Y_1.
4. Mark the points a'_1 and b'_1 on the above projectors such that their distances from X_1Y_1 are equal to the distances of the points a′ and b′ from XY respectively.
5. Join a'_1 and b'_1 to obtain the True Length of the line = **350 mm**.
6. True Inclination θ of the line with HP = Inclination of $a'_1\,b'_1$ with X_1Y_1 = **21°**.
7. Consider an AIP. Repeat the above procedure and draw auxiliary top view on X_2Y_2 parallel to a′ b′.
8. Now line a_1b_1 shows True Length = **350 mm**. True Inclination Φ with VP = **21°**.

POINTS TO REMEMBER

1. We know that the true length and true inclinations of a line can be found by making each of its projections parallel to the reference line XY.
2. ***Change of Position Method:*** In the previous chapter, line inclined to one or both the reference planes is initially assumed to be in simple position parallel to both HP and VP. Projections are then drawn by changing the position of the line from the simple position to the required position. This is know as Change of Position Method.
3. ***Change of Reference Line Method:*** Instead of changing the position of the projection, change of the position of the plane is done. In other words, a new reference line to represent an auxiliary plane is drawn parallel to the projection. The auxiliary projection on the new reference line shows the true length and true inclination of the line.

Problem 14: *(Exercise)*

The end A of a line AB is 30 mm above HP and 12 mm in front of VP. The end B is 16 mm above HP and 40 mm in front of VP. The distance between the end projectors is 60 mm. Draw the projections of the line. Find its true length and true inclinations by auxiliary plane method.

(*Results :* True Length = 67.5 mm, $\theta = 12^{\circ}$ and $\Phi = 24.5^{\circ}$.)

3.6 POINT-VIEW OF A LINE

We know that when a straight line is perpendicular to a reference plane HP or VP, its projection on that plane is a ***point***. Its projection on the other reference plane shows its true length.

That is, the projection of the view of a line showing its true length on an auxiliary plane perpendicular to that view will be a point.

Problem 15: *(Fig. 7)*

The front and top views of a line MN are inclined at an angle of 20° and 30° to the XY line respectively. The distance between the end projectors is 60 mm. The end M is 16 mm above HP and 12 mm in front of VP.

Draw the projections of the line on an auxiliary plane so that in the final view it appears as a point. Also, find the true length of the line. (UQ)

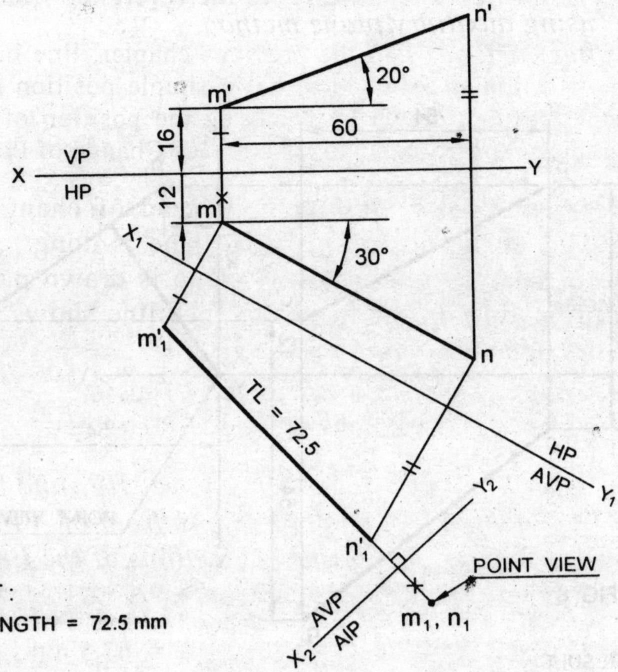

RESULT :

TRUE LENGTH = 72.5 mm

FIG. 7

1. Draw XY line. Draw the front and top views of the line MN using the given data.
2. Draw a new reference line $X_1 Y_1$ parallel to one of the views, say, top view as shown.
3. Draw the auxiliary front view $m_1' n_1'$.
4. Draw another reference line $X_2 Y_2$ perpendicular to $m_1' n_1'$.
5. Draw the auxiliary top view m_1, n_1 which is a *point*.
6. Measure the true length of the line = $m_1' n_1'$ = **72.5 mm.**

Problem 16: *(Exercise)*

The top and front views of a line AB are inclined at an angle of 25° and 45° to the XY line respectively. The end B is 22 mm above and 11 mm in front of VP. The distance between the end projectors is 70 mm.

Draw the projections of the line on an auxiliary plane so that in the final view it appears as a point. Also, find the true length of the line.

Problem 17: *(Fig. 8) The end P of a line PQ is 12 mm in front of VP and 42 mm above HP. The end Q is 54 mm in front of VP and 12 mm above HP. The distance between the end projectors is 54 mm. Determine the shortest distance of the line from XY using auxiliary plane method.* (UQ)

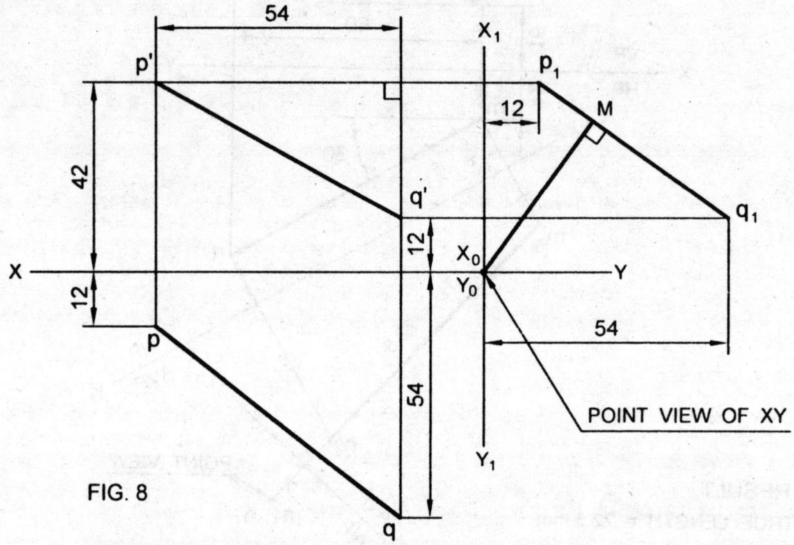

FIG. 8

RESULT :
SHORTEST DISTANCE OF THE LINE FROM XY = X_0M = 41 mm

1. Draw the XY line. Draw the front and top views of the given line.

NOTE : *Measure the distance of the line PQ from XY by drawing an auxiliary view on an Auxiliary Plane, on which the XY line can be represented as a point.*

2. Draw X_1Y_1 perpendicular to XY.
3. Project the auxiliary top view p_1q_1.
4. Point X_0Y_0 is the point view of the XY line.
5. Shortest distance of PQ from XY = Length of the perpendicular drawn from X_0 (or Y_0) to p_1q_1 = X_0M = **41 mm.**

Problem 18: *(Exercise) The end A of a line AB is 40 mm in front of VP and 12 mm above HP. The end B is 16 mm in front of VP and 38 mm above HP. If the distance between the end projectors is 56 mm, draw the projections of the line AB. Determine the shortest distance of the line AB from XY line by using auxiliary plane method.*

(Result : Shortest distance = 38 mm)

*

15

PROJECTIONS OF PLANE SURFACES

1. PLANE SURFACES – Polygonal Surfaces and Circular Lamina

Plane Surfaces or Planes : Any machine or structure is considered to be made up of a number of components which are formed by geometrical surfaces known as *Planes*. A *Plane or Plane Surface,* (also known as *Lamina*), has only two dimensions, *viz.*, length and breadth with negligible thickness. Thickness is neglected since it is too small compared with its length and breadth.

Polygonal Surfaces and Circular Lamina
The plane surfaces may be of polygonal surfaces such as triangular, square, rectangular, polyhedral (pentagonal / hexagonal, etc.) or circular lamina, etc. If sides are equal, it is a *Regular Plane Surface,* otherwise an *Irregular Plane Surface.*

NOTE : 1. ***Plane Surface – Plane – Plane Figure – Lamina – Plate*: All mean the same.**

2. *Do not get confused Plane Surfaces or Planes with Reference Planes or Profile Plane or Auxiliary Planes.*

Traces of Plane Surfaces

Traces: A plane, extended if necessary, will meet the reference planes in lines unless it is parallel to any one of them. These lines are called *Traces of the Plane Surfaces.*

Horizontal Trace (H.T.) of a plane surface is the line in which plane surface meets the HP.

Vertical Trace (V.T.) of a plane surface is the line in which the plane surface meets the VP.

1.1 TYPES OF PLANE SURFACES

1. TYPES OF PERPENDICULAR *PLANES / PLANE SURFACES*

i) Plane surface perpendicular to one reference plane and parallel to the other
ii) Plane surface perpendicular to both the reference planes
iii) Plane surface perpendicular to one reference plane and inclined to the other

2. OBLIQUE *PLANES / PLANE SURFACES*

Plane Surface inclined to both reference planes is termed as Oblique *Plane Surface.*

2. PLANE SURFACES AND THEIR TRACES

2.1 PERPENDICULAR TO HP AND PARALLEL TO VP

Problem 1: *(Fig. 1) A square lamina ABCD of side 40 mm is perpendicular to HP and parallel to VP. Draw its projections and locate its traces.*

1. From the given statement, decide the view to be drawn first.
   Lamina is parallel to VP. When viewed from the front, it is seen in true shape and size. Therefore, **draw the Front View first.**
2. **Front View :** Lamina ABCD is above HP.
   Therefore, draw its Front View **a′ b′ c′ d′** in true size above XY.
3. **Top View :** When the lamina is viewed from the top, edge BC is seen in true length and parallel to VP. Hence draw its Top View **bc** (in true length) parallel to and below XY.
4. **To mark invisible corners in Top View :** Look at Front View in the direction of arrow. **a′** and **d′** are hidden by **b′** and **c′** respectively. Therefore in the Top View, **ad** coincides with **bc** and is marked as **(a)(d)**. Now **bc(d)(a)** is the Top View.

To obtain H.T. and V.T.

5. Extend the plane surface to meet the HP. The extension coincides with the Top View. Hence the Top View **bc (d) (a)** itself is the H.T.
6. Extend the plane surface to meet VP. The plane surface is parallel to VP and it will not meet. Hence, there is no V.T.

2.2 PERPENDICULAR TO VP AND PARALLEL TO HP

Problem 2: *(Fig. 2) A square lamina of side 40 mm is perpendicular to VP and parallel to HP. Draw its projections and find its traces.*

1. When the lamina ABCD is viewed from the top, it is seen in true shape and size. Therefore, **draw the Top View first.**
2. **Top View :** Draw the Top View **abcd** in true size and below XY.
3. **Front View :** View the lamina from front. Edge BC is seen in true length and parallel to HP. Hence draw its Front View **b′ c′** (in true length) parallel to and above XY.
4. **Invisible corners in Front View :** Look at Top View in the direction of arrow shown. **d & a** are hidden in Top View. So mark invisible corners in Front View as **(d′) & (a′)**.
5. **H.T. :** Extend the lamina to meet the HP. It will *not* meet HP and hence *no* H.T.
6. **V.T. :** Extend the lamina to meet the VP. The extension coincides with the Front View. Hence Front View **b′ c′ (d′)(a′)** itself is the V.T.

Problem 3: *(Exercise) A rectangular cardboard ABCD of edges AB = 30 mm and BC = 40 mm is placed such that the edge AB is (i) perpendicular to HP and parallel to VP, (ii) parallel to HP and perpendicular to VP. Draw its projections and traces.*

LAMINA PERPENDICULAR TO ONE REFERENCE PLANE AND PARALLEL TO THE OTHER

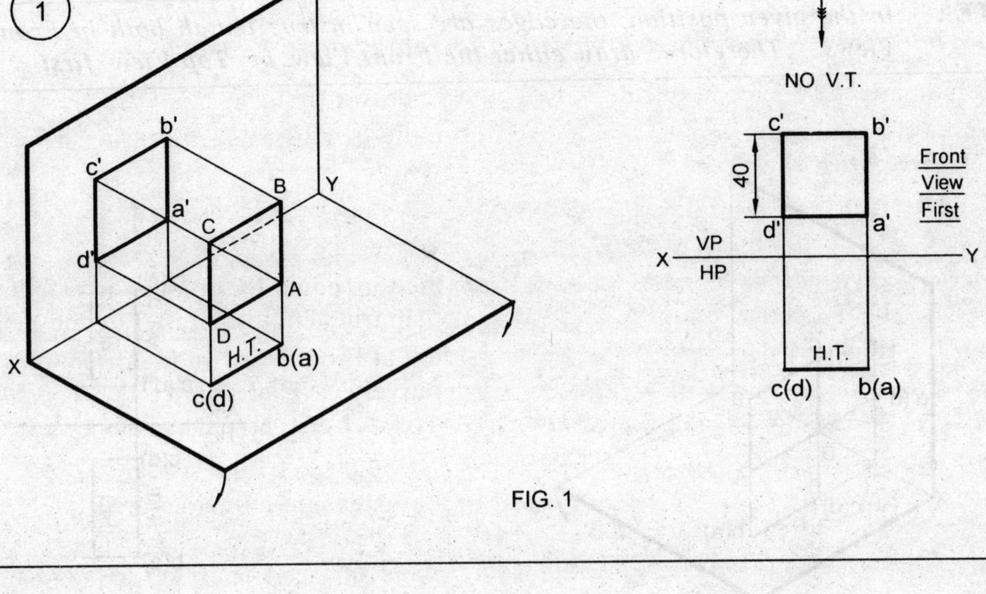

FIG. 1

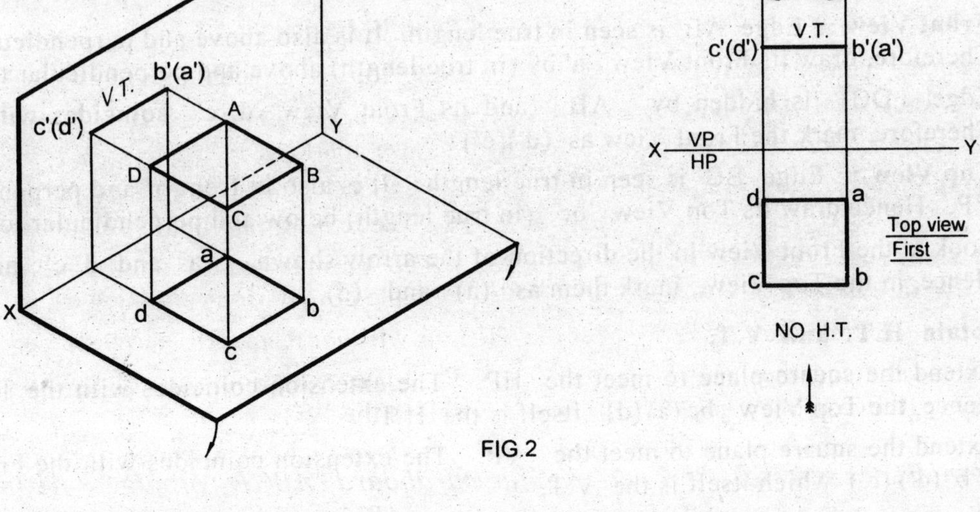

FIG.2

2.3 PERPENDICULAR TO BOTH HP AND VP

Problem 4: *(Fig. 3)* *A thin square plate ABCD of side 40 mm is perpendicular to both HP and VP. Draw its projections and find the traces.*

NOTE : *In the given position, the edges are seen in true length both in Front and Top Views. Therefore,* ***draw either the Front View or Top View first.***

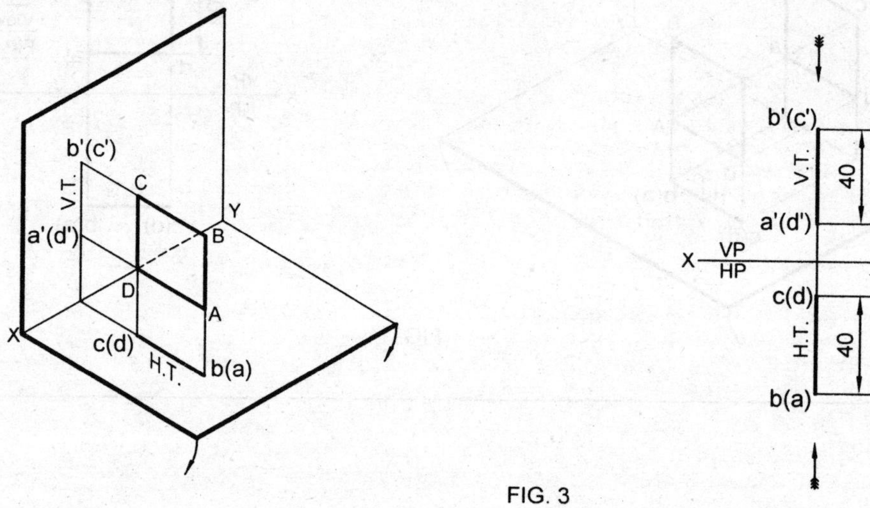

FIG. 3

1. **Front View :** Edge AB is seen in true length. It is also above and perpendicular to HP. Therefore, draw its Front View a′ b′ (in true length) above and perpendicular to XY.
2. Edge DC is hidden by AB and its Front View d′ c′ coincides with a′ b′. Therefore, mark the Front View as (d′)(c′) .
3. **Top View :** Edge BC is seen in true length. It is also in front of and perpendicular to VP. Hence draw its Top View bc (in true length) below and perpendicular to the XY.
4. Look at the Front View in the direction of the arrow shown. a and d′ c′ are hidden. Hence, in the Top View, mark them as (a) and (d).

To obtain H.T. and V.T.

5. Extend the square plane to meet the HP. The extension coincides with the Top View. Hence, the Top View bc (a) (d) itself is the H.T.
6. Extend the square plane to meet the VP. The extension coincides with the Front View a′ b′ (d′) (c′) which itself is the V.T.

Problem 5: *(Exercise)* *A rectangular lamina ABCD of sides 40 mm x 30 mm is perpendicular to both HP and VP. Draw its projections and obtain the traces.*

2.4 PERPENDICULAR TO HP AND INCLINED TO VP

Problem 6: *(Fig. 4) A rectangular lamina of sides 30 mm x 40 mm is perpendicular to HP and inclined at 30° to VP. Draw its projections and obtain the traces.*

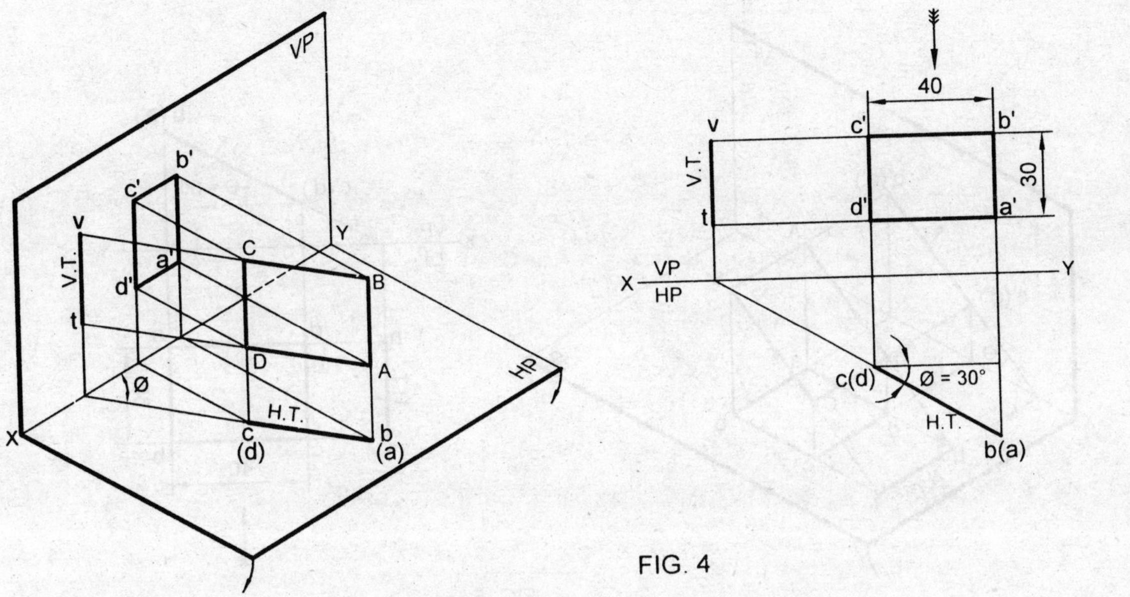

FIG. 4

1. When the lamina (surface) ABCD is viewed from the top, the edge BC is seen in true length. Therefore, **draw the Top View first.**
2. **Top View :** Edge BC is in front of and inclined at an angle of 30° (ϕ) to VP. Hence, draw its Top View bc (in true length) below XY and inclined at 30° to XY.
3. Edge AD is hidden by the edge BC and its Top View (a) (d) coincides with bc. Now bc (a) (d) is the required Top View.
4. **Front View :** View ABCD from front. Front View a′ b′ c′ d′ is seen in reduced size. ABCD is perpendicular to HP. Though ABCD is inclined to VP, edges CD and AB are parallel to VP. Hence draw Front View a′ b′ c′ d′ such that a′ b′ = c′ d′ = AB.
5. **H.T. :** Extend the lamina to meet the HP. The extension coincides with the Top View. Therefore, the Top View bc(a)(d) itself is the H.T.
6. **V.T. :** Extend the lamina to meet the VP at vt which is the required V.T.

Problem 7: *(Exercise) Draw the projections of a pentagonal plane figure of side 28 mm resting with one of its edges on HP such that the plane figure is inclined at 30° to VP and perpendicular to HP. Locate its traces.*

2.5 PERPENDICULAR TO VP AND INCLINED TO HP

Problem 8: *(Fig. 5) A rectangular lamina of sides 30 mm x 40 mm is perpendicular to VP and inclined at 30° to HP. Draw its projections and traces.*

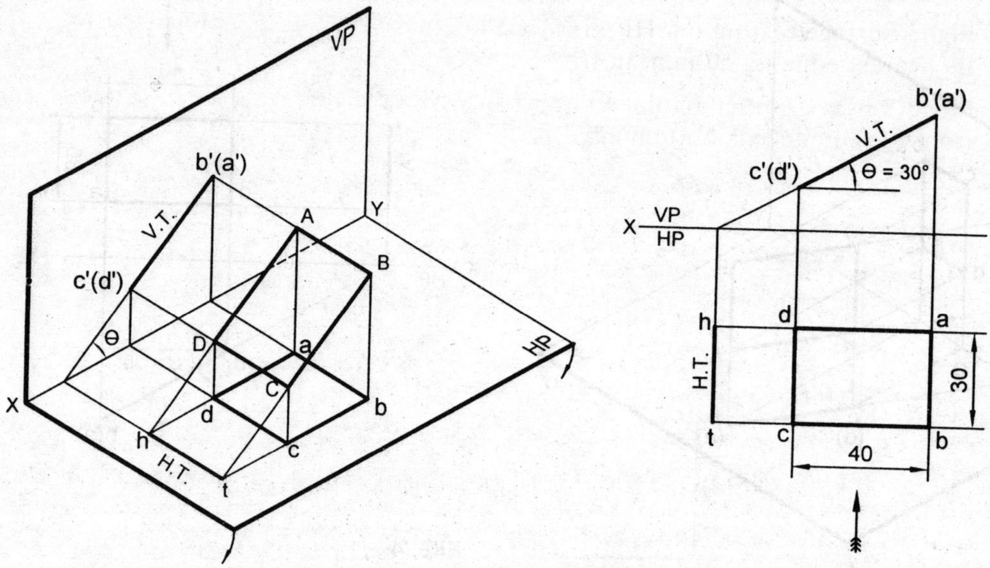

FIG. 5

1. When the lamina ABCD is viewed from front, the edge BC is seen in true length. Therefore, ***draw the Front View first.***
2. **Front View :** Edge BC is above HP and inclined at an angle of 30° (θ) to HP. Hence draw its Front View b′ c′ (in true length) above XY and inclined at 30° to XY.
3. Edge AD is hidden by the edge BC and its Front View a′ d′ coincides with b′ c′.
4. **Top View :** View ABCD from top. Its Top View abcd is seen in reduced size.
5. The lamina is perpendicular to VP. Though the lamina is inclined to HP the edges CD and AB are parallel to HP. Therefore draw the Top View such that ab = cd = AB.
6. **H.T.:** Extend the lamina to meet HP at ht which is the required H.T.
7. **V.T.:** Extend the lamina to meet VP. The extension coincides with the Front View b′ c′ (a′) (d′) which itself is the V.T.

Problem 9: *(Exercise) Draw the projections of a pentagonal plane figure of side 28 mm resting with one of its edges on HP such that the plane figure is inclined at 30° to HP and perpendicular to VP. Locate its traces.*

2.6 REPRESENTATION OF PLANE SURFACES BY H.T. AND V.T.

Problem 10: *(Fig. 6) Represent the following Plane Surfaces by their Traces.*

(i) Plane Surface is perpendicular to both HP and VP with an edge on HP.

(ii) Plane Surface is parallel to and 35 mm above HP and perpendicular to VP.

(iii) Plane Surface is parallel to and 30 mm in front of VP and perpendicular to HP.

(iv) Plane Surface resting on HP, is perpendicular to it. It is inclined at 30° to VP. Its nearest edge is 10 mm in front of VP.

(v) Plane Surface perpendicular to VP, is inclined at 60° to HP. It is 15 mm in front of VP and its bottom edge is 15 mm above HP.

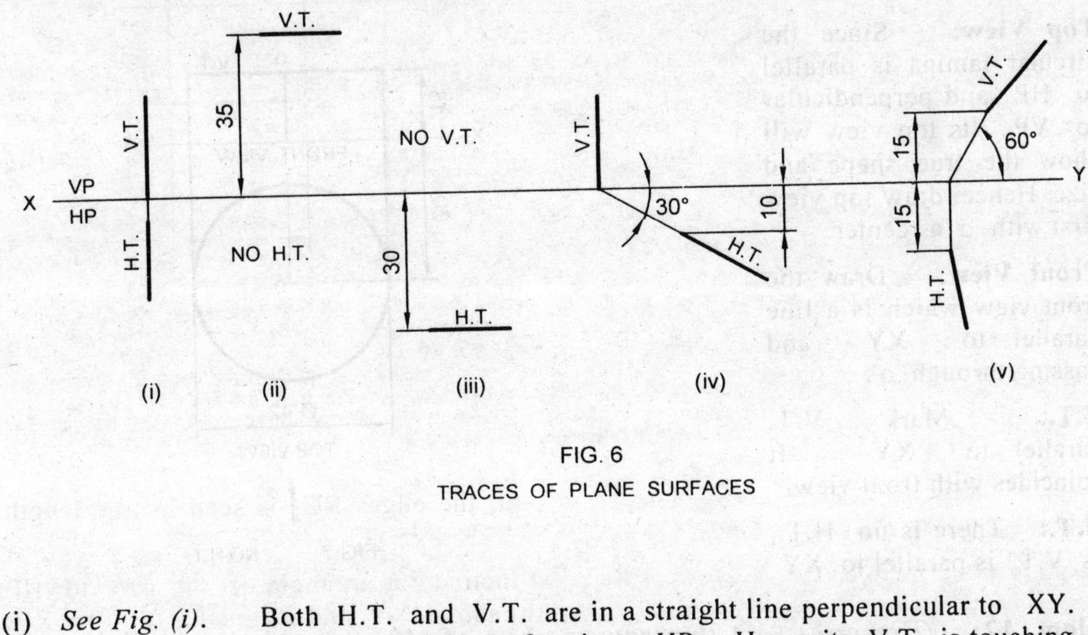

FIG. 6

TRACES OF PLANE SURFACES

(i) *See Fig. (i).* Both H.T. and V.T. are in a straight line perpendicular to XY. An edge of plane surface is on HP. Hence its V.T. is touching XY.

(ii) *See Fig. (ii).* There is no H.T. Its V.T. is a line parallel to XY, 35 mm above it.

(iii) *See Fig. (iii).* H.T. is parallel to XY and 30 mm below it. It has no V.T.

(iv) *See Fig. (iv).* H.T. has its neatest end 10 mm from XY and is inclined at 30° to XY. V.T. is perpendicular to the XY line and touches it. Also, V.T. and H.T. produced, meet at the XY line.

(v) *See Fig. (v).* The nearest end of V.T. is 15 mm above XY and that of H.T. is 15 mm below XY. V.T. is inclined at 60° to XY and H.T. is perpendicular to XY, both meeting on XY.

3. PLANE SURFACES PERPENDICULAR TO ONE REFERENCE PLANE AND PARALLEL TO THE OTHER

The orthographic view of a plane surface on the reference plane parallel to it will show its true shape and size. Hence, draw that view first.

3.1 PLANE SURFACE PERPENDICULAR TO VP AND PARALLEL TO HP

RULE : ***When a plane surface is parallel to HP, top view shows its true shape & size. Therefore, draw the top view first. Project the front view from it.***

Problem 11: *(Fig. 7) A circular lamina of 42 mm diameter has its surface parallel to HP and perpendicular to VP. Its center is 15 mm above HP and 30 mm in front of VP. Draw its projections and show the traces.*

1. **Top View:** Since the circular lamina is parallel to HP and perpendicular to VP, its top view will show the true shape and size. Hence, draw top view first with o as center.
2. **Front View:** Draw the front view which is a line parallel to XY and passing through o'.
3. **V.T.:** Mark V.T. parallel to XY. It coincides with front view.
4. **H.T.:** There is no H.T., as V.T. is parallel to XY.

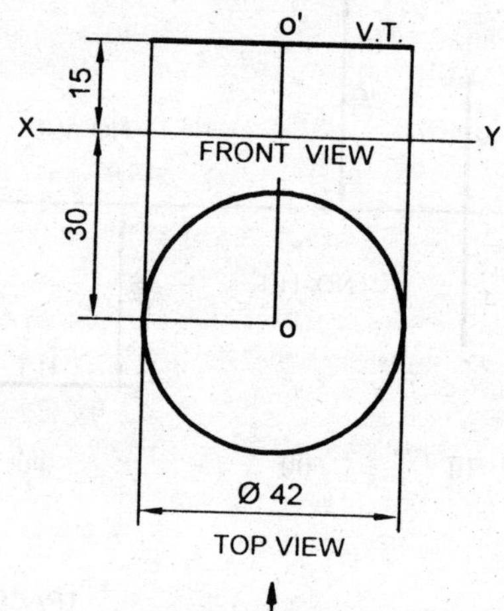

FIG.7 NO H.T.

Problem 12: *(Exercise) A thin square plate of 40 mm side is parallel to HP with one of its sides inclined at an angle of 45° to VP. The plate is 25 mm above HP. Draw its projections. Also show its traces.*

HINT :
1. The square plate is parallel to HP. Therefore, draw the top view first.
2. V.T. coincides with the front view.
3. There is no H.T., since V.T. is parallel to XY.

Problem 13: *(Exercise) An equilateral triangle of 45 mm side has one of its sides inclined at 60° to VP. Its V.T. is parallel to and 30 mm above XY. It has no H.T. Draw the top and front views.*

HINT : V.T. is parallel to XY and there is no H.T. So, the triangle is parallel to HP. Draw top view first such that one side of the triangle is inclined at 60° to XY.

3.2 PLANE SURFACE PERPENDICULAR TO HP AND PARALLEL TO VP

RULE: When a plane surface is parallel to VP, front view shows true shape and size. Hence, draw the front view first and project the top view from it.

Problem 14: *(Fig. 8) A square lamina of 45 mm side has a corner on HP and 25 mm in front of VP. All the sides of the square are equally inclined to HP and parallel to VP. Draw the front and top views. Also, show the traces.*

1. All the sides of the square are parallel to VP. Therefore, the plane surface of the square is also parallel to VP. The front view will show the true shape and size.
2. Draw the front view as a square p′ q′ r′ s′ such that one corner s′ is in XY and all the sides are equally inclined at 45° to XY.
3. Draw the top view p q r s such that it is parallel to XY and 25 mm below it.
4. Top view itself is the H.T. There is no V.T.

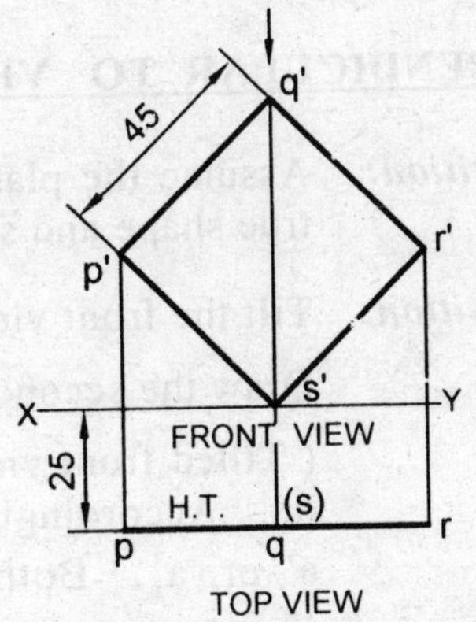

FIG. 8

Problem 15: *(Fig. 9) A regular hexagonal lamina of 22 mm side, rests on one of its sides on HP. It is parallel to and 15 mm away from VP. Draw its projections and show the traces.*

1. Draw the XY line.
2. One of the sides of the hexagon is in HP and the hexagon is parallel to the VP. Hence, draw its front view 1′ 2′ 3′ 4′ 5′ 6′ which will be in true size and shape, having one side 5′ 6′ on XY.
3. Draw the top view 1 2 3 4 5 6 as a line 15 mm below and parallel to XY.
4. Top view is the H.T. There is no V.T.

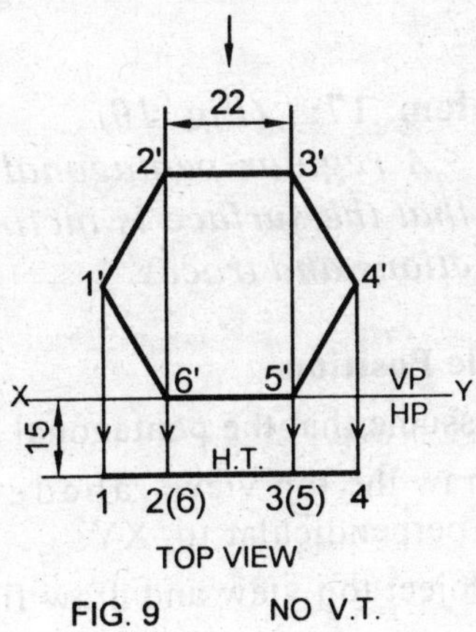

FIG. 9

Problem 16: *(Exercise) A 45° set-square rests on HP on the 90° corner such that two sides of length 160 mm each, containing the corner are equally inclined to HP. Its H.T. is 30 mm below and parallel to XY. Draw the views and locate traces.*

4. PLANE SURFACES PERPENDICULAR TO ONE REFERENCE PLANE AND INCLINED TO THE OTHER

4.1 CHANGE OF POSITION METHOD

Orthographic projections of a plane surface inclined to one reference plane are drawn by changing the position of the plane surface from simple position to the required position in *two* stages, known as the *Change of Position Method.*

4.2 PERPENDICULAR TO VP AND INCLINED TO HP

Simple Position: Assume the plane surface to be parallel to HP. Draw first top view in true shape and size. Draw first front view as a line parallel to XY.

Second Position: Tilt the front view to the given angle to XY.

Draw the second front view.

(Tilted front view may or may not be named as second front view and so on. Accordingly, notations of the tilted front view may be given as either a′ or a_1'. Both the notations are correct.)

Then draw the second top view by projecting all the points from second front view downwards and projecting parallel to XY from first top view.

Intersections of the corresponding vertical and horizontal projectors of a point locate the point.

Join these points in sequence to get second top view.

Problem 17: *(Fig. 10)*

A regular pentagonal plate of side 28 mm is placed with one side on HP such that the surface is inclined at 45° to HP and perpendicular to VP. Draw its projections and traces.

Simple Position

1. Assume that the pentagonal surface is parallel to HP and perpendicular to VP.
2. Draw the top view abcde in true shape and size, (side 28 mm), such that side ae is perpendicular to XY.
3. Project top view and draw first front view a′b′c′(d′)(e′) as a line parallel to XY.

Second Position

4. Tilt first front view about a′(e′) to the given angle of 45° with XY.
5. Redraw it to get second front view a_1' b_1' c_1' (d_1') (e_1').

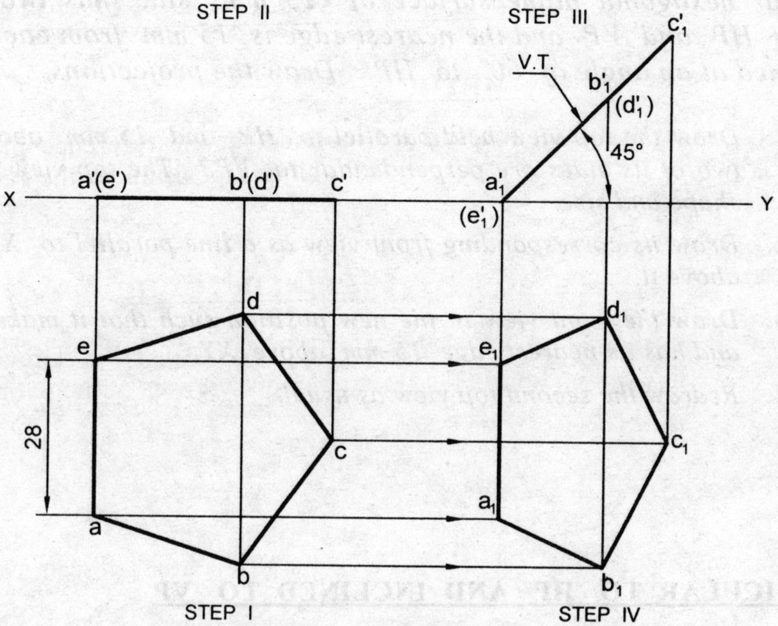

FIG. 10

6. From the first top view, draw horizontal lines through the points a, b, c, d and e.
7. From the second front view, draw vertical lines through the points a_1', b_1', c_1', (d_1') and (e_1').
8. Mark intersecting points of above horizontal and vertical lines as a_1, b_1, c_1, d_1 and e_1.
9. Join $a_1 b_1 c_1 d_1 e_1$ which is the *required second top view*. It is smaller than true size.

V.T. and H.T.

10. V.T. coincides with second front view.
11. Extend the pentagon to meet HP. Line of intersection of pentagon with HP is line pq. pq is the H.T. perpendicular to XY.

Problem 18: *(Exercise)*

Draw the projections of a circular disc of 120 mm diameter resting on HP on a point on the circumference with its surface inclined at 50° to HP and perpendicular to VP. Also, draw the traces.

HINT: A circle has no corners to project one view from another. So divide it into 8/12 parts.

Problem 19: *(Exercise)*

A regular hexagonal plane surface of 25 mm side, has two of its edges parallel to both HP and VP and the nearest edge is 15 mm from each plane. The surface is inclined at an angle of 60° to HP. Draw the projections. (UQ)

HINT :

1. *Draw the top view held parallel to HP and 15 mm above it such that two of its sides are perpendicular to VP. The top view will be in true shape and size.*
2. *Draw its corresponding front view as a line parallel to XY and 15 mm above it.*
3. *Draw the front view in the new position such that it makes 60° to XY and has its nearest edge 15 mm above XY.*
4. *Redraw the second top view as usual.*

4.3 PERPENDICULAR TO HP AND INCLINED TO VP

Simple Position: Assume the plane surface to be parallel to VP.

Draw the first front view in true size and shape.

Project from the first front view and draw the first top view as a line parallel to XY.

Second Position: Tilt the first top view at the given angle to XY the line.

Project the second front view, by drawing vertical projectors through second top view and horizontal projectors through the first front view.

Problem 20: *(Fig. 11)*

A thin circular metal plate of 48 mm diameter, having its plane vertical and inclined at 40° to VP. Its center is 33 mm above HP and 25 mm in front of VP. Draw its projections and locate its traces.

Simple Position

1. Assume the circle to be parallel to VP.
2. Draw first front view as a circle of 48 mm diameter, having its center 33 mm above XY.
3. Divide the circumference of the circle into 12 equal parts as 1, 2, ..., 12.
4. Draw first top view as a line parallel to and 25 mm below XY. Center o coincides with point 4.

Second Position

5. Tilt the circular plate to make 40° with VP. Hence reproduce first top view such that it is inclined at 40° to XY and center o is kept at the same distance of 25 mm below XY.

6. From the first front view, all the points will move along their respective paths parallel to XY to the second front view. To draw the second front view, project all the points upwards from the second top view and horizontally from the first front view as shown.

7. Draw second front view as an ellipse through the points of intersection $1_1'$, $2_1'$, ..., $12_1'$.

8. **H.T. :** H.T. and second top view coincide.

9. **V.T. :** V.T. is the line AB perpendicular to XY. H.T. and V.T. meet XY at C.

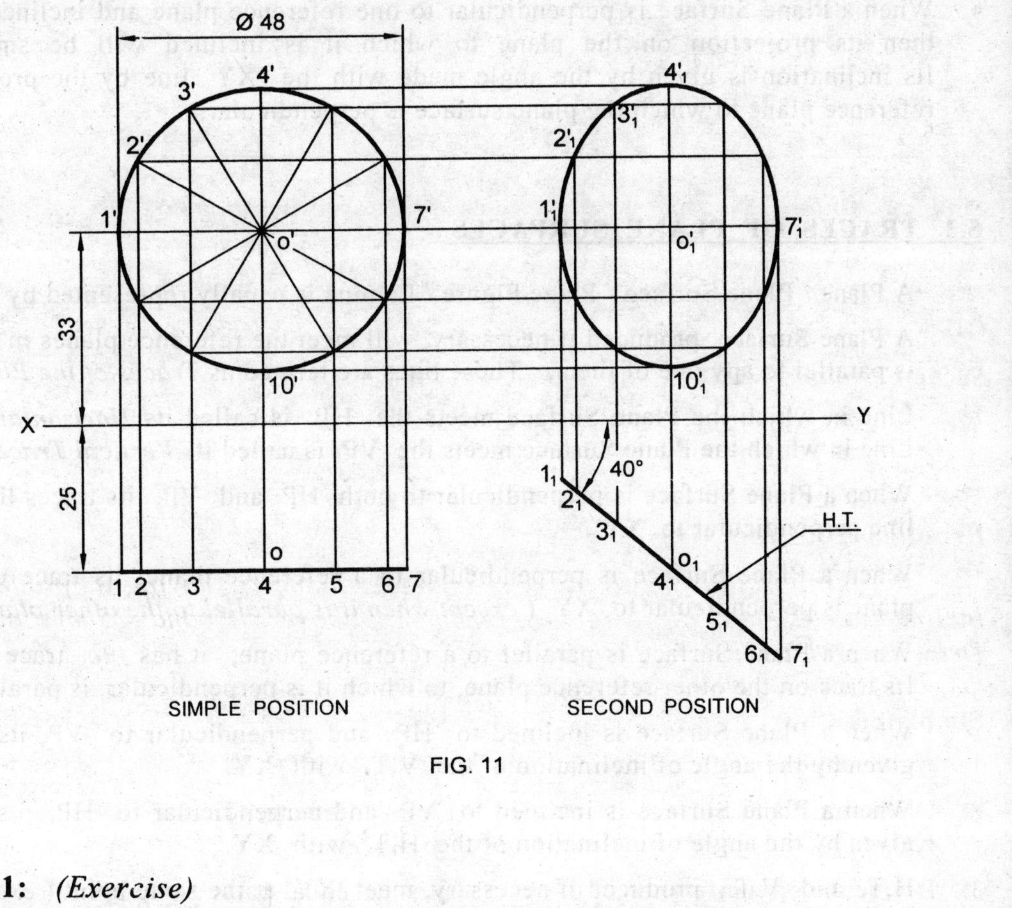

FIG. 11

Problem 21: *(Exercise)*

Draw the orthographic views of a regular hexagonal lamina of 25 mm side, resting on HP on one of its sides with its plane perpendicular to HP and inclined at 45° to VP. Take the nearest corner point 25 mm away from VP. Also show its traces.

5. POINTS TO REMEMBER

5.1 PROJECTIONS OF PLANE SURFACES

- When a Plane Surface is perpendicular to a reference plane, then its projection on that plane will be a straight line.
- When a Plane Surface is parallel to a reference plane, then its projection on that plane will show its true shape and size, *i.e.*,
  - (i) When a Plane Surface is parallel to HP, Top View will show its true shape and size. Hence, the Top View is drawn first and the Front View is projected from it.
  - (ii) When a Plane Surface is parallel to VP, its Front View will show its exact shape and size. Hence, Front View is drawn first and the Top View is projected from it.
- When a Plane Surface is perpendicular to one reference plane and inclined to the other, then its projection on the plane to which it is inclined will be smaller in size. Its inclination is given by the angle made with the XY line by the projection on the reference plane to which the plane surface is perpendicular.

5.2 TRACES OF PLANE SURFACES

- A Plane / Plane Surface / Plane Figure / Lamina is usually represented by its *Traces*.
- A Plane Surface, produced if necessary, will meet the reference planes in lines, unless it is parallel to any one of them. These lines are termed as *Traces of the Plane Surfaces*.
- Line in which the Plane Surface meets the HP is called its *Horizontal Trace* (H.T.). Line in which the Plane Surface meets the VP is called its *Vertical Trace* (V.T.).
- When a Plane Surface is perpendicular to both HP and VP, its traces lie on a straight line perpendicular to XY.
- When a Plane Surface is perpendicular to a reference plane, its trace upon the other plane is perpendicular to XY (*except when it is parallel to the other plane*).
- When a Plane Surface is parallel to a reference plane, it has *no* trace on that plane. Its trace on the other reference plane, to which it is perpendicular, is parallel to XY.
- When a Plane Surface is inclined to HP and perpendicular to VP, its inclination is given by the angle of inclination of the V.T. with XY.
- When a Plane Surface is inclined to VP and perpendicular to HP, its inclination is given by the angle of inclination of the H.T. with XY.
- H.T. and V.T., produced if necessary, meet XY at the same point (*except when both are parallel to XY as in the case of some Oblique Planes*).
- Study of Traces is a pre-requisite for the proper understanding of Section of Solids, since the section planes are generally specified by their Traces.

6. PLANE SURFACES INCLINED TO BOTH HP AND VP – CHANGE OF POSITION METHOD

OBLIQUE PLANES or OBLIQUE PLANE SURFACES are Plane Surfaces which are inclined to both HP and VP.

CHANGE OF POSITION METHOD

When a Plane Surface is inclined to one or both the reference planes, the projections are drawn by the *Change of Position Method*.

When a Plane Surface has its surface inclined to one reference plane and an edge *or* a diagonal *or* a diameter parallel to that plane, inclined to the other reference plane, draw its projections in ***three stages*** as follows:

Simple Position: (***Initial Position***) Assume the plane surface to be parallel to HP and draw its projections. In this position, the First Top View will be in true shape and size and the First Front View will be an edge view, *i.e.*, a straight line parallel to the XY line.

Second Position: Tilt the First Front View so as to make the given angle of inclination with HP and draw the Second Front View.

From this Second Front View, project the second top view. The Second Top View will be smaller than the original plane surface.

Final Position: Obtain the Final Top View, by redrawing the Second Top View such that the given edge or diagonal or diameter is made inclined to XY line at an angle made by the plane surface with the VP. While doing so, the shape of the Final Top View will remain the same though its relative position will change.

The distances of the corners from the XY line, that is, the distances of the corners above the HP, will be the same when the plane surface is turned in top view.

Project the Final Front View by drawing vertical projectors from the Final Top View and horizontal projectors from the Second Front View.

Problem 22: *(Fig. 12)*

A hexagonal lamina of 20 mm side rests on one of its corners on the HP. The diagonal passing through this corner is inclined at 45° to the HP. The lamina is then rotated through 90° such that the top view of this diagonal is perpendicular to the VP and the surface is still inclined at 45° to the HP.

Draw the projections of the lamina. (UQ)

Simple Position

1. Assume the hexagonal lamina ABCDEF to lie on the HP with one of its diagonals AD parallel to the VP.
2. For this position, draw the First Top View abcdef below the XY line.
3. Project its corresponding Front View a' d' on XY.
4. Look at the First Top View in the direction of arrow.
5. Mark hidden corners in the First Front View.

Second Position

6. Diagonal AD of the lamina is inclined at 45° to HP. Hence turn the First Front View a' d' about the corner a' to an angle of 45° to XY to obtain the *Second Front View* $a' d'_1$.
7. Project the corresponding Second Top View $a b_1 c_1 d_1 e_1 f_1$.

NOTE : Second Front and Top views may also be drawn *separately*.

Final Position

8. Then rotate the lamina through 90° such that the Top View $a d_1$ of diagonal AD is perpendicular to the VP.
9. Hence redraw the Second Top View, turning it through 90° such that $a d_1$ is perpendicular to the XY line. $a_2 b_2 c_2 d_2 e_2 f_2$ is the required *Final Top View*.
10. As per given condition, diagonal AD of the lamina still makes an angle of 45° to the HP. Therefore, from the Second Front View draw horizontal lines and from the Final Top View draw the vertical projectors.
11. Complete the Final Front View $a'_2\ b'_2\ c'_2\ d'_2\ e'_2\ f'_2$ as shown.

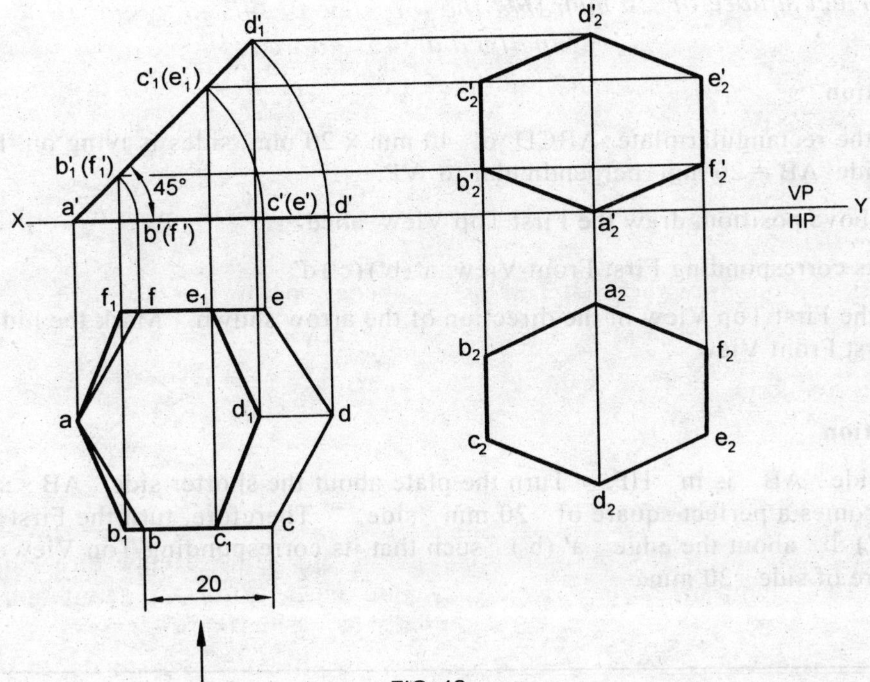

FIG. 12

Problem 23: *(Exercise) A hexagonal sheet metal of 25 mm side rests on the HP such that its surface is inclined at an angle of 45° to the HP and one of its diagonals is inclined at an angle of 45° to the VP. Draw its projections.* (UQ)

HINT :
1. The surface of the sheet metal is to be inclined at 45° to HP and one of its diagonals is to be inclined to the VP. Hence, draw the top and front views of the sheet metal with its surface lying on HP and one of the diagonals perpendicular to VP.
2. Then tilt the sheet on HP with its surface inclined at 45° to HP and draw the top and front views in this position.
3. In this top view, the diagonal perpendicular to VP is in its true length. Hence, redraw the top view such that the diagonal is inclined at 45° to VP.
4. Draw the corresponding front view.

Problem 24: *(Exercise) A thin 30° - 60° set-square of 125 mm longest side, has its longest edge in VP and inclined at 30° to HP. Its surface makes an angle of 40° with VP. Draw its projections.*

HINT : Assume set-square to be in VP such that its hypotenuse is perpendicular to HP.

Problem 25: *(Fig. 13)*

A thin rectangular plate of side 40 mm x 20 mm has its shorter side in the HP and inclined at an angle of 30° to the VP. Project its front view when its top view is a perfect square of 20 mm side. (UQ)

Simple Position

1. Assume the rectangular plate ABCD of 40 mm x 20 mm sides is lying on HP with its shorter side AB = 20 mm perpendicular to VP.
2. For the above position, draw the First Top View abcd.
3. Project its corresponding First Front View a' (b') (c') d'.
4. Look at the First Top View in the direction of the arrow shown. Mark the hidden corners in the First Front View.

Second Position

5. Shorter side AB is in HP. Turn the plate about the shorter side AB until its Top View becomes a perfect square of 20 mm side. Therefore, turn the First Front View a' (b') (c') d' about the edge a' (b') such that its corresponding Top View $a_1\ b_1\ c_1\ d_1$ is a square of side 20 mm.

NOTE : *Inclination of the plate with HP is unknown. Hence first draw the Second Top View $a_1\ b_1\ c_1\ d_1$, a square of 20 mm side.*

Then draw the projectors upwards from the Second Top View.

On these projectors mark the Second Front View $a_1'\ (b_1')\ (c_1')\ d_1'$ *equal to the length of the First Front View = 40 mm as shown.*

Final Position

6. Shorter edge AB is on HP, making an angle of 30° with VP. Hence turn the Second Top View $a_1\ b_1\ c_1\ d_1$ so that the edge $a_1\ b_1$ is at an angle of 30° to XY. $a_2\ b_2\ c_2\ d_2$ is the Final Top View.
7. Project all the points upwards from the Final Top View and also horizontally from the Second Front View.
8. Complete the required Final Front View $a_2'\ b_2'\ c_2'\ d_2'$ as shown.

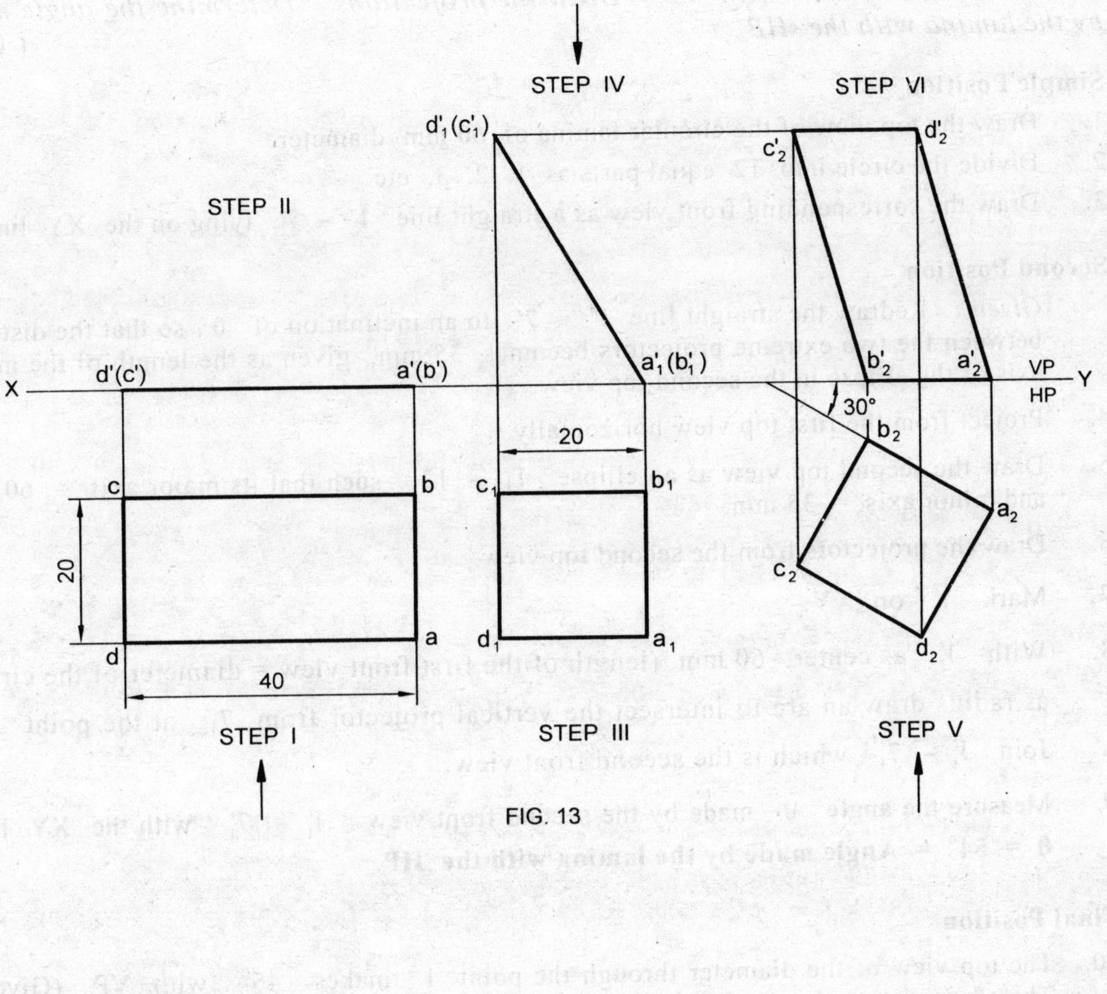

FIG. 13

Problem 26: *(Exercise)*

A rectangular aluminum plate of size 63 mm x 36 mm with negligible thickness, has one of its shorter edges in VP and inclined at 40° to HP. Its front view is a square of sides 36 mm. Draw the top view. (UQ)

HINT : 1. Front view of the plate is a square. Therefore, its surface must be inclined to VP.

2. Assume the plate to be in VP such that its shorter edge is perpendicular to HP.

Problem 27: *(Fig. 14)*

A circular lamina of 60 mm diameter rests on HP on a point 1 on the circumference. The lamina is inclined to HP such that the top view of it is an ellipse of minor axis 35 mm. The top view of the diameter through the point 1 makes an angle of 45° with VP. Draw the projections. Determine the angle made by the lamina with the HP.

(UQ)

Simple Position

1. Draw the top view of the circular lamina of 60 mm diameter.
2. Divide the circle into 12 equal parts as 1, 2, 3, etc.
3. Draw the corresponding front view as a straight line 1′ – 7′ lying on the XY line.

Second Position

Given : Redraw the straight line 1′ – 7′ to an inclination of θ so that the distance between the two extreme projectors becomes 35 mm, given as the length of the minor axis of the ellipse in the second top view.

4. Project from the first top view horizontally.
5. Draw the second top view as an ellipse $1_1 - 12_1$ such that its major axis = 60 mm and minor axis = 35 mm.
6. Draw the projectors from the second top view.
7. Mark $1_1'$ on XY.
8. With $1_1'$ as center, 60 mm (length of the first front view = diameter of the circle) as radius draw an arc to intersect the vertical projector from 7_1 at the point $7_1'$. Join $1_1' - 7_1'$ which is the second front view.
9. Measure the angle θ made by the second front view $1_1' - 7_1'$ with the XY line. **θ = 54° = Angle made by the lamina with the HP.**

Final Position

10. The top view of the diameter through the point 1 makes 45° with VP (Given). Therefore, rotate the axis of the top view $1_1 - 7_1$ such that the final top view $1_2 - 7_2$ makes an angle of 45° with XY.
11. Complete the ellipse of the final top view by redrawing the ellipse of the second top view on the 45° inclined line.
12. Draw the final front view by drawing projectors from the final top view and horizontal lines from the second front view.

Problem 28: *(Fig. 15)* *A semi-circular lamina of 64 mm diameter has its straight edge in VP and inclined at an angle of 45° to HP. The surface of the lamina makes an angle of 30° with VP. Draw the projections.*

27

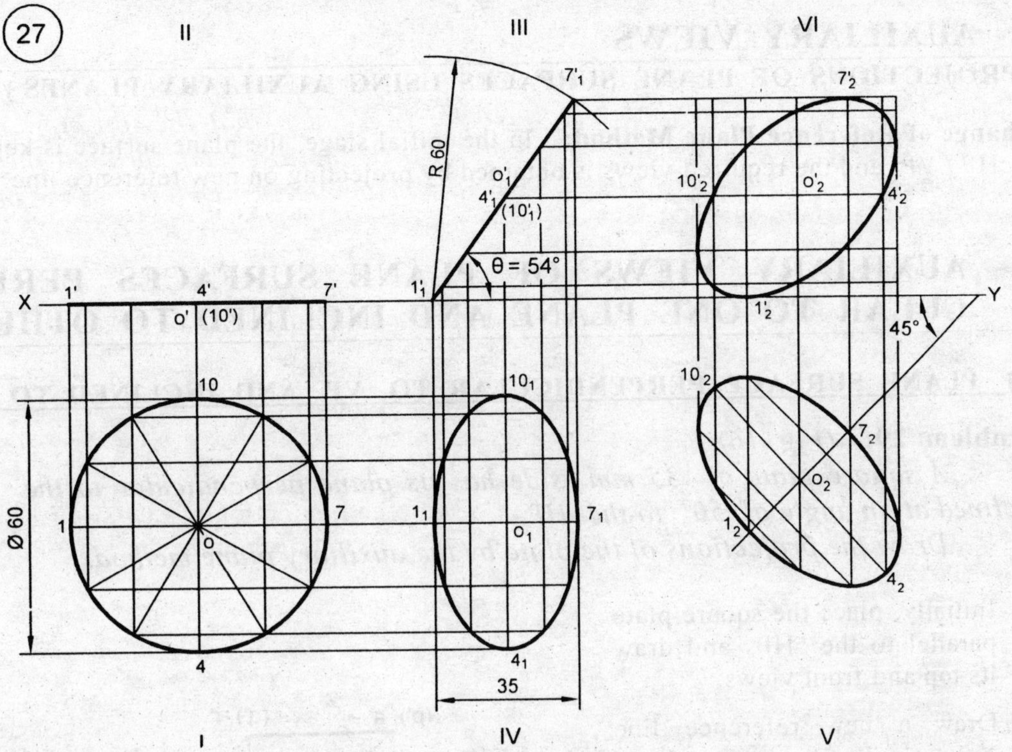

FIG. 14

28

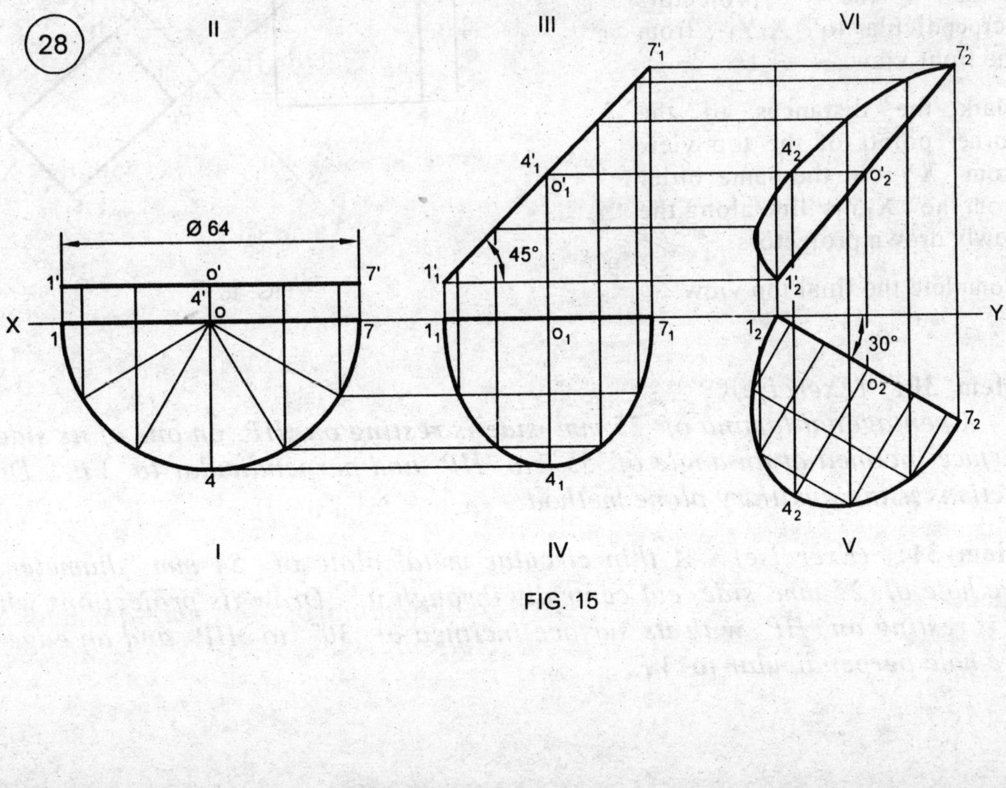

FIG. 15

7. AUXILIARY VIEWS

(PROJECTIONS OF PLANE SURFACES USING AUXILIARY PLANES)

Change of Reference Plane Method: In the initial stage, the plane surface is kept parallel to HP / VP and the required views is obtained by projecting on new reference line X_1Y_1.

8. AUXILIARY VIEWS OF PLANE SURFACES PERPENDICULAR TO ONE PLANE AND INCLINED TO OTHER

8.1 PLANE SURFACE PERPENDICULAR TO VP AND INCLINED TO HP

Problem 29: *(Fig. 16)*

A square plate of 35 mm side has its plane perpendicular to the VP and inclined at an angle of 50° to the HP.

Draw the projections of the plate by the auxiliary plane method.

1. Initially, place the square plate parallel to the HP and draw its top and front views.
2. Draw a new reference line X_1Y_1 inclined at an angle of 50° to XY.
3. Draw the projectors perpendicular to X_1Y_1 from the front view.
4. Mark the distances of the corner points of the top view from XY in the same order from the X_1Y_1 line along the newly drawn projectors.
5. Complete the final top view.

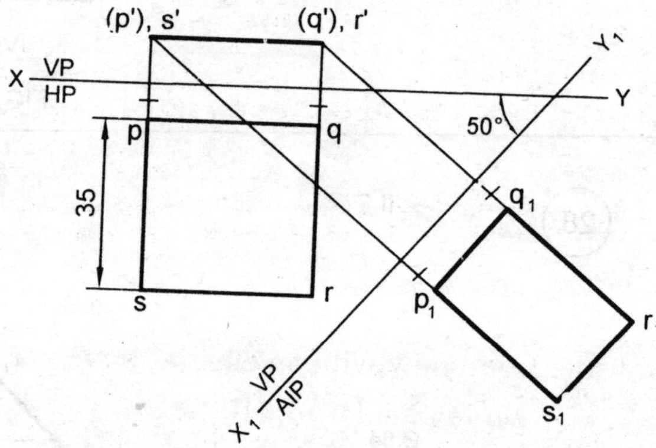

FIG. 16

Problem 30: *(Exercise)*

A pentagonal lamina of 25 mm side is resting on HP on one of its sides with its surface inclined at an angle of 55° to HP and perpendicular to VP. Draw its projections using auxiliary plane method.

Problem 31: *(Exercise)* *A thin circular metal plate of 54 mm diameter has a square hole of 27 mm side, cut centrally through it. Draw its projections when the plate is resting on HP with its surface inclined at 30° to HP and an edge of the square hole perpendicular to VP.*

8.2 PLANE SURFACE PERPENDICULAR TO HP AND INCLINED TO VP

Problem 32: *(Fig. 17) A pentagonal lamina of 24 mm side is resting on HP with one of its edges perpendicular to it and its surface parallel to VP. Draw an auxiliary front view on an auxiliary vertical plane perpendicular to HP and inclined at an angle of 60° to VP.* (UQ)

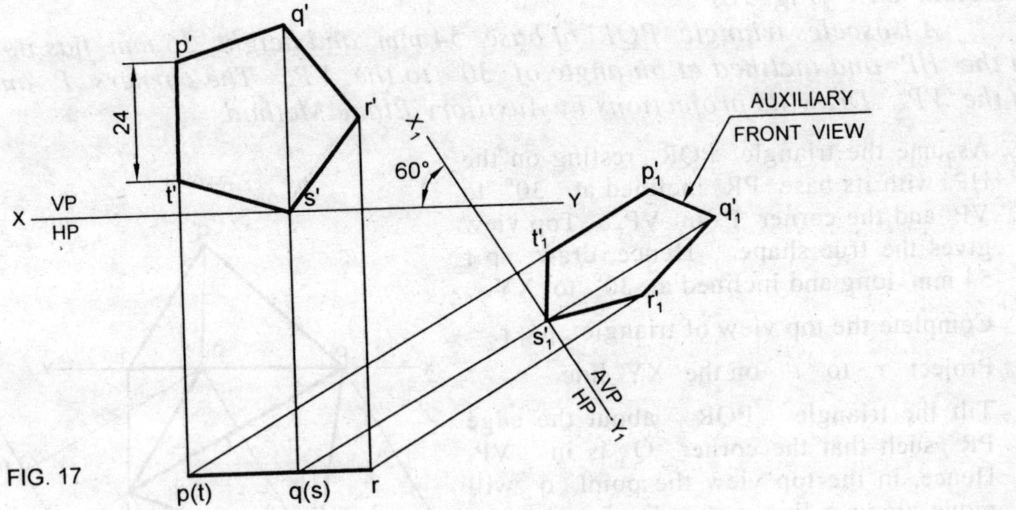

FIG. 17

1. Draw the front view with an edge of the pentagon perpendicular to XY.
2. Lamina is perpendicular to HP. Hence, draw its top view as a line as shown.
3. To draw the auxiliary front view on an AIP perpendicular to HP and inclined at 60° to VP, draw the X_1Y_1 line at an angle of 60° to XY.
4. Draw the projectors perpendicular to X_1Y_1 from all the corners in the top view.
5. On the above projectors, measure the distances of the corners in the front view from XY.
6. Mark these distances from the X_1Y_1 line to obtain the auxiliary front view.

Problem 33: *(Exercise) A circular lamina of 50 mm diameter appears as an ellipse in the front view, its major and minor axes being 50 mm and 30 mm respectively. Draw its top view when the major axis is horizontal.* (UQ)

HINT : The circular lamina is seen as an ellipse in the front view. Hence its surface must be inclined to VP. Assume the lamina to be parallel to the VP and draw its front and top views.

8.3 EDGE VIEW OF A PLANE SURFACE

Auxiliary projections of plane surfaces are used to obtain the following :

- Edge View or Line View of the plane surface
- True shape of the plane surface, in addition to the conventional projections of plane surfaces

A plane surface will be seen as a line when projected on a plane perpendicular to the true length. The projection of this Line View, also known as Edge View of a plane surface on an auxiliary plane parallel to it, will give the true shape and size of the plane surface.

Problem 34: *(Fig. 18)*

A isosceles triangle PQR of base 54 mm and height 36 mm has its base PR in the HP and inclined at an angle of 30° to the VP. The corners P and Q are in the VP. Draw its projections by Auxiliary Plane Method. (UQ)

1. Assume the triangle PQR resting on the HP with its base PR inclined at 30° to VP and the corner P in VP. Top view gives the true shape. Hence, draw p r 54 mm long and inclined at 30° to XY.
2. Complete the top view of triangle p q r.
3. Project r to r′ on the XY line.
4. Tilt the triangle PQR about the edge PR such that the corner Q is in VP. Hence, in the top view the point q will move along a line perpendicular to p r to a point q_1 in XY.
5. To determine the distance of the front view of Q below XY by means of an auxiliary plane, draw a new reference line X_1Y_1 perpendicular to p r.
6. Draw an auxiliary front view as a line q' p_1' showing the Edge View of the triangle. p_1' is the point view of the line p r.
7. When q moves to q_1, q′ will move along the arc drawn with p_1' as center and radius = p_1' q' to q_1' on the line through q_1 drawn perpendicular to X_1Y_1. m q_1' is the required distance.
8. Hence, draw a line perpendicular to the XY through q_1.
9. Mark on it q_2' such that q_1 q_1' = m q_1'.
10. Join p and r′ with q_2'. p q_2' r′ and p q_1 r are the required projections.

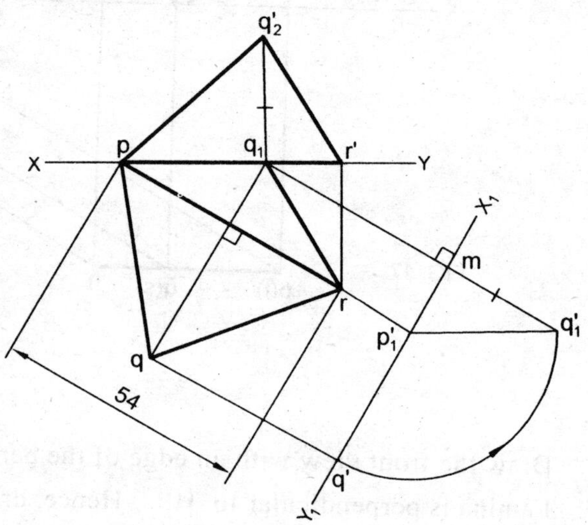

FIG. 18

9. AUXILIARY VIEWS OF PLANE SURFACES INCLINED TO BOTH HP AND VP

Problem 35: *(Fig. 19) A thin rectangular lamina EFGH of 60 mm length and 36 mm width is inclined at an angle of 45° to VP. Its longer edge is making an angle of 30° with HP. Draw the projections by auxiliary plane method.* (UQ)

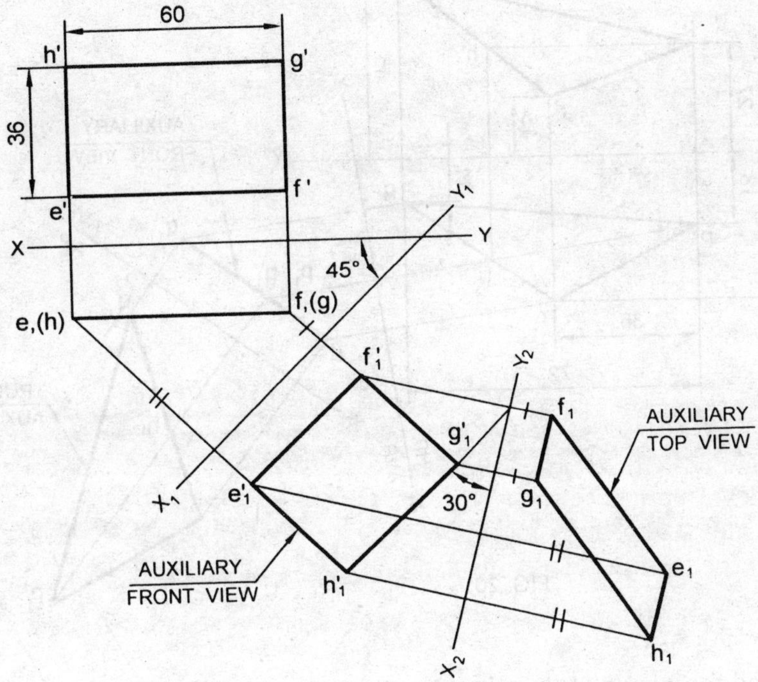

FIG. 19

1. Assume the lamina to be parallel to VP with its longer edge parallel to HP.
2. Draw XY line. Draw the front view e' f' g' h' as a rectangle.
3. Draw the top view e (h) f (g) as a line.
4. Consider an AVP making 45° with XY. Draw the corresponding reference line X_1Y_1.
5. Project and draw the auxiliary front view e_1' f_1' g_1' h_1'.
6. Consider an AIP making 30° with the longer edge g_1' h_1'. Draw reference line X_2Y_2.
7. Project and draw the auxiliary top view e_1 f_1 g_1 h_1.

Problem 36: *(Exercise) A hexagonal G.I. sheet of 25 mm sides is resting on HP with one of its corners touching VP and its surface inclined at 45° to it. One of its edges is inclined at 30° to HP. Obtain its projections.*

Problem 37: *(Fig. 20) A thin triangular sheet has its corners P, Q and R above HP by 27 mm, 45 mm and 15 mm and in front of VP by 18 mm, 11 mm and 36 mm respectively. The projectors through P and Q are 72 mm apart and those through Q and R are 36 mm apart. If the projector through R is between those of P and Q, find the true size of the triangle by auxiliary projection method.* (UQ)

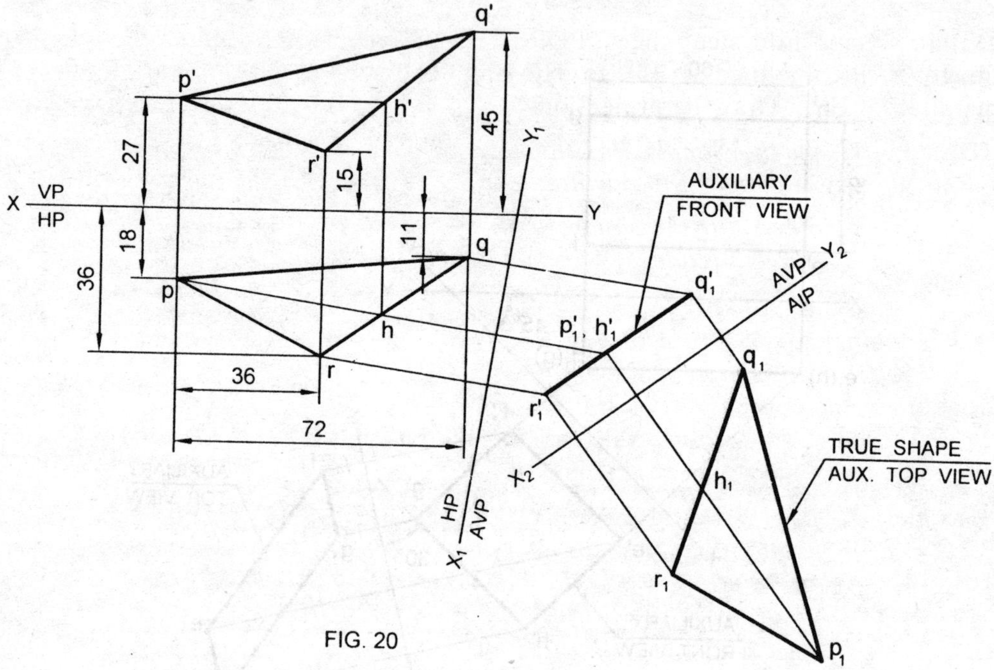

FIG. 20

1. Draw front and top views of the triangle. **To find true shape, setup 2 auxiliary planes.**
2. Draw a horizontal line through any of the corners of the views to cut the opposite side of the triangle. Let p′ h′ be horizontal line drawn in the front view.
3. Draw the same projections of the above line in the top view as p h.
4. Draw X_1Y_1 perpendicular to p h. Draw the auxiliary (second) front view of the triangle as a straight line. That is, the plane of the triangle is perpendicular to AVP.
5. Draw X_2Y_2 parallel to auxiliary (second) front view. Draw auxiliary (second) top view.
6. Since AIP is parallel to the plane of the triangle, the auxiliary (second) top view gives the **True Shape** of the triangle.

Problem 38: *(Exercise) The corners of a quadrilateral PQRS are as follows: P is 25 mm above HP and 50 mm in front of VP. Q is in HP and 80 mm in front of VP. R is 50 mm above HP and 40 mm in front of VP. S is 65 mm above HP and 20 mm in front of VP. The distances between the vertical projectors parallel to XY are as follows. Between P and S is 20 mm, between P and Q is 35 mm and between P and R is 60 mm. Find its true shape.* (UQ)

10. ADDITIONAL PROBLEMS

39. A regular pentagon ABCDE of 25 mm side, has its corner A in HP and the side CD parallel to HP. Draw its projections when its top view is parallel to and 20 mm away from VP. Also show its traces.

40. A thin square mild steel sheet PQRS of 48 mm side has its corner P in HP, its diagonal PR inclined at 45° to HP and the diagonal QS inclined at 30° to VP and parallel to HP. Draw its projections.

HINT : 1. *In the Simple Position, assume the square to be lying in HP with its diagonal PR parallel to VP.*

2. *Then, tilt the square about the corner P so that PR makes 30° with HP, diagonal QS remaining perpendicular to VP and parallel to HP.*

41. A pentagonal plane figure of edges 25 mm is lying on HP with one of its corners touching it such that the plane figure makes 60° with HP. Two of the edges containing the corner on which the plane figure rests make equal inclinations with HP. When the edge opposite to this corner makes 45° with VP, draw the top and front views of the plane figure in this position.

HINT : 1. *The plane figure is resting with a corner on HP such that the two edges containing the corner on which the plane figure rests, make equal inclinations with HP. Hence, draw the top and front views of the plane figure with its surface completely lying on HP.*

2. *The front view will be a line on XY line.*

42. A circular disc of 55 mm diameter rests on HP such that its surface is inclined at 40° to HP. The diameter through the point on which the disc rests on HP appears to be inclined at 30° to VP in the top view. Draw its projections.

HINT : *The circular disc is to be kept with its surface inclined at 40° to HP. Hence draw top and front views when the disc rests with its surface lying on HP.*

43. An isosceles triangle of base 38 mm and sides 58 mm rests with its base on HP and one of its sides lying in VP and the side on which it rests is inclined at 30° to VP. Draw the projections.

HINT : 1. *Surface of the triangle is inclined to VP. Hence, draw its top and front views when placed with its base on HP and the surface parallel to VP.*

2. *Then, tilt the triangle with a corner touching VP and the surface of the triangle inclined at 30° to VP.*

44. A circular lamina of 60 mm diameter is resting on its circumference such that its plane is inclined at 60° to the HP and 30° to the VP. Draw the projections. (UQ)

HINT : 1. Assume the plate to be in VP. Draw the front and top views.
2. Then draw a side view, preferably, left side view.
3. Tilt the plate such that it is inclined at 60° to HP, keeping one end of a diameter in HP and other end in VP. That is, keep a mini-drafter scale to 60° position and set the divider to a length = Diameter of the circle.
4. By trial and error, adjust the position of the set-square by keeping the divider against the edge of the set-square till one leg of the divider is on XY line, while the other leg is on the vertical line.

45. A semi-circular lamina of 60 mm diameter is resting on HP on its diameter AB such that AB is inclined at 45° to VP and the surface of the lamina is making 30° to HP. End A is nearer to VP and is at a distance of 25 mm from it. Draw its projections. (UQ)

HINT : 1. **Initial Position:** Assume the lamina to lie on HP with the diameter AB perpendicular to VP, *i.e.*, XY line.
2. **Second Position:** Keep the surface of the lamina perpendicular to VP and inclined at 30° to HP, still keeping the diameter AB perpendicular to XY.
3. **Final Position:** Diameter AB is made to be inclined at 45° to VP, *i.e.*, XY line.

46. A pentagonal metal sheet of side 30 mm has a square hole of side 20 mm with centrally punched through it. The sheet is resting on HP on one of its sides parallel to VP and the surface of the sheet inclined at 30° to HP. One of the sides of the hole is parallel to the side on which the metal sheet rests. Neglect the thickness of the sheet. Draw its projections.

47. A thin M.S. plate in the form of a regular hexagon of 30 mm side, has a circular hole of 40 mm diameter centrally punched. The plate is resting on HP on one of its edges such that the edge is parallel to VP and the surface of the plate is inclined at 30° to HP. Draw the projections.

48. A regular pentagonal plane figure of 25 mm side has one side AB parallel to VP and making an angle of 40° with HP. Plane of the pentagon makes 35° with the VP. Draw its projections such that the end B of the side AB is 20 mm above the HP and 10 mm in front of VP by (i) Change of Position Method and (ii) Change of Reference Line Method. Compare the solutions.

*

UNIT – III

SYLLABUS

UNIT – III

Orthographic Projections of Solids:
Cylinder, Cone, Prism, Pyramid and Sphere Positions and Axis Inclined to Both the Planes

16

PROJECTIONS OF SOLIDS

1. PROJECTIONS OF SOLIDS

What is a SOLID ?

A *Solid* is defined as an object having three dimensions, *viz.*, length, breadth and height or thickness, measured along the three mutually perpendicular axes. It is bounded by plane faces or curved surfaces or combination of both.

Orthographic Projections of Solids

To represent a solid in the orthographic projection, at least two views are necessary; one view to represent length and height, called *Front View* and the other view to represent length and breadth, called *Top View*. Sometimes the above two views may not be sufficient to explain the details. Hence, a third view called *View from the Right (Right Side View)* or *View from the Left (Left Side View)* becomes necessary.

Multi-view orthographic projections of a solid are obtained by projecting every edge of the solid as well as all the points on the boundary of a curved surface. If these edges or surface boundaries are visible to the observer at infinity in front of plane of projection, the projections are drawn by *thick* lines, otherwise by *short dashed thin* lines for hidden edges.

Importance of the study of Projections of Solids

An engineering component will be formed of a single solid or combination of solids of different shapes. The most common geometrical solids that make engineering components are prisms, pyramids, cylinders, cones, spheres, etc.

For the purpose of drawing, a solid may be considered as a combination of *Points*, *Lines* and *Plane* or *Curved Surfaces*. Orthographic and auxiliary projections of *Points*, *Lines* and *Planes* were already discussed in the previous Chapters. Here, the procedure and conventions discussed earlier are combined and applied together for the projections of *Solids*.

Right Regular Solids

i. If the axis of a solid is *perpendicular* to its base, it is a *Right Solid*.
ii. If all edges of the base / end faces of a *right solid* are of *equal* lengths, it is known as *Right Regular Solid*; otherwise *Irregular Solid*.

2. TYPES OF SOLIDS

Solids are classified into *two* main groups :

1. Polyhedra 2. Solids of Revolution 3. Frustums and Truncated Solids

2.1 POLYHEDRA

Polyhedron: A solid which is bounded by plane surfaces (*i.e.*, faces) is called *Polyhedron*.

Right Regular Polyhedron: If the axis of a polyhedron is perpendicular to the end faces and if all its faces are having the same size and shape, it is a *Right Regular Polyhedron*.

Faces: The regular plane surfaces which form the surfaces of a polyhedron are called *Faces*.

Edges: The lines along which two plane faces or surfaces meet are called *Edges* of the solid.

Right Regular Polyhedra are divided into the following *five* types:

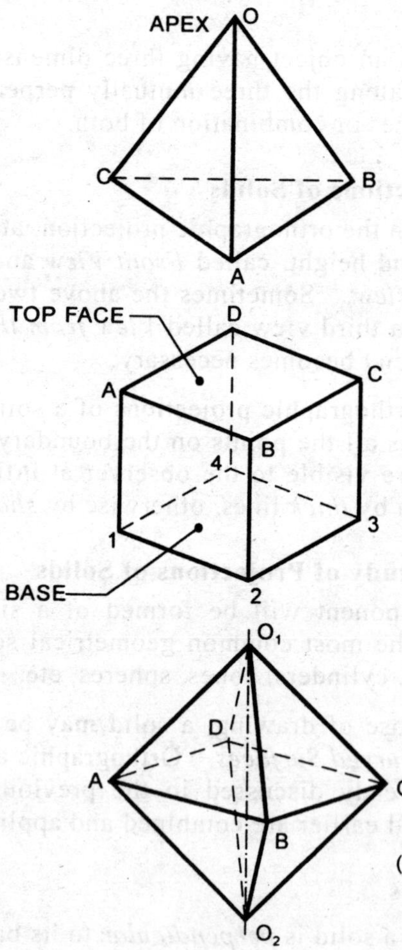

(i) TETRAHEDRON

(ii) CUBE or HEXAHEDRON

(iii) OCTAHEDRON

FIG. 1

1. TETRAHEDRON

See Fig. 1(i).

A tetrahedron has *four* equal equilateral triangular ***Faces***: ABC, OAB, OBC and OCA.

Six ***Edges:*** AB, BC, CA, OA, OB and OC.

O – ***Apex***

2. CUBE

See Fig. 1(ii).

A cube has *six* equal regular squares.

3. OCTAHEDRON

See Fig. 1(iii).

An octahedron has eight equal equilateral triangular ***Faces***, *viz.*, AO_1B, BO_1C, CO_1D, DO_1A, AO_2B, BO_2C, CO_2D and DO_2A.

4. PRISMS

A *Prism* is a polyhedron having two equal and similar faces (called top face and bottom base), parallel to each other and joined by other faces which may be *Rectangles* or *Parallelograms*.

Axis: The imaginary line joining the centers of the faces is called the *Axis*.

NOTE : *A prism is said to be a* ***right regular*** *prism when its axis is perpendicular to its base and all its faces (i.e., lateral surfaces) are equal rectangles.*

Types of Regular Prisms

See Fig. 2.

Prism is named according to the shape of its base as :

Triangular,
Square, Rectangular,
Pentagonal,
Hexagonal, etc.

Nomenclatures of a prism

Axis,
Top face,
Bottom face (Base),
Rectangular faces and
Longer edges.

These are clearly indicated in *Fig. 2.*

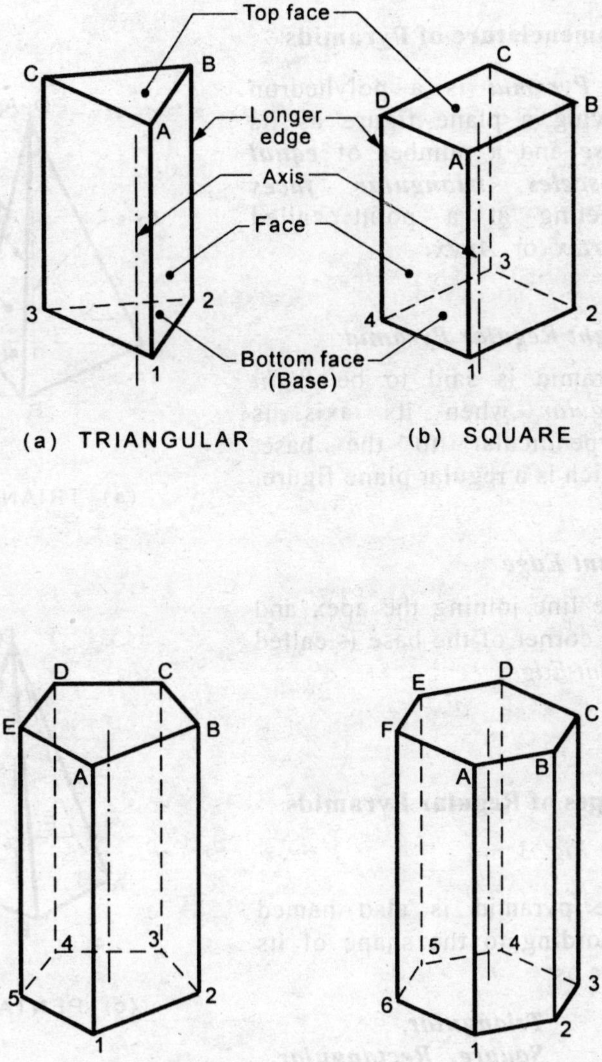

(a) TRIANGULAR (b) SQUARE

(c) PENTAGONAL (d) HEXAGONAL

AB21, etc. - Rectangular faces
A1, etc. - Longer edges

FIG. 2 PRISMS

5. PYRAMIDS

Nomenclature of Pyramids

A *Pyramid* is a polyhedron having a plane figure as its ***base*** and a number of ***equal isosceles triangular faces*** meeting at a point called ***Vertex*** or ***Apex***.

Right Regular Pyramid

Pyramid is said to be *Right Regular* when its axis is perpendicular to the base, which is a regular plane figure.

Slant Edge

The line joining the apex and the corner of the base is called *Slant Edge*.

Types of Regular Pyramids

See Fig. 3.

The pyramid is also named according to the shape of its base as:

Triangular,
Square, Rectangular,
Pentagonal,
Hexagonal, etc.

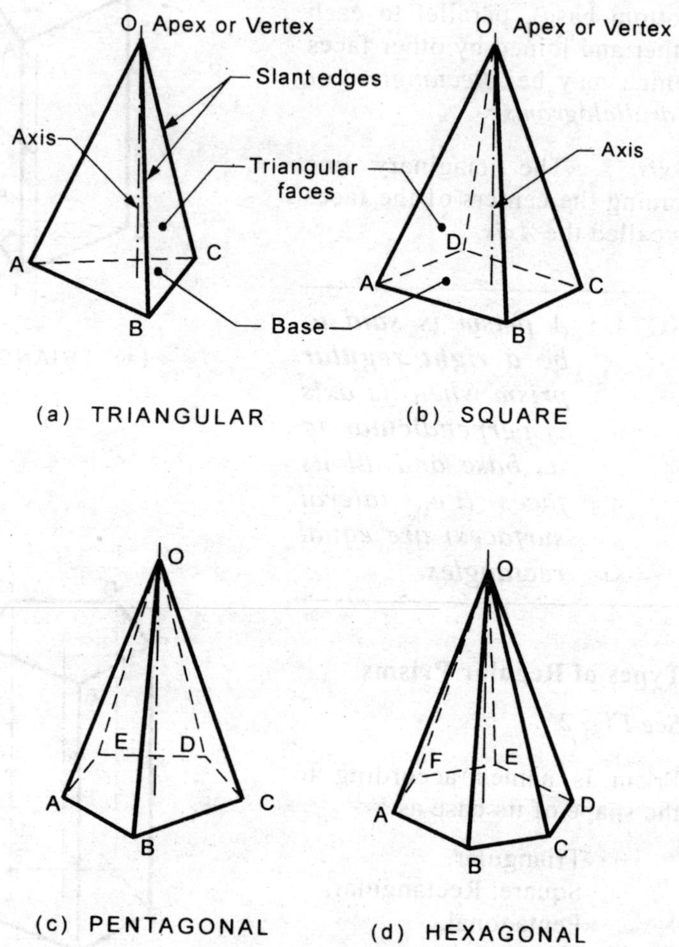

FIG. 3 PYRAMIDS

NOTE : 1. *An* ***Oblique*** *solid has its axis inclined to the base.*

2. *Unless specified as oblique / irregular, a solid is always understood as being* ***'right'*** *and* ***'regular'***

2.2 SOLIDS OF REVOLUTION

Solid of Revolution: *See Fig. 4.* If a plane surface is revolved about one of its edges, the solid generated is called a *Solid of Revolution.* *Examples*: Cylinders, Cones, Spheres, etc.

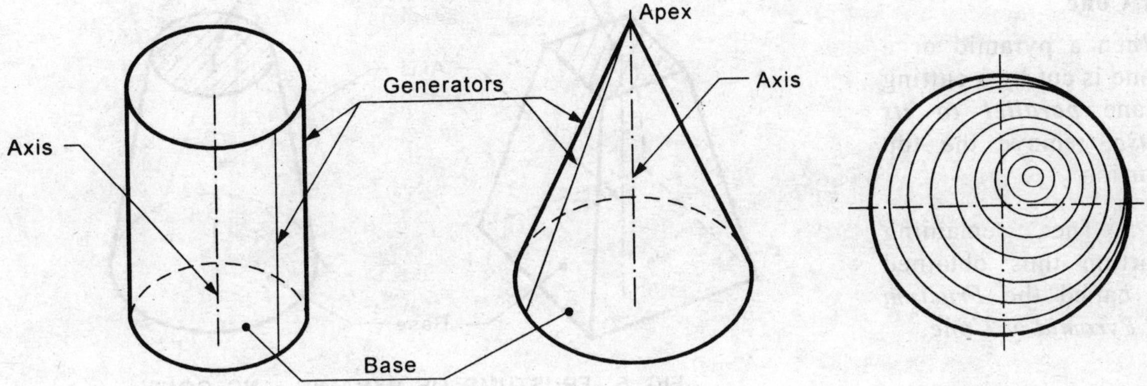

(a) CYLINDER (b) CONE (c) SPHERE

FIG. 4

1. CYLINDER

See Fig. 4(a). A *Right Circular Cylinder* is a solid generated by the revolution of a rectangle about one of its sides which is fixed. It has two circular faces, *i.e.*, top face and bottom base.

Nomenclatures of a Cylinder: Axis, Top face, Bottom face (Base) and Generators.

Axis is the line joining the centers of top face and bottom base. It is perpendicular to base.

2. CONE

See Fig. 4(b). A *Right Cone* is a solid generated by the revolution of a right-angled triangle about one of its perpendicular sides which is fixed. It has a circular base and an apex.

Nomenclatures of a cone: Axis, Apex or Vertex, Base, Generators and Slant Height.

Axis is the line joining the apex with the center of the base.

Generators are straight lines drawn from the apex to the circumference of the base circle.

Slant Height: The length of the generator is the *Slant Height* of the cone.

3. SPHERE

See Fig. 4(c). A *Sphere* is a solid generated by the revolution of a semi-circle about its diameter as its axis. The mid-point of the diameter is the center of the sphere. All points on the surface of the sphere are at equal distance from its center.

2.3 FRUSTUMS AND TRUNCATED SOLIDS

1. FRUSTUMS

See Fig. 5.

Frustums of Pyramid or Cone

When a pyramid or a cone is cut by a cutting plane ***parallel to its base***, remove the top portion.

The remaining portion thus obtained is called the *Frustum of Pyramid or Cone.*

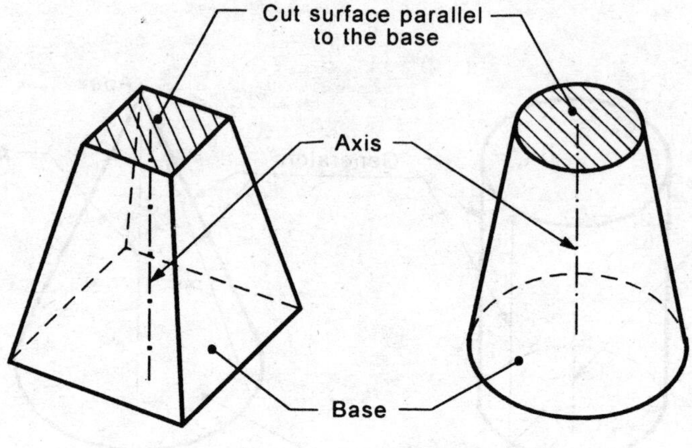

FIG. 5 FRUSTUMS OF PYRAMID AND CONE

2. TRUNCATED SOLIDS

See Fig. 6.

When a solid (prism / cylinder / pyramid / cone) is cut by a cutting plane ***inclined to its base*** (not parallel), remove the top portion.

The remaining portion thus obtained is called the *Truncated Solid.*

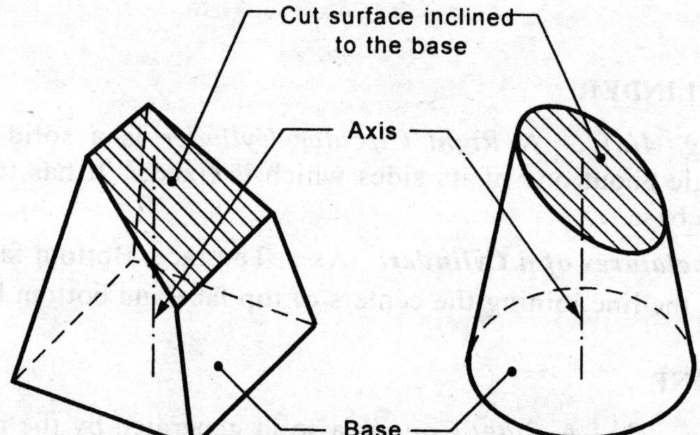

FIG. 6 TRUNCATED PYRAMID AND CONE

NOTE: *Assume the solids to be* ***right*** *and* ***regular*** *unless stated otherwise.*

Oblique Solids

Solids having the axis inclined to their base are known as *Oblique Solids*.
Examples: Oblique pyramids, oblique prisms, oblique cylinders, oblique cones, etc.

3. SOLIDS IN SIMPLE POSITIONS

(i) Axis of the solid perpendicular to HP and parallel to VP
(ii) Axis of the solid perpendicular to VP and parallel to HP
(iii) Axis of the solid parallel to both HP and VP

A solid is said to be in its *Simple Position*, when its axis is perpendicular to one reference plane or parallel to both. When the axis is perpendicular to a reference plane, it is automatically parallel to the other; Also, when the axis is perpendicular to a reference plane, its base will be parallel to that reference plane.

3.1 PROJECTIONS OF SOLIDS IN SIMPLE POSITIONS

We have already studied that when a plane surface is parallel to a reference plane, its projection on that reference plane shows its true shape and size.

RULE : ***Projection of a solid on the reference plane to which its axis is perpendicular, will show the true shape and size of its base.***

- Start with the projection that shows the true shape and size of the base of the solid.
- When the axis of the solid is perpendicular to HP, its base will be parallel to HP. ***The top view shows the true shape and size of the base and hence the top view is to be drawn first*** and the front view projected from it.
- When the axis of the solid is perpendicular to VP, ***the front view shows the true shape and size of the base and hence the front view is to be drawn first***. The top view is projected from it.

3.2 BIS CONVENTIONS

1. A corner of a solid is always denoted by a CAPITAL letter.
2. Projection of the corner on HP is denoted by the lower-case letter in the top view.
3. Projection on VP is denoted by the lower-case letter with a *dash* in the front view.

Invisible Edges

Invisible edges of an object are shown by *lines of dashes*. (Lines of dashes are wrongly called as dotted lines.) To find the visibility or invisibility of the edges, a careful imaginative thinking and visualization is essential. This is based on the following guide lines:

- Outer-most lines of a view are always visible.
- Upper corners / edges / lines of the front view are visible in top view.
- Corners / edges / faces nearer the observer in top view are visible in the front view.
- Corners / edges / faces on the right of front and top views are visible in right side view.
- Corners / edges / faces on the left of front and top views are visible in left side view.

4. PROJECTION OF PRISM, PYRAMID, CYLINDER AND CONE

4.1 AXIS PERPENDICULAR TO HP (Simple Position)

Problem 1: *(Fig. 7) Draw the top and front views of a **cube** of 40 mm side resting with one of its square faces on HP such that one of its vertical faces is parallel to and 10 mm in front of VP.*

1. *See Fig. 7(a).* Look at the pictorial view. Top face ABCD and bottom face 1 2 3 4 are connected by four vertical faces (lateral surfaces) bounded by vertical edges A1, B2, C3 and D4.

From the given statement, decide the view to be drawn first

2. Cube rests with a face on HP. Thus its axis is perpendicular to HP. Hence, the end face is seen in true size in the top view; And **draw the top view first.**

To draw the top view, look down at the cube in the pictorial view from the top

3. Cube rests with a face on HP, *i.e.*, top face ABCD is parallel to HP and an edge CD of a vertical face is parallel to and 10 mm in front of VP. So draw the top view a b c d, a square of 40 mm side with its edge cd parallel to & 10 mm below XY. *See Fig. 7(b).*

4. Bottom face 1 2 3 4 is hidden by the top face ABCD. So in the top view mark the bottom face as (1) (2) (3) (4) such that (1) and a coincide, (2) and b coincide etc. Thus, complete the top view as shown.

To draw the front view

5. Draw the vertical projectors in *thin* lines from the top view.

6. Look at the cube from the front. Bottom face 1 2 3 4 is on HP. Hence draw the front view of the bottom face as 1′ 2′ 3′ 4′ on XY.

7. Height 40 mm is seen in front view. Therefore, draw a horizontal line 40 mm above XY to cut the projectors drawn from a, b, c and d at a′, b′, c′ and d′ respectively.

NOTE : *In the Orthographic views of a solid, corners are first located and they are connected by lines. While doing so, the visible and invisible lines are to be determined. The invisible corners are denoted by letters within brackets.*

To draw the visible and invisible edges in the front view

8. Look at the top view in the direction of the arrow (observer) shown. Face ab (2) (1) is nearer to the observer and is hence visible. So draw a′ b′ 2′ 1′ by *thick line* in front view. That is, existing construction *thin* line is drawn *thick*.

9. Face d c (3) (4) is *invisible* and hence mark it as (d′) (c′) (3′) (4′) in the front view.

10. Invisible edges (4′) (3′), (3′) (c′), (c′) (d′) and (d′) (4′) are hidden by the visible edges 1′ 2′, 2′ b′, b′ a′ and a′ 1′ respectively in the front view and hence *not* shown in dashed lines.

(1)

AXIS PERPENDICULAR TO H.P. AND PARALLEL TO V.P.

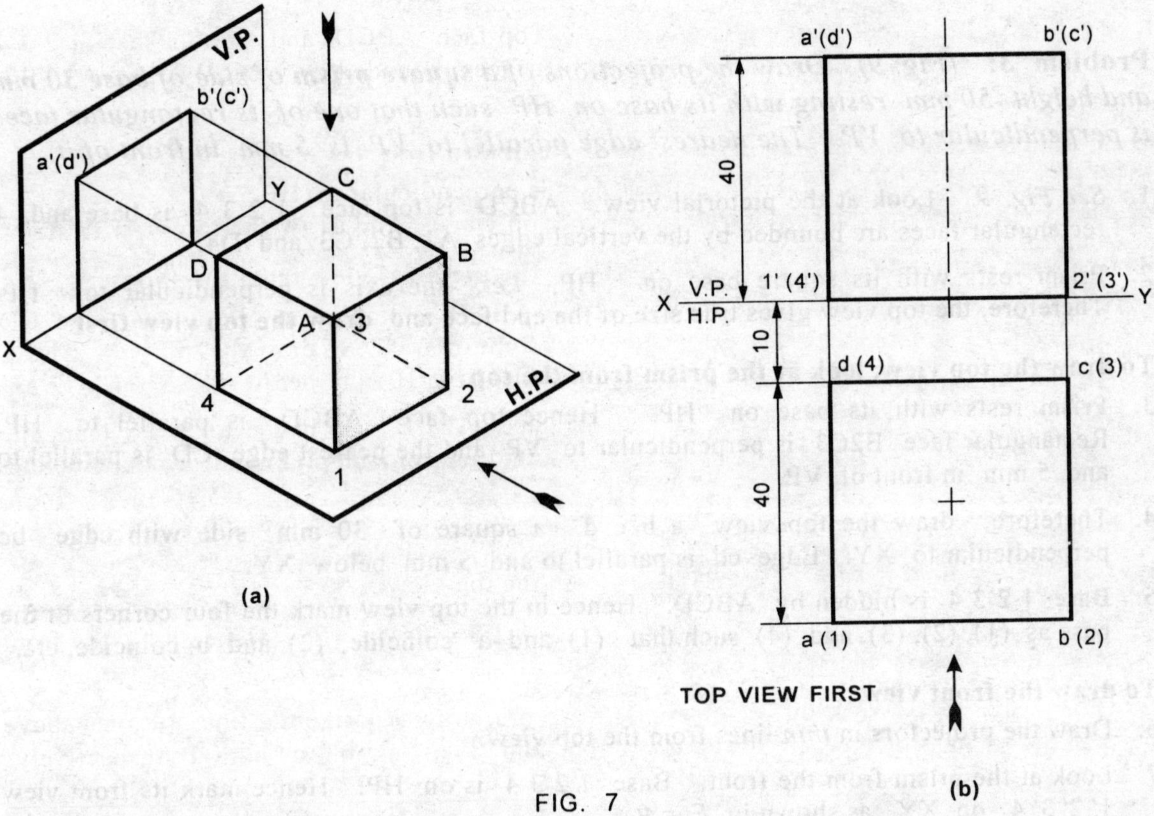

FIG. 7

NOTE :

1. *The view which gives the true shape of the solid shall be drawn first; Other view shall be projected from it.*
2. *When the axis is perpendicular to HP, true size of the end face is seen in top view. Therefore, draw the top view first.*
3. *Then, to mark the invisible corners and edges in the front view, look at the top view in the direction of the arrow shown.*
4. *Mark the axis in the front and top views for all the solids.*

Problem 2: *(Fig. 8) A* ***cube*** *of 40 mm side rests with one of its square faces on HP such that one of its vertical faces is perpendicular to VP. Draw its projections. The nearest edge parallel to VP is 5 mm in front of it.*

See Fig. 8. When face 1AD4 is perpendicular to VP, *face 3CD4 becomes parallel to VP. Now, the Problem is similar to Problem 1.* Therefore, refer the procedure given in Problem 1 and draw the top and front views.

Problem 3: *(Fig. 9) Draw the projections of a* ***square prism*** *of side of base 30 mm and height 50 mm resting with its base on HP such that one of its rectangular faces is perpendicular to VP. The nearest edge parallel to VP is 5 mm in front of it.*

1. *See Fig. 9.* Look at the pictorial view. ABCD is top face, 1 2 3 4 is base and 4 rectangular faces are bounded by the vertical edges A1, B2, C3 and D4.
2. Prism rests with its square base on HP, *i.e.*, the axis is perpendicular to HP. Therefore, the top view gives true size of the end face and **draw the top view first.**

To draw the top view, look at the prism from the top

3. Prism rests with its base on HP. Hence top face ABCD is parallel to HP. Rectangular face B2C3 is perpendicular to VP and the nearest edge CD is parallel to and 5 mm in front of VP.
4. Therefore, draw the top view a b c d, a square of 30 mm side with edge bc perpendicular to XY. Edge cd is parallel to and 5 mm below XY.
5. Base 1 2 3 4 is hidden by ABCD. Hence in the top view mark the four corners of the base as (1), (2), (3) and (4) such that (1) and a coincide, (2) and b coincide, etc.

To draw the front view

6. Draw the projectors in *thin* lines from the top view.
7. Look at the prism from the front. Base 1 2 3 4 is on HP. Hence mark its front view 1′ 2′ 3′ 4′ on XY, as shown in *Fig. 9.*
8. Height 50 mm is seen in the front view. Hence, draw a horizontal line 50 mm above XY to cut the projectors drawn from a, b, c and d at a′, b′, c′ and d′ respectively.

To draw visible and invisible edges in the front view

9. Look at the top view in the direction of arrow shown. The face ab (2) (1) is visible and hence draw a′ b′ 2′ 1′ by continuous thick line in the front view.
10. Face dc (3) (4) is invisible and hence mark this as (d′) (c′) (3′) (4′) in the front view.

    Note that the dashed lines are *not* shown, since they are hidden by thick lines.

NOTE : *Name all the corners / points then and there, i.e., immediately after completing a view, failing which confusion will result.*

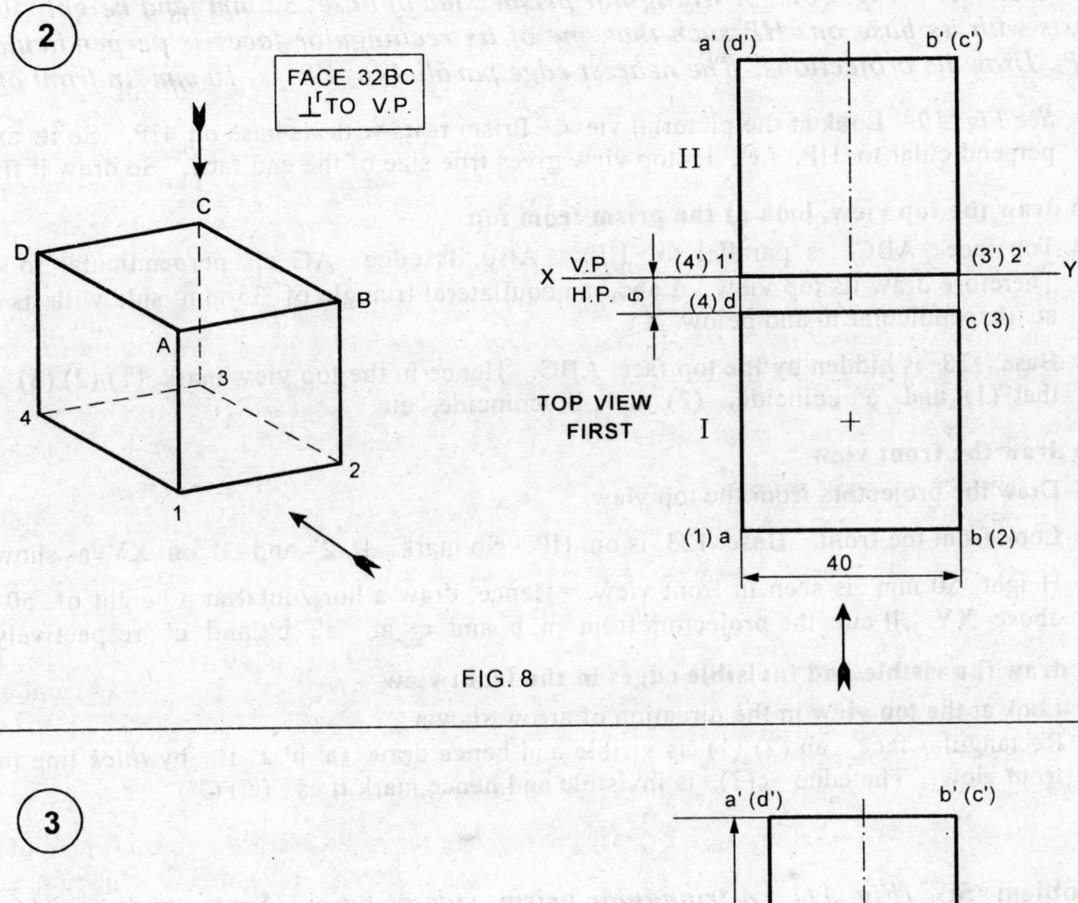
2
FACE 32BC
⊥r TO V.P.
C
D
B
A
3
4
2
1
a' (d')
b' (c')
II
X
V.P.
H.P.
Y
(4') 1'
(3') 2'
5
(4) d
c (3)
TOP VIEW
FIRST
I
(1) a
b (2)
40

FIG. 8

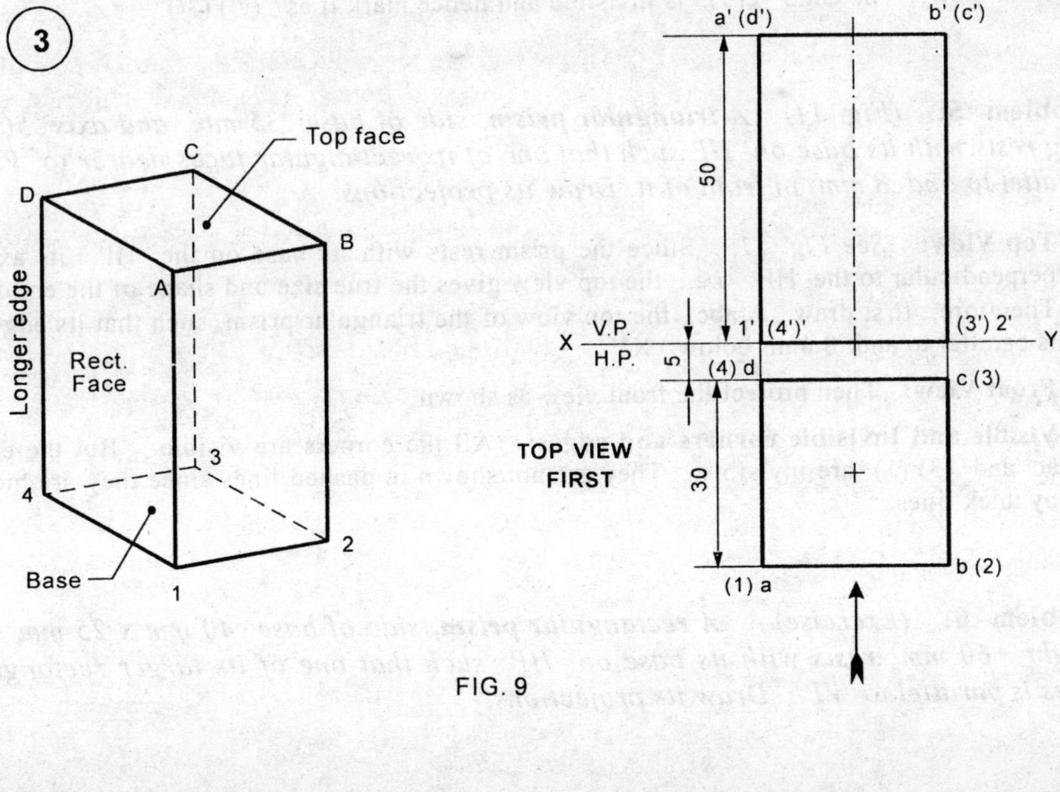
3
Top face
C
D
B
A
Longer edge
Rect.
Face
3
4
2
Base
1
a' (d')
b' (c')
50
X
V.P.
H.P.
Y
1'
(4')'
(3') 2'
5
(4) d
c (3)
TOP VIEW
FIRST
30
b (2)
(1) a

FIG. 9

Problem 4: *(Fig. 10) A **triangular prism**, side of base 35 mm and height 50 mm rests with its base on HP such that one of its rectangular faces is perpendicular to VP. Draw its projections. The nearest edge parallel to VP is 10 mm in front of it.*

1. *See Fig. 10.* Look at the pictorial view. Prism rests with its base on HP. So its axis is perpendicular to HP, *i.e.*, its top view gives true size of the end face. So draw it first.

To draw the top view, look at the prism from top

2. Top face ABC is parallel to HP. Also its edge AC is perpendicular to VP. Therefore draw its top view Δ abc, an equilateral triangle of 35 mm side with its edge ac perpendicular to and below XY.
3. Base 123 is hidden by the top face ABC. Hence in the top view mark (1) (2) (3) such that (1) and a coincide, (2) and b coincide, etc.

To draw the front view

4. Draw the projectors from the top view.
5. Lcok from the front. Base 123 is on HP. So mark 1′ 2′ and 3′ on XY as shown.
6. Height 50 mm is seen in front view. Hence, draw a horizontal at a height of 50 mm above XY. It cuts the projectors from a, b and c at a′, b′ and c′ respectively.

To draw the visible and invisible edges in the front view

7. Look at the top view in the direction of arrow shown.
   Rectangular face ab (2) (1) is visible and hence draw a′ b′ 2′ 1′ by *thick* line in the front view. The edge c(3) is invisible and hence mark it as (c′) (3′).

Problem 5: *(Fig. 11) A **triangular prism**, side of base 35 mm and axis 50 mm long rests with its base on HP such that one of its rectangular faces nearer to VP is parallel to and 8 mm in front of it. Draw its projections.*

1. **Top View:** *See Fig. 11.* Since the prism rests with its base on the HP, its axis is perpendicular to the HP, *i.e.*, the top view gives the true size and shape of the end face. Therefore, first draw Δ abc, the top view of the triangular prism, such that its edge ac is parallel to and 8 mm below XY.
2. **Front View:** Then project the front view as shown.
3. **Visible and Invisible corners and edges:** All the corners are visible. But the edges ac and (3) (1) are invisible. They are not shown in dashed lines since they are hidden by thick lines.

Problem 6: *(Exercise) A **rectangular prism**, side of base 40 mm x 25 mm and height 60 mm rests with its base on HP such that one of its larger rectangular faces is parallel to VP. Draw its projections.*

4

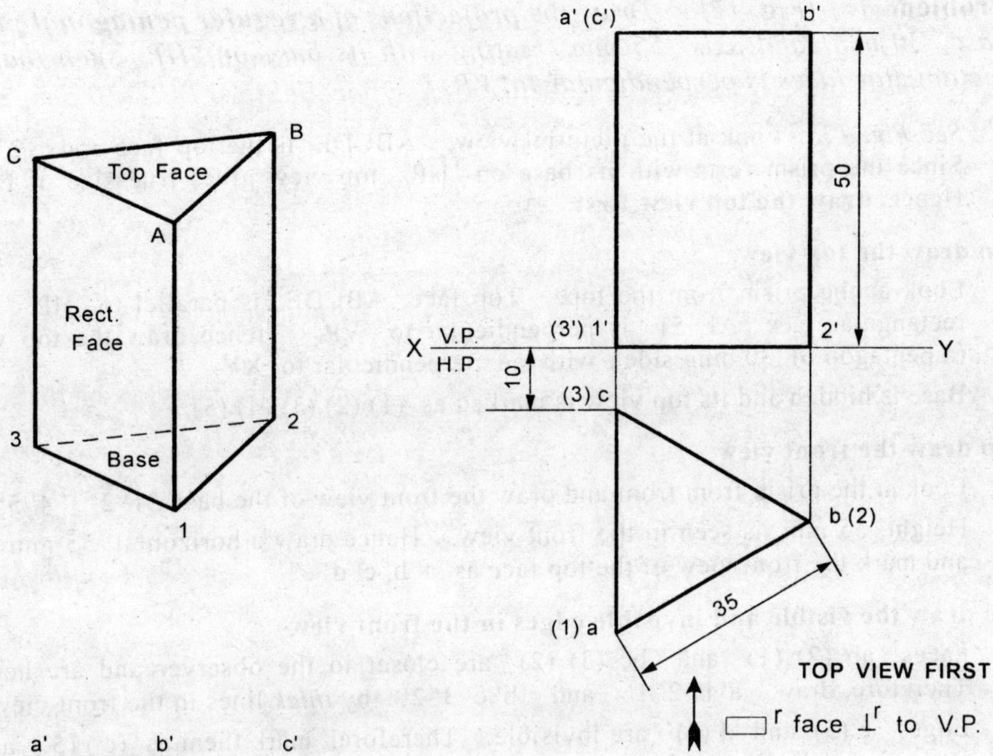

FIG. 10

5

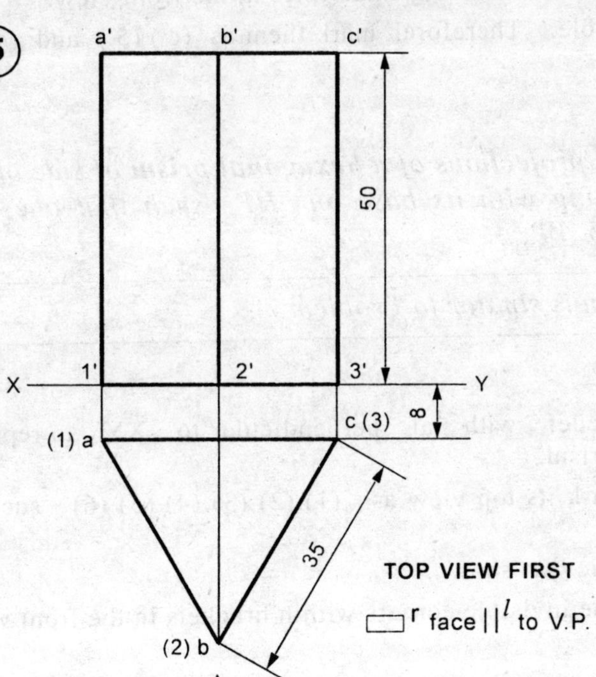

FIG. 11

Problem 7: *(Fig. 12)* *Draw the projections of a* ***regular pentagonal prism*** *side of base 30 mm and axis 55 mm resting with its base on HP such that one of its rectangular faces is perpendicular to VP.*

1. *See Fig. 12.* Look at the pictorial view. ABCDE is the top face and 12345 is base. Since the prism rests with its base on HP, top view gives true size of the end face. Hence, **draw the top view first.**

To draw the top view

2. Look at the prism from the top. Top face ABCDE is parallel to HP. Further the rectangular face AE 51 is perpendicular to VP. Hence draw the top view abcde (a pentagon of 30 mm side) with ae perpendicular to XY.
3. Base is hidden and its top view is marked as (1) (2) (3) (4) (5).

To draw the front view

4. Look at the prism from front and draw the front view of the base 1′ 2′ 3′ 4′ 5′ on XY.
5. Height 55 mm is seen in the front view. Hence draw a horizontal 55 mm above XY and mark the front view of the top face as a′ b′ c′ d′ e′.

To draw the visible and invisible edges in the front view

6. Faces ab (2) (1) and bc (3) (2) are closer to the observer and are hence visible. Therefore, draw a′ b′ 2′ 1′ and b′ c′ 3′ 2′ by *thick* lines in the front view.
7. Edges e (5) and d (4) are invisible. Therefore, mark them as (e′) (5′) and (d′) (4′) in the front view.

Problem 8: *(Fig. 13)* *Draw the projections of a* ***hexagonal prism*** *of side of base 25 mm and height 50 mm resting with its base on HP such that one of its rectangular faces is perpendicular to VP.*

NOTE : *Procedure for this Problem is similar to Problem 7.*

To draw the top view first

1. *See Fig. 13.* Draw a hexagon abcdef with af perpendicular to XY to represent the top view of the top face of the prism.
2. The base is hidden. Therefore, mark its top view as (1) (2) (3) (4) (5) (6) such that (1) and a coincide, etc.
3. **Front view:** Then draw the front view as shown.
4. **Invisible Corners:** Finally mark the invisible corners within brackets in the front view.

Problem 9: *(Exercise)* *Draw the projections of a* ***hexagonal prism****, side of base 25 mm and height 60 mm resting with its base on HP such that one of its rectangular faces is parallel to VP.*

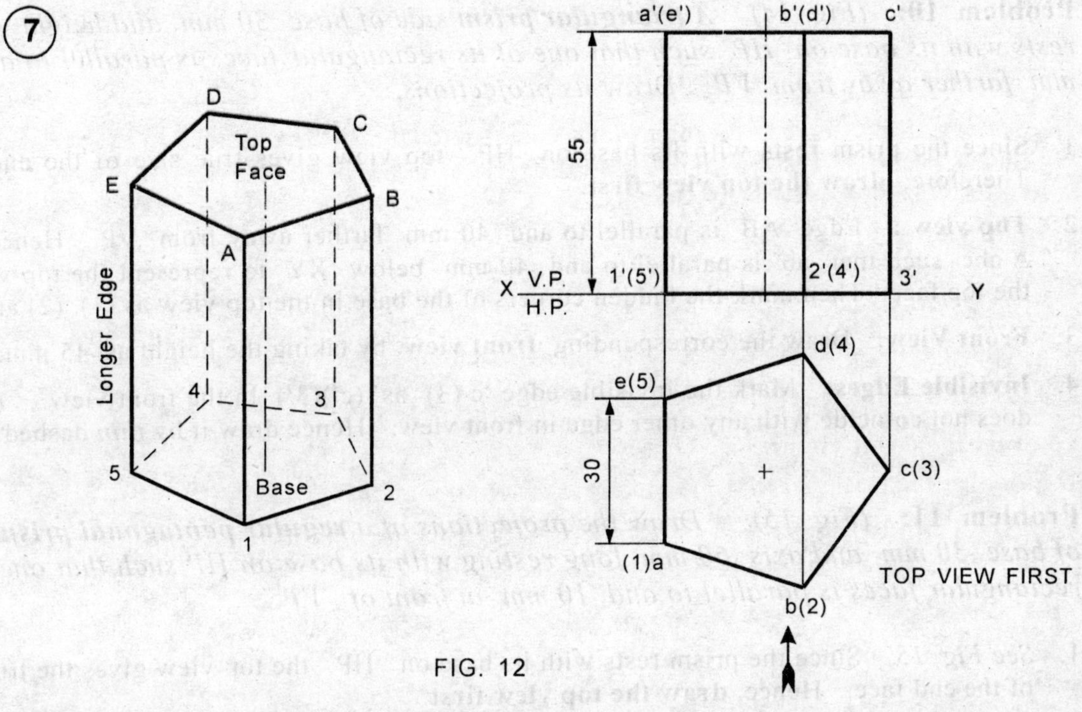
7
D
C
Top
Face
E
B
A
Longer Edge
4
3
5
2
Base
1
a'(e')
b'(d')
c'
55
X
V.P.
H.P.
1'(5')
2'(4')
3'
Y
d(4)
e(5)
30
c(3)
(1)a
b(2)
TOP VIEW FIRST

FIG. 12

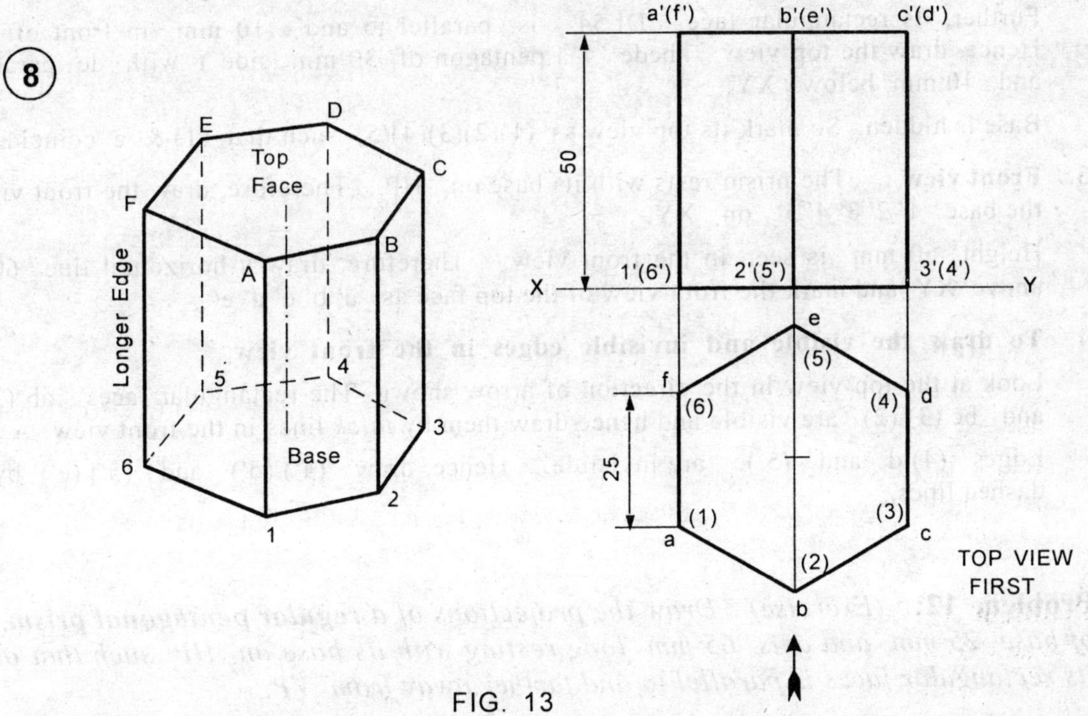
8
E
D
Top
Face
C
F
B
A
Longer Edge
5
4
3
6
Base
2
1
a'(f')
b'(e')
c'(d')
50
X
1'(6')
2'(5')
3'(4')
Y
e
(5)
f
(6)
d
(4)
25
(1)
a
(3)
c
(2)
b
TOP VIEW
FIRST

FIG. 13

Problem 10: *(Fig. 14) A **triangular prism** side of base 30 mm and height 45 mm rests with its base on HP such that one of its rectangular faces is parallel to and 40 mm farther away from VP. Draw its projections.*

1. Since the prism rests with its base on HP, top view gives true size of the end face. Therefore, **draw the top view first.**
2. **Top view :** Edge AB is parallel to and 40 mm farther away from VP. Hence draw Δ abc such that ab is parallel to and 40 mm below XY to represent the top view of the top face. Then mark the hidden corners of the base in the top view as (1), (2) and (3).
3. **Front View:** Draw the corresponding **front view** by taking the height as 45 mm.
4. **Invisible Edges:** Mark the invisible edge c (3) as (c')(3') in the front view. (c')(3') does not coincide with any other edge in front view. Hence draw it by *thin* dashed line.

Problem 11: *(Fig. 15) Draw the projections of a **regular pentagonal prism**, side of base 30 mm and axis 60 mm long resting with its base on HP such that one of its rectangular faces is parallel to and 10 mm in front of VP.*

1. *See Fig. 15.* Since the prism rests with its base on HP, the top view gives the true size of the end face. Hence, **draw the top view first.**
2. **Top view :** Imagine the given position of the prism and look from the top. Prism rests with its base on HP. Hence, the top face ABCDE is parallel to HP.

   Further, its rectangular face DE54 is parallel to and 10 mm in front of VP. Hence, draw the top view abcde (a pentagon of 30 mm side) with de parallel to and 10 mm below XY.

   Base is hidden. So mark its top view as (1)(2)(3)(4)(5) such that (1) & a coincide, etc.
3. **Front view :** The prism rests with its base on HP. Therefore, draw the front view of the base 1' 2' 3' 4' 5' on XY.

   Height 60 mm is seen in the front view. Therefore, draw a horizontal line 60 mm above XY and mark the front view of the top face as a' b' c' d' e'.
4. **To draw the visible and invisible edges in the front view**

   Look at the top view in the direction of arrow shown. The rectangular faces ab (2) (1) and bc (3) (2) are visible and hence draw them by *thick* lines in the front view.

   Edges (4) d and (5)e are invisible. Hence, draw (4') (d') and (5') (e') by *thin* dashed lines.

Problem 12: *(Exercise) Draw the projections of a **regular pentagonal prism**, side of base 25 mm and axis 65 mm long resting with its base on HP such that one of its rectangular faces is parallel to and farther away from VP.*

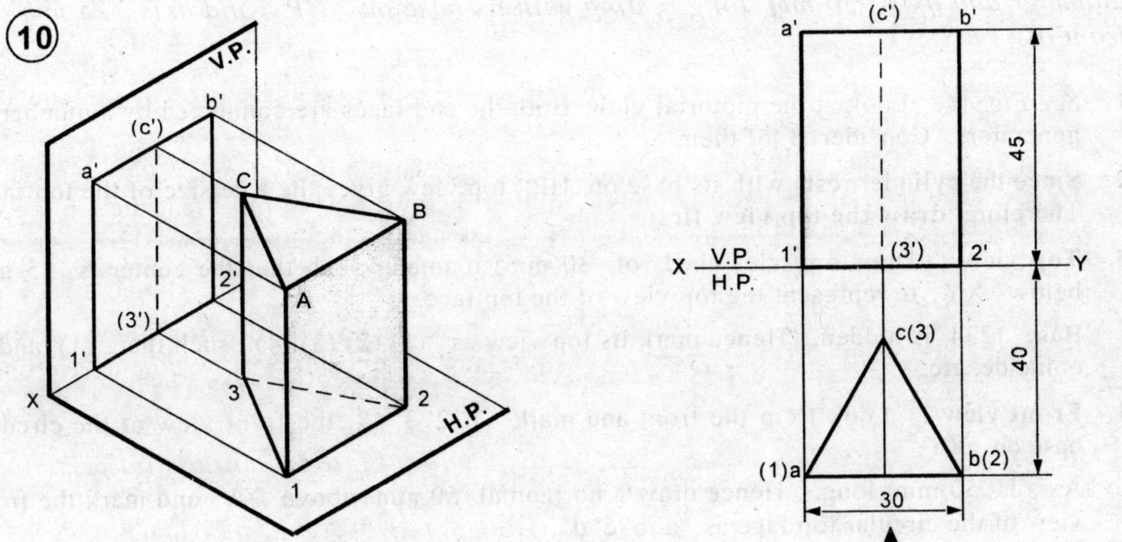

FIG. 14

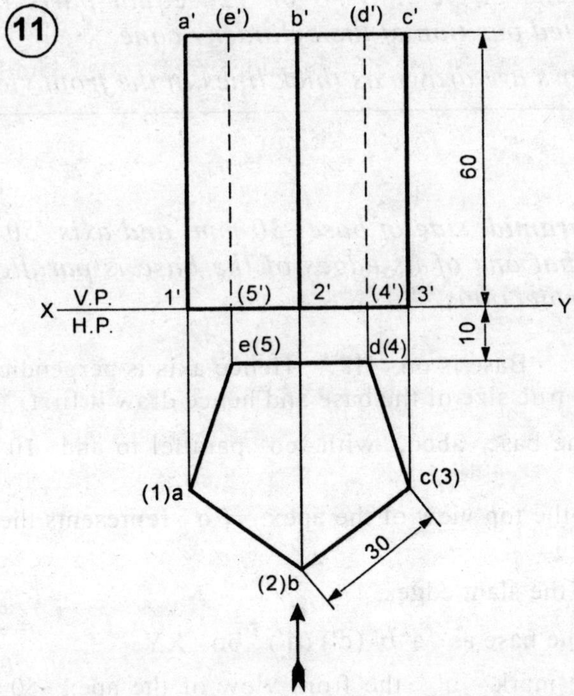

Look at the Top view in the direction of arrow.

∴ (e') (5') and (d') (4') are invisible edges in the front view.

FIG. 15

Problem 13: *(Fig. 16) Draw the projections of a* ***cylinder*** *of base 30 mm diameter and axis 50 mm long resting with its base on HP and axis 25 mm in front of VP.*

1. *See Fig.* 16. Look at the pictorial view. Both the end faces are connected by a number of generators. Consider 8 of them.
2. Since the cylinder rests with its base on HP, top view gives the true size of the top face. Therefore, **draw the top view first.**
3. **Top view :** Draw a circle abcd of 30 mm diameter such that the center is 25 mm below XY to represent the top view of the top face.

   Base 1234 is hidden. Hence mark its top view as (1) (2) (3) (4) such that (1) and a coincide, etc.
4. **Front view :** Look from the front and mark 1′ 2′ 3′ 4′, the front view of the circular base on XY.

   Axis is 50 mm long. Hence draw a horizontal 50 mm above XY and mark the front view of the circular top face as a′ b′ c′ d′.
5. **To draw the visible and invisible generators in the front view :** Generators (1)a, (2)b and (3)c are visible. Hence mark them as 1′ a′, 2′ b′ and 3′ c′ in the front view as shown. The invisible generator d (4) is marked as (d′) (4′).

NOTE : *1. The purpose of dividing the circle into 8 or 12 equal parts is to construct ellipse in the tilted position of the cylinder / cone.*

2. Only the extreme generators are drawn as thick lines in the front view.

Problem 14: *(Fig. 17)* ***A square pyramid****, side of base 30 mm and axis 50 mm long rests with its base on HP such that one of its edges of the base is parallel to and 10 mm in front of VP. Draw its projections.*

1. *See Fig.* 17. Look at the pictorial view. Base is on HP. Hence axis is perpendicular to HP. Thus, the top view gives the true size of the base and hence draw it first.
2. **Top view :** Draw the top view of the base abcd, with cd parallel to and 10 mm below XY.

   Join ac and bd to intersect at o, the top view of the apex. (o represents the top view of the axis also.)

   oa, ob, oc and od are the top view of the slant edges.
3. **Front view :** Draw the front view of the base as a′ b′ (c′) (d′) on XY.

   Length of the axis is 50 mm. Hence mark o′ the front view of the apex 50 mm above XY on the projector from o. Join o′ with b′ (c′) and d′ (a′).

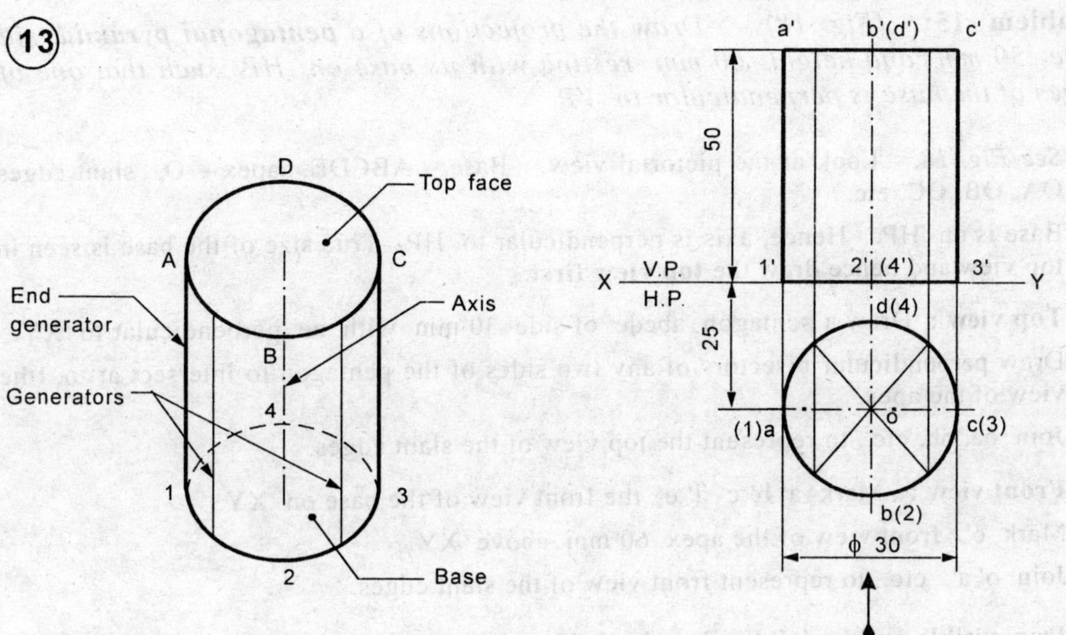
13
D
Top face
A
C
End generator
Axis
B
Generators
4
1
3
2
Base
a'
b'(d')
c'
50
X
V.P.
H.P.
1'
2'(4')
3'
Y
d(4)
25
(1)a
o
c(3)
b(2)
φ 30

FIG. 16

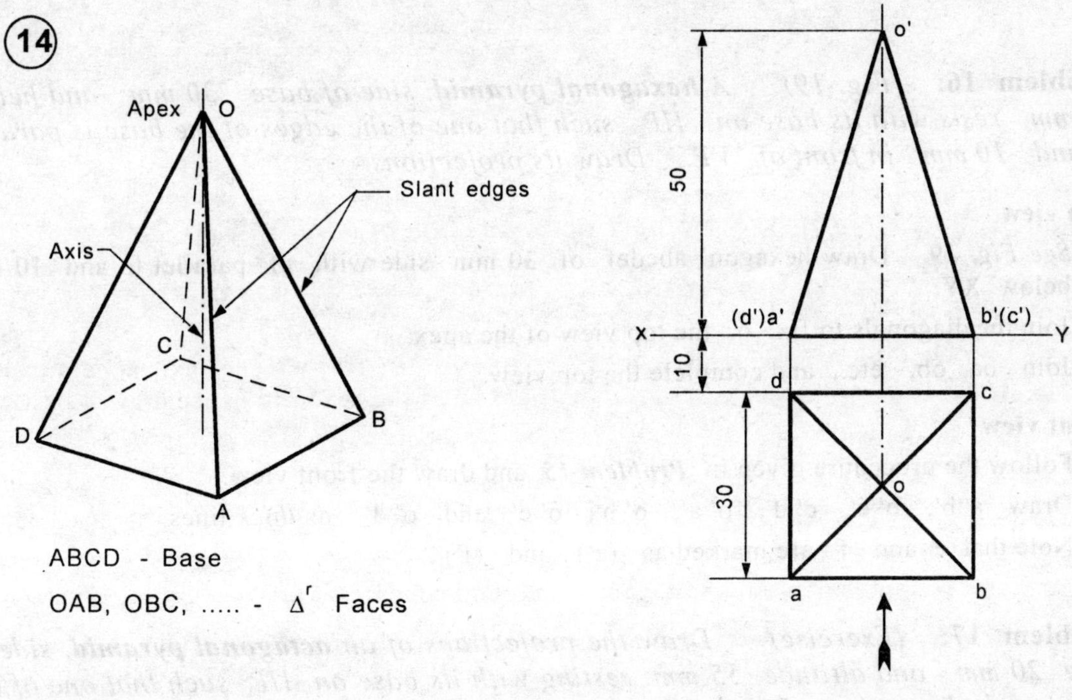
14
Apex O
Slant edges
Axis
C
D
B
A
ABCD - Base
OAB, OBC, ... - Δr Faces
o'
50
X
(d')a'
b'(c')
Y
10
d
c
30
o
a
b

FIG. 17

Problem 15: *(Fig. 18) Draw the projections of a* ***pentagonal pyramid****, side of base 30 mm and height 60 mm resting with its base on HP such that one of the edges of the base is perpendicular to VP.*

1. *See Fig. 18.* Look at the pictorial view. Base – ABCDE, apex – O, slant edges are OA, OB, OC, etc.
2. Base is on HP. Hence, axis is perpendicular to HP. True size of the base is seen in the top view and hence draw the **top view first**.
3. **Top view :** Draw a pentagon abcde of side 30 mm with ae perpendicular to XY.

   Draw perpendicular bisectors of any two sides of the pentagon to intersect at o, the top view of the apex.

   Join oa, ob, etc., to represent the top view of the slant edges.
4. **Front view :** Mark a′ b′ c′ d′ e′ the front view of the base on XY.

   Mark o′, front view of the apex 60 mm above XY.

   Join o′ a′, etc., to represent front view of the slant edges.

To draw visible and invisible edges in the front view

5. ab, bc – edges of base in top view are visible. So draw a′ b′ and c′ d′ as *thick* lines.

   oa, ob and oc are visible and hence draw o′ a′, o′ b′ and o′ c′ as *thick* lines.

   Since ed, dc are not visible, mark them as (e′) (d′) and (d′) (c′).

Problem 16: *(Fig. 19) A* ***hexagonal pyramid****, side of base 30 mm and height 60 mm rests with its base on HP such that one of the edges of the base is parallel to and 10 mm in front of VP. Draw its projections.*

Top view

1. *See Fig. 19.* Draw hexagon abcdef of 30 mm side with ef parallel to and 10 mm below XY.
2. Join the diagonals to fix o, the top view of the apex.
3. Join oa, ob, etc., and complete the top view.

Front view

4. Follow the procedure given in ***Problem 15*** and draw the front view.
5. Draw a′ b′, b′ c′, c′ d′, o′ a′, o′ b′, o′ c′ and o′ d′ as *thick* lines.
6. Note that e and f are marked as (e′) and (f′) .

Problem 17: *(Exercise) Draw the projections of an* ***octagonal pyramid****, side of base 20 mm and altitude 55 mm resting with its base on HP such that one of the edges of the base is parallel to VP.*

15

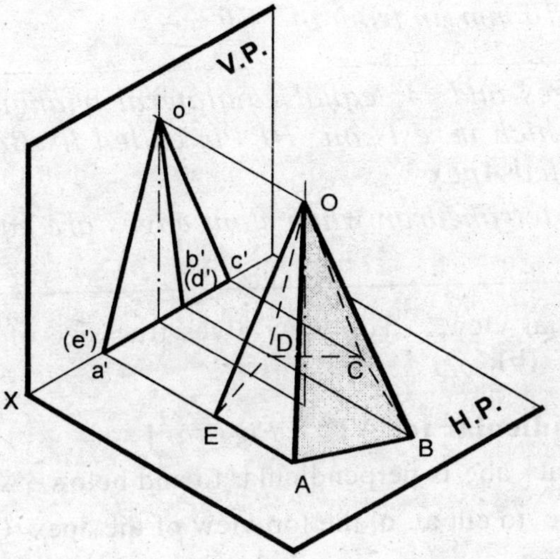

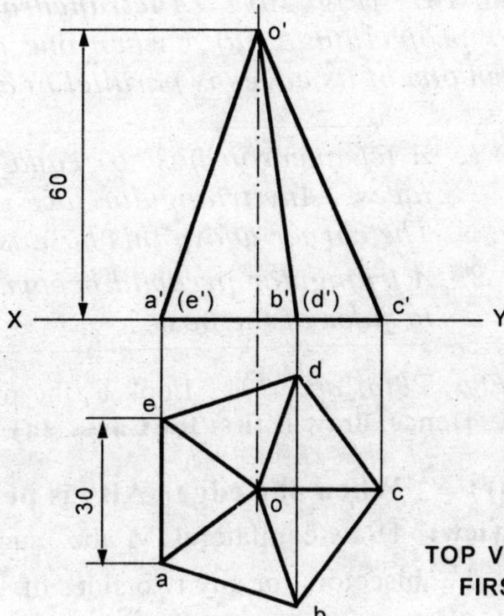

FIG. 18

16

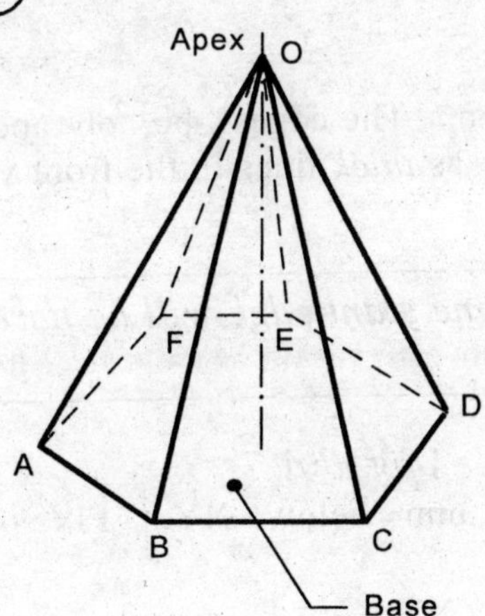

OA, OB, ... - Slant edges

OAB, OBC, ... - Δ^r Faces

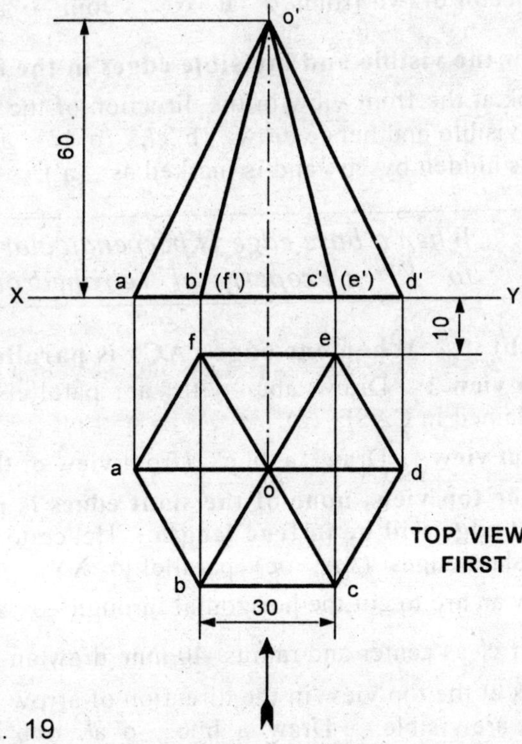

FIG. 19

Problem 18: *(Fig. 20)* *A* ***tetrahedron*** *of side 40 mm rests with its base on HP. Draw its projections* ***(a)*** *when one of its edges is perpendicular to VP and* ***(b)*** *when one of its edges is parallel to and 10 mm in front of VP.*

NOTE : *1. A tetrahedron has 6 equal edges and 4 equal equilateral triangular faces. Any triangular face on which it rests on HP is called its* ***Base****. The corner above this base is called* ***Apex****.*

2. A triangular pyramid becomes a tetrahedron when slant edges are equal to sides of the base.

---

1. *See Figs. 20(a) and (b).* Look at the pictorial view. Top view gives true size of the base. Hence, draw it first for **Cases (a) and (b).**

CASE (a) : When one edge AB is perpendicular to VP [*Fig. (a)*]

2. **Top view:** Draw equilateral Δ abc such that ab is perpendicular to and below XY.
3. Draw ⊥[r] bisectors for any two sides of Δ abc to cut at o, the top view of the apex O.
4. Join oa, ob and oc (top view of the slant edges).
5. **Front view :** Mark a′ b′ c′, the front view of the base on XY. Since ab is perpendicular to XY in the top view, a′ and b′ coincide in the front view.
6. **In the top view, oc is parallel to XY. So its front view will be in true length.** Therefore, with c′ as center and radius 40 mm (one side), draw an arc to cut the projector drawn from o at o. Join o′ b′ and o′ c′.

To draw the visible and invisible edges in the front view

7. Look at the front view in the direction of the arrow shown. The edges bc, ob and oc are visible and hence draw b′ c′, o′ b′ and o′ c′ as *thick* lines in the front view. a is hidden by b and is marked as (a′).

NOTE : *When a base edge is perpendicular to VP, one slant edge will be parallel to VP (Property of Tetrahedron).*

CASE (b) : When one edge AC is parallel to VP [*Fig. (b)*]

1. **Top view :** Draw abc with ac parallel to and 10 mm below XY. Fix o as explained in CASE (a).
2. **Front view :** Draw a′ b′ c′ (front view of the base) on XY.
3. **In the top view, none of the slant edges is parallel to XY. So in the front view no slant edge will be in true length.** Hence to mark o′ turn the top view of any one of the slant edges (*say*, oc) parallel to XY. *i.e.*, with o as center and oc as radius draw an arc to cut the horizontal through o at c_1. Project c_1 and obtain c_1' on XY.
4. With c_1' as center and radius 40 mm draw an arc to cut the projector drawn from o at o.
5. Look at the top view in the direction of arrow shown. All the corners of the base and the apex are visible. Draw a′ b′ c′, o′ a′, o′ b′ and o′ c′ by *thick* lines.

18

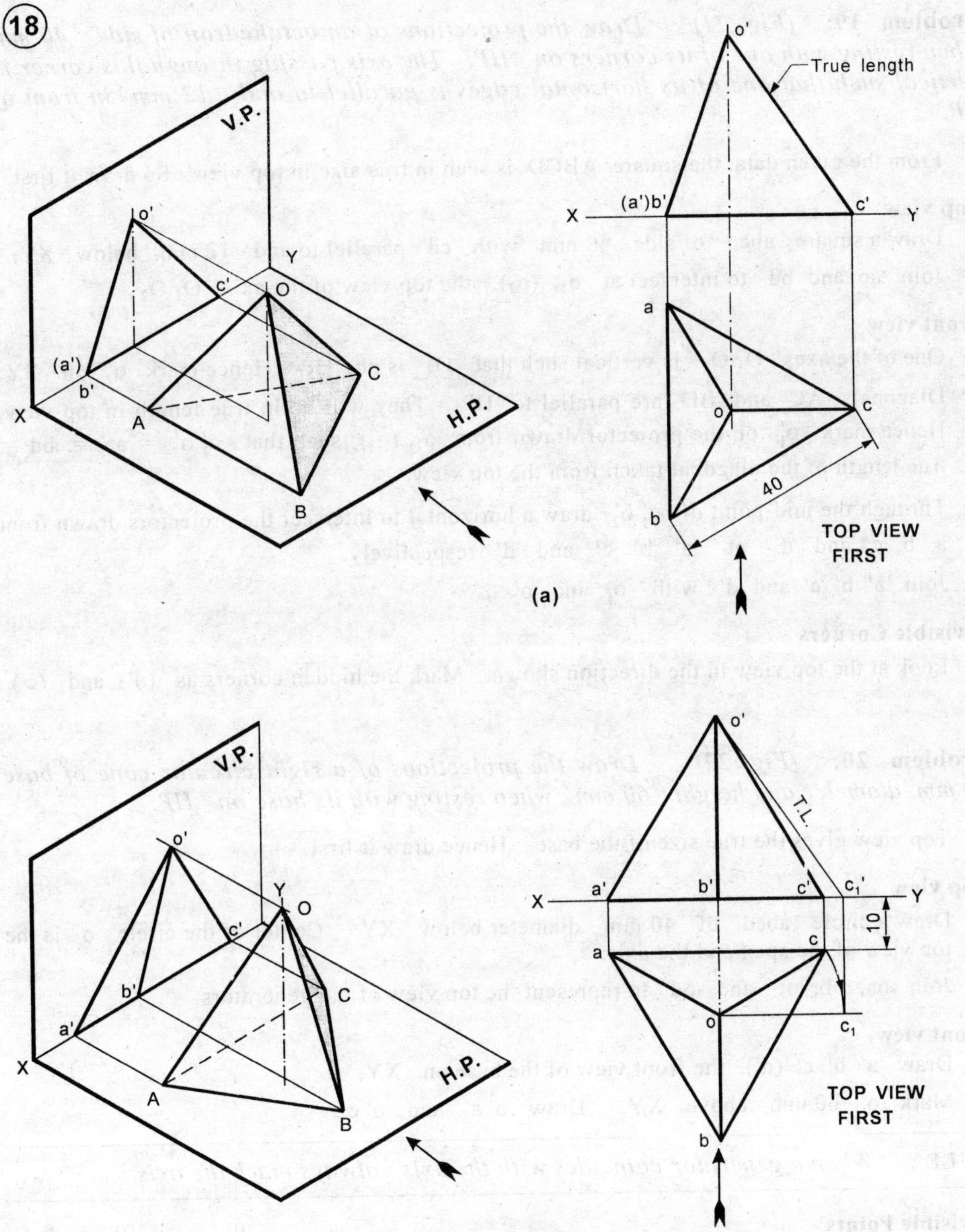
V.P.
H.P.
X
Y
o'
(a')
b'
c'
O
A
B
C
True length
(a')b'
c'
o
a
b
c
40
TOP VIEW
FIRST
(a)
T.L.
a'
b'
c'
c_1'
10
c_1
(b)

FIG. 20

Problem 19: *(Fig. 21)* *Draw the projections of an* ***octahedron*** *of side 40 mm when resting with one of its corners on HP. The axis passing through this corner is vertical such that one of its horizontal edges is parallel to and 12 mm in front of VP.*

1. From the given data, the square ABCD is seen in true size in top view. So draw it first.

Top view

2. Draw a square abcd of side 40 mm with cd parallel to and 12 mm below XY.
3. Join ac and bd to intersect at o_1, (o_2), the top view of the axis $O_1 O_2$.

Front view

4. One of the axes $O_1 O_2$ is vertical such that O_2 is on HP. Hence mark o_2' on XY.
5. Diagonals AC and BD are parallel to HP. They will be in true length in top view. Hence mark o_1' on the projector drawn from o_1, (o_2) such that $o_1' o_2'$ = ac = bd = true length of the diagonal taken from the top view.
6. Through the mid-point of $o_1' o_2'$ draw a horizontal to intersect the projectors drawn from a, b, c and d at a′ b′ c′ and d′ respectively.
7. Join a′ b′ c′ and d′ with o_1' and o_2'.

Invisible Corners

8. Look at the top view in the direction shown. Mark the hidden corners as (d′) and (c′).

Problem 20: *(Fig. 22)* *Draw the projections of a right circular cone of base 40 mm diameter and height 60 mm when resting with its base on HP.*

1. Top view gives the true size of the base. Hence draw it first.

Top view

2. Draw a circle abcd of 40 mm diameter below XY. Center of the circle o is the top view of the apex and the axis.
3. Join oa, ob, oc and od to represent the top view of the generators.

Front view

4. Draw a′ b′ c′ (d′), the front view of the base on XY.
5. Mark o′ 60 mm above XY. Draw o′ a′ and o′ c′.

RULE : *When a generator coincides with the axis, always mark the axis.*

Invisible Points

6. Look at the top view in the direction shown. od is invisible. Hence, in the front view mark the invisible point as (d′). Also o′ (d′) coincides with o′ b′.

(19) OCTAHEDRON

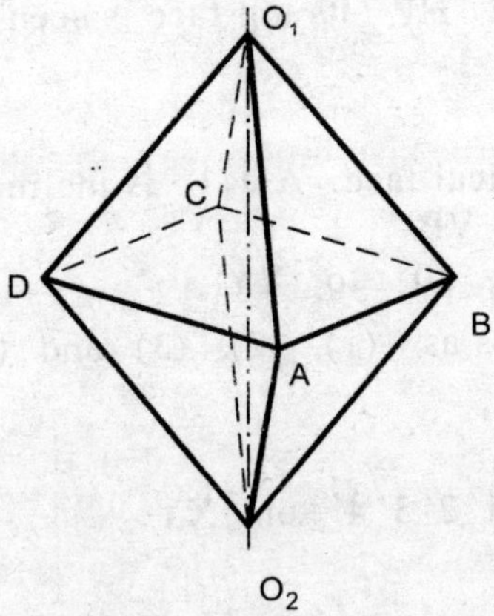

Simple position of an Octahedron means it rests on H.P. with its axis perpendicular to H.P.

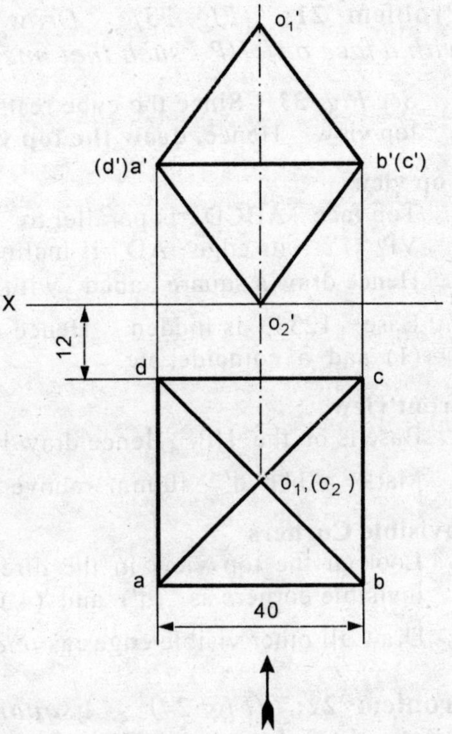

FIG. 21

(20)

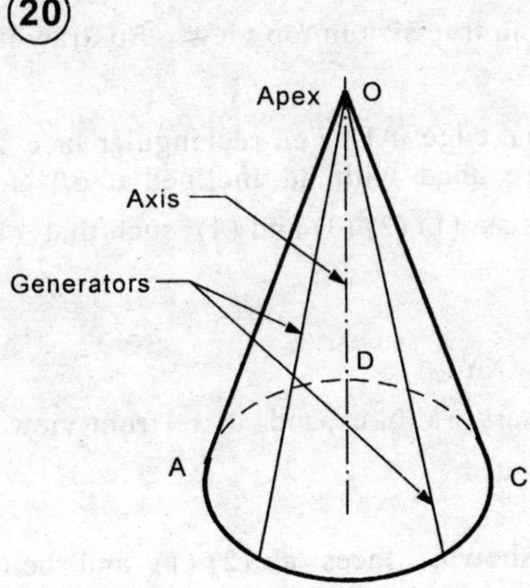

ABCD : Base
OA, OC : End generators
OB, OD, etc : Other generators

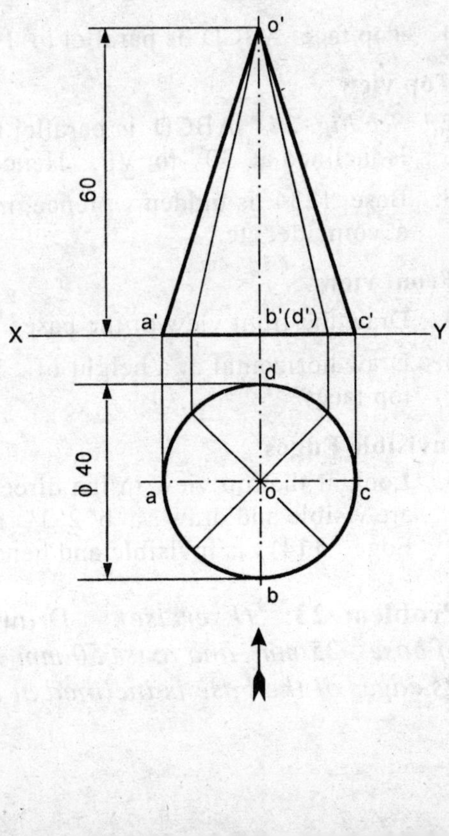

FIG. 22

Problem 21: *(Fig. 23) Draw the projections of a* ***cube*** *of 40 mm side, resting with a face on HP such that one of its vertical faces is inclined at 30° to VP.*

1. *See Fig. 23.* Since the cube rests with a face on HP its top face is seen in true size in top view. Hence, **draw the top view first.**

Top view

2. Top face ABCD is parallel to HP. The vertical face AD41 is inclined at 30° to VP *i.e.*, its edge AD is inclined at 30° to VP.
   Hence draw a square abcd with ad inclined at 30° to XY.
3. Base 1234 is hidden. Hence mark its corners as (1), (2), (3) and (4) such that (1) and a coincide, etc.

Front view

4. Base is on the HP. Hence draw its front view 1′ 2′ 3′ 4′ on XY.
5. Mark a′ b′ c′ d′ 40 mm above XY.

Invisible Corners

6. **L**ook at the top view in the direction of arrow. d (4) is invisible. Hence, mark invisible corners as (d′) and (4′) in the front view and join them by a *thin* dashed line.
7. Draw all other visible edges as *thick* lines.

Problem 22: *(Fig. 24) A* ***square prism****, side of base 35 mm and height 50 mm rests with its base on HP such that one of its rectangular faces is inclined at an angle of 30° to VP. Draw its projections.*

1. Top face ABCD is parallel to HP and is seen in true size in top view. So draw it first.

Top view

2. *See Fig. 24.* ABCD is parallel to HP such that edge AD (*i.e.*, rectangular face AD41) is inclined at 30° to VP. Hence, draw a square abcd with ad inclined at 30° to XY.
3. Base 1234 is hidden. Hence, mark its corners as (1) (2) (3) and (4) such that (1) and a coincide, etc.

Front view

4. Draw the front view of the base 1′ 2′ 3′ 4′ on XY.
5. Draw horizontal at a height of 50 mm and mark a′ b′ c′ and d′ (front view of the top face).

Invisible Edges

6. **L**ook at the top view in the direction of arrow shown; faces ab (2) (1) and bc (3) (2) are visible and draw a′ b′ 2′ 1′ and b′ c′ 3′ 2′ by *thick* lines.
   Edge d (4) is invisible and hence draw (d′) (4′) as *thin* dashed line.

Problem 23: *(Exercise) Draw the projections of a* ***regular hexagonal prism****, side of base 25 mm and axis 50 mm long resting with its base on HP such that one of its edges of the base is inclined at an angle of 20° to VP.*

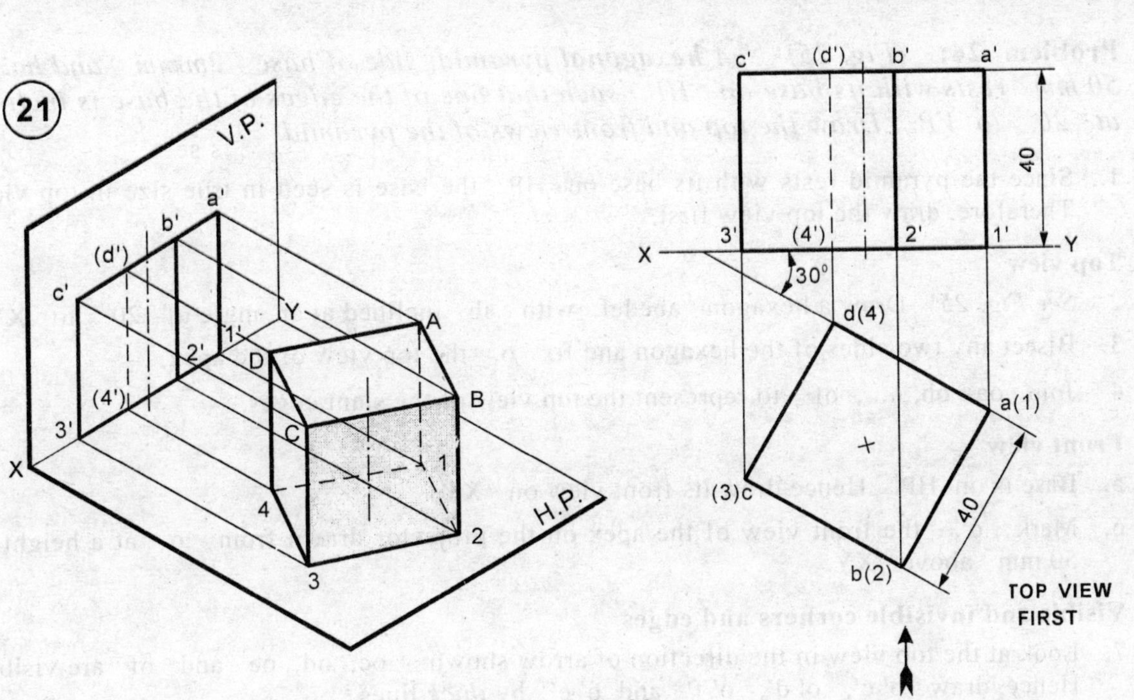
21
V.P.
H.P.
X
Y
a'
b'
c'
(d')
1'
2'
3'
(4')
A
B
C
D
1
2
3
4
40
30°
d(4)
a(1)
(3)c
b(2)
TOP VIEW
FIRST

FIG. 23

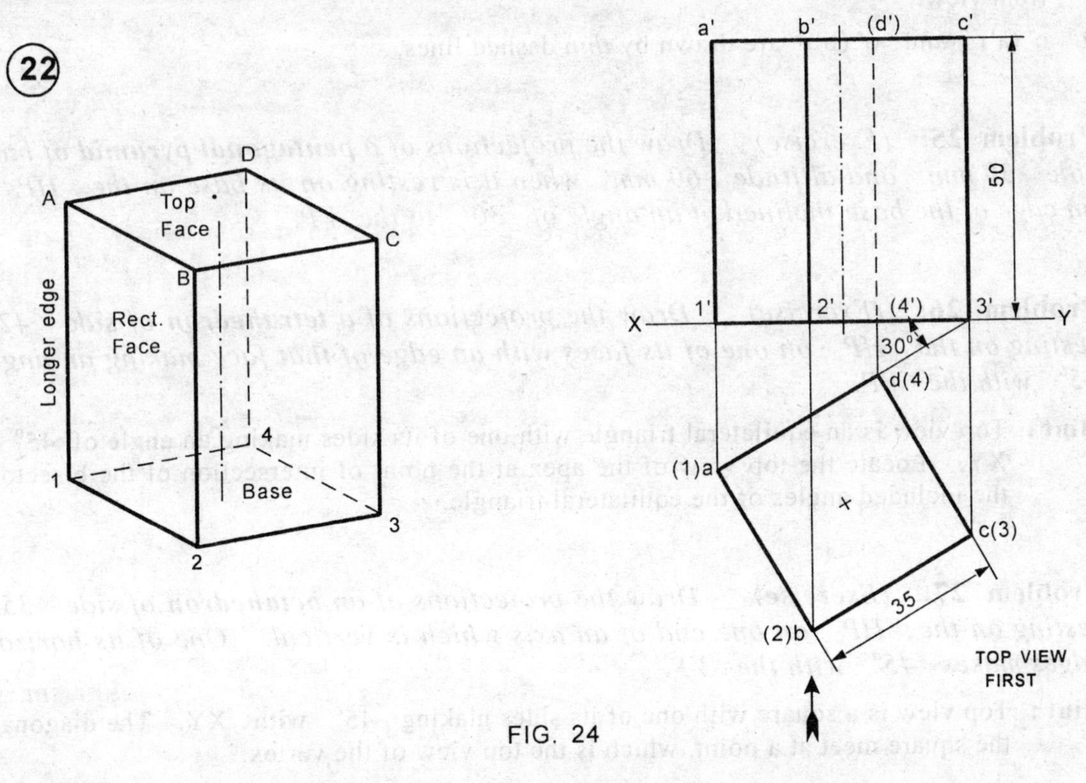
22
A
B
C
D
Top
Face
Rect.
Face
Longer edge
Base
1
2
3
4
a'
b'
(d')
c'
50
X
Y
1'
2'
(4')
3'
30°
d(4)
(1)a
c(3)
(2)b
35
TOP VIEW
FIRST

FIG. 24

Problem 24: *(Fig. 25)* *A* ***hexagonal pyramid****, side of base 25 mm and height 50 mm rests with its base on HP such that one of the edges of the base is inclined at 20° to VP. Draw the top and front views of the pyramid.*

1. Since the pyramid rests with its base on HP, the base is seen in true size in top view. Therefore, draw the top view first.

Top view

2. *See Fig. 25.* Draw a hexagon abcdef with ab inclined at an angle of 20° to XY.
3. Bisect any two sides of the hexagon and fix o, the top view of the apex.
4. Join oa, ob, ..., of to represent the top view of the slant edges.

Front view

5. Base is on HP. Hence draw its front view on XY.
6. Mark o′, the front view of the apex on the projector drawn from o at a height of 50 mm above XY.

Visible and invisible corners and edges

7. Look at the top view in the direction of arrow shown. oc, od, oe and of are visible. Hence, draw o′ c′, o′ d′, o′ f′ and o′ e′ by *thick* lines.
8. oa and ob are invisible. Hence mark the invisible corners as (a′) and (b′) in the front view.
9. o′ (a′) and o′ (b′) are drawn by *thin* dashed lines.

Problem 25: *(Exercise)* *Draw the projections of a* ***pentagonal pyramid*** *of base of side 30 mm and altitude 60 mm when it is resting on its base on the HP with an edge of the base inclined at an angle of 30° to the VP.*

Problem 26: *(Exercise)* *Draw the projections of a* ***tetrahedron*** *of side 42 mm resting on the HP on one of its faces with an edge of that face making an angle of 45° with the VP.*

Hint : Top view is an equilateral triangle with one of its sides making an angle of 45° with XY. Locate the top view of the apex at the point of intersection of the bisectors of the included angles of the equilateral triangle.

Problem 27: *(Exercise)* *Draw the projections of an* ***octahedron*** *of side 35 mm resting on the HP on one end of an axis which is vertical. One of its horizontal edges makes 45° with the VP.*

Hint : Top view is a square with one of its sides making 45° with XY. The diagonals of the square meet at a point, which is the top view of the vertex.

24

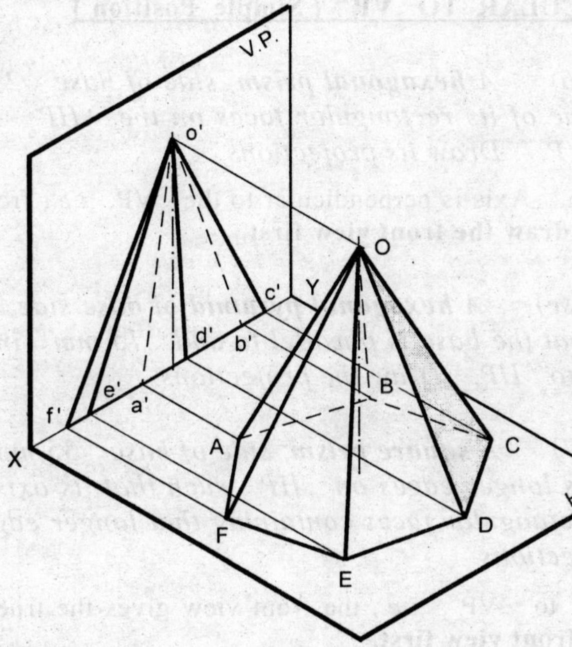
V.P.
o'
O
c'
Y
d'
b'
e'
a'
B
f'
A
C
X
H.P.
F
D
E

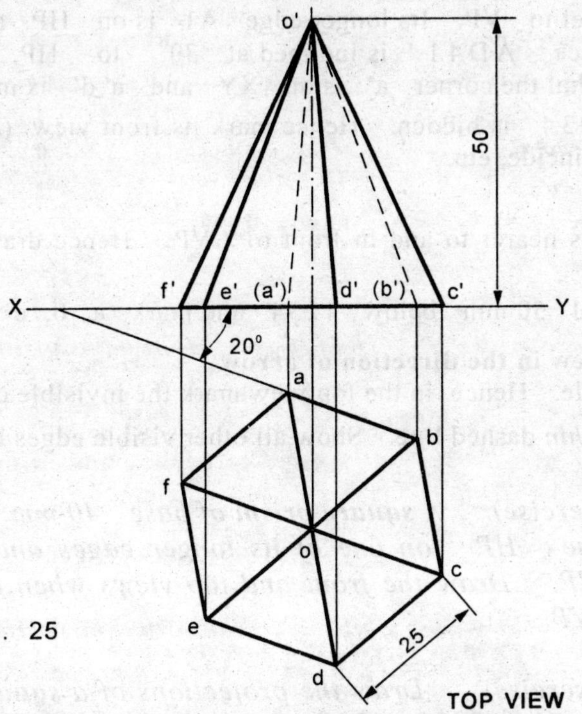
o'
50
f'
e'
(a')
d'
(b')
c'
X
Y
20°
a
b
f
o
c
e
25
d
TOP VIEW
FIRST

FIG. 25

4.2 AXIS PERPENDICULAR TO VP (Simple Position)

Problem 28: *(Fig. 26) A* ***hexagonal prism****, side of base 25 mm and axis 60 mm long, lies with one of its rectangular faces on the HP such that the axis is perpendicular to the VP. Draw its projections.*

See Fig. 26 for Solution. Axis is perpendicular to the VP, *i.e.*, front view gives the true size of the base. Hence, **draw the front view first.**

Problem 29: *(Exercise) A* ***hexagonal pyramid*** *of base side 25 mm and height 65 mm is kept such that the base is parallel to and 15 mm in front of VP with an edge perpendicular to HP. Draw the projections.*

Problem 30: *(Fig. 27) A* ***square prism****, side of base 35 mm and axis 50 mm long lies with one of its longer edges on HP such that its axis is perpendicular to VP. Also one of its rectangular faces containing that longer edge is inclined at 30° to HP. Draw its projections.*

1. Axis is perpendicular to VP , *i.e.*, the front view gives the true size of the end face. Therefore, **draw the front view first.**

Front view

2. ABCD is parallel to VP. Its longer edge A1 is on HP and perpendicular to VP. Its rectangular face AD41 is inclined at 30° to HP. Hence, draw a square a′ b′ c′ d′ such that the corner a′ is on XY and a′ d′ is inclined at 30° to XY.
3. *See Fig. 27.* 1234 is hidden. Hence mark its front view (1′) (2′) (3′) (4′) such that (1′) and a′ coincide, etc.

Top view

4. Face 1234 is nearer to and in front of VP. Hence draw its top view below and **nearer to XY.**
5. Draw a horizontal 50 mm below 1234 and mark a, b, c and d.

Look at the front view in the direction of arrow

6. (1′) a′ is invisible. Hence, in the top view mark the invisible corners as (1) and (a).
7. Join (1) (a) by *thin* dashed line. Show all other visible edges by *thick* lines in top view.

Problem 31: *(Exercise) A* ***square prism*** *of base 40 mm edge and axis 70 mm long is lying on the HP on one of its longer edges and with its faces equally inclined to the HP. Draw the front and top views when the axis of the prism is perpendicular to VP.*

Problem 32: *(Exercise) Draw the projections of a* ***square pyramid*** *of base 40 mm side and axis 60 mm long, has its base in the VP. One edge of the base is inclined at 30° to the HP and a corner contained by that edge is on the HP.*

28

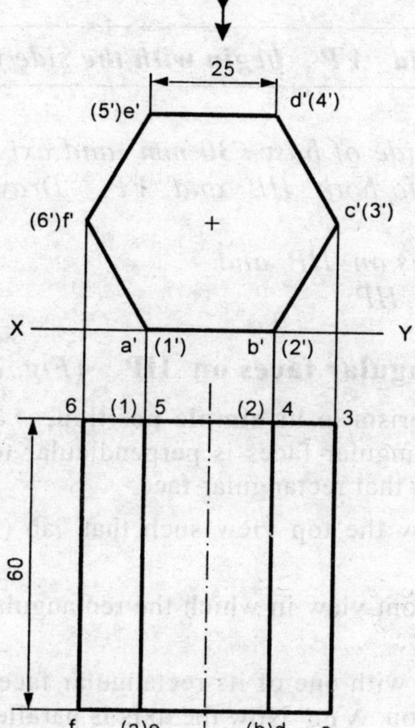

FIG. 26

30

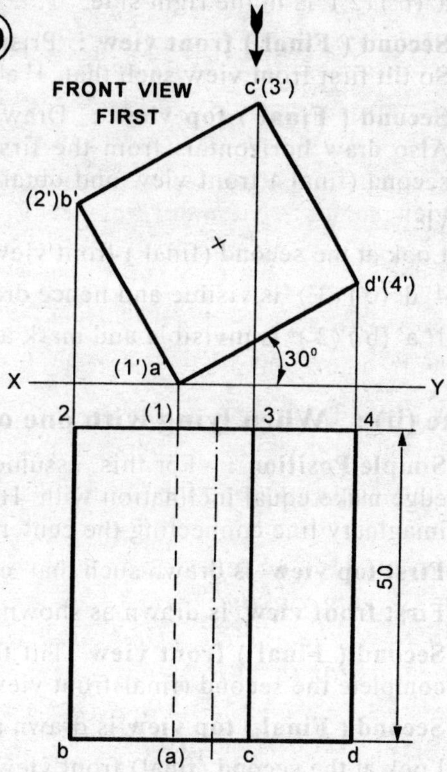

FIG. 27

4.3 AXIS PARALLEL TO BOTH HP AND VP (Simple Position)

RULE: ***As the axis is parallel to both HP and VP, begin with the side view.***

Problem 33: *(Fig. 28)* *A* ***square prism****, side of base 30 mm and axis 45 mm long lies on HP such that its axis is parallel to both HP and VP. Draw the top and front views of the prism, when*

i) it lies with one of its rectangular faces on HP and

ii) it lies with one of its longer edges on HP. (UQ)

Case (i) : When lying with one of its rectangular faces on HP *(Fig. a)*

1. **Simple Position :** For this, first assume the prism in its **simple position**, *i.e.*, resting with its base on HP such that one of its rectangular faces is perpendicular to VP to enable to tilt the prism on the right side to lie on that rectangular face.
2. **First top view :** For the simple position draw the top view such that ab (2) (1) is perpendicular to XY and on the right side.
3. **First front view :** Project the corresponding front view in which the rectangular face 1′ a′ (b′) (2′) is in the right side.
4. **Second (Final) front view :** Prism has to lie with one of its rectangular faces on HP. So tilt first front view such that 1′ a′(b′)(2′) is on XY. Now the axis is parallel to XY.
5. **Second (Final) top view :** Draw the projectors from the second (final) front view. Also draw horizontals from the first top view to cut the respective projectors from the second (final) front view and obtain the corresponding points of the second (final) top view.
6. Look at the second (final) front view in the direction of arrow.

   4′ d′ (c′) (3′) is visible and hence draw 4_1 d_1 c_1 3_1 as *thick* line.

   1′ a′ (b′) (2′) is invisible and mark as (1_1) (a_1) (b_1) (2_1).

Case (ii) : When lying with one of its longer edges on HP *(Fig. b)*

1. **Simple Position :** For this, assume that the two rectangular faces containing the longer edge make equal inclination with HP. So while drawing the top view in simple position, imaginary line connecting the center of the base with its corner should be parallel to VP.
2. **First top view** is drawn such that c(3) is on the right side and o3 is parallel to VP.
3. **First front view** is drawn as shown.
4. **Second (Final) front view** Tilt the first front view such that 3′ c′ lies on XY and complete the second (final)front view.
5. **Second (Final) top view** is drawn as explained in ***Case (I)***.
6. Look at the second (final) front view in the direction of arrow shown.

   Mark the invisible corners as (3_1) and (c_1).

(33)

AXIS PARALLEL TO BOTH H.P. AND V.P.

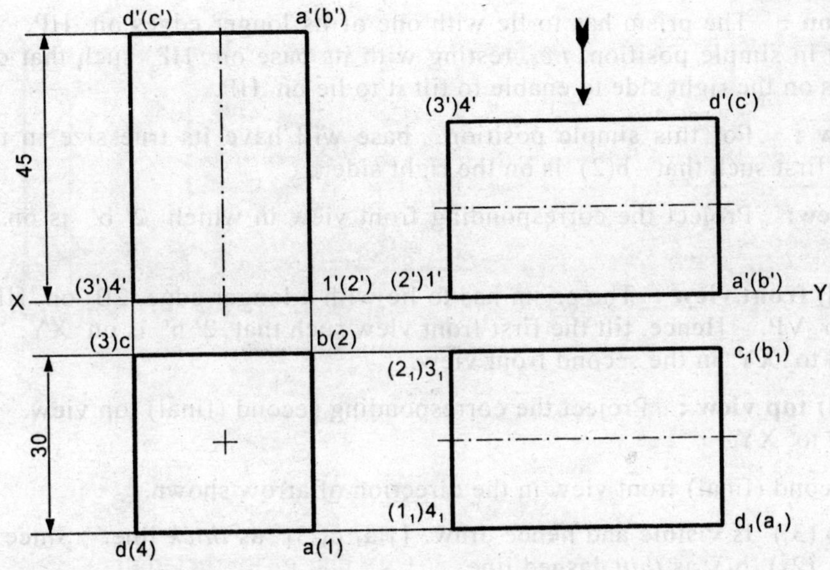

Case I

FIG. 28 (a)

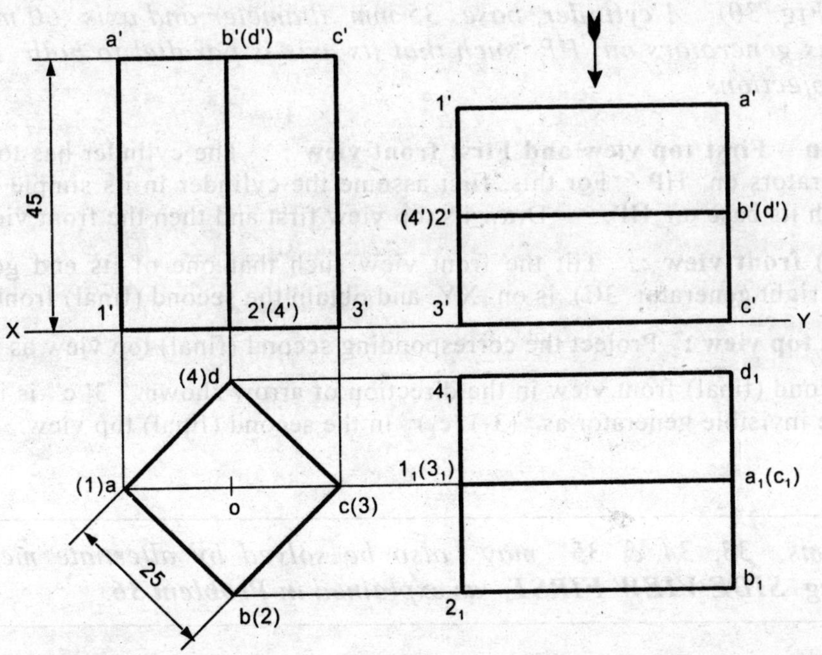

Case II

FIG. 28 (b)

Problem 34: *(Fig. 29)* *A* ***triangular prism****, side of base 35 mm and height 60 mm lies with one of its longer edges on HP such that its axis is parallel to both HP and VP. Draw its projections.* (UQ)

1. **Simple Position :** The prism has to lie with one of its longer edges on HP. For this, first assume it in simple position, *i.e.*, resting with its base on HP such that one of its longer edges is on the right side to enable to tilt it to lie on HP.
2. **First top view :** For this simple position, base will have its true size in top view. Hence draw it first such that b(2) is on the right side.
3. **First front view:** Project the corresponding front view in which 2′ b′ is on the right side.
4. **Second (Final) front view :** The prism has to lie with a longer edge 2B on HP and its axis parallel to VP. Hence, tilt the first front view such that 2′ b′ is on XY. Now the axis is parallel to XY in the second front view.
5. **Second (Final) top view :** Project the corresponding second (final) top view. Now the axis is parallel to XY.
6. Look at the second (final) front view in the direction of arrow shown.

   Face 1′ a′ (c′) (3′) is visible and hence draw 1_1 a_1 c_1 3_1 as *thick* line. Since 2′ b′ is invisible draw (2_1) (b_1) as *thin* dashed line.

Problem 35: *(Fig. 30)* *A* ***cylinder****, base 35 mm diameter and axis 60 mm long lies with one of its generators on HP such that its axis is parallel to both HP and VP. Draw its projections.*

1. **Simple Position – First top view and First front view :** The cylinder has to lie with one of its generators on HP. For this, first assume the cylinder in its simple position, *i.e.*, resting with its base on HP. Draw its top view first and then the front view.
2. **Second (Final) front view :** Tilt the front view such that one of its end generators (preferably the right generator 3C) is on XY and obtain the second (final) front view.
3. **Second (Final) top view :** Project the corresponding second (final) top view as shown.
4. Look at the second (final) front view in the direction of arrow shown. 3′ c′ is invisible. Hence mark the invisible generator as (3_1) (c_1) in the second (final) top view.

NOTE : *Problems 33, 34 & 35 may also be solved by alternate method by drawing* ***SIDE VIEW FIRST****, as explained in* ***Problem 36****.*

34

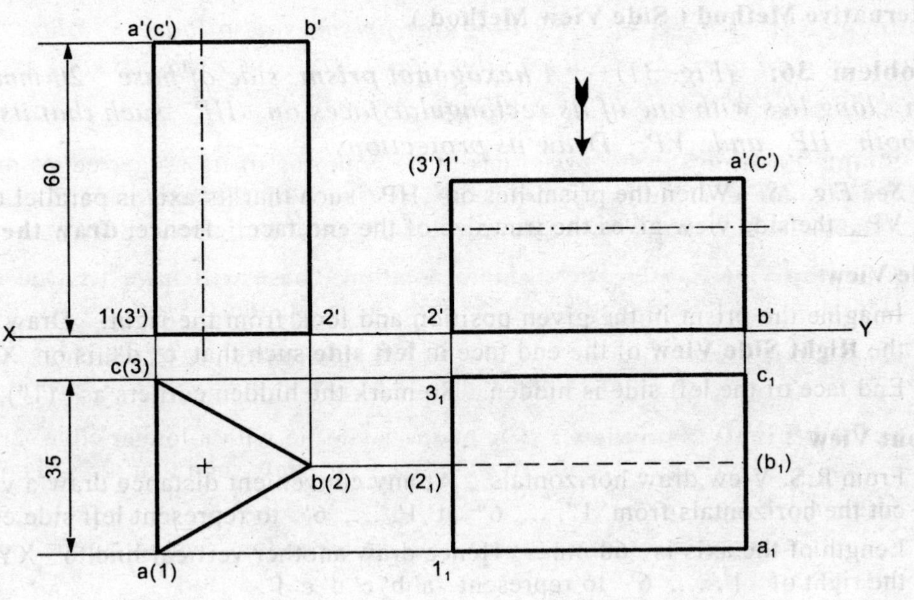

(i) Simple Position (ii) Final Position

FIG. 29

35

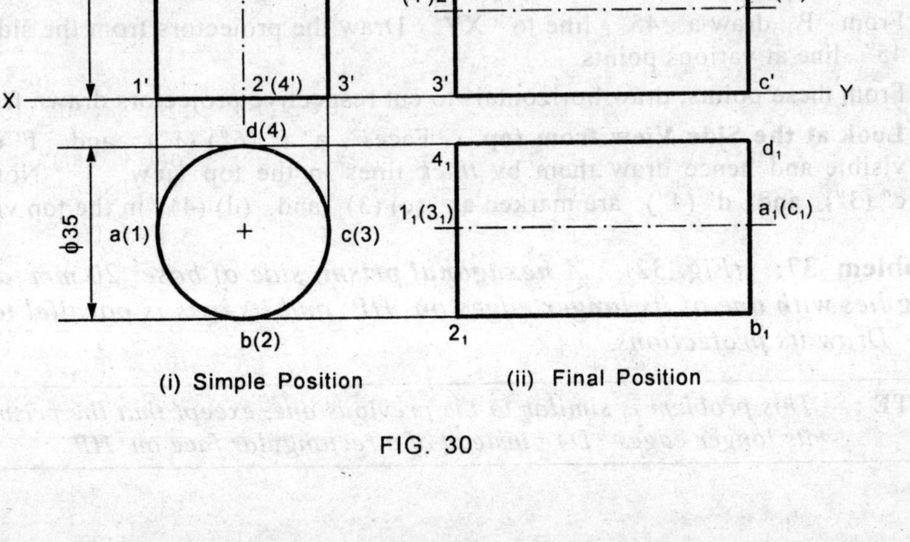

(i) Simple Position (ii) Final Position

FIG. 30

Alternative Method (Side View Method)

Problem 36: *(Fig. 31)* *A **hexagonal prism**, side of base 20 mm and axis 60 mm long lies with one of its rectangular faces on HP such that its axis is parallel to both HP and VP. Draw its projections.*

1. *See Fig. 28.* When the prism lies on HP such that its axis is parallel to both HP and VP , the side view gives the true size of the end face. Hence, **draw the side view first.**

Side View

2. Imagine the prism in the given position and look from the right. Draw a″ b″ c″ d″ e″ f″, the **Right Side View** of the end face in **left side** such that c″ d″ is on XY.
3. End face of the left side is hidden. So mark the hidden corners as (1″), (2″), ..., (6″).

Front View

4. From R.S. View draw horizontals. At any convenient distance draw a vertical to XY to cut the horizontals from 1″, ..., 6″ at 1′, ..., 6′ to represent left side end face 123456.
5. Length of the axis is 60 mm. Hence draw another vertical line to XY at 60 mm to the right of 1′, ..., 6′ to represent a′ b′ c′ d′ e′ f′.

Look at the Side View from left

6. The faces a″ b″ (2″) (1″) and b″ c″ (3″) (2″) are visible. Hence draw a′ b′ 2′ 1′ and b′ c′ 3′ 2′ by *thick* line.
7. Faces f″ e″ (5″) (6″) and e″ d″ (4″) (5″) are invisible and are marked as (f′) (e′) (5′) (6′) and (e′) (d′) (4′) (5′).

Top view

8. Draw the projectors from the front view. Draw X_1Y_1 at P perpendicular to XY.

NOTE : *Distances of all corners of the prism in the top view from XY should be = the respective distances of all the corners in the side view from X_1Y_1.*

---

9. From P draw a 45° line to XY. Draw the projectors from the side view to cut the 45° line at various points.
10. From these points, draw horizontals to cut respective projectors drawn from front view.
11. **Look at the Side View from top.** Faces a″ f″ (6″) (1″) and f″ e″ (5″) (6″) are visible and hence draw them by *thick* lines in the top view. Note that the edges c″ (3″) and d″ (4″) are marked as (c) (3) and (d) (4) in the top view.

Problem 37: *(Fig. 32)* *A **hexagonal prism**, side of base 20 mm and axis 60 mm long lies with one of its longer edges on HP and its axis is parallel to both HP and VP. Draw its projections.*

NOTE : *This problem is similar to the previous one, except that the prism lies with one of its longer edges D4 instead of a rectangular face on HP.*

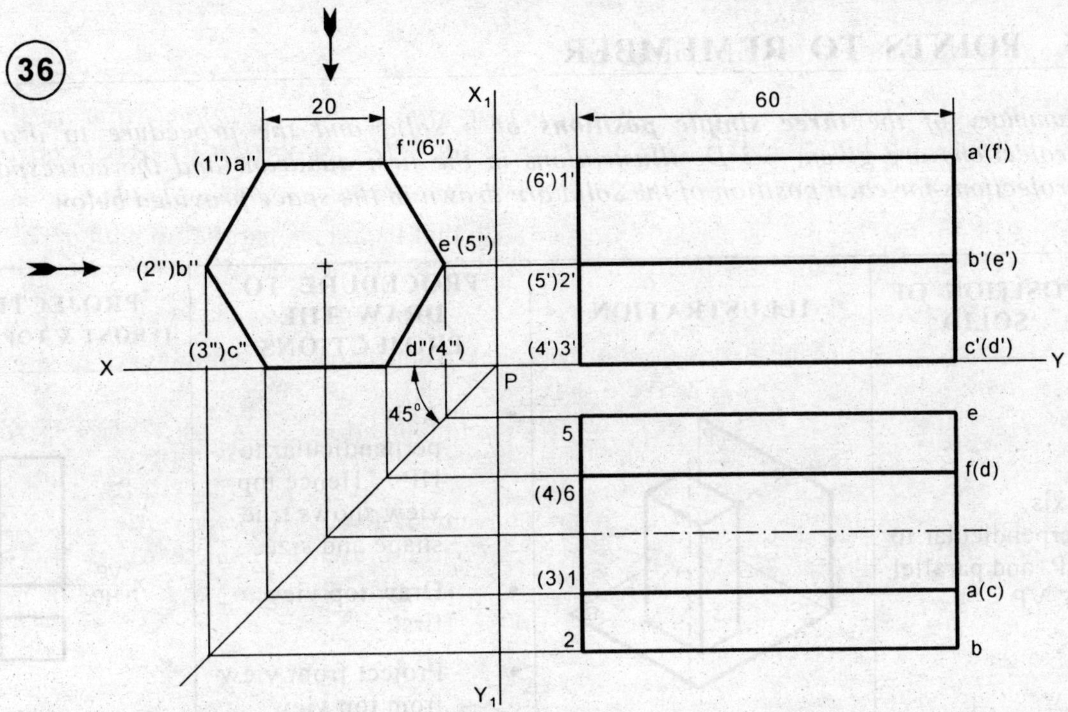
36
20
60
X1
Y1
X
Y
P
45°
(1")a"
f"(6")
(2")b"
e'(5")
(3")c"
d"(4")
(6')1'
(5')2'
(4')3'
a'(f')
b'(e')
c'(d')
5
(4)6
(3)1
2
e
f(d)
a(c)
b

FIG. 31

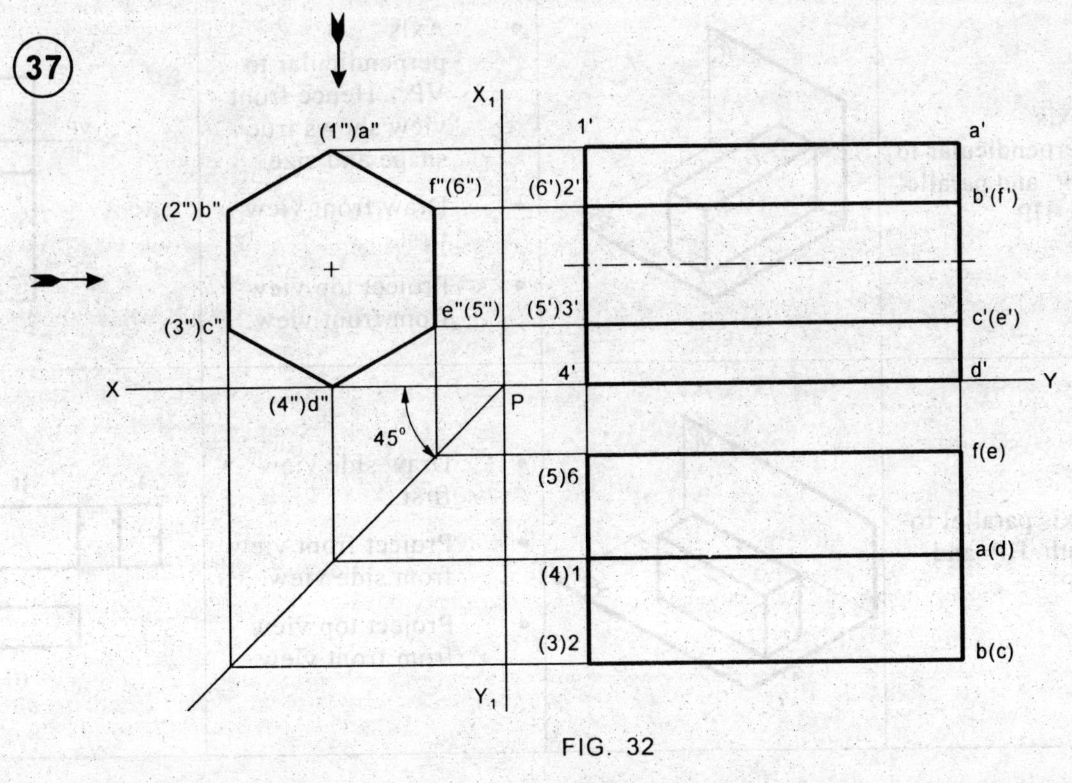
37
X1
Y1
X
Y
P
45°
(1")a"
f"(6")
(2")b"
e"(5")
(3")c"
(4")d"
1'
(6')2'
(5')3'
4'
a'
b'(f')
c'(e')
d'
(5)6
(4)1
(3)2
f(e)
a(d)
b(c)

FIG. 32

5. POINTS TO REMEMBER

Summary of the ***three simple positions*** *of a solid and the procedure to draw the projections are given. 3-D illustrations in the first quadrant and the corresponding projections for each position of the solid are drawn in the space provided below:*

POSITION OF SOLID	ILLUSTRATION	PROCEDURE TO DRAW THE PROJECTIONS	PROJECTIONS (FRONT & TOP VIEWS)
Axis perpendicular to HP and parallel to VP	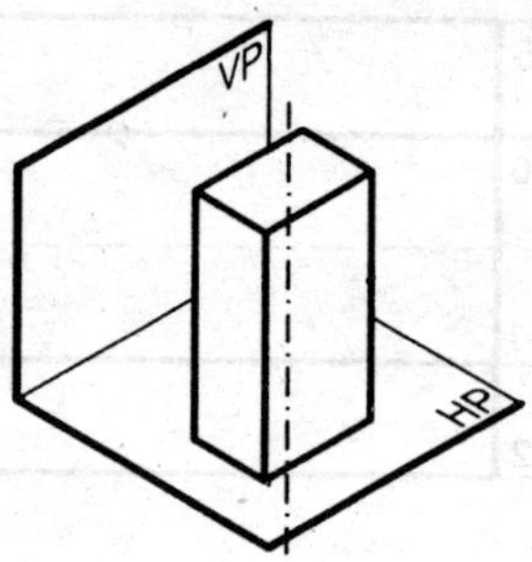	• Axis perpendicular to HP. Hence top view shows true shape and size. • Draw top view first. • Project front view from top view.	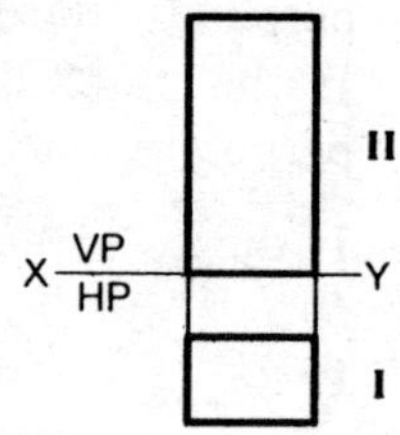
Axis perpendicular to VP and parallel to HP	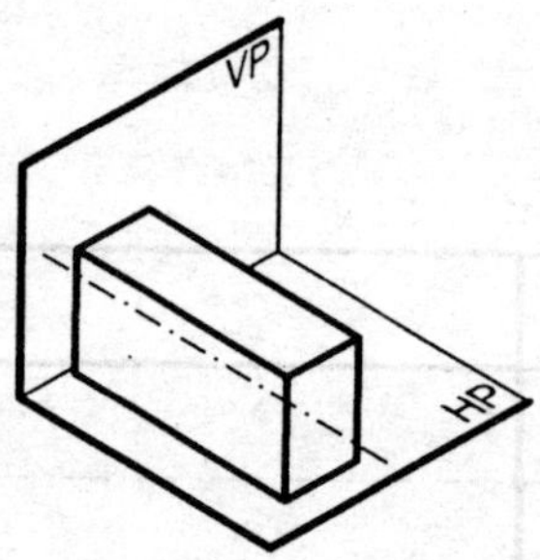	• Axis perpendicular to VP. Hence front view shows true shape and size. • Draw front view first. • Project top view from front view.	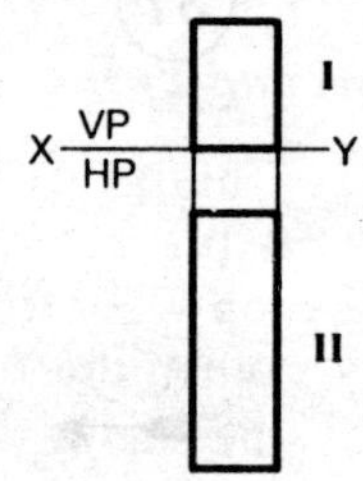
Axis parallel to both HP and VP	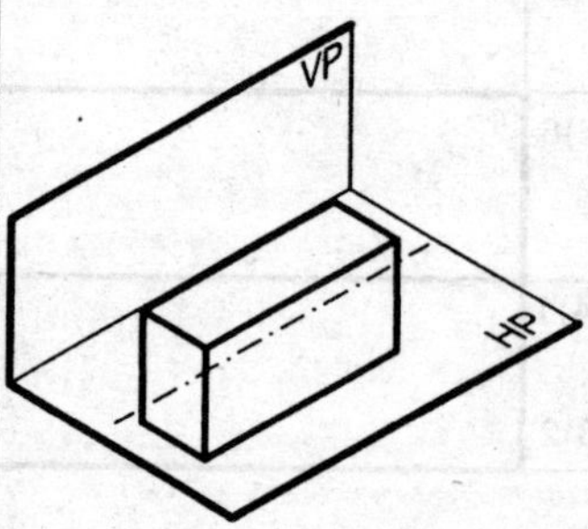	• Draw side view first. • Project front view from side view. • Project top view from front view.	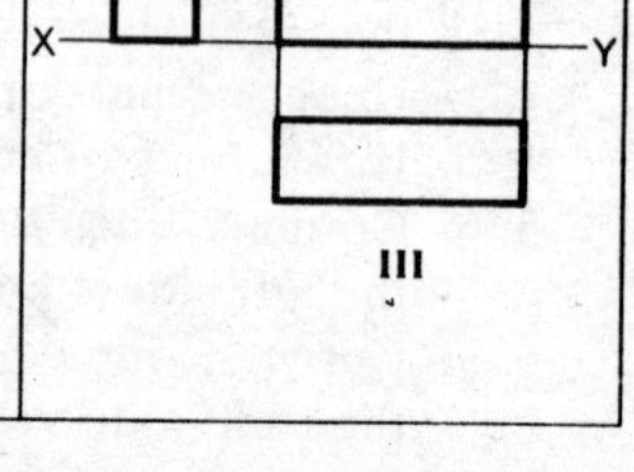

6. CHANGE OF POSITION METHOD

6.1 AXIS OF SOLID INCLINED TO ONE PLANE AND PARALLEL TO OTHER

When the axis of a solid is inclined to one reference plane and parallel to the other, its projections, namely, front and top views are drawn in *two* stages as explained below:

Simple Positions

Assume the solid to be kept in *simple position*, *i.e.*, its axis perpendicular to HP or VP.

- If the axis is inclined to HP, assume axis to be perpendicular to HP in simple position.
- If the axis is inclined to VP, assume axis to be perpendicular to VP in simple position.

- If the solid has an *edge* of its base in HP or parallel to the HP, keep that edge perpendicular to VP in simple position.
- If the solid has an *edge* of its base in VP or parallel to VP, keep that edge perpendicular to HP in simple position.

- If the solid has a *corner* of its base in HP, keep the sides of the base containing that corner equally inclined to VP in simple position.
- If the solid has a *corner* of its base in VP, keep the sides of the base containing that corner equally inclined to HP in simple position.

6.2 AXIS OF SOLID INCLINED TO HP AND PARALLEL TO VP

Simple Position: Assume the solid to be kept with its axis perpendicular to HP. In this simple position, draw the top view first and project the front view from it.

Second Position: Draw second (final) projections by one of the following *two* methods:

1. **Change of Position Method**
   Tilt the front view to the required position (*i.e.*, inclined to HP) and project the second top view from the tilted front view.
   When the front view is tilted, the distances of the corners of the solid from VP remain the same. In other words, the distances of the corners from XY will be the same both in the first top view and the second top view. Therefore, draw horizontal lines from the corners in the first top view to locate the same below XY in the second top view.
2. **Change of Reference Line Method (Refer Next Chapter)**

HOW TO DRAW VISIBLE AND INVISIBLE EDGES ?

1. Outermost lines or edges in any view are always visible.
2. Corners, lines or edges and faces nearer the observer in the top view will be visible in the front view.
3. Upper corners and lines or edges in the front view are visible in the top view.
4. If two lines or edges cross each other, one of them represents invisible edge.
5. Corners, lines or edges and faces on the left of the front or top view will be visible in the left side view.
6. Corners, lines or edges and faces on the right of the front or top view will be visible in the right side view.

7. AXIS OF SOLID INCLINED TO ONE PLANE AND PARALLEL TO OTHER

7.1 AXIS OF SOLID INCLINED TO HP AND PARALLEL TO VP

Problem 38: *(Fig. 33)* *A* ***pentagonal prism****, side of base 25 mm and axis 50 mm long, rests with one of its edges on HP such that the base containing that edge makes an angle of 30° to HP and its axis is parallel to VP. Draw its projections.*

Simple Position

1. When the prism has to rest with **one of its edges (*i.e.*, the base edge)** on HP, first assume the prism in simple position such that its base is on HP and one of the edges of the base is ⊥ʳ to VP, preferably on the right side for tilting it to the given final position.
2. For this simple position, **draw the Top View first.**
3. Now in the top view, (3)(4) is perpendicular to XY and on the right side.
4. Draw the front view, in which the edge 3′ (4′) appears as a point.

Second (Final) Front View

5. Prism has to rest with one of its edges on HP and its base makes 30° to HP. Hence, tilt the first front view such that 3′ (4′) is on XY and its base is inclined at 30° to XY. Thus obtain the second (final) front view.

Second (Final) Top View

6. Draw the projectors from the final front view.
7. Draw horizontal lines from the first top view to cut the corresponding projectors from final front view. Thus obtain the final top view. The axis is parallel to XY in this view.

To draw the visible and invisible edges in the final top view

8. Draw the boundary lines of this view by *thick* lines.
9. Look at the final front view in the direction of the arrow shown.
10. Edges a′ 1′, a′ b′ and a′ (e′) are visible. Hence draw $a_1 1_1$, $a_1 b_1$ and $a_1 e_1$ as *thick* lines.
11. 3′ (4′) is invisible. Hence draw $(3_1)(4_1)$ as *thin dashed* lines.
12. In a view, the line connecting a visible point and an invisible point should be shown by *thin* dashed line. Hence join $(3_1) 2_1$, $(3_1) c_1$, $(4_1) 5_1$ and $(4_1) d_1$ by *dashed* lines.

Problem 39: *(Exercise)* *A* ***hexagonal prism****, side of base 24 mm and axis 55 mm long, rests with one of its edges on HP such that the base containing that edge makes an angle of 45° to HP and its axis is parallel to VP. Draw its projections.*

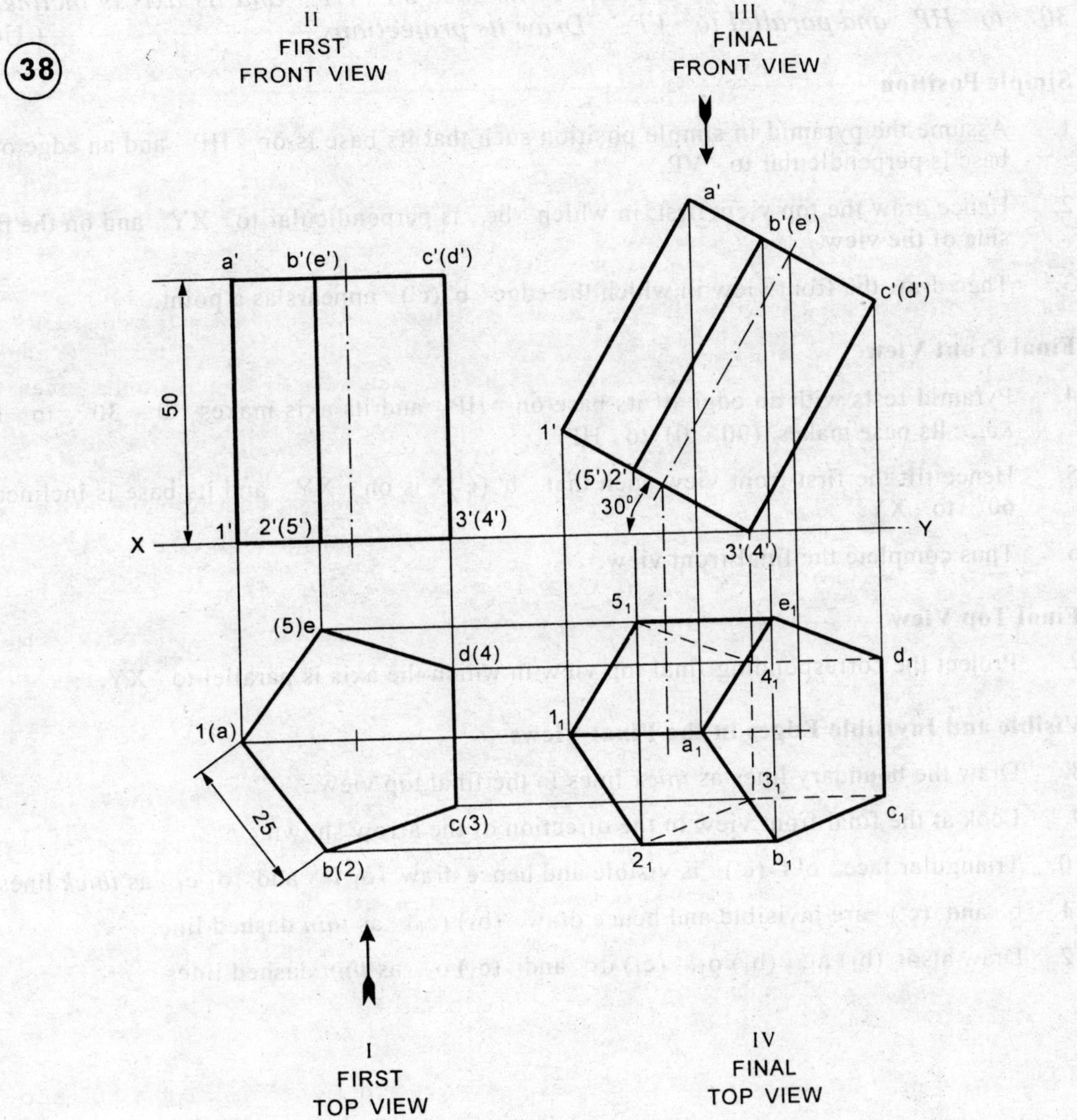

FIG. 33

NOTE : *In an orthographic view of a solid, the line connecting an invisible corner and a visible corner should be showr. by a dashed line.*

Problem 40: *(Fig. 34) A* ***hexagonal pyramid****, side of base 25 mm and axis 50 mm long, rests with one of the edges of its base on HP and its axis is inclined at 30° to HP and parallel to VP. Draw its projections.* (UQ)

Simple Position

1. Assume the pyramid in simple position such that its base is on HP and an edge of its base is perpendicular to VP.

2. Hence draw the top view first, in which bc is perpendicular to XY and on the right side of the view.

3. Then draw the front view in which the edge b′ (c′) appears as a point.

Final Front View

4. Pyramid rests with an edge of its base on HP and its axis makes $\theta = 30^\circ$ to HP, *i.e.*, its base makes $(90 - \theta)$ to HP.

5. Hence tilt the first front view such that b′ (c′) is on XY and its base is inclined at 60° to XY.

6. Thus complete the final front view.

Final Top View

7. Project the corresponding final top view in which the axis is parallel to XY.

Visible and Invisible Edges in the Final Views

8. Draw the boundary lines as *thick* lines in the final top view.

9. Look at the final front view in the direction of the arrow shown.

10. Triangular face o′ f′ (e′) is visible and hence draw $o_1 f_1$ and $o_1 e_1$ as *thick* lines.

11. b′ and (c′) are invisible and hence draw (b_1) (c_1) as *thin* dashed line.

12. Draw also (b_1) a_1, (b_1) o_1, (c_1) d_1 and (c_1) o_1 as *thin* dashed lines.

Problem 41: *(Exercise)*

A ***square pyramid****, side of base 30 mm and height 65 mm, rests with one of the edges of its base on HP such that its base makes 30° to HP. Draw its projections.*

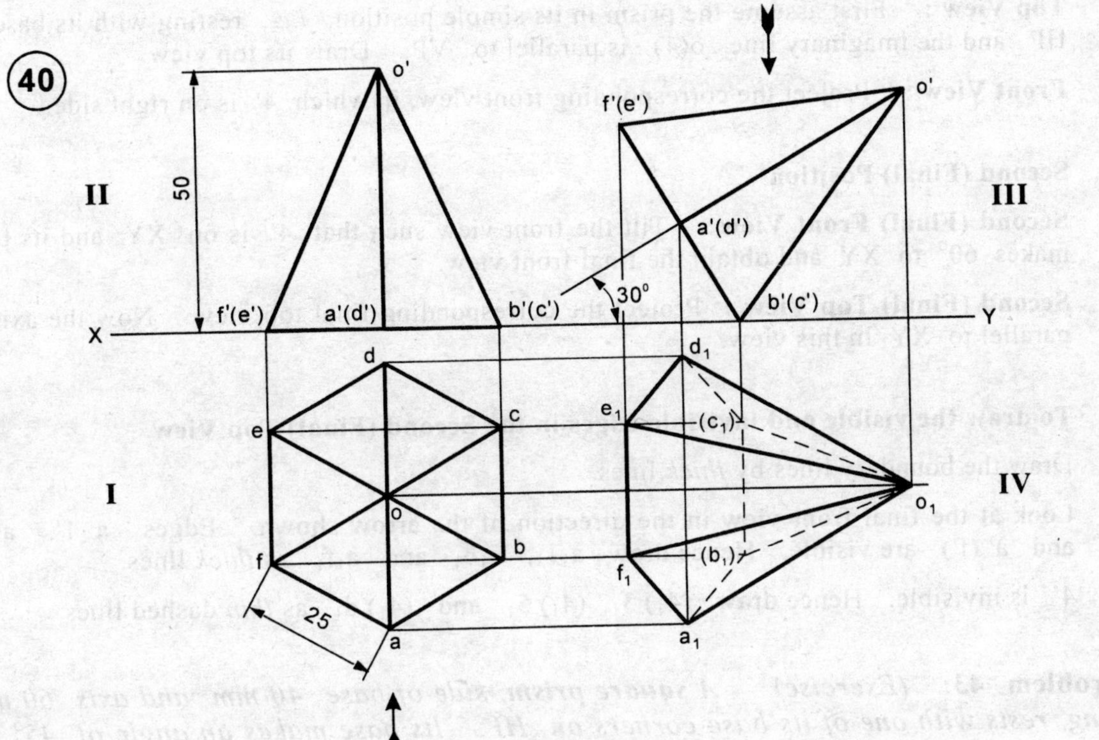

FIG. 34

Problem 42: *(Fig. 35)* *A* ***hexagonal prism****, side of base 25 mm and axis 50 mm long rests with one of its base corners on HP such that its base makes an angle of 60° to HP and its axis is parallel to VP. Draw its projections.* (UQ)

1. Simple Position

NOTE : *When the prism has to rest with one of its base corners on HP, assume that the two edges of the base containing that corner make equal inclinations with HP. So while drawing the top view of the prism in simple position, the imaginary line connecting the center of the base with the corner should be parallel to VP.*

Top View : First assume the prism in its simple position, *i.e.*, resting with its base on HP and the imaginary line o(4) is parallel to VP. Draw its top view.

Front View : Project the corresponding front view, in which 4′ is on right side.

2. Second (Final) Position

Second (Final) Front View : Tilt the front view such that 4′ is on XY and its base makes 60° to XY and obtain the final front view.

Second (Final) Top View : Project the corresponding final top view. Now the axis is parallel to XY in this view.

3. To draw the visible and invisible edges in the Second (Final) Top View

Draw the boundary lines by *thick* lines.

Look at the final front view in the direction of the arrow shown. Edges a′ 1′, a′ b′ and a′ (f′) are visible. Hence draw $a_1 1_1$, $a_1 b_1$ and $a_1 f_1$ as *thick* lines.

4′ is invisible. Hence draw $(4_1)\,3_1$, $(4_1)\,5_1$ and $(4_1)\,d_1$ as *thin* dashed lines.

Problem 43: *(Exercise) A **square prism**, side of base 40 mm and axis 60 mm long, rests with one of its base corners on HP. Its base makes an angle of 45° to HP and its axis is parallel to VP. Draw its projections.*

Problem 44: *(Fig. 36) A **pentagonal pyramid**, side of base 25 mm and axis 55 mm long, lies with one of its slant edges on HP such that its axis is parallel to VP. Draw its projections.*

(UQ)

1. First assume the pyramid to rest with its base on HP such that one of its corners D is on the right side and an imaginary line connecting the corner with the center of the base is parallel to VP. For this position draw the top view such that d is on right side and the imaginary line is parallel to XY. The imaginary line coincides with slant edge od in the top view.
2. Now project the corresponding front view.
3. Complete the final front and top views.
4. When seen in the direction of arrow, the base and apex are visible. Slant edge o′ d′ is invisible. Hence, draw the slant edge $o_1\,d_1$ by dotted line in the final top view.

Problem 45: *(Exercise) A **square pyramid** of side of base 50 mm and height 75 mm lies on HP on one of its triangular faces with its axis parallel to VP. Draw its projections.*

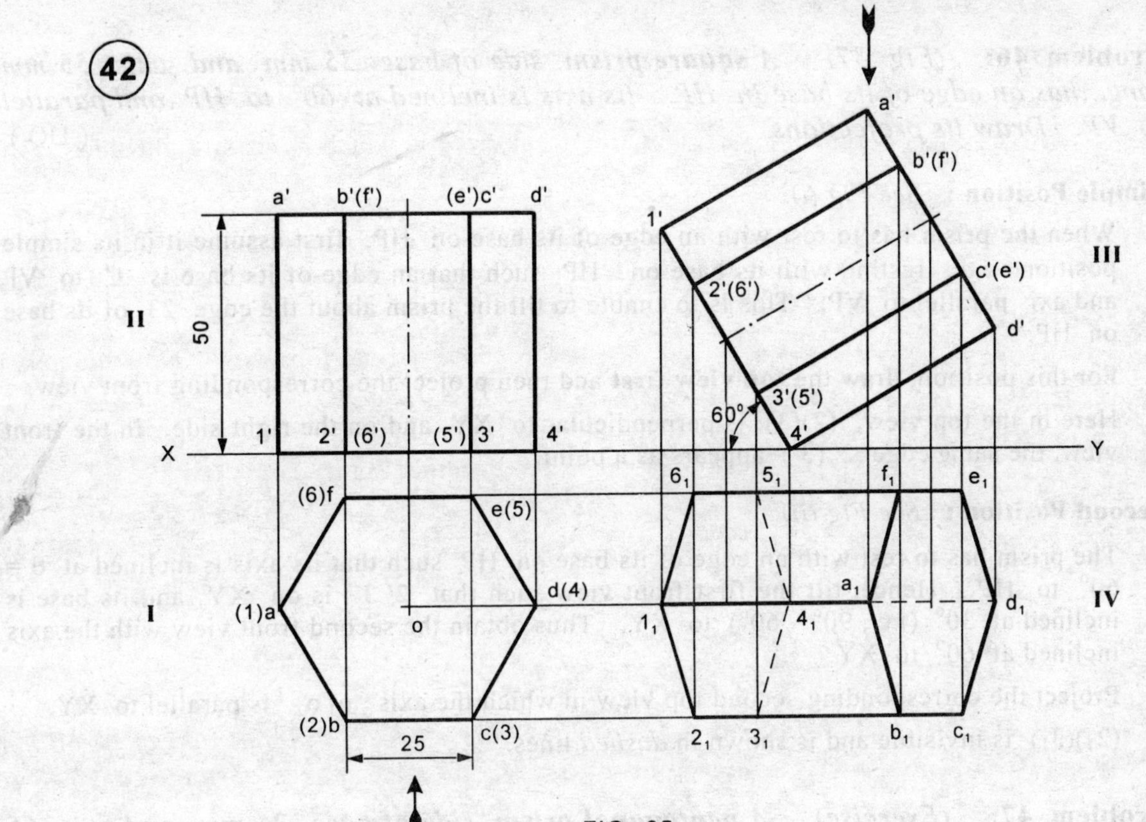
42
II
I
III
IV
a'
b'(f')
(e')c'
d'
50
1'
2'
(6')
(5')
3'
4'
X
Y
(6)f
e(5)
d(4)
(1)a
o
(2)b
c(3)
25
a'
b'(f')
c'(e')
d'
1'
2'(6')
3'(5')
4'
60°
6₁
5₁
f₁
e₁
a₁
d₁
1₁
4₁
2₁
3₁
b₁
c₁

FIG. 35

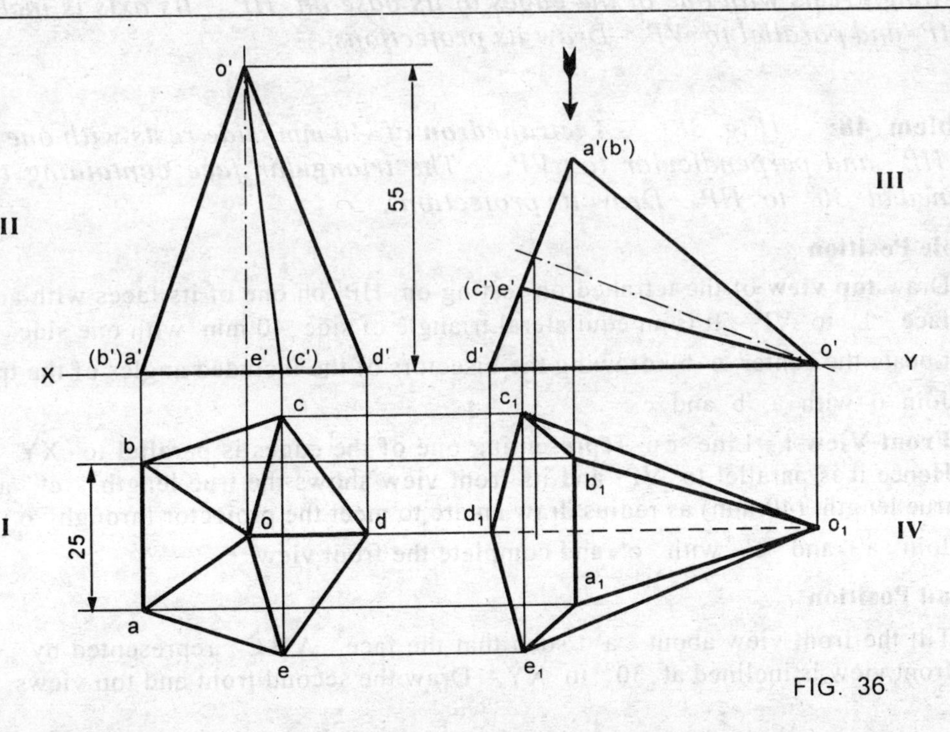
44
II
I
III
IV
o'
55
(b')a'
e'
(c')
d'
X
Y
c
b
o
d
25
a
e
a'(b')
(c')e'
d'
o'
c₁
b₁
d₁
o₁
a₁
e₁

FIG. 36

Problem 46: *(Fig. 37)* *A* ***square prism****, side of base 35 mm and axis 55 mm long, has an edge of its base in HP. Its axis is inclined at 60° to HP and parallel to VP. Draw its projections.* (UQ)

Simple Position : *See Fig (i).*

1. When the prism has to rest with an edge of its base on HP, first assume it in its simple position, *i.e.*, resting with its base on HP such that an edge of its base is $\perp^r$ to VP and axis parallel to VP. This is to enable to tilt the prism about the edge 23 of its base on HP.
2. For this position, draw the **top view first** and then project the corresponding front view.
3. Here in the top view, (2)(3) is perpendicular to XY and on the right side. In the front view, the same edge 2′ (3′) appears as a point.

Second Position : *See Fig (ii).*

4. The prism has to rest with an edge of its base on HP such that its axis is inclined at θ = 60° to HP. Hence, tilt the first front view such that 2′ 3′ is on XY and its base is inclined at 30° (*i.e.*, 90° – 60°) to XY. Thus obtain the second front view with the axis inclined at 60° to XY.
5. Project the corresponding second top view in which the axis $o_1 o_2$ is parallel to XY.
6. $(2_1)(3_1)$ is invisible and is shown in *dashed* lines.

Problem 47: *(Exercise)* *A* ***pentagonal prism****, side of base 25 mm and axis 65 mm long, rests with one of the edges of its base on HP. Its axis is inclined at 30° to HP and parallel to VP. Draw its projections.*

Problem 48: *(Fig. 38)* *A* ***tetrahedron*** *of 40 mm side rests with one of its edges on HP and perpendicular to VP. The triangular face containing that edge is inclined at 30° to HP. Draw its projections.* (UQ)

Simple Position

1. Draw **top view** of the tetrahedron resting on HP on one of its faces with an edge of the face $\perp^r$ to VP. It is an equilateral triangle of side 40 mm with one side $\perp^r$ to XY.
2. Locate the center o by drawing the bisectors of the included angles of the triangle.
3. Join o with a, b and c.
4. **Front View :** Line c o representing one of the edges is parallel to XY in top view. Hence it is parallel to VP and its front view shows the true length. c′ as center and true length (40 mm) as radius draw an arc to meet the projector through o at o′.

   Join a′ and c′ with o′ and complete the front view.

Second Position

5. Tilt the front view about a′ such that the face A B C represented by a′ c′ in the front view is inclined at 30° to XY. Draw the second front and top views.

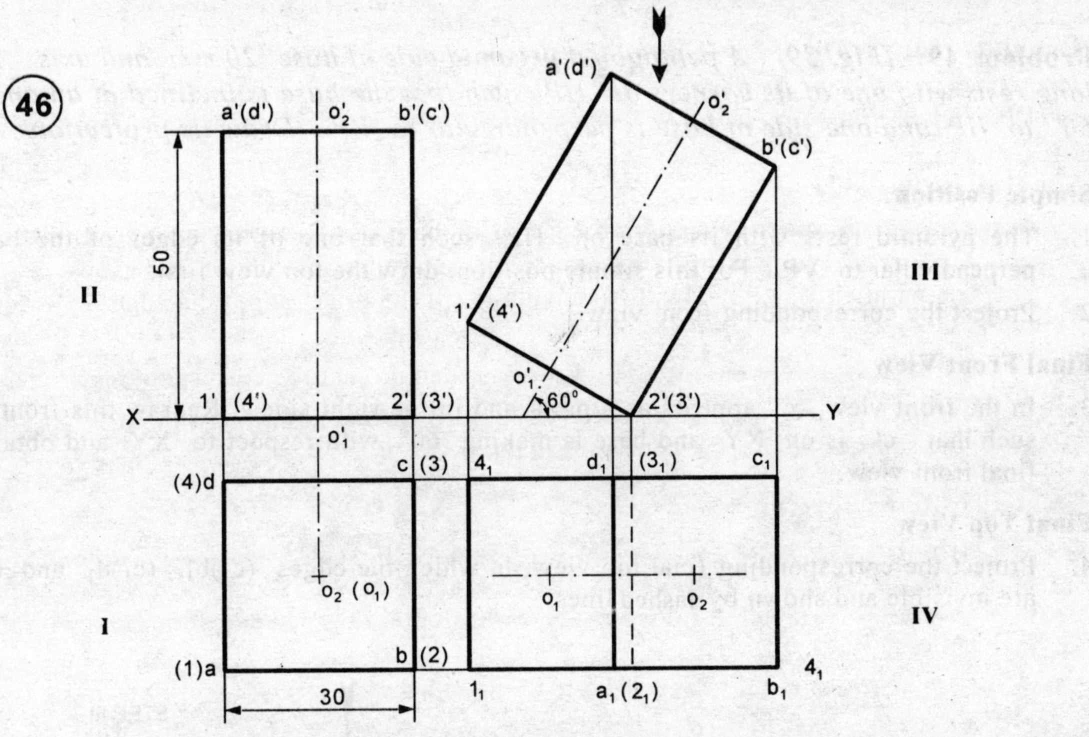

(i) Simple Position (ii) Second Position

FIG. 37

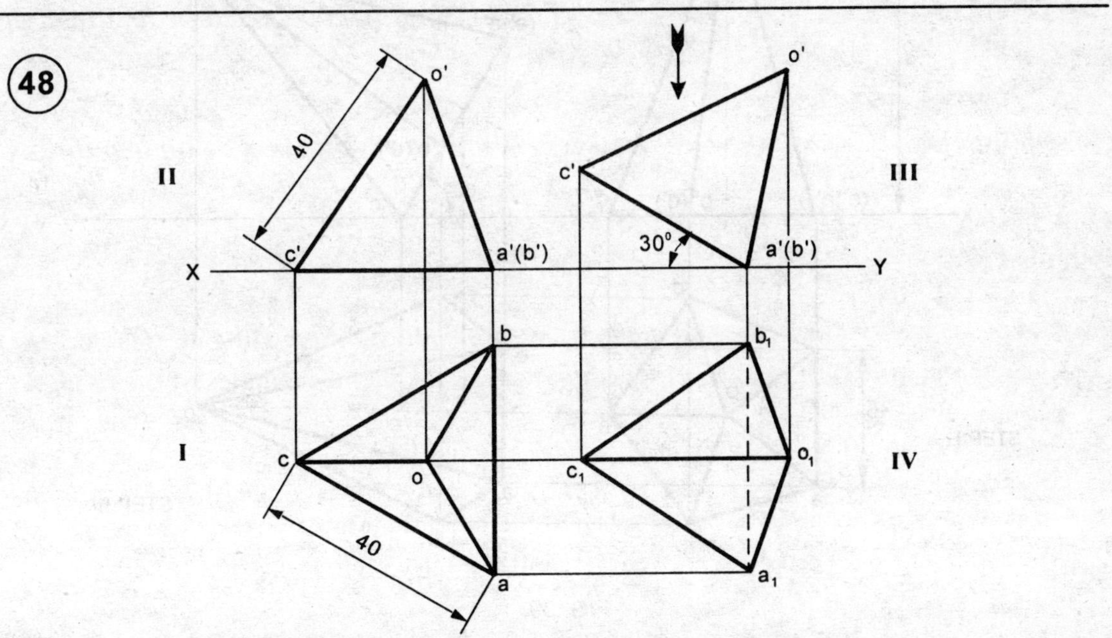

(i) Simple Position (ii) Second Position

FIG. 38

Problem 49: *(Fig. 39) A pentagonal pyramid side of base 20 mm and axis 45 mm long rests with one of its corners on HP such that the base is inclined at an angle of 60° to HP and one side of base is perpendicular to VP. Draw its projections.* (UQ)

Simple Position

1. The pyramid rests with its base on HP such that one of its edges of the base is perpendicular to VP. For this simple position, draw the top view first.
2. Project the corresponding front view.

Final Front View

3. In the front view, c′ appears as a point and in the right side. Redraw this front view such that c′ is on XY and base is making 60° with respect to XY and obtain the final front view.

Final Top View

4. Project the corresponding final top view, in which the edges $(c_1)b_1$, $(c_1)d_1$ and $(c_1)o_1$ are invisible and shown by dashed lines.

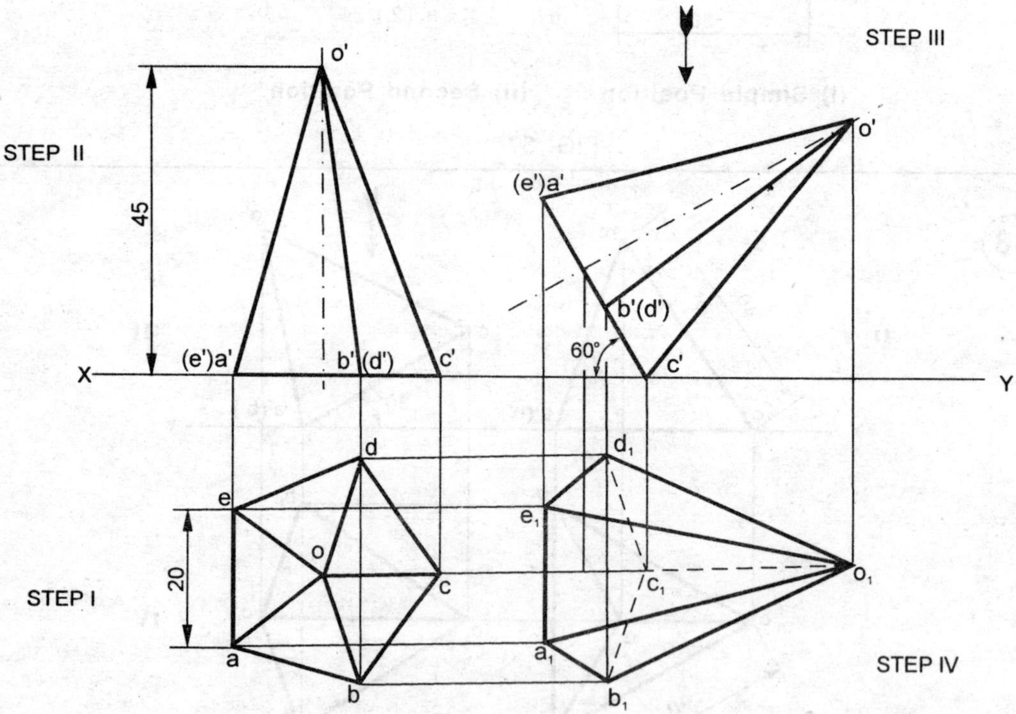

FIG. 39

Problem 50: *(Fig. 40) A hexagonal pyramid side of base 25 mm, axis 50 mm long lies with one of its triangular faces on the HP and its axis is parallel to the VP. Draw its projections.* (UQ)

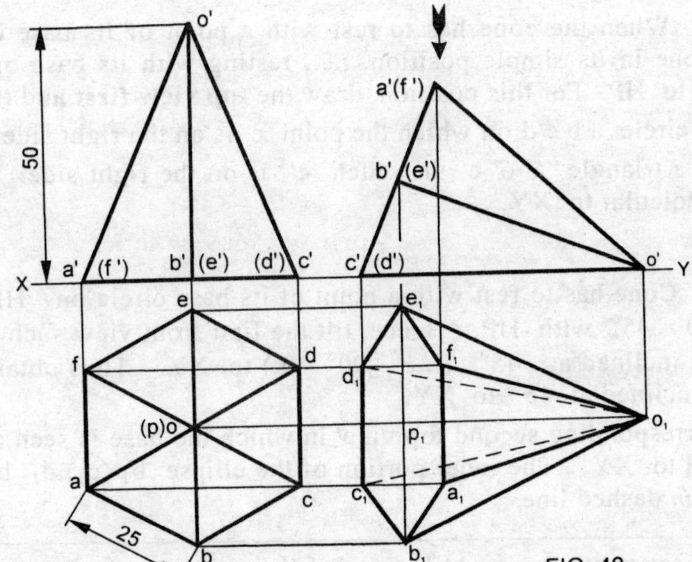

FIG. 40

Problem 51: *(Fig. 41) A pentagonal pyramid, side of base 30 mm and axis 60 mm long is lying with one of its triangular faces on the HP and axis parallel to the VP. Draw its projections.* (UQ)

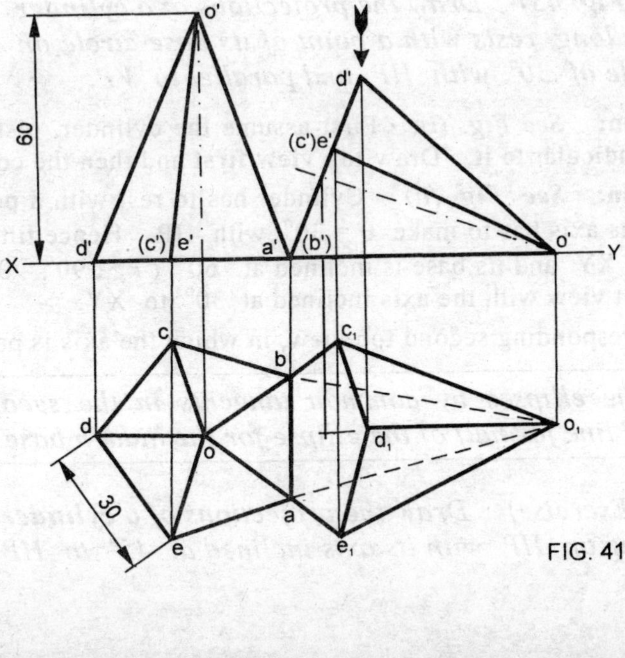

FIG. 41

Problem 52: *(Fig. 42) Draw the projections of a* ***cone****, base 30 mm diameter and axis 50 mm long, resting on HP on a point of its base circle with the axis making an angle of 45° with HP and parallel to VP.*

Simple Position

1. *See Fig. (i).* When the cone has to rest with a point of its base circle on HP, first assume the cone in its simple position, *i.e.*, resting with its base on HP and the axis perpendicular to HP. For this position, draw the **top view first** and then the front view.
2. Top view is a circle a b c d in which the point c is on the right side.
3. Front view is a triangle a′ o′ c′ in which c′ is on the right side. Also the axis of the cone is perpendicular to XY.

Second Position

4. *See Fig. (ii).* Cone has to rest with a point of its base circle on HP, such that its axis has to make $\theta = 45°$ with HP. Hence, tilt the first front view such that c′ is on XY and its base is inclined at 45°, (*i.e.*, $90° - \theta$) to XY. Thus obtain second front view with the axis inclined at 45° to XY.
5. Project the corresponding second top view in which the base is seen as an ellipse and the axis is parallel to XY. The inner portion of the ellipse b_1 (c_1) d_1 between the tangents is shown as *thin* dashed line.

NOTE : *These tangents need not be any of the particular generators that have been chosen, i.e., need not be $o_1 b_1$ or $o_1 d_1$.*

Problem 53: *(Exercise) A* ***cone*** *of base 50 mm diameter and axis 65 mm long, lies with one of its generators on HP and axis parallel to VP. Draw its projections.*

Problem 54: *(Fig. 43) Draw the projections of a* ***cylinder****, base 30 mm diameter and axis 40 mm long, rests with a point of its base circle on HP such that the axis is making an angle of 30° with HP and parallel to VP.*

1. **Simple position:** *See Fig. (i).* First assume the cylinder, resting with its base on HP and axis perpendicular to it. Draw **top view first** and then the corresponding front view.
2. **Second position:** *See Fig. (ii).* Cylinder has to rest with a point of its base circle on HP such that its axis has to make $\theta = 30°$ with HP. Hence tilt the front view such that point 3′ is on XY and its base is inclined at 60° (*i.e.*, $90° - \theta$) to XY. Thus obtain the second front view with the axis inclined at 30° to XY.

   Project the corresponding second top view, in which the axis is parallel to XY.

NOTE : *Join the ellipses by common tangents in the second top view. Draw dashed line for half of the ellipse for the hidden base.*

Problem 55: *(Exercise) Draw the projections of a* ***cylinder*** *35 mm diameter and 50 mm long, lying on HP with its axis inclined at 45° to HP and parallel to VP.*

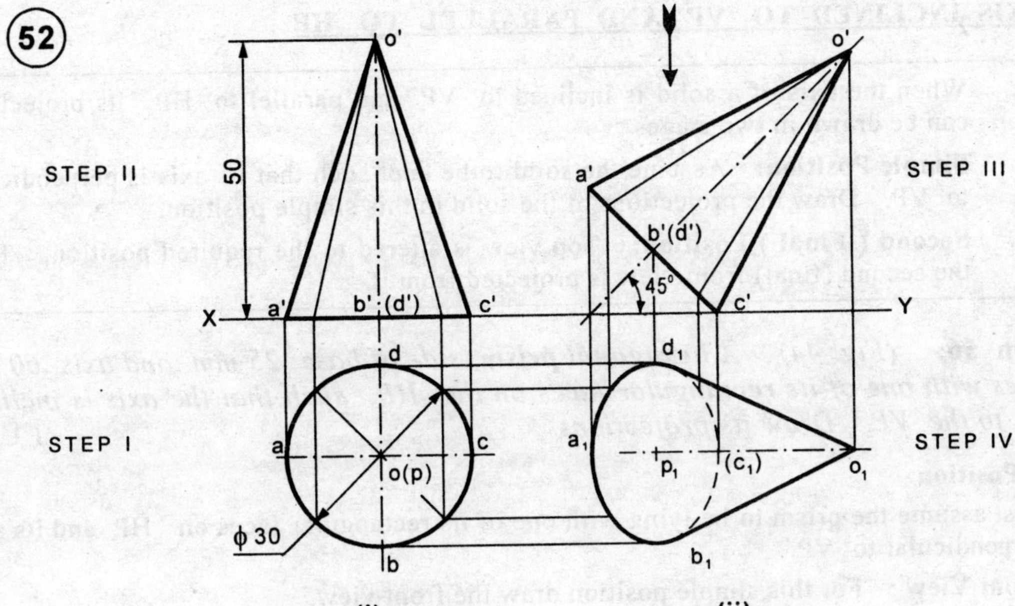

(i) Simple Position

(ii) Second Position

FIG. 42

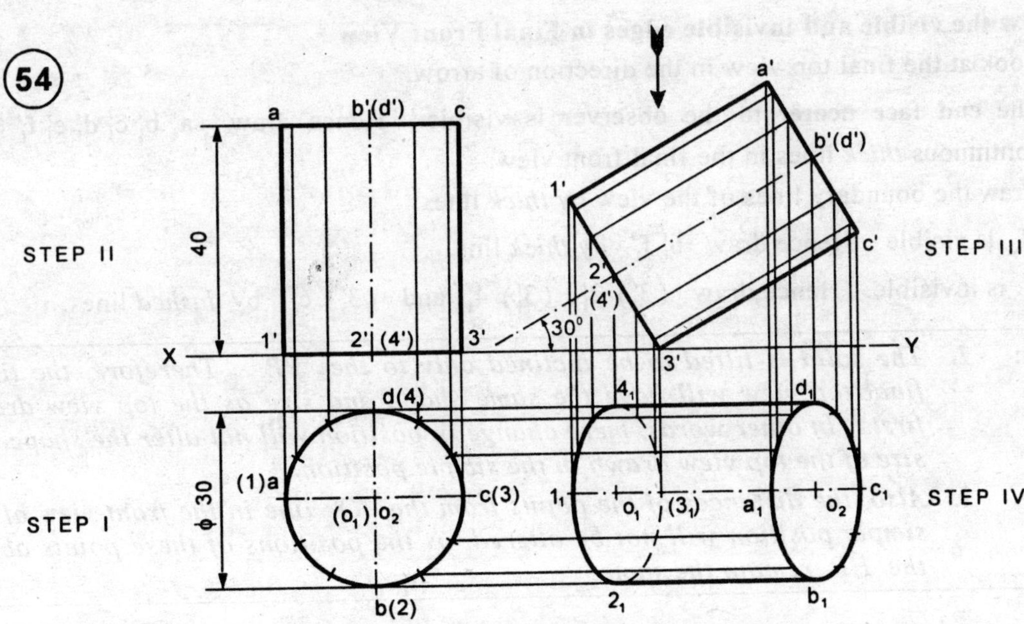

(i) Simple Position

(ii) Second Position

FIG. 43

7.2 AXIS INCLINED TO VP AND PARALLEL TO HP

NOTE : When the axis of a solid is inclined to VP and parallel to HP, its projections can be drawn in two stages :

Simple Position: Assume the solid to be kept such that its axis is perpendicular to VP. Draw the projections of the solid in this simple position.

Second (Final) Position: Top view is altered to the required position. Then the second (final) front view is projected from it.

Problem 56: *(Fig. 44) A **hexagonal prism**, side of base 25 mm and axis 60 mm long, lies with one of its rectangular faces on the HP, such that the axis is inclined at 45° to the VP. Draw its projections.* (UQ)

Simple Position

1. First assume the prism to be lying with one of its rectangular faces on HP and its axis perpendicular to VP.
2. **Front View :** For this simple position draw the front view.
3. **Top View :** Draw the corresponding top view projecting from the front view.
4. **Second (Final) Top View :** Redraw top view such that axis is inclined at 45° to XY.
5. **Second (Final) Front View :** Project the corresponding final front view, in which axis is parallel to XY.

To draw the visible and invisible edges in Final Front View

6. Look at the final top view in the direction of arrow.
7. The end face nearer to the observer is visible. Hence draw $a_1'\ b_1'\ c_1'\ d_1'\ e_1'\ f_1'$ by continuous *thick* lines in the final front view.
8. Draw the boundary lines of the view by *thick* lines.
9. 6f is visible. Hence draw $6_1'\ f_1'$ by *thick* line.
10. 3 is invisible. Hence draw $(3_1')\ 2_1'$, $(3_1')\ 4_1'$ and $(3_1')\ c_1'$ by *dashed* lines.

NOTE : *1. The solid is tilted to be inclined only to the VP. Therefore, the tilted final top view will have the same shape and size as the top view drawn first. In other words, mere change in position will not alter the shape and size of the top view drawn in the simple position.*

2. Also, the distances of the points from the XY line in the front view of the simple position will not be altered, as the positions of these points above the HP remain the same.

Problem 57: *(Fig. 45) Draw the projections of a pentagonal prism of 30 mm side of base and 65 mm long lying on one of its longer edges on the HP with one rectangular face perpendicular to the HP such that the axis makes 60° with VP.*

56

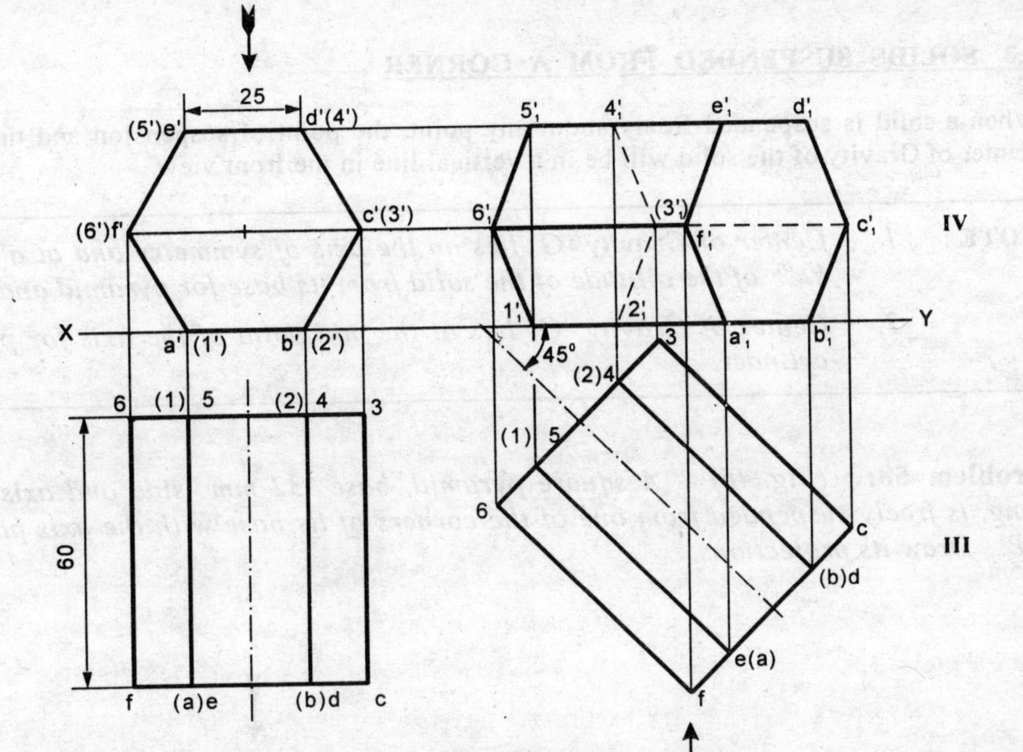

FIG. 44

57

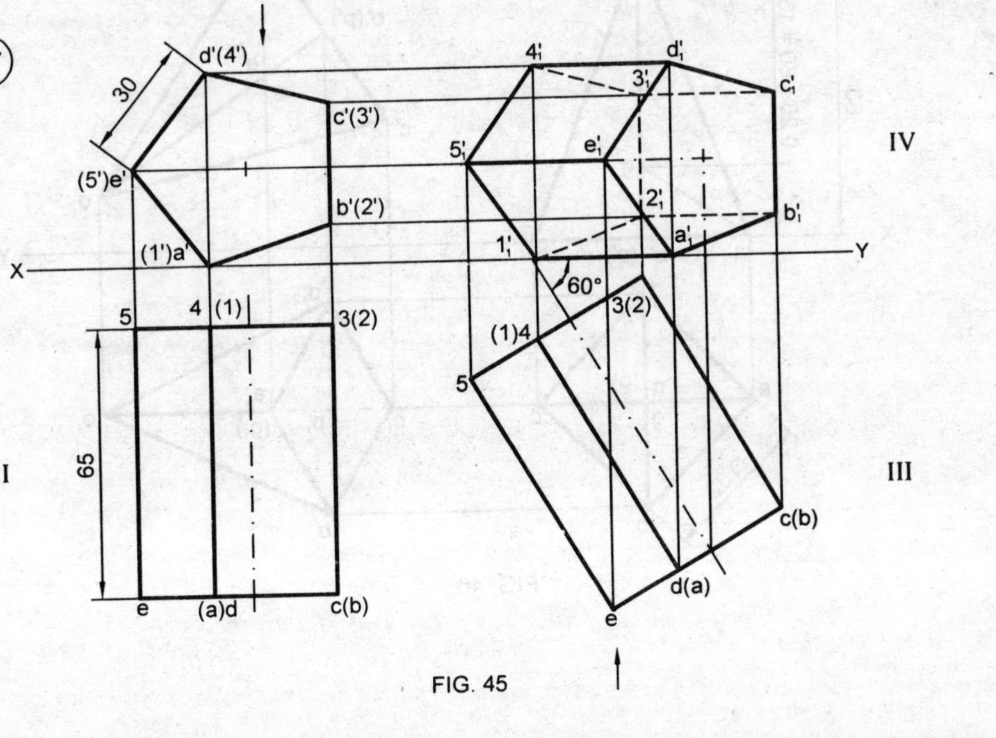

FIG. 45

7.3 SOLIDS SUSPENDED FROM A CORNER

When a solid is suspended freely about any point, the point of suspension and the point of Center of Gravity of the solid will be in a vertical line in the front view.

NOTE : 1. *Center of Gravity G lies on the axis of symmetry and at a height of ¼ th of the altitude of the solid from its base for pyramid and cone.*

2. *Center of Gravity G lies at the mid point of the axis for prism and cylinder.*

Problem 58: *(Fig. 46) A square pyramid, base 32 mm side and axis 60 mm long, is freely suspended from one of the corners of its base with the axis parallel to VP. Draw its projections.* (UQ)

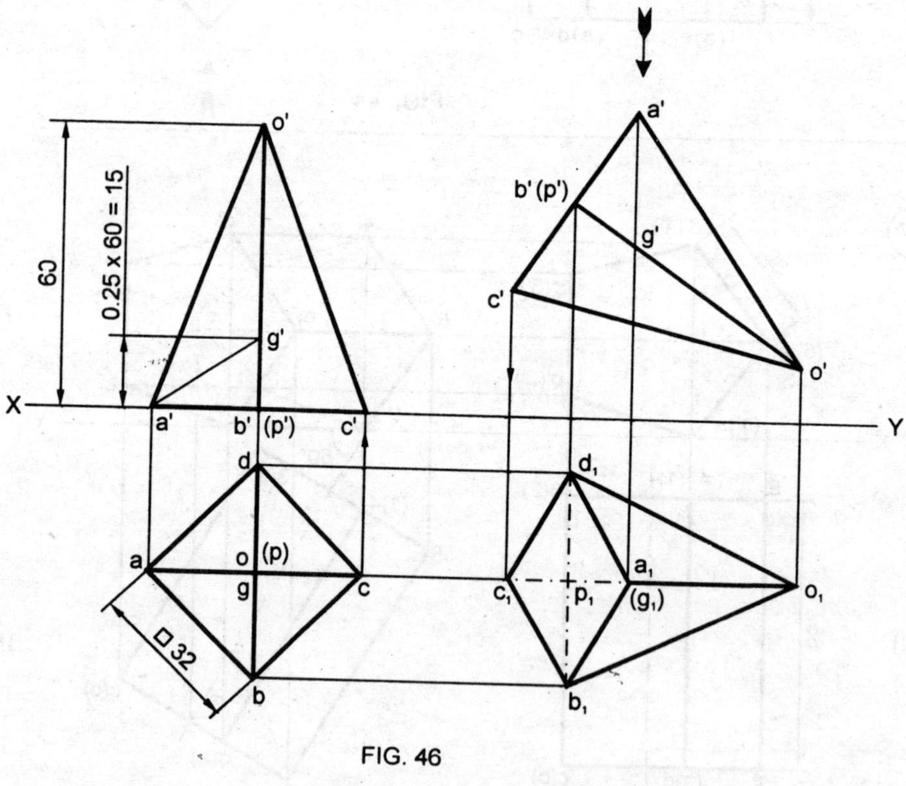

FIG. 46

1. Assume the pyramid to be suspended from the corner A of its base with its axis parallel to VP.

Initial Position

2. Initially, keep the pyramid such that its base is on HP and the line joining A (corner from which it is suspended) with the center of gravity G, parallel to VP. Therefore, g will coincide with the top view o of the axis.
3. Now draw the top view as a square a b c d such that a g, *i.e.*, a o is parallel to XY.
4. Draw the corresponding front view.
5. Mark g′ at a distance equal to ¼ of the height of the axis from XY.
6. Join a′ and g′.
7. Tilt the front view such that a′ g′ is perpendicular to XY.
8. Project the corresponding top view, in which the axis o_1 p_1 of the pyramid still remaining parallel to VP.

Problem 59: *(Exercise)*

A pentagonal prism of side of base 26 mm and height 58 mm is freely suspended from a corner such that the axis of the prism is parallel to VP. Draw its projections. (UQ)

Problem 60: *(Exercise)*

A triangular prism of 30 mm side of base and height 65 mm is suspended freely from one of its corners such that its axis is parallel to VP. Draw its projections.

Problem 61: *(Exercise)*

A hexagonal pyramid of base 25 mm side and height 60 mm is freely suspended from a corner of its base. The projection of the axis on HP is parallel to VP. Draw the projections.

Problem 62: *(Exercise)*

A cylinder of diameter of base 50 mm and height 60 mm is suspend freely from a point on its circular rim. The projection of the axis on HP is parallel to the XY line. Draw the projections of the cylinder. (UQ)

8. POINTS TO REMEMBER

POSITION OF SOLID	ILLUSTRATION	PROCEDURE TO DRAW THE PROJECTIONS	PROJECTIONS (FRONT & TOP VIEWS)
Axis parallel to VP and inclined to HP	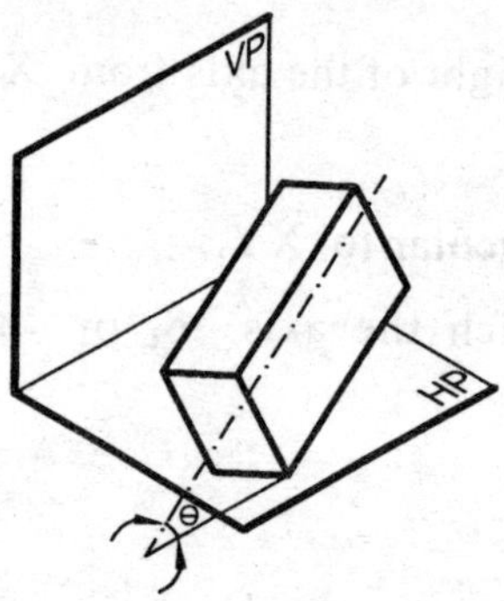	**Simple Position** • Draw front and top views in simple position with axis perpendicular to HP. **Second Position** • Draw second front view in the tilted position. • Project the second top view from the second front view.	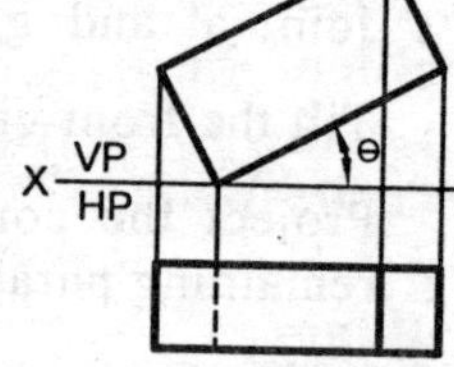 Second Position
Axis parallel to HP and inclined to VP	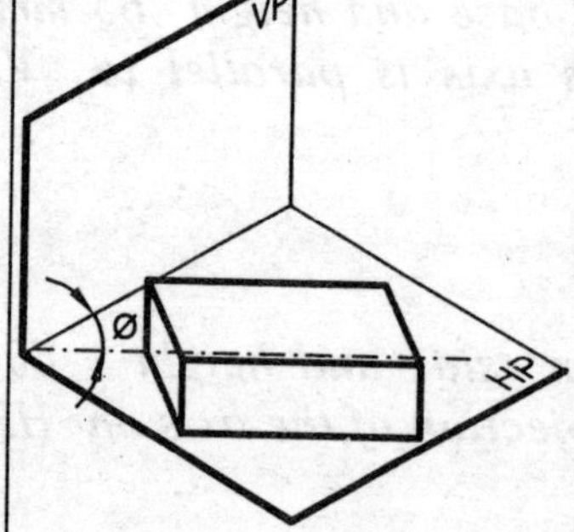	**Simple Position** • Draw the front and top views in simple position with axis perpendicular to VP. **Second Position** • Draw second top view in the tilted position. • Project the second front view from the second top view.	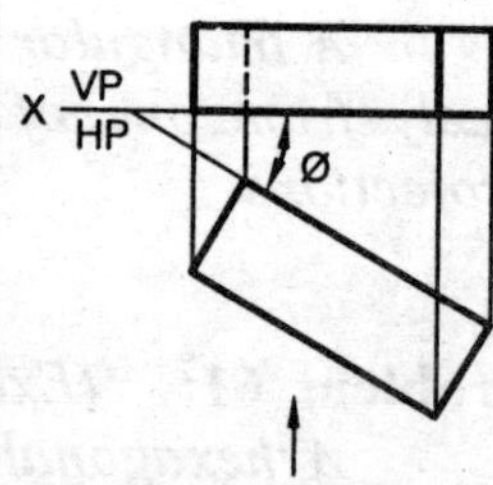 Second Position

9. ADDITIONAL PROBLEMS – I

Problem 63: A cube of 45 mm sides has one of its body diagonal horizontal and parallel to VP with one of its edges touching it. Draw its projections.

Problem 64: The body diagonal of a cube is 70 mm long. Draw the projections of the cube when the body diagonal is perpendicular to HP.

Problem 65: Draw the projections of a square prism of 28 mm side of base and height 55 mm rests with one of its corners on HP such that one of the base edges containing the corner on which the prism rests is inclined at 30° to HP, the axis is inclined at 25° to HP and parallel to VP. Draw its projections on HP and VP.

Problem 66: A pentagonal prism of 30 mm side of base and height 60 mm is resting on a corner of its base on HP with a longer edge containing that corner is inclined at 45° to HP and parallel to VP. Draw its projections. (UQ)

Problem 67: A regular pentagonal prism lies with its axis inclined at 60° to HP and parallel to VP. The prism is 70 mm long and has a face width of 25 mm. The highest rectangular face is perpendicular to VP. The prism leans to the left. Draw the top and front views.

Problem 68: A right regular triangular pyramid, base 35 mm side and axis 60 mm rests with one of its inclined lateral edges on HP such that the two triangular faces containing the inclined edge on which it rests make equal inclinations with HP. The projection of the axis on the HP is parallel to VP. Draw the projections. (UQ)

Problem 69: A hexagonal pyramid of 25 mm side of base and 50 mm high rests with one of its corners of the base on HP with its axis inclined at 45° to HP. The two base edges containing the corner of which it rests, made equal inclinations with HP. Draw its projections on VP and HP.

Problem 70: A right circular cone of base diameter 50 mm and height 60 mm is placed such that one diameter AB of the base is inclined at 45° to HP and the other diameter CD of the base is parallel to both HP and VP. The diameters AB and CD are perpendicular to each other. Draw the projections of the cone. (UQ)

Problem 71: A cube of edges 40 mm is hung by a string attached to one of its corners. Draw its projections when two of its edges containing the corner from which it is hung are equally inclined to VP.

Problem 72: A square prism of side of base 30 mm and axis 55 mm long lies on HP on one of its longer edges with its faces equally inclined to HP. Draw its projections when its axis is inclined at an angle of 60° to VP.

10. AXIS OF SOLID INCLINED TO BOTH HP AND VP

The projections of a solid with its axis inclined to both HP and VP are drawn in ***three stages*** as follows : (This is an extension of the procedure followed for solids with axis inclined to one reference plane.)

Stage I : Keep the axis of the solid in Simple Position (i.e., perpendicular to one of the reference planes), suitable to reach the final required position. Then draw the first set of views.

Stage II: Tilt the solid so as to bring the axis inclined to one of the reference planes and parallel to the other. Then draw the second set of views.

Stage III : Tilt the solid further so as to bring the axis inclined to both the reference planes. Then draw the third and final set of views.

Change of Position Method is similar to the one explained in the projections of plane surfaces inclined to both the planes.

Problem 73: *(Fig. 47) A square prism, side of base 30 mm and axis 50 mm long, has its axis inclined at 60° to HP. It has an edge of its base in the HP and inclined at 45° to VP. Draw its projections.* (UQ)

Simple Position : *See Fig (i).*

1. When the prism has to rest with an edge of its base on H.P. first assume it in its simple position, i.e., resting with its base on H.P. such that an edge of its base is perpendicular to V.P. This is to enable to tilt the prism about the edge 23 of its base on H.P. For this position, draw the top view first and then project the corresponding front view.
2. Here in the top view, (2)(3) is perpendicular to XY and on the right side. In the front view, the same edge 2′ (3′) appears as a point.

Second position : *See Fig (ii).*

3. The prism has to rest with an edge of its base on H.P. such that its axis is inclined at $\theta = 60^\circ$ to H.P. So tilt the first front view such that 2′ 3′ is on XY and its base is inclined at 30° (i.e., 90° – 60°) to XY. Thus obtain the second front view.
4. Project the corresponding second top view in which $(2_1)(3_1)$ is perpendicular to XY.

Final position : *See Fig (iii).*

5. As per the given condition, again the edge $(2_1)(3_1)$ has to make an inclination of 45° with V.P., keeping the inclination of the axis with H.P. constant. Hence, the shape and size of the second top view will remain the same, but its position alone will change.

   Therefore tilt the second top view such that (2_1) (3_1) is inclined at 45° to and below XY. Thus obtain the final top view.

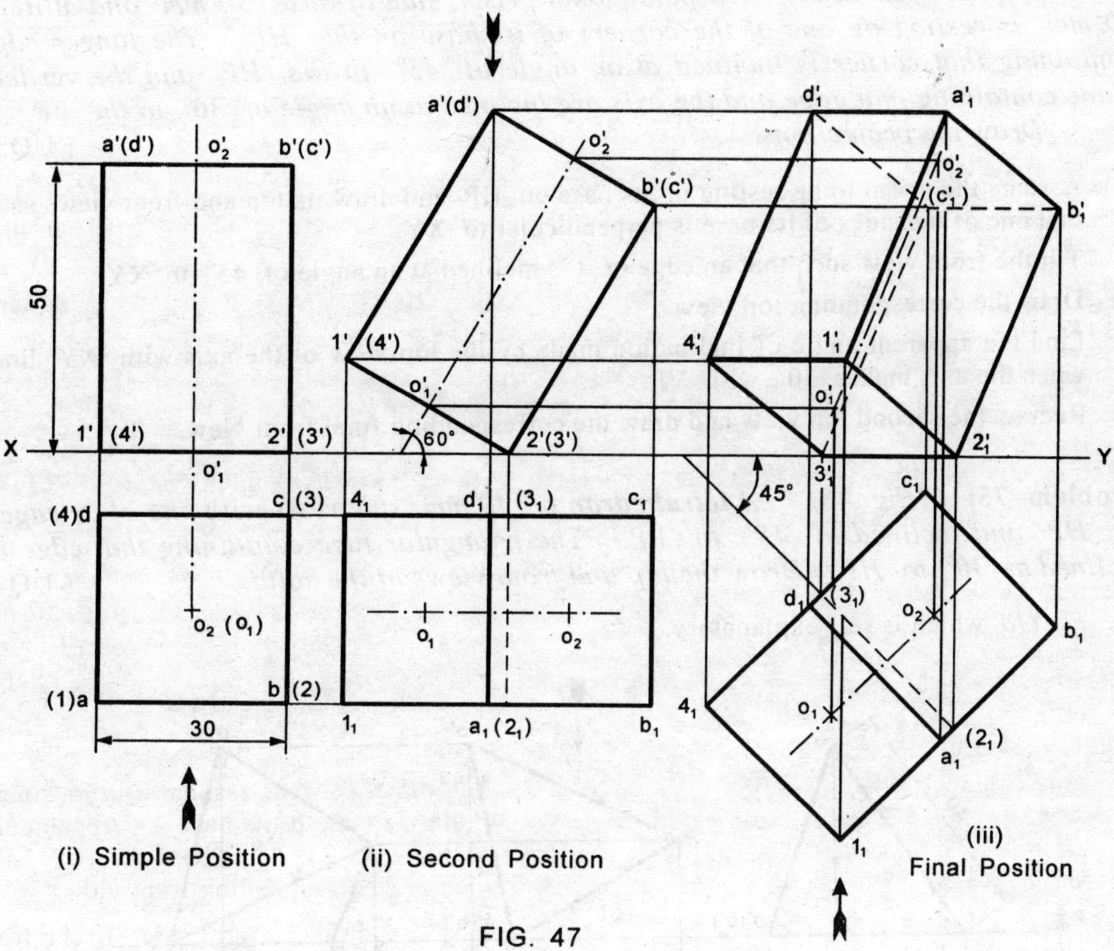

FIG. 47

6. In the final front view height of all the corners from XY will remain the same as in the second front view. Therefore draw projectors from the final top view and horizontals from the second front view and obtain the final front view.

To draw the visible and invisible edges in the final front view

7. First draw the boundary lines of the final front view by *thick* lines.
8. Look at the final top view in the direction of arrow. $1_1\,a_1$, $1_1\,(2_1)$ and $1_1\,4_1$ are visible. Hence, draw $1_1'\,a_1'$, $1_1'\,2_1'$ and $1_1'\,4_1'$ by *thick* lines.
9. c_1 is invisible. Hence, mark it as (c_1') in the final front view.
10. $c_1(3_1)$, c_1b_1 and c_1d_1 are invisible. Hence, draw $(c_1')\,3_1'$, $(c_1')\,b_1'$ and $(c_1')\,d_1'$ by *dashed* lines.

Problem 74: *(Exercise)* *A pentagonal prism, side of base 30 mm and altitude 60 mm is resting on one of the corners of its base on the HP. The longer edge containing that corner is inclined at an angle of 45° to the HP and the vertical plane containing that edge and the axis are inclined at an angle of 30° to the VP. Draw the projections.* (UQ)

1. Assume the prism to be resting on its base on HP and draw its top and front views such that one of the sides of its base is perpendicular to XY.
2. Tilt the front view such that an edge of it is inclined at an angle of 45° to XY.
3. Draw the corresponding top view.
4. Find the apparent angle of inclination made by the top view of the axis with XY line, when the axis makes 30° with VP.
5. Redraw the second top view and draw the corresponding final front view.

Problem 75: *(Fig. 48)* ***A tetrahedron*** *of 40 mm side rests with one of its edges on HP and inclined at 45° to VP. The triangular face containing that edge is inclined at 30° to HP. Draw the top and front views of the solid.* (UQ)

See *Fig. 48* which is self-explanatory.

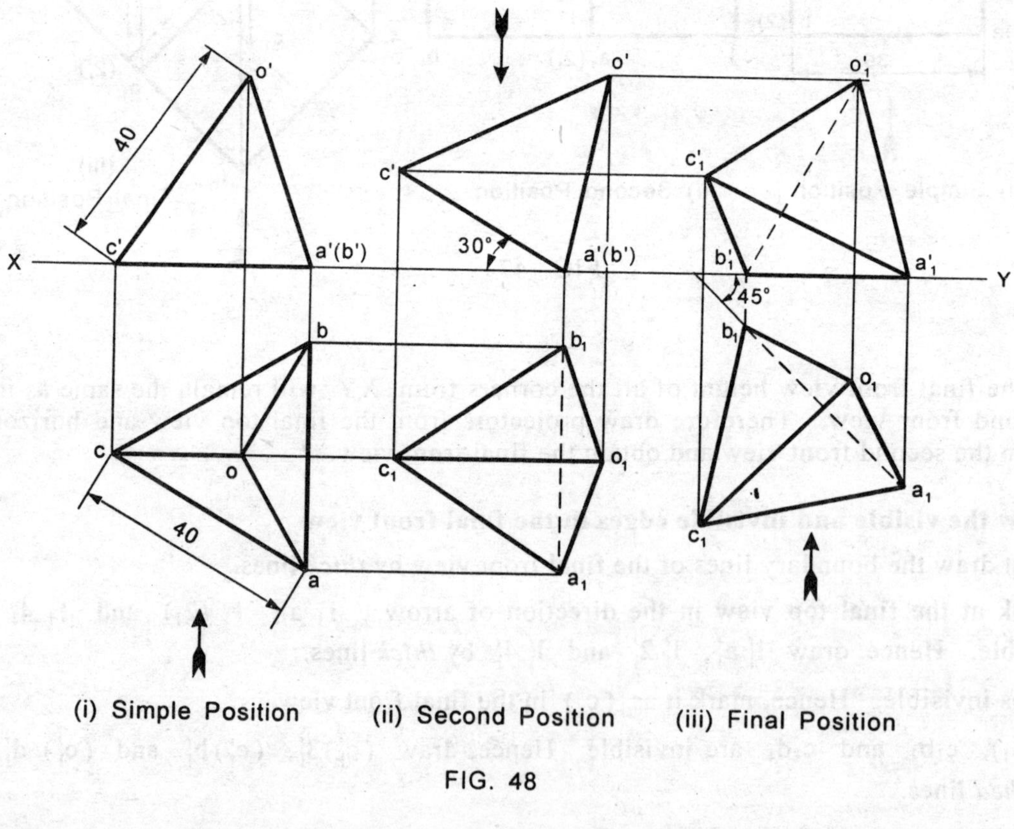

FIG. 48

Problem 76: *(Exercise) Draw the projections of a pentagonal pyramid of 25 mm side and height 55 mm, when it is resting on one of its base edges.*

(i) The axis is making an angle of 30° with HP and the top view of the axis making 45° with VP.

(ii) The axis is making an angle of 30° with HP and 45° with VP. (UQ)

1. Draw the projections of the pyramid in the simple position, keeping the base on HP and one of the base edges perpendicular to VP.
2. Keep the axis inclined at an angel of 30° to XY and draw the second front view.
3. Project horizontally from the top view and vertically from the second front view to obtain the second top view.

Case (i)

4. Axis of the top view is inclined at 45° to XY. Hence redraw the second top view such that the axis of the view is inclined at 45° to XY. Thus the third top view is drawn.
5. Project horizontally from the second front view and vertically from the third top view to obtain the third front view.

Case (ii)

6. Angle of inclination of axis with VP is 45°. Hence, to reproduce the top view, find the apparent angle (angle seen in the projection) which is always greater than the true angle.
7. Redraw the second top view to obtain the fourth top view.
8. Project horizontally from the second front view and vertically from the fourth top view to obtain the fourth front view.

Problem 77: *(Fig. 49) Draw the front view and top view of an octahedron of edges 30 mm when one face is in the HP and one edge of that face makes 45° with the VP.* (UQ)

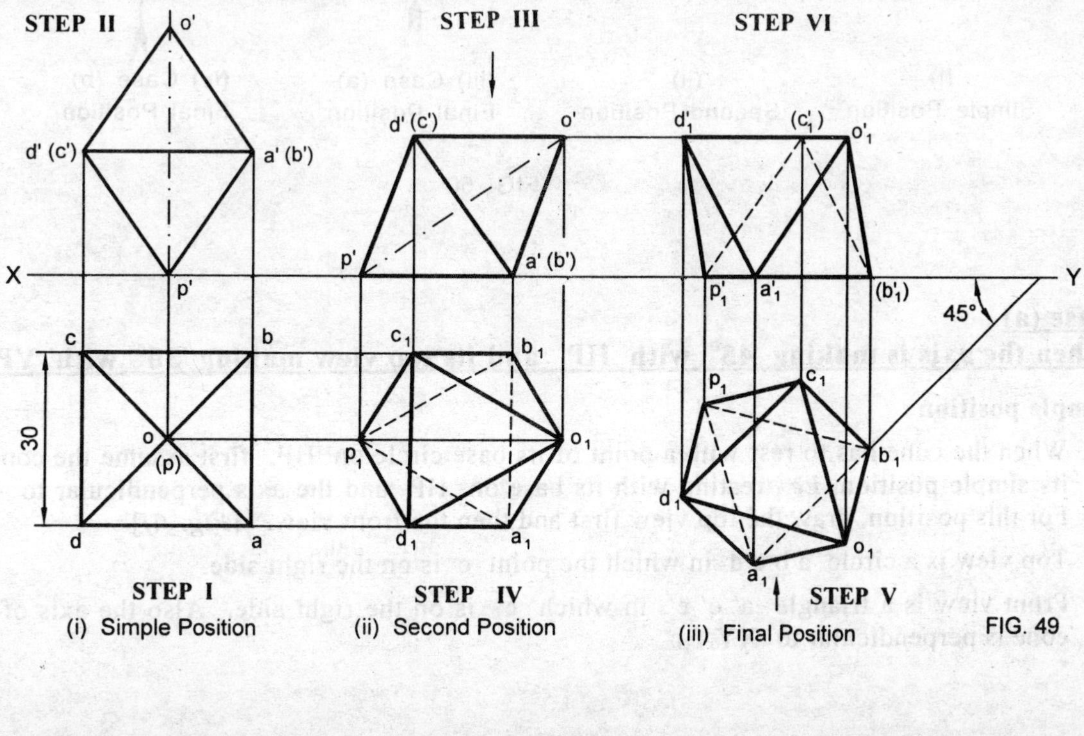

(i) Simple Position (ii) Second Position (iii) Final Position

FIG. 49

Problem 78: *(Fig. 50) Draw the projections of a* **cone**, *base 30 mm diameter and axis 50 mm long, resting on HP on a point of its base circle with*

(a) the axis making an angle of 45° with HP and its top view making an angle of 30° with VP

(b) the axis making 45° with HP and 30° with VP. (UQ)

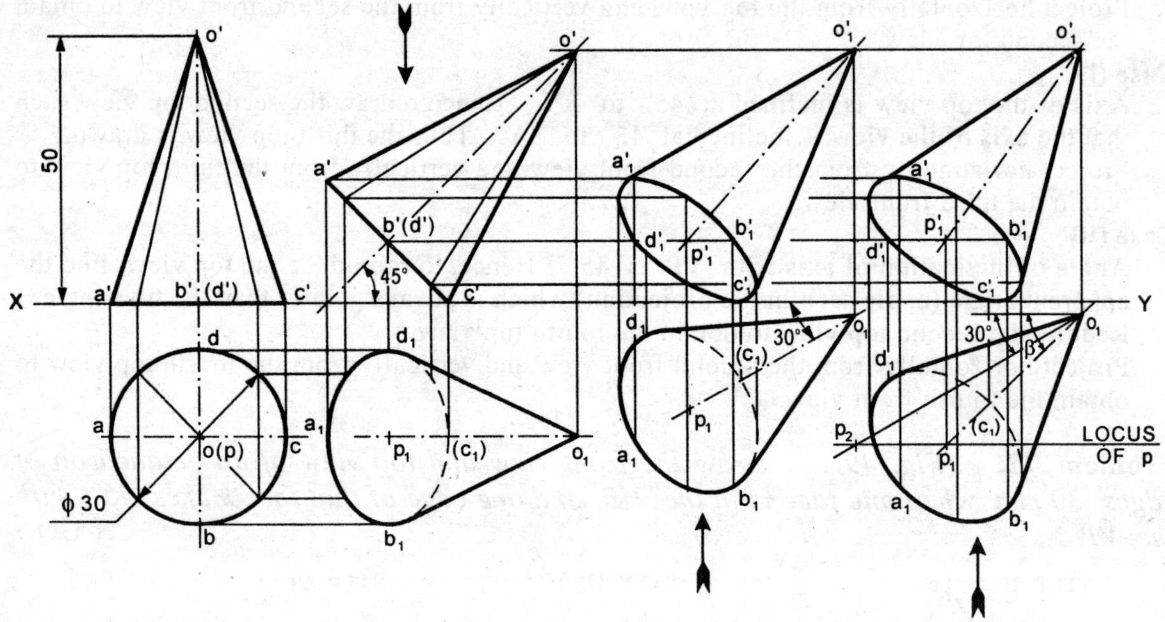

(i) Simple Position

(ii) Second Position

(iii) Case (a) Final Position

(iv) Case (b) Final Position

FIG. 50

Case (a)

When the axis is making 45° with HP and its top view making 30° with VP

Simple position

1. When the cone has to rest with a point of its base circle on HP, first assume the cone in its simple position, *i.e.*, resting with its base on HP and the axis perpendicular to HP. For this position, draw the top view first and then the front view. [*Fig. (i)*]
2. Top view is a circle a b c d in which the point c is on the right side.
3. Front view is a triangle a' o' c' in which c' is on the right side. Also the axis of the cone is perpendicular to XY.

Second position

4. The cone has to rest with a point of its base circle on HP, such that its axis has to make $\theta = 45^\circ$ with HP. Hence tilt the first front view such that c' is on XY and its base is inclined at an angle of 45°, (*i.e.*, $90 - \theta$) to XY. Thus obtain the second front view. [*Fig. (ii)*]

5. Project the corresponding second top view in which the base is seen as an ellipse and the axis is parallel to XY. The inner portion of the ellipse b_1 (c_1) d_1 between the tangents is shown as dashed line.

NOTE : *These tangents need not be any of the particular generators that have been chosen, i.e., need not be* $o_1 b_1$ *or* $o_1 d_1$.

Final position

6. As per the given data, the top view of the axis has to make 30° with VP, keeping the inclination of the axis with HP constant. So the shape and size of the second top view will remain the same, but its position alone will change.

   Therefore tilt the second top view such that $o_1 p_1$, the top view of the axis is inclined at 30° to XY and obtain the final top view. Then project the corresponding final front view. [*Fig. (iii)*]

7. Look at the final top view in the direction of arrow. Base is closer than the apex. So draw the base completely by thick line in the final front view. Also from o_1' draw two tangents to the ellipse.

Case (b)

When the axis is making 45° with HP and 30° with VP

1. Repeat steps 1 to 5 in case (a) and obtain the second top view.

2. **Final position**

   Since the axis makes 45° with HP and 30° with VP, its top view will make an apparent angle β greater than 30° with XY (as per the concept of the projections of straight line inclined to both HP and VP)

3. **To determine β**

   From o_1 draw a line inclined at 30° to XY and mark $o_1 p_2$ on it such that it is equal to the true length of the axis = 50 mm. With o_1 as center and $o_1 p_1$ (length of the axis in the top view) as radius draw an arc to cut the locus of p at p_1. Now the angle made by $p_1 o_1$ with XY is β (> 30°).

4. Redraw the second top view with respect to $p_1 o_1$ which makes an angle of β with XY. Project the corresponding final front view. [*Fig. (iv)*]

Problem 79: *(Fig. 51) A cone 40 mm base and axis 50 mm long touches the VP on a point of its base circle. The axis is inclined at 30° to VP and the front view of the axis is inclined at 45° to HP. Draw its projections.* (UQ)

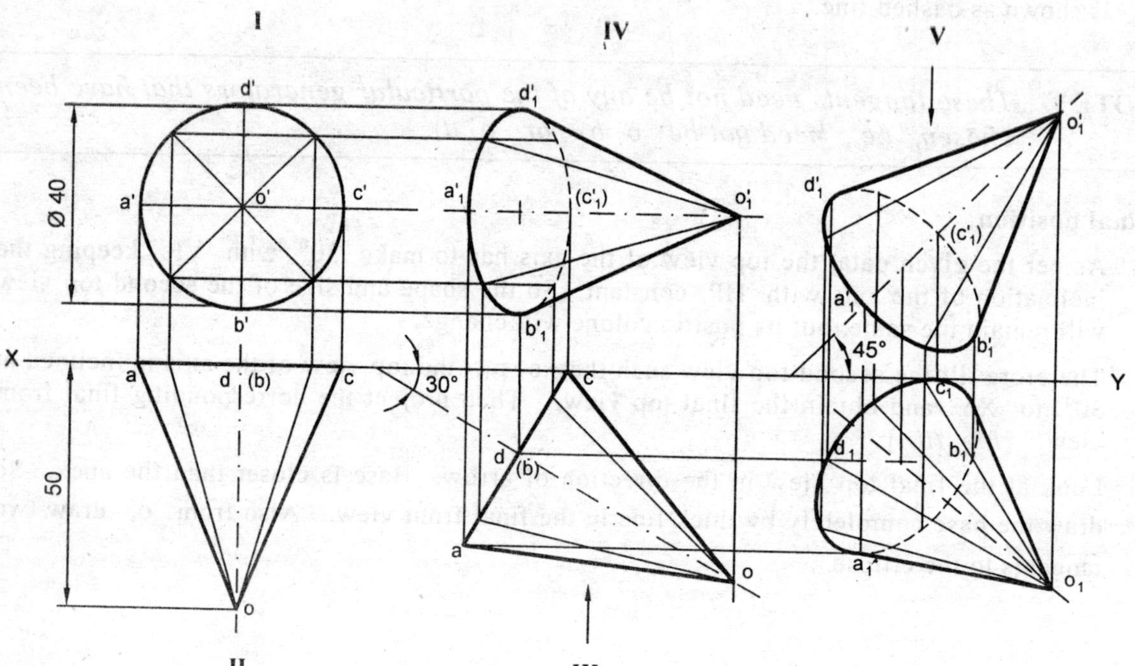

FIG. 51

Problem 80: *(Fig. 52) Draw the projections of a cylinder, base 30mm diameter and axis 40 mm long, rests with a point of its base circle on HP such that the axis is making an angle of 30° with HP and its top view perpendicular to VP.*

Simple position : See *Fig. (i)*.

1. First assume the cylinder, resting with its base on HP and axis perpendicular to it. Draw top view and the corresponding front view.

Second position : See *Fig. (ii)*.

2. The cylinder has to rest with a point of its base circle on HP such that its axis makes $\theta = 30^{\circ}$ with HP. Hence tilt the front view such that the point 3' is on XY and its base is inclined at 60° (*i.e.*, $90^{\circ} - \theta$) to XY. Thus obtain the second front view.
3. Project the corresponding second top view, in which axis is parallel to XY.

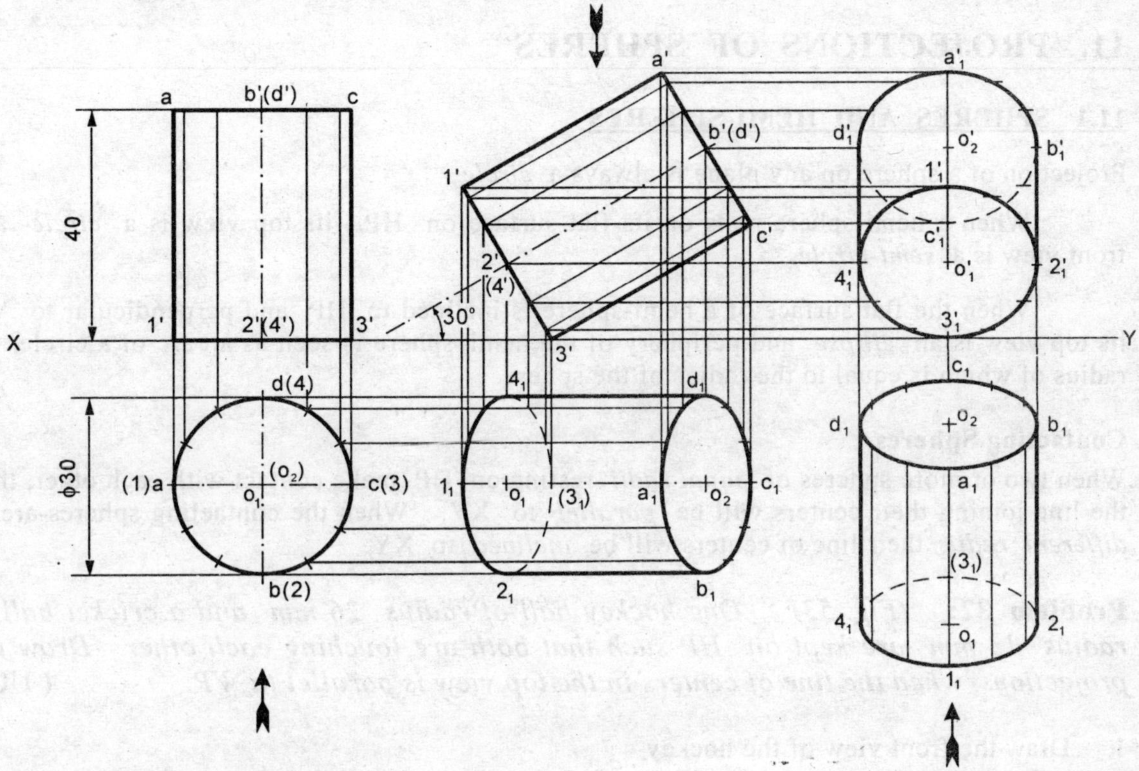

(i) Simple Position (ii) Second Position (iii) Final Position

FIG. 52

Final position

4. As per the given condition the cylinder is to be tilted such that top view of the axis is perpendicular to XY, keeping the inclination of the axis with HP constant. Hence the shape and size of the second top view will remain the same, but its position alone will change.

5. Therefore tilt the second top view such that o_1 o_1, the top view of the axis is perpendicular to XY and obtain the final top view.

6. Then project the final front view as in *Fig. (iii)*.
Look at the top view in the direction of arrow. 1_1 2_1 (3_1) 4_1 is nearer to the observer. Hence draw the curve $1'_1$ $2'_1$ $3'_1$ $4'_1$ by thick line.

7. b_1 c_1 d_1, the portion of the top face is invisible in final top view. Hence draw the curve $b'_1(c'_1)d'_1$ by dashed line in the final front view.

Problem 81: *(Exercise) A cylinder of diameter 30 mm and axis height 60 mm lying on the ground on a point of its base circle such that the axis is inclined at 45° to the HP and the plane containing the axis makes an angle of 30° with the VP. Draw the projections of the cylinder.*

11. PROJECTIONS OF SPHERES

11.1 SPHERES AND HEMI-SPHERES

Projection of a sphere on any plane is always a *circle.*

When a hemi-sphere rests on its flat surface on HP, its top view is a *circle* and front view is a *semi-circle.*

When the flat surface of a hemi-sphere is inclined to HP and perpendicular to VP, its top view is an *ellipse* and periphery of the hemi–sphere is seen as an arc of a circle, the radius of which is equal to the radius of the sphere.

Contacting Spheres

When two or more spheres of *equal radii* resting on HP make contact with each other, then the line joining their centers will be *parallel* to XY. When the contacting spheres are of *different radii,* their line of centers will be *inclined* to XY.

Problem 82: *(Fig. 53) One hockey ball of radius 26 mm and a cricket ball of radius 13 mm are kept on HP such that both are touching each other. Draw the projections, when the line of centers in the top view is parallel to VP.* (UQ)

1. Draw the front view of the hockey ball (ϕ 52 mm).
2. To draw the cricket ball, draw a locus line 13 mm above HP for its center.
3. Cut an arc on it with p′ as center and 39 (= 26 + 13) mm radius.
4. With q′ as center draw the front view of the cricket ball.
5. Since the line joining the centers of spheres is parallel to VP, draw a line for it in top view.
6. Draw the projectors to obtain p and q.
7. Complete the top view as shown.

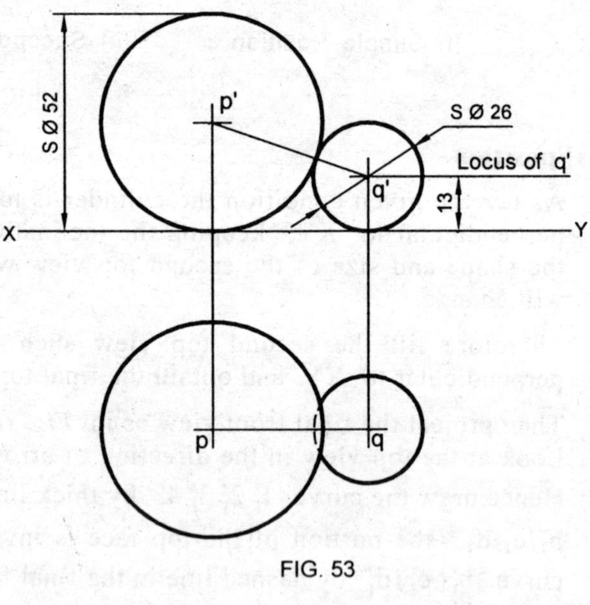

FIG. 53

Problem 83: *(Exercise) Two spheres of radii 24 mm and 36 mm are lying on the HP touching each other, such that the line joining the centers makes an angle of 45° with VP. Draw the projections.*

Problem 84: *(Fig. 54)* *A hemispherical bowl of 48 mm diameter is resting on HP with its flat surface inclined at 45° to HP and perpendicular to VP. Draw the projections of the bowl.* (UQ)

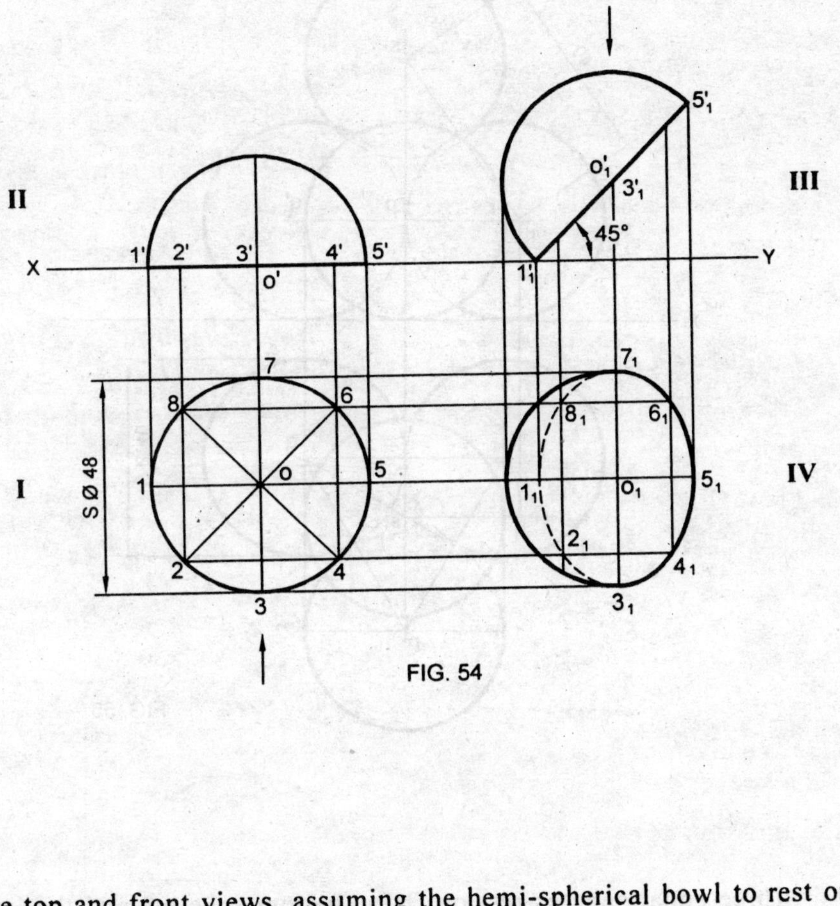

FIG. 54

1. Draw the top and front views, assuming the hemi-spherical bowl to rest on HP on its flat surface.
2. Tilt the front view such that the flat surface is inclined at 45° to XY.

Problem 85: *(Exercise)*

An apple of 50 mm diameter is cut at two ends. The cut surfaces are flat and circular in shape of 35 mm and 25 mm diameters. The cut surfaces are parallel to each other. Draw the projections of the apple when one of the cut surfaces is perpendicular to VP and inclined at 35° to HP.

Problem 86: *(Fig. 55) Four equal spheres of diameters 40 mm are arranged in a pyramidal form such that three spheres are resting on HP. The line joining the centers of two of them is parallel to VP. Each sphere is in contact with the other three. Draw the projections of the sphere assembly.* (UQ)

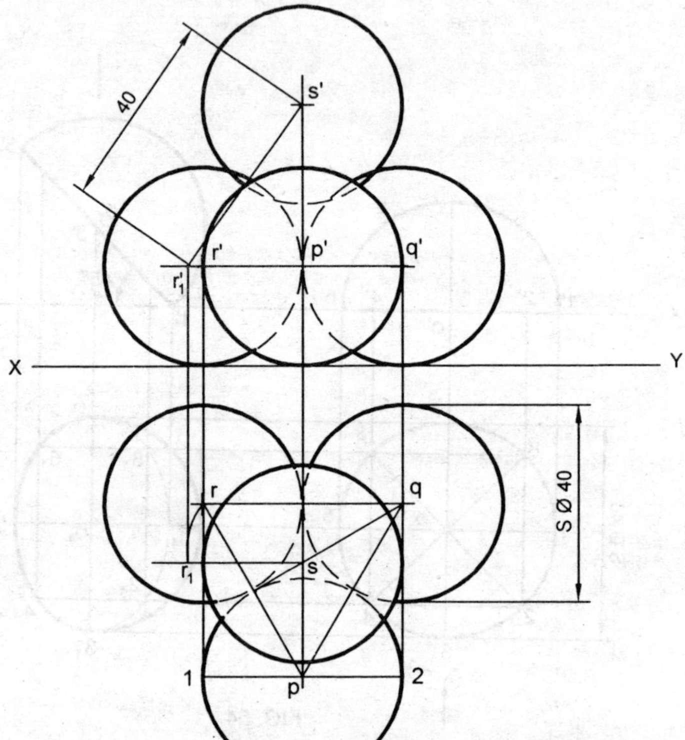

FIG. 55

1. *Given :* Three equal spheres rest on HP (bottom spheres) and the line joining the centers of two of them is parallel to VP.

   Hence, the line joining the centers of the two spheres is parallel to XY.

2. Looking from the top, the centers p, q and r of the three bottom spheres lying on HP form an equilateral triangle of sides = (20 + 20) mm = 40 mm.

   Hence, draw an equilateral triangle p q r of side 40 mm with one side parallel to XY in the top view.

3. p, q and r as centers, draw three circles of 40 mm diameter to represent the three bottom spheres.

4. Draw vertical projectors from the top view to obtain the front view.
5. In the front view mark the centers p′, q′ and r′ such that they lie on a line parallel to and 20 mm above XY.
6. Draw the front view as three circles with p′, q′ and r′ as center and 20 mm radius.

Fourth (Top) Sphere

7. Now, place the fourth sphere on the top of the three spheres.
8. In the top view, the center s of the top sphere is nothing but the center of the equilateral triangle p q r. With center s and radius 20 mm, draw the top view of the top sphere as a circle as shown.
9. Mark the hidden portions of the three bottom spheres, hidden by the top sphere.
10. Draw the projector through s to locate center s′ of the top sphere in the front view.

To locate s′

NOTE : **True Distance** between the center of the top sphere and that of any one of the bottom spheres = Sum of the radii = (20 + 20) mm = 40 mm. But, the lines sp, sq and sr are not parallel to XY in the top view. Hence their front views will not show their true lengths.

---

11. Therefore, to locate the position of s′, make one of the lines, say, sp is turned to sp_1 parallel to XY.
12. Project p_1 to p_1' on a line drawn through p′ parallel to XY (which is the locus of p′).
13. p_1' as center and 40 mm as radius, draw an arc to intersect the vertical projector through s at the point s′. Now s′ is the center of the top sphere in the front view.
14. s′ as center and 20 mm as radius, draw a circle to represent the top sphere.
15. In the front view, a portion of the upper circle is hidden. Also, the circles with centers p′ and q′ are partly hidden. The bottom sphere R nearer to the observer is seen fully in the front view.

Problem 87: *(Exercise)*

Draw the projections of three equal spheres of 40 mm diameters, resting on HP touching each other. The line of centers of two of them is inclined at an angle of 30° to VP. A fourth sphere of diameter 44 mm is kept over the three spheres in a pyramidal form touching the bottom three spheres.

12. COMBINATION OF SOLIDS

Problem 88: *(Fig. 56) A hexagonal headed bolt of 27 mm diameter has its cylindrical portion 58 mm long. The hexagonal head is of side 27 mm, having a thickness of 18 mm. Draw the projections of the bolt, when it is lying on HP with an edge of the head perpendicular to VP and axis inclined at 60° to HP.* (UQ)

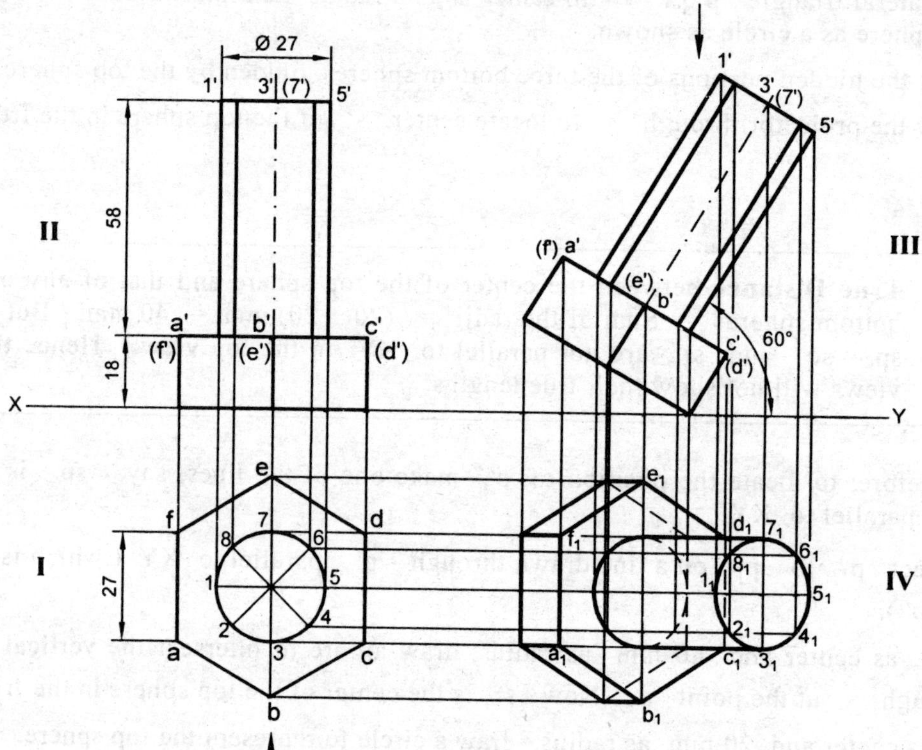

FIG. 56

1. Draw the top view first with an edge of the head perpendicular to XY.
2. Project the front view.
3. Tilt the front view such that the axis of the bolt is inclined at an angle of 60° to XY.
4. Now project the final top view as shown.

Problem 89: *(Exercise) A rivet has a cylindrical stem of 20 mm diameter and 50 mm length. The rivet head has hemispherical shape of 30 mm diameter. The rivet is lying on HP with its head touching VP and the axis making 45° with VP. Draw its projections.*

13. ADDITIONAL PROBLEMS – II

Problem 90: A pentagonal prism of side of base 28 mm and altitude 58 mm rests on the HP on one of its base edges such that the base is inclined at an angle of 30° to HP and the top view of the axis is inclined at an angle of 60° to VP. Draw the projections.

Problem 91: Draw the projections of a hexagonal prism, side of base 30 mm and axis 60 mm long, resting with one of the edges of its base on HP. Its axis is inclined at 30° to the HP and the top view of the axis at 45° to VP.

Problem 92: An equilateral triangular pyramid of side of base 34 mm and height 68 mm is lying on HP on a slant face such that the nearest slant edge containing the resting face parallel to VP. Draw the projections.

Problem 93: A square pyramid of side of base 40 mm and altitude 60 mm is lying on HP on one of its triangular faces with the plane containing the axis inclined at an angle of 45° to the VP, such that the apex is far away from VP. Draw the projections.

Problem 94: A hexagonal pyramid, side of base 25 mm and axis 50 mm long, rests with one of the corners of its base on HP. Its axis is inclined at 30° to HP and 45° to VP. Draw its projections.

Problem 95: A frustum of cone of base diameter 54 mm, top face diameter 27 mm and altitude 50 mm is lying on HP on one of its generators with the plane containing it is inclined at an angle of 40° to VP. Draw the projections.

Problem 96: A cylindrical block of diameter 75 mm and thickness 25 mm, has a hexagonal hole of 25 mm side, cut centrally through its flat faces. Draw the projections of the block, when its flat faces are vertical and inclined at 30° to VP and a face of the hexagonal hole is parallel to HP.

Problem 97: A pile of spheres are formed by placing a sphere of 40 mm diameter over three other spheres of 27 mm diameter lying on HP so that each one touches the other two. The lines joining the centers of any two of the bottom spheres with the other bottom sphere make equal inclinations with VP and being nearer to the observer. Draw the projections of the spheres.

HINT

The centers of the three spheres lying on HP, such that each one touches the other two, when connected, forms an *equilateral triangle.*

Problem 98: A 55 mm diameter spherical paper weight has a 40 mm diameter of flat base. It is placed on HP such that its flat base is perpendicular to HP and inclined at 40° to VP. Draw its projections.

Problem 99: Three equal spheres of 25 mm diameter are placed centrally in a hemispherical bowl of 75 mm diameter, which is lying on its curved surface on HP. Each sphere touches the other two.

Draw the top and front views of the bowl with the spheres, when the line joining the centers of two spheres is perpendicular to VP. Find the height of the centers of the spheres above HP.

(UQ)

HINT

Here, the centers of the three spheres will be at a vertical of an equilateral triangle of side = 12.5 + 12.5 + 25 mm. Draw an equilateral triangle with a side perpendicular to XY in the top view and draw the three circles. Project the front view.

*

17

AUXILIARY PROJECTIONS OF SOLIDS

1. CHANGE OF REFERENCE LINE METHOD (AUXILIARY PROJECTION METHOD)

We have already discussed the Auxiliary Projections of Points and Straight Lines in the Chapter 14 and of Plane Surfaces in the Chapter 15. The method adopted is Change of Reference Line Method, otherwise known as Auxiliary Projection Method.

This Chapter deals with the auxiliary projections of solids when (i) the axis of the solid is inclined to one plane and parallel to the other, and (ii) the axis of the solid is inclined to both the reference planes, so as to obtain the auxiliary views of the solids.

2. AXIS OF SOLID INCLINED TO ONE PLANE AND PARALLEL TO OTHER

When axis is inclined to one plane and parallel to other, draw its projections in ***two stages*** :

Stage I – Simple Position

In the first stage, assume the solid to be in simple position, *i.e.*, its axis perpendicular to one of the planes. (That is, if the axis is inclined to HP, assume the solid to be perpendicular to HP. Similarly, if the axis is inclined to VP, assume the solid to be perpendicular to VP.)

NOTE :

1. *If the solid has an edge of its base in HP or parallel to HP, keep that edge perpendicular to VP. If the edge of the base is in VP or parallel to VP, keep that edge perpendicular to HP.*
2. *If the solid has a corner of its base on HP, keep the sides of the base containing that corner equally inclined to VP. If the solid has a corner of its base in VP, keep the sides of base containing that corner inclined to HP.*

Stage II

- Change of Position Method: Discussed in the previous Chapter 16.
- Change of Reference Line or Auxiliary Plane Method: In this method, introduce an auxiliary plane according to the required conditions. Project the required final views.

2.1 AUXILIARY VIEWS OF PRISM, PYRAMID, CYLINDER AND CONE

Problem 1: *(Fig. 1) A hexagonal prism, side of base 20 mm and axis 48 mm long, rests with its base on HP such that an edge of the base is parallel to VP. Draw the projection of the prism on an Auxiliary Inclined Plane (A.I.P.) which makes an angle of 60° with the HP.* (UQ)

Simple Position

1. Prism rests with its base on HP and an edge of the base is parallel to VP. For this position, draw top and front views.

Auxiliary Top View on A.I.P.

2. A.I.P. is inclined at 60° to HP. Hence draw a new reference line X_1Y_1 at an angle of 60° to XY to project the Auxiliary Top View.
3. From all corners of the prism in the front view, draw projectors perpendicular to X_1Y_1.
4. Measure the distances of all the corners from XY in the first top view and transfer the same with reference to X_1Y_1 on the corresponding projectors. Thus locate all the points in the auxiliary top view.

To mark visible and invisible edges in the Auxiliary Top View

5. Look at the front view in the direction of arrow shown.
6. End face nearer to the observer is visible. Hence draw $a_1 b_1 c_1 d_1 e_1 f_1$ by *continuous thick lines* in Auxiliary Top View.
7. Draw the boundary lines of the view by *thick* lines.
8. a'1' is visible and hence draw $a_1 1_1$ by *thick* line.
9. 4' is invisible in the front view. Hence draw the edges $(4_1)\ 3_1$, $(4_1)\ 5_1$ and $(4_1)\ d_1$ as dashed lines in the Auxiliary Top View.

Problem 2: *(Exercise) A pentagonal prism, side of base 25 mm and axis 50 mm long, rests with one of its shorter edges on HP such that the base containing that edge makes an angle of 30° to HP and its axis is parallel to VP. Draw the projections.*

Problem 3: *(Fig. 2)*

Draw the projections of a cube of 30 mm side, resting in the HP on one of its corners with a solid diagonal perpendicular to the HP.

See Fig. 2 for Solution.

Problem 4: *(Exercise) Draw the projections of a cube of 35 mm side, when it rests on HP on one of its corners and a face containing that corner is inclined at an angle of 30° to HP and perpendicular to VP.*

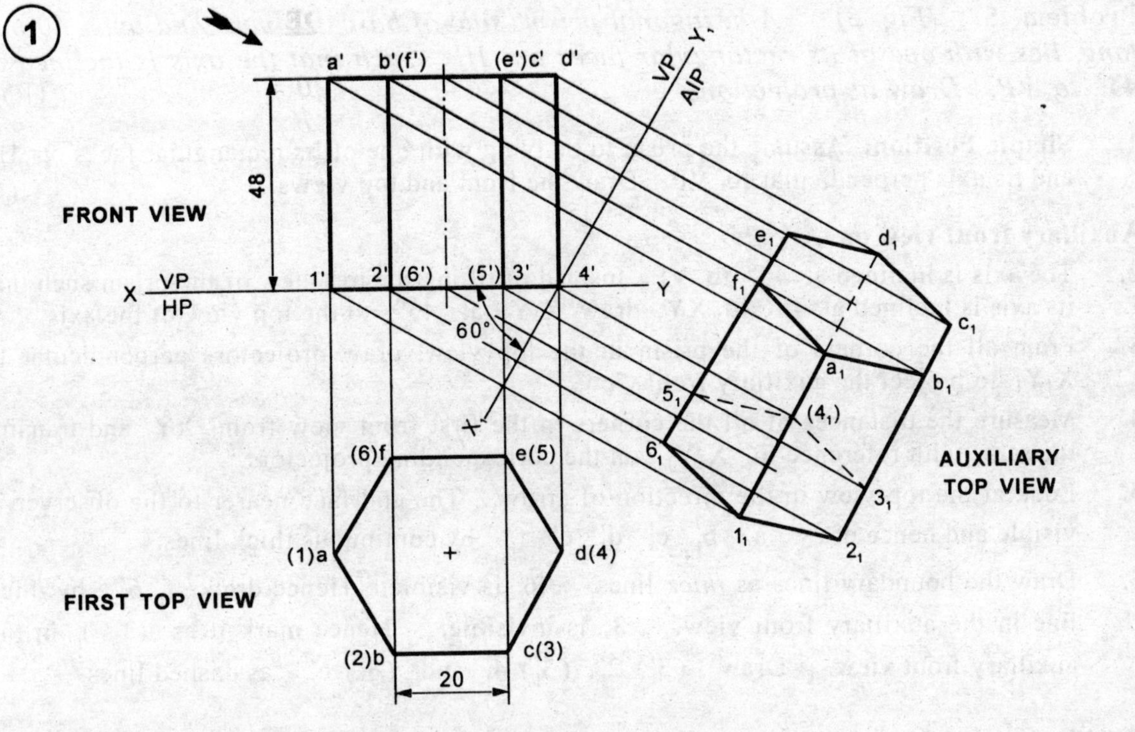

FIG. 1 CHANGE OF REFERENCE LINE METHOD

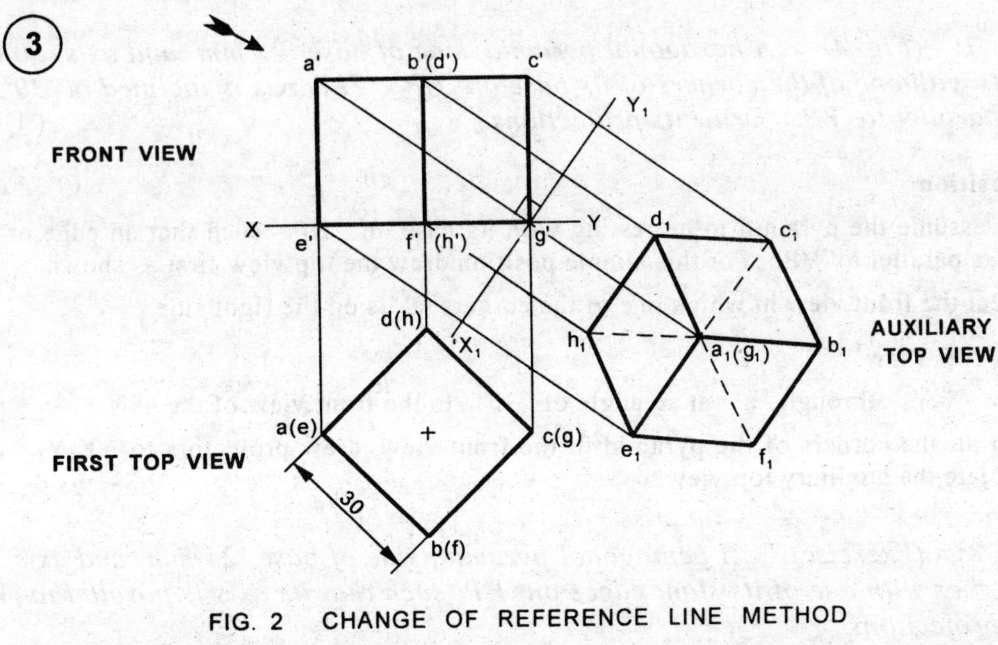

FIG. 2 CHANGE OF REFERENCE LINE METHOD

Problem 5: *(Fig. 3)* *A hexagonal prism, side of base 25 mm and axis 60 mm long, lies with one of its rectangular faces on HP, such that the axis is inclined at 45° to VP. Draw its projections.* (UQ)

1. **Simple Position:** Assume the prism to be lying with one of its rectangular faces on HP and its axis perpendicular to VP. Draw the front and top views.

Auxiliary front view on A.V.P.

2. The axis is inclined at 45° to VP. Instead of tilting the top view of the prism such that its axis is inclined at 45° to XY, draw X_1Y_1 at 45° to the top view of the axis.
3. From all the corners of the prism in the top view, draw projectors perpendicular to X_1Y_1 to project the auxiliary front view.
4. Measure the distances of all the corners in the first front view from XY and transfer the same with reference to X_1Y_1 on the corresponding projectors.
5. Look at the top view in the direction of arrow. The end face nearer to the observer is visible and hence draw a_1' b_1' c_1' d_1' e_1' f_1' by continuous thick lines.
6. Draw the boundary lines as *thick* lines. f6 is visible. Hence draw f_1' $6_1'$ by thick line in the auxiliary front view. 3 is invisible. Hence mark it as ($3_1'$) in the auxiliary front view. Draw ($3_1'$) $2_1'$, ($3_1'$) $4_1'$ and ($3_1'$) c_1' as dashed lines.

Problem 6: *(Exercise)* *A hexagonal prism, side of base 22 mm and axis 58 mm long, lies with one of its rectangular faces on HP, such that the axis is inclined at an angle of 60° to VP. Draw its projections.*

Problem 7: *(Fig. 4)* *A hexagonal pyramid, side of base 25 mm and axis 50 mm long, rests with one of the corners of its base on HP. The axis is inclined at 30° to HP and parallel to VP. Draw its projections.* (UQ)

Simple Position

1. First assume the pyramid to be resting with its base on HP such that an edge of the base is parallel to VP. For this simple position draw the top view first as shown.
2. Project the front view in which one of the corners a' is on the right side.

Auxiliary Top View

3. Draw X_1Y_1 through a' at an angle of 30° to the front view of the axis.
4. From all the corners of the pyramid in the front view, draw projectors to X_1Y_1 and complete the auxiliary top view.

Problem 8: *(Exercise)* *A pentagonal pyramid, side of base 25 mm and axis 55 mm long, lies with one of its slant edges on HP such that its axis is parallel to VP. Draw its projections.*

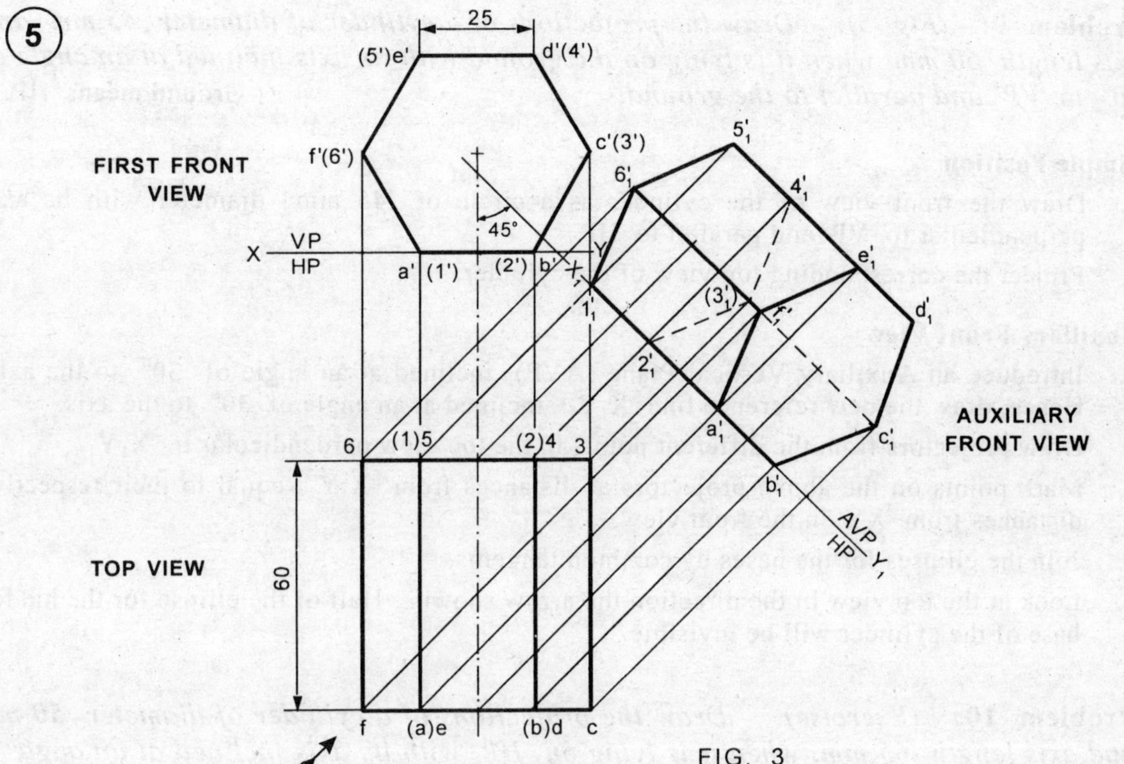
5
25
(5')e'
d'(4')
c'(3')
f'(6')
FIRST FRONT VIEW
45°
X
VP
HP
Y
a'
(1')
(2')
b'
5'1
6'1
4'1
e'1
(3'1)
f'1
d'1
1'1
2'1
a'1
c'1
b'1
AUXILIARY FRONT VIEW
AVP
HP
Y1
6
(1)5
(2)4
3
TOP VIEW
60
f
(a)e
(b)d
c

FIG. 3

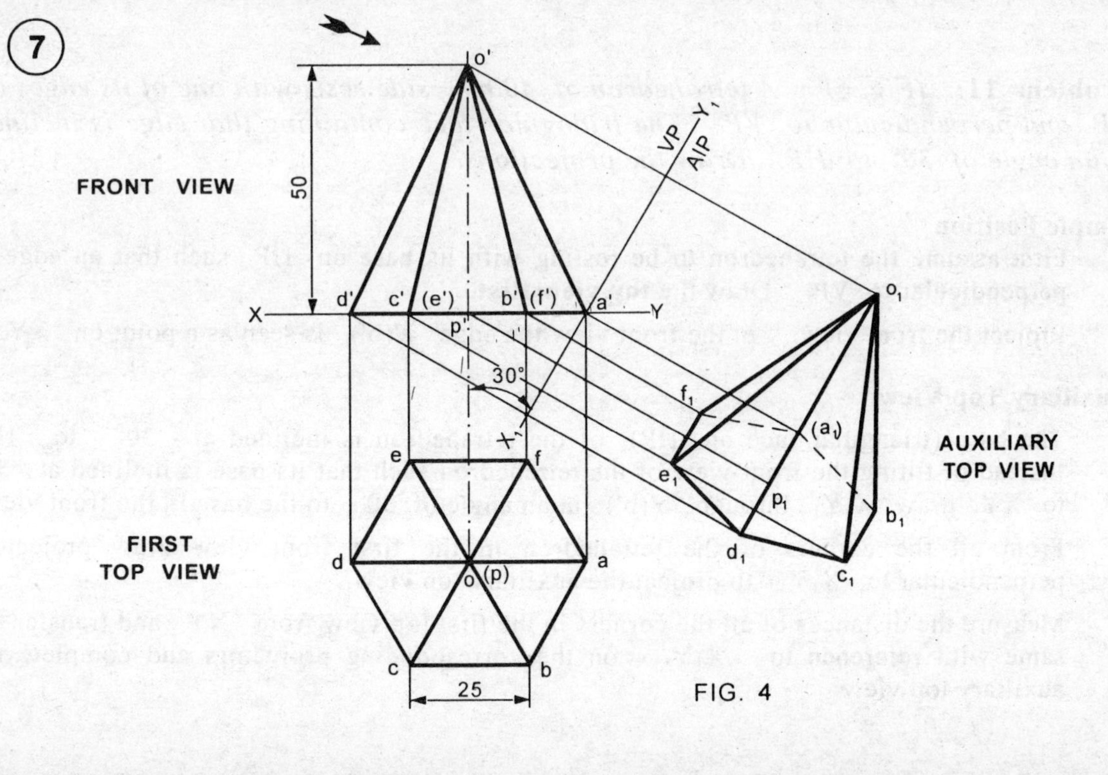
7
o'
FRONT VIEW
50
VP
AIP
Y1
X
d'
c'
(e')
b'
(f')
a'
Y
p'
30°
X1
o1
f1
(a1)
e1
p1
b1
d1
c1
AUXILIARY TOP VIEW
e
f
FIRST TOP VIEW
d
o (p)
a
c
b
25

FIG. 4

Problem 9: *(Fig. 5) Draw the projections of a cylinder of diameter 45 mm and axis length 60 mm when it is lying on the ground with its axis inclined at an angle of 30° to VP and parallel to the ground.* (Ground means HP.)

Simple Position

1. Draw the front view of the cylinder as a circle of 45 mm diameter with its axis perpendicular to VP and parallel to HP.
2. Project the corresponding top view of the cylinder.

Auxiliary Front View

3. Introduce an Auxiliary Vertical Plane (AVP) inclined at an angle of 30° to the axis. Hence, draw the new reference line X_1Y_1 inclined at an angle of 30° to the axis.
4. Draw projectors from the different points in the top view perpendicular to X_1Y_1.
5. Mark points on the above projectors at distances from X_1Y_1 equal to their respective distances from XY in the front view.
6. Join the ellipses for the bases by common tangents.
7. Look at the top view in the direction the arrow shown. Half of the ellipse for the hidden base of the cylinder will be invisible.

Problem 10: *(Exercise) Draw the projections of a cylinder of diameter 50 mm and axis length 65 mm when it is lying on HP with its axis inclined at an angle of 45° to VP and parallel to HP.*

Problem 11: *(Fig. 6) A tetrahedron of 40 mm side rests with one of its edges on HP and perpendicular to VP. The triangular face containing that edge is inclined at an angle of 30° to HP. Draw the projections.* (UQ)

Simple Position

1. First assume the tetrahedron to be resting with its base on HP such that an edge is perpendicular to VP. Draw the **top view first**.
2. Project the front view. In the front view the edge a'(b') is seen as a point on XY.

Auxiliary Top View

3. The base (triangular face on HP) of the tetrahedron is inclined at 30° to HP. Instead of tilting the front view of the tetrahedron such that its base is inclined at 30° to XY, draw X_1Y_1 through a'(b') at an angle of 30° to the base in the front view.
4. From all the corners of the tetrahedron in the first front view draw projectors perpendicular to X_1Y_1 to project the auxiliary top view.
5. Measure the distances of all the corners in the first top view from XY and transfer the same with reference to X_1Y_1 on the corresponding projectors and complete the auxiliary top view.

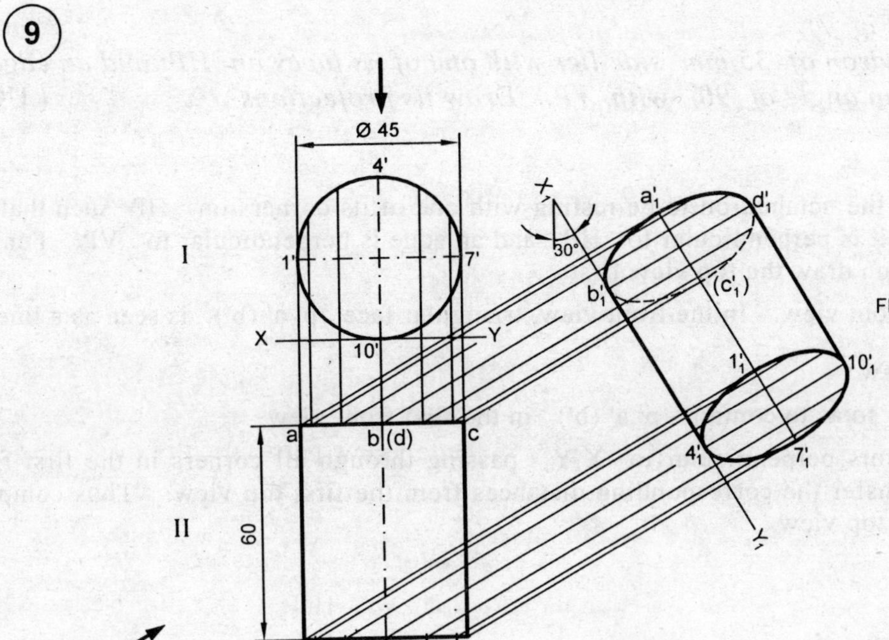
9
Ø 45
4'
I
1'
7'
10'
X
Y
a'1
d'1
b'1
(c'1)
III
AUXILIARY
FRONT VIEW
30°
1'1
10'1
4'1
7'1
a
b (d)
c
II
60
1
4 (10)
7

FIG. 5

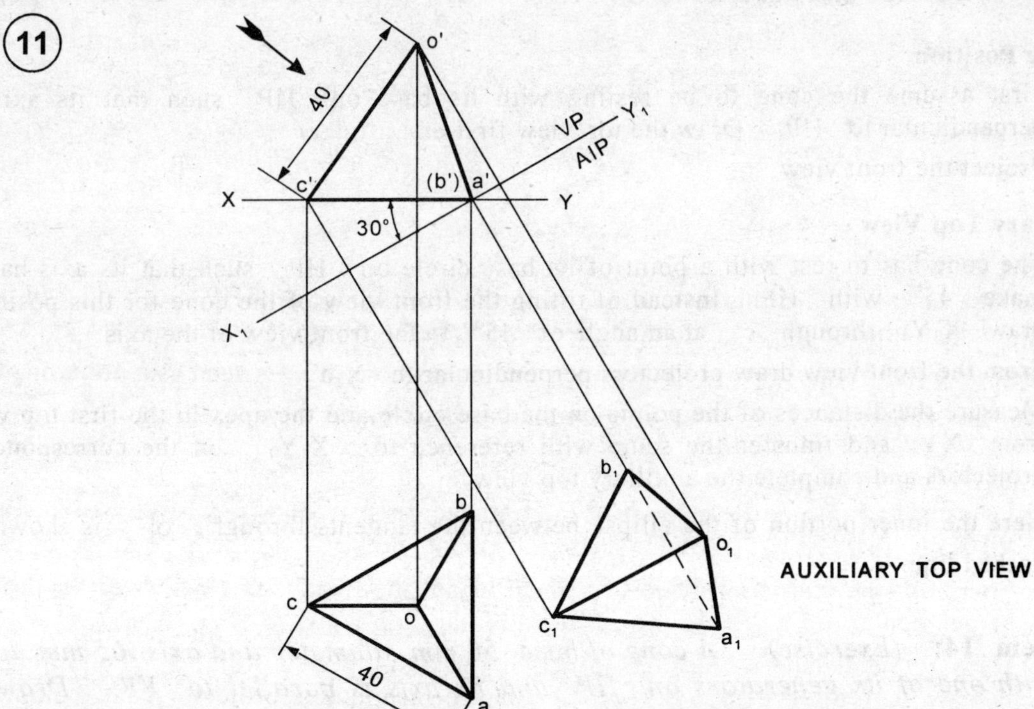
11
o'
40
VP
AIP
Y1
X
c'
(b') a'
Y
30°
X1
b1
o1
c1
a1
b
c
o
a
40
AUXILIARY TOP VIEW

FIG. 6

Problem 12: *(Fig. 7)*

An octahedron of 35 mm side lies with one of its faces on HP and an edge of that face makes an angle of 90° with VP. Draw its projections. (UQ)

Simple Position

1. First assume the octahedron to be resting with one of its corners on HP such that the axis through it is perpendicular to HP and an edge is perpendicular to VP. For this simple position draw the top view first.
2. Project the front view. In the front view, triangular face p' a' (b') is seen as a line.

Auxiliary Top View

3. Draw X_1Y_1 so as to contain p' a' (b') in the first front view.
4. Draw projectors perpendicular to X_1Y_1 passing through all corners in the first front view and transfer the corresponding distances from the first top view. Thus complete the auxiliary top view.

Problem 13: *(Fig. 8)*

Draw the projections of a cone, base 30 mm diameter and axis 50 mm long, when it is resting on HP on a point of its base circle with the axis making an angle of 45° with HP and parallel to VP. (UQ)

Simple Position

1. First assume the cone to be resting with its base on HP such that its axis is perpendicular to HP. Draw the top view first.
2. Project the front view.

Auxiliary Top View

3. The cone has to rest with a point of its base circle on HP such that its axis has to make 45° with HP. Instead of tilting the front view of the cone for this position, draw X_1Y_1 through c' at an angle of 45° to the front view of the axis.
4. From the front view draw projectors perpendicular to X_1Y_1.
5. Measure the distances of the points on the base circle and the apex in the first top view from XY and transfer the same with reference to X_1Y_1 on the corresponding projectors and complete the auxiliary top view.
6. Here the inner portion of the ellipse between the tangents through o_1' is shown by dashed line.

Problem 14: *(Exercise)* *A cone of base 50 mm diameter and axis 65 mm long, lies with one of its generators on HP and its axis is parallel to VP. Draw its projections.*

(12)

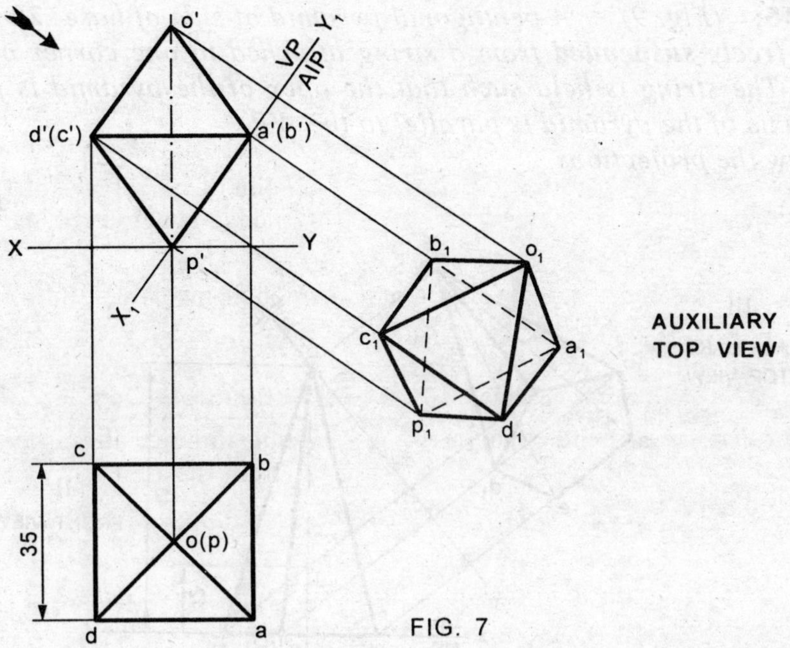

FIG. 7

(13)

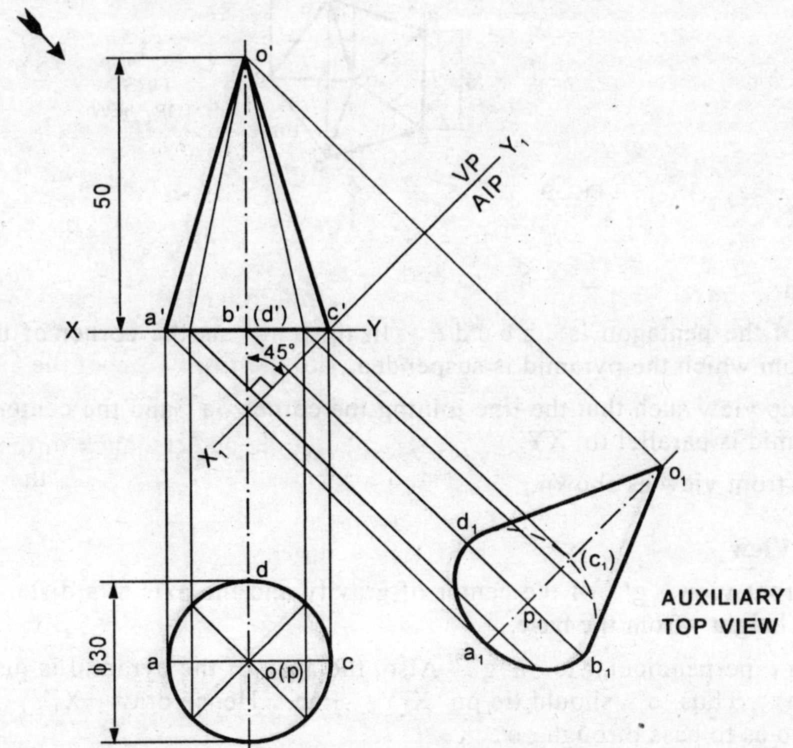

FIG. 8

Problem 15: *(Fig. 9) A pentagonal pyramid of side of base 20 mm and altitude 48 mm is* ***freely suspended*** *from a string attached to one corner of the base of the pyramid. The string is held such that the apex of the pyramid is just touching the HP. The axis of the pyramid is parallel to the VP.*

Draw the projections. (UQ)

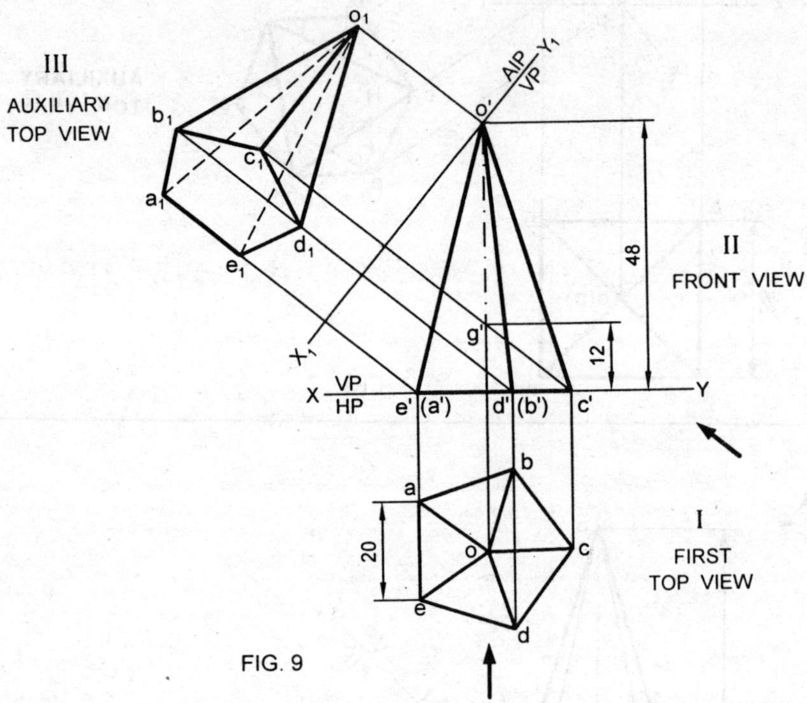

FIG. 9

Simple Position

1. Top view of the pentagon is a b c d e. In this, a is the corner of the base of the pyramid from which the pyramid is suspended.
2. Draw the top view such that the line joining the corner a and the center of gravity g of the pyramid is parallel to XY.
3. Project the front view as shown.

Auxiliary Top View

4. Mark the front view g' of the center of gravity and the axis at a distance of 12 mm (= 1 / 4 th height) from the base.
5. Draw X_1Y_1 perpendicular to a' g'. Also, the apex of the pyramid is just touching the HP (Given). Thus o' should lie on X_1Y_1 line. Hence draw X_1Y_1 perpendicular to a' g' so as to pass through o'.
6. Project the Auxiliary Top View as shown.

Problem 16: *(Exercise)* *A hexagonal pyramid of side of base 25 mm and height 55 mm is* ***suspended freely*** *from a corner such that the axis is parallel to VP. Draw its projections.*

Problem 17: *(Fig. 10)*

A ***frustum*** *of a square pyramid, base side 40 mm, top side 22 mm and altitude 53 mm has its bases equally inclined to VP and its axis is parallel to VP. It is tilted such that its axis is inclined at an angle of 30° to HP.*

Draw the projections of the solid. (UQ)

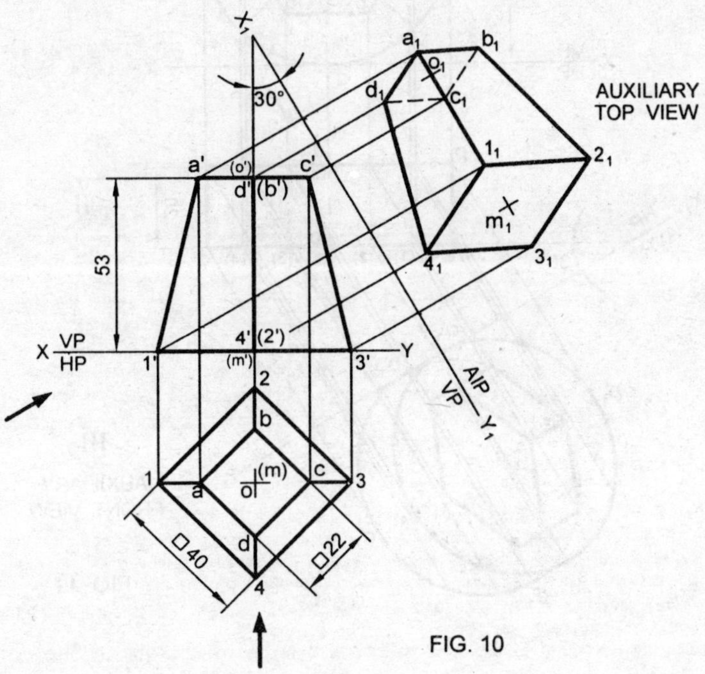

FIG. 10

Problem 18: *(Exercise)*

A ***frustum*** *of a pentagonal pyramid, base 32 mm side, top 26 mm side and altitude 58 mm is tilted about a corner of the base till its axis is parallel to the ground and inclined at an angle of 45° to VP. Top view of the frustum is nearer to VP.*

Draw the projections of the solid. (UQ)

2.2 COMBINATION OF SOLIDS

Problem 19: *(Fig. 11) A cylindrical block of 60 mm diameter and 20 mm thick has a hexagonal hole of 20 mm side cut centrally through its flat faces. Draw the top and front views of the block when its flat faces are vertical and inclined at an angle of 30° to VP and the two parallel faces of the hole are parallel to HP.* (UQ)

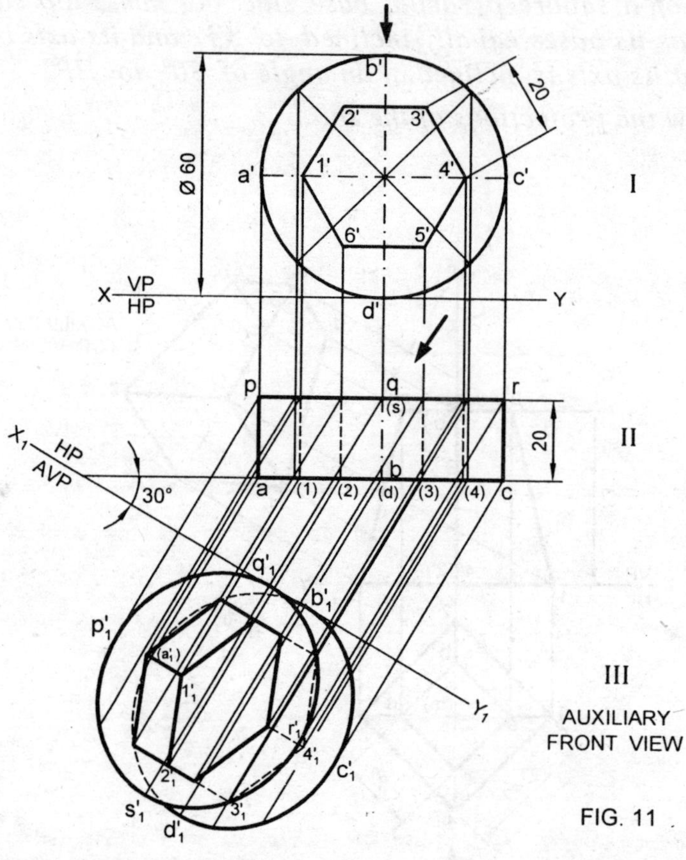

FIG. 11

1. **Simple Position:** *See Fig. 11.* Assume the cylindrical block to rest on HP with its axis perpendicular to VP and an edge of the hexagonal hole is parallel to HP. Draw the front and top views.
2. **Second Position:** Draw X_1Y_1 representing an AVP making 30° with the flat faces of the block in the top view. Project the Auxiliary Front View as shown.

Problem 20: *(Exercise) A hexagonal nut of side 40 mm and height 40 mm has a hole of diameter 40 mm drilled centrally through its ends. Draw its projections when it is resting on one of its corners of the base on the ground with its axis inclined at 60° to HP and two of its faces parallel to VP.* (UQ)

Problem 21: *(Fig. 12)*

A cube of 40 mm side is resting on HP with its adjacent vertical faces equally inclined to VP. A hexagonal pyramid of 16 mm base and altitude 48 mm is placed centrally on the top of the cube.

Draw the projections of the **combination**, *if the axes are collinear and a pair of edges of the base of the pyramid is parallel to VP.*

Draw an auxiliary top view on an auxiliary inclined plane perpendicular to VP and inclined at 45° to HP. (UQ)

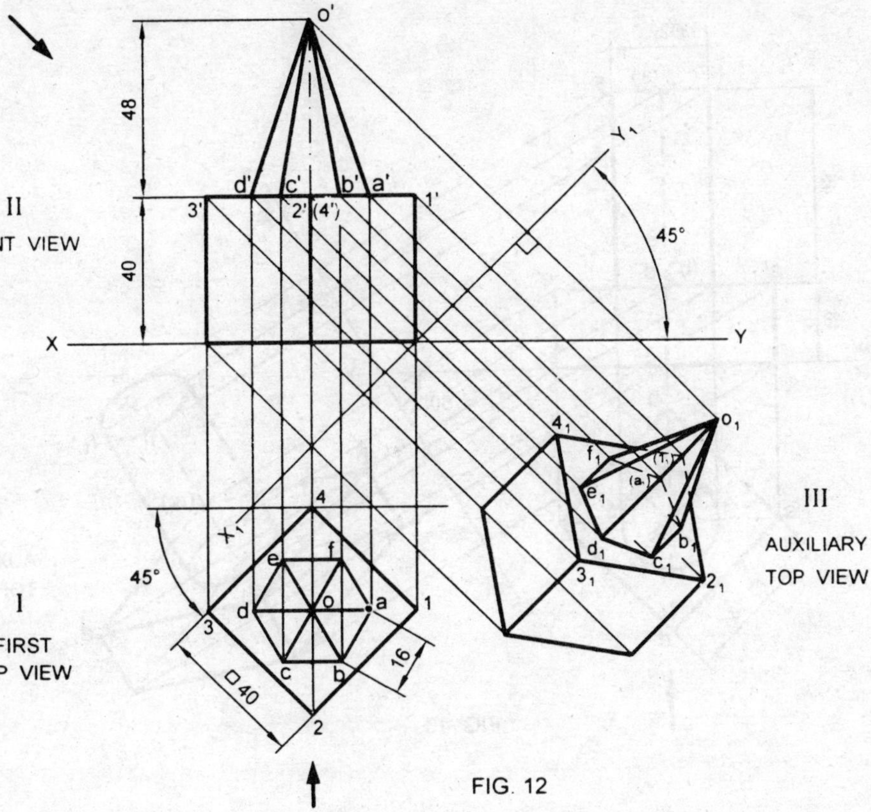

FIG. 12

Problem 22: *(Exercise)* *A right regular hexagonal prism of side of base 20 mm and axis 24 mm long is surmounted coaxially by a hexagonal pyramid of side of base 20 mm and axis 40 mm long.*

Draw the projections of the **combination** *when resting on HP on one of the edges of the base of the prism perpendicular to VP with the axis inclined at 30° to HP.* (UQ)

Problem 23: *(Fig. 13)* *A square-headed bolt, without chamfer, has 22.5 mm diameter and cylindrical portion has 45 mm length. The square head has 36 mm side and 18 mm thickness. Draw the projections when the bolt is resting upon its head on the ground, such that the axis is vertical and the rectangular faces are equally inclined to VP.*

Add a top view on an Auxiliary Inclined Plane inclined at 60°. to HP. (UQ)

NOTE : *The square-headed bolt is a combination of a square prism and a cylinder.*

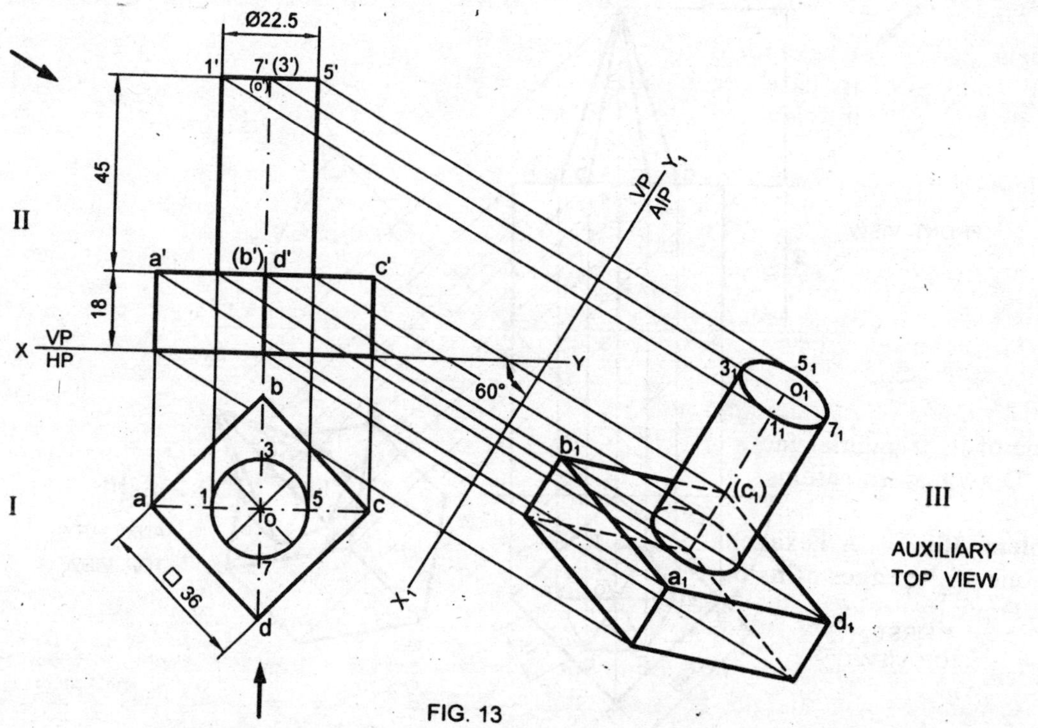

FIG. 13

Problem 24: *(Exercise)* *A hexagonal-headed bolt, without chamfer, is 22 mm in diameter and the cylindrical portion is 50 mm long. The side of the hexagonal head is 22 mm having a thickness of 18 mm. The bolt is placed with a horizontal edge of the head on HP such that it is perpendicular to VP and the axis of the bolt is inclined at 40° to HP.*

Draw the projections of the bolt, neglecting its threads. (UQ)

3. ADDITIONAL PROBLEMS - I

☞ All the problems given in the previous Chapter 16 – Projections of Solids can also be solved by the Change of Reference Line Method. Then, compare the solutions obtained by the Change of Position Method.

☞ Similarly, all the problems given in this Chapter can also be solved by the Change of Position Method. Then, compare the solutions obtained by the Change of Reference Line Method.

Problem 25: A square prism, side of base 30 mm and axis 50 mm long, has an edge of its base in HP. Its axis is inclined at 60° to HP and parallel to VP. Draw its projections.

Problem 26: A hexagonal prism of side of base 26 mm and axis length 52 mm rests on HP on one of its base edges with its axis inclined at 60° to HP and parallel to VP. Draw its top and front views. (UQ)

Problem 27: A cube of 35 mm side rests with one of its edges on HP and one of the faces containing that edge is inclined at 30° to HP. The edge on which the cube rests is parallel to VP. Draw its projections. (UQ)

Problem 28: Draw the projections of a pentagonal prism, base 25 mm side and axis 50 mm long, resting on one of its rectangular faces on HP, with the axis inclined at 45° to VP.

Problem 29: A square pyramid of 50 mm side of base and 50 mm length of axis rests on one of its triangular faces on HP and having a slant edge containing that face parallel to VP. Draw the projections.

Problem 30: A hexagonal pyramid, side of base 25 mm and axis 50 mm long, rests with one of the edges of its base on HP and its axis is inclined at 30° to HP and parallel to VP. Draw its projections of the pyramid.

Problem 31: Draw the projections of a rectangular pyramid of 40 mm x 25 mm sides of base and 40 mm height, rests on one of the longer edges of its base on HP and the triangular face containing that edge is perpendicular to both HP and VP.

Problem 32: A hexagonal pyramid of side of base 24 mm and altitude 60 mm is resting on HP on one of its base edges such that the triangular face containing that edge is perpendicular to both HP and VP. Draw its projections.

Problem 33: Draw the projections of a cylinder, base 30 mm diameter and axis 40 mm long, rests with a point of its base circle on HP such that the axis is making an angle of 30° with HP and parallel to VP.

Problem 34: A cylindrical disc, 50 mm diameter and 26 mm thick, has its flat face inclined at 35° to V.P. Draw its projections.

Problem 35: Draw the projections of a cylinder of 40 mm diameter and altitude 60 mm, when it is resting on HP with its axis inclined at 45° to HP and parallel to VP.

Problem 36: A solid square pyramid of side of base 33 mm and altitude 44 mm is resting on HP on one of its base edges with its axis inclined at an angle of 45° to HP and parallel to VP. Draw its projections. (UQ)

Problem 37: A hexagonal pyramid has an altitude of 50 mm and side of base 25 mm. It rests with one of its sides of the base on HP such that the triangular face containing that side is perpendicular to HP. Draw the projections. (UQ)

Problem 38: A cone, base 40 mm diameter and axis 50 mm long, touches VP on a point of its base circle. Its axis is inclined at 30° to VP and parallel to HP. Draw its projections.

Problem 39: A frustum of a cone, base 54 mm diameter, top 36 mm diameter and height 50 mm rests on one of its generators such that a plane containing the axis and that generator makes an angle of 45° with VP. Draw its projections. (UQ)

Problem 40: A hollow cylinder of outside diameter 65 mm, thickness of wall 12 mm and height 65 mm rests on HP with its axis inclined at an angle of 50° to VP and parallel to HP. Draw its projections.

Problem 41: A thin lamp shade in the form of a frustum of cone of base diameter 200 mm, top face diameter 80 mm and height of the axis 150 mm is lying on its side on HP and the axis parallel to VP. Draw its projections.

Problem 42: A hopper made up of thin sheet metal is in the form of an inverted frustum of a cone. Its top is 450 mm diameter, bottom 230 mm diameter and height 230 mm. A circular ring of 50 mm height is welded to the frustum at the bottom and another circular ring of 60 mm height is welded to the frustum at the top. Draw its projections when the axis makes 45° with HP.

Problem 43: A pentagonal prism of 22 mm side of base and altitude 58 mm is freely suspended from a corner such that the axis is parallel to VP. Draw its projections.

Problem 44: A cylindrical slab of 60 mm diameter has a 30 mm square hole axially cut in it. The cylindrical slab stands on HP with its axis making an angle of 60° to HP. One of the faces of the square hole is parallel to VP. Draw the projections of the solid if its thickness is 20 mm. (UQ)

4. AXIS INCLINED TO BOTH THE HP AND THE VP

4.1 AUXILIARY VIEWS

More than two front / top views are necessary to describe the shape and size of solids inclined with either HP or VP or both. Thus in order to show the true shape and size of such solids, it is necessary to project additional views called auxiliary views on the planes other than the principle planes.

4.2 PROJECTION OF SOLID WITH AXIS INCLINED TO BOTH HP AND VP

The projections of a solid with its axis inclined to both HP and VP are drawn in three stages as follows :

Stage I : Simple Position
Stage II : Second Position – Axis Inclined to One Plane and Parallel to the Other
Stage III : Final Position

The second and final positions may be obtained either by the change of positions of the solid (that is, change of position of views), called Change of Position Method or by the change of reference lines, called Change of Reference Line Method.

4.3 COMPARISON OF CHANGE OF REFERENCE LINE METHOD AND CHANGE OF POSITION METHOD

1. Change of Position Method is more time-consuming, especially when the solid has curved surfaces or too many corners or edges. Therefore, Change of Reference Line Method is preferable.
2. In the Change of Position Method, the reproduction of the views in the new position have to be carried out. This may result in certain degree of inaccuracy in the drawings.
3. However, take sufficient care in transferring the distances of various points or corners from the original reference line, while adopting the Change of Reference Line Method.
4. New reference line should pass through the corner, edge or face on which the solid is resting in the second or final position. Any line parallel to this reference line may be drawn to avoid overlapping of views already drawn.

Problem 45: *(Fig. 14) A square prism, side of base 30 mm and axis 50 mm long, has its axis inclined at 60° to HP. It has an edge of its base in the HP and inclined at 45° to VP. Draw its projections.*

Simple Position

1. When the prism has to rest with an edge of its base on HP, first assume it in its simple position, *i.e.*, resting with its base on HP, such that an edge of its base is perpendicular to VP. For this position draw the top view first and then the front view.

Auxiliary Top View

2. The axis is inclined at 60° to HP. Instead of tilting the front view of the prism such that its axis is inclined at 60° to XY, draw X_1Y_1 through 2'(3') at 60° to the front view of the axis.

NOTE : *Here X_1Y_1 represents the intersection of VP and A.I.P.*

---

3. From all the corners of the prism in the first front view draw projectors perpendicular to X_1Y_1 to project the auxiliary top view.
4. Measure the distances of all the corners in the first top view from XY and transfer the same with reference to X_1Y_1 on the corresponding projectors.
5. Look at the first front view in the direction of arrow shown. The edge 2' (3') is invisible. Hence draw (2_1) (3_1) by dashed line in the auxiliary top view. Here this edge is perpendicular to VP.

Auxiliary front view

6. The edge 23 is inclined at 45° to VP (given). Draw X_2Y_2 at an angle of 45° to (2_1) (3_1) to project the auxiliary front view.

NOTE : *Here X_2Y_2 represents the intersection of A.I.P. and A.V.P.*

---

7. From all the corners in auxiliary top view draw projectors perpendicular to X_2Y_2.
8. Measure the distances of all the corners in the first front view from X_1Y_1 and transfer the same with reference to X_2Y_2 on the corresponding projectors.

To mark the visible and invisible edges in the Auxiliary Front View

9. Look at the auxiliary top view in the direction of arrow shown. The rectangular face 1_1 a_1 b_1 (2_1) is visible. Hence draw $1_1'$ a_1' b_1' $2_1'$ as continuous *thick* lines in the auxiliary front view.
10. Draw the boundary lines $2_1'$ $3_1'$, $3_1'$ $4_1'$, $4_1'$ d_1', d_1' a_1' as *thick* lines. 1_1 4_1 is nearer to the observer and hence draw $1_1'$ $4_1'$ as thick line. Since c_1 is invisible draw (c_1') b_1', (c_1') d_1', (c_1') $3_1'$ as dashed lines.

45

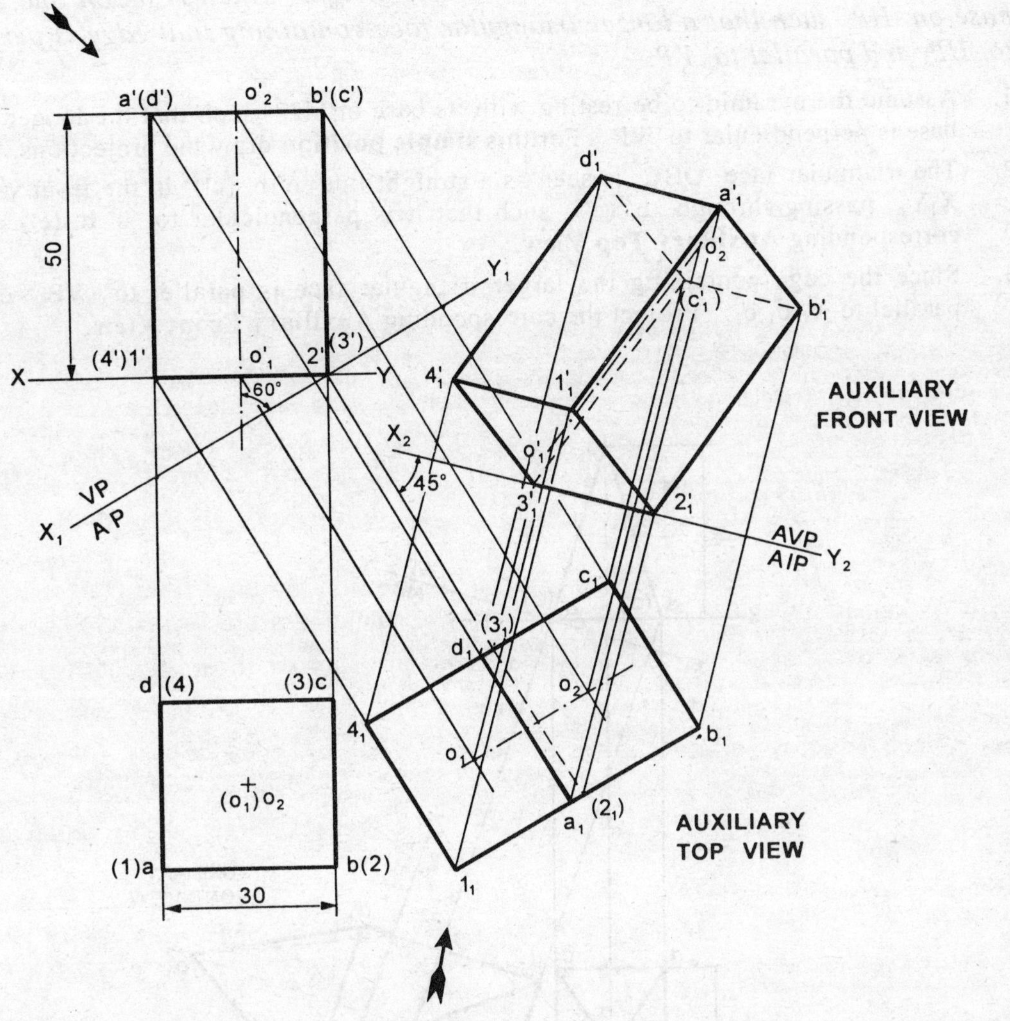

FIG. 14

Problem 46: *(Exercise)*

Draw the projections of a hexagonal prism base of 30 mm side and axis 50 mm when resting on one of its longest edges on HP and axis inclined at 40° to VP. One of the rectangular faces containing the longer edge on which the prism rests is inclined at 30° to HP.

Problem 47: *(Fig. 15) Draw the projections of a rectangular pyramid of 40 mm x 25 mm sides of the base and 40 mm height, which rests on one edge of the base on HP such that a larger triangular face containing that edge is perpendicular to HP and parallel to VP.* (UQ)

1. Assume the pyramid to be resting with its base on HP such that the longer edge of the base is perpendicular to VP. For this **simple position** draw the projections.
2. The triangular face OBC is seen as a straight line o′ b′ (c′) in the front view. Draw X_1Y_1 passing through b′ (c′) such that it is perpendicular to o′ b′ (c′). Draw the corresponding **Auxiliary Top View**.
3. Since the edge containing the larger triangular face is parallel to VP, draw X_2Y_2 parallel to $c_1\ o_1\ b_1$. Project the corresponding **Auxiliary Front View**.

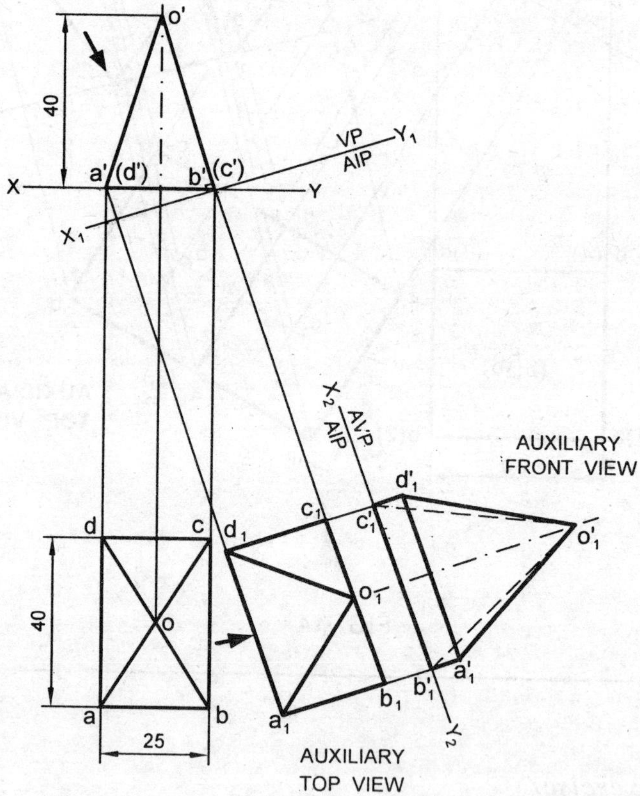

FIG. 15

Problem 48: *(Exercise) A square pyramid of base edges 35 mm and height 50 mm has its base on HP with one of the base edges inclined at 30° to VP. Draw the top and front views. Obtain an auxiliary top view from the front view on an auxiliary plane inclined at 45° to HP.*

Problem 49: *(Fig. 16) A hexagonal pyramid, side of base 25 mm and axis 50 mm long, rests with one of the corners of its base on HP. The axis is inclined at 30° to HP and 45° to VP. Draw its projections.*

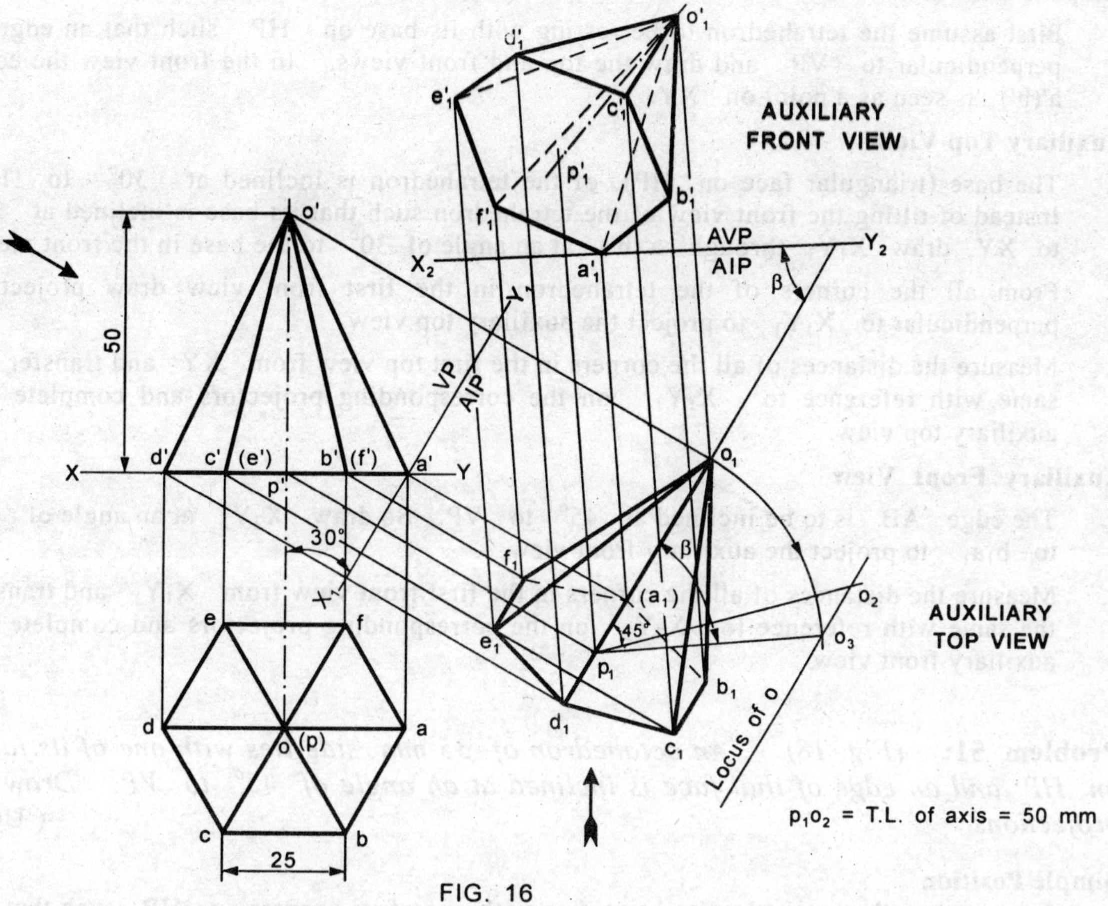

FIG. 16

1. **Simple Position :** First assume the pyramid to be resting with its base on HP such that an edge of the base is parallel to VP. Draw the projections.
2. **Auxiliary Top View:** Draw X_1Y_1 through a′ at 30° to the front view of the axis. From all the corners of the pyramid in the front view, draw projectors to X_1Y_1 and complete the auxiliary top view.
3. **Auxiliary Front View:** Since the axis is inclined at an angle of 30° to H.P. and 45° to V.P., its top view will make an apparent angle β (> 45°) with X_2Y_2. Find β as shown in the auxiliary top view.

   Draw X_2Y_2 at an angle of β to the auxiliary top view of the axis and complete the auxiliary front view as shown.

Problem 50: *(Fig. 17) A tetrahedron of 40 mm side rests with one of its edges on HP and inclined at 45° to VP. The triangular face containing that edge is inclined at 30° to HP. Draw the top and front views of the solid.* (UQ)

Simple Position

1. First assume the tetrahedron to be resting with its base on HP such that an edge is perpendicular to VP and draw the top and front views. In the front view the edge a'(b') is seen as a point on XY.

Auxiliary Top View

2. The base (triangular face on HP) of the tetrahedron is inclined at 30° to HP. Instead of tilting the front view of the tetrahedron such that its base is inclined at 30° to XY, draw X_1Y_1 through a'(b') at an angle of 30° to the base in the front view.

3. From all the corners of the tetrahedron in the first front view draw projectors perpendicular to X_1Y_1 to project the auxiliary top view.

4. Measure the distances of all the corners in the first top view from XY and transfer the same with reference to X_1Y_1 on the corresponding projectors and complete the auxiliary top view.

Auxiliary Front View

5. The edge AB is to be inclined at 45° to VP. So draw X_2Y_2 at an angle of 45° to b_1a_1 to project the auxiliary front view.

6. Measure the distances of all the corners in the first front view from X_1Y_1 and transfer the same with reference to X_2Y_2 on the corresponding projectors and complete the auxiliary front view.

Problem 51: *(Fig. 18) An octahedron of 35 mm side lies with one of its faces on HP and an edge of that face is inclined at an angle of 45° to VP. Draw its projections.* (UQ)

Simple Position

1. First assume the octahedron to be resting with one of its corners on HP such that the axis through it is perpendicular to HP and an edge is perpendicular to VP. For this simple position draw the projections. In the front view the triangular face p' a' (b') is seen as a line.

Auxiliary Top View

2. Draw X_1Y_1 so as to contain p' a' (b') in the first front view. Draw projectors perpendicular to X_1Y_1 passing through all corners in the first front view and transfer the corresponding distances from the first top view. Thus complete the auxiliary top view.

Auxiliary Front View

3. Draw X_2Y_2 passing such that it is making an angle of 45° to b^1a^1 and complete the auxiliary front view as shown.

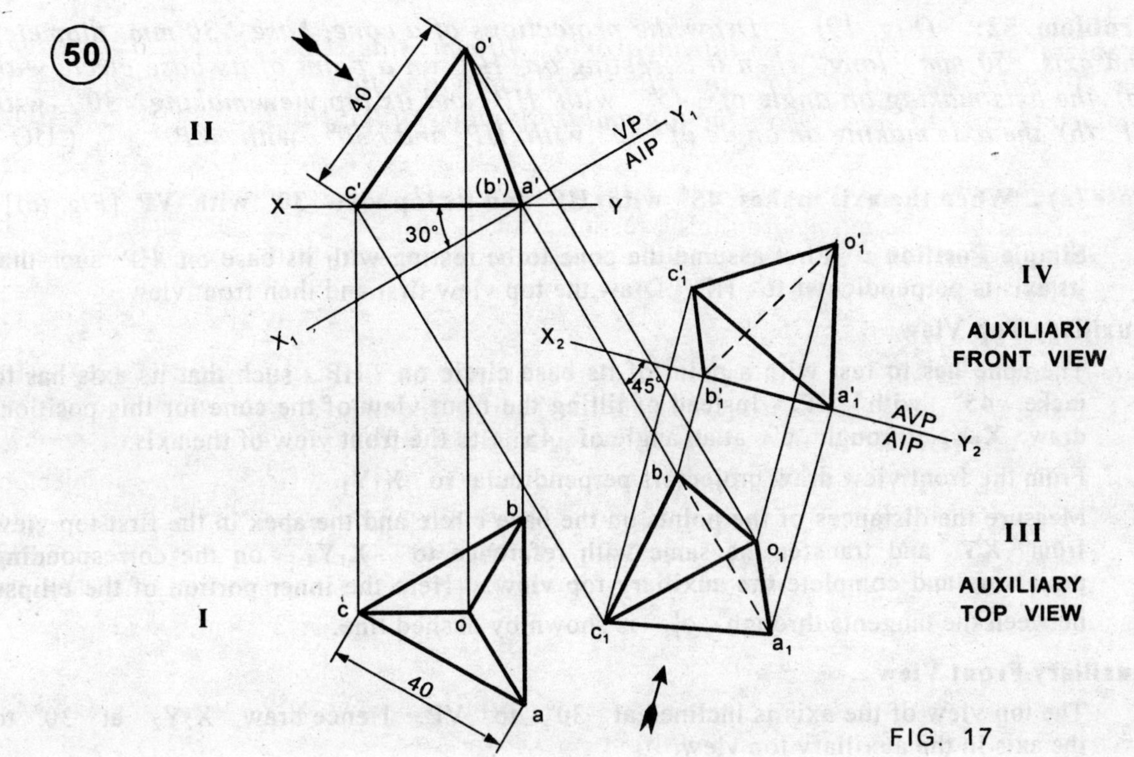
50
II
I
IV
III
AUXILIARY FRONT VIEW
AUXILIARY TOP VIEW
VP
AIP
AVP
AIP
30°
45°
40
40

FIG. 17

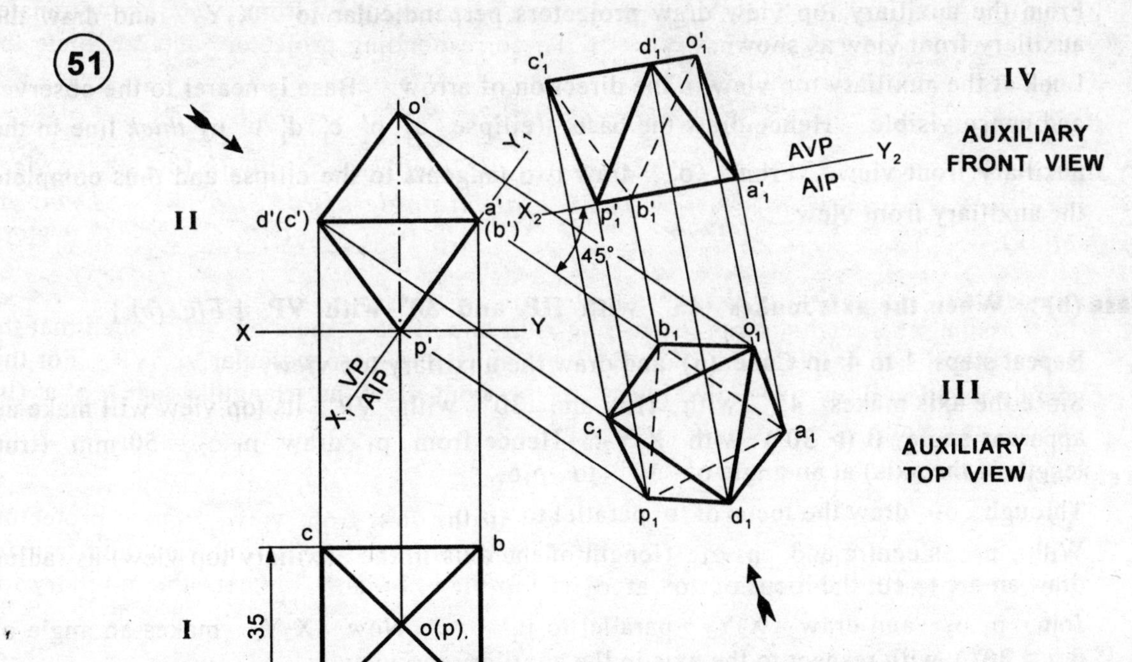
51
II
I
IV
III
AUXILIARY FRONT VIEW
AUXILIARY TOP VIEW
AVP
AIP
VP
AIP
45°
35

FIG. 18

Problem 52: *(Fig. 19) Draw the projections of a cone, base 30 mm diameter and axis 50 mm long, when it is resting on HP on a point of its base circle with (a) the axis making an angle of 45° with HP and its top view making 30° with VP (b) the axis making an angle of 45° with HP and 30° with VP.* (UQ)

Case (a) : When the axis makes 45° with HP and its top view 30° with VP [*Fig. (a)*]

1. **Simple Position :** First assume the cone to be resting with its base on HP such that its axis is perpendicular to HP. Draw the top view first and then front view.

Auxiliary Top View

2. The cone has to rest with a point of its base circle on HP such that its axis has to make 45° with HP. Instead of tilting the front view of the cone for this position, draw X_1Y_1 through c′ at an angle of 45° to the front view of the axis.
3. From the front view draw projectors perpendicular to X_1Y_1.
4. Measure the distances of the points on the base circle and the apex in the first top view from XY and transfer the same with reference to X_1Y_1 on the corresponding projectors and complete the auxiliary top view. Here the inner portion of the ellipse between the tangents through o_1' is shown by dashed line.

Auxiliary Front View

5. The top view of the axis is inclined at 30° to VP. Hence draw X_2Y_2 at 30° to the axis in the auxiliary top view.
6. From the auxiliary top view draw projectors perpendicular to X_2Y_2 and draw the auxiliary front view as shown.
7. Look at the auxiliary top view in the direction of arrow. Base is nearer to the observer and hence visible. Hence draw the base (ellipse a_1' b_1' c_1' d_1') by *thick* line in the auxiliary front view. From o_1' draw two tangents to the ellipse and thus complete the auxiliary front view.

Case (b) : When the axis makes 45° with HP and 30° with VP [*Fig. (b)*]

1. Repeat steps 1 to 4 in **Case (a)** and draw the auxiliary top view.
2. Since the axis makes 45° with HP and 30° with VP, its top view will make an apparent angle β (> 30°) with X_2Y_2. Hence from p_1 draw $p_1\ o_2$ = 50 mm (true length of the axis) at an angle of 30° to p_1o_1.
3. Through o_2 draw the locus of o parallel to p_1o_1.
4. With p_1 as centre and $p_1\ o_1$ (length of the axis in the auxiliary top view) as radius draw an arc to cut the locus of o at o_3.
5. Join $p_1\ o_3$ and draw X_2Y_2 parallel to it. Now X_2Y_2 makes an angle of β (> 30°) with respect to the axis in the auxiliary top view.
6. Draw projectors perpendicular to X_2Y_2 from the auxiliary top view and **complete the** auxiliary front view.

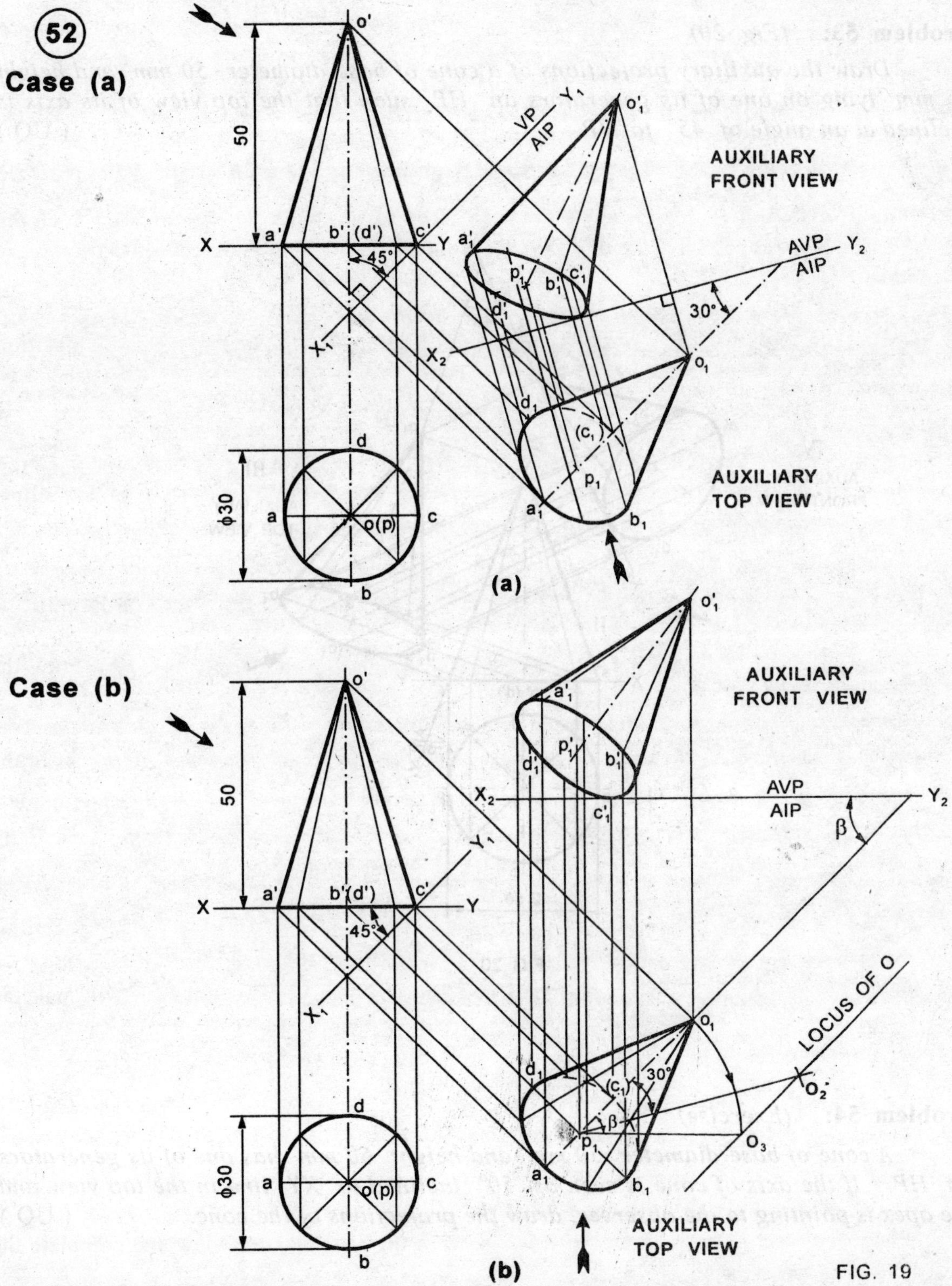
52
Case (a)
o'
50
a'
b' (d')
c'
X
Y
45°
X1
VP
AIP
Y1
o'1
a'1
p'1
b'1
c'1
d'1
AUXILIARY FRONT VIEW
AVP
AIP
Y2
30°
X2
o1
d1
(c1)
p1
a1
b1
AUXILIARY TOP VIEW
d
a
c
o(p)
b
ϕ30
Case (b)
o'
50
a'
b'(d')
c'
X
Y
45°
X1
Y1
o'1
a'1
p'1
d'1
b'1
c'1
AUXILIARY FRONT VIEW
X2
AVP
AIP
Y2
β
LOCUS OF O
o1
d1
(c1)
30°
β
p1
a1
b1
O2
O3
AUXILIARY TOP VIEW
d
a
c
o(p)
b
ϕ30

(a)

(b)

FIG. 19

Problem 53: *(Fig. 20)*

Draw the auxiliary projections of a cone of base diameter 50 mm and height 75 mm lying on one of its generators on HP such that the top view of its axis is inclined at an angle of 45° to VP. (UQ)

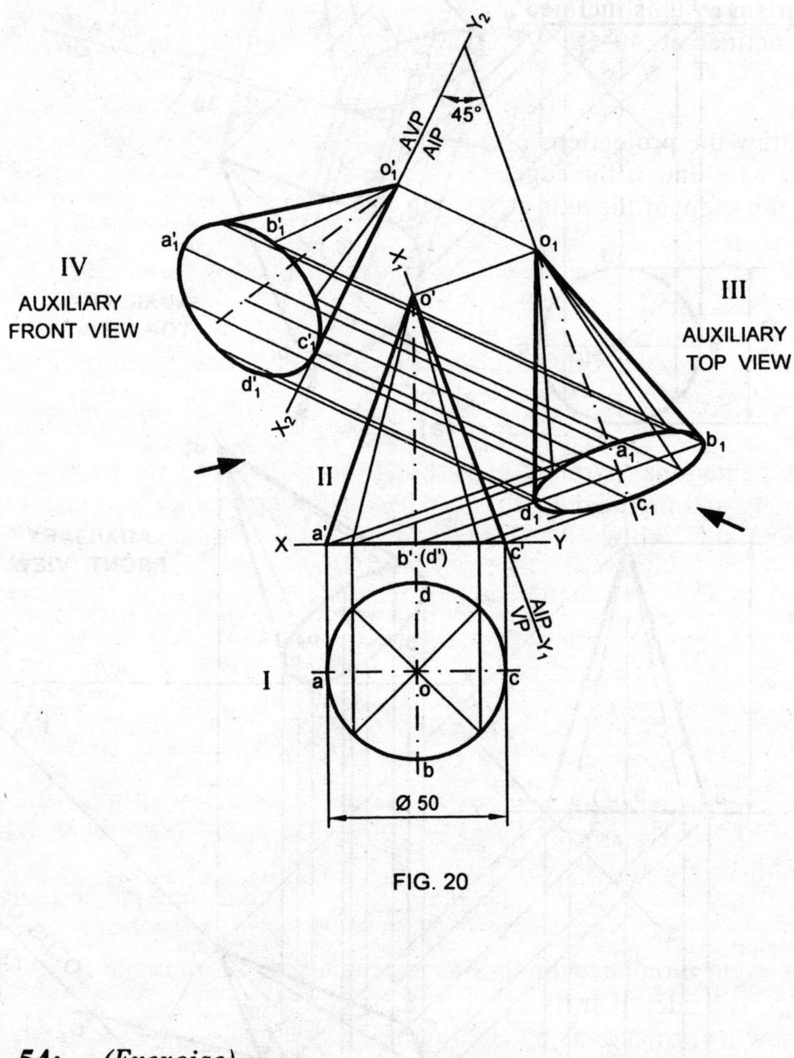

FIG. 20

Problem 54: *(Exercise)*

A cone of base diameter 50 mm and height 60 mm has one of its generators on HP. If the axis of cone is seen as 50° inclined to XY line in the top view and the apex is pointing to the observer, draw the projections of the cone. (UQ)

5. ADDITIONAL PROBLEMS – II

Problem 55: A pentagonal prism of 25 mm side of base and height 50 mm rests with one of its longer edges on HP such that, one of the rectangular faces containing the longer edge on which the prism rests is inclined at 30° to HP and is nearer to the observer. The axis of the prism is inclined at 40° to VP. Draw the top and front views. (UQ)

Problem 56: Draw the projections of a hexagonal prism, side of base 30 mm and axis 60 mm long, resting with one of the edges of its base on HP such that its axis is inclined at 30° to HP and the top view of the axis at 45° to VP.

Problem 57: A hexagonal pyramid with side of base 20 mm and height 60 mm is resting on a side such that the axis makes an angle of 30° with both the planes. Draw the projections.

Problem 58: A pentagonal pyramid of base 24 mm side and axis 64 mm long has an edge of the base in VP and inclined at 30° to HP, while the triangular face containing that edge makes an angle of 45° with VP. Draw the projections of the pyramid.

Problem 59: A pentagonal pyramid, side of base 25 mm and axis 55 mm long, has its axis inclined at 45° to HP. It has an edge of its base in the HP and inclined at 30° to the VP. Draw its projections.

Problem 60: A square pyramid, base 35 mm side and axis 50 mm long, is freely suspended from one of the corners of its base. Draw its projections, when the axis as a vertical plane makes an angle of 45° with VP. (UQ)

Problem 61: A right circular cylinder of base diameter 40 mm and length of axis 50 mm, rests on its base rim such that its axis is 45° inclined to HP and the top view is 60° inclined to VP. Draw its projections by auxiliary plane method. (UQ)

Problem 62: Draw the projections of the frustum of a cone whose base diameter is 80 mm, top diameter is 40 mm and axis height is 50 mm, if it is lying on HP on one of its generators. Assume that the plane containing the axis and the generator makes an angle of 45° to VP. (UQ)

Problem 63: A triangular prism, base 40 mm side and axis 50 mm long, is lying on HP on one of its rectangular faces with the axis perpendicular to VP. A cone, base 40 mm diameter and axis 50 mm long, is resting on HP and is leaning centrally on a face of the prism, with its axis parallel to VP.

Draw the projections of the solids and project another front view on a reference line making 60° with XY. (UQ)

HINT: First draw the cone with its base on HP and find the length of its generator.

Problem 64: A glass spherical paper weight has flat base and spherical top. The spherical diameter is 54 mm and the flat base diameter is 40 mm. It is resting on HP such that its flat base is inclined at 30° to VP and perpendicular to HP. Draw its projections.

Problem 65: A brass spherical flower-vase with flat top and spherical bottom, is placed over a hollow square prism of outer base edge 50 cm, inner base edge 36 cm and height 20 cm. The greatest diameter of the flower-vase is 56 cm and flat top face is of diameter 40 cm. Draw its projections when the top face is horizontal. Draw an auxiliary top view of the combination of solids on a plane inclined at 40° to HP.

HINT

Draw the top view. The bottom of the spherical flower-vase sits partially into the hollow square prism. Locate the centre of sphere in the front view by drawing arcs with the two intersecting points of the prism and sphere as centers and radius of the sphere as radius.

Problem 66: A G.I. bucket, 300 mm diameter at the top and 225 mm diameter at the bottom has a circular ring 225 mm diameter and 50 mm width attached at the bottom. Total height of the bucket is 300 mm. Draw the projections of the bucket when its axis is inclined at 60° to the HP and as a vertical plane makes as angle of 45° with the VP. Assume the thickness of the G.I. sheet to be negligible.

Problem 67: An equilateral triangle of 40 mm side represents the front view of a cone standing on its base. It is tilted until its axis makes an angle of 30° with HP and the top view of the axis inclined at an angle of 50° to VP. Draw the front view of the cylinder.

UNIT – IV

SYLLABUS

UNIT – IV

Isometric Projections of Lines, Planes and Simple Solids

18

ISOMETRIC PROJECTION

1. PICTORIAL PROJECTIONS

In orthographic projection, two or more views are drawn to describe an object completely and accurately in appearance and size. To visualize the shape of the whole object in its 3-D form, all the two or three orthographic views of the object have to be read simultaneously and a conclusion is to be drawn *mentally* about its shape.

Each of the orthographic views shows only two dimensions. Therefore, such a drawing can convey information only to those who are trained in reading the multi-view orthographic drawings. This difficulty is overcome by resorting to pictorial type of projection. It displays all the three dimensions of an object in one and the same view.

Engineers often find that they must prepare pictorial drawings, showing three or more number of faces in one view to convey the technical information to technically lay persons. Pictorial drawings are obtained by projecting an object only on one plane of projection.

Uses

1. Pictorial drawing conveys the shape of the object in such a way that even a lay man who is not familiar with the art of reading a drawing, can easily visualize its shape.
2. Pictorial drawings are used for machine / structural details, architectural buildings, piping diagrams, furniture design, maintenance manuals illustrating the methods of assembling and disassembling of machines and structures, advertisement hoardings, preparing catalogs of products, etc.
3. It is always used by engineers in the design and development of new or complicated parts, the shapes of which are difficult to understand from the multi-view orthographic projections.

Limitations

1. Pictorial drawings may require more time for execution than orthographic drawings.
2. These may have a distorted and unreal appearance.
3. Also, these are difficult to dimension. Some lines on a pictorial drawing cannot be directly measured.

1.1 TYPES OF PICTORIAL PROJECTIONS

1. **Axonometric Projection**

   It is a pictorial projection projected on VP by placing the object with its three mutually perpendicular surfaces inclined to the VP. The projectors are parallel to each other and perpendicular to the VP.

2. **Oblique Projection**

   It is produced by parallel projectors that make an angle other than 90° with the plane of projection. In doing so, an oblique projection is not an orthographic view.

3. **Perspective Projection**

   It shows an object as it appears instead of its true shape and size.

1.2 TYPES OF AXONOMETRIC PROJECTION

Because an object can be placed in an unlimited number of positions relative to the Axonometric Plane of Projection, an infinite number of axonometric views may be drawn. These views will vary in proportions, lengths, sizes of angles, etc. For practical reasons, a few of these positions are classified into the following types:

1. **Isometric** 2. **Dimetric** 3. **Trimetric**

2. PRINCIPLES OF ISOMETRIC PROJECTION

What is Isometric Projection ?

Iso (in *Greek*) means ***equal*** ; ***metric*** means ***measure***. Hence ***Isometric Projection*** means ***a system of projection of equality of measure***.

An *Isometric Projection* is a pictorial projection in which three dimensions of a solid are not only shown in one view, but also their dimensions can be measured from it directly.

In Isometric Projection, the principal axes makes equal angles with the plane of projection and are therefore foreshortened equally. In ***Diametric*** and ***Trimetric Projections***, the edges are unequally foreshortened, making it difficult to represent circular features of an object. Hence these methods are *not* used in practice.

Orthographic Projection of a Cube

The principle involved in drawing an Isometric Projection can best be explained by drawing *Orthographic Projection of a Cube*, resting on HP on one of its corners with a *solid diagonal* (Solid Diagonal is an imaginary line joining one of the corners at the top and the diametrically opposite corner at the bottom) perpendicular to HP.

Isometric Projection

Fig. 1(b) shows the top view of the cube, obtained from orthographic projection in *Fig. 1(a)*. It gives a fair idea of the pictorial form of the cube and is called its *Isometric Projection*.

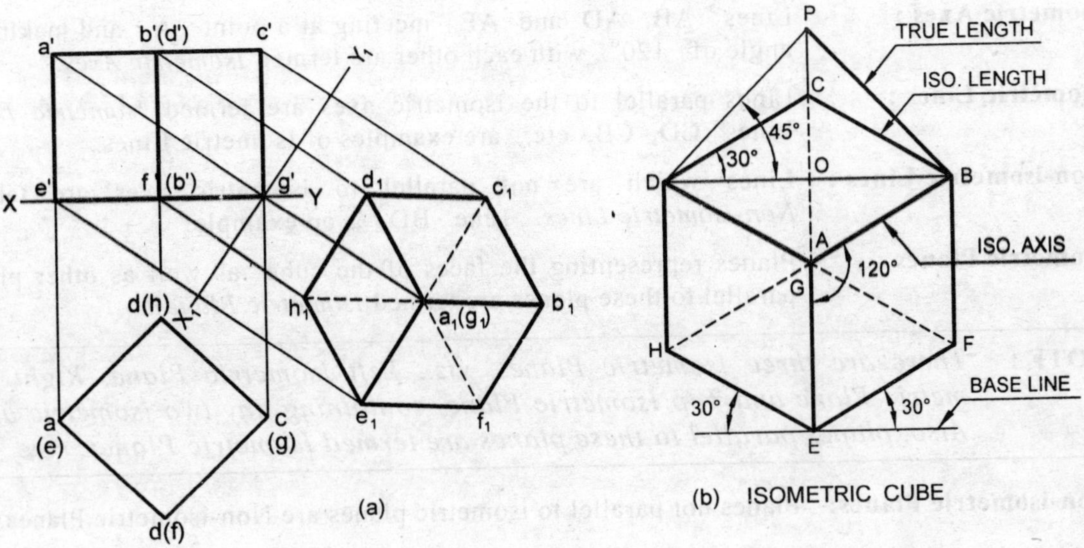

FIG. 1

RULE : **To distinguish Isometric Projection from Orthographic Projection, use CAPITAL LETTERS for the corners in the Isometric Projection.** *See Fig. (b).* ABCDEFGH is the Isometric Projection of the Cube.

Refer Fig. 1(b)

- All the edges and faces of the cube are equally inclined to the plane of projection. Hence the square faces are seen as similar and equal rhombuses.
- The three edges / sides of the cube AB, AD and AE are meeting at the point A. These edges are mutually perpendicular to each other in the solid.
- Above three edges form 90° in the Orthographic Projection in *Fig. 1(a)* and are now seen to include an angle of 120° with each other in Isometric Projection in *Fig. 1(b)*.
- Edges AB and AD make 30° with the horizontal. Edge AE is vertical.
- The three edges AB, AD and AE meeting at A and forming a solid right angle are also equally inclined to VP and therefore they are *equally foreshortened.*

  All the other edges like CB and BF are parallel to any of the lines AB, AD and AE and are also *equally foreshortened.* These edges being *equally foreshortened*, leads us to the problem of selecting an ***Isometric Scale***.
- BD and AC represent equal diagonals of the cube, but are not equally shortened in Isometric Projection. Diagonal DB of the top face of the cube is parallel to the VP. Hence it retains its true length in Isometric. However, diagonal AC is shortened.

2.1 TERMINOLOGY

Isometric Axes : Lines AB, AD and AE meeting at a point A and making an angle of 120° with each other are termed *Isometric Axes.*

Isometric Lines : Lines parallel to the isometric axes are termed *Isometric Lines.* Lines CD, CB, etc., are examples of Isometric Lines.

Non-isometric Lines : Lines which are not parallel to isometric axes are termed *Non-isometric Lines.* Line BD is an example.

Isometric Planes : Planes representing the faces of the cube as well as other planes parallel to these planes are termed *Isometric Planes.*

NOTE : *There are three Isometric Planes, viz., Left Isometric Plane, Right Isometric Plane and Top Isometric Plane, containing any two isometric axes; Also, planes parallel to these planes are termed Isometric Planes.*

Non-isometric Planes: Planes not parallel to isometric planes are Non-isometric Planes.

3. ISOMETRIC SCALE

Isometric projection is drawn using *Isometric Scale*, which converts true lengths into isometric lengths which are *foreshortened.* (Edges of an object are foreshortened means that the edges are caused to appear as if shortened.)

Ratio of Isometric Length to True Length

See Fig. 1(b). Construct a square DPBQ. DB is its diagonal and has true length. Top face of the cube is represented by rhombus DCBA in isometric. True length DP is shown by foreshortened length DC in isometric.

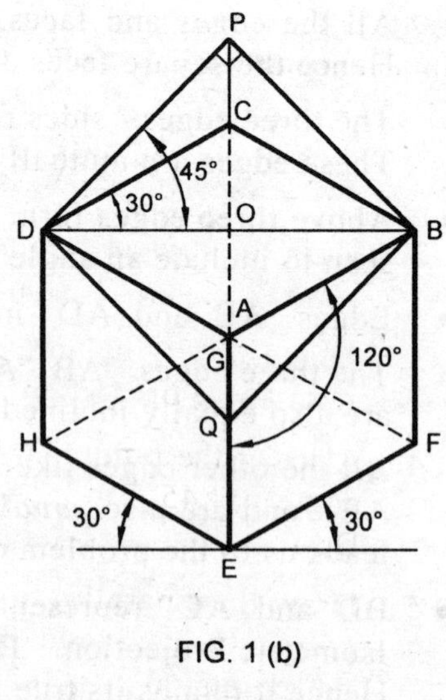

FIG. 1 (b)

In the right angle triangle PDO, $\dfrac{DP}{DO} = \dfrac{1}{\cos 45^\circ} = \sqrt{2}$

In the right angle triangle CDO, $\dfrac{DC}{DO} = \dfrac{1}{\cos 30^\circ} = \dfrac{2}{\sqrt{3}}$

$$\therefore \frac{\text{Isometric Length}}{\text{True Length}} = \frac{DC}{DP} = \frac{2}{\sqrt{3}} \times \frac{1}{\sqrt{2}}$$

$$= \mathbf{0.8165} = \mathbf{0.82} \text{ (say)}$$

Construction of Isometric Scale

Method I

1. *See Fig. 2.* Draw a horizontal line AB.
2. From A draw a line AC at 45° to represent actual or true length and another line AD at 30° to AB to measure isometric length.
3. On AC mark the points 0, 1, 2, etc., to represent actual lengths.
4. From these points draw verticals to meet AD at 0′, 1′, 2′, etc. The length A1′ represents the isometric scale length of A1 and so on.
5. In Isometric Projection, actual (true) length of object is foreshortened *i.e.*, in the Isometric Projection, all the edges of an object along the direction of the three isometric axes are foreshortened to *0.82 times* their actual / true lengths.

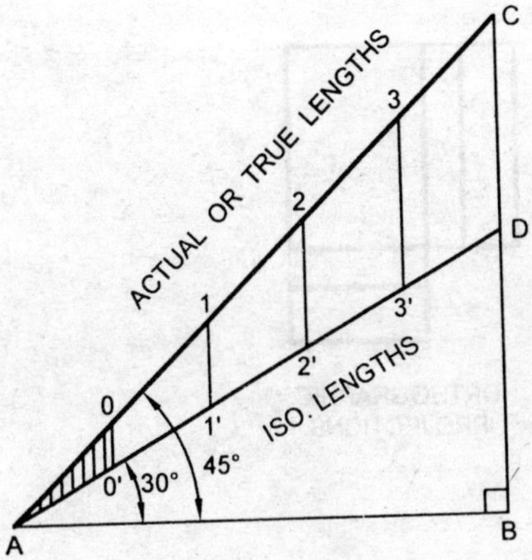

FIG. 2 ISOMETRIC SCALE

Method II

See Fig. 3. It shows **simplified** Isometric Scale construction.

1. Draw a horizontal line AC equal to true length.
2. Mark the division points of true length on AC as 0, 1, 2 and 3.
3. At the point A, draw a line inclined at an angle of 15° to AC.
4. Also, at the point C draw a line at an angle of 45° to CA. Both meet at D.
5. Draw lines parallel to CD through the points 0, 1, 2 and 3 to meet AD at the points 0′, 1′, 2′ and 3′. The line AD gives the isometric lengths.

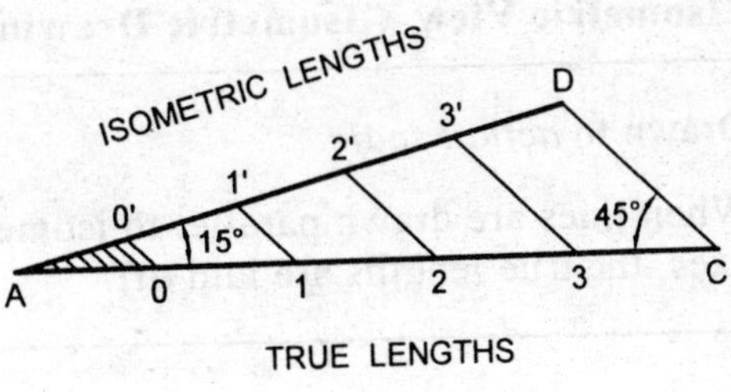

FIG. 3 ISOMETRIC SCALE

(PREFERRED)

4. ISOMETRIC VIEW

Isometric View, also known as Isometric Drawing is drawn with True Scale. Isometric Projection is drawn using Isometric Scale.

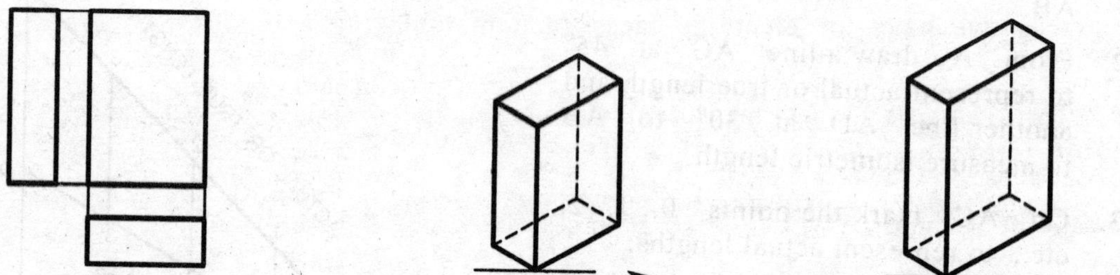

ORTHOGRAPHIC PROJECTIONS (a) ISOMETRIC PROJECTIONS (b) ISOMETRIC VIEW (c)

FIG. 4

See Fig. 4. *Fig. (a)* shows the orthographic projections of a rectangular prism.

Fig. (b) shows its *Isometric Projection drawn to the Isometric Scale.*

Fig. (c) shows its *Isometric View drawn to the actual scale.*

NOTE : 1. *Arrows in Figs. 4(b) and 4(c) indicate the direction in which the object is viewed for its front view.*

2. *The front view is so chosen that it shows the largest dimensions and most shape revealing contour.*

Isometric View (Isometric Drawing)	Isometric Projection
Drawn to *actual scale.* When lines are drawn parallel to isometric axes, the true lengths are laid off.	Drawn to *isometric scale.* When lines are drawn parallel to isometric axes, the lengths are ***foreshortened to 0.82 times*** the actual lengths.

Isometric Projection is drawn by reducing each dimension to Isometric Scale. It is time consuming and distances between any two points cannot be measured with actual scale. Hence, the general practice by the engineers is to draw Isometric View which is exactly similar in shape of Isometric Projection. Also, it is easy to draw and direct measurement of actual dimensions is possible.

5. CONVENTIONS

1. For drawing isometric, assume the object to be placed on HP only.

2. To draw isometric projection / view, view the object such that the front – right edge or the left edge becomes the nearest to the observer. Note that while representing the faces, we have to select the faces such that maximum details of the objects are visible.

3. ***Measurements:*** All measurements should be made along the isometric lines or isometric axes only.

4. ***Vertical Edges:*** All the vertical edges of the object are to be shown vertical in the isometric.

5. ***Horizontal Edges:*** Horizontal edges of the object parallel to the isometric axes are drawn at an angle of 30° to the horizontal.

6. ***Circles:*** Circles are represented as ellipses in isometric.

7. ***Hidden Edges:*** Hidden edges need not be shown in isometric, unless specified otherwise for locating a corner or an edge or a face.

8. ***Construction Lines:*** Always retain the construction lines after completing the isometric.

9. ***Isometric Scale:*** Draw isometric projection using isometric scale, unless specified otherwise to draw isometric view or isometric drawing.

10. ***Different Positions of Isometric Axes***

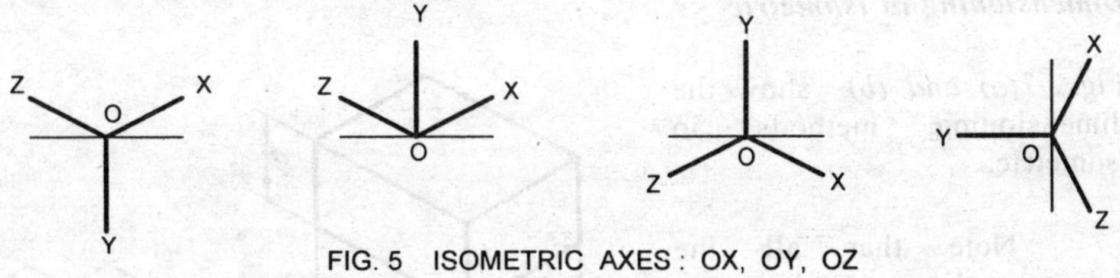

FIG. 5 ISOMETRIC AXES : OX, OY, OZ

The three isometric axes may be arranged in any one of the positions shown in *Fig. 5*.

11. ***Isometric and Non-Isometric Lines***

Lines drawn parallel to the isometric axes are called Isometric Lines. Lines which are not parallel to any of the isometric axes are called Non-Isometric Lines and hence they are not foreshortened in the same proportion as the isometric lines. Also, they should not be drawn at an angle of 30°. Non-isometric lines are drawn by locating their ends and then joining them.

12. ***Inclined Edges:*** Inclined edges which are not parallel to the isometric axes should never be drawn at the given inclination in the isometric. Draw these inclined edges by first locating the end points in isometric and then join them.

13. **Angles in Isometric**

Angles cannot be set-off directly in isometric. To draw an isometric angle, it is necessary to locate the end points of the lines which are subtending the angle.

14. ***Isometric and Non-Isometric Planes***

The three Isometric Axes form the three Isometric Planes, *viz.*, Left, Right and Top. There are three Isometric Planes containing any two isometric axes.

Planes parallel to these isometric planes are also termed as Isometric Planes. The planes which are not parallel to any of the isometric planes are termed as Non-Isometric Planes.

See Fig. 6. Isometric Planes are marked as 1. Non-isometric Planes are marked as 2.

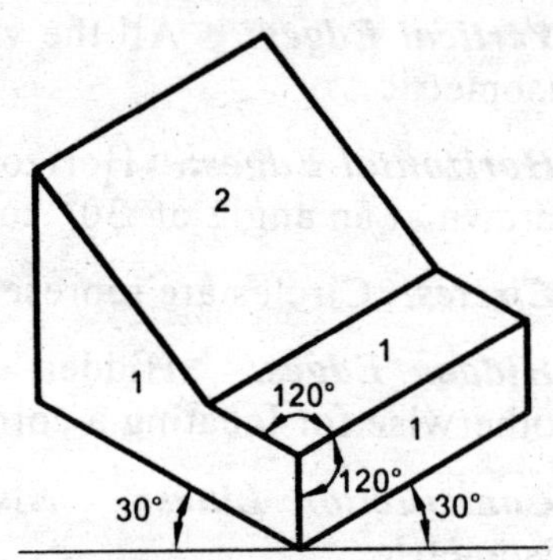

FIG. 6 ISOMETRIC PLANES - 1
NON-ISOMETRIC PLANES - 2

15. ***Dimensioning in Isometric***

Figs. 7(a) and (b) show the dimensioning methods in isometric.

Note that all the dimension lines are drawn in line with the vertical or sloping lines at 30°. No other lines are allowed.

The strokes of the letters and numerals are to be parallel to the isometric axes.

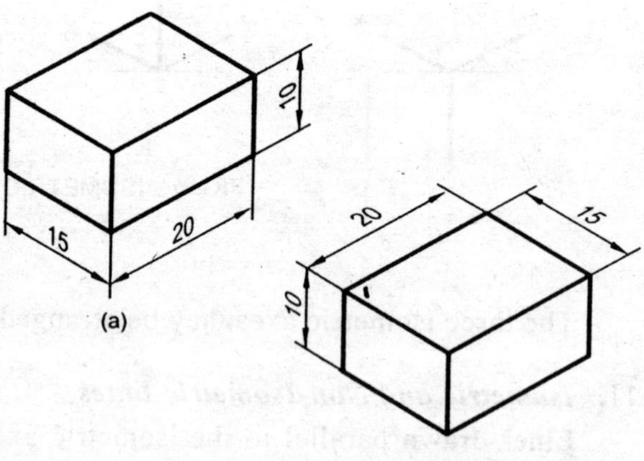

FIG. 7

6. ISOMETRIC PROJECTION / VIEW OF LINES

6.1 ISOMETRIC LINES

Lines drawn parallel to the isometric axes are called Isometric Lines.

Problem 1: *(Fig. 8)*

Draw the isometric projection and isometric view of a horizontal line AB measuring 50 mm length.

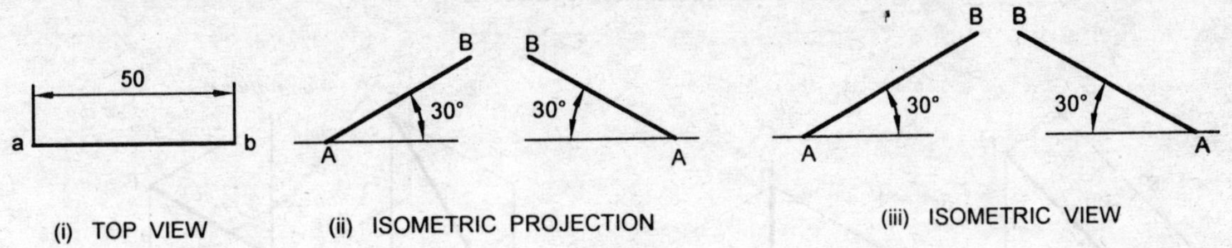

FIG. 8

Problem 2: *(Fig. 9)*

Draw the isometric projection and isometric view of a vertical line CD measuring 50 mm length.

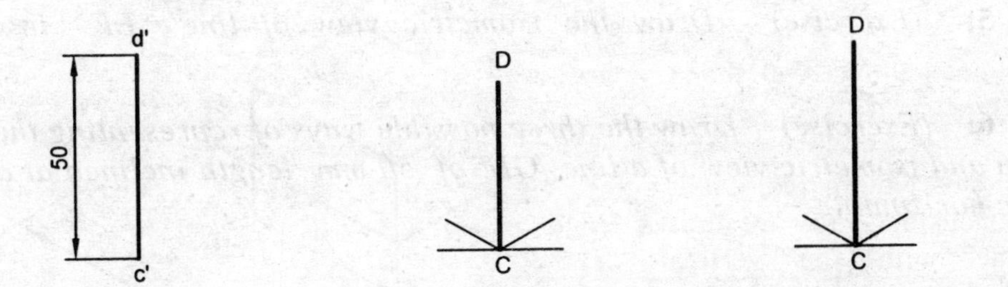

FIG. 9

Problem 3: *(Exercise)* *Draw the three possible ways of representing the isometric projection and isometric view of a line GH of 48 mm length.*

(Three possible ways means the two horizontal and one vertical positions of the line.)

6.2 NON-ISOMETRIC LINES

Lines which are not parallel to any of the isometric axes are called Non-Isometric Lines. Since the non-isometric lines are not parallel to the isometric axes, they are not foreshortened in the same proportion as the isometric lines. Also, they should not be drawn at an angle of 30°. Non-isometric lines are drawn by locating their ends and then joining them.

Problem 4: *(Fig. 10)*

Draw the three possible ways of representing the isometric projection of a line EF of 50 mm length inclined at an angle of 45° to the horizontal.

See Fig. 10. EF is a non-isometric line. Its isometric projection is drawn by locating its ends and then joining them in the three isometric planes.

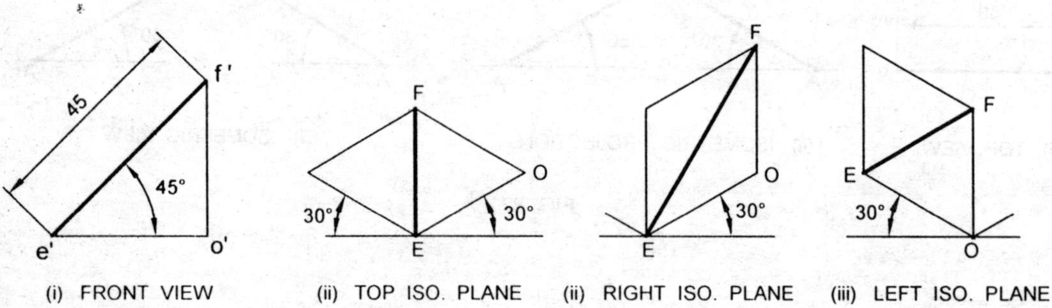

FIG. 10 ISOMETRIC PROJECTION

Problem 5: *(Exercise) Draw the isometric view of line EF in the above Problem.*

Problem 6: *(Exercise) Draw the three possible ways of representing the isometric projection and isometric view of a line GH of 50 mm length inclined at an angle of 60° to the horizontal.*

6.3 ISOMETRIC OF ARCS

Problem 7: *(Fig. 11) Draw the isometric projection of a 90° arc of a circle of radius 40 mm shown in Fig. (i) in (a) vertical and (b) horizontal positions.*

1. *See Fig. (i).* The 90° arc pq is tangential to lines pr and qr at p and q.
2. *See Figs. (ii), (iii) & (iv).* Locate the points R, P and Q in the isometric projection.
3. At P and Q draw perpendiculars to isometric lines PR and QR to meet at O_1.
4. O_1 as center and O_1P (= O_1Q) as radius draw the arcs in the isometric projection.

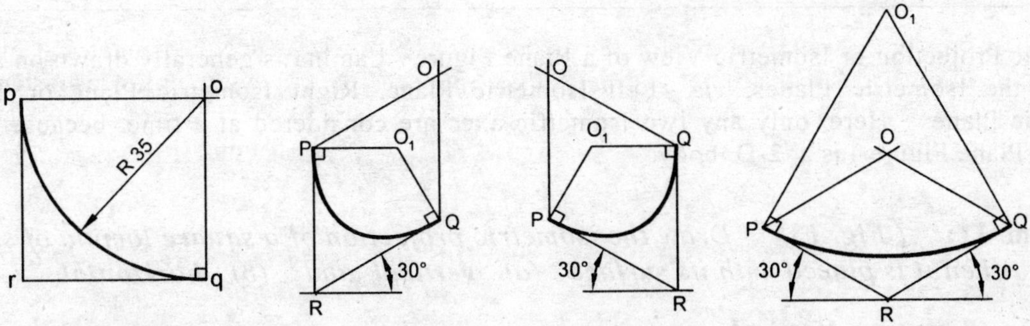

(i) TOP VIEW (ii) RIGHT ISO. PLANE (iii) LEFT ISO. PLANE (iv) HORIZONTAL I. P.

FIG. 11 ISOMETRIC PROJECTION OF ARC

Problem 8: *(Exercise) Draw the isometric view of the above Problem.*

6.4 ISOMETRIC OF CURVES

Problem 9: *(Fig. 12) Draw the isometric projection of an irregular curve EF shown in Fig. 12(i).*

1. Enclose the given curve in a rectangle PQRS.
2. Mark points 1, 2, 3, etc., on the curve.
3. Draw horizontals and verticals through these points.
4. Mark their coordinates as (x_1, y_1), (x_2, y_2), (x_3, y_3), etc. Mark these coordinates by using superimposed running dimensioning in the horizontal and vertical directions.
5. Draw the isometric projection of the rectangle PQRS.
6. Mark the distances (x_1, y_1), (x_2, y_2), (x_3, y_3), etc. on it.
7. Through x_1, x_2, etc., draw lines parallel to SP.
8. Through y_1, y_2, etc., draw lines parallel to SR to meet the above lines at 1, 2, etc.
9. Join 1, 2, 3, etc., which gives the required isometric curve EF.

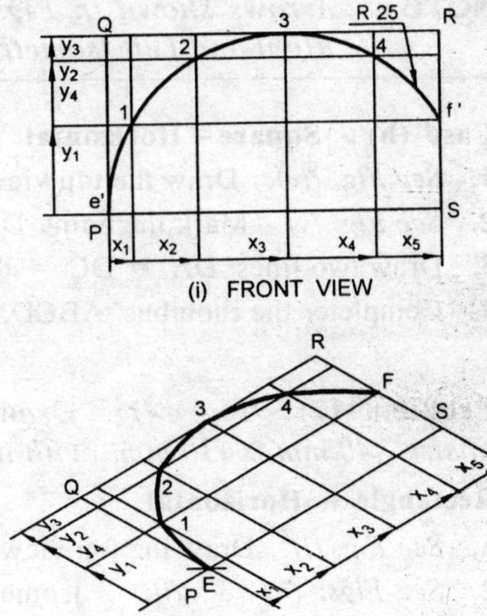

FIG. 12 ISO. PROJECTION OF CURVE

Problem 10: *(Exercise) Draw the isometric view of the above Problem.*

7. ISOMETRIC PROJECTION OF PLANE FIGURES

Isometric Projection or Isometric View of a Plane Figure / Lamina is generally drawn on any one of the Isometric Planes, *viz.*, Left Isometric Plane, Right Isometric Plane or Top Isometric Plane. Here, only any two isometric axes are considered at a time, because the object (Plane Figure) is a 2-D body.

Problem 11: *(Fig. 13) Draw the isometric projection of a* ***square*** *lamina of side 35 mm when it is placed with its surface (a)* ***vertical*** *and (b)* ***horizontal.***

Case (a) : Square – Vertical

1. *See Fig. (i).* Draw the front view a' b' c' d' of the square of side 35 mm.

NOTE : *In isometric projection / view, draw the vertical lines vertical and the horizontal lines inclined at an angle of 30° to the horizontal.*

---

2. *See Fig. (ii).* Mark point D on a horizontal line.
3. Draw a vertical line DA = d' a' = 35 mm.
4. Again through D, draw a line DC = d' c' inclined at 30° to the horizontal.
5. Complete rhombus ABCD, the required isometric projection in the Left Isometric Plane.
6. *See Fig. (iii).* Also, draw isometric projection in the direction of the other sloping axis.

NOTE : *Arrows shown in Figs. (ii) & (iii) indicate the directions of viewing in Right and Left Isometric Planes respectively, to obtain given front view.*

Case (b) : Square – Horizontal

1. *See Fig. (iv).* Draw the top view abcd of the square of 35 mm side.
2. *See Fig. (v).* Mark the point D on a horizontal line.
3. Draw two lines DA = DC = da = dc = 35 mm inclined at 30° to the horizontal.
4. Completer the rhombus ABCD, which is the required isometric projection.

Problem 12: *(Fig. 14) Draw the isometric projection of a* ***thin rectangular sheet*** *of size 40 mm x 30 mm, with its surface* ***horizontal.***

Rectangle – Horizontal

1. *See Fig. (i).* Draw the top view efgh of the rectangle 40 mm x 30 mm as shown.
2. *See Figs. (ii) & (iii).* Isometric projections of the rectangle are drawn with the side HG looking from the left as well as looking from the right.

Problem 13: *(Exercise) Draw the isometric projection / view of a* ***thin rectangular sheet*** *of size 40 mm x 30 mm, with its surface* ***vertical.***

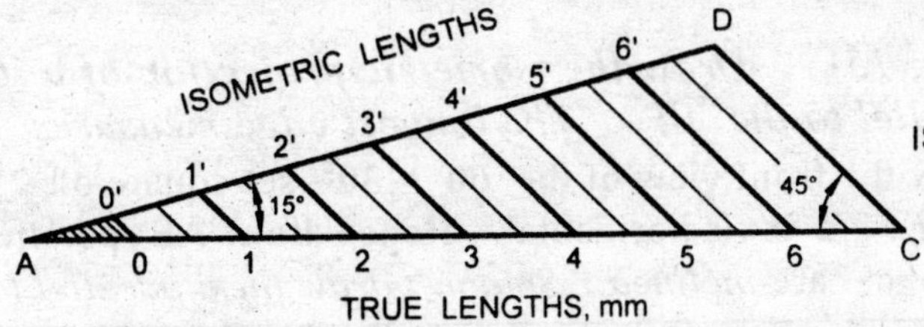

ISOMETRIC SCALE

11

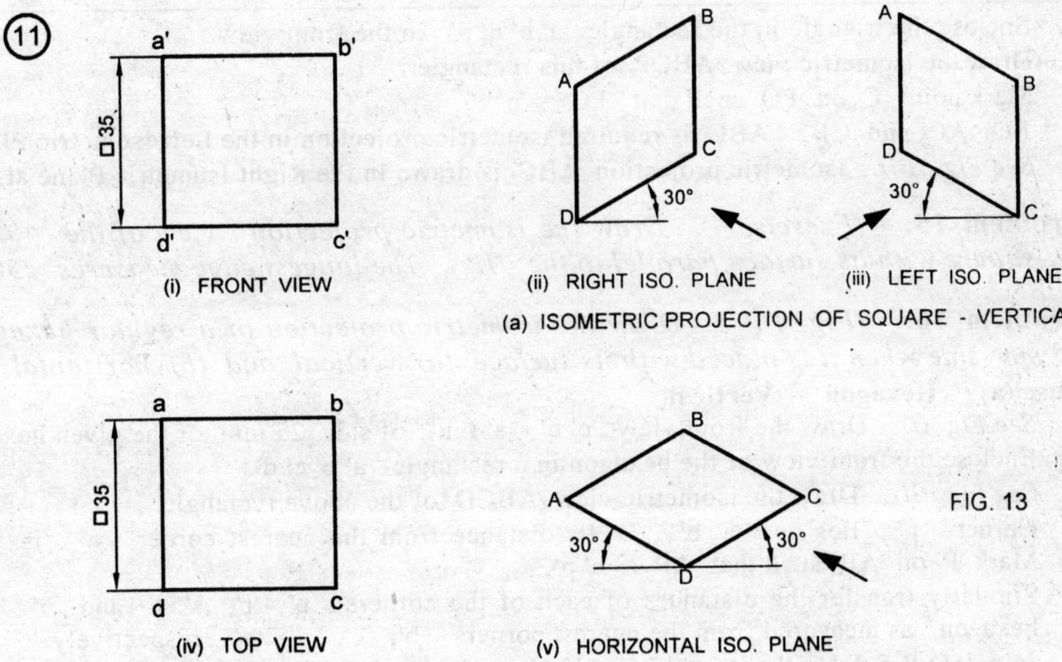

(i) FRONT VIEW (ii) RIGHT ISO. PLANE (iii) LEFT ISO. PLANE

(a) ISOMETRIC PROJECTION OF SQUARE - VERTICAL

FIG. 13

(iv) TOP VIEW (v) HORIZONTAL ISO. PLANE

(b) ISOMETRIC PROJECTION OF SQUARE - HORIZONTAL

12

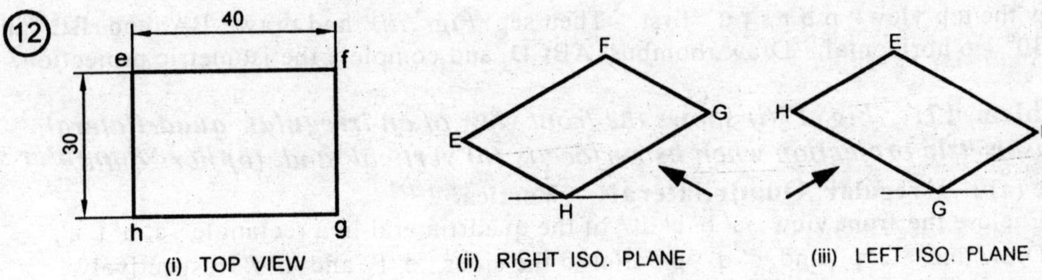

(i) TOP VIEW (ii) RIGHT ISO. PLANE (iii) LEFT ISO. PLANE

ISOMETRIC PROJECTION OF RECTANGLE - HORIZONTAL

FIG. 14

NOTE : Use True Scale for drawing Isometric View / Isometric Drawing.

Problem 14: *(Fig. 15) Draw the isometric projection of a 60° - 30°* ***set-square*** *with its surface parallel to the VP. The longest edge measures 250 mm.*

1. *See Fig. (i).* Draw the front view of the 60° – 30° set-square of 250 mm base.
2. *See Figs. (i) & (ii).* a′ b′ is horizontal. Hence draw AB parallel to a sloping axis.
3. Sides a′ c′ and b′ c′ are *inclined. So don't draw them parallel to any isometric axis.*

NOTE: *In isometric, draw an angle by fixing positions of its ends on isometric lines.*

4. Enclose the triangle in the rectangle a′ b′ q′ p′ in the front view.
5. Draw the isometric view ABQP of this rectangle.
6. Mark point C on PQ such that PC = p′ c′.
7. Join AC and CB. ABC is required isometric projection in the Left Isometric Plane.
8. *See Fig. (iii).* Isometric projection ABC is drawn in the Right Isometric Plane also.

Problem 15: *(Exercise) Draw the isometric projection / view of the 60° - 30°* ***set-square*** *with its surface parallel to the HP. The longest edge measures 250 mm.*

Problem 16: *(Fig. 16) Draw the isometric projection of a regular* ***hexagon*** *of 25 mm side when it is placed with its surface (a) vertical and (b) horizontal.*

Case (a) : Hexagon – Vertical

1. *See Fig. (i).* Draw the front view p′ q′ r′ s′ t′ u′ of side 25 mm of the given hexagon.
2. Enclose the front view of the hexagon in a rectangle a′ b′ c′ d′.
3. *See Fig. (ii).* Draw the isometric view ABCD of the above rectangle.
4. Corner p′ lies on a′ b′. Its distance from the nearest corner a′ is a′ p′. Mark P on AB such that AP = a′ p′.
5. Similarly transfer the distances of each of the corners q′, r′, s′, t′ and u′ of the hexagon as measured from the nearest corners b′, c′, d′, a′ respectively.
6. Join P Q R S T U, the isometric projection of hexagon on Left Iso. Plane when vertical.

Exercise: *Draw the isometric projection of the above hexagon on the Right Isometric Plane.*

Case (b) : Hexagon – Horizontal

Draw the top view p q r s t u first. Then see *Fig. (iii)* and draw BA and BC inclined at 30° to horizontal. Draw rhombus ABCD and complete the isometric projection.

Problem 17: *Fig. 17(i) shows the front view of an* ***irregular quadrilateral.*** *Draw the isometric projection when its surface is (a) vertical and (b) horizontal.*

Case (a) : Irregular Quadrilateral – Vertical

1. Enclose the front view a′ b′ c′ d′ of the quadrilateral in a rectangle a′ b′ f′ e′.
2. Draw lines c′ p′ and c′ q′ parallel to the sides e′ f′ and b′ f′ respectively.
3. *See Fig. (ii).* Draw the isometric projection ABFE of the rectangle a′ b′ f′ e′.
4. Fix the point D in EF as explained in Problem 8.
5. Draw the iso. projection CPFQ of the rectangle c′ p′ f′ q′ as shown to fix the point C.
6. Draw the quadrilateral ABCD, the required isometric projection on Left Isometric Plane.

Case (b) : Irregular Quadrilateral – Horizontal

Draw top view abcd first. Draw the isometric projection in horizontal position [*Fig. (iii)*].

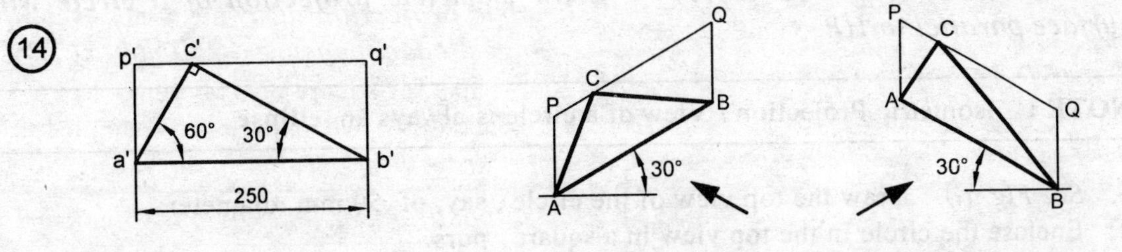

(i) FRONT VIEW (ii) RIGHT ISO. PLANE (iii) LEFT ISO. PLANE

FIG. 15 ISOMETRIC PROJECTION OF SET-SQUARE

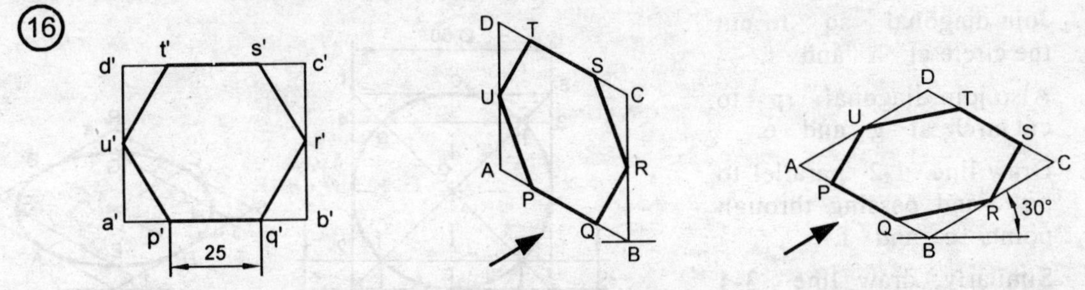

(i) FRONT VIEW (ii) SURFACE VERTICAL ON L.P (iii) SURFACE HORIZONTAL

FIG. 16 ISOMETRIC PROJECTION OF HEXAGON

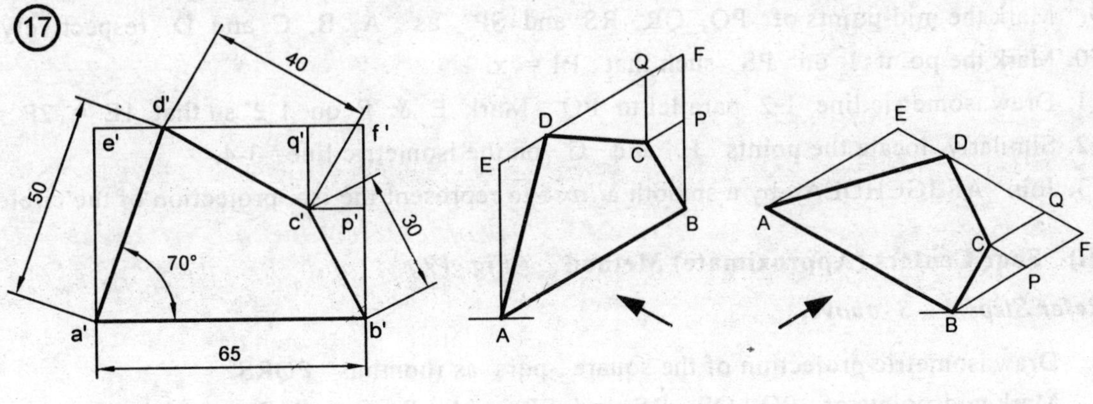

(i) FRONT VIEW (ii) SURFACE VERTICAL (iii) SURFACE HORIZONTAL

FIG. 17 ISOMETRIC PROJECTION OF QUADRILATERAL

Problem 18: *(Figs. 18 & 19)* *Draw isometric projection of a* ***circle*** *with its surface parallel to HP.*

NOTE : Isometric Projection / View of a circle is always an **ellipse**

---

1. *See Fig. (i).* Draw the top view of the circle , say, of 50 mm diameter.
2. Enclose the circle in the top view in a square pqrs.
3. Mark mid-points of pq, qr, rs and sp as a, b, c and d respectively.

Two Methods of Construction of Ellipse are as follows :

(i) Method of Points (*Fig. 18*)

4. Join diagonal sq to cut the circle at h and f.
5. Also join diagonal rp to cut circle at g and e.
6. Draw line 1-2 parallel to pq and passing through points e and f.
7. Similarly, draw line 3-4 parallel to sr and passing through h and g.
8. Draw the isometric projection of the square pqrs as rhombus PQRS.

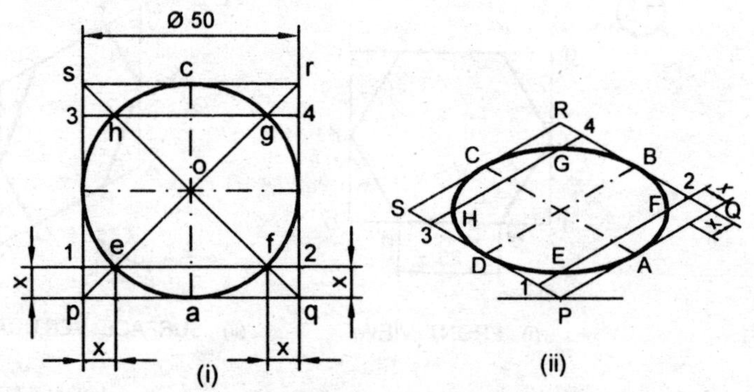

FIG. 18 METHOD OF POINTS

9. Mark the mid-points of PQ, QR, RS and SP as A, B, C and D respectively.
10. Mark the point 1 on PS such that P1 = x.
11. Draw isometric line 1-2 parallel to PQ. Mark E & F on 1-2 so that 1E = 2F = x.
12. Similarly, locate the points H and G on the isometric line 3-4.
13. Join AFBGCHDEA by a smooth *ellipse* to represent the iso. projection of the circle.

(ii) Four Centers (Approximate) Method *(Fig. 19)*

Refer Steps 1 – 3 above.

4. Draw isometric projection of the square pqrs as rhombus PQRS.
5. Mark mid-points of PQ, QR, RS and SP as A, B, C and D respectively.
6. Join P with B and C, the mid-points of opposite sides of the rhombus.
7. Join R with A and D.
8. P as center and PC as radius draw an arc CB.
9. R as center and RA as radius draw an arc AD.

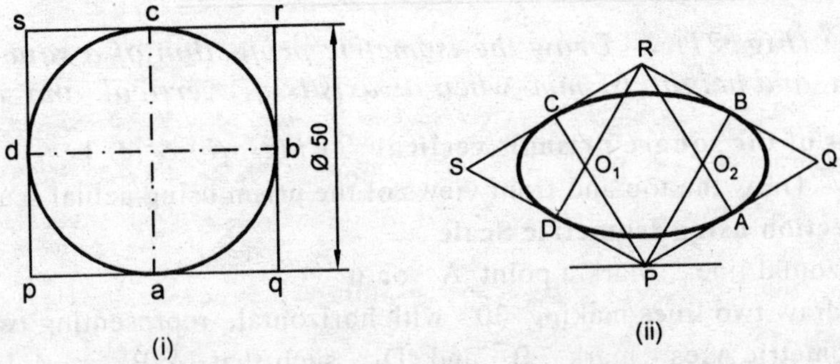

FIG. 19 FOUR CENTERS METHOD

10. PC and RD cut at O_1. O_1 as center and O_1D as radius draw an arc DC.
11. Similarly, with O_2 (intersection of RA and PB) as center and O_2B as radius draw an arc BA. Thus complete the *ellipse*.

NOTE : ***Method of Points*** *gives an accurate ellipse, but it is time consuming.* ***Four Centers Method*** *gives only an approximate ellipse.* ***But, due to the ease in construction and to avoid the labour of sketching freehand ellipse, Four Centers Method is preferable.***

Problem 19: *(Fig. 20) Draw the isometric projection of a* ***circle*** *of 50 mm diameter with its surface parallel to VP.*

Fig. 20 shows the isometric projection of the circle with its surface parallel to VP as an ellipse obtained in the rhombus on Left Isometric Plane.

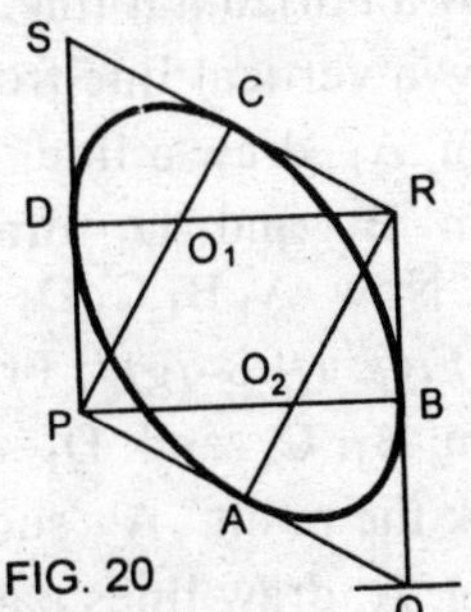

FIG. 20
ELLIPSE ON LIP

Problem 20: *(Exercise) Draw the isometric projection of the above* ***circle*** *with its surface parallel to VP such that the ellipse is on Right Isometric Plane.*

Problem 21: *(Exercise) Draw the isometric projection of a* ***semi-circle*** *of radius of 28 mm when its surface is parallel to HP.*

HINT: First consider the semicircle as a full circle and draw the corresponding ellipse in the isometric projection. Then, retain the required portion of the semicircle and its corresponding semi-ellipse in the projection.

8. ISOMETRIC PROJECTION OF SOLIDS

8.1 CUBE, SQUARE AND RECTANGULAR PRISMS

Problem 22: *(Fig. 21)* *Draw the isometric projection of a* ***square prism*** *of side of base 35 mm and height 65 mm when its axis is (i) vertical and (ii) horizontal.*

Case (i) : Axis of the square prism is vertical : [*Figs. (b) & (c)*]

1. *See Fig. (a).* Draw the top and front views of the prism using actual scale.

Isometric Projection using Isometric Scale

2. Draw a horizontal line. Mark a point A_1 on it.
3. From A_1 draw two lines making 30° with horizontal, representing two isometric axes.
4. On these isometric axes, mark B_1 and D_1 such that $A_1B_1 = A_1D_1$ = Side of the square, measured from the Isometric Scale using the divider.
5. From B_1 draw a line parallel to A_1D_1. From D_1 draw a line parallel to A_1B_1. Both meet at C_1. *Rhombus* $A_1B_1C_1D_1$ is the *isometric projection of the base.* *[Fig. (b)]*
6. *See Fig. (c).* At A_1 erect vertical to represent the third axis.
7. Mark the point A on it such that A_1A = Height of the prism in Isometric Scale.
8. From B_1, C_1 and D_1 draw vertical lines.
9. From A draw lines parallel to A_1B_1 and A_1D_1 and obtain the points B and D on the respective vertical lines.
10. From B and D draw lines parallel to AD and AB respectively to meet at C.
11. All the edges of top face, vertical edges DD_1, AA_1, BB_1 and base edges D_1A_1, A_1B_1 are visible in the isometric projection. Hence draw them as *thick* lines as shown.

Case (ii) : Axis of the square prism is horizontal : [*Figs. (d), (e), (f) & (g)*]

1. Draw a horizontal line. Mark a point A_1 on it.
2. Draw a vertical line from A_1. Mark D_1 on it such that A_1D_1 = 35 mm in Iso. Scale.
3. From A_1 draw a line 30° to the horizontal. Mark B_1 on it such that $A_1B_1 = A_1D_1$.
4. From B_1 and D_1 draw lines parallel to A_1D_1 and A_1B_1 respectively. Both meet at C_1. Now $A_1B_1C_1D_1$ represents the end face of the prism as shown in *Figs. (d) & (e).*
5. *See Figs. (f) & (g).* From A_1 draw a 30° line on the other side to represent third axis.
6. From B_1, C_1 and D_1 draw lines parallel to the third axis.
7. Mark the point A such that A_1A = Height of the prism in Isometric Scale.
8. From A draw lines parallel to A_1B_1 and A_1D_1 to obtain B and D respectively.
9. From B and D draw lines parallel to AD and AB to obtain C.
10. Draw all the visible edges in the isometric projection as *thick* lines.

NOTE : In isometric projection / view, hidden lines are not generally shown.

Problem 23: *(Exercise)* *Draw the isometric projection of a* ***square prism*** *of 40 mm side of base and 60 mm altitude when its axis is (i) vertical and (ii) horizontal.*

Problem 24: *(Exercise)* *Draw the isometric projection of a* ***cube*** *of side 45 mm.*

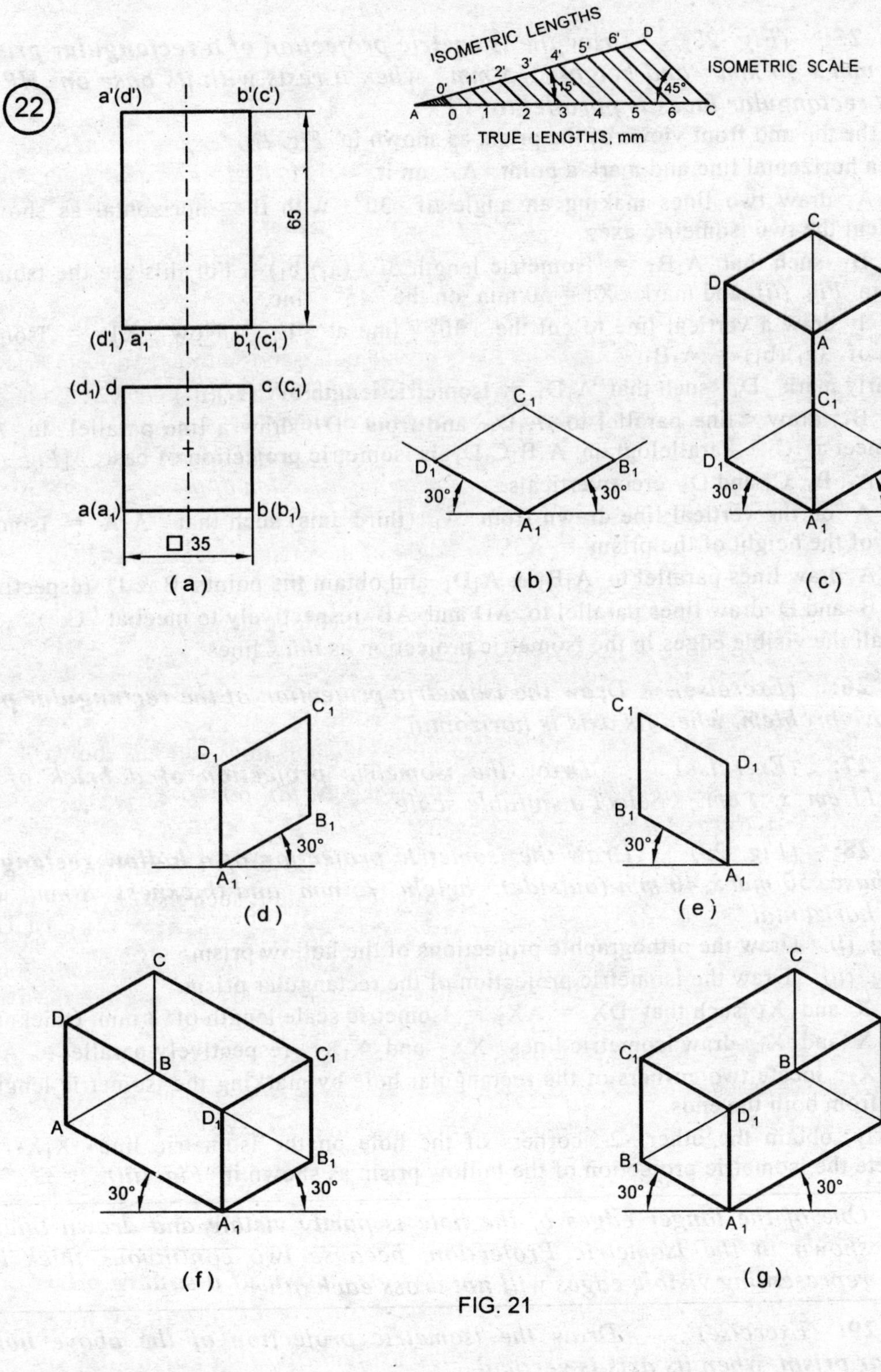
22
ISOMETRIC LENGTHS
ISOMETRIC SCALE
TRUE LENGTHS, mm
A 0 1 2 3 4 5 6 C
0' 1' 2' 3' 4' 5' 6' D
15°
45°
a'(d')
b'(c')
65
(d'1) a'1
b'1 (c'1)
(d1) d
c (c1)
a(a1)
b(b1)
□ 35
(a)
(b)
(c)
(d)
(e)
(f)
(g)
30°

FIG. 21

Problem 25: *(Fig. 22)* *Draw the isometric projection of a* ***rectangular prism*** *of base 50 mm x 40 mm and height 75 mm, when it rests with its base on HP and one of its rectangular faces is parallel to VP.*

1. Draw the top and front views of the prism as shown in *Fig. (i)*.
2. Draw a horizontal line and mark a point A_1 on it.
3. From A_1 draw two lines making an angle of 30° with the horizontal as shown to represent the two isometric axes.
4. Mark B_1 such that A_1B_1 = Isometric length of $(a_1)(b_1)$. For this see the Isometric Scale in *Fig. (ii)* and mark X1 = 50 mm on the 45° line.
5. From 1 draw a vertical line to cut the 30° line at 1′. Now X1′ = Isometric length of $(a_1)(b_1) = A_1B_1$.
6. Similarly mark D_1 such that A_1D_1 = Isometric length of $(a_1)(d_1)$ = X2′.
7. From B_1 draw line parallel to A_1D_1 and from D_1 draw a line parallel to A_1B_1. Both meet at C_1. Parallelogram $A_1B_1C_1D_1$ is isometric projection of base. [*Fig. (iii)*]
8. From A_1, B_1, C_1 and D_1 erect verticals.
9. Mark A on the vertical line drawn from A_1 (third axis) such that A_1A = Isometric length of the height of the prism = X3′.
10. From A draw lines parallel to A_1B_1 & A_1D_1 and obtain the points B & D respectively.
11. From B and D draw lines parallel to AD and AB respectively to meet at C.
12. Draw all the visible edges in the isometric projection as *thick* lines.

Problem 26: *(Exercise)* *Draw the isometric projection of the* ***rectangular prism*** *in the above problem, when its axis is horizontal.*

Problem 27: *(Exercise)* *Draw the isometric projection of a* ***brick*** *of size 22 cm x 11 cm x 4 cm. Select a suitable scale.*

Problem 28: *(Fig. 23)* *Draw the isometric projection of a* ***hollow rectangular prism*** *of base 50 mm x 40 mm (outside), height 75 mm and thickness 8 mm when its axis is horizontal.* (UQ)

1. *See Fig. (i).* Draw the orthographic projections of the hollow prism.
2. *See Fig. (ii).* Draw the isometric projection of the rectangular prism.
3. Mark X and X_1 such that DX = AX_1 = Isometric scale length of 8 mm (thickness).
4. From X and X_1 draw isometric lines XX_2 and X_1X_3 respectively parallel to AB.
5. On XX_2 locate two corners of the rectangular hole by marking the isometric length of 8 mm from both the ends.
6. Similarly, obtain the other 2 corners of the hole on the isometric line X_1X_3 and complete the isometric projection of the hollow prism as shown in *Fig. (iii)*.

NOTE : *One of the longer edges of the hole is partly visible and drawn thick as shown in the Isometric Projection, because two continuous thick lines representing visible edges will* ***not*** *cross each other.*

Problem 29: *(Exercise)* *Draw the isometric projection of the above* ***hollow rectangular prism****, when its axis is vertical.*

25

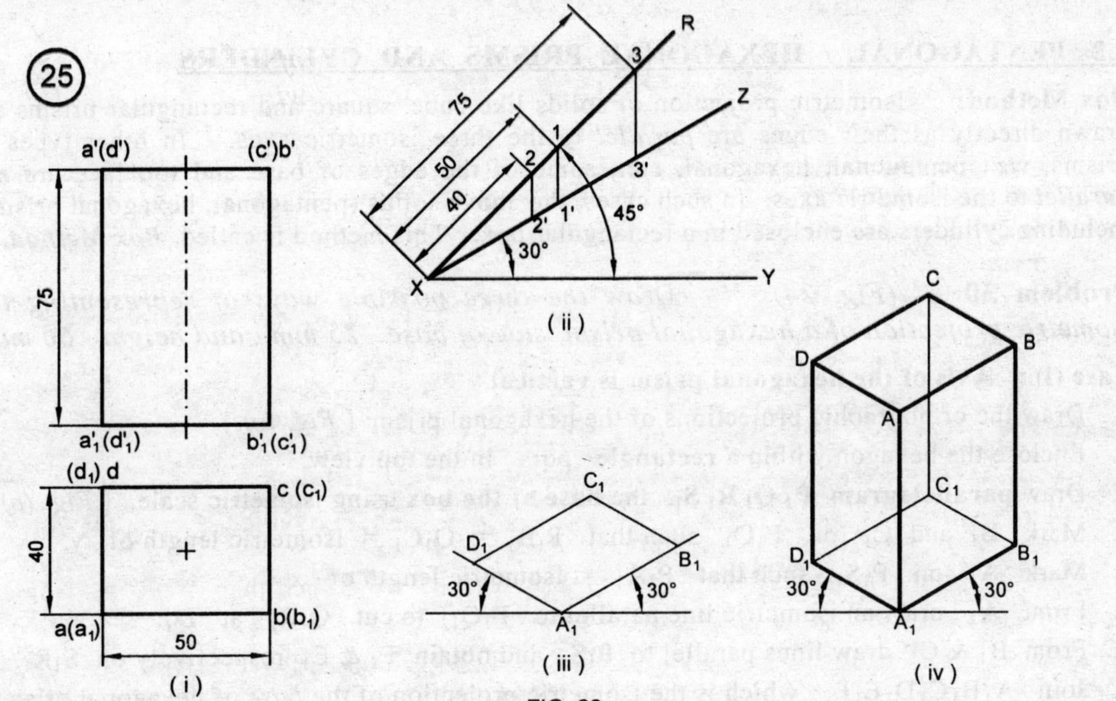

FIG. 22

28

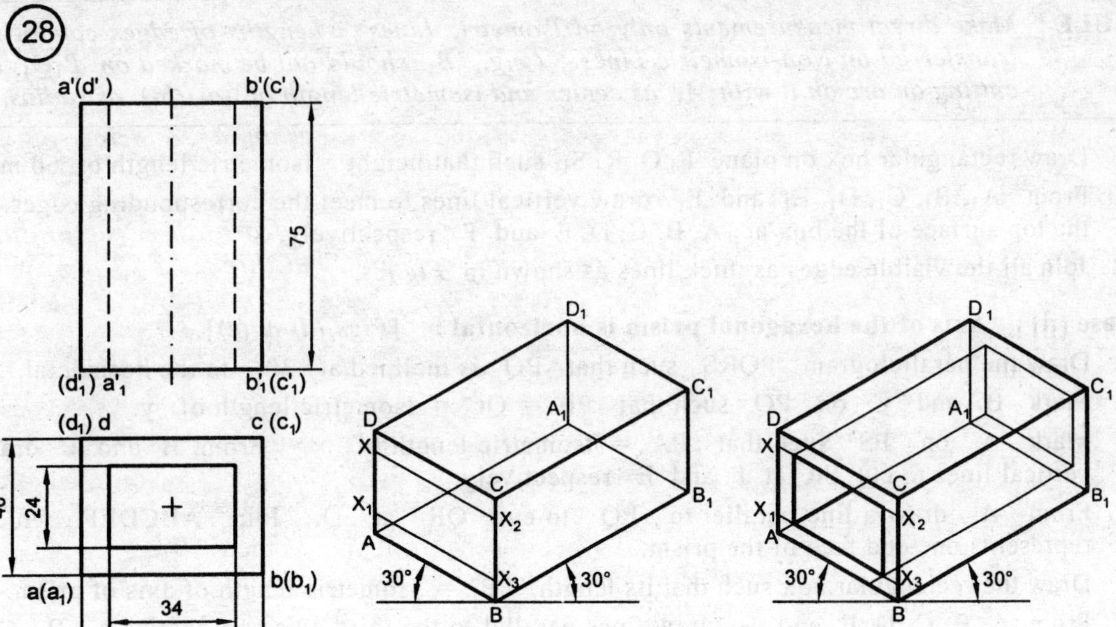

FIG. 23

8.2 PENTAGONAL / HEXAGONAL PRISMS AND CYLINDERS

Box Method: Isometric projection of solids like cube, square and rectangular prisms are drawn directly as their edges are *parallel* to the three isometric axes. In other types of prisms, *viz.*, pentagonal, hexagonal, etc., some of the edges of base and top face are ***not parallel*** to the isometric axes. In such cases, the above solids (pentagonal, hexagonal prisms) including cylinders are enclosed in a rectangular box. This method is called ***Box Method.***

Problem 30: *(Fig. 24) Draw the three possible ways of representing the isometric projection of a* ***hexagonal prism****, side of base 25 mm and height 60 mm.*

Case (i) : Axis of the hexagonal prism is vertical

1. Draw the orthographic projections of the hexagonal prism [*Fig. (a)*].
2. Enclose the hexagon within a **rectangle** pqrs in the top view.
3. Draw **parallelogram $P_1 Q_1 R_1 S_1$**, the **base of the box** using isometric scale. [*Fig. (b)*]
4. Mark B_1 and C_1 on P_1Q_1 such that $P_1B_1 = Q_1C_1$ = Isometric length of y.
5. Mark A_1 on P_1S_1 such that P_1A_1 = Isometric length of x.
6. From A_1 draw an isometric line parallel to P_1Q_1 to cut Q_1R_1 at D_1.
7. From B_1 & C_1 draw lines parallel to P_1S_1 and obtain F_1 & E_1 respectively on S_1R_1.
8. Join $A_1B_1C_1D_1E_1F_1$ which is the isometric projection of the *base* of hexagonal prism.

RULE : *Make direct measurements only on Isometric Lines. Lengths of edges cannot be transferred on Non-isometric Lines. (e.g., B_1 should not be marked on P_1Q_1 by cutting an arc on it with A_1 as center and isometric length of $(a_1)(b_1)$ as radius.)*

---

9. Draw rectangular box on plane $P_1 Q_1 R_1 S_1$ such that height = Isometric length of 60 mm.
10. From A_1, B_1, C_1, D_1, E_1 and F_1 draw vertical lines to meet the corresponding edges of the top surface of the box at A, B, C, D, E and F respectively.
11. Join all the visible edges as thick lines as shown in *Fig.(c)*.

Case (ii) : Axis of the hexagonal prism is horizontal : *[Figs.(d) & (e)]*

1. Draw the parallelogram PQRS such that PQ is inclined at 30° to the horizontal.
2. Mark B and C on PQ such that PB = QC = Isometric length of y.
3. Mark A on PS such that PA = Isometric length of x. From B and C draw vertical lines to cut SR at F and E respectively.
4. From A draw a line parallel to PQ to cut QR at D. Join ABCDEF which represents one end face of the prism.
5. Draw the rectangular box such that its length PP_1 = Isometric length of axis of prism.
6. From A, B, C, D, E and F draw lines parallel to the third axis and obtain A_1, B_1, C_1, D_1, E_1 and F_1 on the corresponding edges of the box.
7. Then complete the isometric projections as shown in *Figs. (d) & (e)*.

Problem 31: *(Exercise) Draw the three possible ways of representing the isometric projection of a* ***pentagonal prism****, side of base 30 mm and height 60 mm.*

30

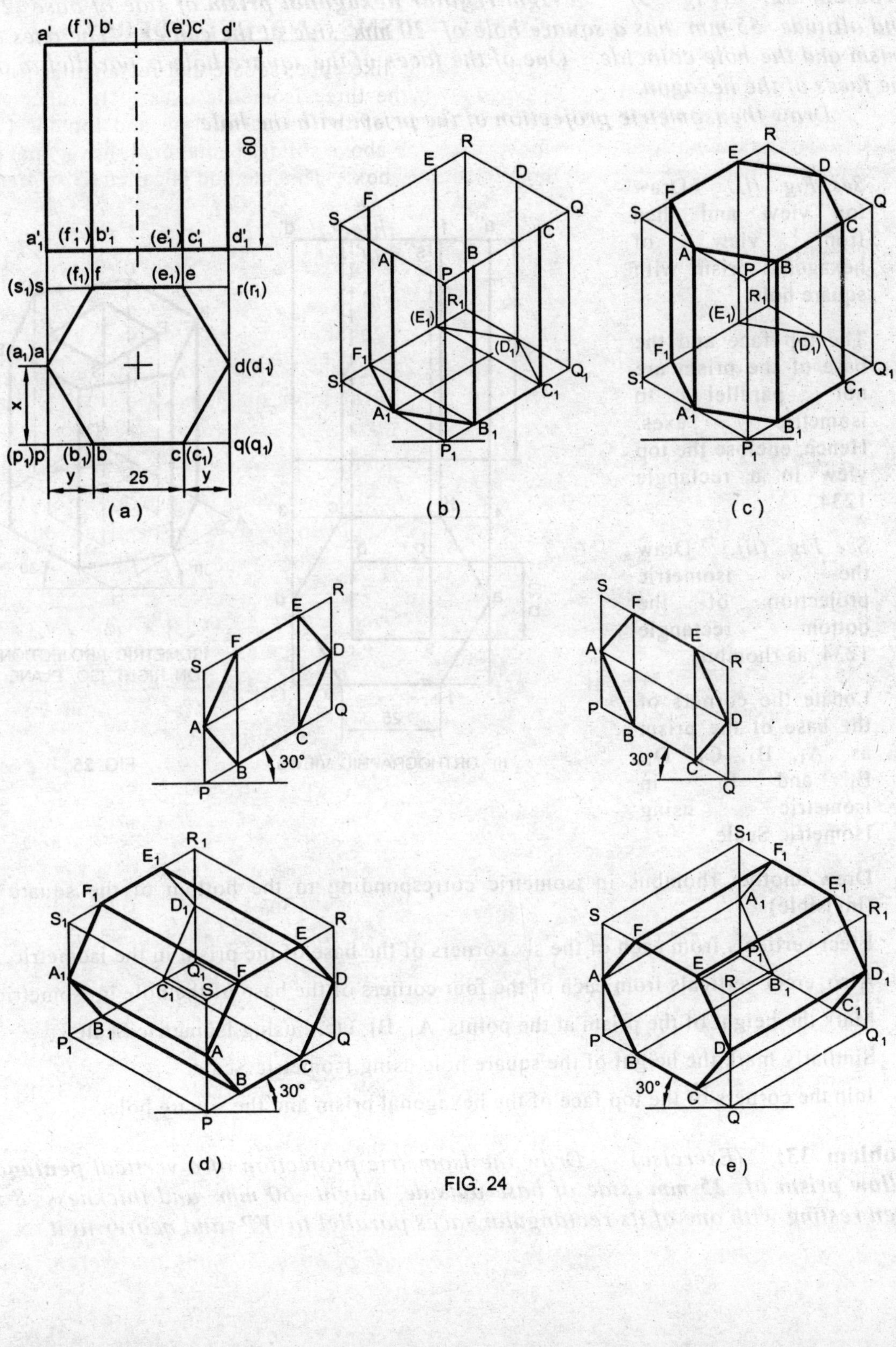
a'
(f ') b'
(e')c'
d'
60
a'1
(f'1) b'1
(e'1) c'1
d'1
(s1)s
(f1) f
(e1) e
r(r1)
(a1)a
d(d1)
x
(p1)p
(b1) b
c (c1)
q(q1)
y
25
y
(a)
R
E
D
F
S
Q
C
B
A
P
R1
(E1)
(D1)
F1
S1
Q1
C1
A1
B1
P1
(b)
(c)
30°
(d)
(e)

FIG. 24

Problem 32: *(Fig. 25) A right* ***regular hexagonal prism*** *of side of base 25 mm and altitude 55 mm has a square hole of 20 mm side at the center. The axes of the prism and the hole coincide. One of the faces of the square hole is parallel to one of the faces of the hexagon.*

Draw the isometric projection of the prism with the hole. (UQ)

1. *See Fig. (i).* Draw top view and then front view of hexagonal prism with square hole.
2. The top face and the base of the prism are not parallel to isometric axes. Hence, enclose the top view in a rectangle 1234.
3. *See Fig. (ii).* Draw the isometric projection of the bottom rectangle 1234 as rhombus.
4. Locate the corners of the base of the prism as A_1, B_1, C_1, D_1, E_1 and F_1 in isometric using Isometric Scale.

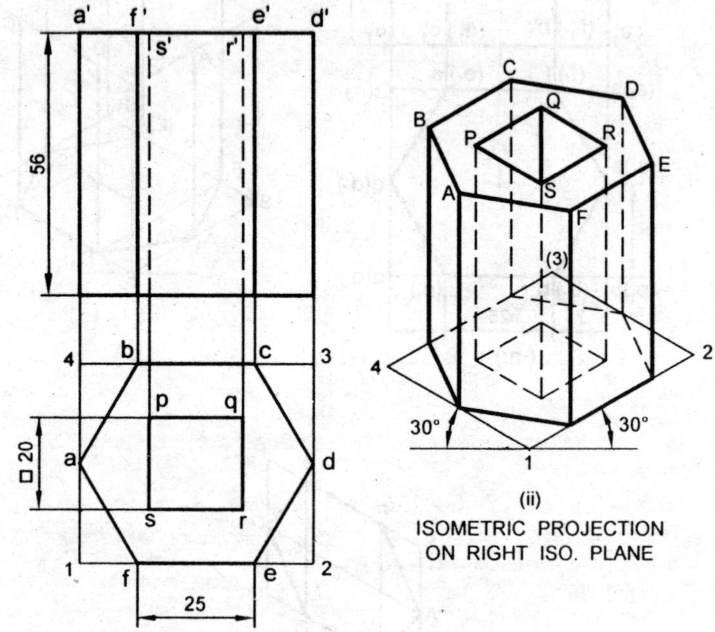

(i) ORTHOGRAPHIC VIEWS

(ii) ISOMETRIC PROJECTION ON RIGHT ISO. PLANE

FIG. 25

5. Draw another rhombus in isometric corresponding to the bottom of the square hole (invisible).
6. Erect verticals from each of the six corners of the base of the prism in the isometric.
7. Also, erect verticals from each of the four corners of the base of the hole in isometric.
8. Mark the height of the prism at the points A_1, B_1, etc., using Isometric Scale.
9. Similarly mark the height of the square hole using Isometric Scale.
10. Join the corners of the top face of the hexagonal prism and the square hole.

Problem 33: *(Exercise) Draw the isometric projection of a vertical* ***pentagonal hollow prism*** *of 25 mm side of base outside, height 60 mm and thickness 8 mm, when resting with one of its rectangular faces parallel to VP and nearer to it.*

Problem 34: *(Fig. 26) A right hexagonal prism of side of base 24 mm and axis 56 mm long is lying on one of the corner of the base. Its axis is inclined at an angle of 30° to HP. Draw the orthographic projections and the isometric projection of the solid.* (UQ)

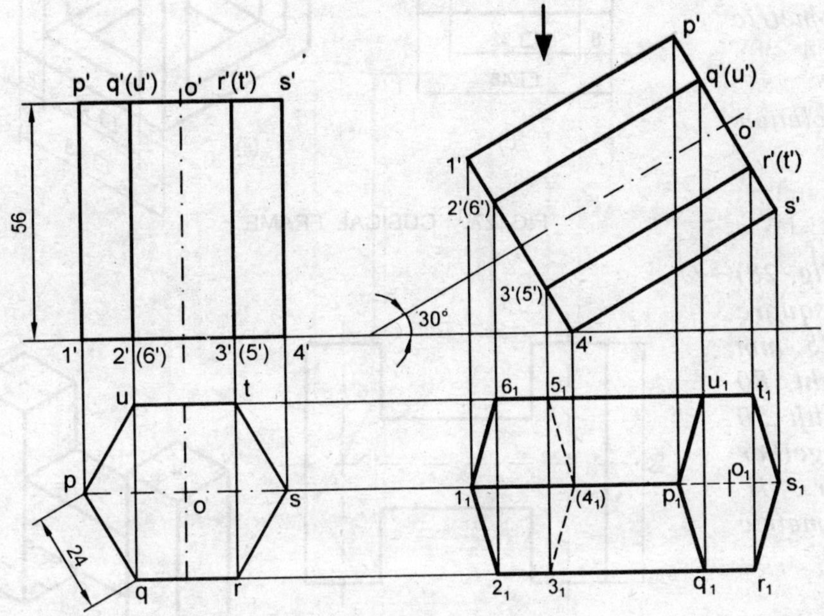

(i)

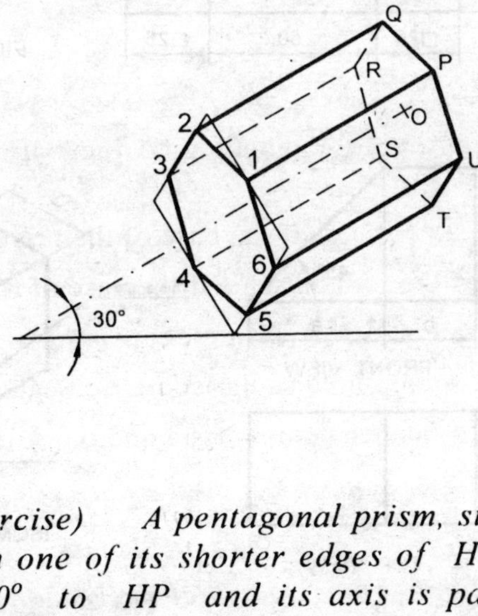

ISO. PROJECTION OF AUXILIARY TOP VIEW

(ii)

FIG. 26

Problem 35: *(Exercise) A pentagonal prism, side of base 25 mm and axis 50 mm long rests, with one of its shorter edges of HP such that the base containing that edge makes 30° to HP and its axis is parallel to VP. Draw both the orthographic projections and the isometric projection of the solid.*

Problem 36: *(Fig. 27)* *Fig. (i) shows the orthographic projection of a cubical frame having the same view in all the three directions. Draw its isometric projection.*

See Fig. 27(ii) for Solution.

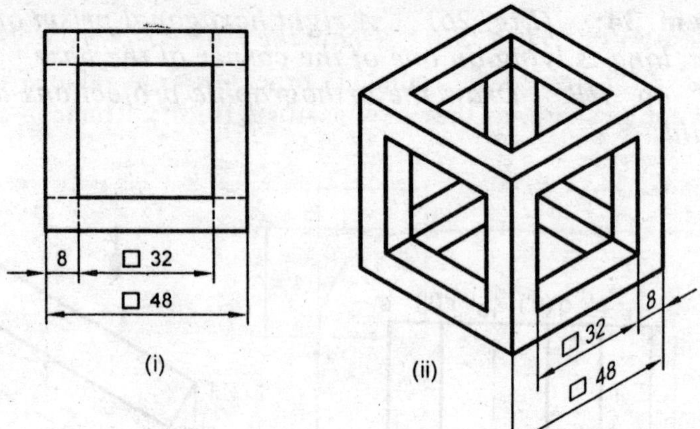

FIG. 27 CUBICAL FRAME

Problem 37: *(Fig. 28)* *Three square rods of 25 x 25 mm cross section, height 80 mm and inside width 50 mm are nailed together to form the letter H. Draw its isometric projection.*

See Fig. 28 for Solution.

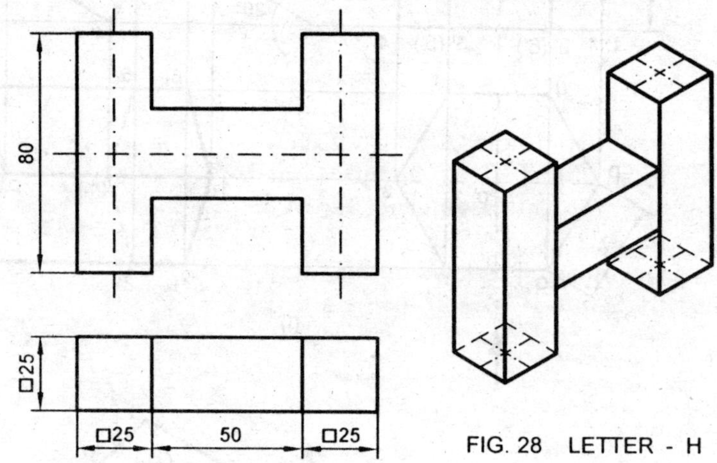

FIG. 28 LETTER - H

Angle in Isometric

Problem 38: *(Fig. 29)* Draw the isometric projection of a taper block with 45° inclination.

1. Do not set-off angles directly in isometric.
2. Translate the angle to linear measurements along the isometric axes.
3. Locate the end points of the lines which subtend the angle.

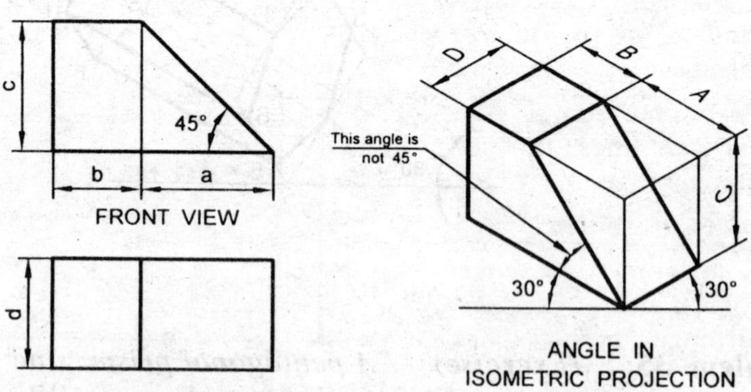

TOP VIEW

FIG. 29

Problem 39: *(Fig. 30) Draw the isometric projection of the* ***Curved Solid*** *shown.*

(Have you observed the rounded corners of your table ? Corners of any machine element should not be sharp, as they may easily chip-off. Rounded corner has fillet radius as shown.)

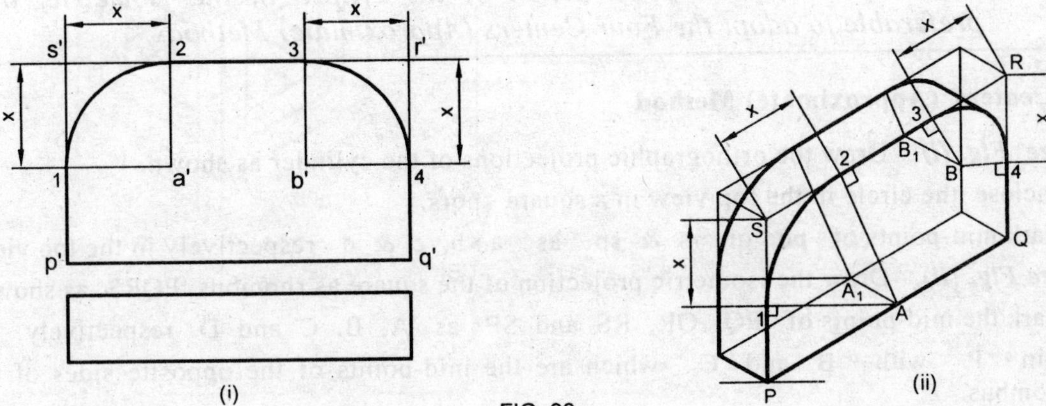

FIG. 30

1. *See Fig. (i).* Enclose the front view of the solid in a box.
2. Mark the points 1, 2, 3 and 4 at a distance of x from the top corners of the box.
3. *See Fig. (ii).* In isometric, fix A at the intersection of the perpendiculars drawn from 1 and 2.
4. A as center and A1 as radius, draw the arc 1 – 2.
5. Fix the point B and draw the arc 3 - 4.
6. Similarly fix the centers A_1 and B_1 and draw the arcs at the rear side of the solid.

Problem 40: *(Fig. 31) Draw the isometric projection of an* ***irregular curved solid*** *shown.*

1. Enclose the front view in a rectangle.
2. Mark any convenient number of points on the curve.
3. Draw horizontal and vertical lines through these points.
4. Draw an isometric box.
5. Transfer the coordinates of the different points from the orthographic view to the isometric.
6. Join the points by smooth curve.

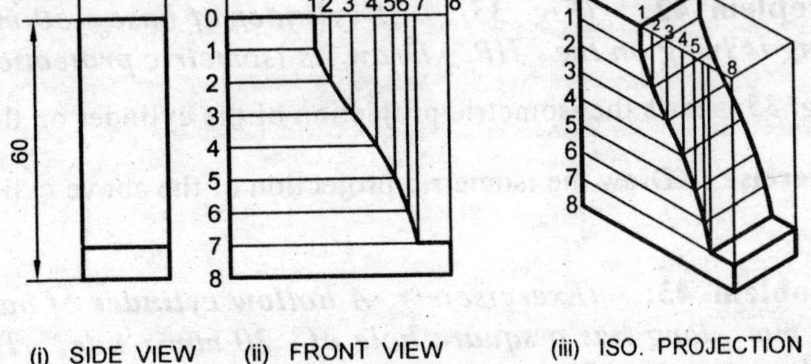

FIG. 31

Problem 41: *(Fig. 32) Draw the isometric projection of a* ***cylinder*** *of base 50 mm diameter and 70 mm height when it rests with its base on HP.*

NOTE : *For easy and quick construction of the ellipse in the isometric, it is preferable to adopt the Four Centers (Approximate) Method.*

Four Centers (Approximate) Method

1. ***See Fig. (i).*** Draw the orthographic projections of the cylinder as shown.
2. Enclose the circle in the top view in a square pqrs.
3. Mark mid-points of pq, qr, rs & sp as a, b, c & d respectively in the top view.
4. ***See Fig. (ii).*** Draw the isometric projection of the square as rhombus PQRS as shown.
5. Mark the mid-points of PQ, QR, RS and SP as A, B, C and D respectively.
6. Join P with B and C which are the mid-points of the opposite sides of the rhombus.
7. Similarly, join R with A and D.
8. With P as center and PC as radius draw an arc CB.
9. Similarly, R as center and RA as radius draw an arc AD.
10. Lines PC and RD intersect at O_1.
11. With O_1 as center and O_1D as radius draw an arc DC.
12. Similarly, with O_2 (intersection of RA and PB) as center and O_2B as radius draw an arc BA. Thus complete the ellipse to represent the base circle. [*Fig. (ii)*].
13. ***See Fig. (iii).*** Draw the isometric projection of the box, taking its height = Height of the cylinder in isometric length.
14. Construct another ellipse on the top surface of the box.
15. Draw two common tangents to the ellipses and complete the isometric projection of the cylinder.

Problem 42: *(Fig. 33) A* ***cylinder*** *of base 50 mm diameter and axis 70 mm long is lying on the HP. Draw its isometric projection when the axis is horizontal.*

Fig. 33 shows the isometric projection of the cylinder on the Right Isometric Plane.

Exercise : Draw the isometric projection of the above cylinder on the Left Iso. Plane.

Problem 43: *(Exercise) A* ***hollow cylinder*** *of base 50 mm diameter and axis 70 mm long has a square hole of 20 mm side. The axes of the cylinder and the hole coincide. Draw the three possible ways of representing the isometric projection of the solid.*

(UQ)

41

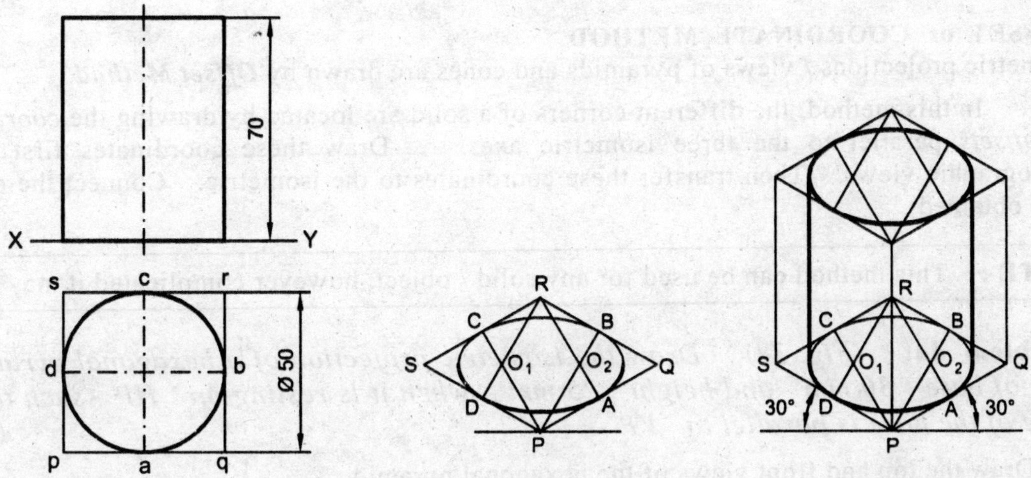

(i) (ii) (iii)

FIG. 32

42

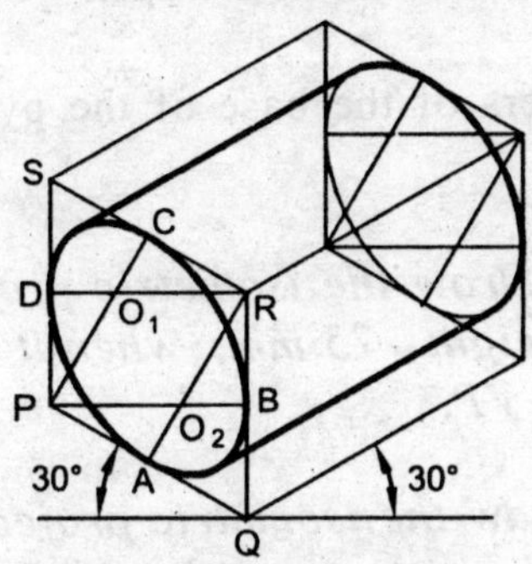

FIG. 33

ISOMETRIC PROJECTION OF CYLINDER
ON RIGHT ISOMETRIC PLANE

NOTE :
1. Use BOX METHOD for the isometric projection / view of cylinders. For this, enclose the cylinder in a square prism.
2. Use FOUR CENTER METHOD for the construction of Ellipse.
3. The invisible part of the base of the cylinder need not be shown.

8.3 PYRAMIDS AND CONES

OFFSET or COORDINATE METHOD

Isometric projections / views of pyramids and cones are drawn by ***Offset Method***.

In this method, the different corners of a solid are located by drawing the *coordinates* or *offsets* parallel to the three isometric axes. Draw these coordinates first in the orthographic views. Then transfer these coordinates to the isometric. Connect the corners thus obtained.

NOTE : This method can be used for any solid / object, however complicated it may be.

Problem 44: *(Fig. 34) Draw the isometric projection of a hexagonal pyramid of side of base 30 mm and height 75 mm , when it is resting on HP such that an edge of the base is parallel to VP.* (UQ)

1. Draw the top and front views of the hexagonal pyramid.
2. Enclose the hexagon in a rectangle pqrs in the top view.
3. Draw the isometric projection of the base of the pyramid in the parallelogram PQRS.
4. Note that FC is an isometric line on which O_1 lies. Hence mark O_1 on the **isometric line FC** such that $FO_1 = y$.
5. From O_1 erect a vertical line and mark the apex O such that O_1O = length of the axis = 75 mm.
6. Join O with all the corners of the base of the pyramid and complete the isometric projection.

Problem 45: *(Exercise) Draw the isometric projection of a triangular pyramid of side of base 35 mm and height 75 mm, when it is resting on HP such that an edge of the base is parallel to VP.*

Problem 46: *(Fig. 35) Draw the isometric projection of a* ***cone*** *of base 40 mm diameter and height 58 mm when it rests with its base on HP. (i.e., Axis is vertical)*

1. Draw the orthographic projections of the cone.
2. Enclose the circle in the top view in a square pqrs.
3. Draw the rhombus PQRS using the Isometric Scale.
4. Draw the base of the cone as an ellipse by ***Four Centers (Approximate) Method***.
5. From B and C draw lines parallel to isometric axes and obtain O_1 (***Offset Method***).
6. From O_1 draw a line parallel to the third axis.
7. On this line mark O such that O_1O = isometric length of the height of the cone.
8. From O draw 2 tangents to the ellipse and complete the isometric projection of the cone.

Problem 47: *(Exercise) Draw the isometric .projection of the cone in the previous Problem when its axis is horizontal.*

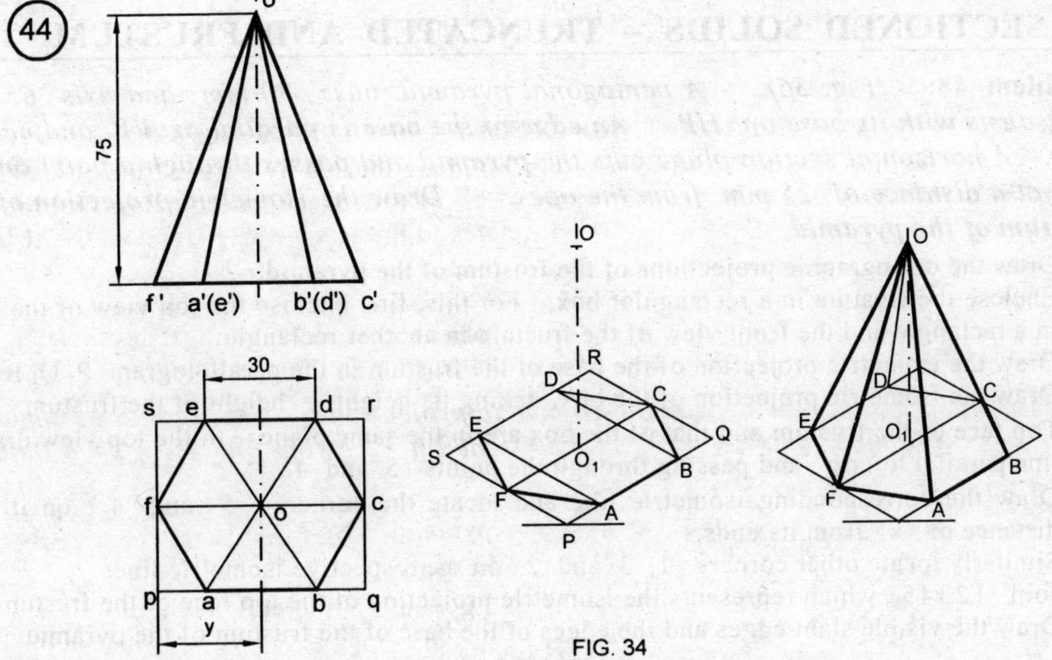
44
o'
75
f' a'(e') b'(d') c'
30
s e d r
f o c
p a b q
y
O
R
D C
E Q
S O1 B
F A
P
O
D C
E O1 B
F A

FIG. 34

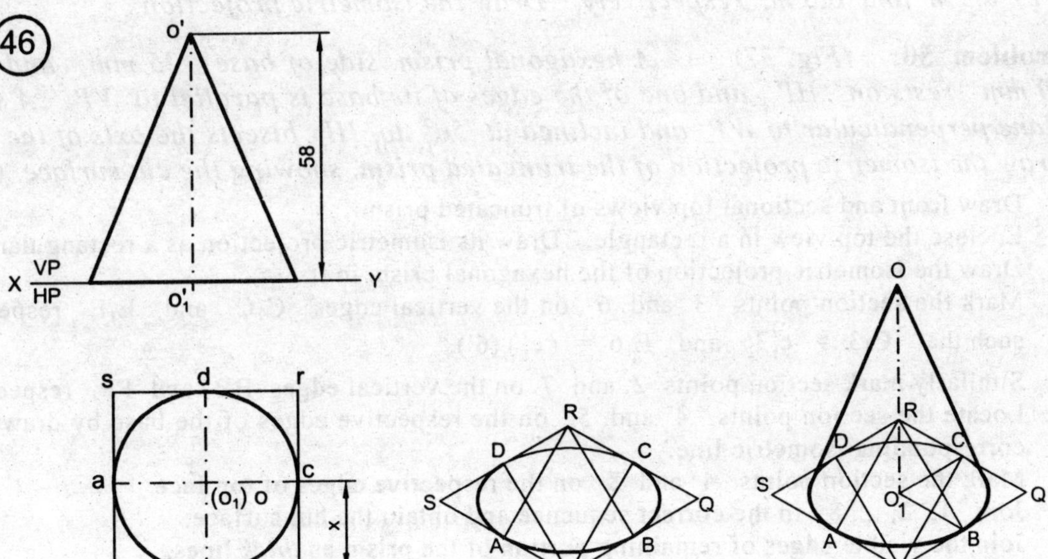
46
o'
58
X VP HP o1' Y
s d r
a c
(o1) o
x
p b q
x
Ø 40
R
D C
S Q
A B
P
O
R
D C
S O1 Q
A B
P

FIG. 35

9. SECTIONED SOLIDS – TRUNCATED AND FRUSTUM

Problem 48: *(Fig. 36) A pentagonal pyramid, base 30 mm and axis 65 mm long, rests with its base on HP. An edge of the base is parallel to VP and nearer to it. A horizontal section plane cuts the pyramid and passes through a point on the axis at a distance of 25 mm from the apex. Draw the isometric projection of the* ***frustum of the pyramid.*** (UQ)

1. Draw the orthographic projections of the frustum of the pyramid.
2. Enclose the frustum in a rectangular box. For this, first enclose the top view of the base in a rectangle and the front view of the frustum in another rectangle.
   Draw the isometric projection of the base of the frustum in the parallelogram $P_1 Q_1 R_1 S_1$.
   Draw the isometric projection of the box, taking its height = height of the frustum.
3. Top face of the frustum and that of the box are in the same plane. In the top view draw a line parallel to pq and passing through the points 5 and 4.
4. Draw the corresponding isometric line and locate the corners 5 and 4 on it at a distance of x from its ends.
5. Similarly locate other corners 1, 3 and 2 on the respective isometric lines.
6. Join 12345 which represents the isometric projection of the top face of the frustum.
7. Draw the visible slant edges and the edges of the base of the frustum of the pyramid.

Problem 49: *(Exercise) A masonry pillar is in the form of a* ***frustum of a hexagonal pyramid.*** *The pillar is 2 m high and the sides of its base and top face are 0.5 m and 0.3 m respectively. Draw the isometric projection.* (UQ)

Problem 50: *(Fig. 37) A hexagonal prism, side of base 25 mm and height 50 mm rests on HP and one of the edges of its base is parallel to VP. A section plane perpendicular to VP and inclined at 50° to HP bisects the axis of the prism. Draw the isometric projection of the* ***truncated prism****, showing the cut surface.* (UQ)

1. Draw front and sectional top views of truncated prism.
2. Enclose the top view in a rectangle. Draw its isometric projection as a rectangular box.
3. Draw the isometric projection of the hexagonal prism in it.
4. Mark the section points 3 and 6 on the vertical edges C_1C and E_1E respectively such that $C_13 = c_1'3'$ and $E_16 = (e_1')(6')$.
5. Similarly mark section points 2 and 7 on the vertical edges BB_1 and FF_1 respectively.
6. Locate the section points 4 and 5 on the respective edges of the base by drawing the corresponding isometric line.
7. Mark the section points 1 and 8 on the respective edges of top face.
8. Join 1, 2, ..., 8 in the correct sequence and obtain the cut surface.
9. Join the visible edges of remaining portion of the prism as *thick* lines.

Problem 51: *(Exercise) A* ***square prism****, side of base 30 mm and axis 65 mm long, rests with its base on HP and one of its rectangular faces is inclined at 30° to VP. A section plane perpendicular to VP and inclined at 60° to HP cuts the prism at a point on the axis 40 mm from the base. Draw the isometric projection of the prism, showing the cut surface.* (UQ)

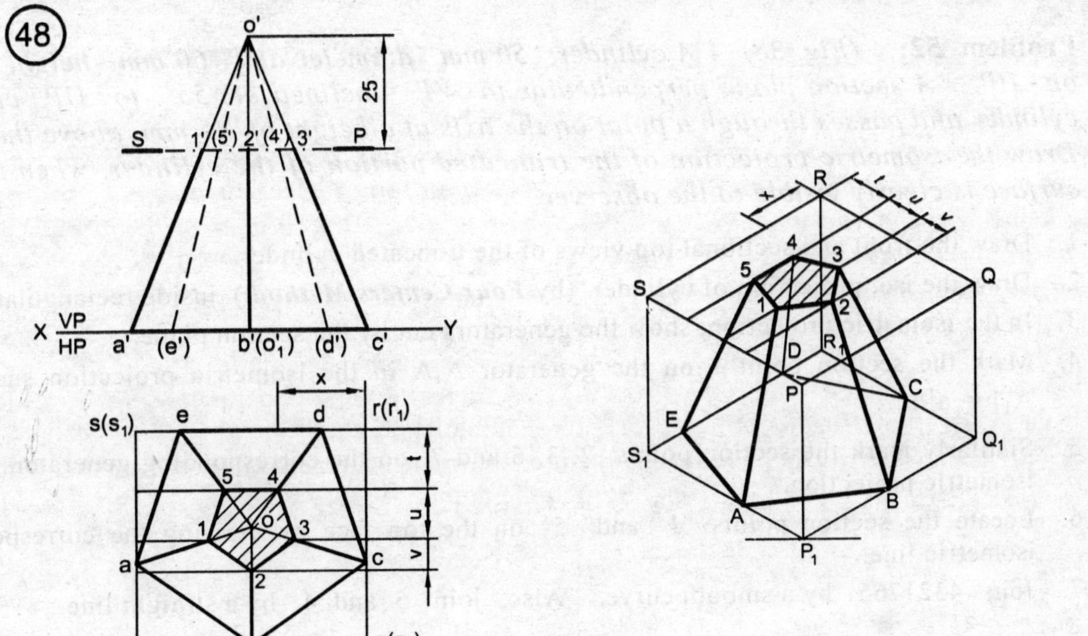
48
o'
25
S
1'
(5')
2'
(4')
3'
P
X
VP
HP
a'
(e')
b'(o'₁)
(d')
c'
Y
x
s(s₁)
e
d
r(r₁)
5
4
o
1
3
a
c
2
t
u
v
p(p₁)
b
q(q₁)
R
t
u
v
S
Q
5
4
3
1
2
D
R₁
P
E
C
S₁
Q₁
A
B
P₁

FIG. 36

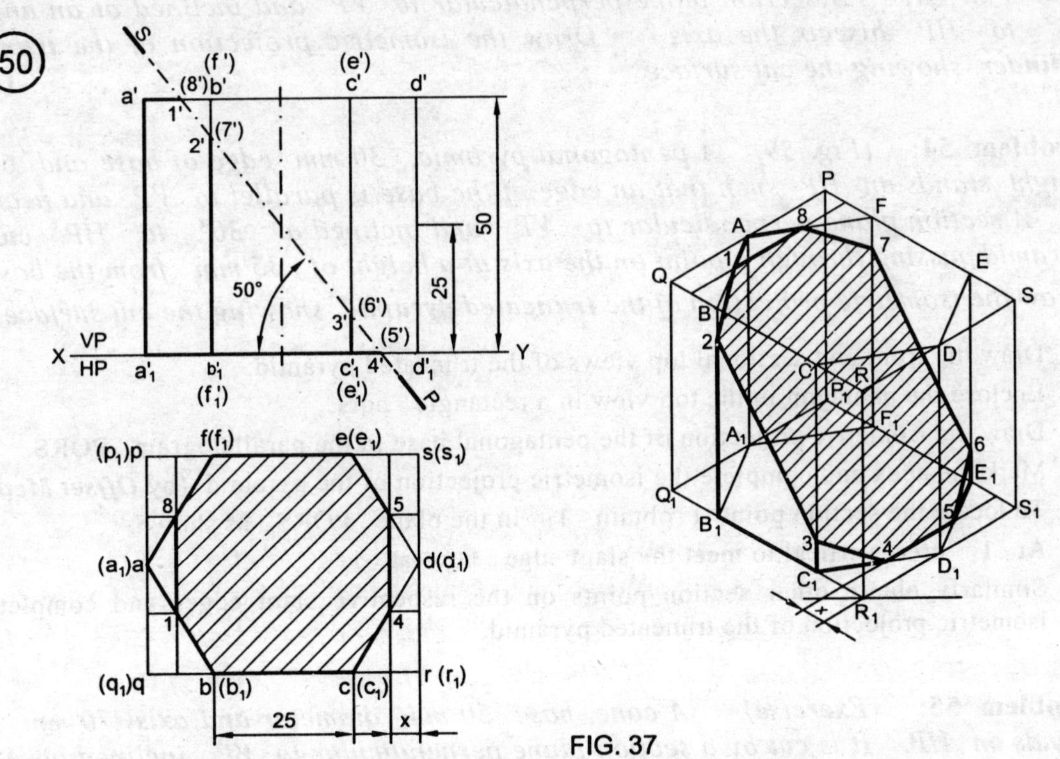
50
S
(f ')
a'
(8')
b'
1'
2'
(7')
(e')
c'
d'
50
25
50°
(6')
3'
(5')
4'
X
VP
HP
a'₁
b'₁
(f'₁)
c'₁
(e'₁)
d'₁
Y
P
(p₁)p
f(f₁)
e(e₁)
s(s₁)
8
5
(a₁)a
d(d₁)
1
4
(q₁)q
b(b₁)
c(c₁)
r(r₁)
25
x
P
F
A
8
7
1
Q
E
S
B
2
D
C
R
P₁
F₁
A₁
6
E₁
Q₁
S₁
B₁
5
3
4
C₁
D₁
R₁
x

FIG. 37

Problem 52: *(Fig. 38) A cylinder 50 mm diameter and 60 mm height stands on HP. A section plane perpendicular to VP, inclined at 55° to HP cuts the cylinder and passes through a point on the axis at a height of 45 mm above the base. Draw the isometric projection of the* ***truncated portion of the cylinder****, when the cut surface is clearly visible to the observer.* (UQ)

1. Draw the front and sectional top views of the truncated cylinder.
2. Draw the iso. projection of cylinder (by ***Four Centers Method***) inside rectangular box.
3. In the isometric projection, show the generators cut by the section plane.
4. Mark the section point 1 on the generator A_1A in the isometric projection such that $A_1 1 = a_1' 1'$.
5. Similarly mark the section points 2, 3, 6 and 7 on the corresponding generators in the isometric projection.
6. Locate the section points 4 and 5 on the top face by drawing the corresponding isometric line.
7. Join 4321765 by a smooth curve. Also, join 5 and 4 by a straight line.

Problem 53: *(Exercise) A cylinder of 60 mm diameter and 70 mm height, stands on HP. A section plane perpendicular to VP and inclined at an angle of 45° to HP bisects the axis. Draw the isometric projection of the* ***truncated cylinder****, showing the cut surface.* (UQ)

Problem 54: *(Fig. 39) A pentagonal pyramid, 30 mm edge of base and 65 mm height, stands on HP such that an edge of the base is parallel to VP and nearer to it. A section plane perpendicular to VP and inclined at 30° to HP cuts the pyramid passing through a point on the axis at a height of 35 mm from the base. Draw the isometric projection of the* ***truncated pyramid****, showing the cut surface.* (UQ)

1. Draw the front and sectional top views of the truncated pyramid.
2. Enclose the pentagon in the top view in a rectangle pqrs.
3. Draw the isometric projection of the pentagonal base in the parallelogram PQRS.
4. Mark the apex and complete the isometric projection of the pyramid (by ***Offset Method***).
5. To locate the section point 1 obtain 1_1 in the plane PQRS as shown.
6. At 1_1 erect vertical to meet the slant edge OA at 1.
7. Similarly obtain other section points on the respective slant edges and complete the isometric projection of the truncated pyramid.

Problem 55: *(Exercise) A cone, base 50 mm diameter and axis 70 mm long, stands on HP. It is cut by a section plane perpendicular to VP, inclined at 45° to HP and passing through a point on the axis 35 mm above the base. Draw the isometric projection of the* ***truncated cone****, showing the cut surface.* (UQ)

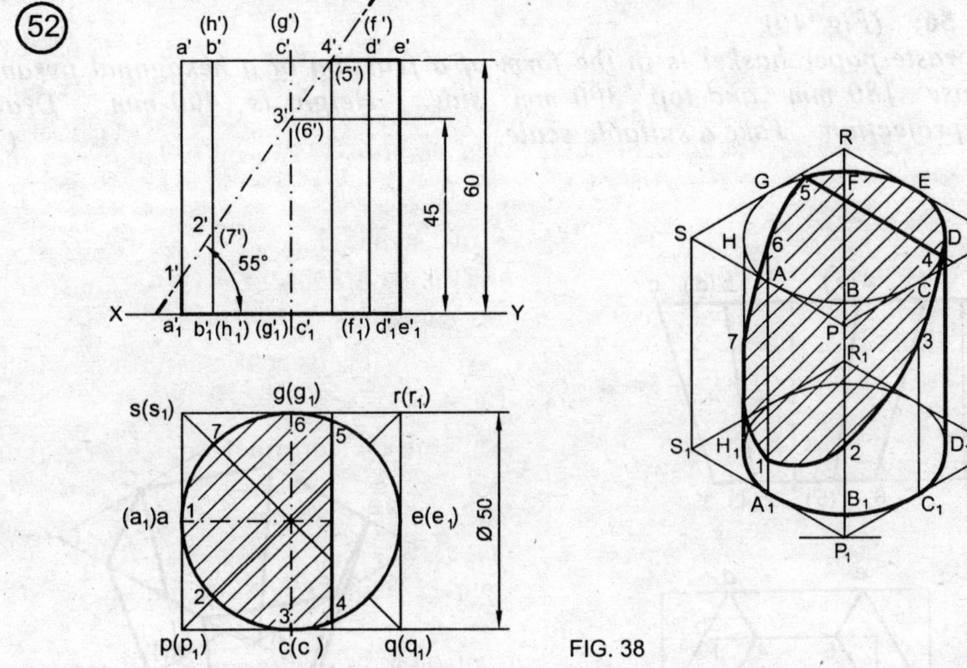
52
(h') (g') (f ')
a' b' c' 4' d' e'
(5')
3'
(6')
60
45
2'
(7')
1'
55°
X Y
a'_1 $b'_1(h'_1)$ $(g'_1)c'_1$ (f'_1) $d'_1e'_1$
$s(s_1)$ $g(g_1)$ $r(r_1)$
7 6 5
$(a_1)a$ 1 $e(e_1)$
Ø 50
2 3 4
$p(p_1)$ c(c) $q(q_1)$
R
G 5 F E
S H 6 D Q
4
A B C
P
7 R_1 3
S_1 H_1 1 2 D_1 Q_1
A_1 B_1 C_1
P_1

FIG. 38

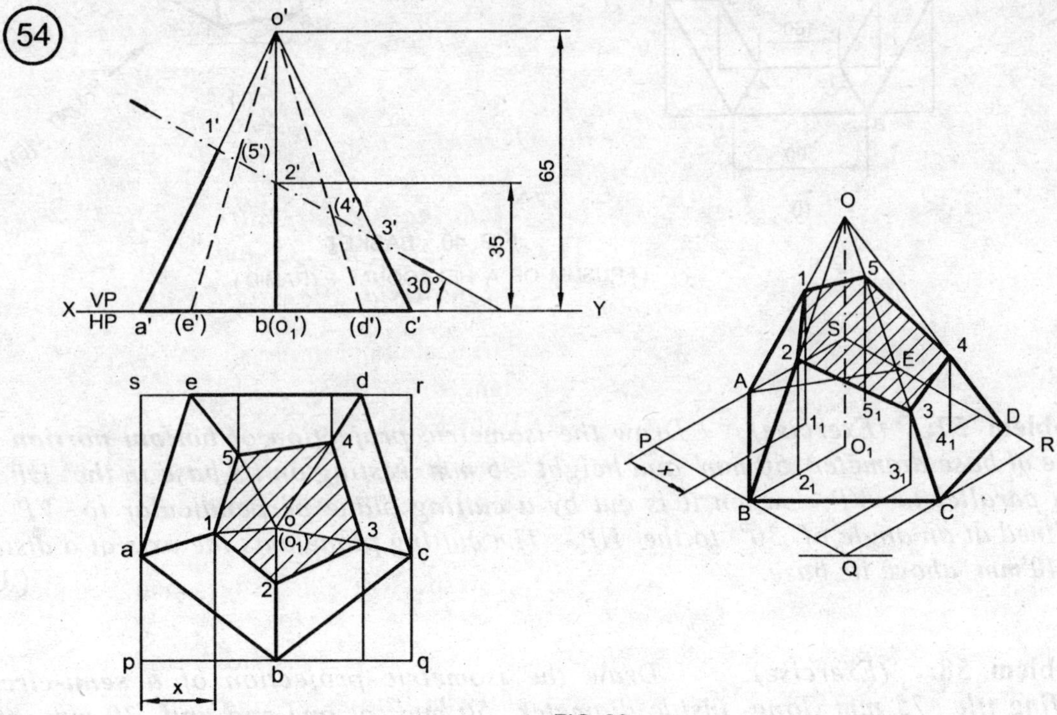
54
o'
1'
(5')
2'
(4')
3'
65
35
30°
X VP HP Y
a' (e') $b(o_1')$ (d') c'
s e d r
5 4
1 o 3
(o_1)
a c
2
p b q
x
O
1 5
S
2 E 4
A
5_1 3 D
P 1_1 4_1 R
O_1
2_1 3_1
B C
Q

FIG. 39

Problem 56: *(Fig. 40)*

A waste paper basket is in the form of a ***frustum*** *of a hexagonal pyramid of side of base 180 mm and top 300 mm side. Height is 400 mm. Draw the isometric projection. Take a suitable scale.* (UQ)

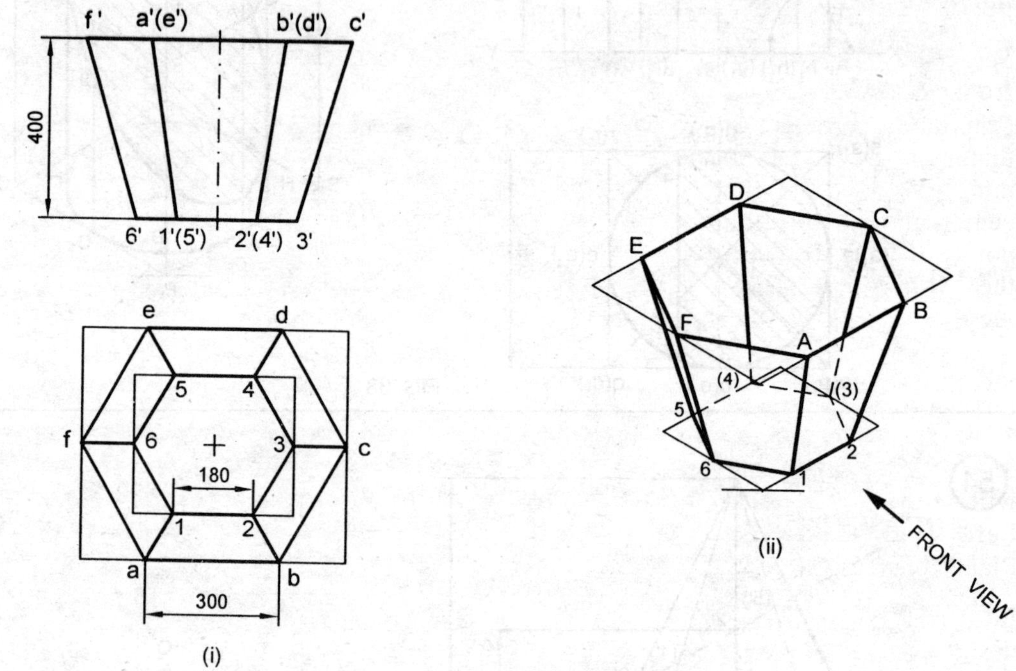

FIG. 40 BASKET

(FRUSUM OF A HEXAGONAL PYRAMID)

Problem 57: *(Exercise)* *Draw the isometric projection of* ***bottom portion of a cone*** *of base diameter 50 mm and height 75 mm resting on its base in the HP and axis parallel to VP, when it is cut by a cutting plane perpendicular to VP and inclined at an angle of* 30° *to the HP. The cutting plane cuts the axis at a distance of 40 mm above its base.* (UQ)

Problem 58: *(Exercise)* *Draw the isometric projection of a* ***semi-circular roofing*** *tile 75 mm long, inside diameter 50 mm at one end and 30 mm at the other end. Thickness of the tile is 10 mm.* (UQ)

10. ISOMETRIC PROJECTION OF SPHERES

The orthographic view of a sphere seen from any direction is a circle of diameter equal to the diameter of the sphere. ***Isometric Projection of a sphere is also a circle of diameter equal to the actual diameter of the sphere.*** We shall prove this concept in the *following Problem*:

Problem 59: *(Fig. 41) Draw the isometric projection of a sphere of 50 mm diameter, resting on the HP.*

1. *See Fig. (i).* Draw front and top views of sphere as circles of diameter D = Sϕ 50 mm. o′ is the center and p′ is the point of contact with the HP in the front view.
2. *See Fig. (ii).* Assume three Section Planes 1 1 1 1, 2 2 2 2 and 3 3 3 3 coinciding with the Isometric Left, Right and Top Planes respectively.
3. In all the above cases, the shape of the sections are circles of diameter D.

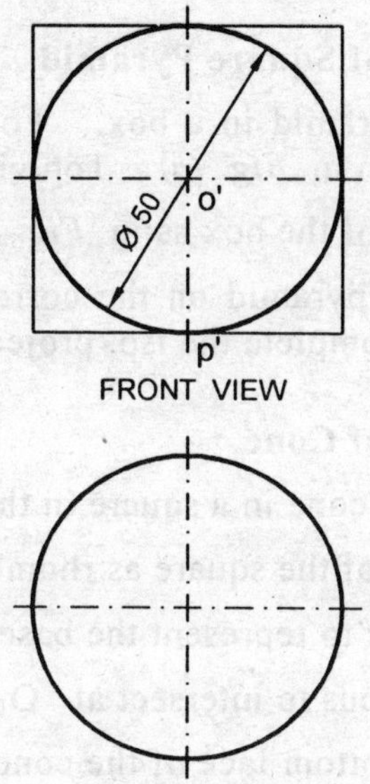

(i)

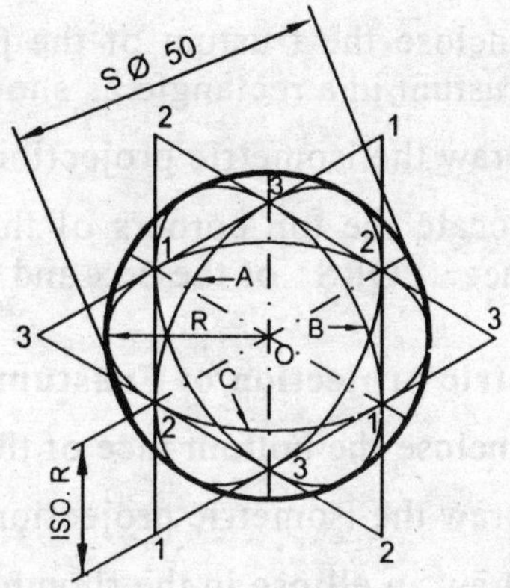

(ii)

FIG. 41
ISOMETRIC PROJECTION OF SPHERE

4. Draw the isometric projections of these three circles obtained on the three isometric planes as three ellipses, *i.e.*, O as center and Major diameter = True diameter of sphere = D, draw three ellipses, *i.e.*, A and B on LIP and RIP and C in horizontal position.
5. In all the above three cases, Major Diameter = D, since the distance of outer-most points on the ellipse from the center O = D/2. Hence, in Isometric Projection, distances of all the points on the surface of a sphere from its center are = Radius of the sphere.
6. Circle drawn enveloping these ellipses represents the isometric projection of the sphere. Hence Isometric Projection of a sphere is a circle of diameter = True diameter of sphere.
7. However, the distance of the center O of the sphere from its point of contact P with HP is equal to the Isometric Radius of the sphere = OP = 0.82 (D/2).

RULE : Use Isometric Scale for isometric projections of spheres or having spherical parts.

11. COMBINATION OF SOLIDS (COMPOUND SOLIDS)

Problem 60: *(Fig. 42)* ***A combination of solids*** *is formed as follows: A frustum of a cone has its top and bottom diameters 35 mm and 50 mm respectively and altitude 53 mm. It rests on the top face of a frustum of a square pyramid. Sides of the top and bottom faces of the pyramid are 58 mm and 70 mm respectively. Height is 22 mm.*

Draw the isometric projection of the combination of solids. (UQ)

Isometric projection of Frustum of Square Pyramid

1. Enclose the frustum of the pyramid in a box. For this, enclose the front view of the frustum in a rectangle as shown in *Fig. (a)*. Top view is self-enclosed.
2. Draw the isometric projection of the box as in *Fig. (b)*.
3. Locate the top corners of the pyramid on the corresponding isometric lines on the top face PQRS of the box and complete the iso. projection of the frustum of the pyramid.

Isometric projection of Frustum of Cone

4. Enclose the bottom face of the cone in a square in the top view in *Fig. (a)*.
5. Draw the isometric projection of the square as rhombus on top face of the box.
6. Draw an ellipse in the rhombus to represent the base of the cone.
7. Join the diagonals of the rhombus to intersect at O_1.
8. From the center O_1 of the bottom face of the cone, erect a vertical and mark O such that OO_1 = Altitude of the frustum of the cone = 53 mm.
9. Through O draw two isometric lines AC and BD such that AC = ac and BD = bd.
10. Through A, B, C and D, draw isometric lines to obtain a rhombus.
11. Inside the rhombus, draw the isometric projection of the top face of the cone (*top ellipse*).
12. Draw *two common tangents* to ellipses and complete the isometric projection of the frustum of cone as shown.

Problem 61: *(Fig. 43)* *Draw the isometric projection of a funnel consisting of a cylinder and a frustum of a cone. The diameter of the cylinder is 20 mm and the top diameter of the funnel is 68 mm. The heights of the frustum and cylinder are each equal to 40 mm.* (UQ)

See Fig. 43 for Solution.

Problem 62: *(Exercise)* *Draw the isometric projection of a bucket of top diameter 250 mm and bottom diameter 160 mm with a circular ring of 50 mm width attached at the bottom. Total height of the bucket is 320 mm.* (UQ)

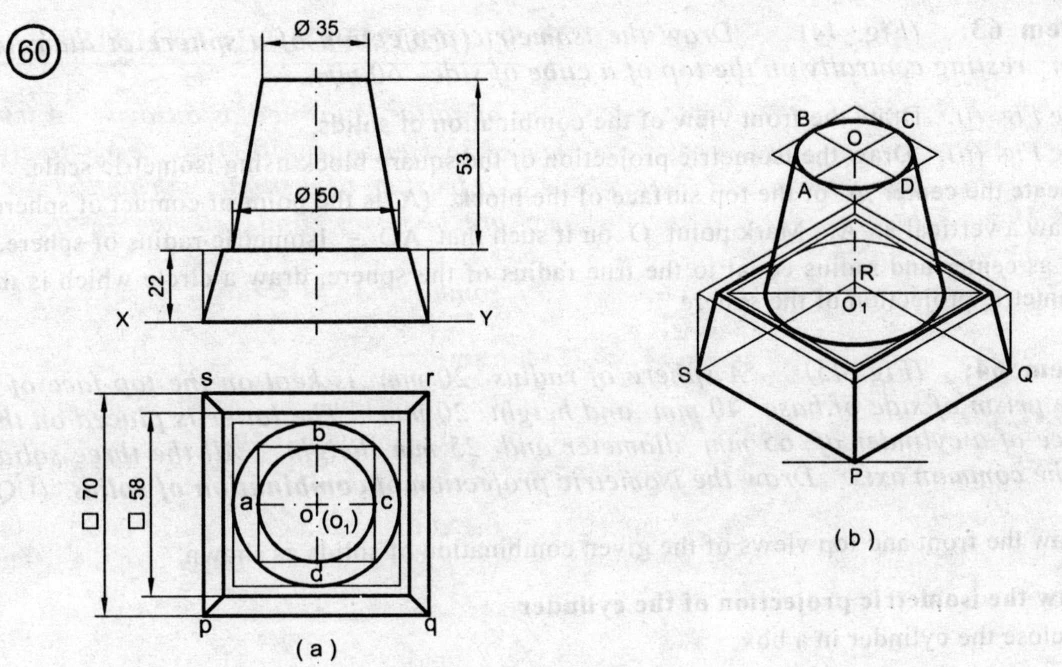
60
Ø 35
53
Ø 50
22
X
Y
s
r
b
a
o
(o1)
c
d
□ 70
□ 58
p
q
(a)
B
C
O
A
D
R
O1
S
Q
P
(b)

FIG. 42

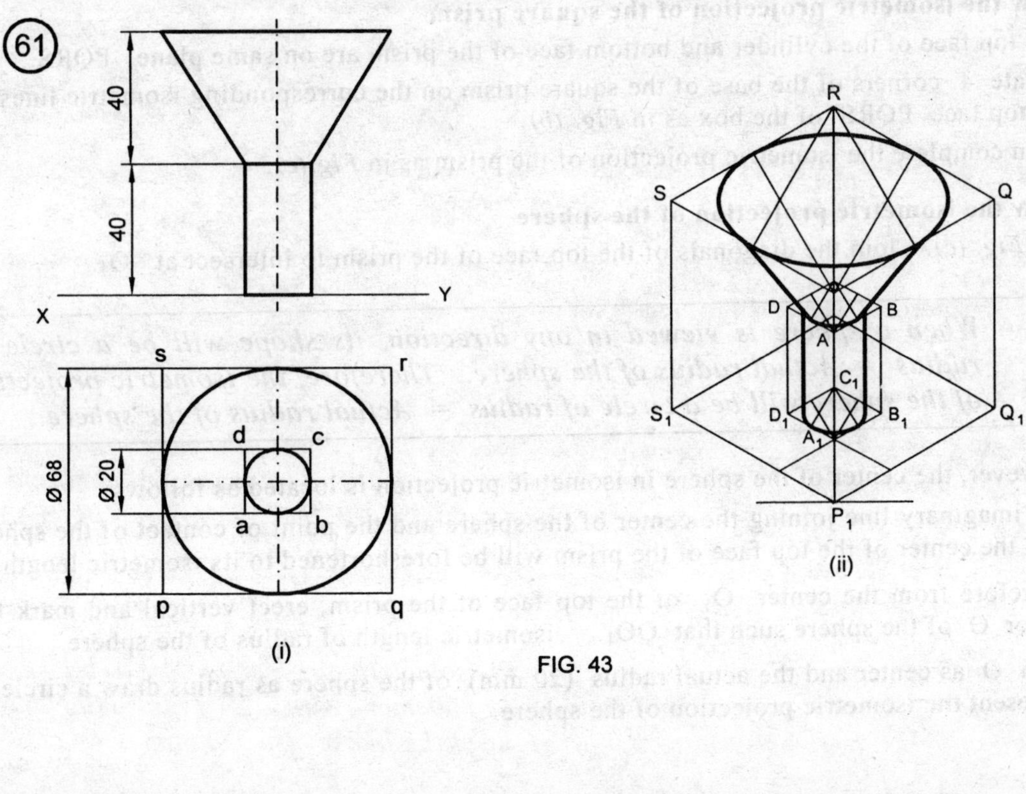
61
40
40
X
Y
s
r
d
c
Ø 20
Ø 68
a
b
p
q
(i)
R
S
Q
D
B
A
C1
S1
D1
B1
Q1
A1
P1
(ii)

FIG. 43

Problem 63: *(Fig. 44)* *Draw the isometric projection of a* ***sphere*** *of diameter 50 mm resting centrally on the top of a* ***cube*** *of side 60 mm.*

1. *See Fig. (i).* Draw the front view of the combination of solids.
2. *See Fig. (ii).* Draw the isometric projection of the square block using isometric scale.
3. Locate the center A of the top surface of the block. (A is the point of contact of sphere)
4. Draw a vertical at A. Mark point O on it such that AO = Isometric radius of sphere.
5. O as center and radius equal to the true radius of the sphere, draw a circle which is the isometric projection of the sphere.

Problem 64: *(Fig. 45)* *A sphere of radius 20 mm is kept on the top face of a square prism of side of base 40 mm and height 20 mm. The latter is placed on the top face of a cylinder of 65 mm diameter and 25 mm height. All the three solids have the common axis. Draw the isometric projection of* ***combination of solids***. (UQ)

1. Draw the front and top views of the given combination of solids as shown.

To draw the isometric projection of the cylinder

2. Enclose the cylinder in a box.
3. Draw the isometric projection of the box.
4. Draw the isometric projection of the cylinder inside the box by ***Four Centers Method***.

To draw the isometric projection of the square prism

5. The top face of the cylinder and bottom face of the prism are on same plane PQRS.
6. Locate 4 corners of the base of the square prism on the corresponding isometric lines on the top face PQRS of the box as in *Fig. (b)*.
7. Then complete the isometric projection of the prism as in *Fig. (c)*.

To draw the isometric projection of the sphere

8. *See Fig. (c).* Join the diagonals of the top face of the prism to intersect at O_1.

NOTE: *When a sphere is viewed in any direction, its shape will be a circle of radius = Actual radius of the sphere. Therefore, the isometric projection of the sphere will be a circle of radius = Actual radius of the sphere.*

---

9. However, the center of the sphere in isometric projection is located as follows:
   The imaginary line joining the center of the sphere and the point of contact of the sphere with the center of the top face of the prism will be foreshortened to its isometric length.
10. Therefore from the center O_1 of the top face of the prism, erect vertical and mark the center O of the sphere such that OO_1 = isometric length of radius of the sphere.
11. With O as center and the actual radius (20 mm) of the sphere as radius draw a circle to represent the isometric projection of the sphere.

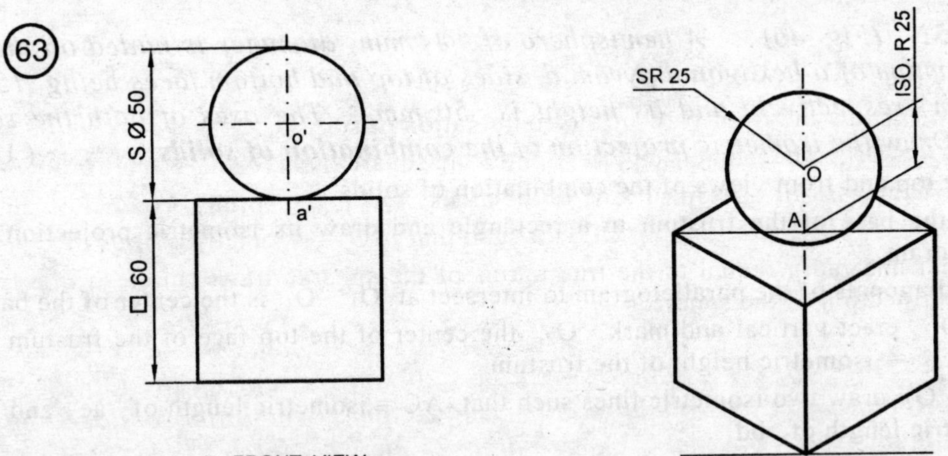

(i) FRONT VIEW

(ii) ISOMETRIC PROJECTION

FIG. 44

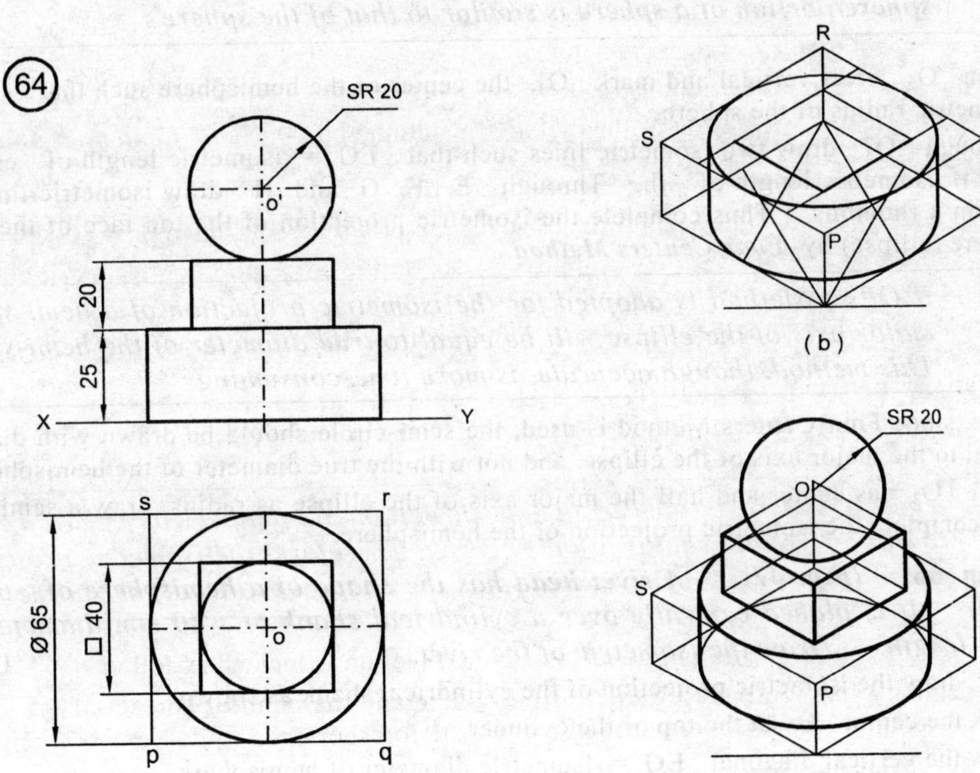

(a)

(b)

(c) ISOMETRIC PROJECTION

FIG. 45

Problem 65: *(Fig. 46) A hemisphere of 44 mm diameter is nailed on the top face of a frustum of a hexagonal pyramid, sides of top and bottom faces being 15 mm and 30 mm respectively and its height is 50 mm. The axes of both the solids coincide. Draw the isometric projection of the* ***combination of solids.*** (UQ)

1. Draw the top and front views of the combination of solids.
2. Enclose the base of the frustum in a rectangle and draw its isometric projection as a parallelogram.
3. Join the diagonals of the parallelogram to intersect at O_1. O_1 is the center of the base.
4. From O_1 erect vertical and mark O_2, the center of the top face of the frustum such that O_1O_2 = isometric height of the frustum.
5. Through O_2 draw two isometric lines such that AC = isometric length of ac and BD = Isometric length of bd.
6. Through A, B, C and D draw isometric lines and obtain the parallelogram as shown and complete the isometric projection of the top face.

NOTE : *Procedure to obtain the isometric projection of the surface of a hemi-sphere/portion of a sphere is similar to that of the sphere.*

---

7. From O_2 erect vertical and mark O_3, the center of the hemisphere such that O_2O_3 = Isometric radius of the sphere.
8. Through O_3 draw two isometric lines such that EG = isometric length of eg and FH = isometric length of fh. Through E, F, G and H draw isometric lines and obtain a rhombus. Thus complete the isometric projection of the top face of the hemi-sphere (ellipse) by ***Four Centers Method.***

NOTE : *If Offset Method is adopted for the isometric projection of a hemi-sphere, major axis of the ellipse will be equal to true diameter of the hemi-sphere. This method, though accurate, is more time-consuming.*

Here, since Four Centers Method is used, the semi-circle should be drawn with diameter equal to the major axis of the ellipse, and not with the true diameter of the hemisphere.

9. With O_3 as center and half the major axis of the ellipse as radius, draw a semi-circle and complete the isometric projection of the hemisphere.

Problem 66: *(Fig. 47)* ***A rivet head*** *has the shape of a hemisphere of 60 mm diameter. It is placed centrally over a cylindrical shank of 40 mm diameter and 60 mm length. Draw the isometric of the rivet.* (UQ)

1. First, draw the isometric projection of the cylindrical shank as shown.
2. Mark the center O at the top of the cylinder.
3. Draw the vertical diagonal EG = Isometric diameter of hemisphere.
4. Draw the sides EF and EH at an angle of 30° to the horizontal.
5. Draw the horizontal diagonal FH and complete the rhombus EFGH and the ellipse to represent the bottom flat surface of the hemisphere.
6. O as center and true radius of the sphere, draw a semi-circle and complete the isometric.

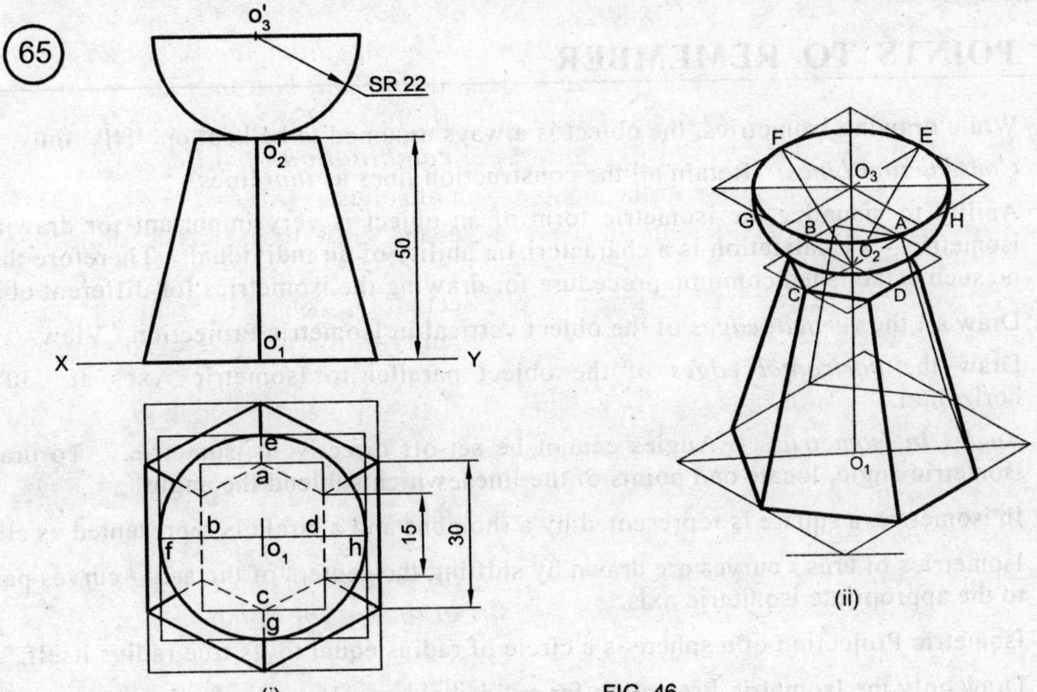
65
O'_3
SR 22
O'_2
50
O'_1
X
Y
e
a
b
d
15
30
f
O_1
h
c
g
(i)
F
E
O_3
G
B
A
H
O_2
C
D
O_1
(ii)

FIG. 46

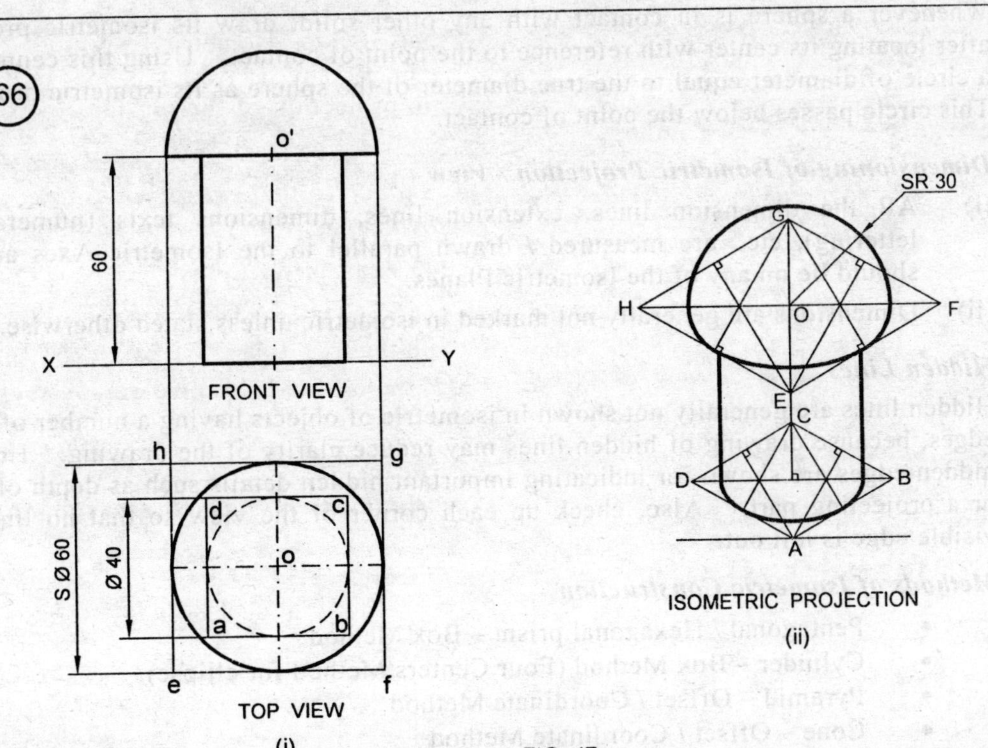
66
o'
60
X
Y
FRONT VIEW
h
g
d
c
S Ø 60
Ø 40
o
a
b
e
f
TOP VIEW
(i)
SR 30
G
H
O
F
E
C
D
B
A
ISOMETRIC PROJECTION
(ii)

FIG. 47

12. POINTS TO REMEMBER

1. While drawing isometrics, the object is always assumed to be kept on HP only.
2. *Construction Lines:* Retain all the construction lines as *thin* lines.
3. Ability to visualize the isometric form of an object is very important for drawing an isometric. Visualization is a characteristic ability of an individual. Therefore there is no such established common procedure for drawing the isometrics for different objects.
4. Draw all the *vertical edges* of the object vertical in Isometric Projection / View.
5. Draw the *horizontal edges* of the object parallel to Isometric Axes at 30° to horizontal.
6. *Angles in Isometric :* Angles cannot be set-off directly in isometric. To draw an isometric angle, locate end points of the lines, which subtend the angle.
7. In isometric, a square is represented by a rhombus and a circle is represented as ellipse.
8. Isometrics of arcs / curves are drawn by shifting the centers of the arcs / curves parallel to the appropriate isometric axis.
9. Isometric Projection of a sphere is a circle of radius equal to its true radius itself.
10. Draw only the Isometric Projection for solids having sphere or spherical parts.
11. Whenever a sphere is in contact with any other solid, draw its isometric projection after locating its center with reference to the point of contact. Using this center, draw a circle of diameter equal to the true diameter of the sphere as its isometric projection. This circle passes below the point of contact.
12. ***Dimensioning of Isometric Projection / View***

    (i) All the dimension lines, extension lines, dimensions texts (numerals and lettering), etc., are measured / drawn parallel to the Isometric Axes and they should lie on any of the Isometric Planes.

    (ii) Dimensions are generally not marked in isometric unless stated otherwise.
13. ***Hidden Lines***

    Hidden lines are generally not shown in isometric of objects having a number of hidden edges, because drawing of hidden lines may reduce clarity of the drawing. However, hidden edges are shown for indicating important hidden details such as depth of a hole or a projecting part. Also, check up each corner of the view so that no line for a visible edge is left out.
14. ***Methods of Isometric Construction***

    - Pentagonal / Hexagonal prism – Box Method
    - Cylinder – Box Method (Four Centers Method for ellipse)
    - Pyramid – Offset / Coordinate Method
    - Cone – Offset / Coordinate Method
    - Frustum of Pyramid / Cone – Box Method (easier)
    - Truncated Pyramid – Offset Method

13. ADDITIONAL PROBLEMS

67. Draw the isometric projection of a hollow cube whose outer edges are 44 mm and inner edges are 33 mm.

68. Draw the isometric projection of a pentagonal pyramid of side of base 32 mm and axis 60 mm long resting with its base on HP and a rectangular face is parallel to VP.

69. A hexagonal prism, side of base 25 mm and height 65 mm has a square hole of side 12 mm cut through it. The axes of the hole and the prism coincide. Also a face of the hole is parallel to a face of the prism. Draw the isometric projection of the solid.

70. The external and internal diameters of a cylindrical bush are 55 mm and 35 mm. Length is 100 mm. Draw the isometric projection when its axis is horizontal. (UQ)

71. Draw the isometric projection of a vertical pentagonal hollow prism of 28 mm side of base outside, altitude 68 mm and thickness 10 mm, when placed with one of its rectangular faces parallel to VP and nearer to it.

72. A flower vase is in the form of a frustum of a pentagonal pyramid of base side 25 mm, top side 40 mm and height 65 mm. Draw the isometric projection. (UQ)

73. A plate washer of diameter 60 mm and thickness 6 mm has a hexagonal hole of side 20 mm at its center. Draw the isometric projection of the washer, such that the curved surface is lying on HP. (UQ)

74. A hexagonal prism of side of base 27 mm and altitude 63 mm rests on one of the corners of the base. Its axis is inclined at an angle of 30° to HP. Draw its projections and the corresponding isometric projection.

75. A box made of 20 mm thick wooden planks has the overall dimensions of 200 mm x 100 mm x 150 mm including the lid. The lid is of dimensions 200 mm x 100 mm x 40 mm outside. Draw the isometric projection when the lid is open through 120°.

76. The outside dimensions of a box made of 20 mm thick planks are 500 mm x 300 mm x 300 mm height. The depth of the lid on the outside is 100 mm. Draw the isometric projection of the box to a suitable scale when the lid is 90° open. (UQ)

HINT: *Draw the side view of the box first. Note that the thickness of the wooden plank for the box and the lid should be deducted for drawing the inside lines.*

77. A memorial is in the form of a cube of side 50 cm and thereafter tapers into a frustum of a square pyramid whose top surface is a square of 30 cm side. Total height of the memorial is 80 cm. Draw its isometric projection. (UQ)

78. Draw the isometric projection of the frustum of an octagonal pyramid of 20 mm sides at the top and 30 mm sides at the bottom and 65 mm height. The two sides of the base make equal inclinations with VP. (UQ)

79. A pentagonal prism, side of base 30 mm and height 70 mm, is resting upon HP on its base, keeping one of its rectangular faces perpendicular to VP. A section plane 45° inclined to HP, bisects the axis of the prism.

Draw the isometric projection of the truncated prism showing the cut surface. (UQ)

80. A hexagonal pyramid, side of base 30 mm and axis 65 mm long, rests with its base on HP and an edge of the base is perpendicular to VP. It is cut by a section plane perpendicular to VP, inclined at 30° to HP and passing through a point on the axis 20 mm below the apex.

Draw the isometric projection of the truncated pyramid, showing the sectioned surface. (UQ)

81. A cylindrical slab 50 mm in diameter and 45 mm thick is surmounted by a cube of 30 mm edge. A square pyramid of altitude 45 mm and side of 25 mm rests on the top of the cube with the side of its base parallel to the sides of the top of the cube. The axes of the solids are collinear. Draw the isometric projection. (UQ)

82. A hollow hexagonal prism, side of base 30 mm and axis 65 mm long, has a square hole of side 20 mm through it such that axis of the hole coincides with that of the prism. The prism rests with its base on HP and one of its rectangular faces (which is parallel to one of the rectangular faces of the square hole) is perpendicular to VP. It is cut by a section plane perpendicular to VP, inclined at 45° to HP and passing through a point on the axis 20 mm from its top end.

Draw the isometric projection of the truncated prism, showing the cut surface. (UQ)

83. On a square prism of sides 65 mm and height 30 mm, another square prism of 35 mm sides and height 35 mm is placed vertically and symmetrically. Draw the isometric projection of this setup. (UQ)

84. Draw the iso. projection of a hexagonal prism of side of base 25 mm and height 55 mm, on the top of which is placed a cone of base diameter 50 mm and height 24 mm.

85. A frustum of a cone has its top and bottom diameters 25 mm and 50 mm respectively and the altitude is 55 mm. It is resting vertically on a cylindrical block of 75 mm diameter and 30 mm thick such that both the solids have the common axis. Draw the isometric projection of the combination of solids.

86. A combination of solids is formed as follows: A hollow cylinder of 50 mm and 35 mm outside and inside diameters and height 60 mm stands vertically on a square prism of 65 mm side and 35 mm height.

Draw the isometric projection. (UQ)

87. A glass bottle consists of a cylinder of 60 mm diameter and 40 mm height with a cylindrical neck of 20 mm diameter and 20 mm height. The bottle is fitted with a cork of 18 mm diameter which projects by 5 mm. Draw its isometric projection. (UQ)

88. A cylindrical disc of 20 mm diameter and altitude 20 mm is placed vertically at the center of the rectangular face of a hexagonal prism of 28 mm sides and axis 60 mm. Draw the isometric projection of the combination. (UQ)

89. A tapering pillar of 100 mm bottom diameter and 50 mm top diameter holds a 100 mm diameter sphere at its top center. The height of the pillar is 100 mm. Draw the isometric projection of the combination of solid. (UQ)

90. A compound solid consists of a frustum of a cone and a hexagonal prism. The frustum of the cone is kept centrally on the top of the prism. The frustum of the cone has the bottom base diameter 45 mm, top face diameter 27 mm and length of the axis is 72 mm. The sides of the hexagonal prism are 32 mm and the axis is 22 mm long.

Draw the isometric projection of the compound solid. (UQ)

91. A stool has a square top of 400 mm side and 30 mm thick. It has three vertical cylindrical legs of diameter 40 mm and height 500 mm symmetrically placed with their centers lying on a pitch circle of 360 mm diameter, to support the top. A hollow cylinder of 300 mm outer diameter, 200 mm inner diameter and height 150 mm is placed centrally over the stool with its axis vertical.

Draw the isometric projection. (UQ)

92. A combination of solids is formed by a vertical cylinder of 75 mm diameter and altitude 30 mm, a vertical square prism of 40 mm sides and altitude 30 mm and a vertical square pyramid of 30 mm sides and altitude 45 mm. The square prism is resting on the flat face of the cylinder and the square pyramid is placed on the vertical square prism. All the three solids are coaxial.

Draw the isometric projection. (UQ)

UNIT – V

SYLLABUS

UNIT – V

Conversion of Orthographic Views into Isometric Views and Vice-Versa

19

CONVERSION OF ORTHOGRAPHIC VIEWS INTO ISOMETRIC VIEWS

1. VISUALIZATION OF OBJECTS FROM ORTHOGRAPHIC VIEWS

Reading of Orthographic Views

Reading of orthographic views is a reversal of the process of making drawings. *Reading* a drawing is defined as the process of recognizing and applying the principles of orthographic projections to interpret the shape of an object from the orthographic views.

Visualization of Objects from Orthographic Views

For a beginner, it may be difficult to visualize an object from 2 or 3 orthographic views. There is no systematic procedure for analyzing complex objects. The ability of visualizing an object from its orthographic views will certainly improve only by practice.

How to visualize an object from orthographic views ?

- You must know the meaning of various types of Lines, Areas, etc., in orthographic views.
- Then, you have to exercise the mental process of interpreting them.
- Obtain a general idea about the overall shape of the object.
- Start reading the major individual features of the object by referring all related views.
- Preference may be given to familiar shapes and read them completely.
- Then start reading unfamiliar and complex features.
- Note the relationship between individual parts and their way of joining.

Note that the entire drawing of an object cannot be read at a glance. A drawing must be read, referring to each part systematically, back and forth from one view to the other if a single orthographic view is not sufficient to identify a part completely. While visualizing an object, you should think of a 3D object and not a 2D flat projection.

The basics of both orthographic and isometric drawings are important to make effective isometric sketches. Just as learning the grammar of English does not make one a creative writer, so also will the training in sketching not make one a creative designer.

2. PRINCIPLES OF ISOMETRIC VIEWS

1. *Isometrics give the true shape of the object, while perspective and oblique views do not.* **Isometric views are therefore generally adopted by engineers in practice.**

2. ***Front Side of the Isometric View***
   Front side of isometric view may be taken towards left or towards right to show the maximum details clearly.

3. ***Horizontal Edges***
   Horizontal edges of the object are *parallel* to Isometric Axes at 30° to the horizontal.

4. ***Vertical Edges***
   All vertical edges of the object are *vertical* in Isometric View.

5. ***Inclined Edges***
   Draw inclined edges by first locating end points in Isometric & then join.

6. ***Circle***
   Isometric view of a circle is always an ellipse.

7. ***Semi-circle***
   To sketch the isometric view of semi-circle, consider it as a full circle and sketch the ellipse. Then, retain only the required portion of the semi-circle.

8. ***Angles in Isometric***
   Angles cannot be set-off directly in isometric. To draw an isomeric angle, locate the end points of the lines, which subtend the angle.

9. ***Hidden Lines***
   Hidden lines are generally not shown in isometric views of objects having a number of hidden edges, because drawing of hidden lines may reduce clarity of the drawing. However, hidden edges are shown for indicating important hidden details such as depth of a hole or a projecting part.

   Also, check up each corner of the isometric view so that no line for a visible edge is left out.

10. ***Dimensioning***
    All the dimension lines, dimensions texts (numerals and lettering), etc., are measured / drawn parallel to the Isometric Axes and they should lie on any of the Isometric Planes.

    NOTE, if any, should be written so that it can be read from the bottom.

    Dimensions are generally not marked in isometrics unless stated otherwise.

3. STEP-BY-STEP PROCEDURE :
CONVERSION OF ORTHOGRAPHIC VIEWS INTO ISOMETRIC VIEWS

Problem 1 : *Fig. 1(i) shows the front and left side views of an ANGLE PLATE. Draw its isometric (pictorial) view Step-by-Step.*

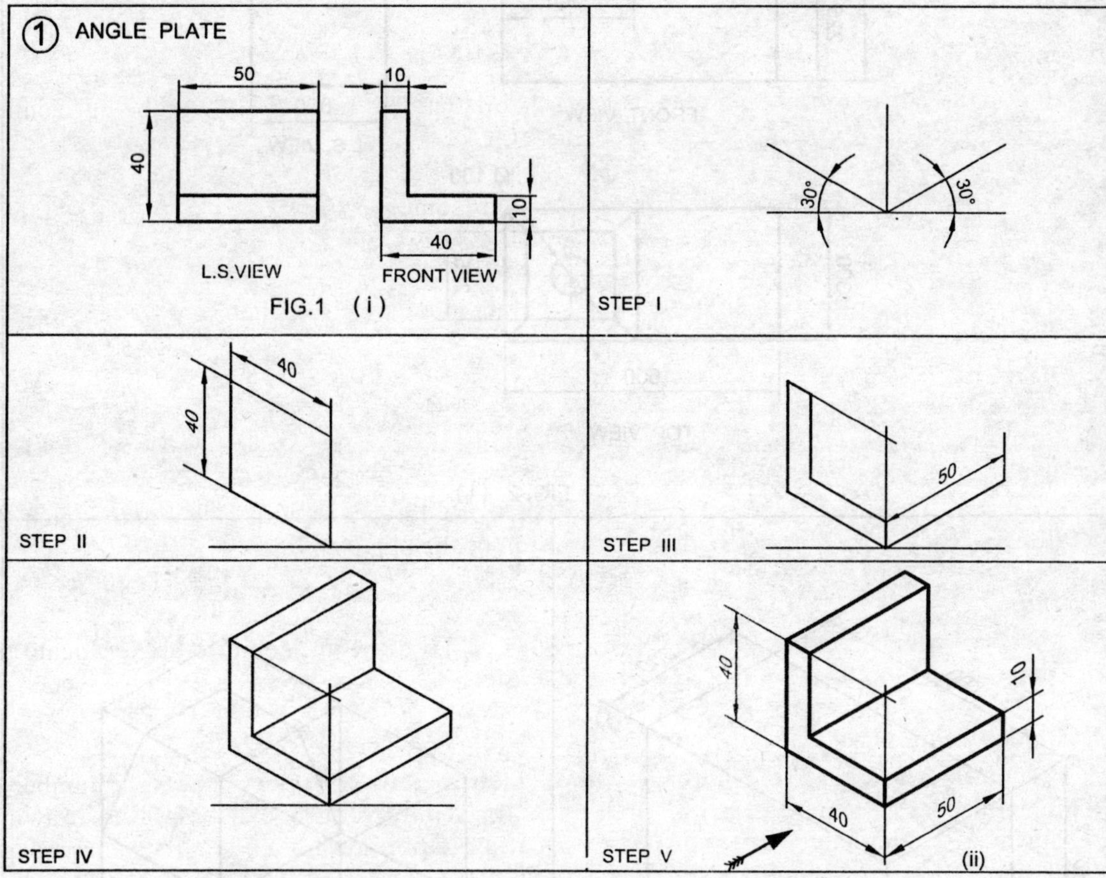

STEP I	:	Draw the three isometric axes in thin lines.
STEP II	:	Draw a rhombus of 40 mm side to represent an isometric plane containing two isometric axes.
STEP III	:	Draw the front view (L-shape) in this plane.
STEP IV	:	From each corner of the 'L' draw lines parallel to the third axis. Mark length 50 mm on these lines and complete the isometric view. [Fig. (ii)]
STEP V	:	Darken the isometric view. Add the dimensions, if necessary.

Problem 2: MIXIE - BASE (UQ)

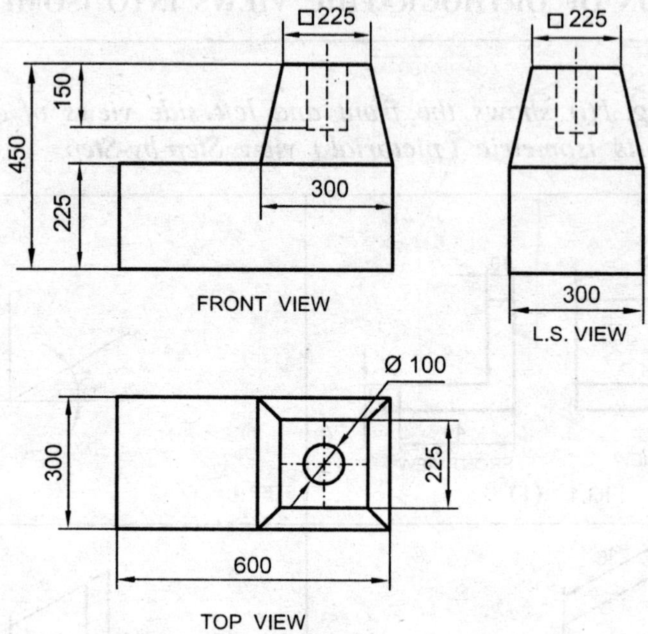

FIG. 2 (i)

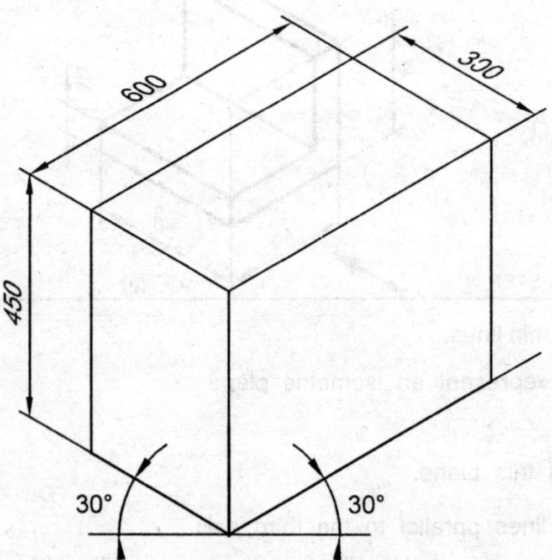

STEP I

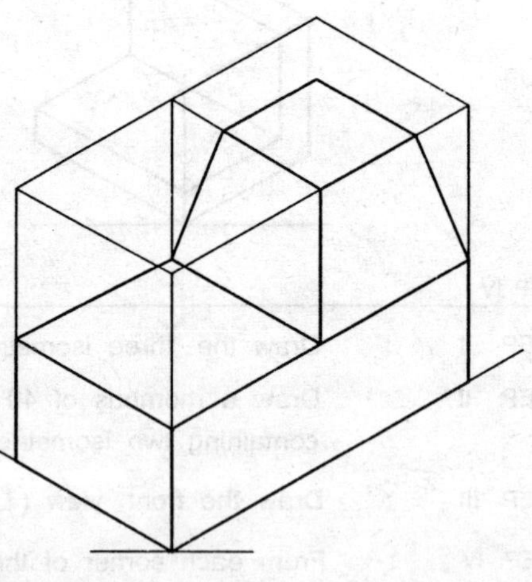

STEP II

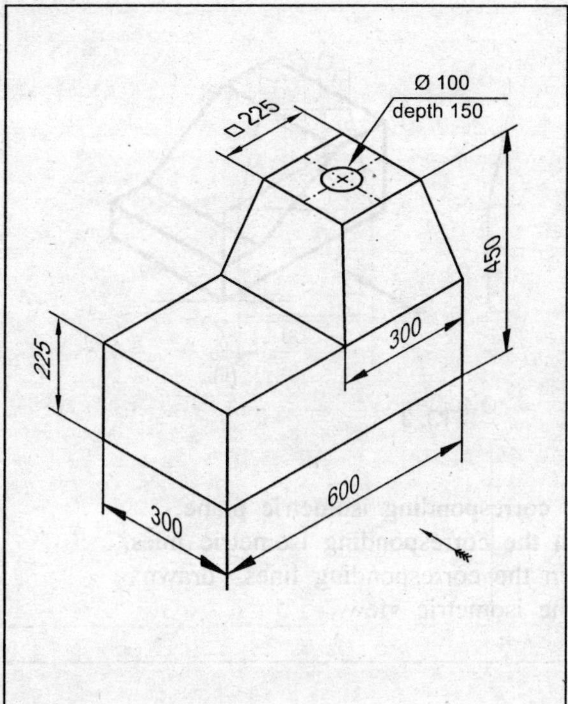

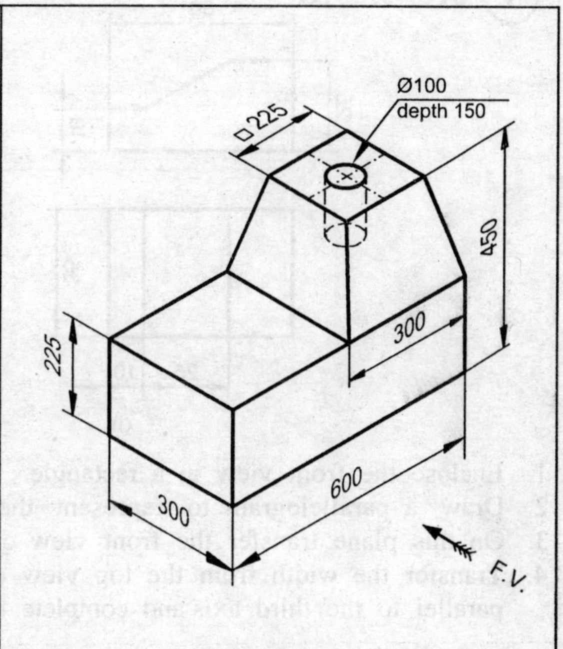

FIG. 2 (ii)

STEP I : Isometric Drawing starts with the three isometric axes. BOX METHOD is preferable. Hence, draw a box in thin lines along the isometric axes for the overall length, width and height of the object.

STEP II : Mark the various points at appropriate distances of each feature along the isometric axes. Draw lightly, the lines corresponding to the points marked as above.

STEP III : Draw the ellipse to represent the hole of 100 mm diameter.

STEP IV : Darken the required lines of all the features in the Isometric View.

NOTE: For complicated objects, the cylindrical features should be first drawn as prisms using BOX METHOD. Centers of holes are located and blocked in.

(3) STOP BLOCK

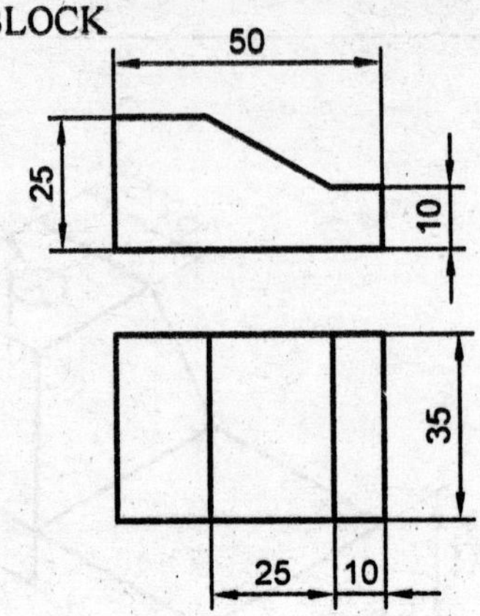

(i)

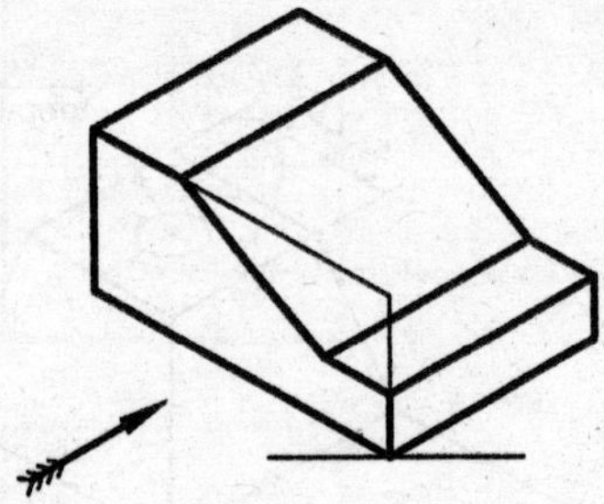

(ii)

FIG. 3

1. Enclose the front view in a rectangle.
2. Draw a parallelogram to represent the corresponding isometric plane.
3. On this plane, transfer the front view on the corresponding isometric lines.
4. Transfer the width from the top view on the corresponding lines, drawn parallel to the third axis and complete the isometric view.

(4) CAVITY BLOCK

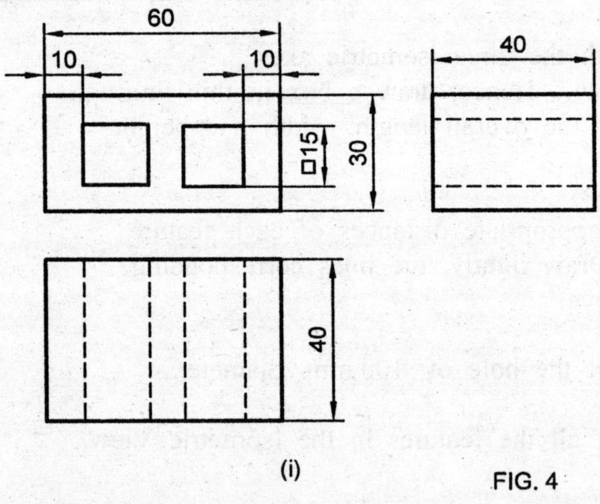

(i)

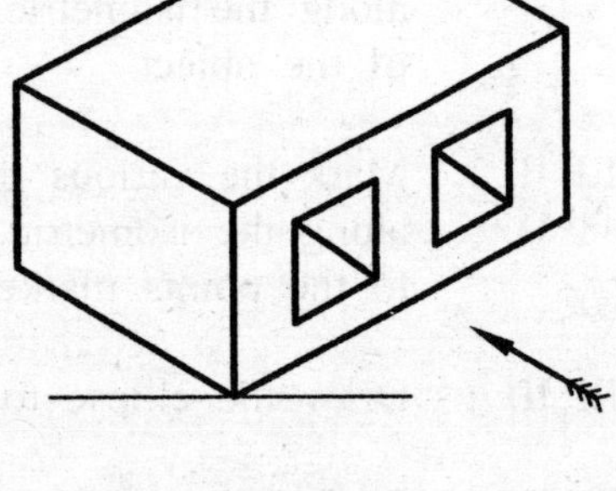

(ii)

FIG. 4

1. Draw the front view in an isometric plane.
2. Mark the corners of square holes on the respective isometric lines.
3. From various corners, draw lines parallel to the third axis.
4. Mark the thickness from top or side view and complete the isometric view.

(5) SQUARE BLOCK WITH HOLE

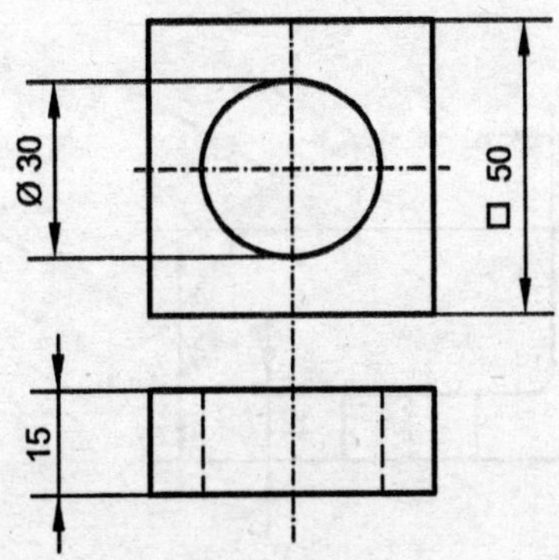

FIG. 5 (i)

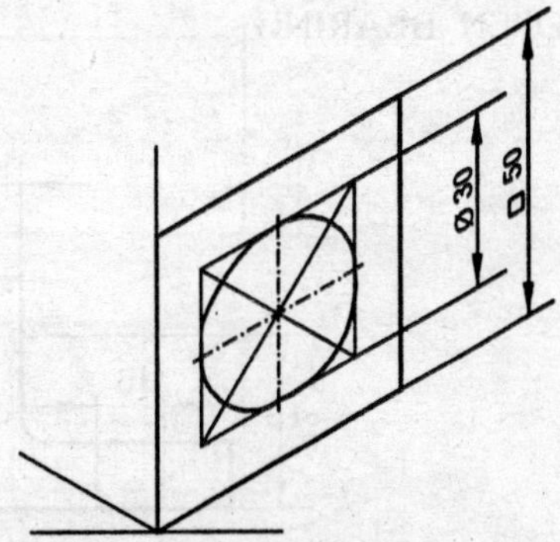

STEP I

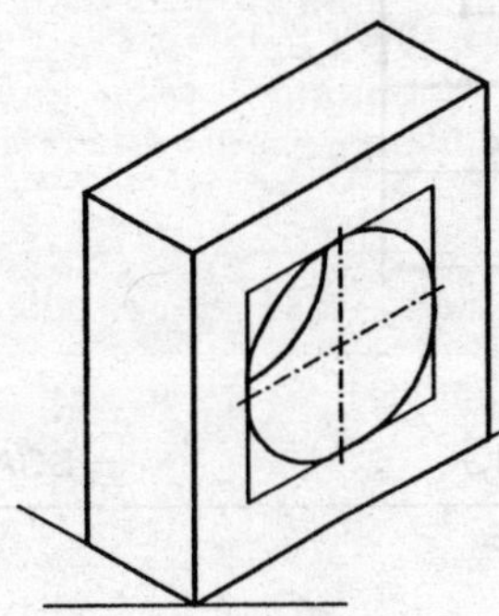

STEP II

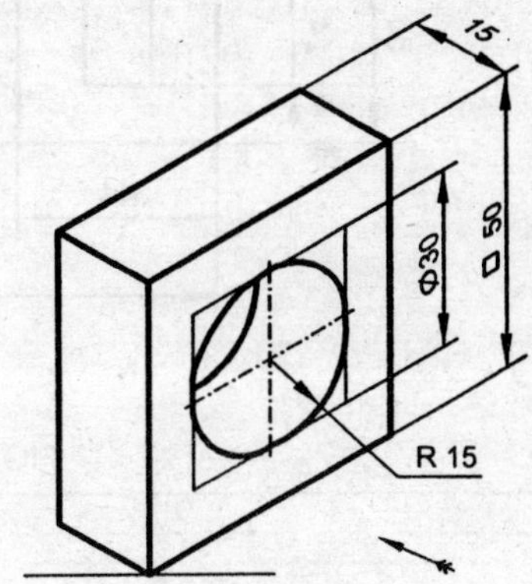

(ii)

STEP III

NOTE: In the isometric view, draw the hole as two parallel ellipses corresponding to the front and rear circles of the hole.

STEP I : Draw the rhombus in the isometric to represent the square in the front view. Draw the ellipse nearer to the observer as shown.

STEP II : Thickness of the block is 15 mm. Obtain the centers for drawing the rear ellipse by shifting the center of the front ellipse backwords along the isometric axes by a distance of 15 mm.

STEP III : Retain only the visible portion of the rear ellipse.

(6) OPEN BEARING

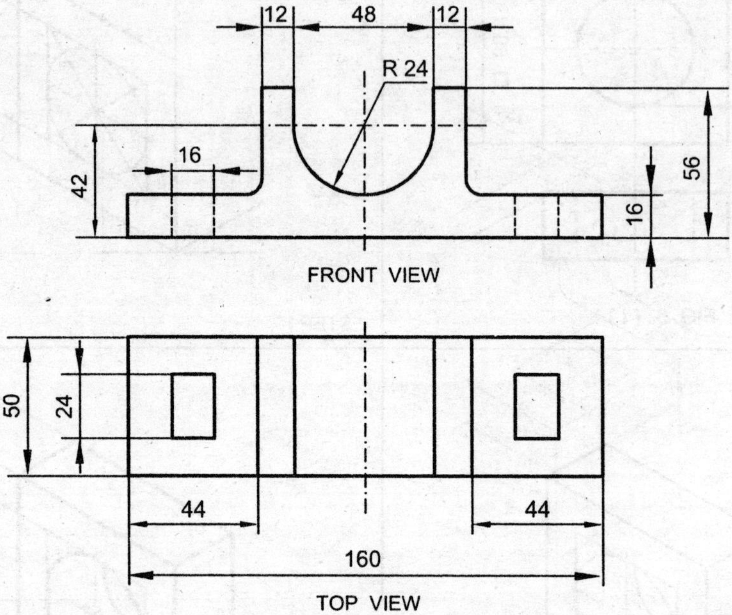

FIG.6 (i) OPEN BEARING SCALE 2:1

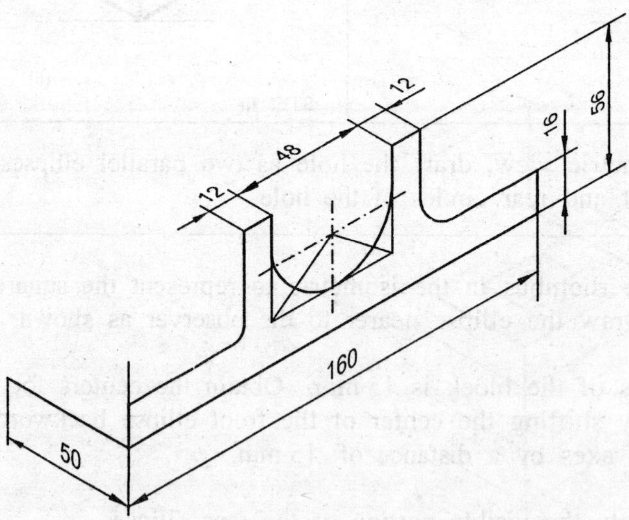

STEP I

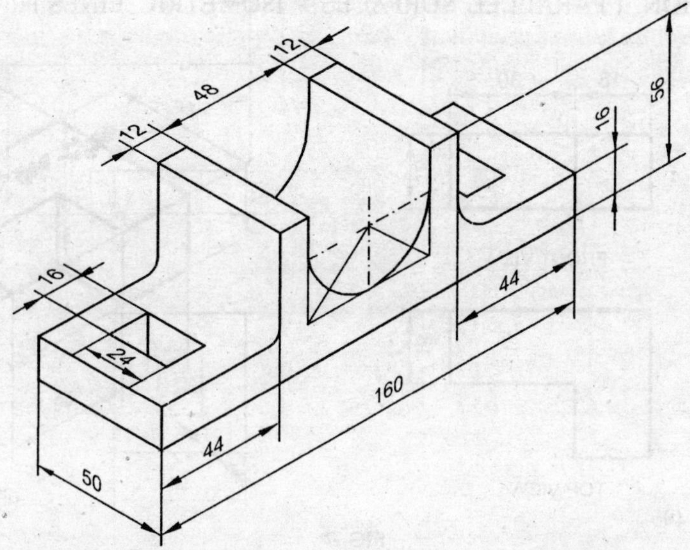

STEP II

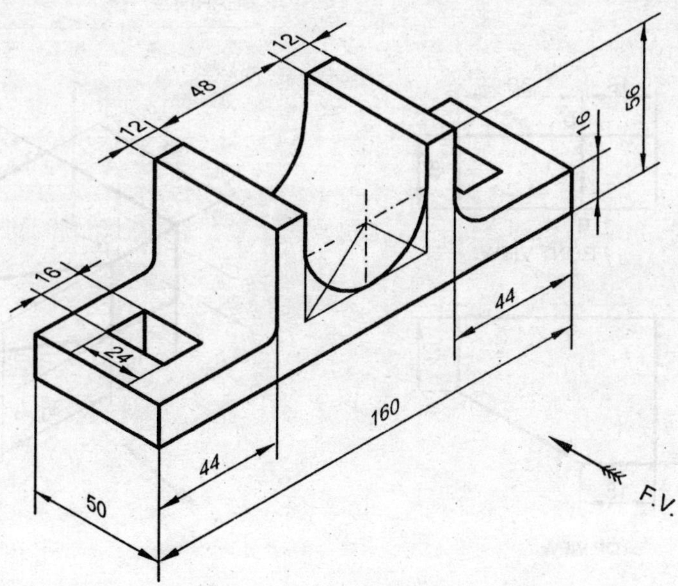

STEP III

(ii) OPEN BEARING

4. TYPICAL PRACTICAL PROBLEMS

(7) L-SECTION (PARALLEL SURFACES - ISOMETRIC LINES)

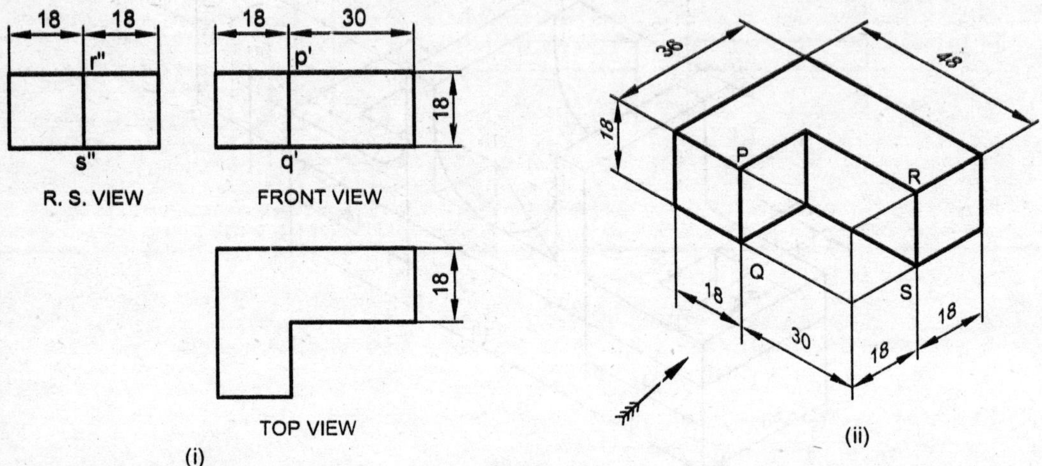

FIG. 7

(8) WEDGE PIECE (OBLIQUE SURFACE - NON-ISOMETRIC LINES)

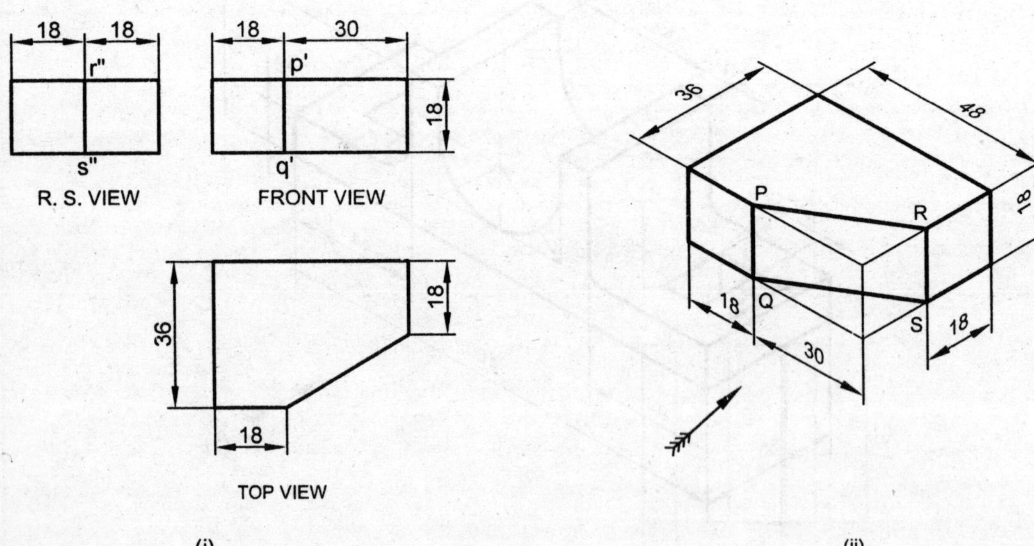

FIG. 8

9 I - BEAM

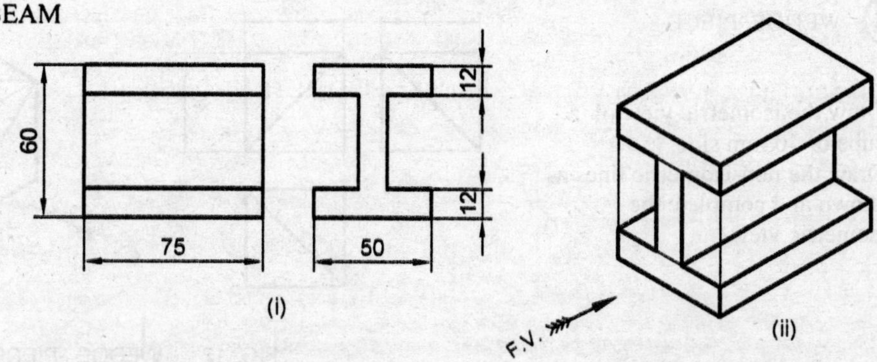

FIG. 9 I - BEAM

10 DOOR - STEPS

(Fig. 10) Draw the isometric view of DOOR-STEPS having three steps of 22 cm tread and 15 cm rise. The steps measure 75 cm lengthwise.

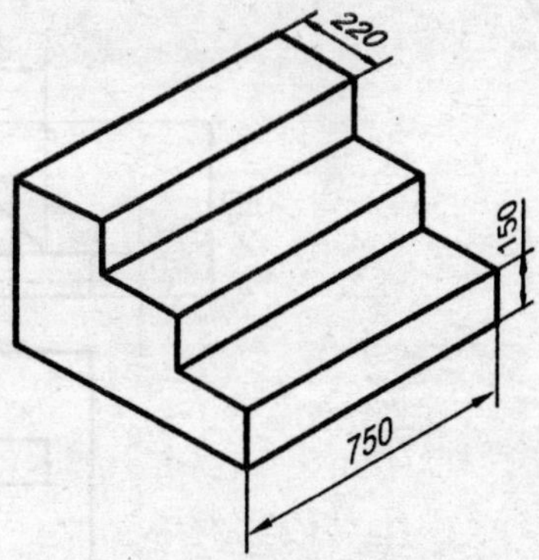

FIG. 10 DOOR - STEPS
SCALE - 10 : 1

11 V - BLOCK

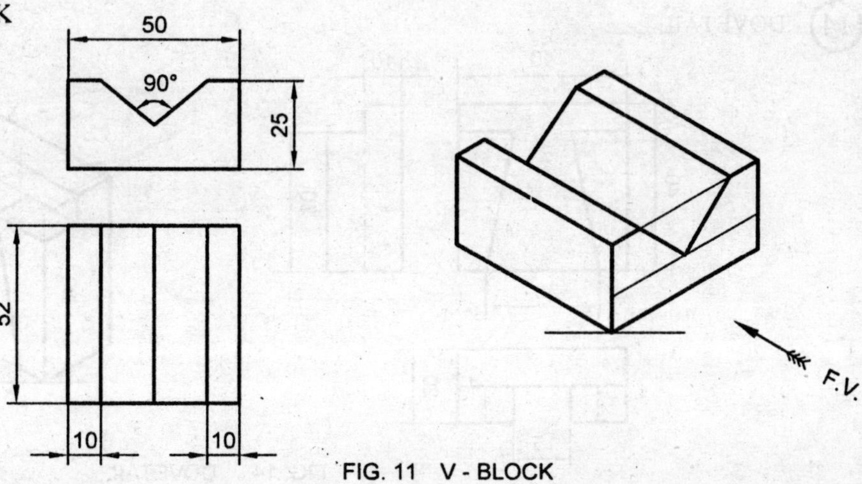

FIG. 11 V - BLOCK

(12) WEDGE PIECE

Draw the isometric view of a cube of 40 mm side.
Draw the non-isometric lines as shown and complete the isometric view.

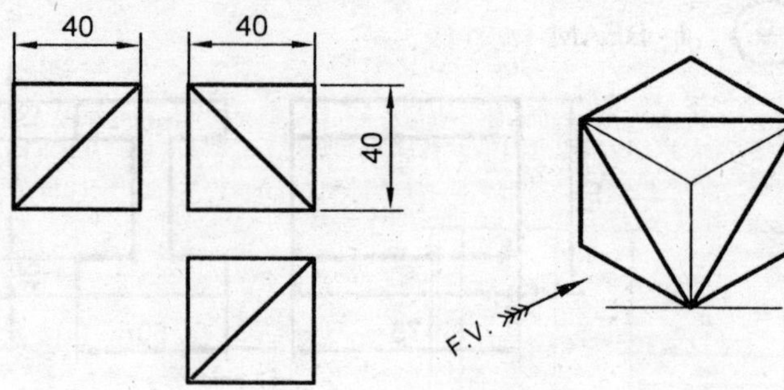

FIG. 12 WEDGE PIECE

(13) RIBBED ANGLE PLATE

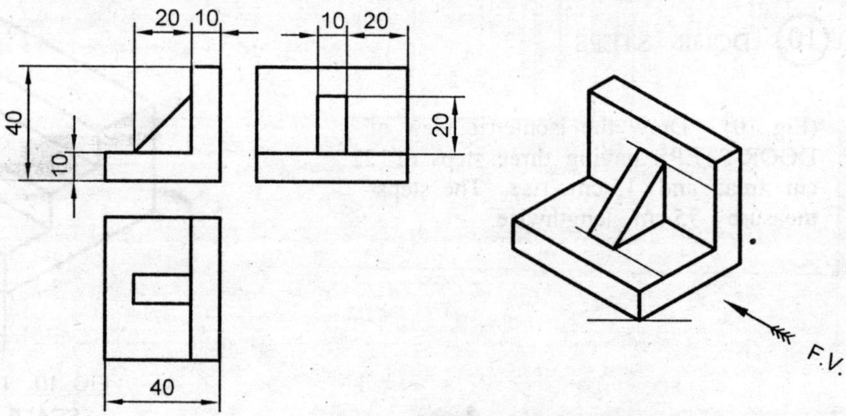

FIG. 13 RIBBED ANGLE PLATE

(14) DOVETAIL

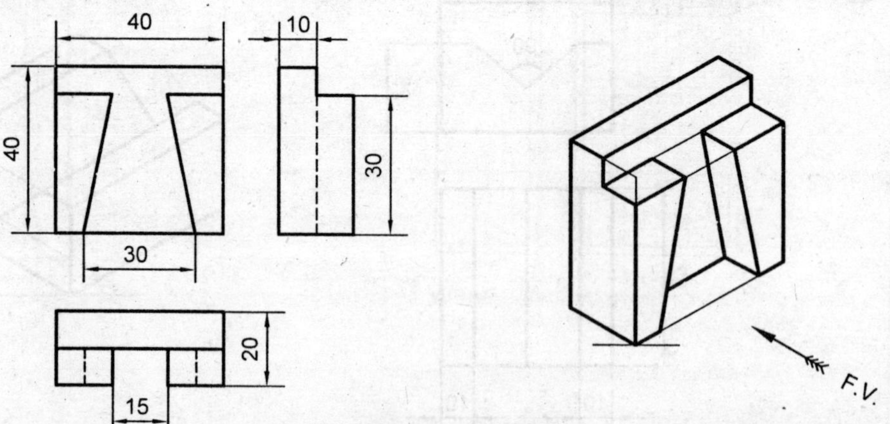

FIG. 14 DOVETAIL

(15) CLAMP (UQ)

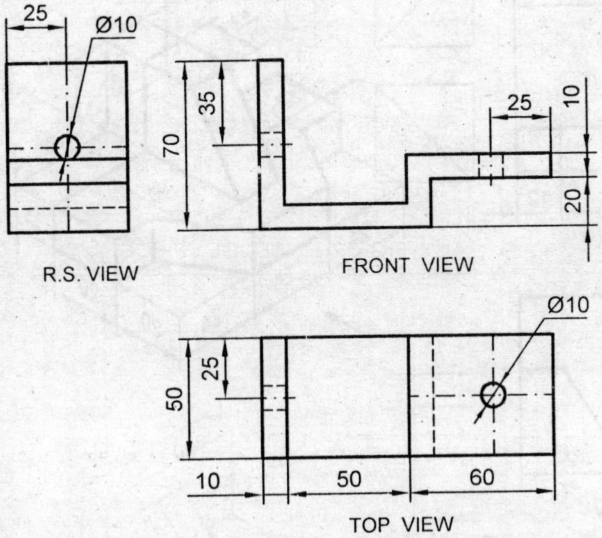

(i)

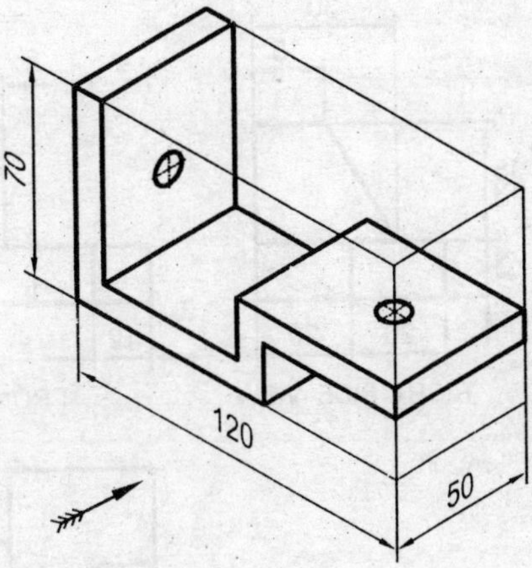

(ii)

FIG. 15

(16) WOODEN BOOK - RACK

The outside dimensions of a WOODEN BOOK-RACKRACK are 44 cm x 25 cm x 68 cm. It contains three shelves at equal height. Thickness of the teak plank is 20 mm. Draw its isometric view. (UQ)

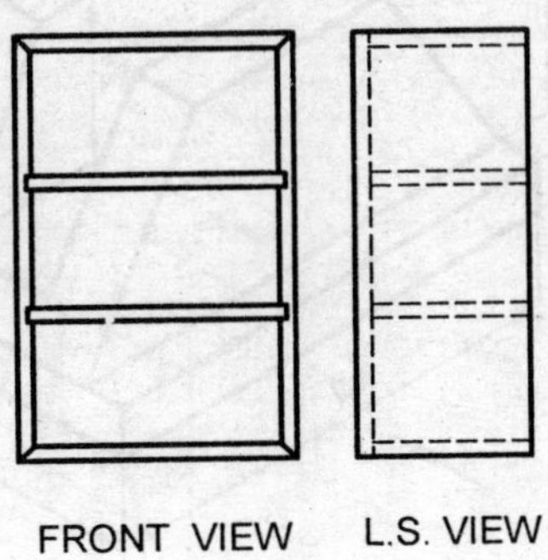

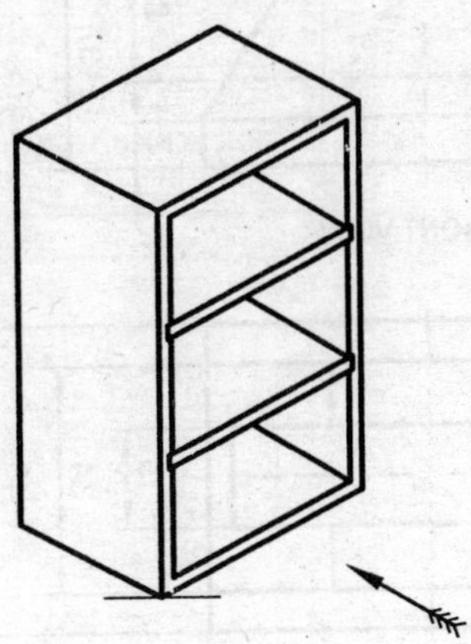

FIG. 16

TAKE A SUITABLE SCALE. ADD THE DIMENSIONS.

(17) MACHINE BLOCK

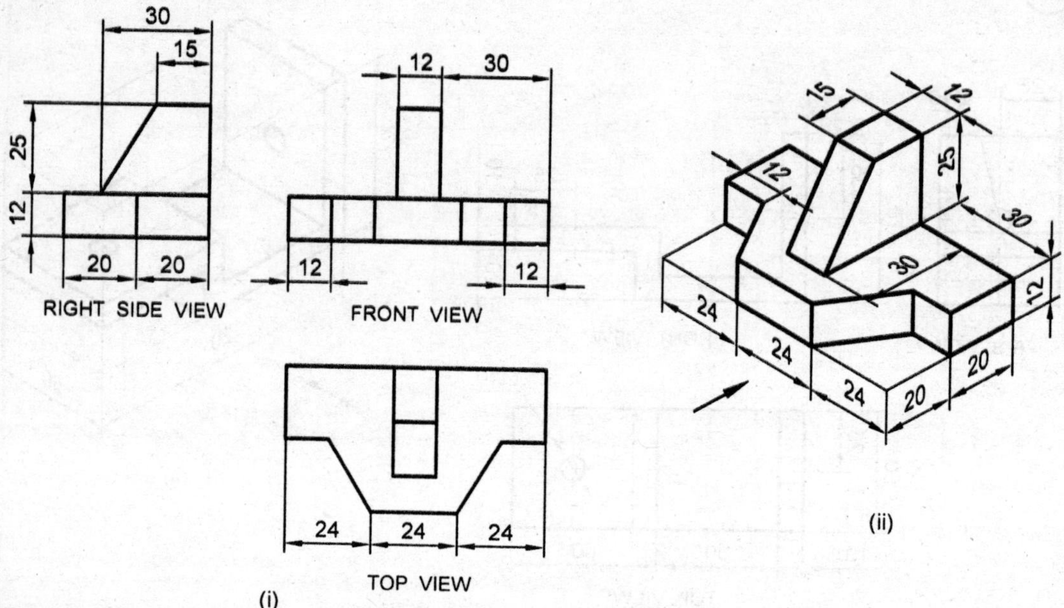

FIG. 17

(18) SLIDING BLOCK

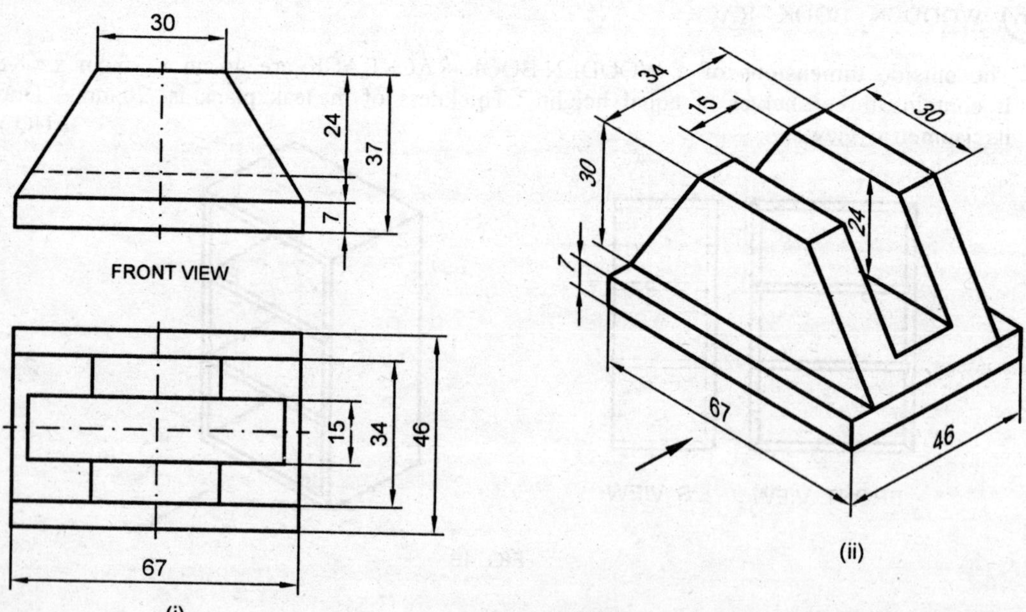

FIG. 18

(19) END STOP

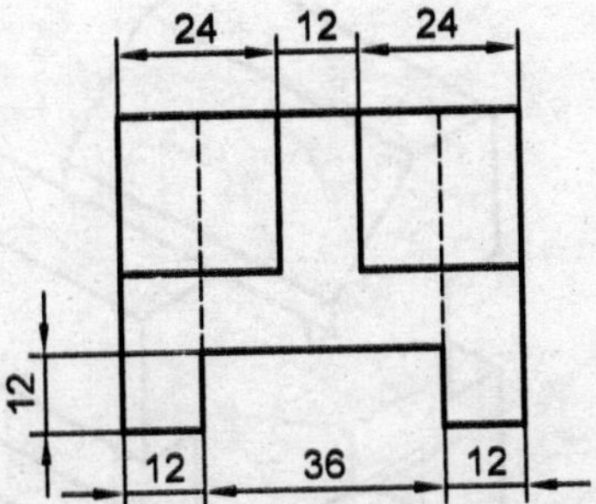

F.V.

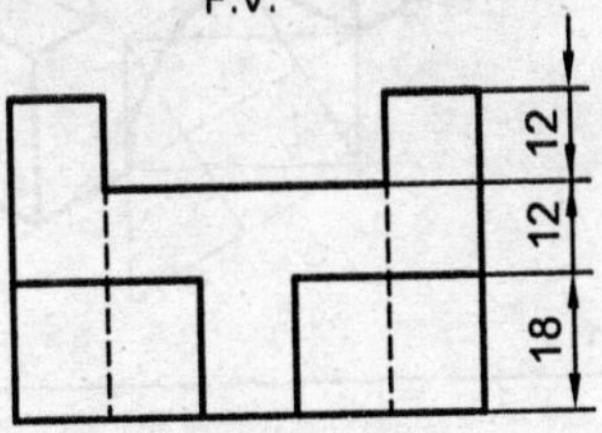

T.V.

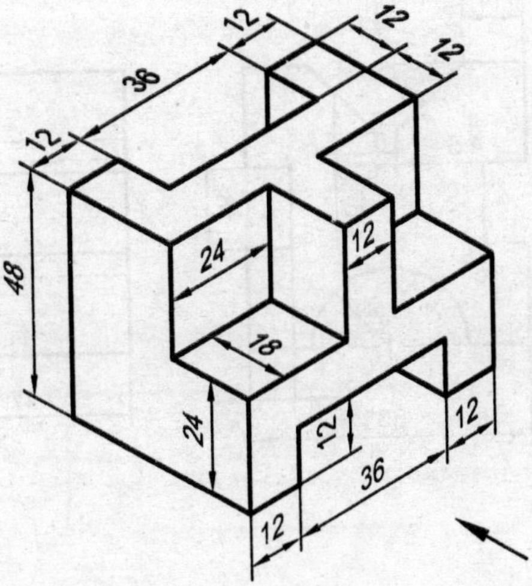

FIG. 19

(20) OFFICE BUILDING

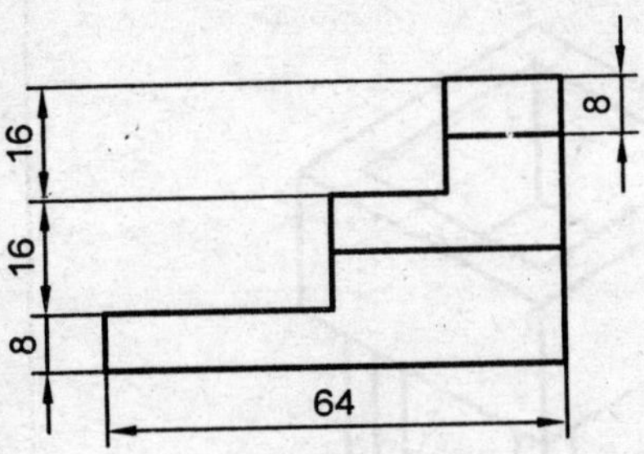

FRONT VIEW

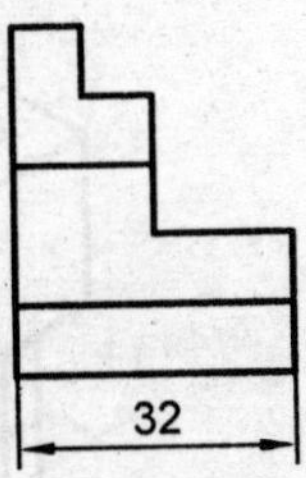

LEFT SIDE VIEW

(i)

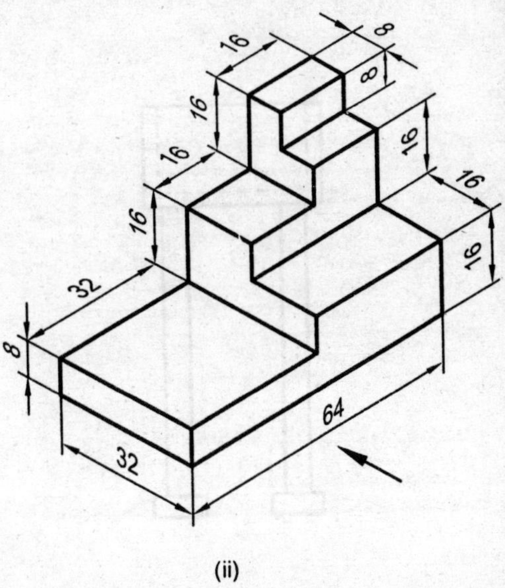

(ii)

FIG. 20

(21) MODIFIED V - BLOCK (UQ)

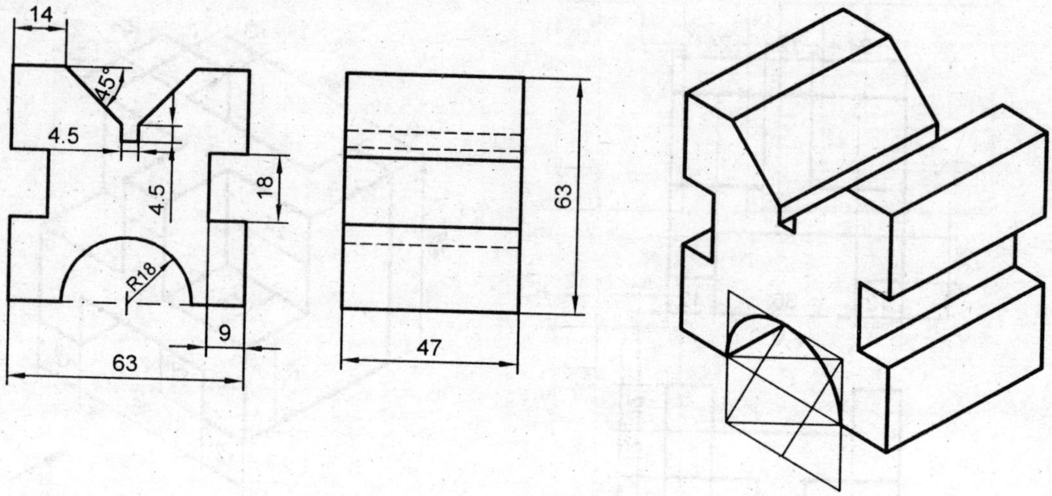

FIG. 21

(22) OVERHEAD TANK

Draw the isometric view of a square OVERHEAD TANK open at the top and supported on four stanchions. Assume suitable dimensions. (UQ)

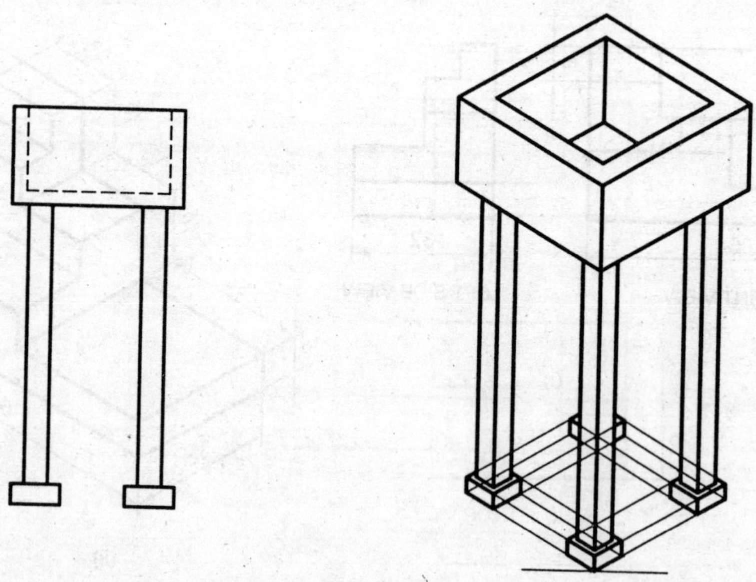

FIG. 22

(23) ADJUSTOR BLOCK

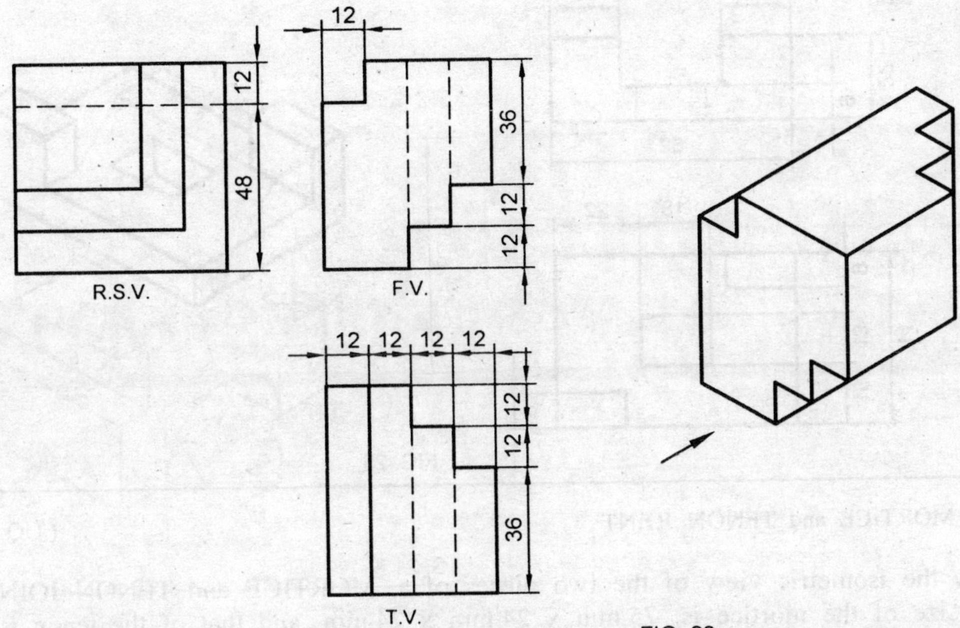

FIG. 23

(24) SUPREME COURT

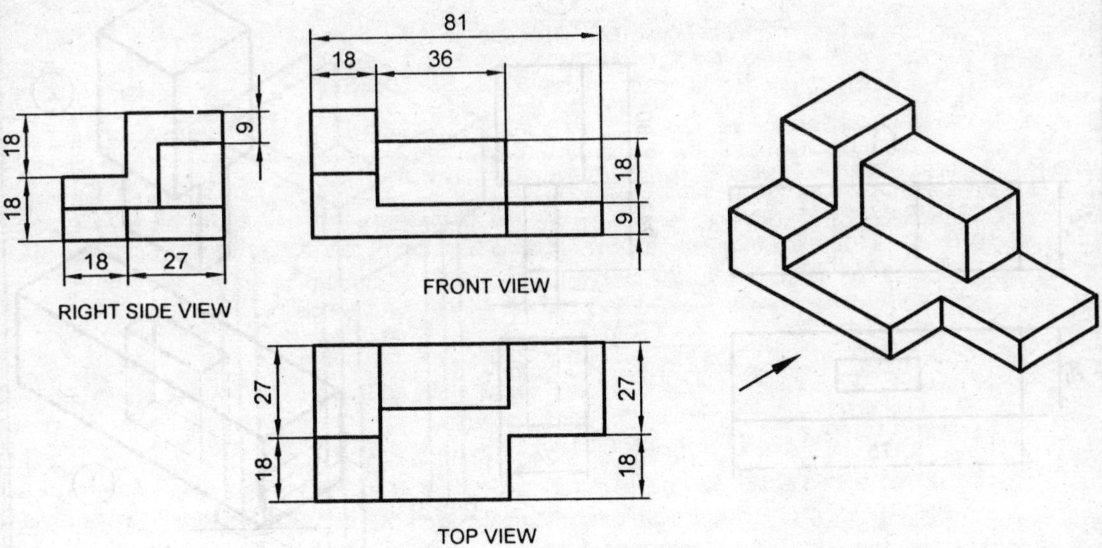

FIG. 24

(25) CENTER GUIDE

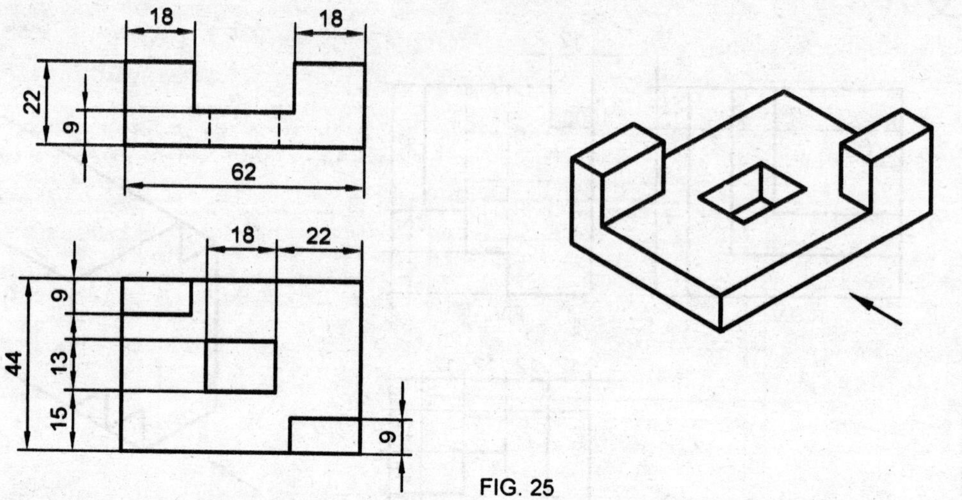

FIG. 25

(26) MORTiCE and TENON JOINT

(UQ)

Draw the isometric view of the two pieces of a MORTICE and TENON JOINT. The size of the mortice is 75 mm x 24 mm x 24 mm and that of the tenon is 24 mm x 24 mm x 8 mm. The two pieces are to be shown separately in a position ready for fitting.

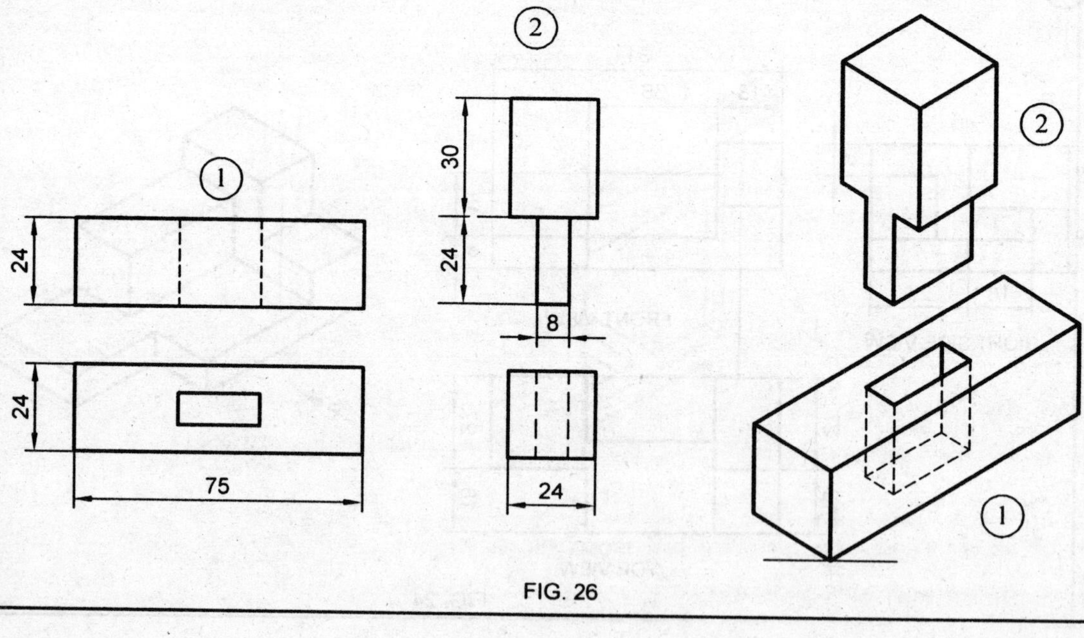

FIG. 26

(27) WOODEN TABLE (UQ)

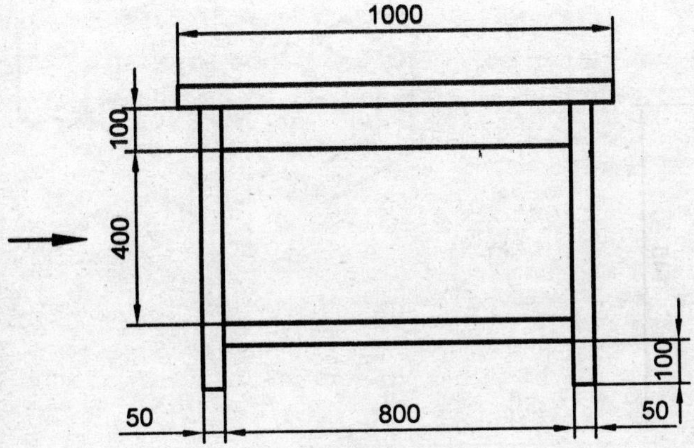

FRONT VIEW

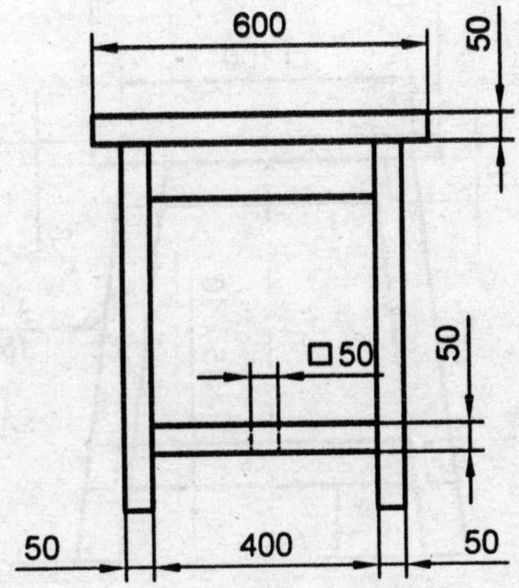

LEFT SIDE VIEW

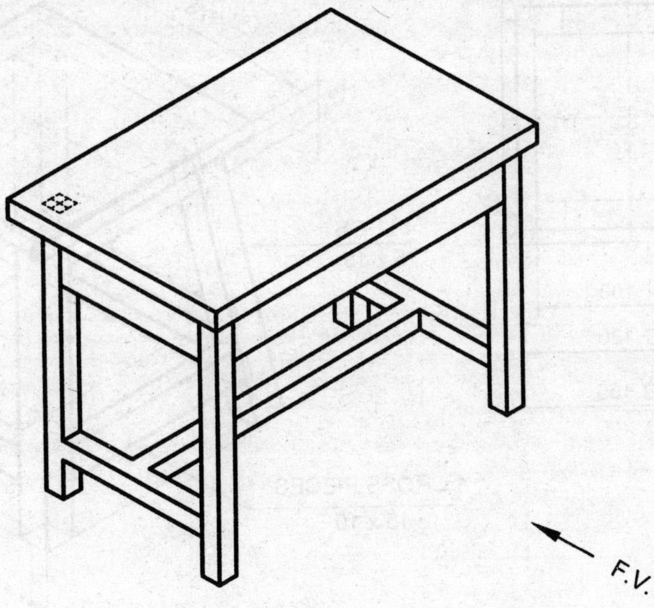

ISOMETRIC VIEW

FIG. 27

(28) SQUARE STOOL

(UQ)

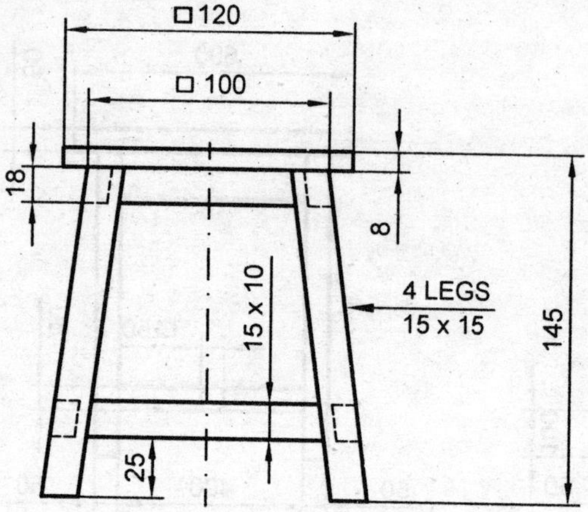

FRONT VIEW

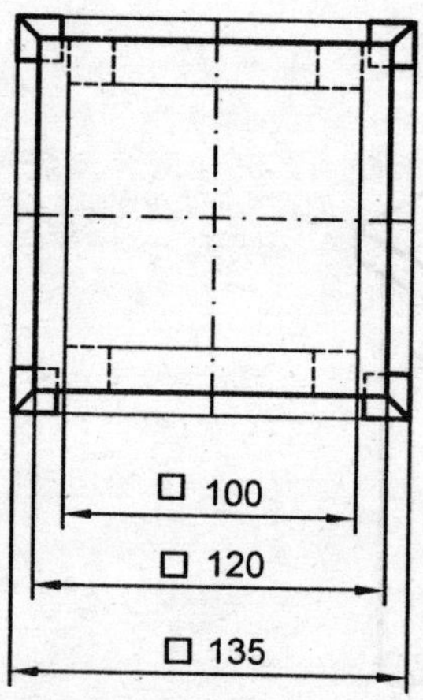

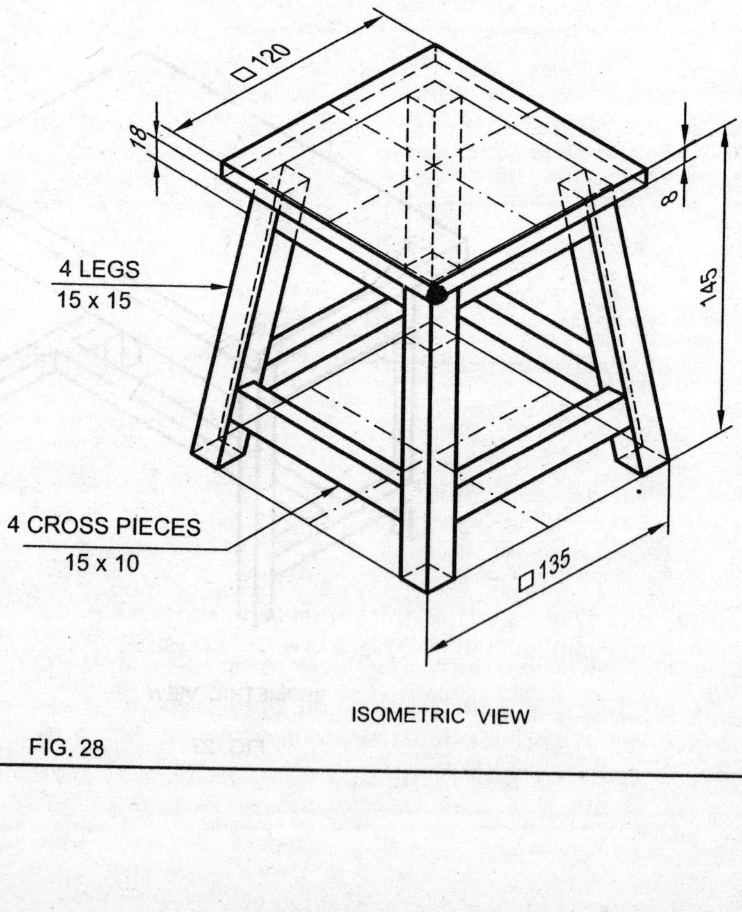

ISOMETRIC VIEW

FIG. 28

(29) BOX (LID IS 90° OPEN) (UQ)

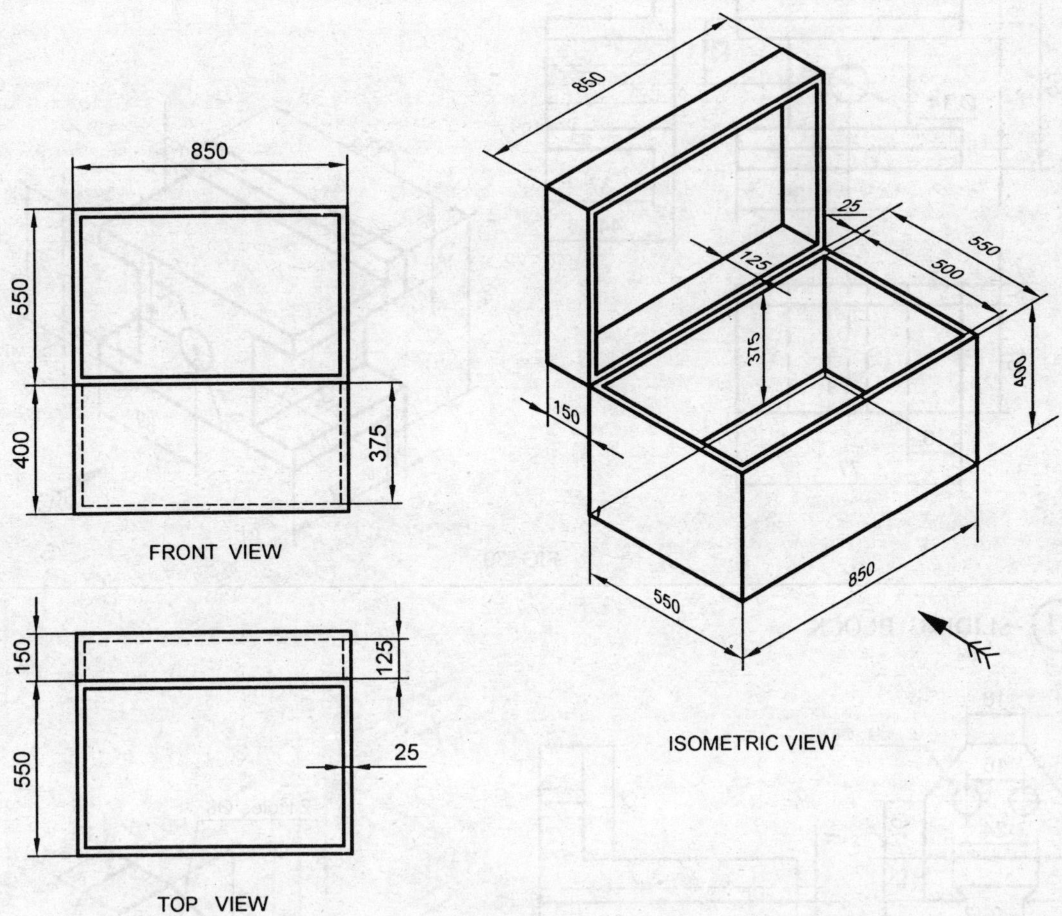

FIG. 29

NOTE : Deduct plank thickness at top and bottom for the inside of the box and the lid.

Exercises : (a) Sketch the Left Side View.
(b) Sketch the Isometric View of the BOX, when the lid is 120° open.

(30) STEEL CASTING

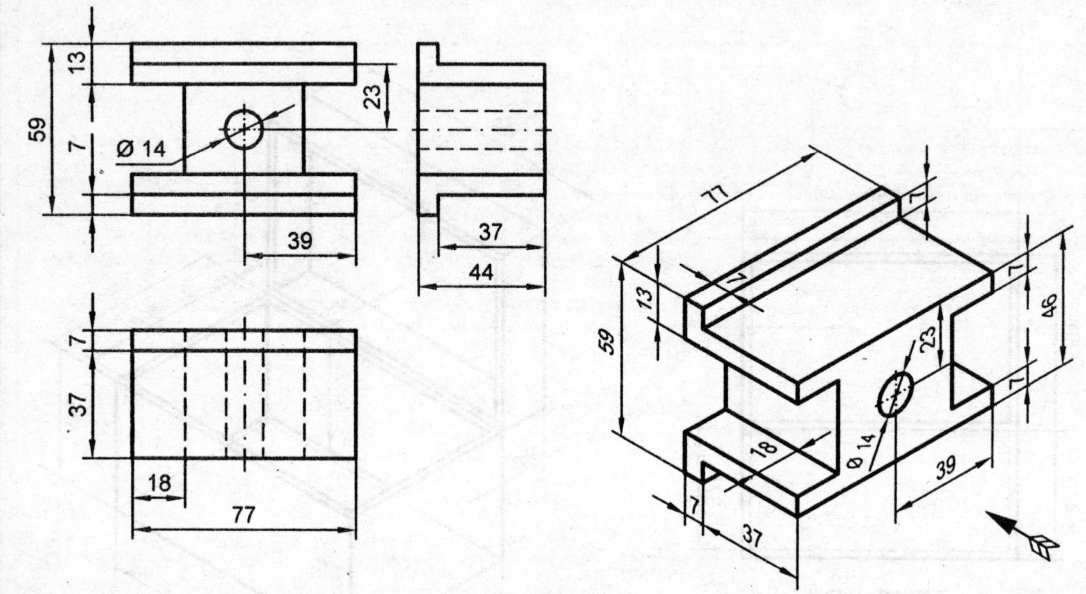

FIG. 30

(31) SLIDING BLOCK

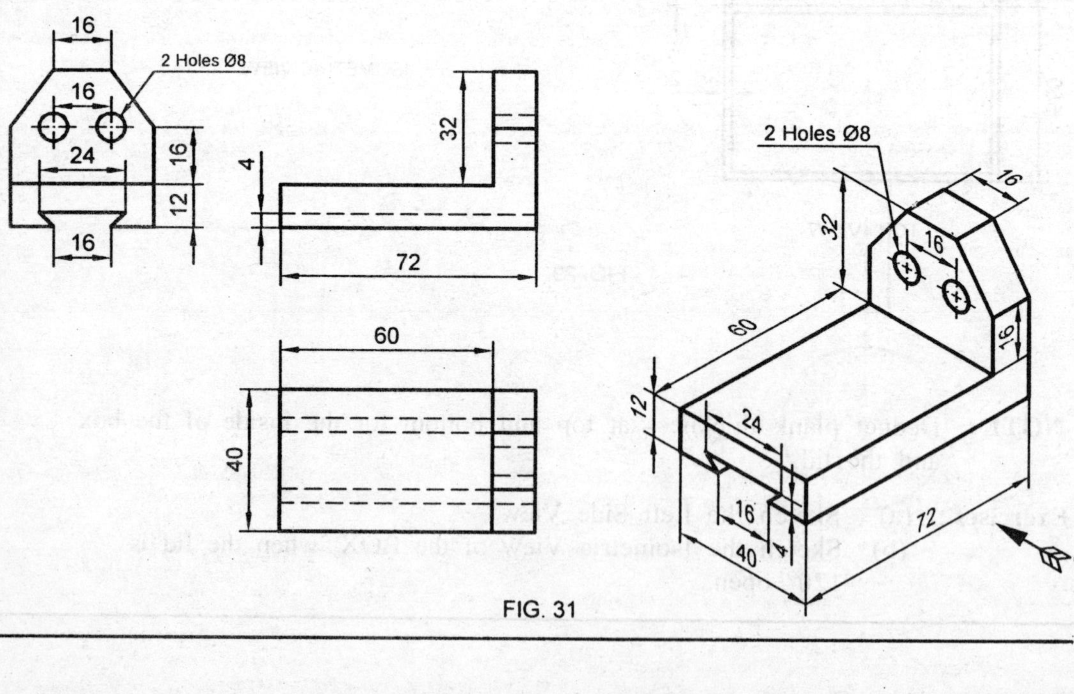

FIG. 31

(32) SIMPLE BRACKET

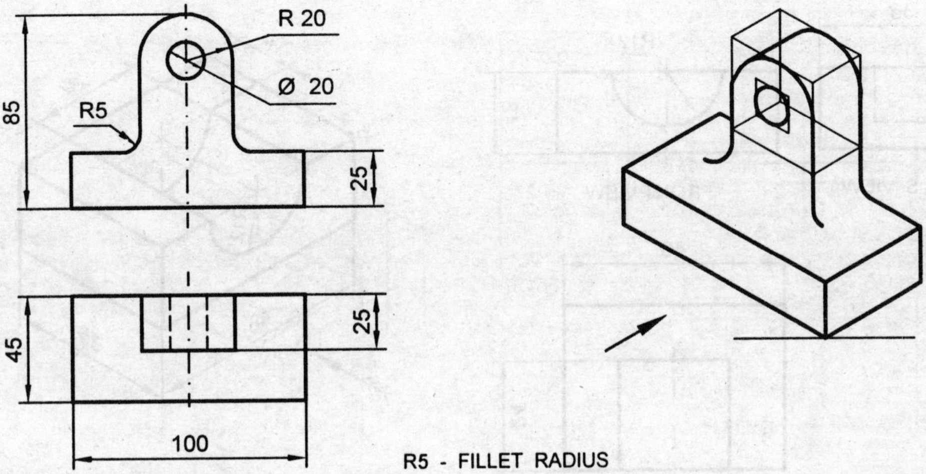

R5 - FILLET RADIUS

FIG. 32

(33) VERTICAL SHAFT WITH A CIRCULAR DISC

A circular disc of 100 mm diameter and 25 mm thick is mounted at the top of a vertical shaft of 25 mm diameter and 120 mm long. Draw the isometric view.

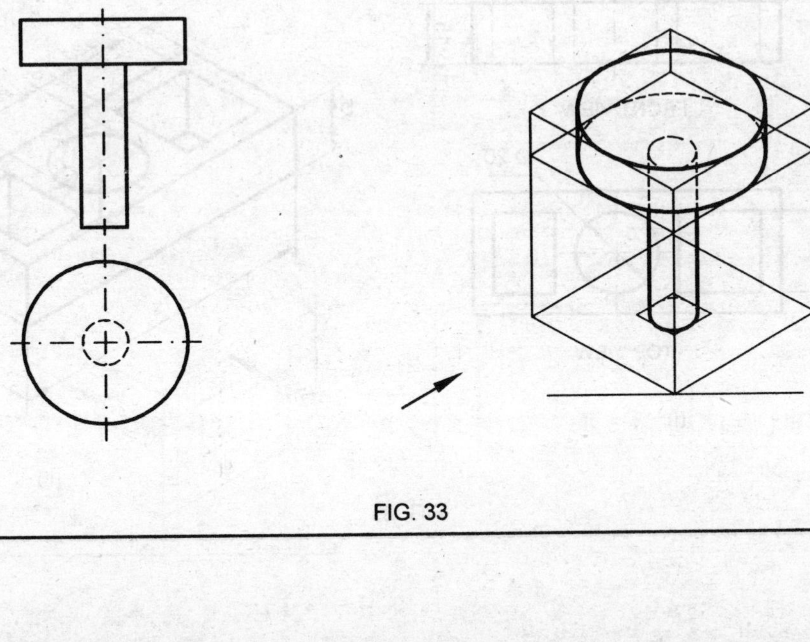

FIG. 33

(34) SEMI - CIRCULAR BLOCK

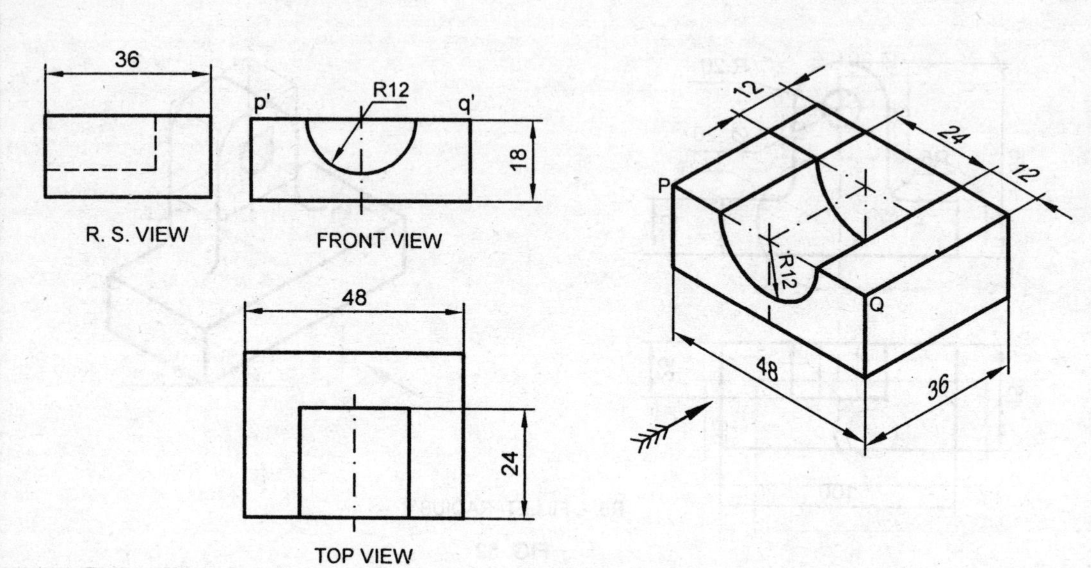

FIG. 34

(35) C.I. CASTING

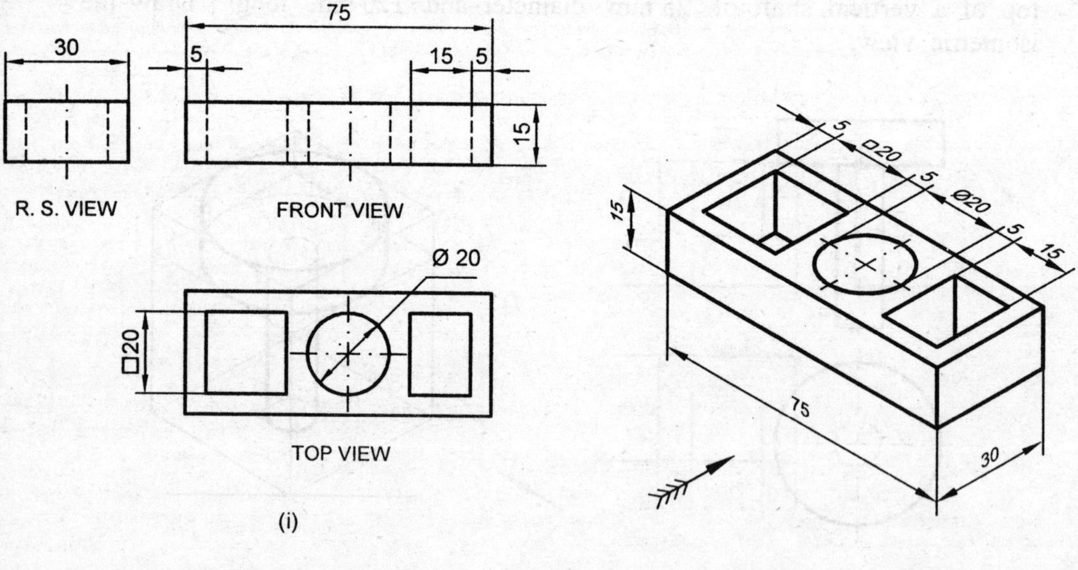

FIG. 35

(36) INK BOTTLE (UQ)

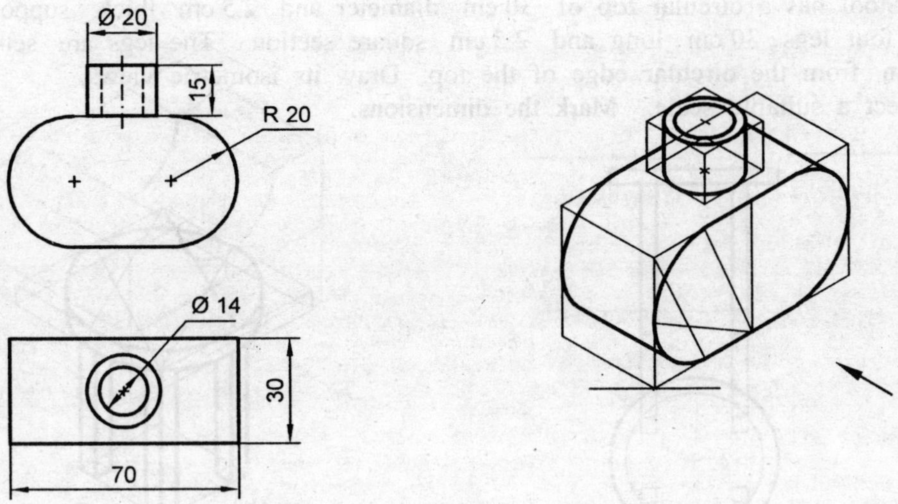

FIG. 36

(37) CUT BLOCK (UQ)

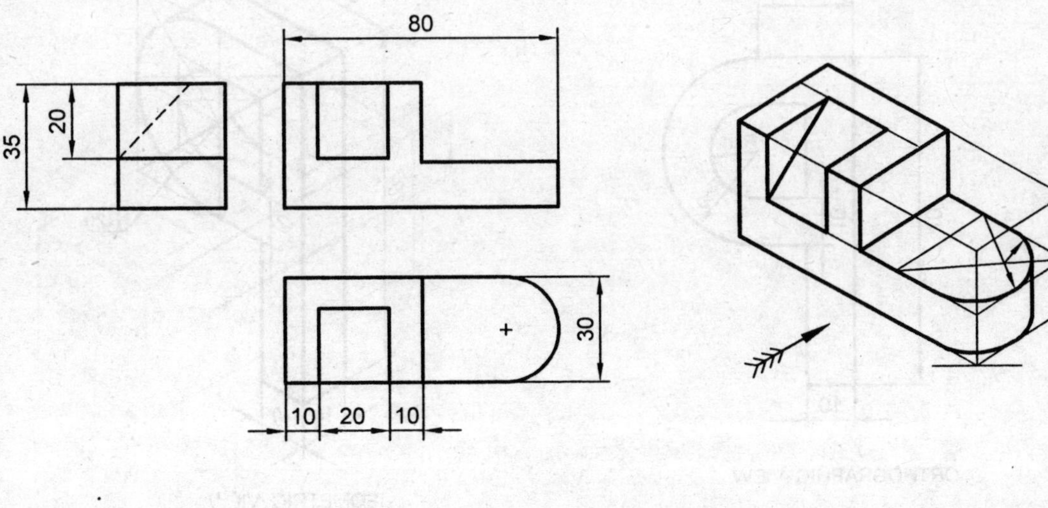

FIG. 37

(38) STOOL

(UQ)

A Stool has a circular top of 30 cm diameter and 2.5 cm thick, supported by four legs 30 cm long and 2.5 cm square section. The legs are set-in 4 cm from the circular edge of the top. Draw its isometric view. Select a suitable scale. Mark the dimensions.

FIG. 38

(39) LETTER - P

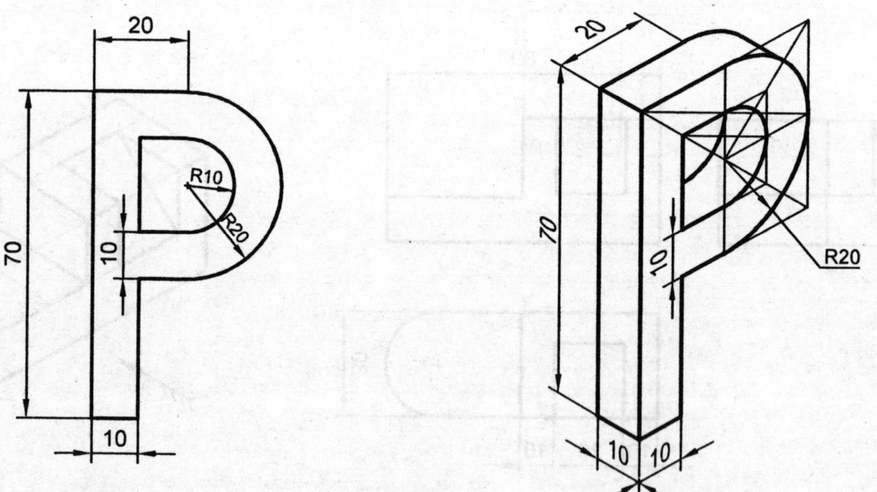

ORTHOGRAPHIC VIEW

ISOMETRIC VIEW

FIG. 39

(40) SQUARE - HEADED BOLT

Draw the isometric view of a SQUARE - HEADED BOLT 30 mm diameter and 95 mm long with a square neck of 25 mm thickness. The bolt head is of 50 mm square and 25 mm thickness. Assume that the bolt head has no chamfer.

(UQ)

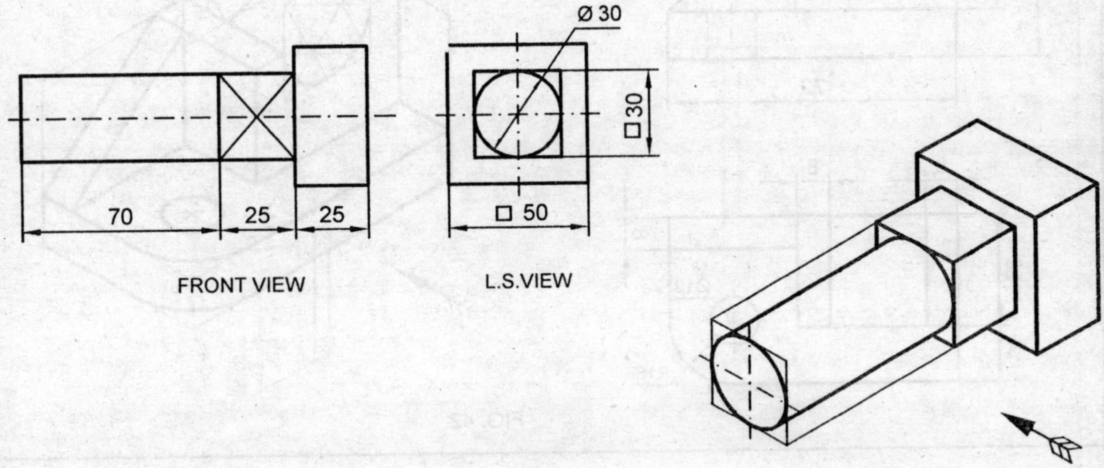

FIG. 40

(41) GLAND

(UQ)

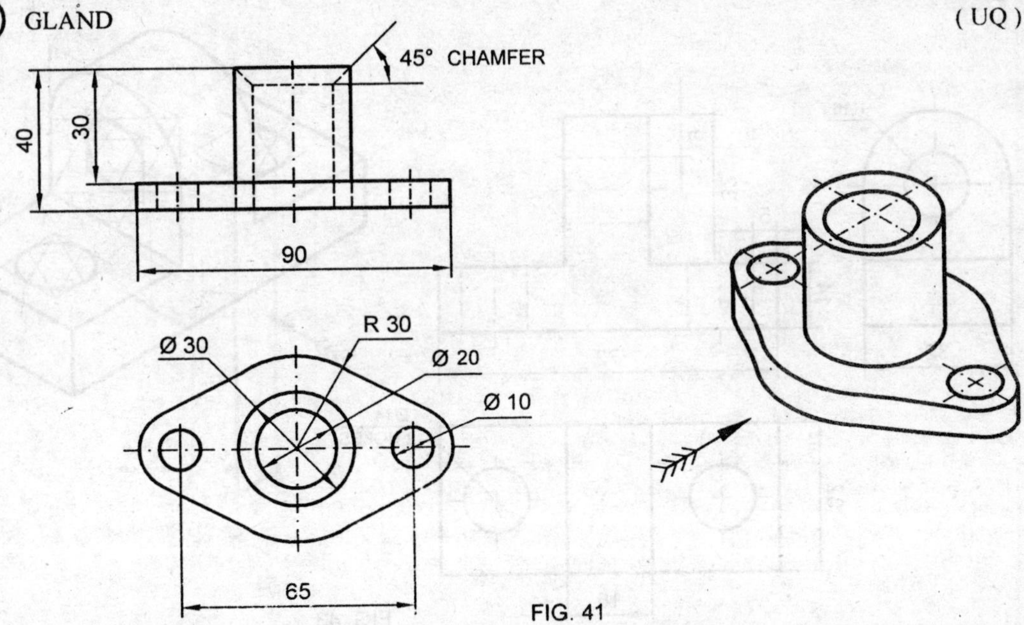

FIG. 41

(42) ROD SUPPORT (UQ)

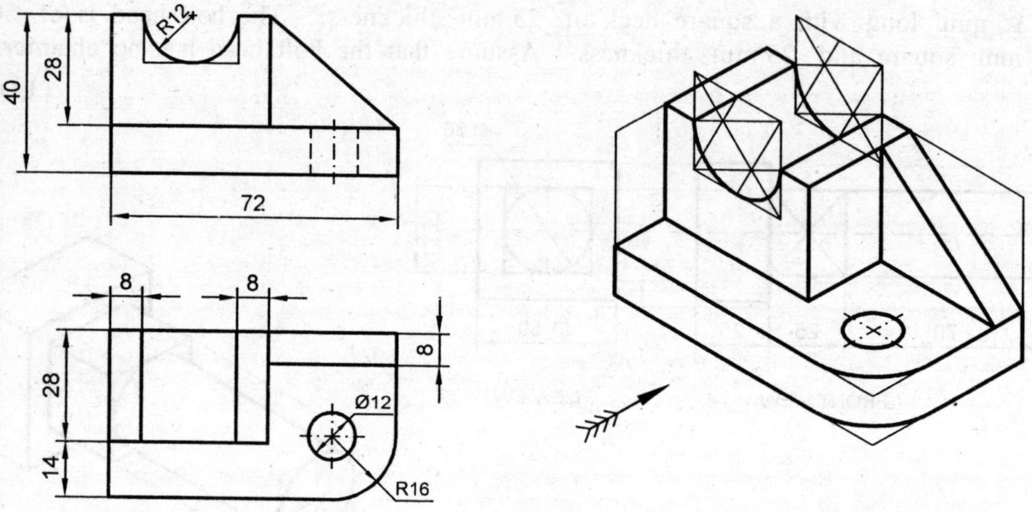

FIG. 42

(43) TOOL HOLDER (UQ)

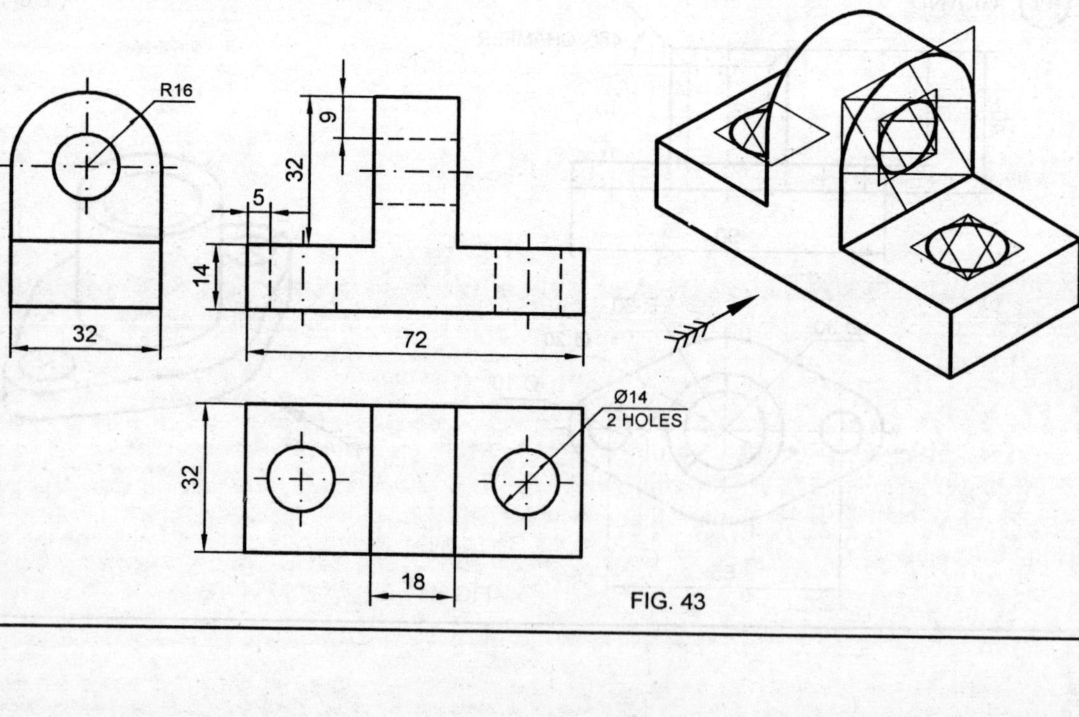

FIG. 43

5. EXERCISES

(44) CHANNEL

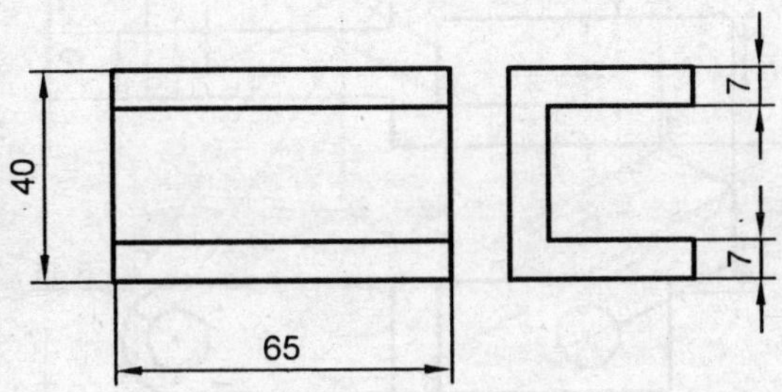

FIG. 44

(45) ELECTRO - MAGNET

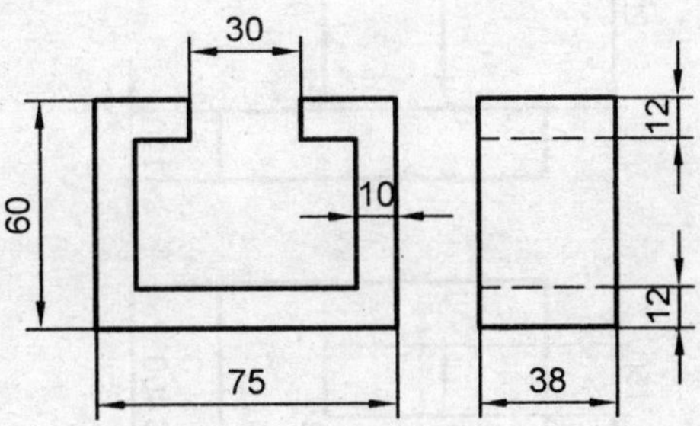

FIG. 45

(46) WOODEN STEPS

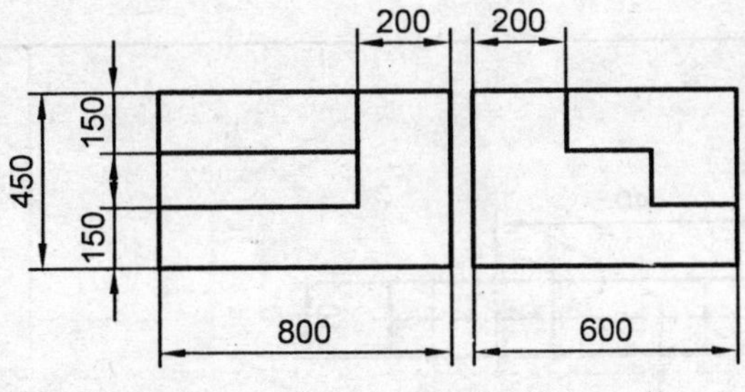

FIG. 46

(47) TEE BRACKET

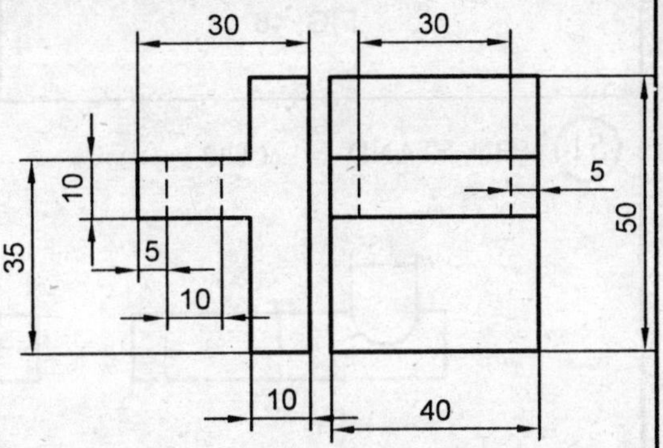

FIG. 47

(48)

A glass bottle consists of a cylinder of 60 mm diameter and 40 mm height with a cylindrical neck of 20 mm diameter and 20 mm height. The bottle is fitted with a cork of 18 mm diameter which projects by 5 mm. Draw its isometric view. (UQ)

(49) MILE STONE

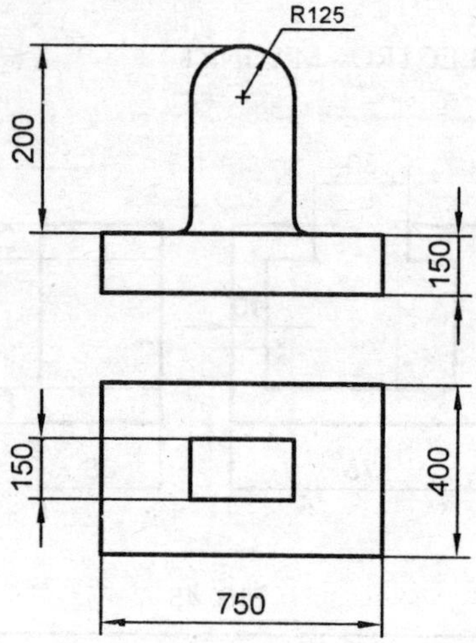

FIG. 48

(50) FORK (UQ)

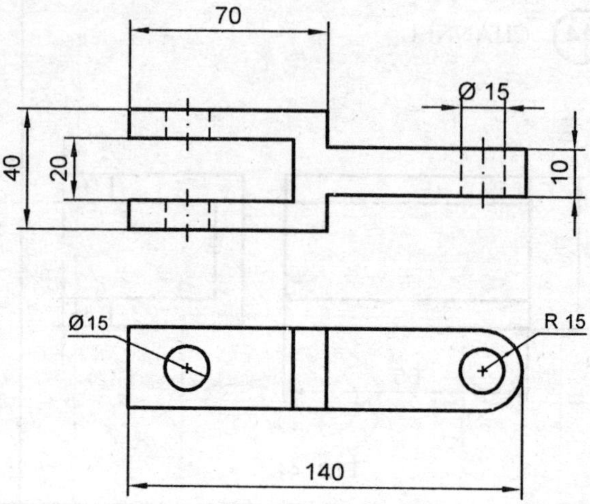

FIG. 49

(51) PEN STAND (UQ)

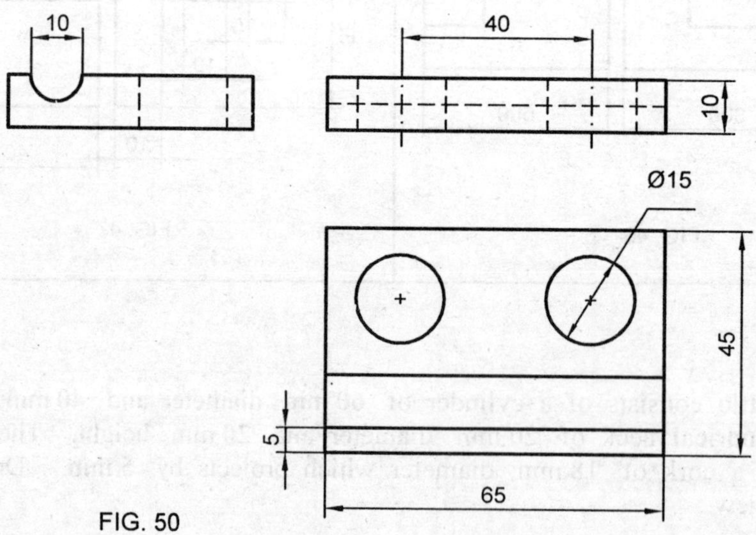

FIG. 50

(52) RIBBED SHAFT BRACKET

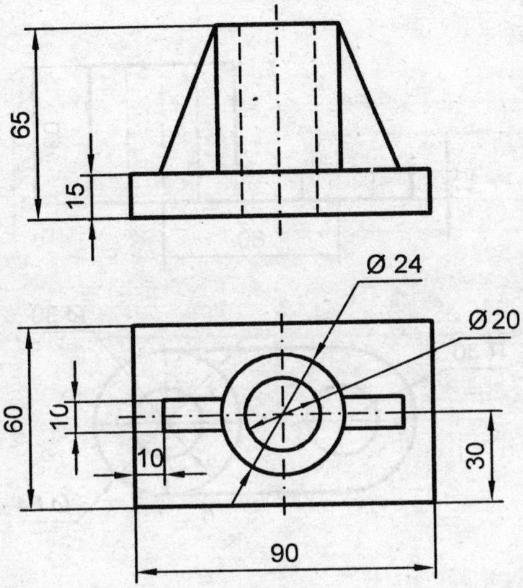

FIG. 51

(53) HINGE

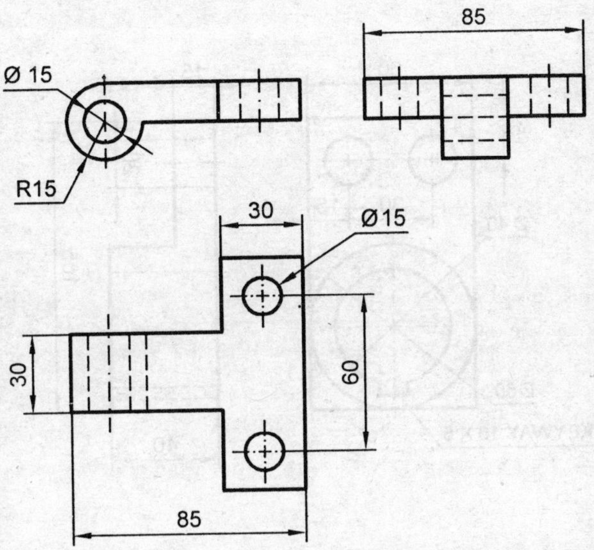

FIG. 52

(54) TRAY

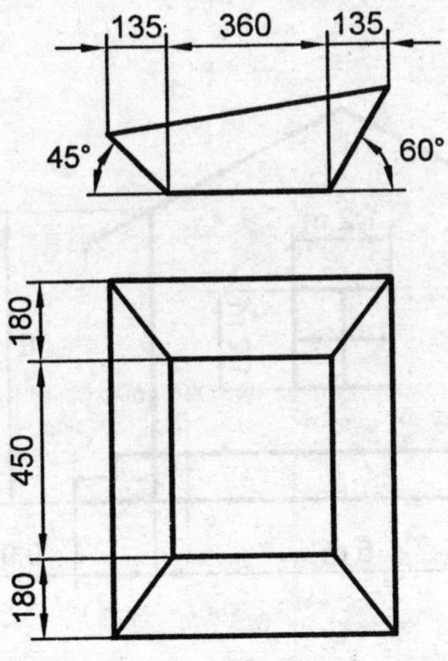

FIG. 53

(55) HOLDER

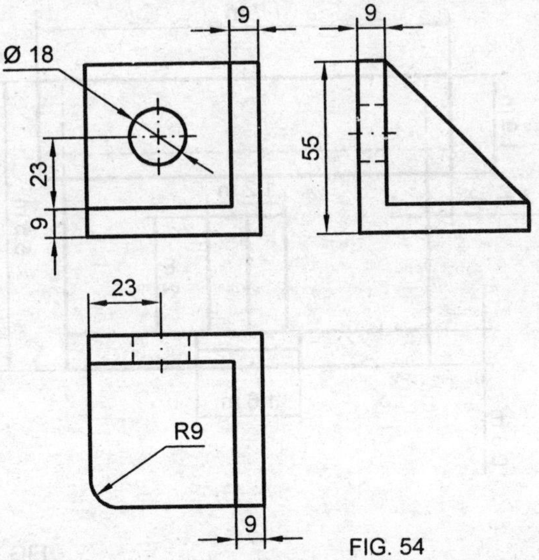

FIG. 54

56 HORIZONTAL SHAFT SUPPORT

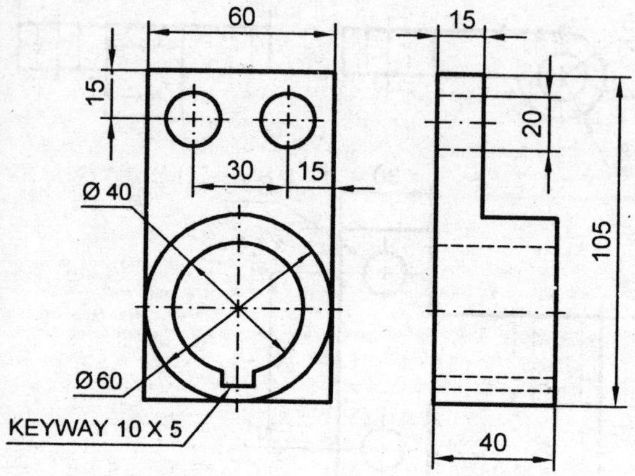

FIG. 55

57 VERTICAL SHAFT SUPPORT

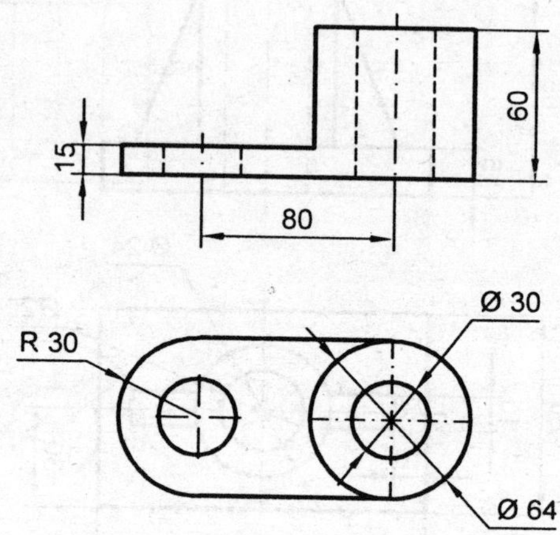

FIG. 56

58 PUMP HOUSE

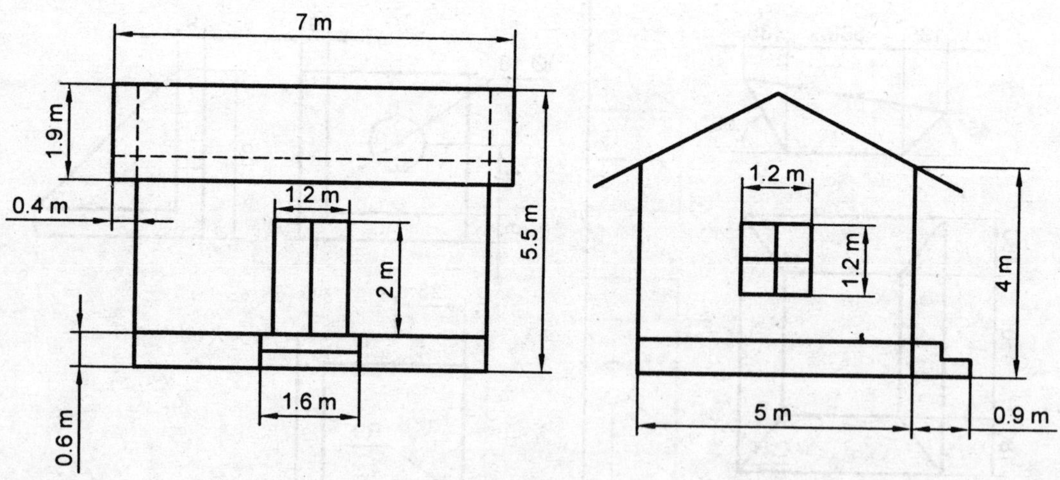

FIG. 57

(59) ADJUSTING ARM (UQ)

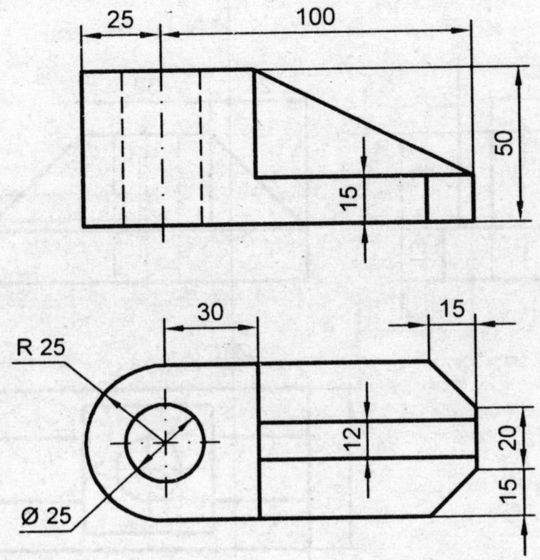

FIG. 58

(60) ROD ANCHOR (UQ)

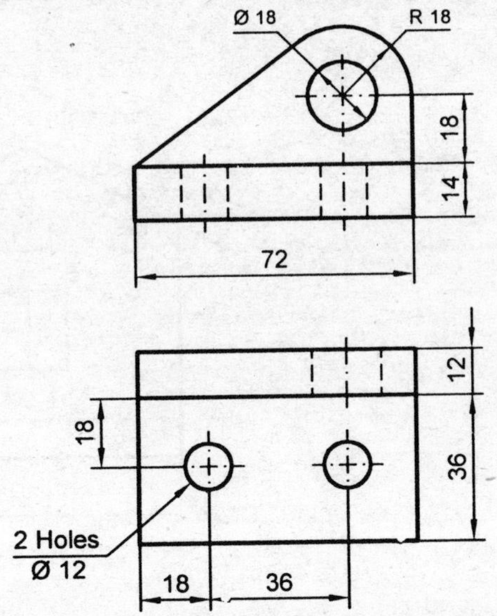

FIG. 59

(61) TOOL POST (UQ)

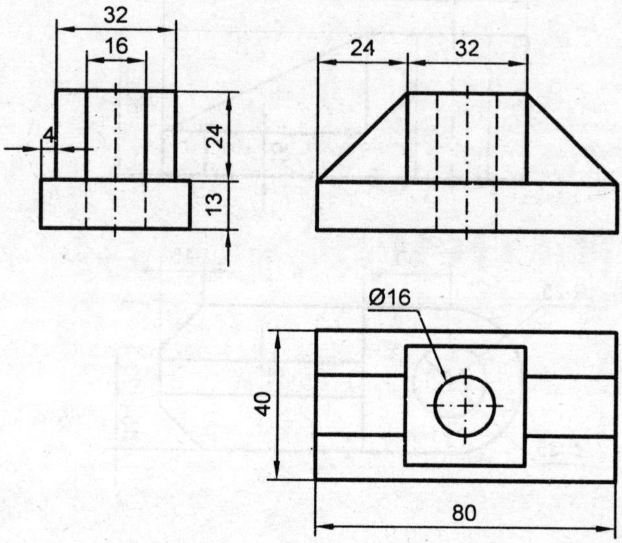

FIG. 60

20

CONVERSION OF ISOMETRIC VIEWS INTO ORTHOGRAPHIC VIEWS

1. ISOMETRIC VIEWS

Pictorial Views

Pictorial view shows the shapes of the front, top and a side face of an object. Therefore, it is easier to visualize its overall shape and develop the ability to sketch orthographic views from its pictorial view. Reading a pictorial sketch is rather easy and fast compared to orthographic sketches, because the object can be easily visualized. However, *pictorial views cannot give full dimensional details and true shapes*. Therefore, in engineering profession, pictorial views are used only to show the overall shape of objects.

Isometric Views

Pictorial views may be classified as: (i) Isometric, (ii) Oblique and (iii) Perspective.

NOTE: ***Isometric views, in which proportions can be easily judged, are generally adopted by engineers in practice.***

An isometric view conveys a natural representation of the object by showing the front, top and side faces in the same view – much as one would see an object in the world. This technique is especially useful when *communicating* the appearance of design. Orthographic views are more appropriate for *communicating* specific sizes and manufacturing data.

2. ORTHOGRAPHIC VIEWS

Orthographic projection is a ***multi-view projection*** by which the exact shape of an object can be represented by two or three separate views produced on the planes of projection that are at right-angles to each other. Each view shows the true shape and size of the object for a particular view direction and these views all together describe the object completely.

Engineers use orthographic views, *viz.*, Front, Top and Left or Right Side Views for describing the shape and size of 3D objects on a 2D drawing sheet. The objects include all types of machine parts, civil structures, electrical, electronic and instrumentation components, etc.

3. CONVERSION OF ISOMETRIC VIEWS INTO ORTHOGRAPHIC VIEWS

In orthographic views, the surfaces of the object except those lying inclined to the planes of projection are shown in their true shapes. Square feature in orthographic view is shown as rhombus in isometric view; rectangular shape as parallelogram and circular feature as ellipse in isometric view. Therefore, while drawing the orthographic views from isometric views, you have to visualize the true shape of the surfaces of the object. The three surfaces, *viz.*, top, right and left and the details on them are read one by one to visualize the object.

Conversion of an isometric view into orthographic views requires sound knowledge of the principles of isometric and orthographic projections and a little bit of imagination, to visualize the true shapes, *i.e.*, square / rectangle or circle / semi-circle / arc, etc., of the surfaces of the object.

3.1 PRINCIPLE OF VISUALIZATION

The ability to read and write engineer's language is a basic requirement for all engineers.

Reading: *Reading* a drawing / sketch is defined as the process of applying the principles of projection. Reading an isometric view is rather easy and fast compared to orthographic sketches, because the object can be easily visualized.

Visualization: *Visualization* is the understanding part of the reading of the sketches. Visualization is actually a mental process by which the details furnished in an isometric or orthographic sketch is translated to understand the shape of the object.

To understand the shape of an object from its isometric view, break the object into primary geometrical shapes like prisms, pyramids, cylinders, cones, etc. These shapes may be additions in the form of projections or subtractions in the form of cavities. The entire object is obtained by adding or subtracting the above geometrical shapes. Assemble them mentally to obtain the final shape.

NOTE :

1. *An isometric view does not show the real shapes of its surfaces. Hidden parts are also not clearly shown. All these have to be visualized.*
2. *Look at the isometric view in the direction of any one of the three isometric axes. Only two of the three overall dimensions, viz., Length / Width / Height or Thickness will be visible.*
3. *While reading an isometric view, assume a hidden part of a symmetrical object to be similar to the corresponding visible part.*
4. *If depth of a hole is not given, assume it to be a* ***through hole****. Through hole is a hole which goes completely through the object. If it is not a through hole, it is called* ***blind hole****, depth of which is to be indicated.*
5. ***Fillet:*** *Fillet is a rounded internal corner on a component. Sharp corners are difficult to cast and are also weak. Assume suitable radii, 3–6 mm for the fillet.*

4. CONVENTIONS

1. In Isometric View, only the overall shape of the object can be visualized. But the Orthographic View gives the true dimensions or sizes of the object.
2. Top view and front view are always in line ***vertically***.
   Front view and side view are always in line ***horizontally***.
3. Length of the top view is the ***same*** as length of the front view.
   Height of the side view is the ***same*** as the height of the front view.
   Width of the top view is the ***same*** as width of the side view.
4. Each view gives ***two*** dimensions, *i.e.*, front view gives length and height, top view gives length and width and side view gives height and width.
5. ***Lines and Points:*** Surfaces of the object, parallel to the direction of vision will be seen as ***lines*** in orthographic views. Also, edges of the object, parallel to the direction of vision will be seen as ***points*** in orthographic views.
6. ***True Shape and Size:*** If a surface is parallel to a plane (*i.e.*, normal to the direction of viewing), its projection on that plane will show its ***true shape and size***.
7. ***Plane Surfaces inclined to the direction of view:*** Plane surfaces inclined to the direction of viewing will appear in sizes smaller than true sizes in orthographic view.
8. ***Square and Rectangular Surfaces:*** Rhombus and parallelogram in the isometric views are square and rectangular surfaces respectively in the orthographic views.
9. ***Curved Surface:*** Curved surface in the isometric view of an object when viewed normal to it, appears as rectangle in the orthographic view.
10. ***Circle:*** Ellipse in isometric view should be visualized as a circle in orthographic view.
11. ***Semi-circle or Arc of a circle:*** Semi-circle or Arc of a circle in orthographic view is shown as semi-ellipse or the portion of an ellipse in the isometric view.
12. Dimensions which are parallel to the direction of vision will ***not*** be seen.

4.1 PRECEDENCE OF LINES

Lines such as visible line, hidden line and center line (axis) at times coincide on the sketches. It is necessary to know which line should be shown under such circumstances.

1. When a visible line coincides with a hidden line (*i.e.*, dashed line), sketch visible line.
2. When a visible line coincides with a center line (axis), sketch the visible line and extend the center line (axis) beyond the outlines of the view.
3. When a section plane line coincides with a center line (axis), show the center line and sketch the section plane line outside the outlines of the object at the ends of the center line by *thick* dashes.
4. When a dashed line (invisible line) coincides with the center line (axis), sketch the dashed line.

5. VISUALIZATION OF OBJECTS FROM ISOMETRIC VIEWS

Engineering drawings of 3D objects are prepared on 2D drawing sheets by drawing ***Points***, ***Lines*** and ***Areas / Surfaces***. Therefore, the meaning of ***Points***, ***Lines*** and ***Areas / Surfaces*** should be clearly known to understand the graphic language.

Identification of Surfaces: The concept of *Identification of Surfaces* of an object for its visualization followed by the preparation of the orthographic views is discussed below:

5.1 REPRESENTATION OF A POINT

Point: An edge of an object, when perpendicular to HP or VP, its orthographic projection on that plane is a *Point*.

5.2 REPRESENTATION OF A LINE

Line: A *Line* on an orthographic view may be used to represent the following:

1. Edge view of the plane surface of an object
2. Intersection of two plane, curved or spherical surfaces of an object
3. Boundary of the surfaces of an object

5.2.1 Edge View of Plane Surface of an Object

If a plane surface is perpendicular to VP or HP, its projection on that plane is a ***Line***.

Example: *See Fig. 1.* If a plane surface ABEF of an object is perpendicular to VP and parallel to HP, its projection on VP is a *line*. *This line represents the following:*

(i) Plane surface ABEF in Front View,

(ii) Edge AB.

5.2.2 Intersection of Two Plane, Curved or Spherical Surfaces of an Object

If a surface is intersecting with another surface, a ***line*** is formed. The intersecting surfaces may be *plane, curved* or *spherical* in shape. *Examples:*

See Fig. 1. If a plane surface ABCD is intersecting with another plane surface ABEF, a *line* AB is formed.

If a curved surface, *say,* the surface of a cylinder, is intersecting with its plane surface, the line of intersection is a *curved line* (here, *a circle*).

If a spherical surface, *say,* the surface of a hemi-sphere, is intersecting with its plane surface, the line of intersection is a *curved line* (here, *a circle*).

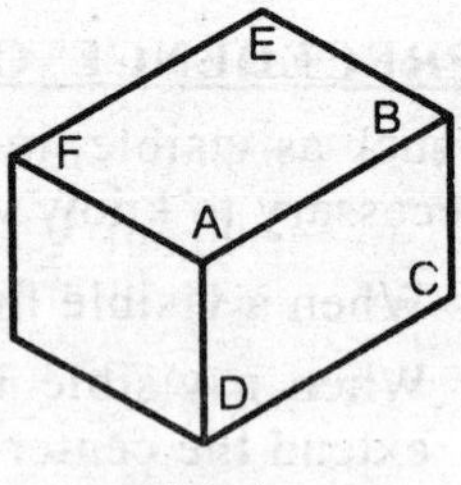

FIG. 1

5.2.3 Boundary of the Surfaces of an Object

Boundary of a plane surface A B C D is its outer *edges* AB, BC, CD and DA. Boundary of a circular plane is a *circle*.

5.3 REPRESENTATION OF AN AREA

An area on an orthographic view represents the projected surface area of a part of an object. The meaning of an area in an orthographic view can be understood only after analyzing the corresponding part in the other view or views.

5.3.1 Parallel Surfaces (Area parallel or perpendicular to HP / VP)

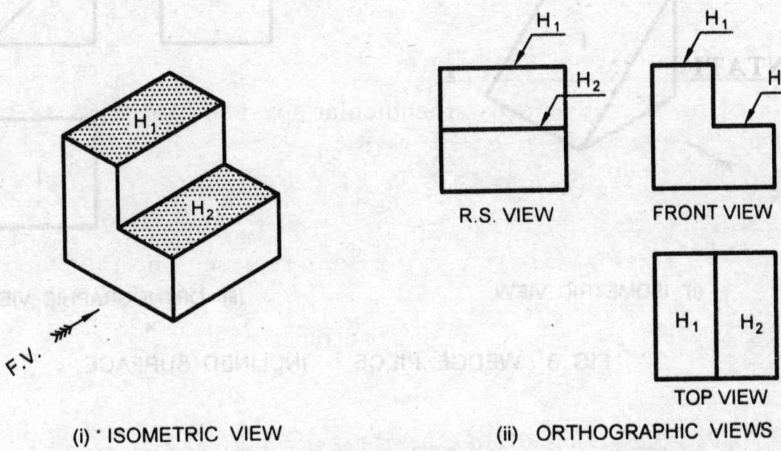

FIG. 2 L-BLOCK - PARALLEL SURFACES

RULE: 1. If a surface of an object is parallel to HP / VP, the projection of the surface on that plane to which it is parallel will have its true size and shape.

2. Projection of a surface normal to the plane of projection is represented by a straight line.

See Fig. 2. In the Pictorial View of the L-BLOCK, surfaces H_1 and H_2 are horizontal and are perpendicular to VP. Hence,

(i) in the Top View, horizontal surfaces H_1 and H_2 will have ***true shape*** and ***size***.

(ii) in the Front View, the horizontal ***surfaces*** H_1 and H_2 are indicated by horizontal ***lines*** H_1 and H_2.

(iii) in the Right Side View, the horizontal ***surfaces*** H_1 and H_2 are indicated by horizontal ***lines*** H_1 and H_2, as they are perpendicular to the Left Profile Plane.

NOTE: 1. *Parallelogram and rhombus in the isometric view are rectangle and square surfaces respectively in the orthographic views.*

2. *Lines / Edges of the object parallel to the direction of viewing will be seen as points in the orthographic views.*

5.3.2 Inclined Surface (Area Inclined to VP or HP)

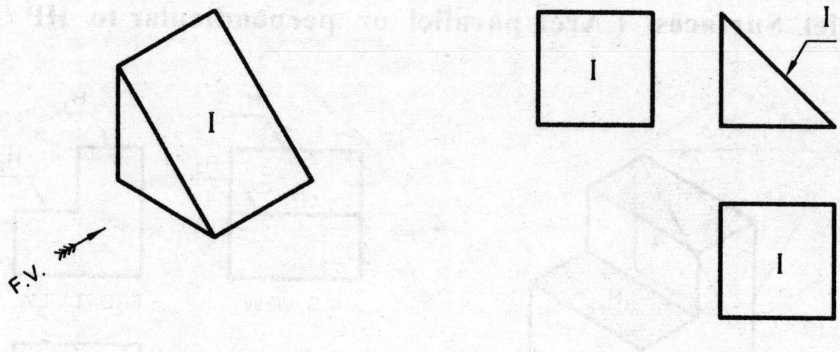

(i) ISOMETRIC VIEW

(ii) ORTHOGRAPHIC VIEWS

FIG. 3 WEDGE PIECE - INCLINED SURFACE

RULE: 1. If a surface of an object is perpendicular to either HP or VP and inclined to the other plane, the projection of the surface to which it is perpendicular will be a *straight line*.

2. Projection of the surface on the HP or VP to which it is inclined will appear *foreshortened*.

See Fig. 3.

In the Pictorial View of the WEDGE PIECE, the surface I is inclined to HP. Hence,

(i) in the Front View, this inclined ***surface*** is indicated by an inclined ***line*** I as it is inclined to HP.

(ii) in the Top View, this inclined surface will not have the true shape and size, but *foreshortened* and is indicated as I.

(iii) in the Right Side View, this inclined surface will *not* have the true shape and size, but *foreshortened* and is indicated as I.

5.3.3 Oblique Surface (Area Inclined to VP, HP and PP)

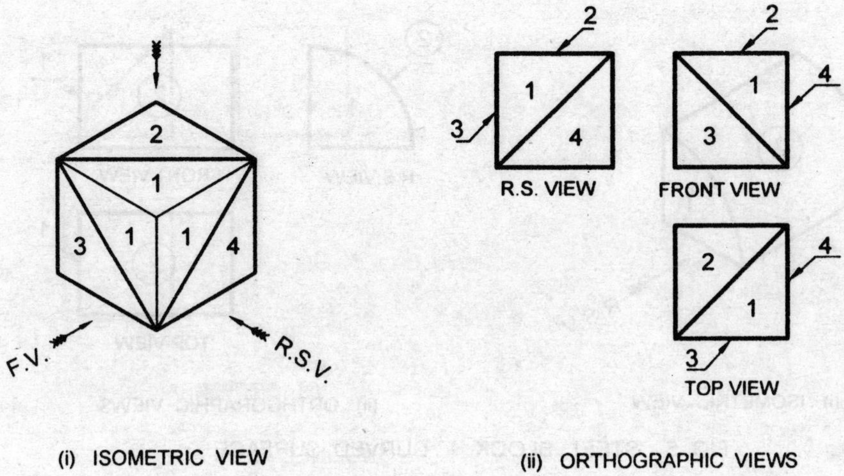

FIG. 4 WEDGE BLOCK - OBLIQUE SURFACE

What is an Oblique Surface ?

If a surface of an object is inclined to the three reference planes, the surface is called as *Oblique Surface*.

RULE: **An oblique surface is inclined to all the three planes, *viz.*, VP, HP and PP. It is seen in all the three views, *viz.*, Front, Top and Side Views in *foreshortened* form.**

See Fig. 4.

It shows the Isometric View of a WEDGE BLOCK having oblique surface. Oblique surface is seen in all the three views, viz., Front, Top and Side Views in foreshortened form.

5.3.4 Area Inclined to HP and VP and Perpendicular to PP

RULE: **If the surface is inclined to both HP and VP and perpendicular to PP, its Front and Top Views will get *foreshortened* while the Side View is a *straight line*.**

5.3.5 Curved Surface (Curved Area Perpendicular to VP or HP)

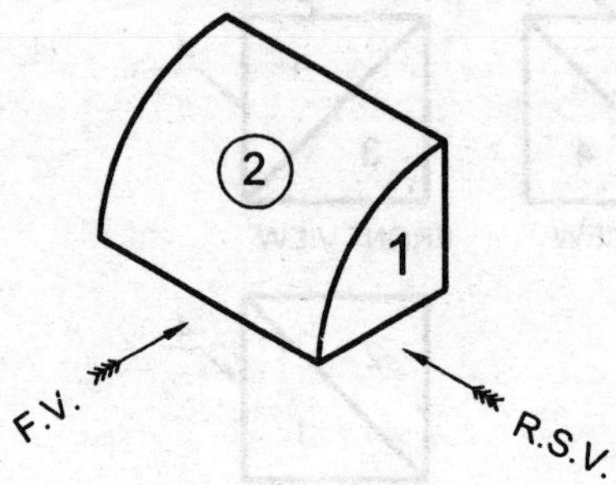

(i) ISOMETRIC VIEW

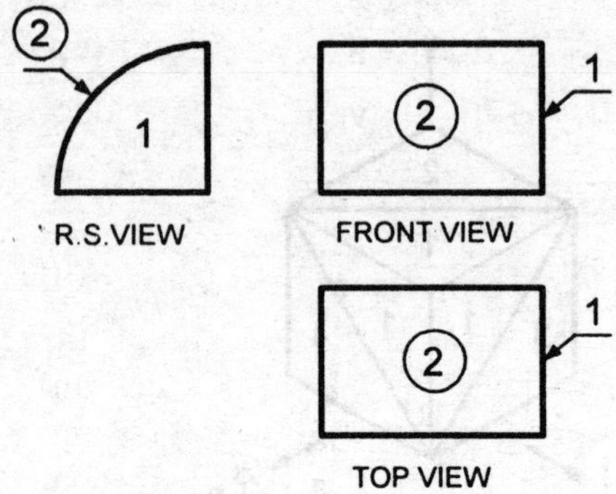

(ii) ORTHOGRAPHIC VIEWS

FIG. 5 STEEL BLOCK - CURVED SURFACE

RULE: 1. Curved surface of an object, when viewed normal to it, appears as a *rectangle* in the orthographic view.

2. If a curved surface of an object is perpendicular to HP or VP, its projection on that plane is represented by *curved line (arc)*, while its projection on the other plane is *foreshortened* area.

Fig. 5 shows the Isometric View of a STEEL BLOCK having curved surface. Identify the visible surfaces and name them using numerals or alphabets as explained below:

Surface 1 is perpendicular to HP. Hence, its true shape is seen in the Right Side View. But it is seen as *lines* in Front and Top Views. These lines are indicated by the corresponding numeral 1.

Curved surface 2 , when viewed normal to it, is seen as *rectangles* in Front and Top Views, but seen as an *arc* in the Right Side View. The curved surface 2 is indicated by the corresponding numeral 2 enclosed in circles.

NOTE: ***Visualization of Circles and Arcs of circles in Orthographic Views :***

1. Circle in an orthographic view is shown as an ellipse in an isometric view.

2. Arc of a circle in an orthographic view is shown as the portion of an ellipse in the isometric view.

---

6. STEP-BY-STEP PROCEDURE : CONVERSION OF ISOMETRIC VIEWS TO ORTHOGRAPHIC VIEWS

Visualization of the isometric views may be started from simple objects. Complex objects may be visualized after sufficient practice. While preparing orthographic views from isometric views, follow a systematic **Step-by-Step** procedure as explained below:

Remember ! BIS conventions are also to be observed *in toto.*

ISOMETRIC VIEW

1. Study the Isometric View of the given object and its details carefully.

   An isometric view shows three mutually perpendicular planes, *i.e.*, Left, Right and Top. The three faces on these three planes and details on them are read one-by-one to visualize the object. Thus, each part of the object is identified, thereby all the visible and invisible surfaces can be identified.

2. **SELECTION OF ORTHOGRAPHIC VIEWS**

   Front View

   Choose the front view by seeing the object in a direction in which its length, the major shape and maximum details are seen. Further, place the object so as to obtain minimum number of hidden lines in the front view.

   Indicate the direction from which the object is to be seen for its front view by means of an arrow. Mark this arrow parallel to a sloping axis. Obtain other views by looking in the directions parallel to other two isometric axes.

   Other Orthographic Views

   Decide the necessary *other views* to be sketched.

   ONE VIEW DRAWING : A single view is sufficient for simple objects, if a qualifying *Note* is added. For example, a cylinder is drawn as a rectangle and the symbol ϕ for diameter is added for the dimension.

   TWO VIEW DRAWINGS : The object having both the right-hand side and the left-hand side of symmetrical shape, requires only two views. Careful analysis of the object will enable you to determine whether a top or a side view is necessary after the front view is selected.

   THREE VIEW DRAWINGS : *Irregular, unsymmetrical* and *complex objects* generally require three views for the complete shape description.

Rules for Selection of Views

- *Minimum hidden lines:* Select the view such that minimum hidden lines are necessary to show the internal details.
- *Cannon ball blunder:* Select the least number of views for complete description of the object. Drawing more views than necessary is familiarly called the *Cannon-ball blunder*. It derives its name from drawing three views of a cannon ball.

ORTHOGRAPHIC VIEWS

3. ***Overall Dimensions:*** Determine the overall dimensions of the required orthographic views, *i.e.*, Length L, Width W and Height H (or Thickness T) from the given isometric view.

4. ***Scale:*** Select a suitable scale, if necessary.

5. ***Block-in Views:*** Draw rectangular blocks (*in thin lines*) for the overall dimensions of each view, keeping sufficient space between the views for easy and legible placing of dimensions, Note, etc. on the views. This is known as *Block-in views.*

6. ***Center Lines:***
   (i) Mark centers for circles in the front view. Draw center lines passing through these centers for the axes of symmetry. When a cylindrical part / hole is seen as a rectangle, draw only one center line for its axis; when seen as a circle draw two axes perpendicular to each other.

   (ii) Extend the above center lines in the other orthographic views.

7. Refer : **CONVENTIONS** in Section 4 and **VISUALIZATION OF OBJECTS FROM ISOMETRIC VIEWS** in Section 5.

   Draw the details of the object (*in thin lines*) ***simultaneously*** in all the views by drawing projectors in the following order :

   (i) Circles and Arcs of circles,
   (ii) Visible lines of the general shape of the object,
   (iii) Minor details of the object such as lines, small fillet curves, etc.

8. Darken all the visible lines and circles / arcs forming the orthographic views.

9. Mark the invisible lines, if any.

NOTE: *Dashed lines (Hidden lines) should be drawn correctly, as per the BIS conventions of orthographic projections.*

---

10. Mark the dimensions, distributing them appropriately in all the views. Also, add Note, etc., at the appropriate views.

11. Erase the construction lines and projection lines.

Problem 1 : (Fig. 6) I-BEAM (PARALLEL SURFACES)

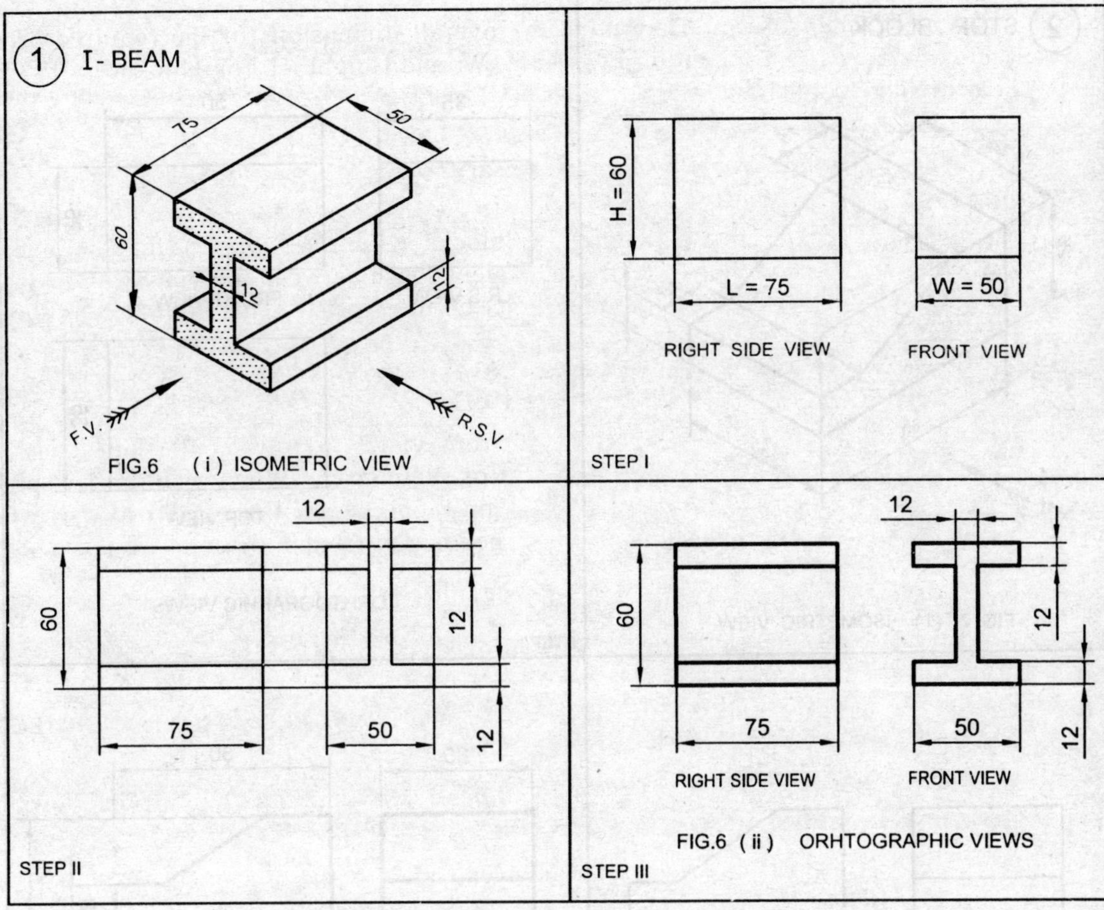

FIG.6 (i) ISOMETRIC VIEW

FIG.6 (ii) ORHTOGRAPHIC VIEWS

Isometric View :
1. Decide the required views, i.e., Front and Right Side Views. Top View is not required.
2. Mark their directions in the Pictorial View.
3. Measure the overall Length L, Width W and Height H of the object.

STEP I :
4. Using L, W and H, mark spaces for the Front and Right Side Views in the form of rectangles (in thin lines).

STEP II :
5. Draw the details simultaneously in both Front and Right Side Views by drawing the corresponding projectors (in thin lines).

STEP III :
6. Darken all the visible lines. Erase the projection / construction lines.
7. Mark the dimensions.

Problem 2 : (Fig.7) STOP BLOCK (INCLINED SURFACE)

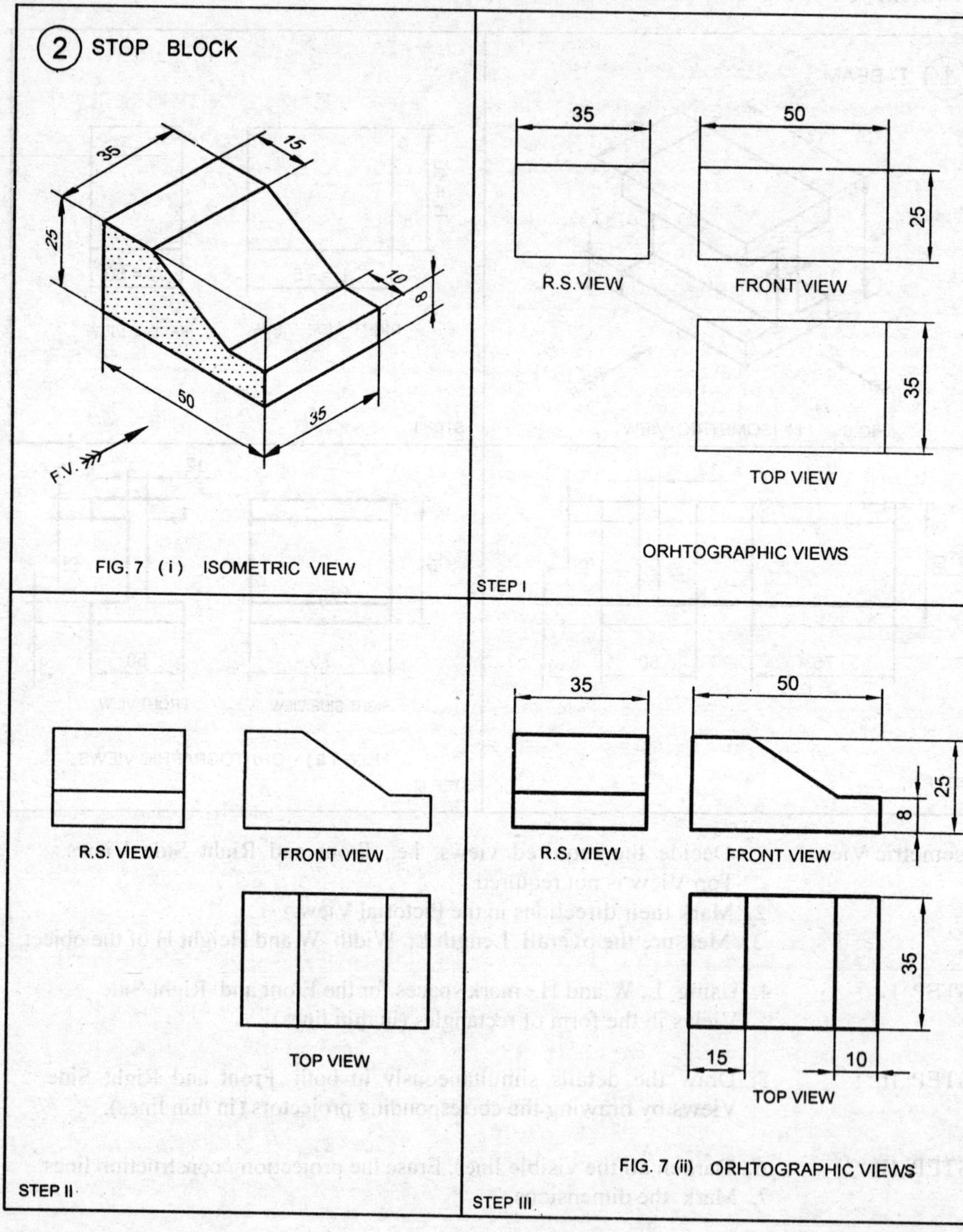

FIG. 7 (i) ISOMETRIC VIEW

FIG. 7 (ii) ORHTOGRAPHIC VIEWS

Problem 3 : (Fig.8) WEDGE PIECE (OBLIQUE SURFACE)

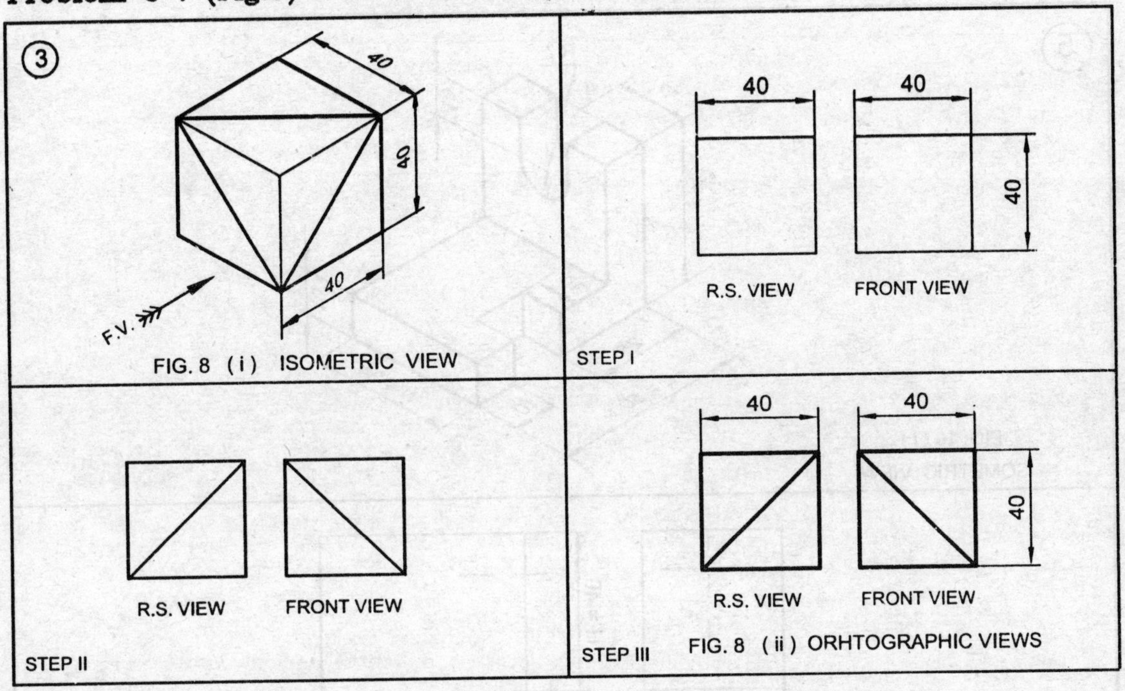

FIG. 8 (i) ISOMETRIC VIEW

FIG. 8 (ii) ORHTOGRAPHIC VIEWS

Problem 4 : (Fig.9) STEEL BLOCK (CURVED SURFACE)

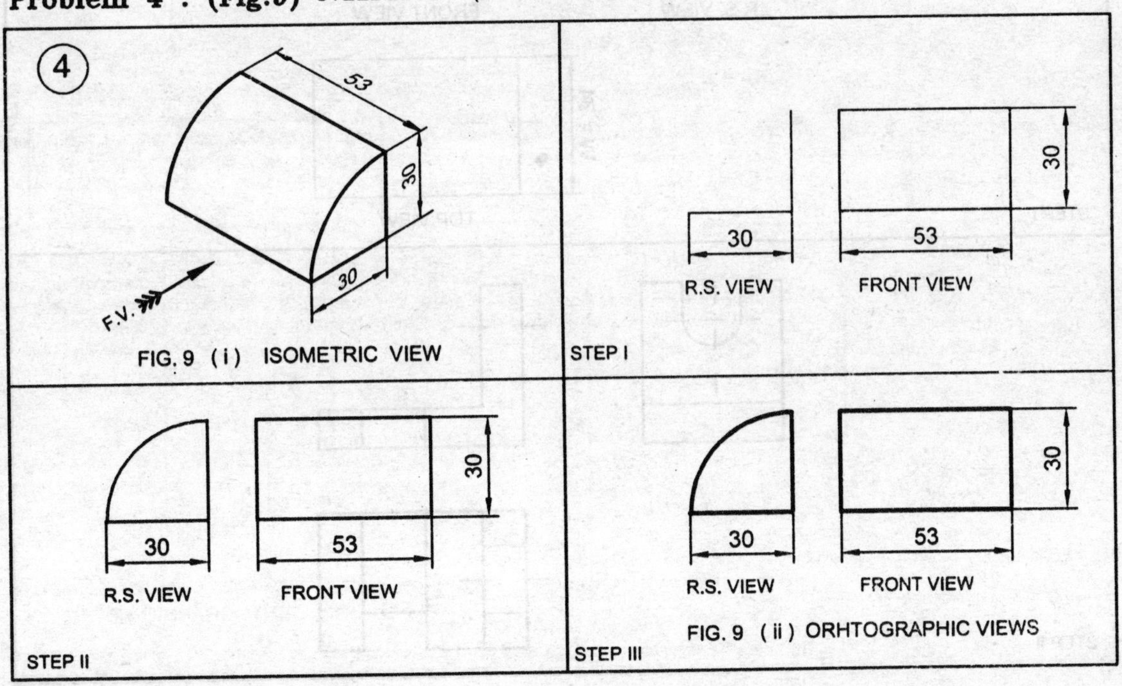

FIG. 9 (i) ISOMETRIC VIEW

FIG. 9 (ii) ORHTOGRAPHIC VIEWS

Problem 5 : (Fig.10) GUIDE BLOCK (UQ)

5

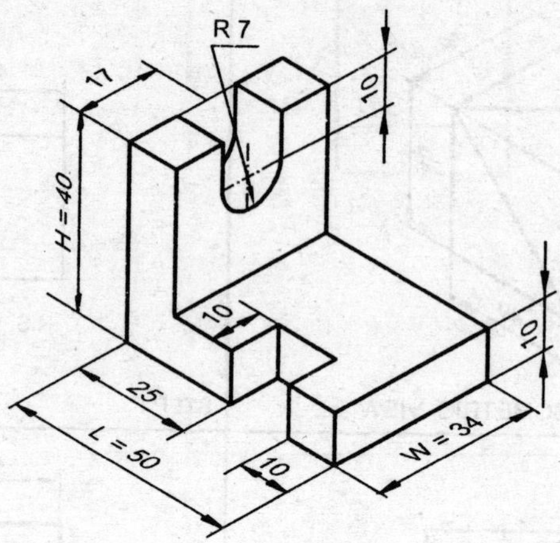

FIG. 10 (i)
ISOMETRIC VIEW

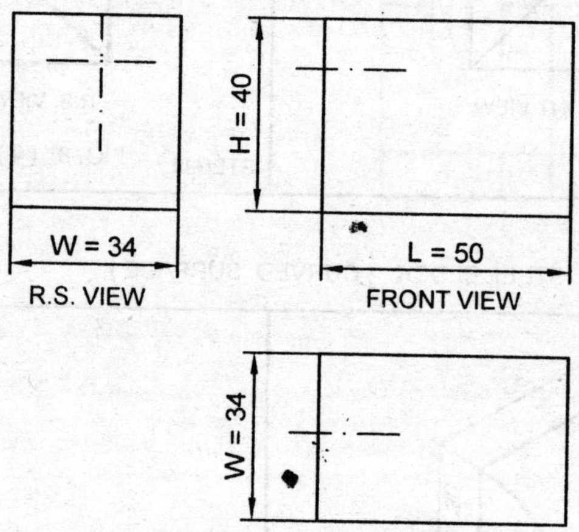

TOP VIEW

STEP I

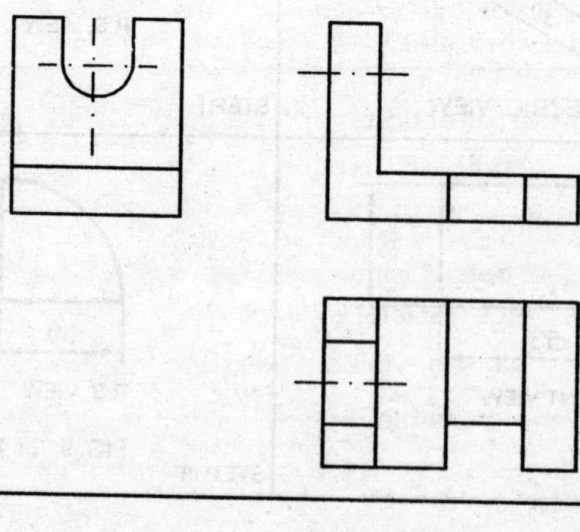

STEP II

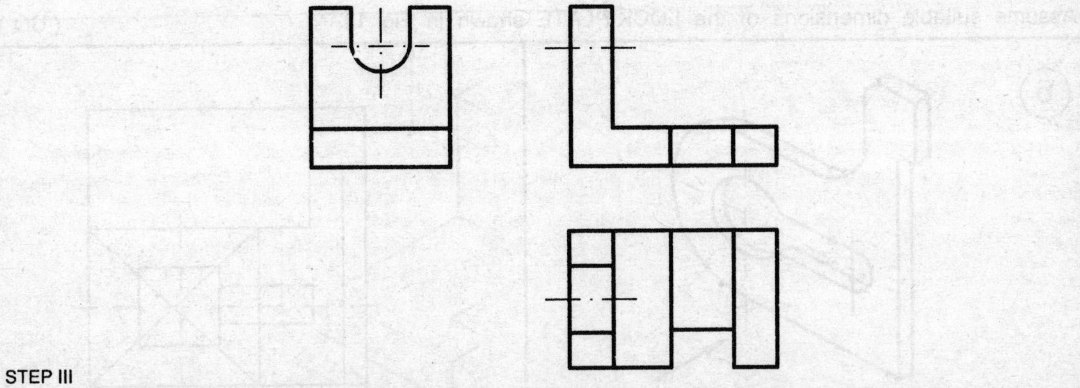

STEP III

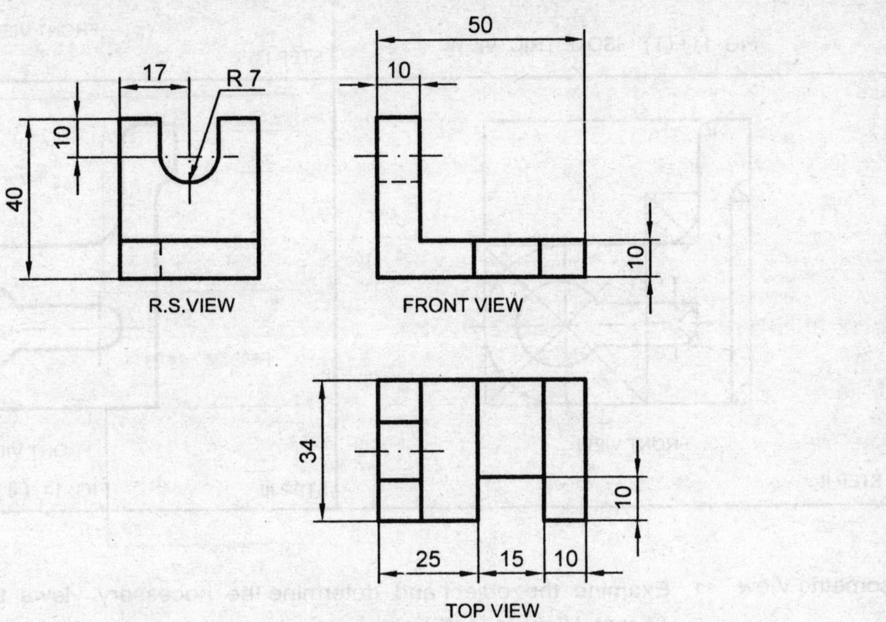

STEP IV

FIG 10 (ii) ORHTOGRAPHIC VIEWS

Pictorial View : 1. Mark the directions of Front, Top and Right Side Views.

STEP I : 2. Draw rectangles using the overall dimensions L, W and H.

: 3. Mark centre lines (axes) at the appropriate places.

STEP II : 4. Draw the details simultaneously in all 3 views in the following order : Circles / Arcs, Visible lines of major shape of the object, Minor details.

STEP III : 5. Darken all the visible lines and curves.

STEP IV : 6. Mark the invisible lines. Add the Jimensions.

Problem 6 : (Fig. 11) LOCK PLATE

Assume suitable dimensions of the LOCK PLATE shown in Fig. 11. (UQ)

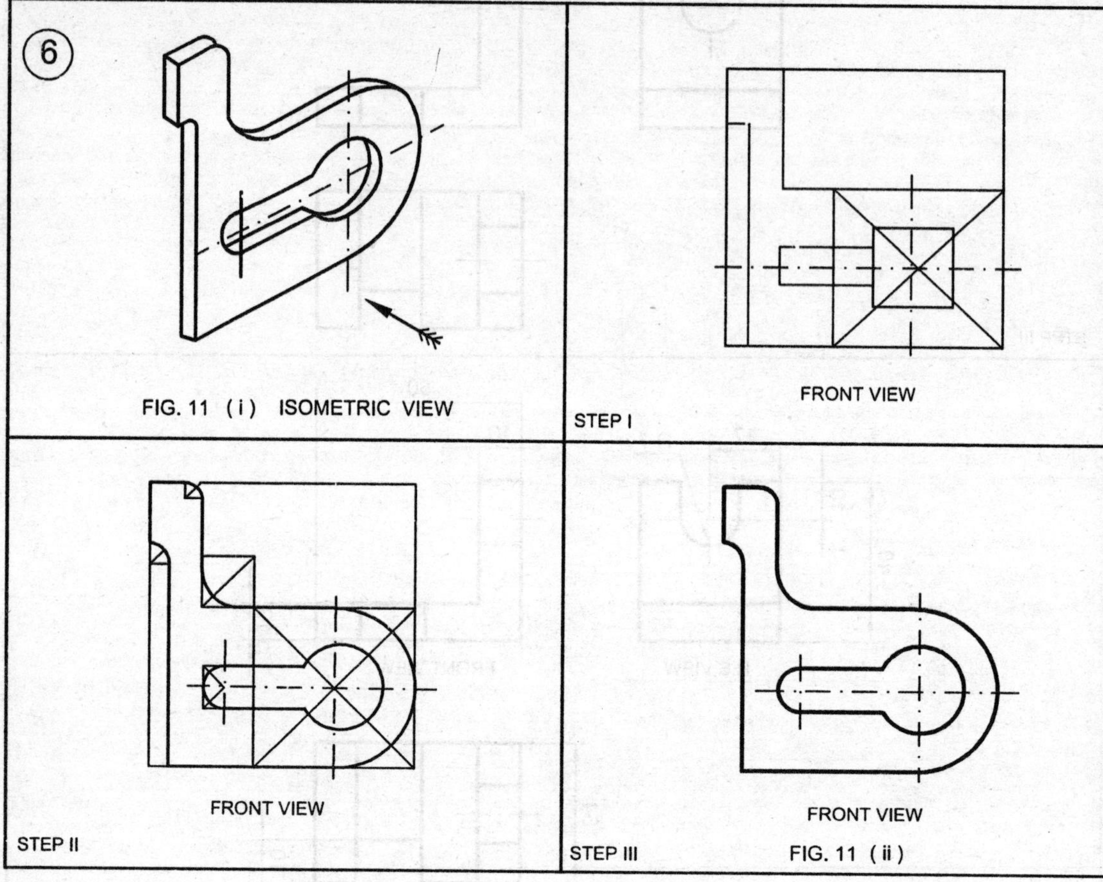

Isometric View : 1. Examine the object and determine the necessary views to be drawn. (Front View is sufficient.)

: 2. Mark suitable dimensions of the Lock Plate in the Isometric View.

STEP I : 3. Draw the overall size of the front view as rectangular blocks. Draw the center lines.

STEP II : 4. Draw the outline of each larger main feature, including circles and arcs lightly. This is the most important part of the drawing.

STEP III : 5. Complete the view by darkening all the lines and arcs.

: 6. Mark the dimensions.

Problem 7 : (Fig.12) SQUARE - HEADED BOLT (UQ)

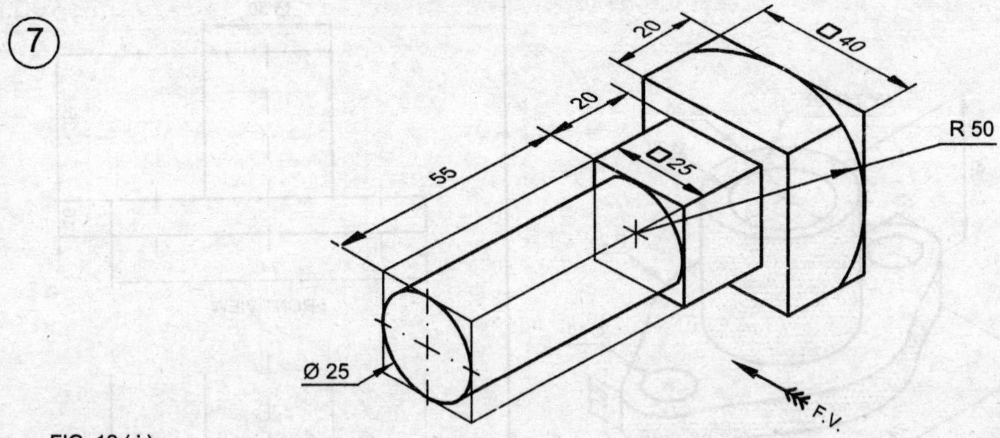

FIG. 12 (i)
ISOMETRIC VIEW

TOP VIEW NOT NEEDED

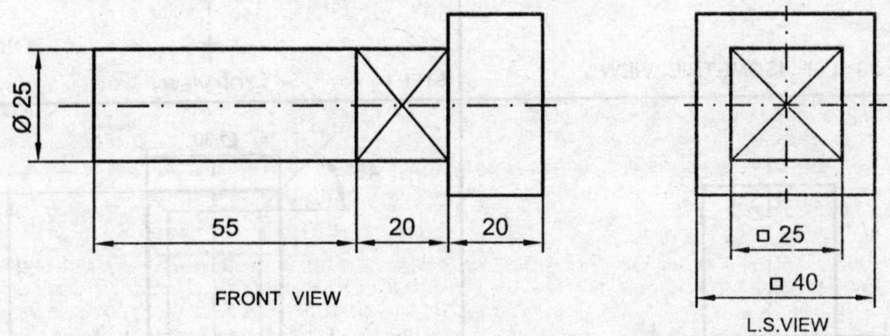

STEP I

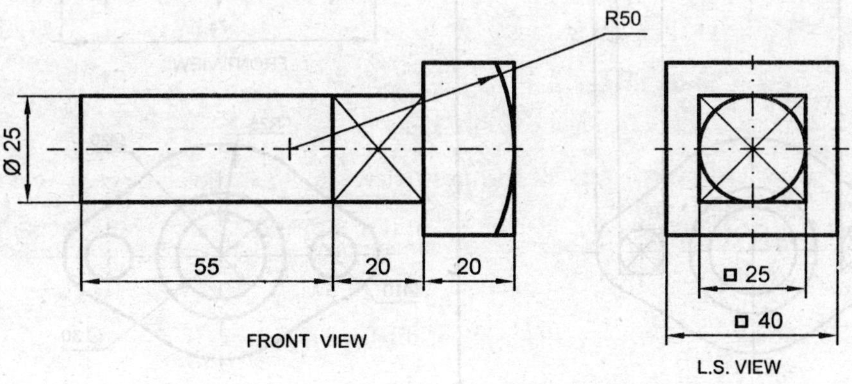

FIG. 12 (ii) ORHTOGRAPHIC VIEWS

STEP Ii

Problem 8: (Fig.13) GLAND

(8)

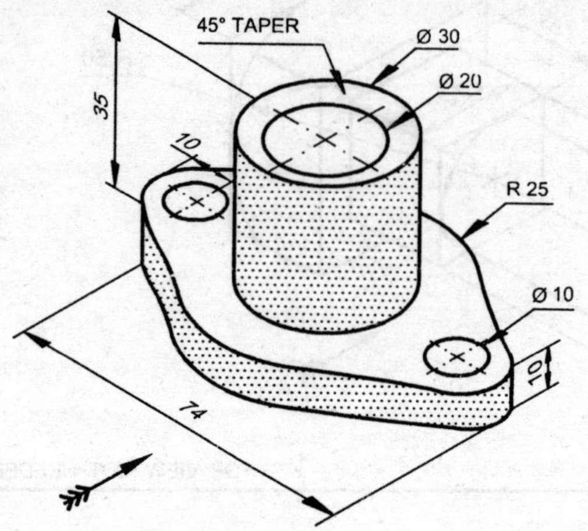

FIG. 13 (i) ISOMETRIC VIEW

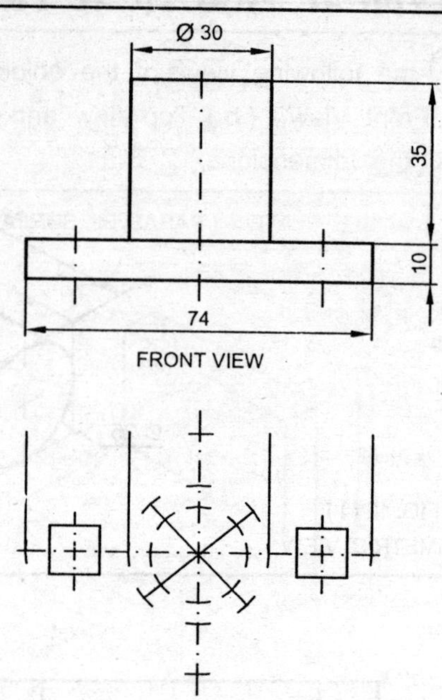

STEP I

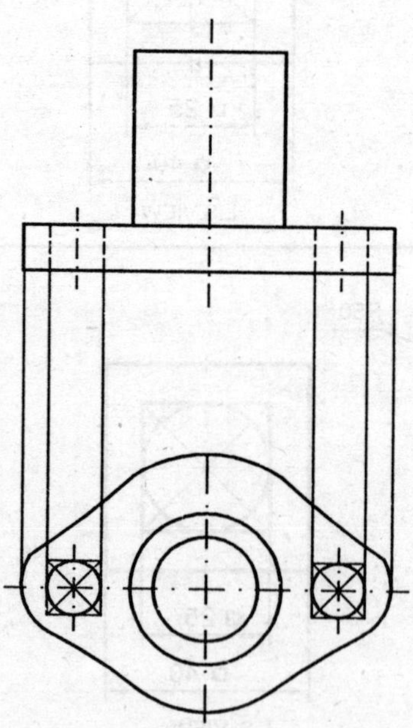

STEP II

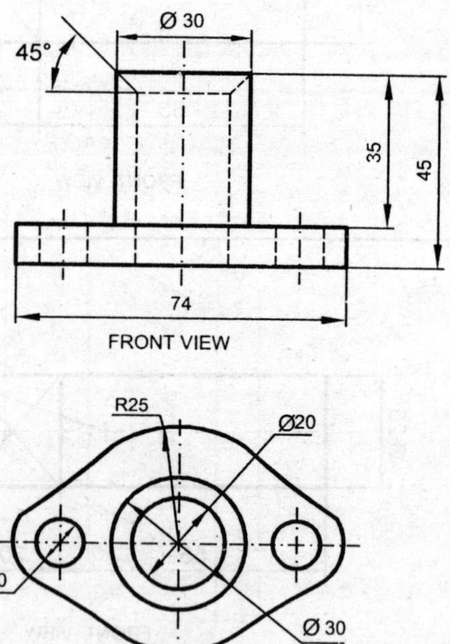

STEP III FIG. 13 (ii) ORHTOGRAPHIC VIEWS

7. TYPICAL PRACTICAL PROBLEMS

Draw the following views of the objects shown in Figs. 14 to 32 :

(a) Front View , (b) Top View and (c) Side View as viewed from the side available for view. Mark the dimensions.

(9) ANGLE PLATE : (PARALLEL SURFACES)

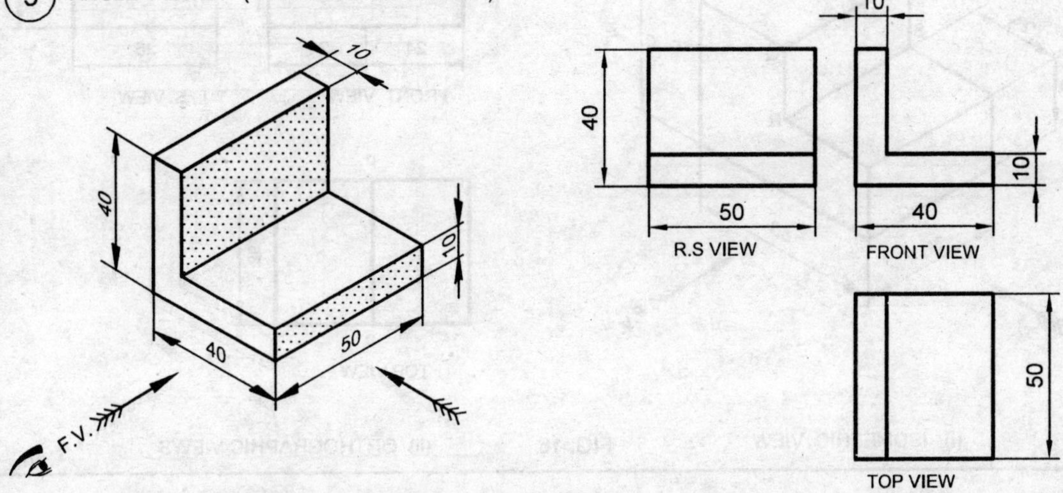

(i) ISOMETRIC VIEW (ii) ORTHOGRAPHIC VIEWS

FIG. 14

(10) CONCRETE SLAB : (PARALLEL SURFACES)

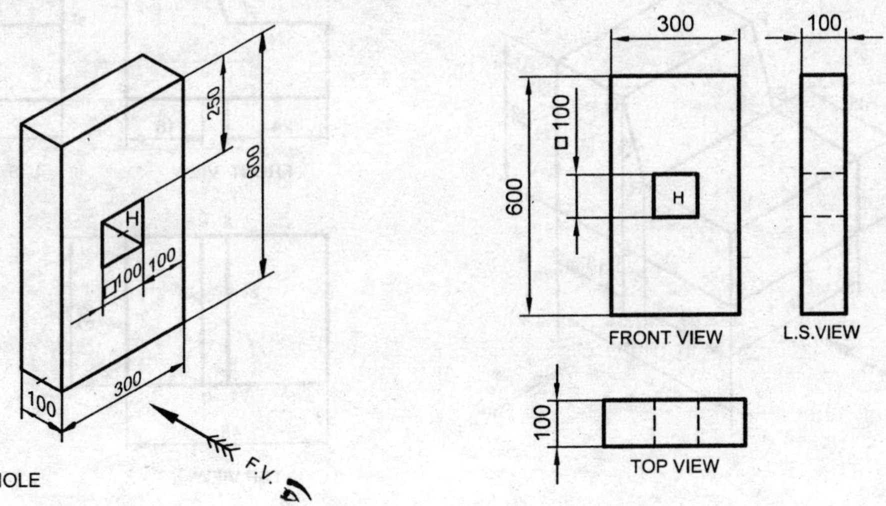

H - THROUGH HOLE

(i) ISOMETRIC VIEW (ii) ORTHOGRAPHIC VIEWS

FIG. 15

(11) L - BLOCK (PARALLEL SURFACES)

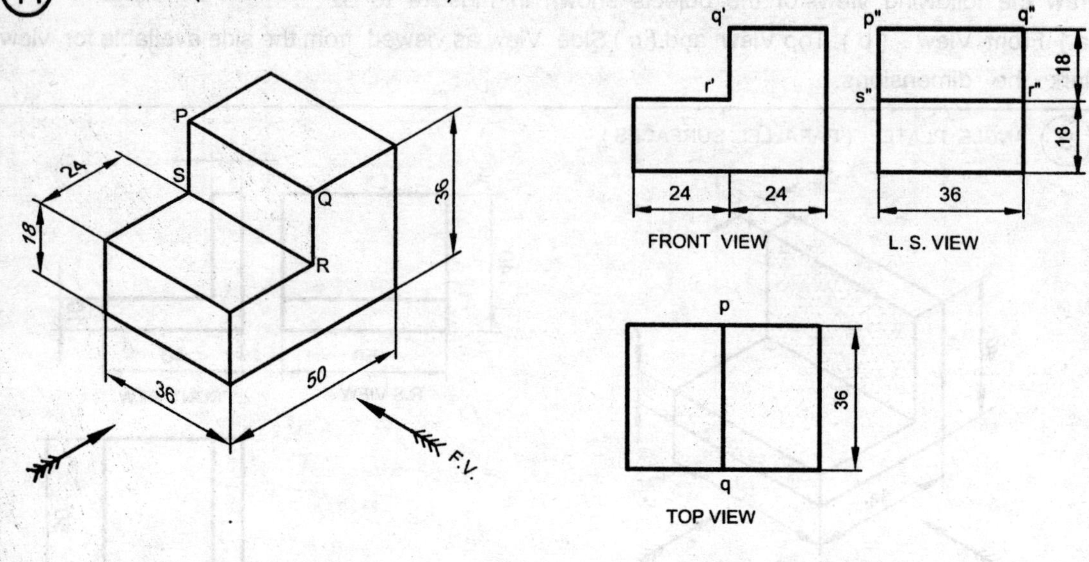

(i) ISOMETRIC VIEW FIG. 16 (ii) ORTHOGRAPHIC VIEWS

(12) TAPER BLOCK (INCLINED SURFACE)

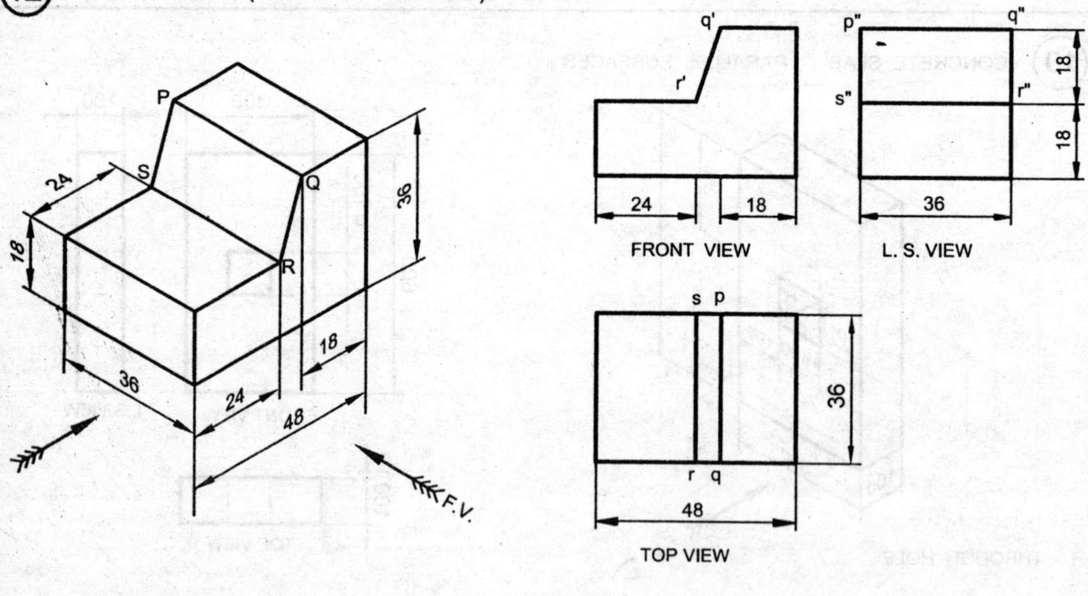

(i) ISOMETRIC VIEW (ii) ORTHOGRAPHIC VIEWS

FIG. 17

(13) RIBBED ANGLE PLATE: (INCLINED SURFACE)

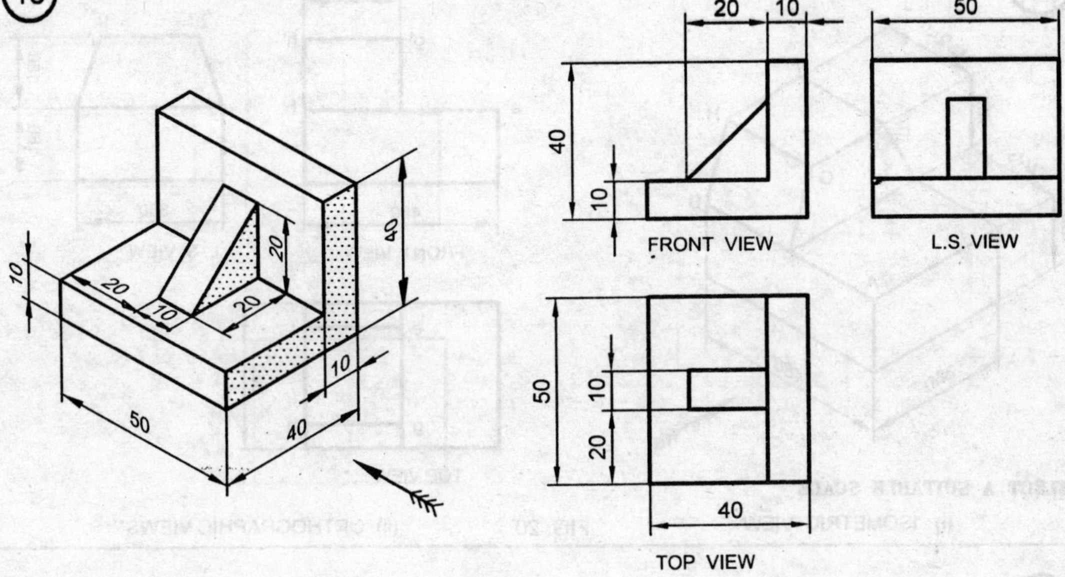

(i) ISOMETRIC VIEW FIG.18 (ii) ORTHOGRAPHIC VIEWS

(14) DOVETAIL BLOCK : (INCLINED SURFACES)

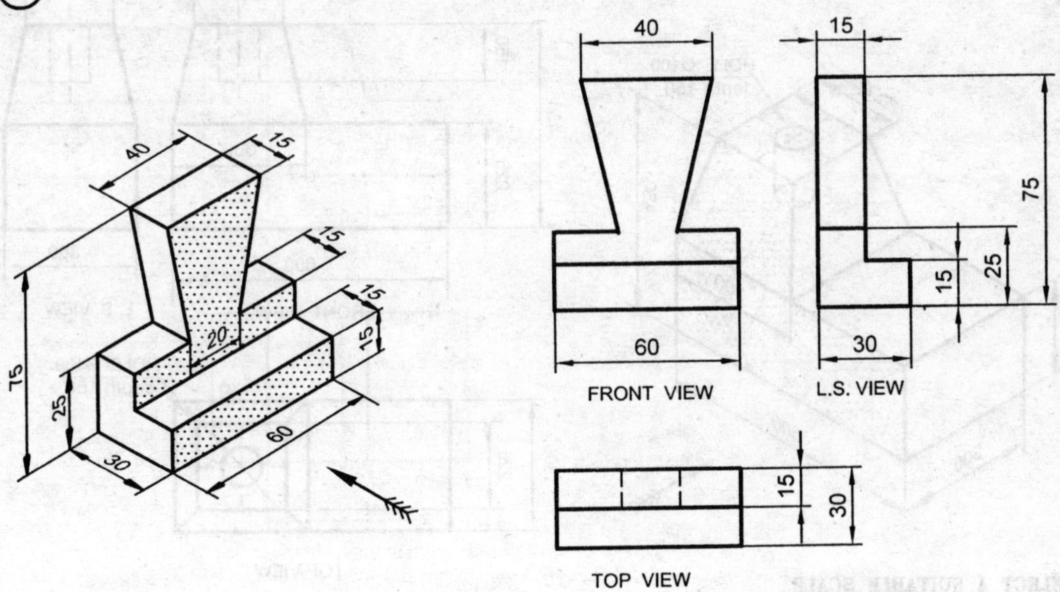

(i) ISOMETRIC VIEW FIG. 19 (ii) ORTHOGRAPHIC VIEWS

15 VICTORY STAND

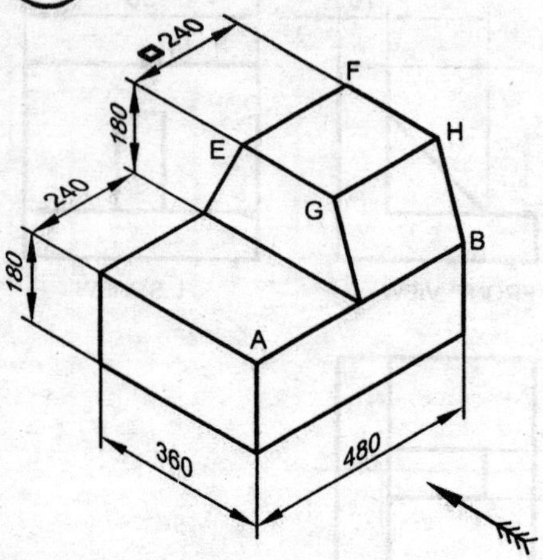

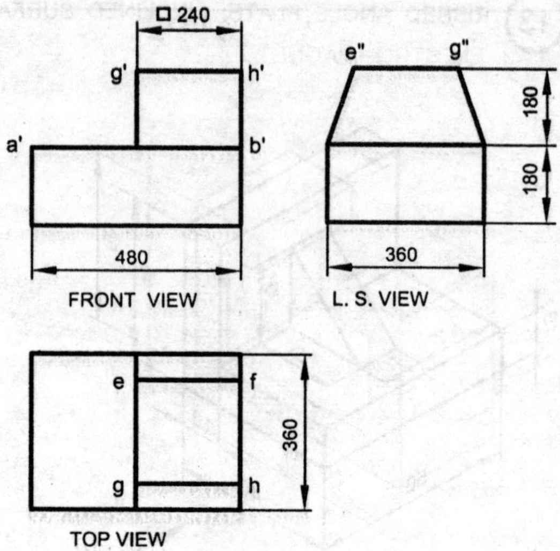

SELECT A SUITABLE SCALE

(i) ISOMETRIC VIEW FIG. 20 (ii) ORTHOGRAPHIC VIEWS

16 MIXIE - BASE (UQ)

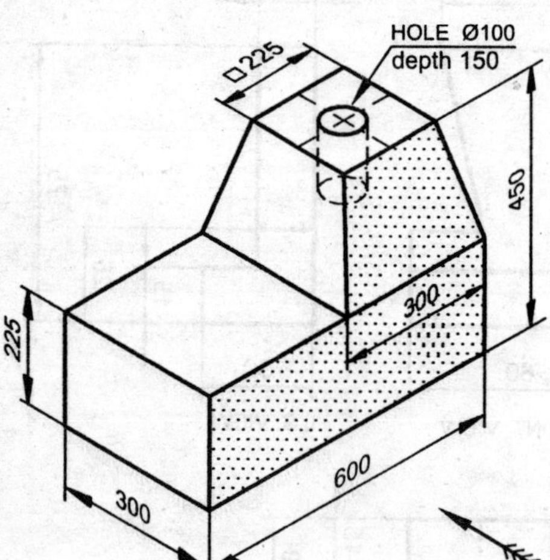

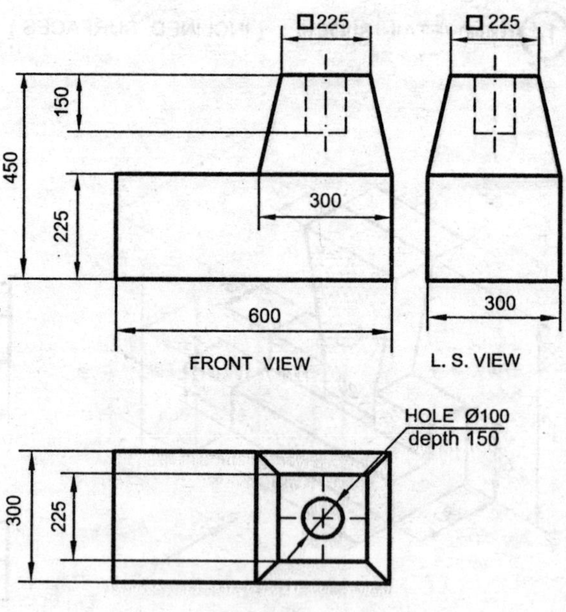

SELECT A SUITABLE SCALE

(i) ISOMETRIC VIEW FIG. 21 (ii) ORTHOGRAPHIC VIEWS

(17) ELECTRO-MAGNET

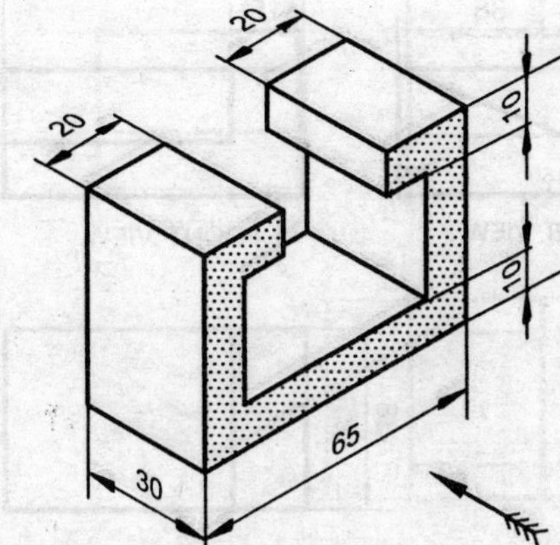

(i) ISOMETRIC VIEW

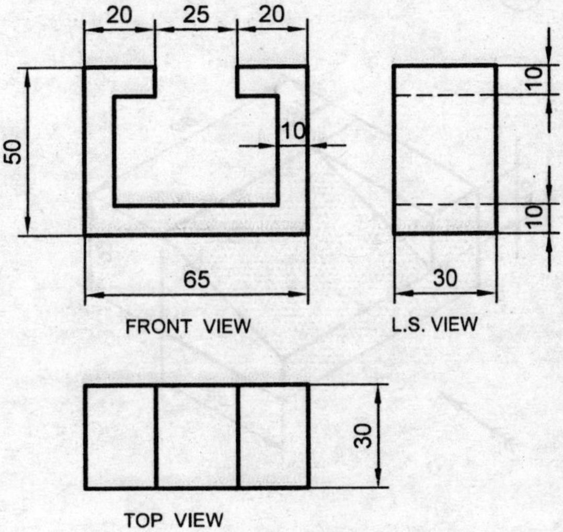

(ii) ORTHOGRAPHIC VIEWS

FIG. 22

(18) CAVITY BLOCK

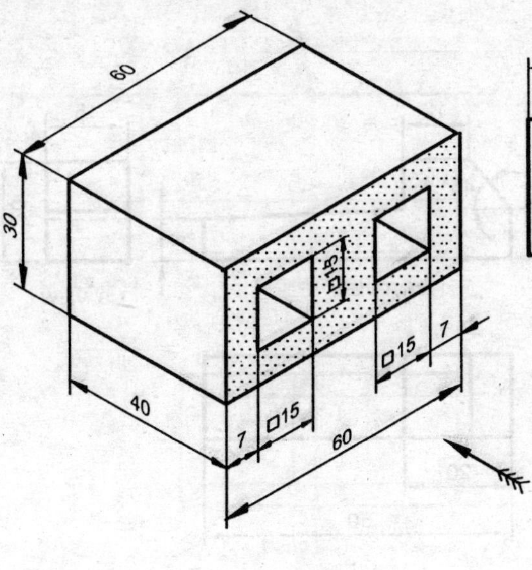

(i) ISOMETRIC VIEW

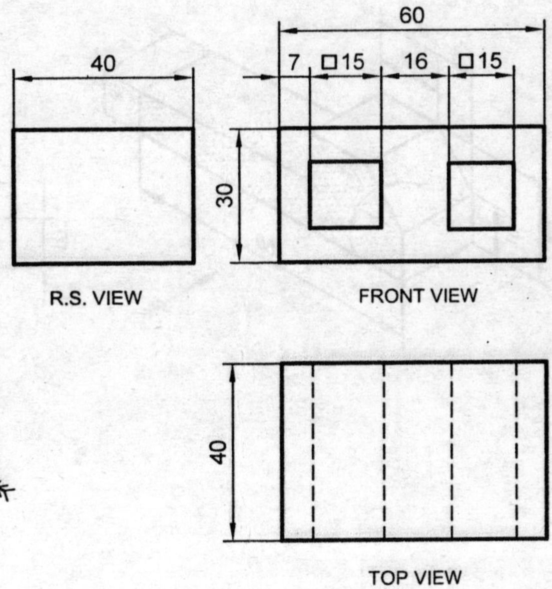

(ii) ORTHOGRAPHIC VIEWS

FIG. 23

(19) STEEL BLOCK

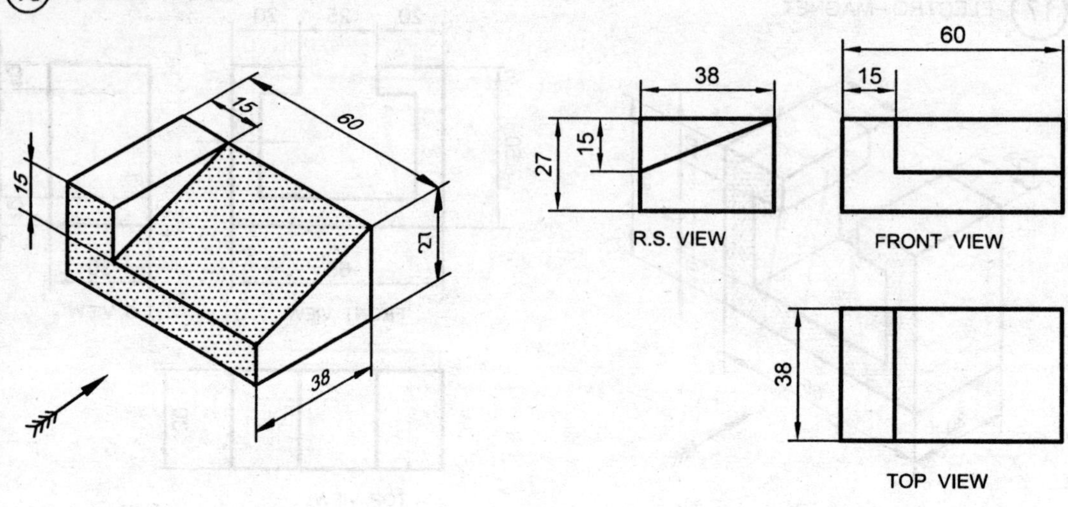

(i) ISOMETRIC VIEW FIG. 24 (ii) ORTHOGRAPHIC VIEWS

(20) GIB - HEADED KEY (UQ)

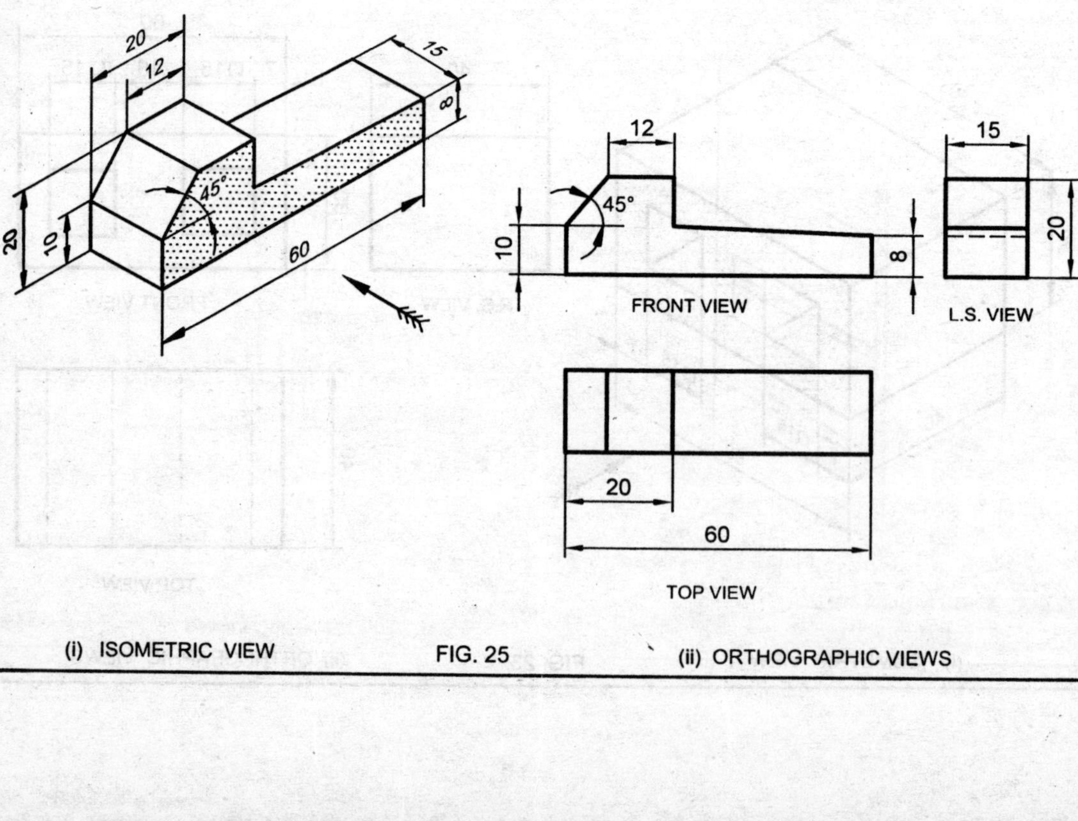

(i) ISOMETRIC VIEW FIG. 25 (ii) ORTHOGRAPHIC VIEWS

(UQ)

21 DOVETAIL BLOCK

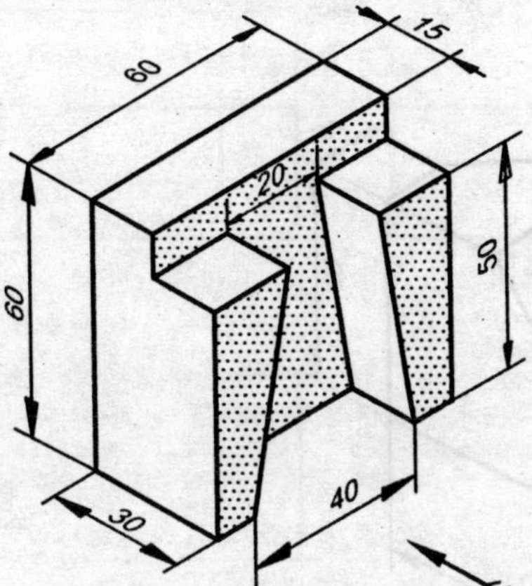

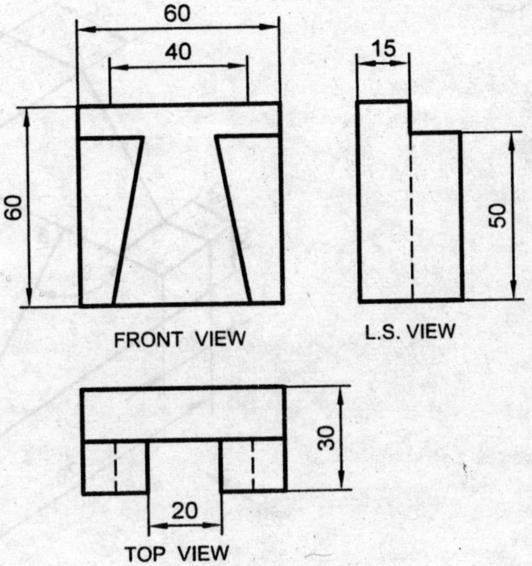

(i) ISOMETRIC VIEW FIG. 26 (ii) ORTHOGRAPHIC VIEWS

(UQ)

22 WOODEN BOOK - RACK

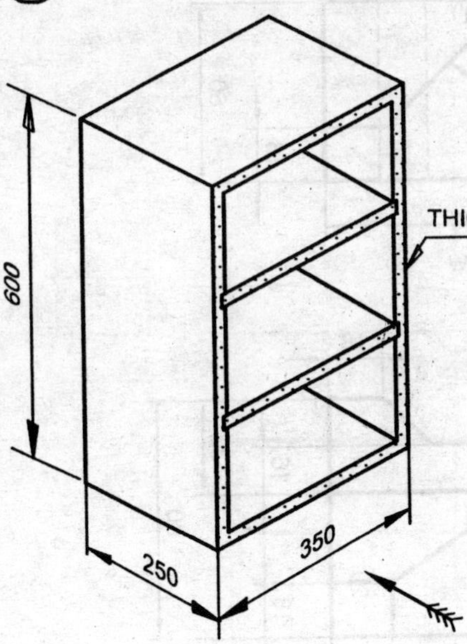

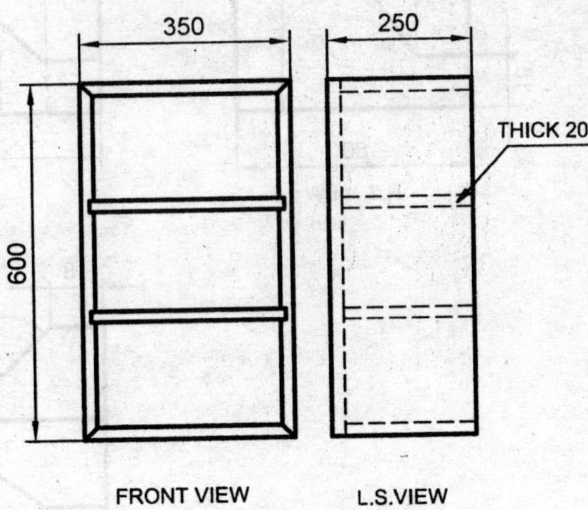

SELECT A SUITABLE SCALE

(i) ISOMETRIC VIEW FIG. 27 (ii) ORTHOGRAPHIC VIEWS

(23) 90° - GUIDE BRACKET

(UQ)

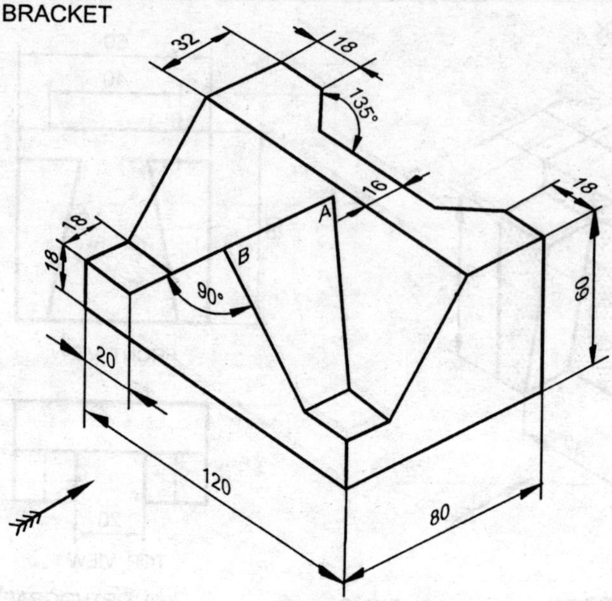

(i) ISOMETRIC VIEW

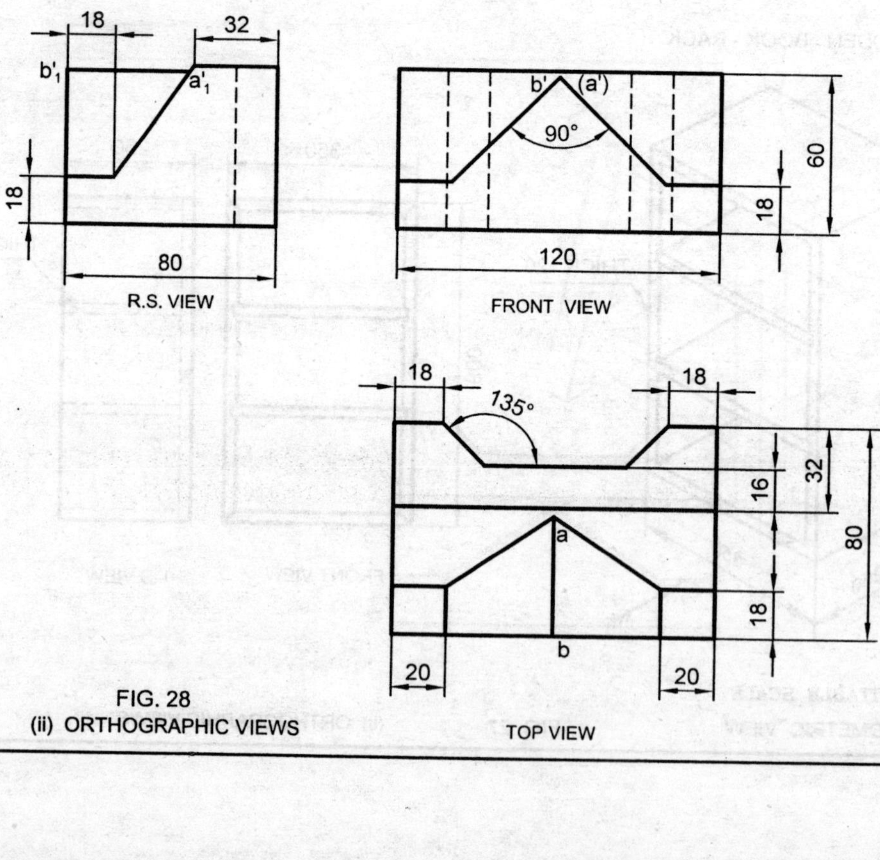

FIG. 28

(ii) ORTHOGRAPHIC VIEWS

24 TRAY (UQ)

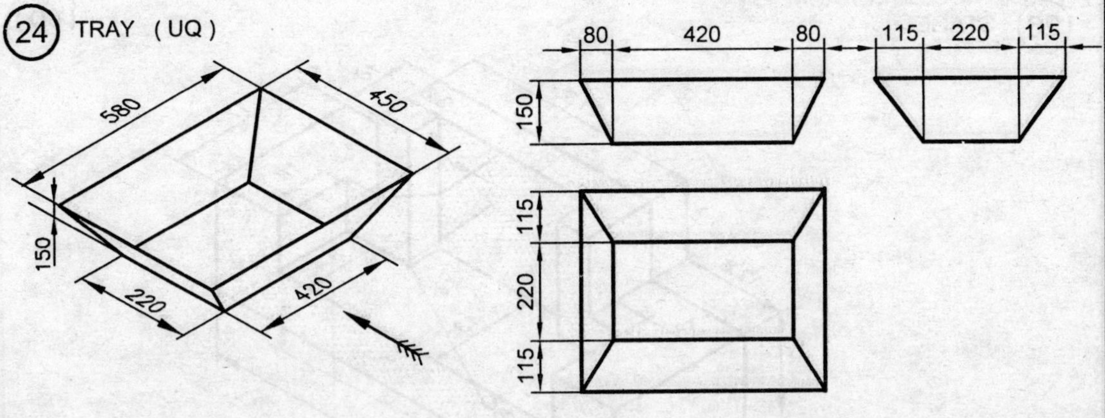

(i) ISOMETRIC VIEW FIG. 29 (ii) ORTHOGRAPHIC VIEWS

25 SQUARE STOOL

(UQ)

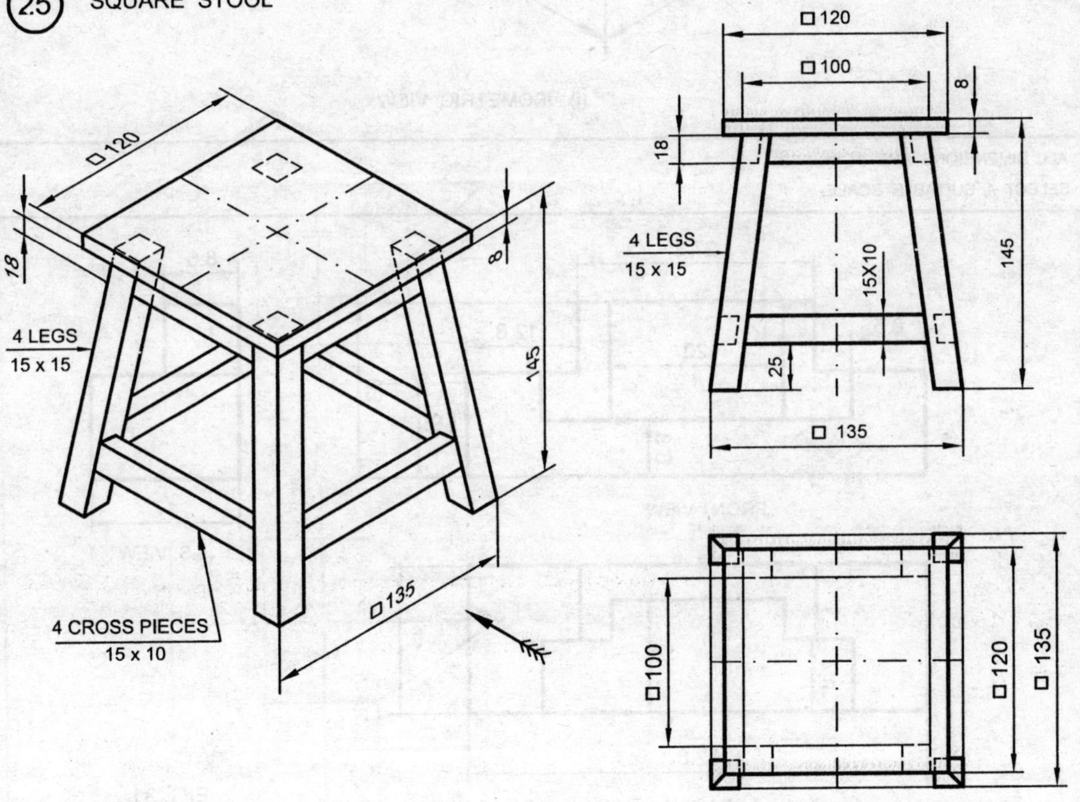

(i) ISOMETRIC VIEW FIG. 30 (ii) ORTHOGRAPHIC VIEWS

(26) STAGE

(UQ)

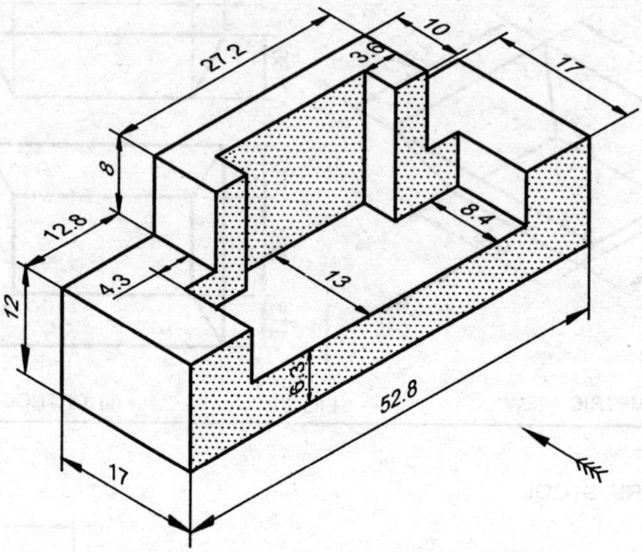

(i) ISOMETRIC VIEW

ALL DIMENSIONS ARE IN m.

SELECT A SUITABLE SCALE.

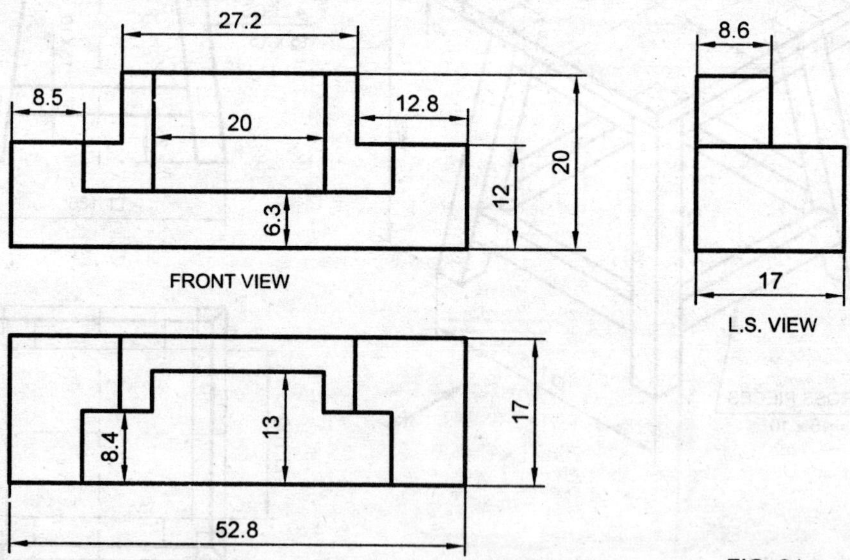

FIG. 31

(ii) ORTHOGRAPHIC VIEWS

(27) PUMP HOUSE

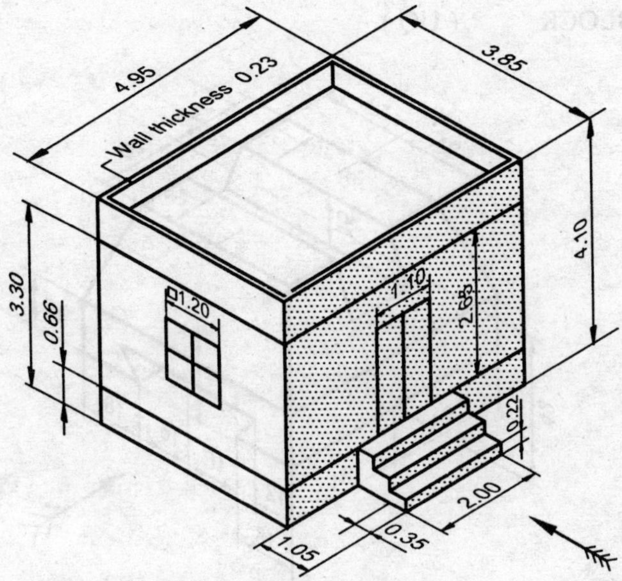

(i) ISOMETRIC VIEW

ALL DIMENSIONS ARE IN m Only. SELECT A SUITABLE SCALE.
DRAW THE TOP VIEW AT THE SILL LEVEL OF THE WINDOW.
MARK THE DIMENSIONS IN THE ORTHOGRAPHIC VIEWS.

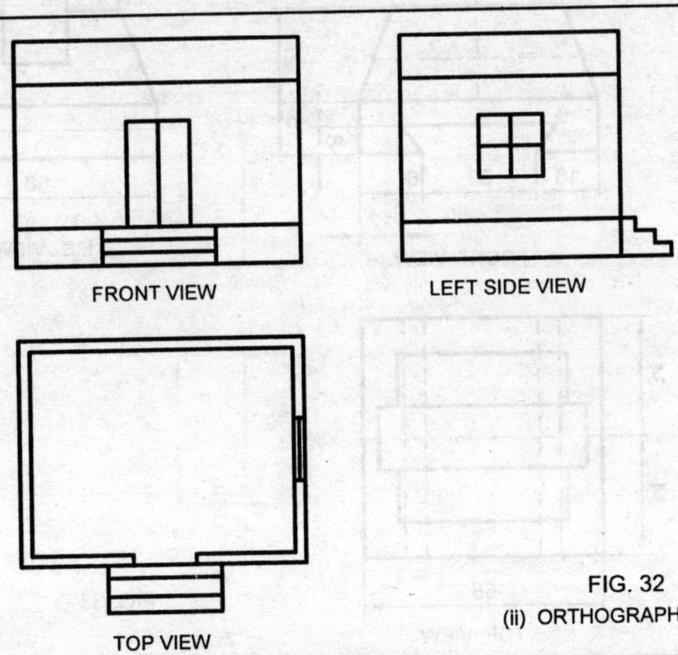

FIG. 32
(ii) ORTHOGRAPHIC VIEWS

(28) TAPER BLOCK (UQ)

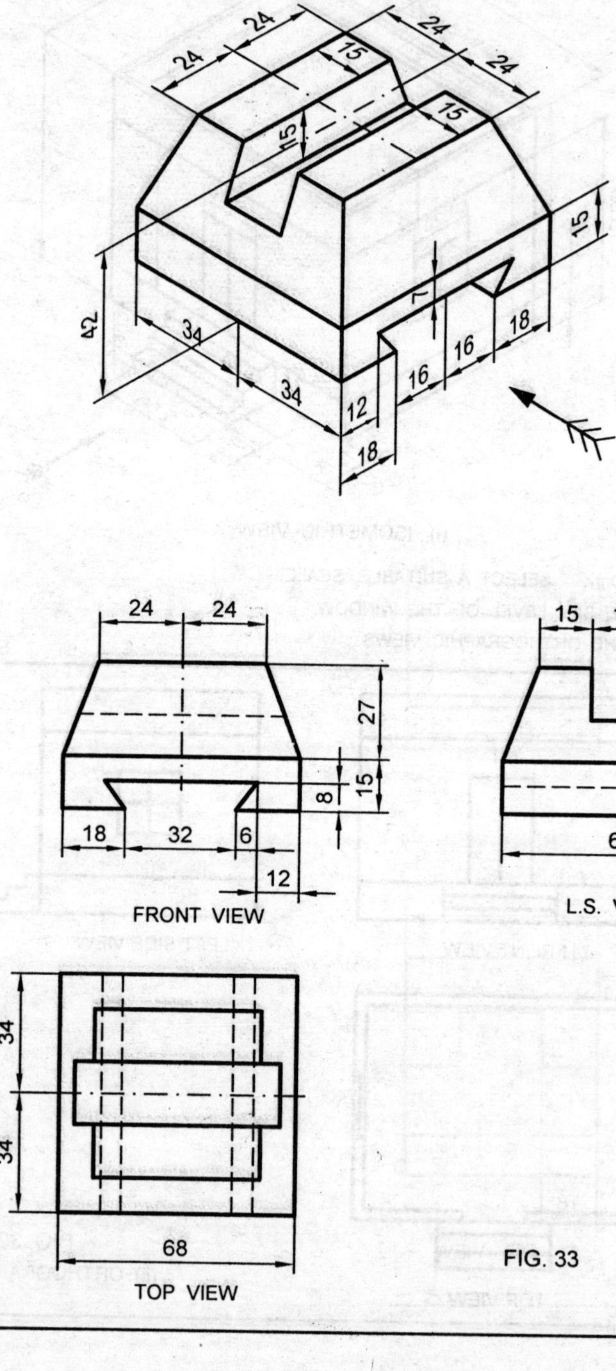

FIG. 33

(29) BASE BLOCK (UQ)

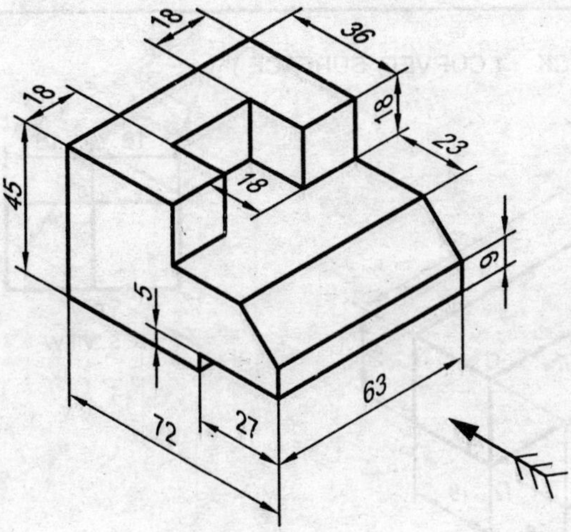

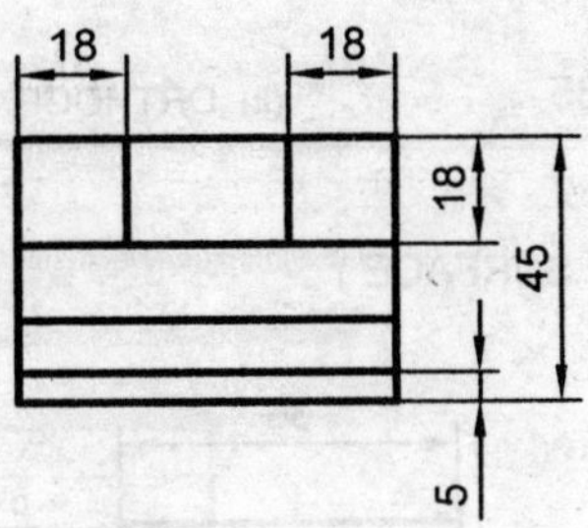

FRONT VIEW

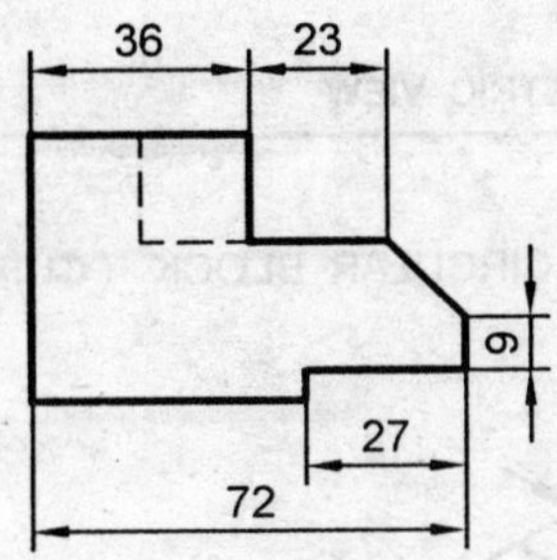

LEFT SIDE VIEW

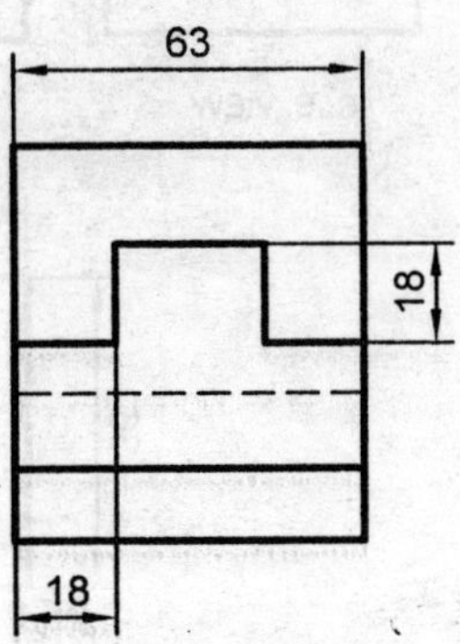

TOP VIEW

FIG. 34

7.1 CURVED OBJECTS

(30) CURVED BLOCK (CURVED SURFACE)

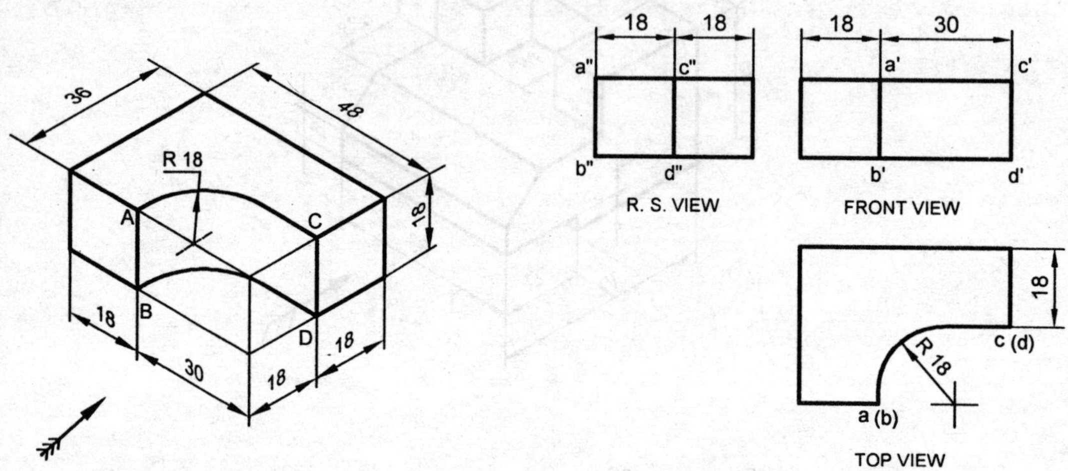

(i) ISOMETRIC VIEW FIG. 35 (ii) ORTHOGRAPHIC VIEWS

(31) SEMI - CIRCULAR BLOCK (CURVED SURFACE)

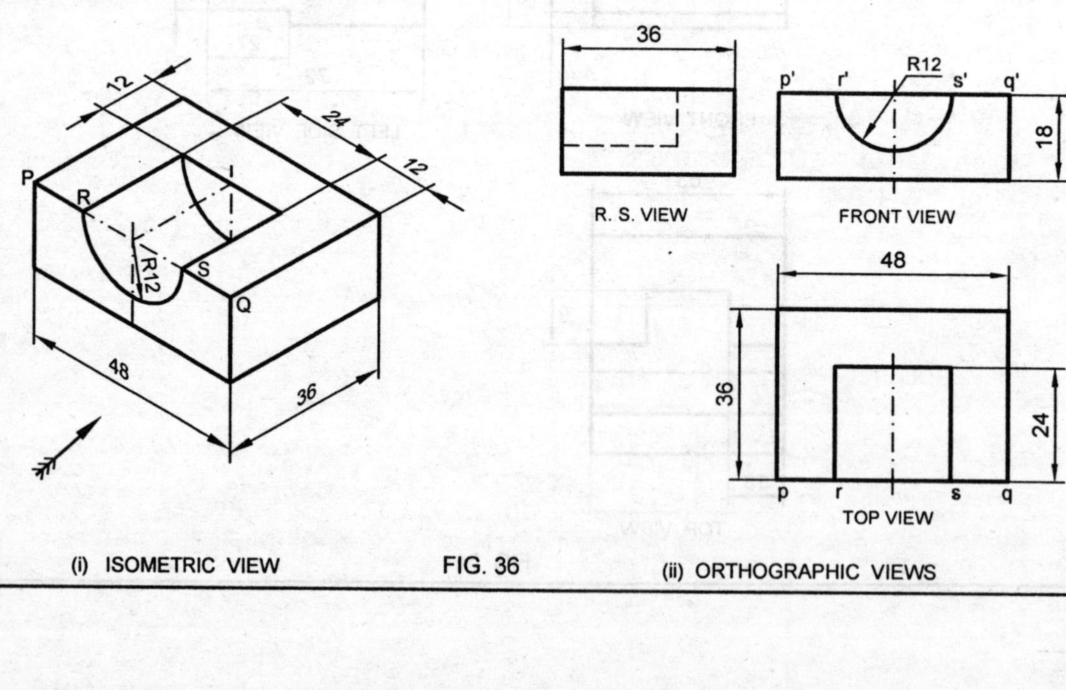

(i) ISOMETRIC VIEW FIG. 36 (ii) ORTHOGRAPHIC VIEWS

32 STOPPER

(UQ)

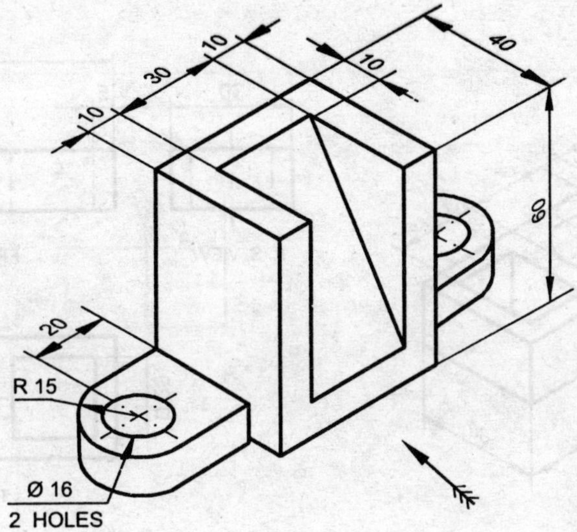

(i) ISOMETRIC VIEW

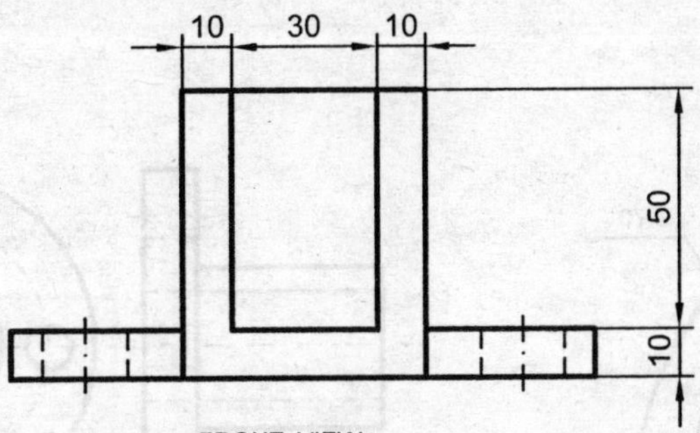

FRONT VIEW

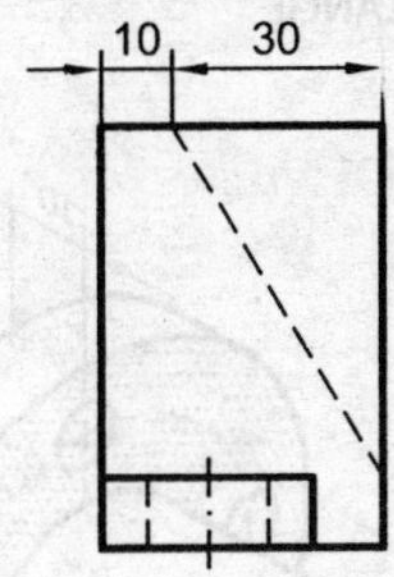

L.S. VIEW

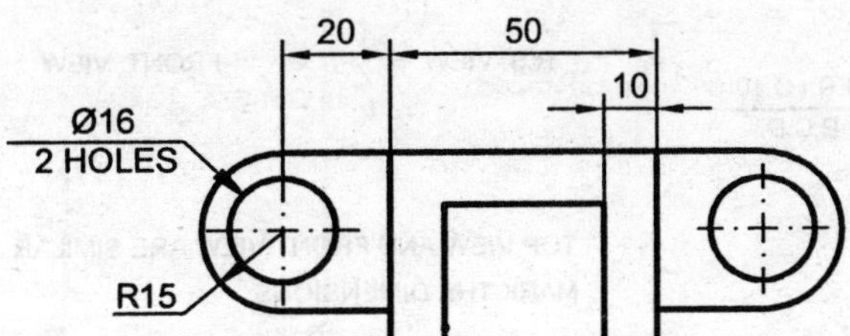

TOP VIEW

FIG. 37

(ii) ORTHOGRAPHIC VIEWS

(33) C.I. CASTING

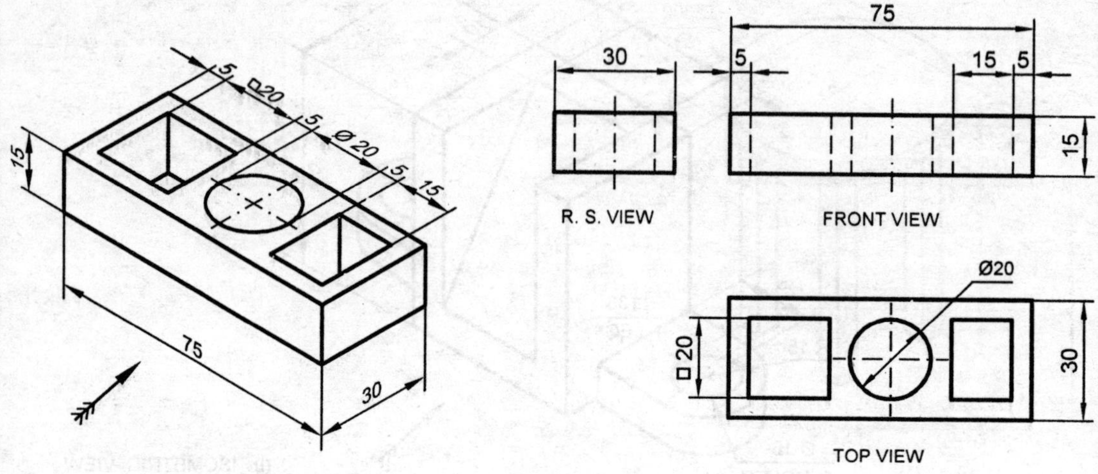

(i) ISOMETRIC VIEW FIG. 38 (ii) ORTHOGRAPHIC VIEWS

(34) FLANGE

(UQ)

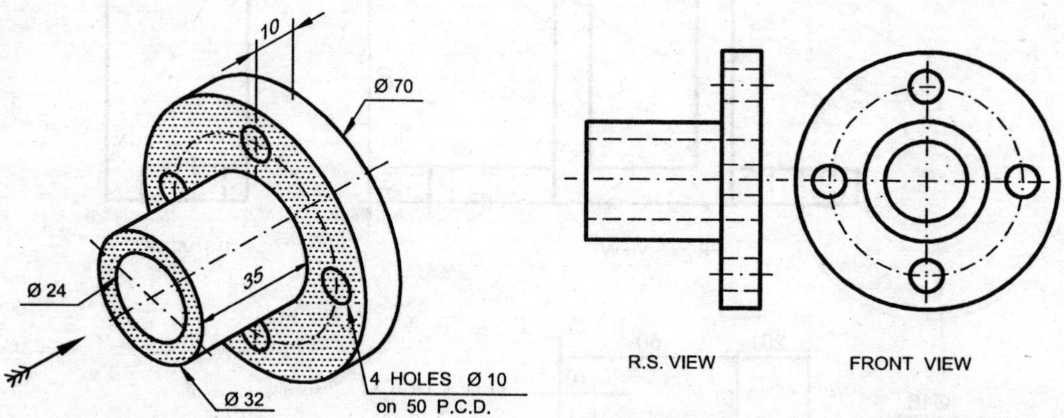

TOP VIEW AND FRONT VIEW ARE SIMILAR.

MARK THE DIMENSIONS.

(i) ISOMETRIC VIEW FIG.39 (ii) ORTHOGRAPHIC VIEWS

35 CIRCULAR STOOL

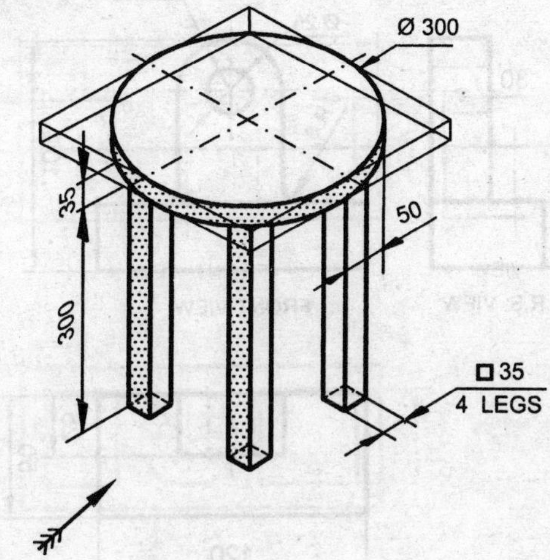

(i) ISOMETRIC VIEW

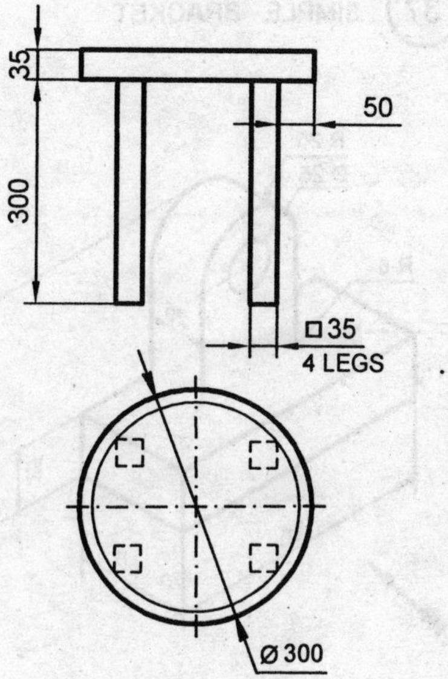

FIG. 40

(ii) ORTHOGRAPHIC VIEWS

36 FORK

(UQ)

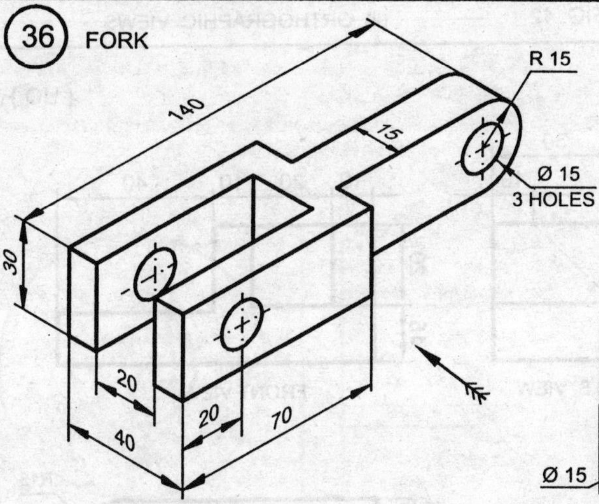

(i) ISOMETRIC VIEW

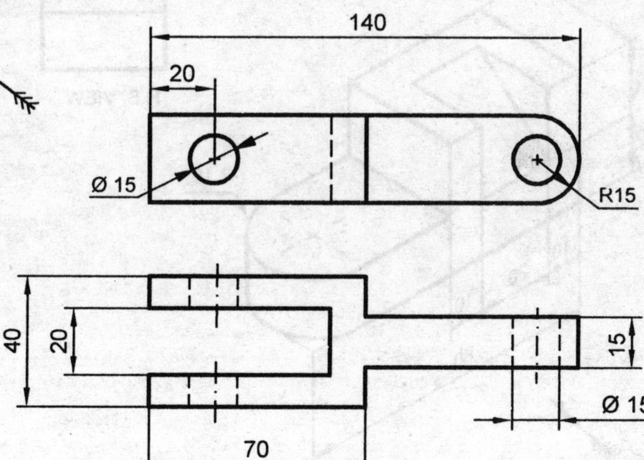

DRAW THE LEFT SIDE VIEW.

FIG. 41

(ii) ORTHOGRAPHIC VIEWS

(37) SIMPLE BRACKET

(UQ)

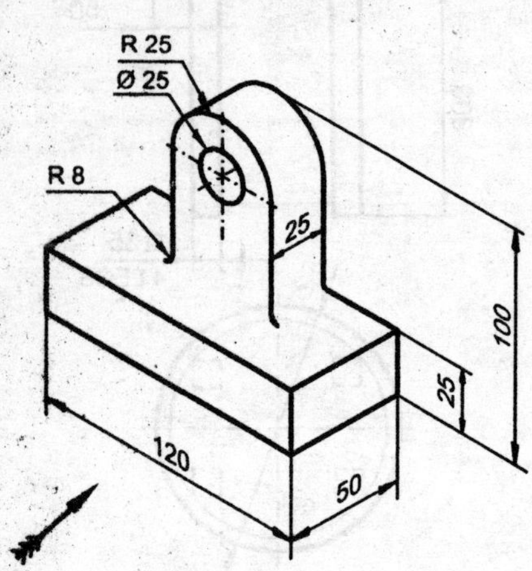

(i) ISOMETRIC VIEW

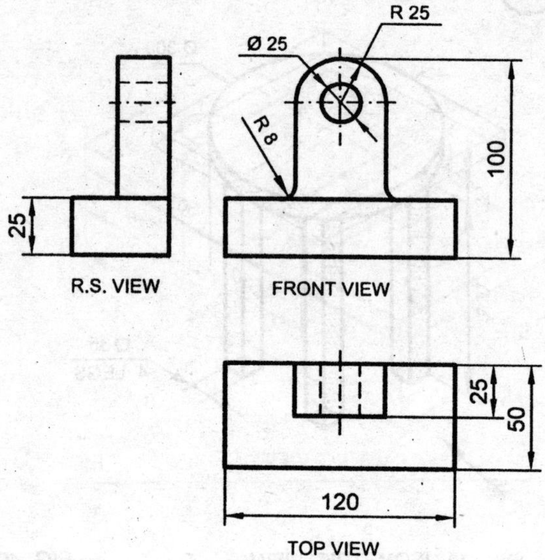

(ii) ORTHOGRAPHIC VIEWS

FIG. 42

(38) CUT BLOCK

(UQ)

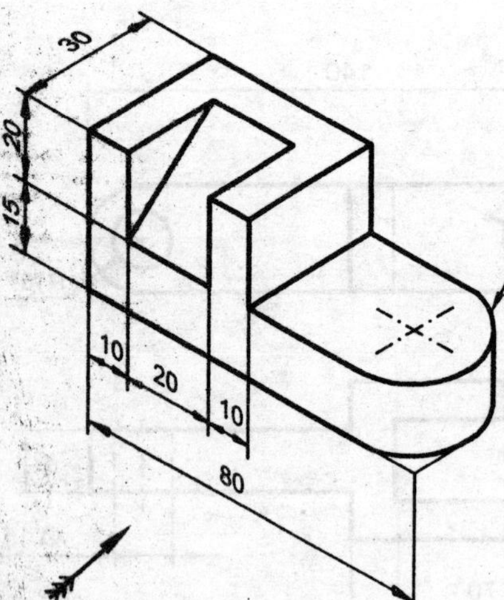

(i) ISOMETRIC VIEW

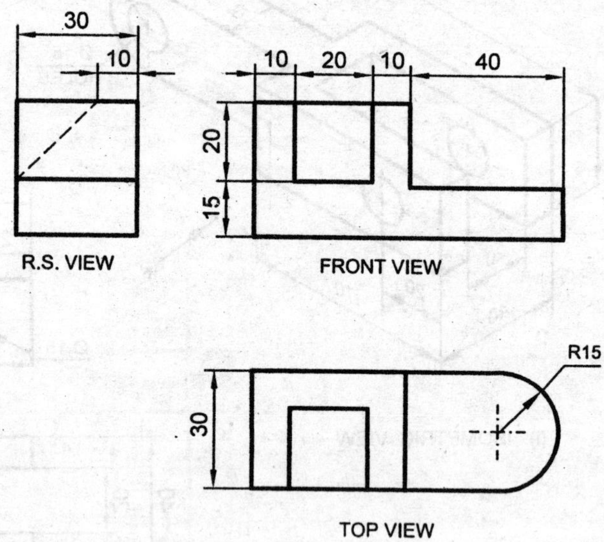

(ii) ORTHOGRAPHIC VIEWS

FIG. 43

39 OPEN BEARING (UQ)

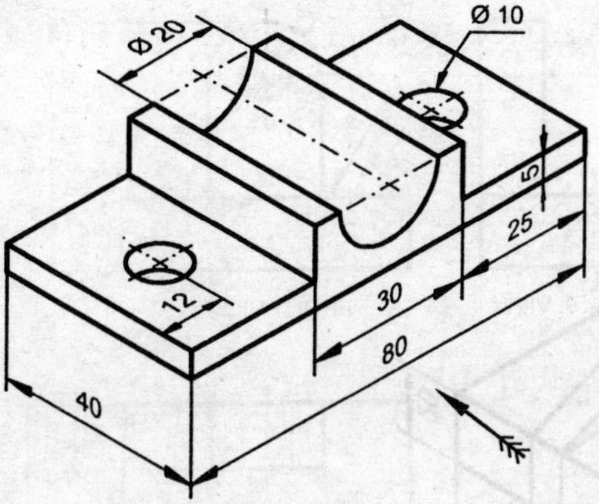

(i) ISOMETRIC VIEW

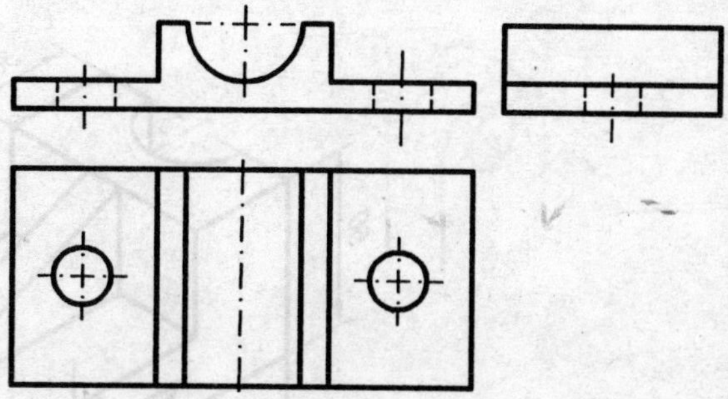

FIG. 44

(ii) ORTHOGRAPHIC VIEWS

40 JIG PLATE (UQ)

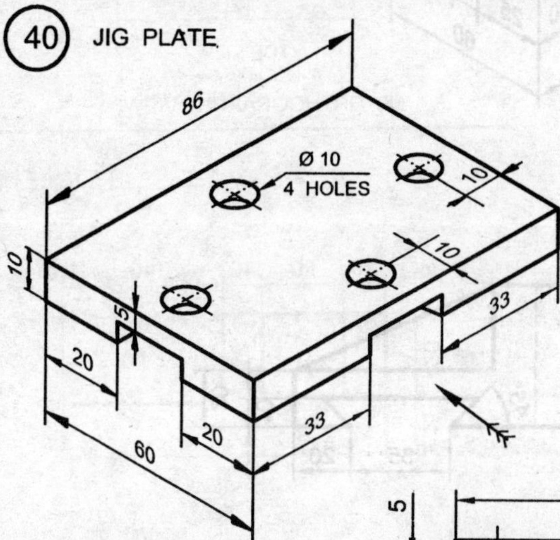

(i) ISOMETRIC VIEW

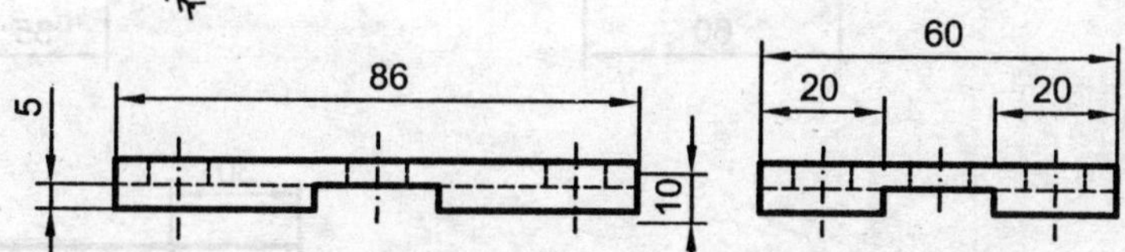

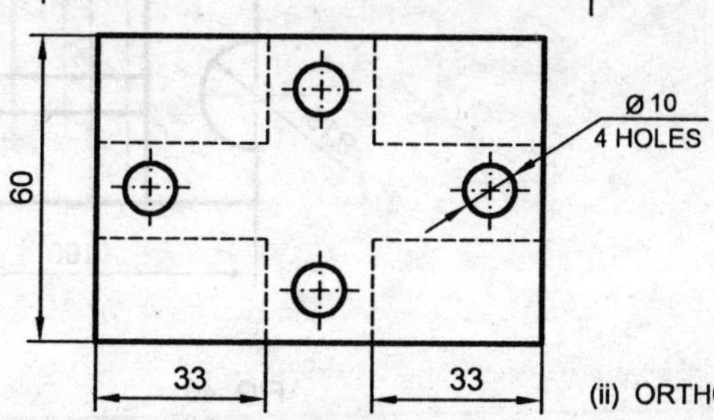

FIG. 45

(ii) ORTHOGRAPHIC VIEWS

(41) GUIDE BLOCK (UQ)

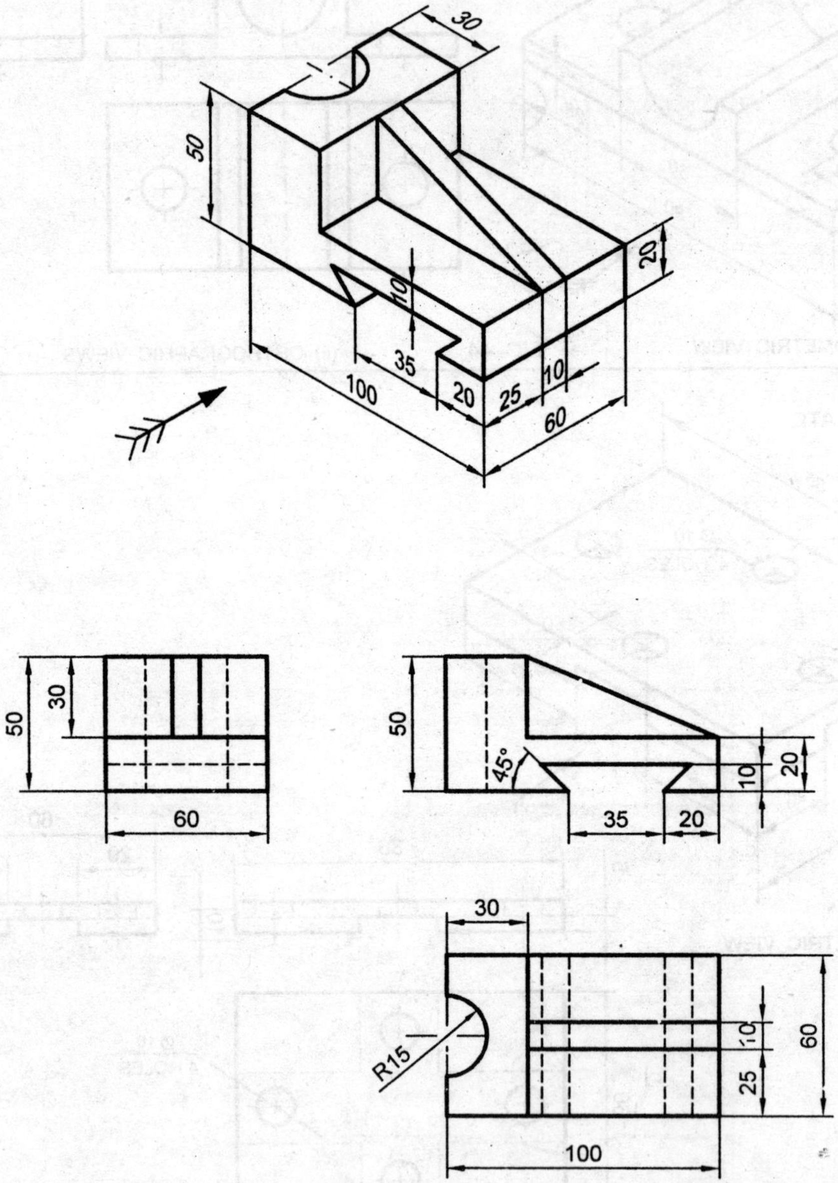

FIG. 46

(42) CYLINDRICAL BLOCK (UQ)

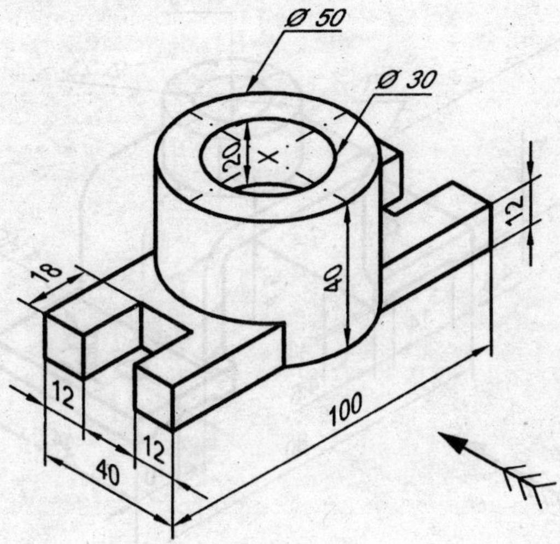

(i)

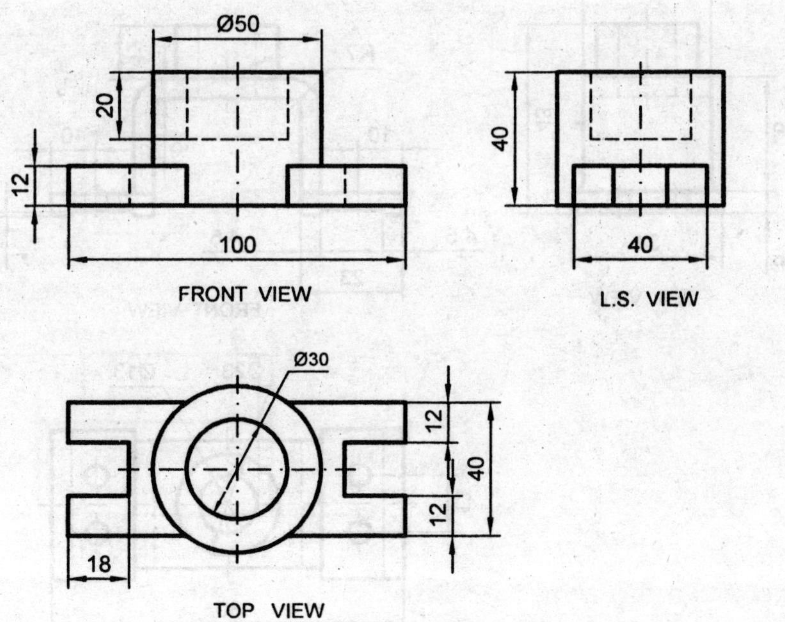

FIG. 47 (ii)

(43) C - FRAME (UQ)

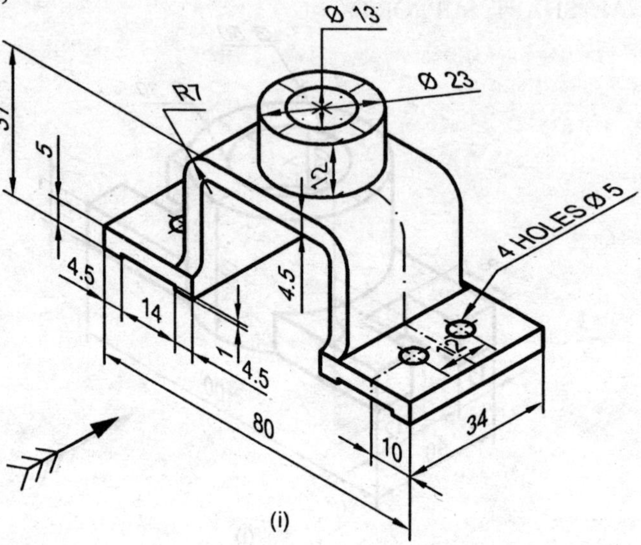

(i)

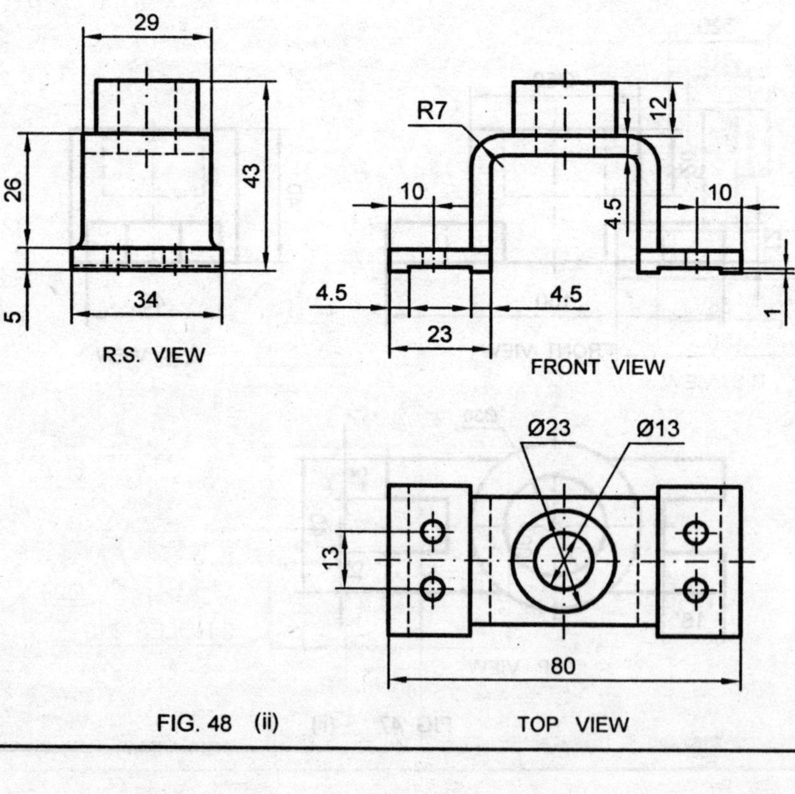

FIG. 48 (ii)

(44) VERTICAL SHAFT SUPPORT (UQ)

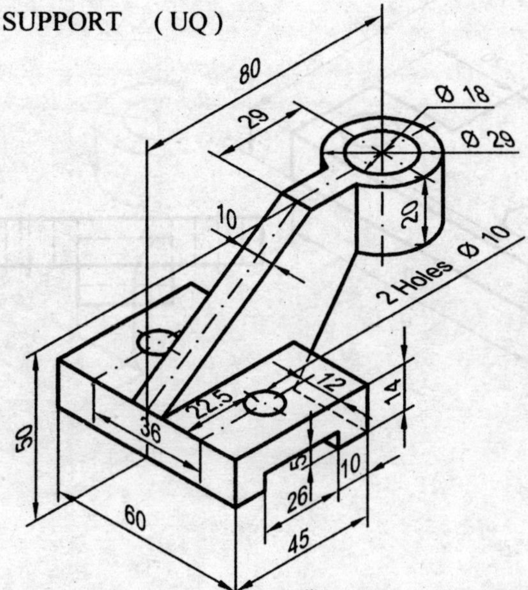

(i)

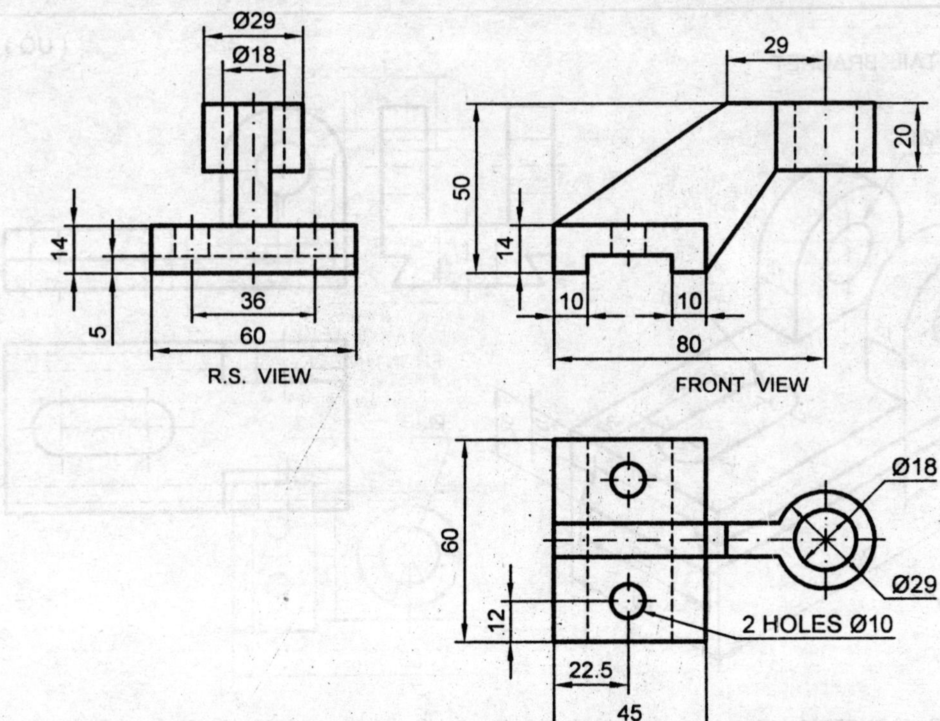

FIG. 49 (ii) TOP VIEW

(UQ)

45 HINGE

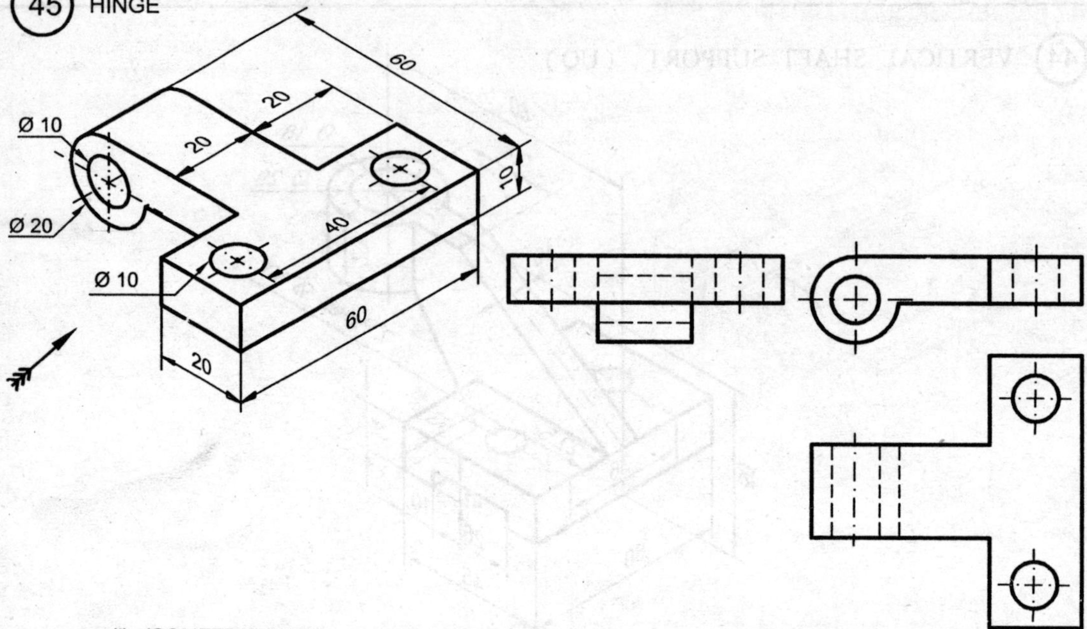

(i) ISOMETRIC VIEW FIG. 50 (ii) ORTHOGRAPHIC VIEWS

(UQ)

46 DOVETAIL BRACKET

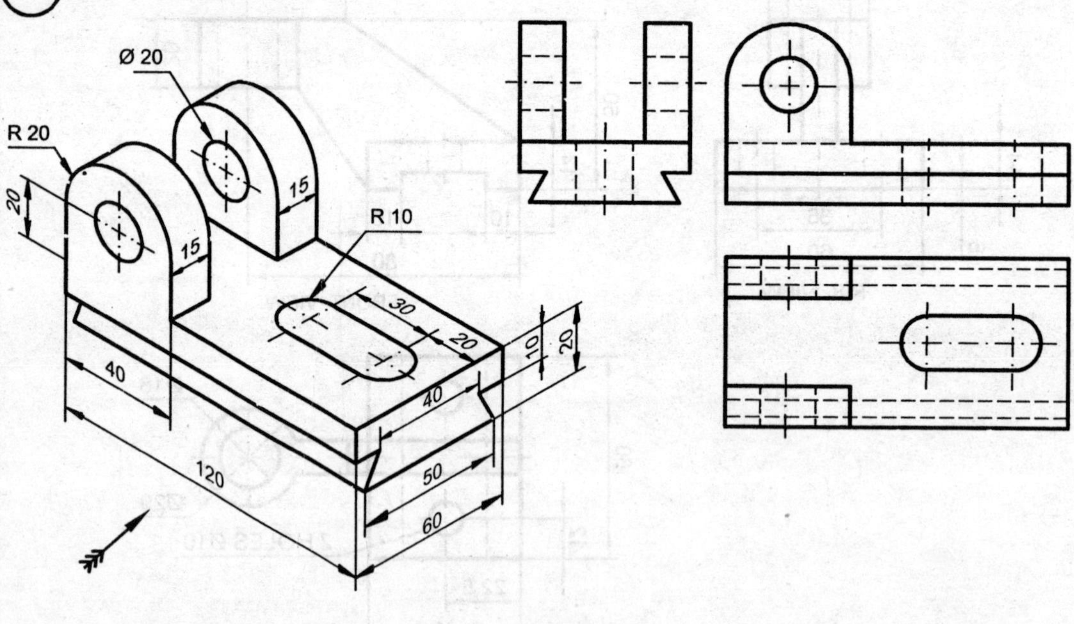

(i) ISOMETRIC VIEW FIG. 51 (ii) ORTHOGRAPHIC VIEWS

(UQ)

•47 V - BLOCK WITH CIRCULAR ROD

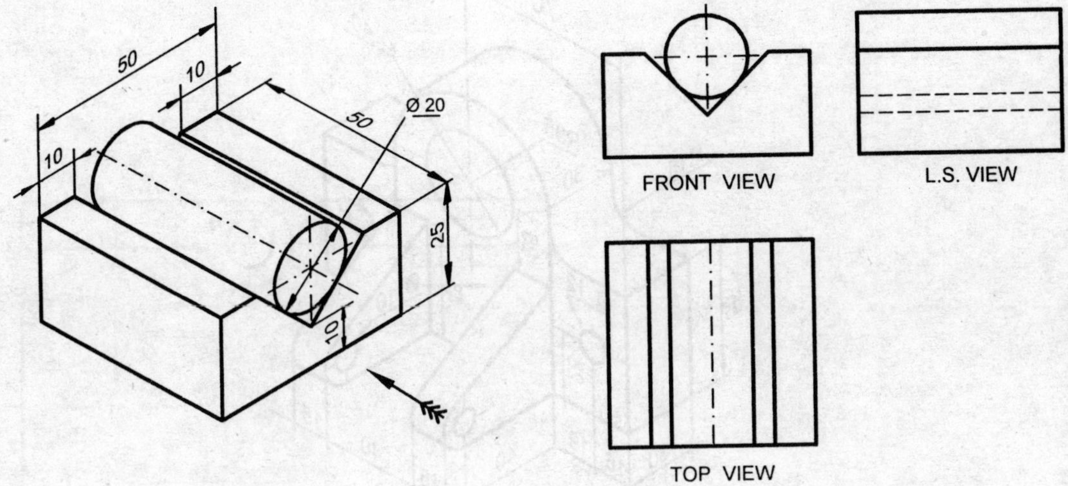

(i) ISOMETRIC VIEW FIG. 52 (ii) ORTHOGRAPHIC VIEWS

(UQ)

48 BEARING BLOCK

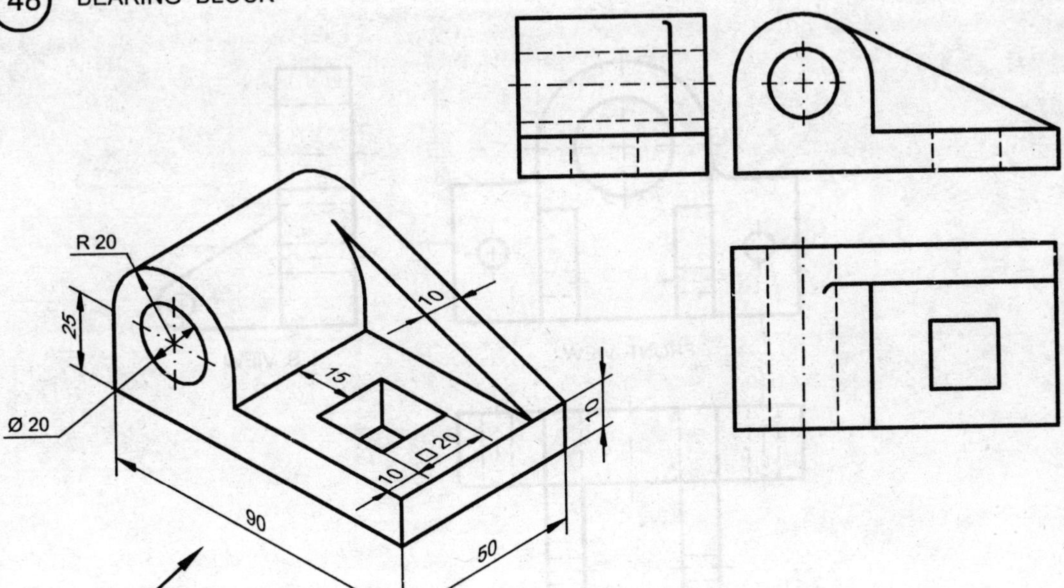

(i) ISOMETRIC VIEW FIG. 53 (ii) ORTHOGRAPHIC VIEWS

(49) ADJUSTOR ARM (UQ)

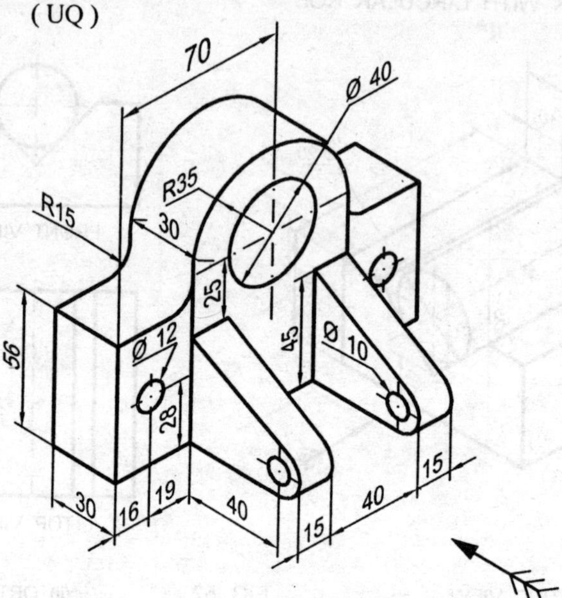

(i)

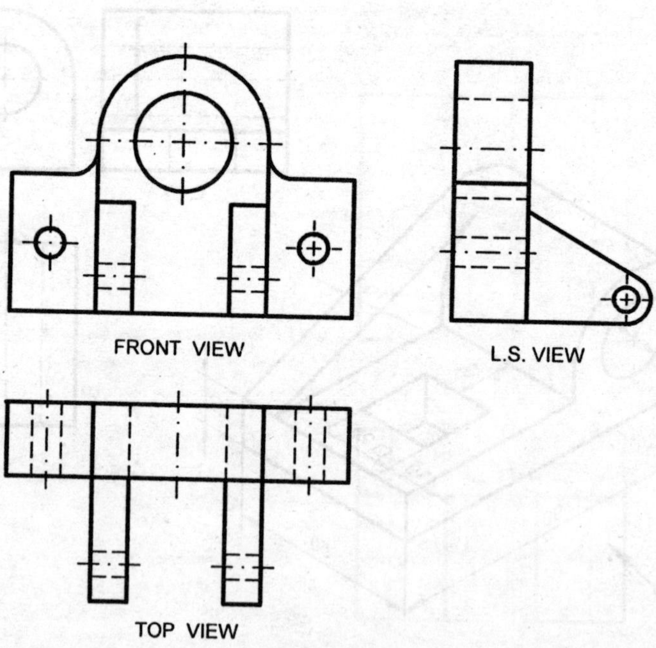

FIG. 54 (ii)

(50) CORNER STOP (UQ)

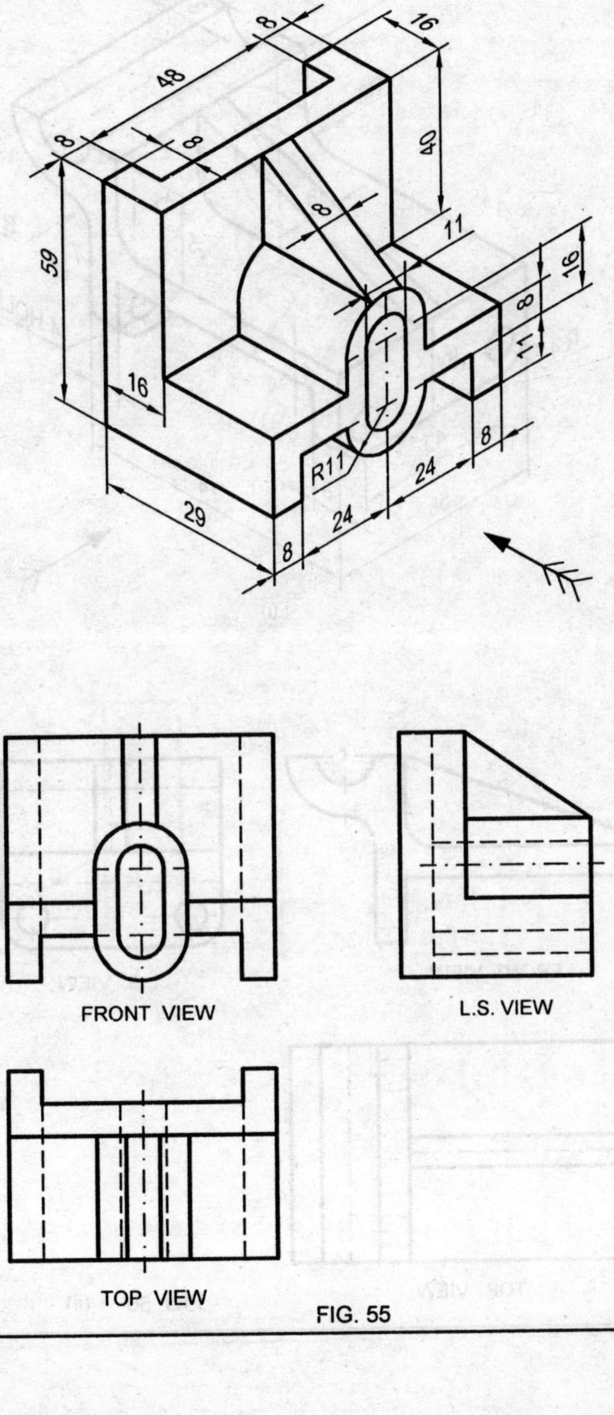

FIG. 55

51 ROD RUPPORT (UQ)

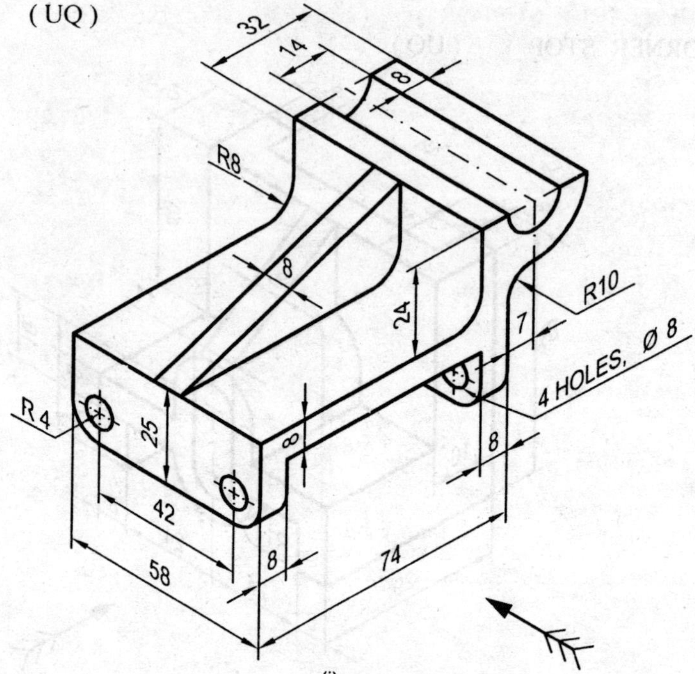

(i)

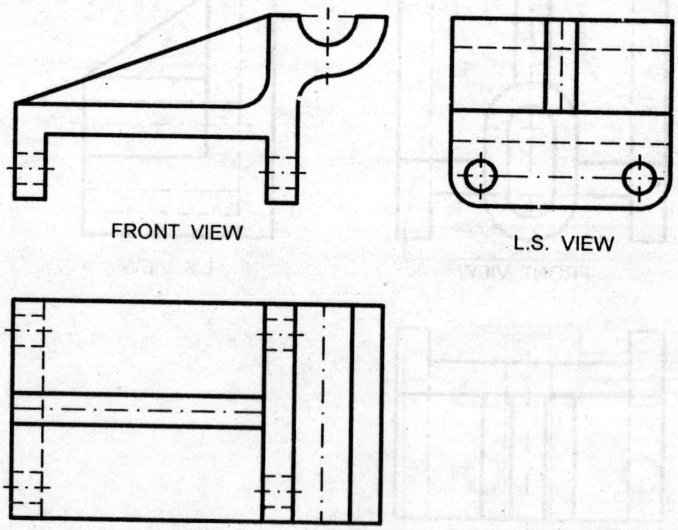

FIG. 56 (ii)

(52) ADJUSTABLE SUPPORT (UQ)

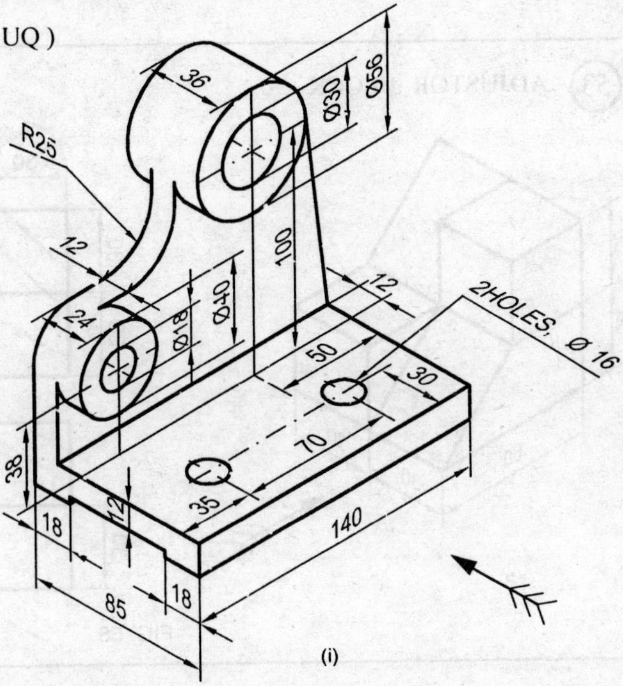

(i)

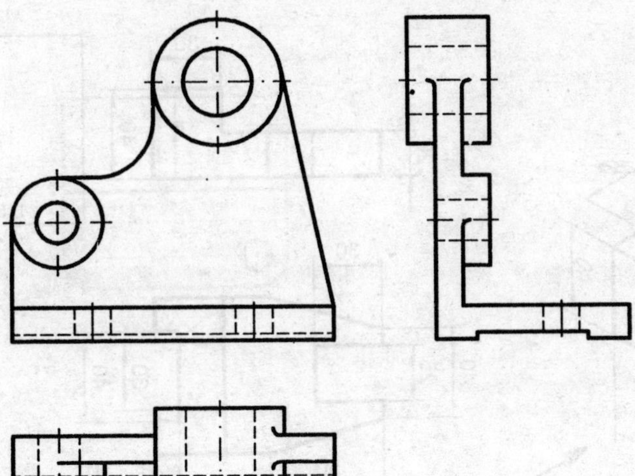

FIG. 57 (ii)

7.2 MISSING VIEWS

(53) ADJUSTOR BLOCK

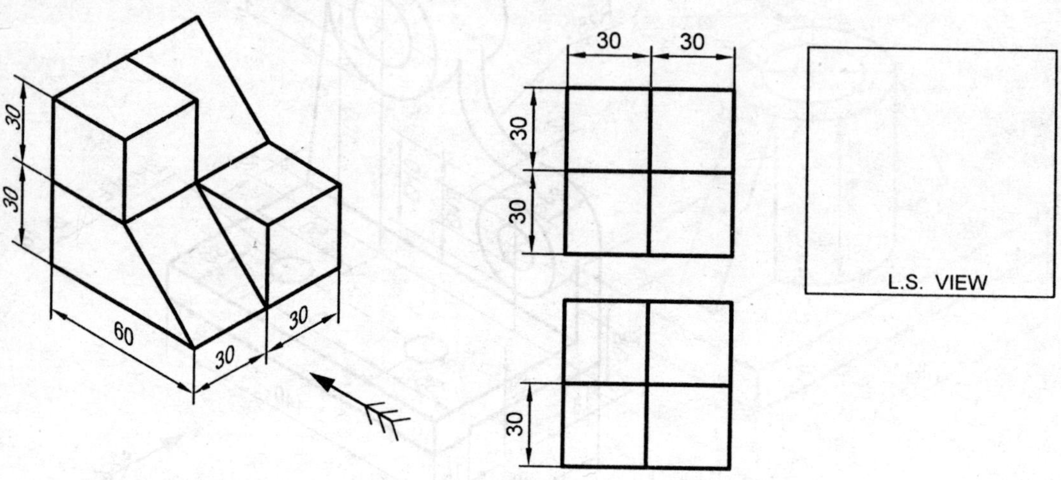

FIG. 58

(54) SLOTTED BLOCK (UQ)

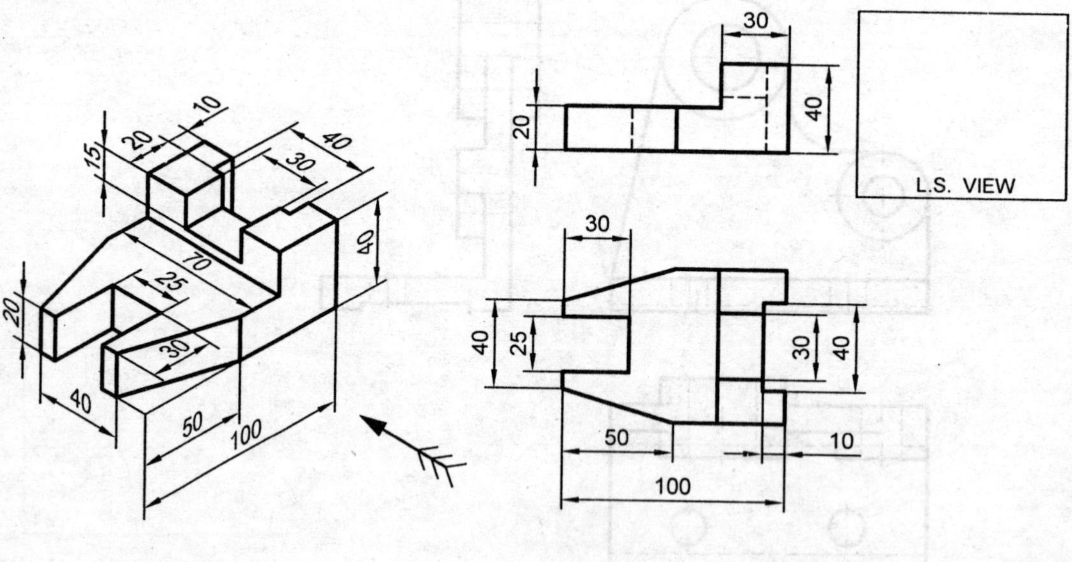

FIG. 59

55 END BRACKET

(UQ)

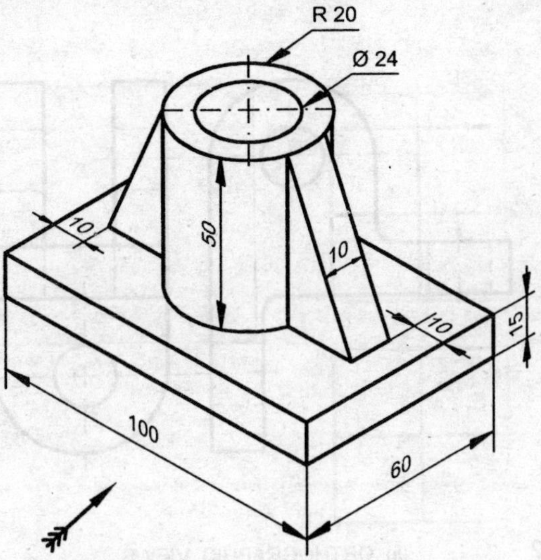

(i) ISOMETRIC VIEW

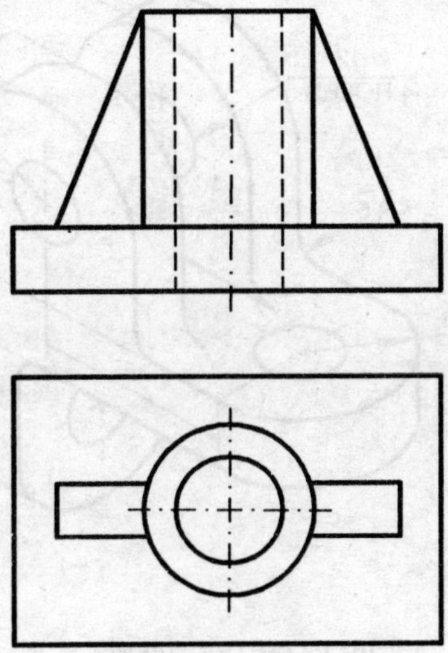

(ii) ORTHOGRAPHIC VIEWS

FIG. 60

56 MACHINE CLAMP

(UQ)

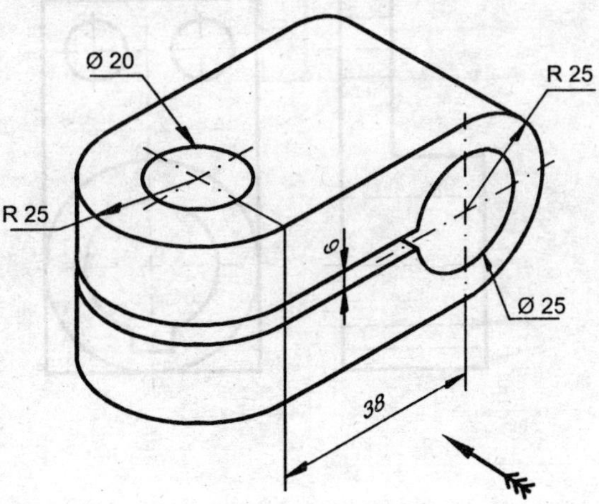

(i) ISOMETRIC VIEW

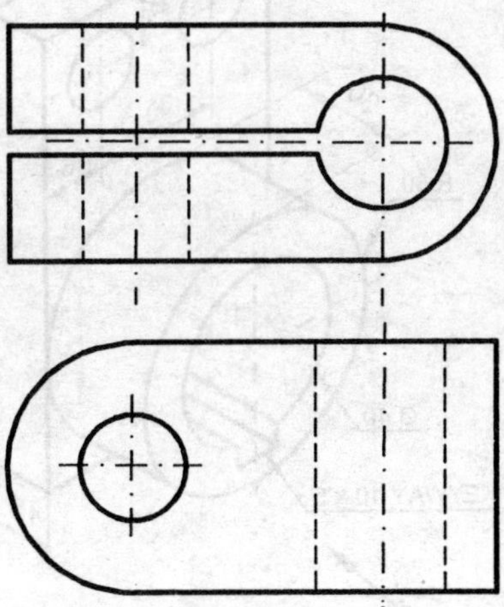

(ii) ORTHOGRAPHIC VIEWS

FIG. 61

57 BRACKET

(UQ)

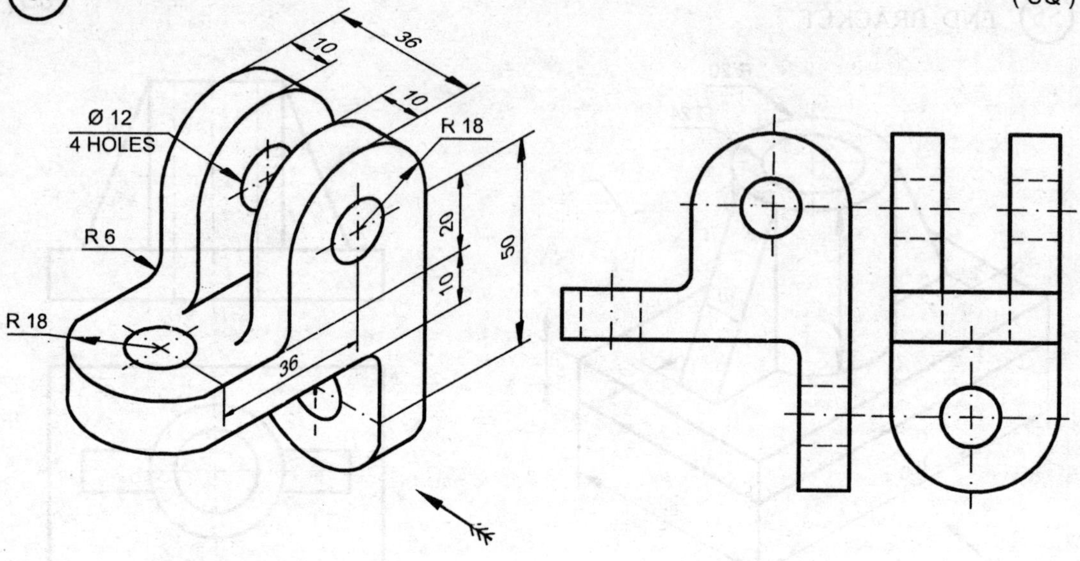

(i) ISOMETRIC VIEW

FIG. 62

(ii) ORTHOGRAPHIC VIEWS

58 HORIZONTAL SHAFT SUPPORT

(UQ)

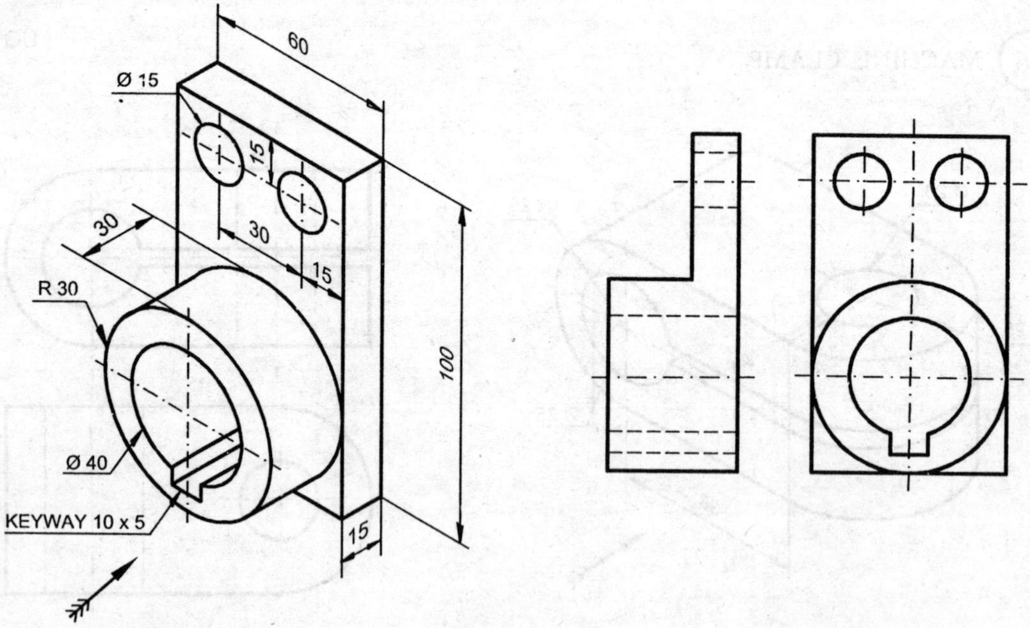

(i) ISOMETRIC VIEW

FIG. 63

(ii) ORTHOGRAPHIC VIEWS

59 INK BOTTLE

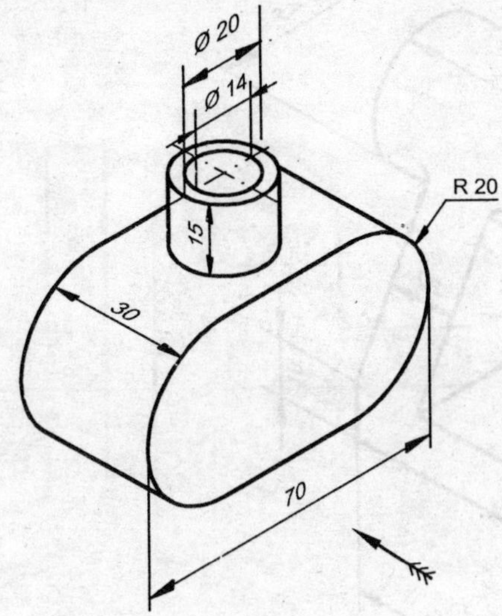

(i) ISOMETRIC VIEW

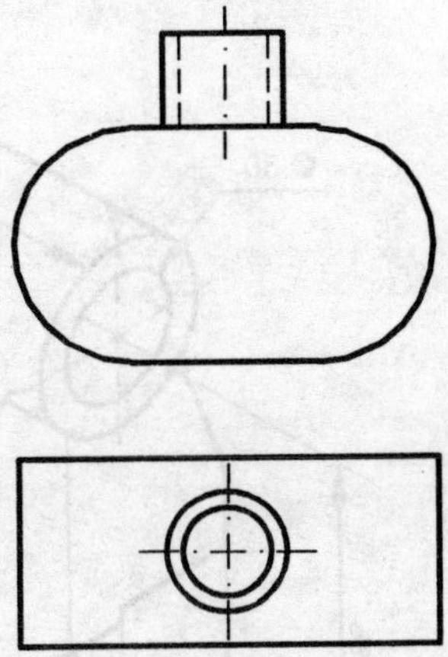

(ii) ORTHOGRAPHIC VIEWS

FIG. 64

(UQ)

60 SWAGE BLOCK

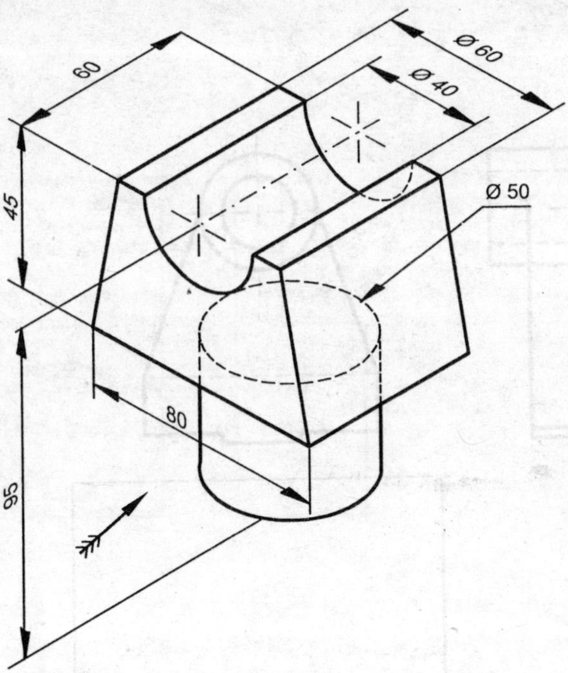

(i) ISOMETRIC VIEW

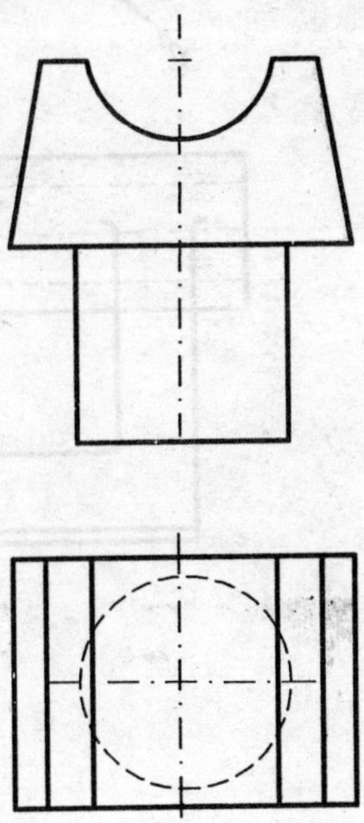

(ii) ORTHOGRAPHIC VIEWS

FIG. 65

(61) TAILSTOCK (UQ)

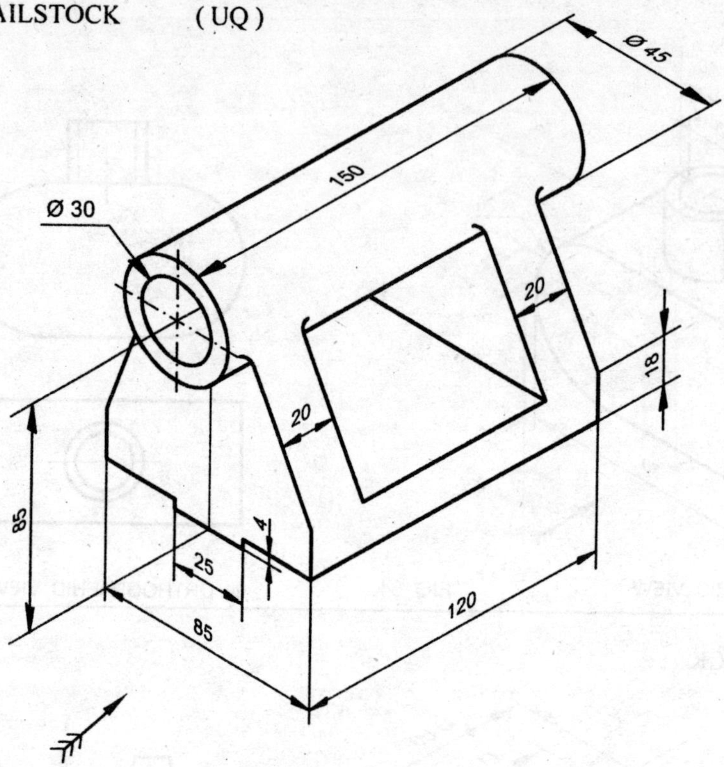

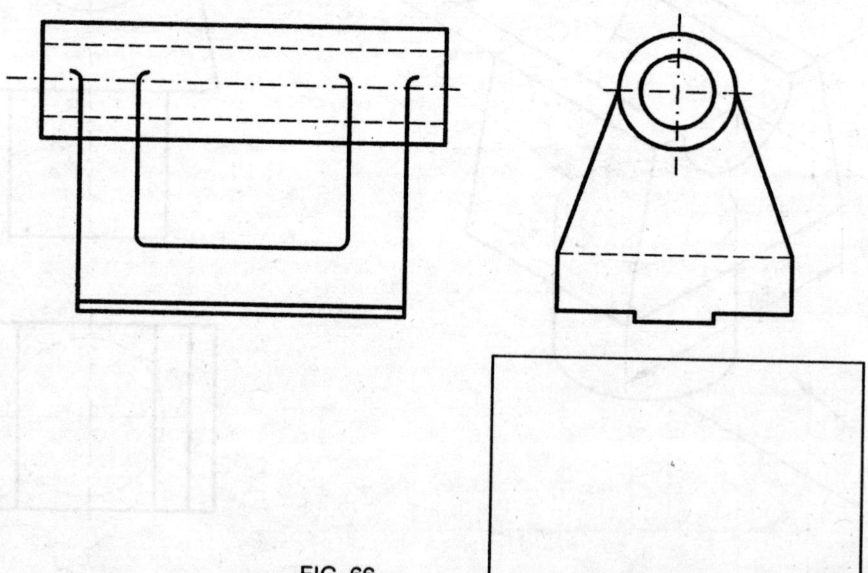

FIG. 66

(62) SHAFT SUPPORT (UQ)

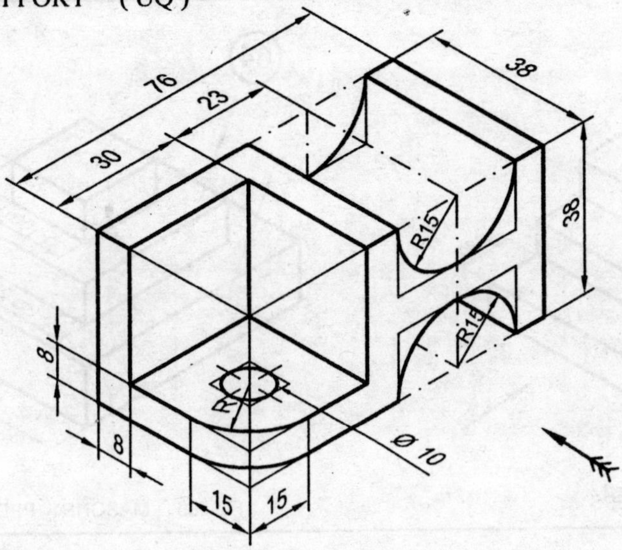

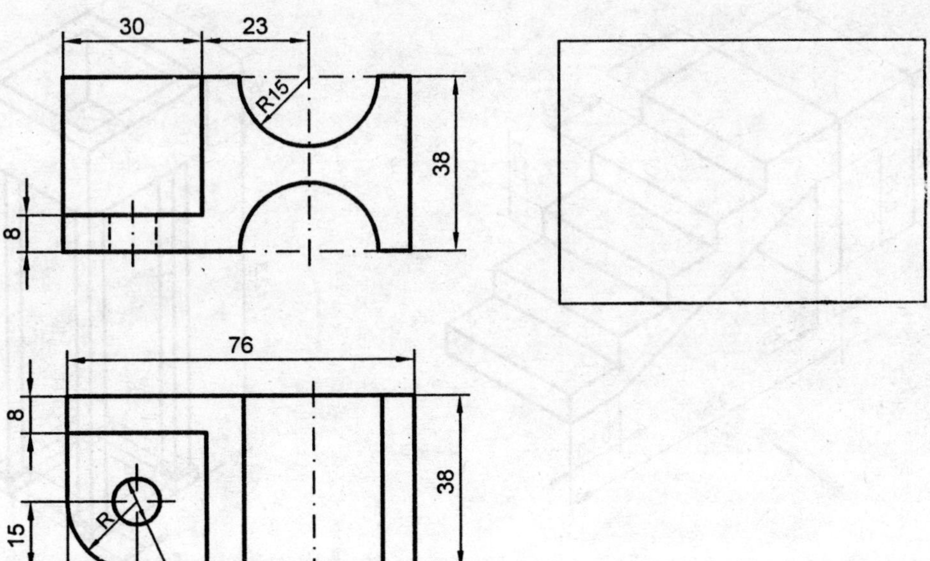

FIG. 67

8. EXERCISES

(63)

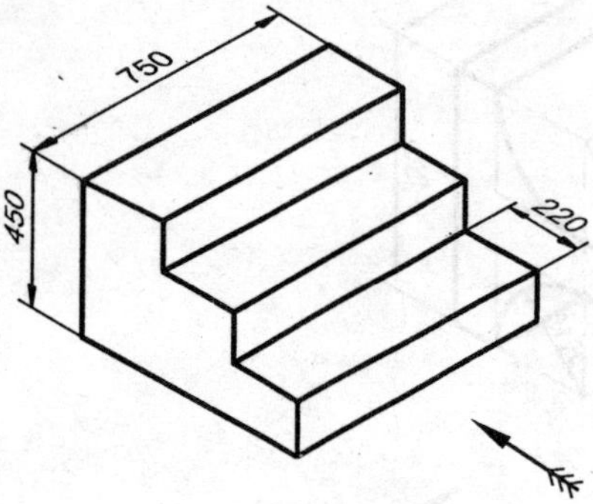

FIG. 68 DOOR STEPS

(64)

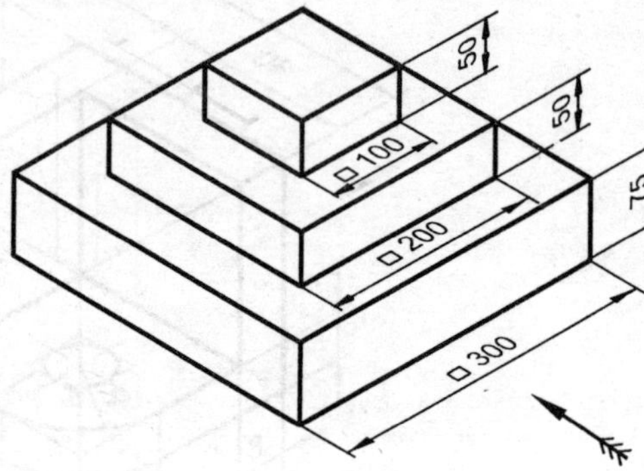

FIG. 69 MASONRY PEDESTAL

(65)

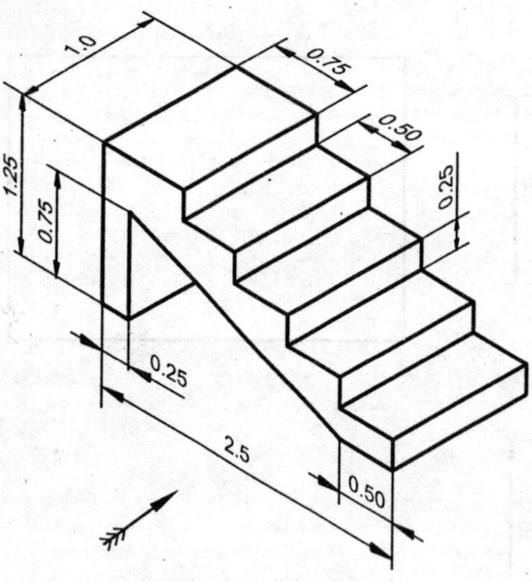

FIG. 70 STAIRCASE

ALL DIMENSIONS ARE IN m.

(66)

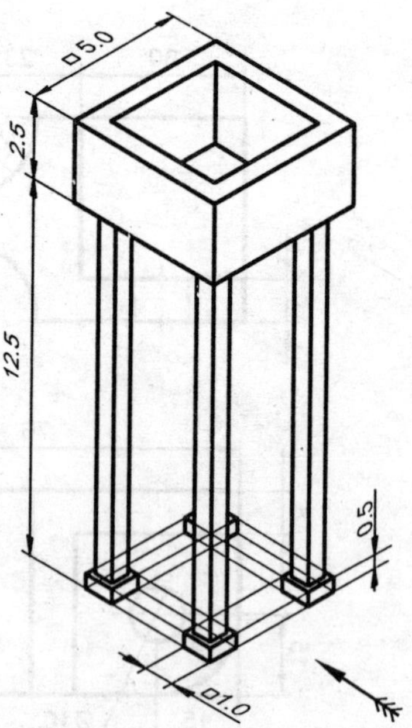

FIG. 71 WATER TANK

(67)

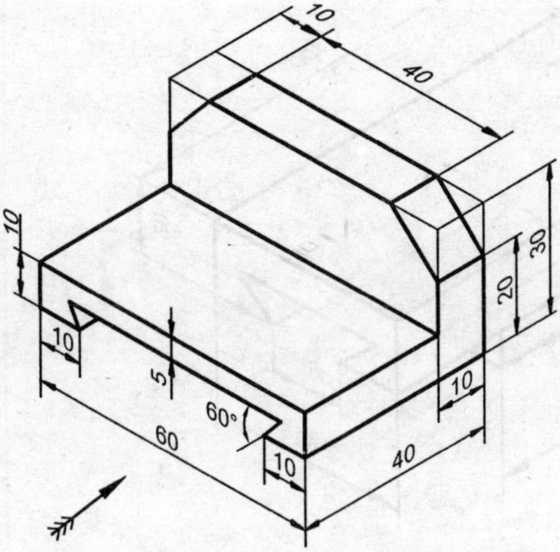

FIG. 72 DOVETAIL GUIDE

(68)

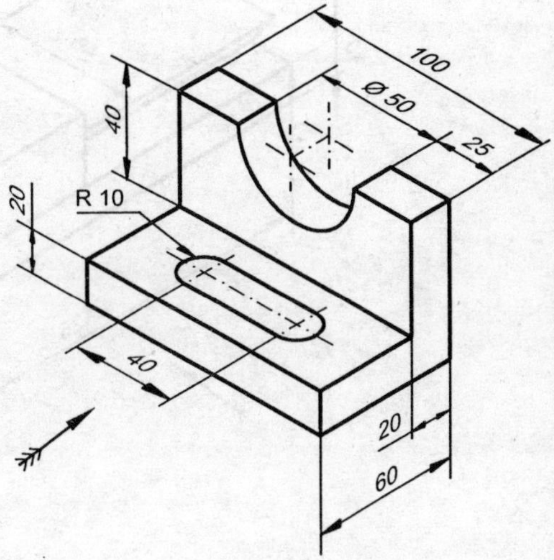

FIG.73 ANGULAR BLOCK

(69)

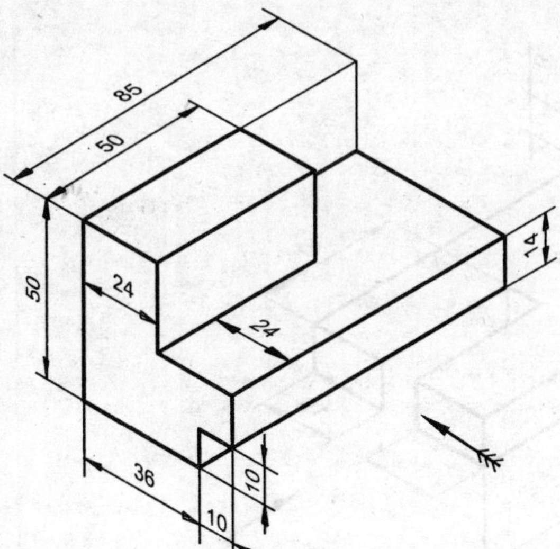

FIG. 74 STOP BLOCK

(70)

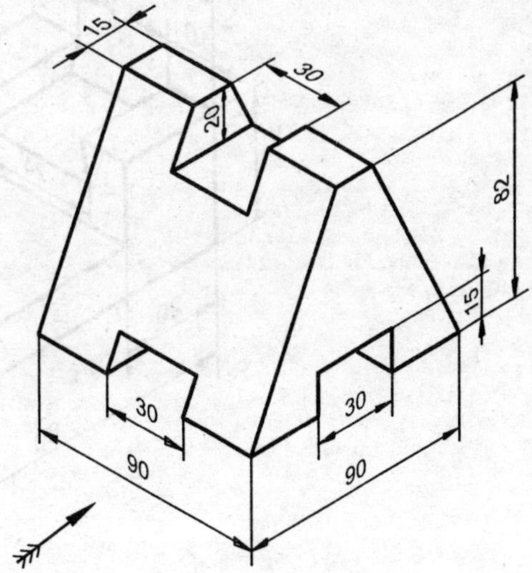

FIG. 75 H - BLOCK (UQ)

71

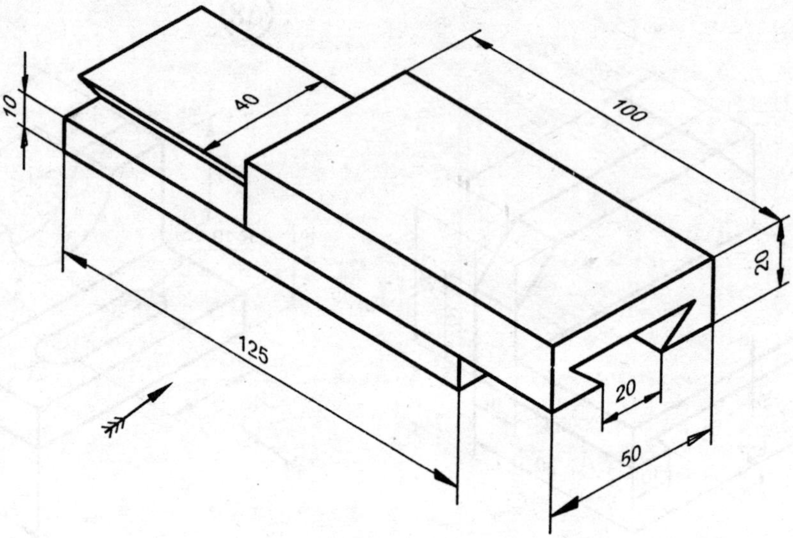

FIG. 76 DOVETAIL SLIDE (UQ)

72

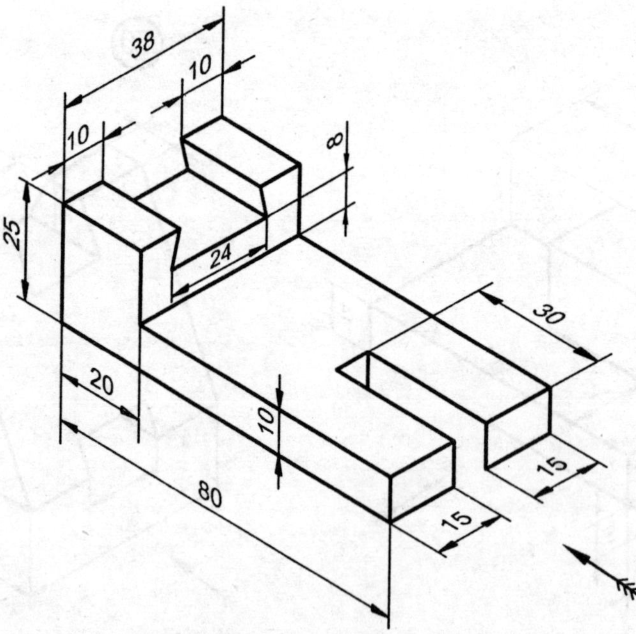

FIG. 77 DOVETAIL STOP (UQ)

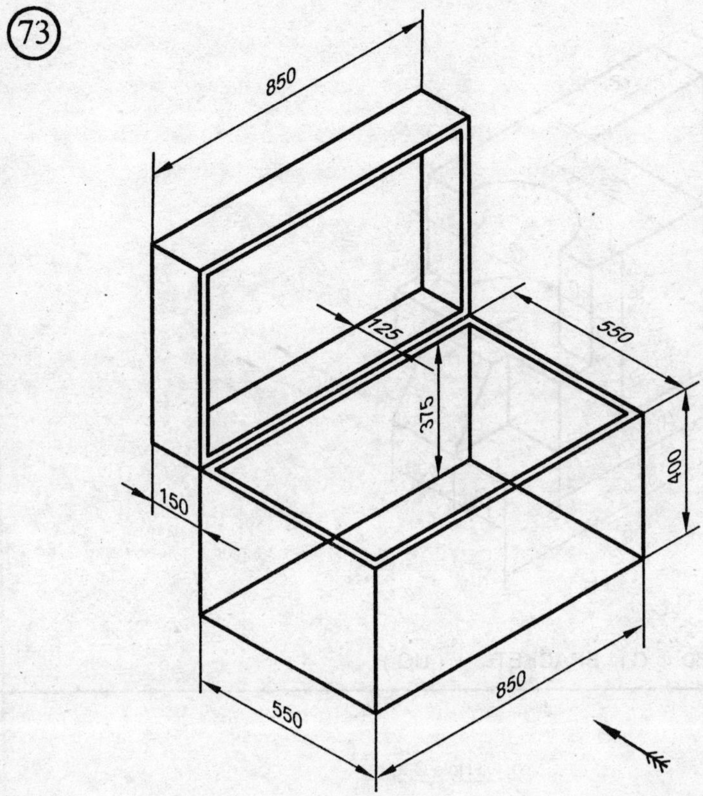

FIG. 78

BOX (LID IS 90° OPEN) (UQ)

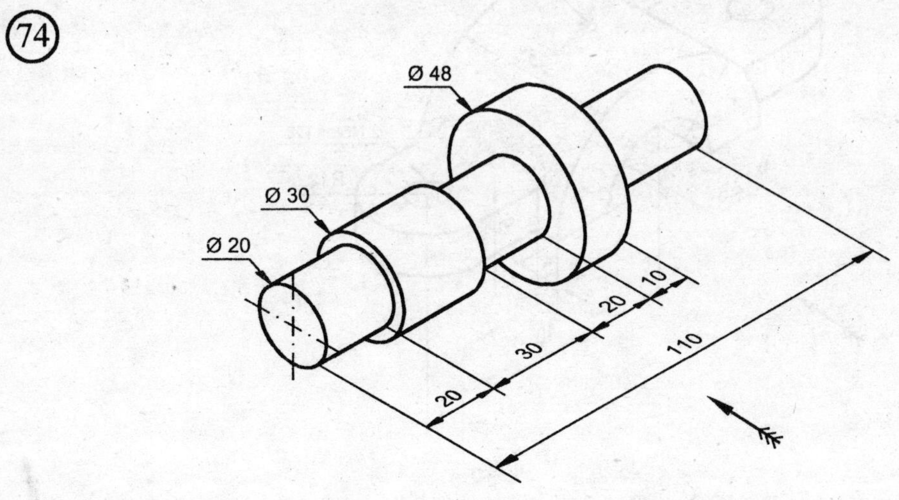

FIG. 79 SHAFT WITH MOUNTINGS

(75)

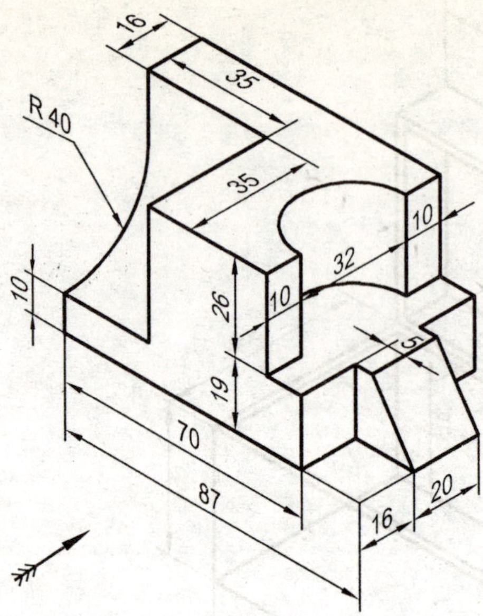

FIG. 80 C.I. BRACKET (UQ)

(76)

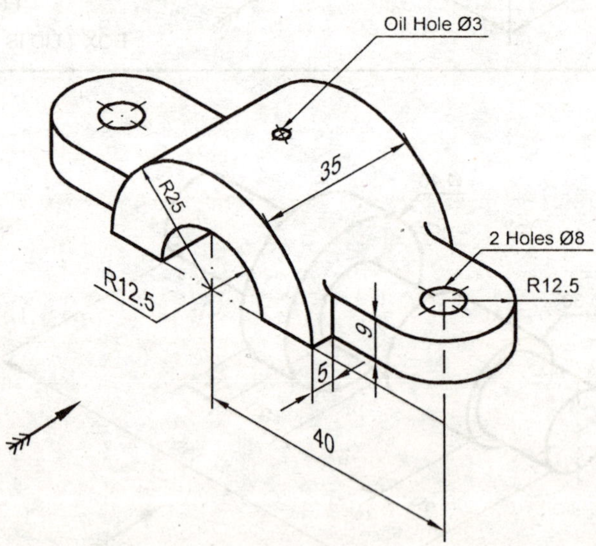

FIG. 81 BEARING CAP (UQ)

(77)

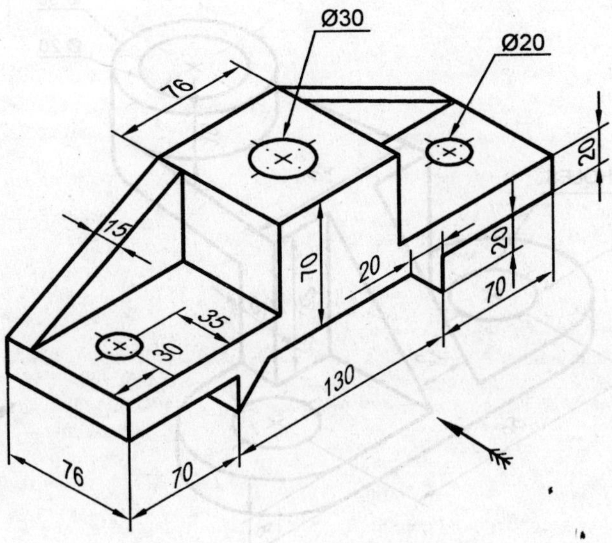

FIG. 82 SLIDING FRAME (UQ)

(78)

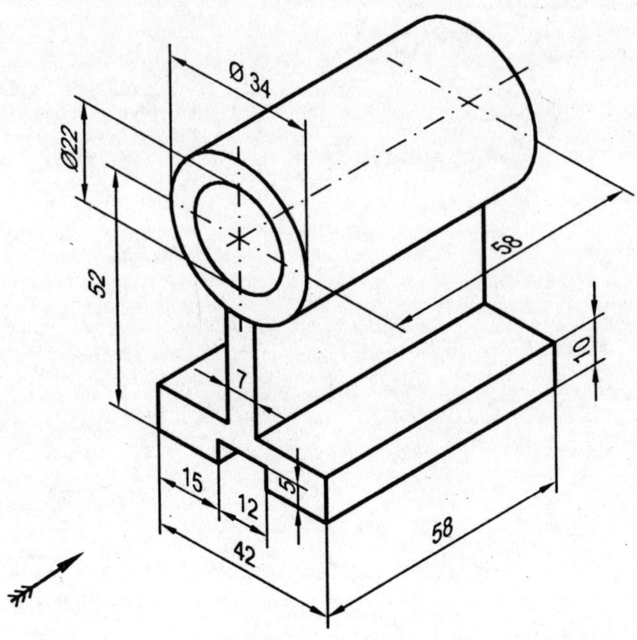

FIG. 83 DRUM WITH MOUNTING (UQ)

(79)

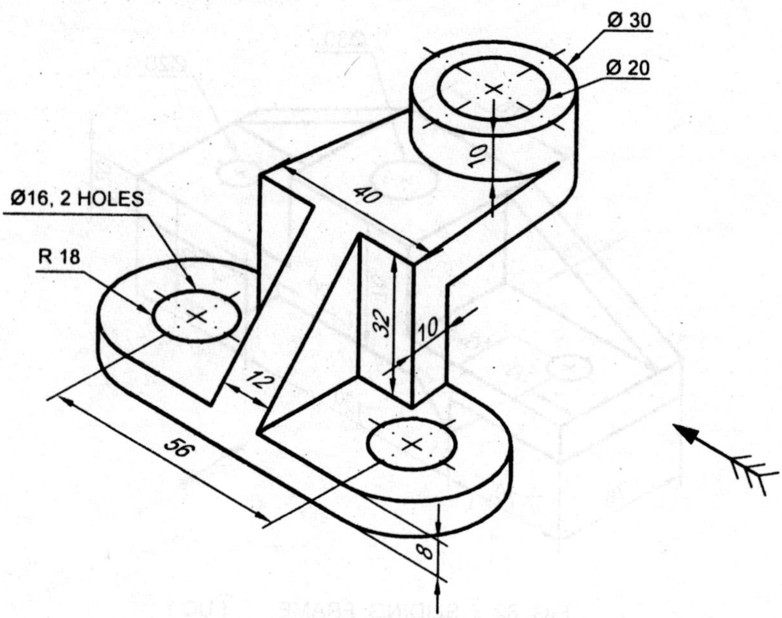

FIG. 84 SHAFT SUPPORT (UQ)

UNIT – VI

SYLLABUS

UNIT – VI

Sections of Solids – Developments of Surfaces – Intersection of Surfaces – Perspective Projection

21

SECTIONS OF SOLIDS

1. WHY IS SECTIONING NECESSARY?

Section Plane (S.P.) ***or*** **Cutting Plane**

In the orthographic views, the interior hollow portions and invisible outer edges / faces are represented by *dashed lines*. When the interior of the object is complex, these dashed lines are numerous. Therefore, the drawing is difficult to interpret.

To overcome this difficulty, it is imagined that the object is being cut through *completely* or *partially* by an *imaginary plane* called *Cutting* or *Section Plane*. Portion of the object between the S.P. and the observer is *assumed to be removed* to show the invisible and interior details of the object clearly. Show the S.P. in the view in which it is seen as a straight line, i.e., S.P. is shown by its Trace.

Sectioning: The imaginary process of cutting or sectioning the object is called *Sectioning*.

Cut Surface: Surface obtained by cutting an object by S.P. is called *Cut Surface* or *Section*.

BIS Representation of Section Plane Line: S.P. is conventionally represented as a long-dashed dotted *narrow* line and is drawn *wide* at the ends, bends and the changes of direction. The direction of viewing is perpendicular to the S.P. It is shown by arrows resting on the *thick* ends. The cut surface of the object is shown by a series of *narrow* lines (Type 01.1) called *Section Lines*, inclined at 45° to the axis or to the main outline of section.

2. HATCHING – BIS CONVENTIONS

The process of drawing section lines is called *Hatching*.

Spacing of Lines: Spacing between hatching lines should be in proportion to the size of the hatched areas. Minimum spacing between hatching lines should not be less than 0.7 mm.

See Fig. 1. Hatching is done by using continuous narrow (*thin*) lines (Type 01.1) at an angle of 45° to the principle outlines or line of symmetry of the sections as shown.

See Fig. 2. Separate areas of a section of the same component shall be hatched in an identical manner. The hatching of adjacent components shall be carried out with different directions or spacings as shown.

See Fig. 3. Hatching is interrupted when it is not possible to place inscriptions, say, S outside the hatched area as shown.

FIG. 1

FIG. 2

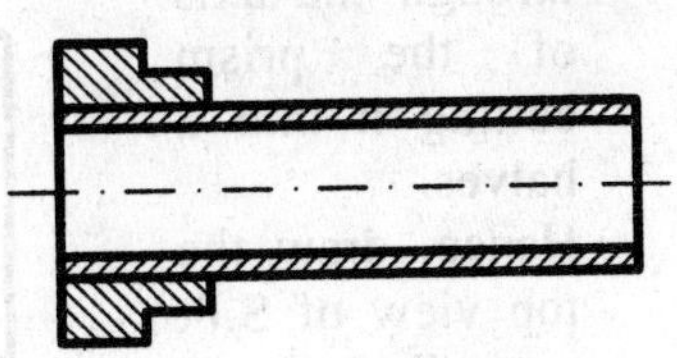

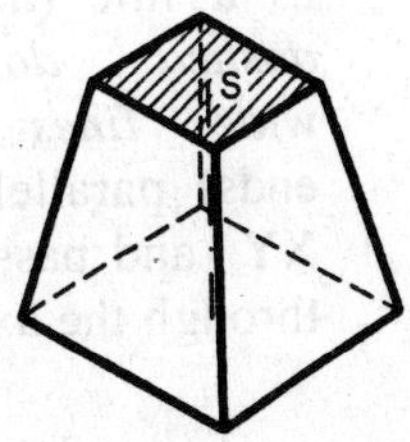

FIG. 3

Note: 1. *If the boundary of the object itself is inclined at 45° to the horizontal, section lines can be drawn at 45° to the main axis of the object.*

2. *Section lines should be equally spaced. Depending on the area to be hatched, spacing normally varies from 1.0 to 3.0 mm.*

Sectional View: It is the drawing showing that part of the object which is between the plane of projection and the S.P. It may be either a Sectional Front View or Sectional Top View.

True Shape of Section: If a solid is sectioned by a S.P. and it is projected to a plane parallel to it, the shape of the section obtained will be exactly the section exposed by the S.P. This shape is called *True Shape of Section*.

Apparent Shape of Section: If the S.P. is inclined to the plane of projection, the shape obtained will not be the true shape. Such a shape is called *Apparent Shape of Section*.

2.2 TYPES OF SECTION PLANES

(i) S.P. Perpendicular to HP and Parallel to VP
(ii) S.P. Perpendicular to VP and Parallel to HP
(iii) S.P. Perpendicular to VP and Inclined to HP
(iv) S.P. Perpendicular to HP and Inclined to VP
(v) S.P. Perpendicular to both HP and VP

3. SECTIONS OF PRISMS AND CUBES

Problem 1: *(Fig. 4)* ***A rectangular prism****, side of base 40 mm x 25 mm and height 60 mm, rests with its base on HP such that one of its larger rectangular faces is parallel to VP. A section plane perpendicular to HP and parallel to VP cuts the prism into two equal halves. Draw its top view and sectional front view.*

1. Draw the top and front views of the prism by *thin* lines. Name all the corners.
2. S.P. is ⊥' to HP, parallel to VP and passing through the axis of the prism cutting it into 2 halves. Hence, draw the top view of S.P. as a line (*long-dashed dotted wide lines* at ends) parallel to XY and passing through the axis.

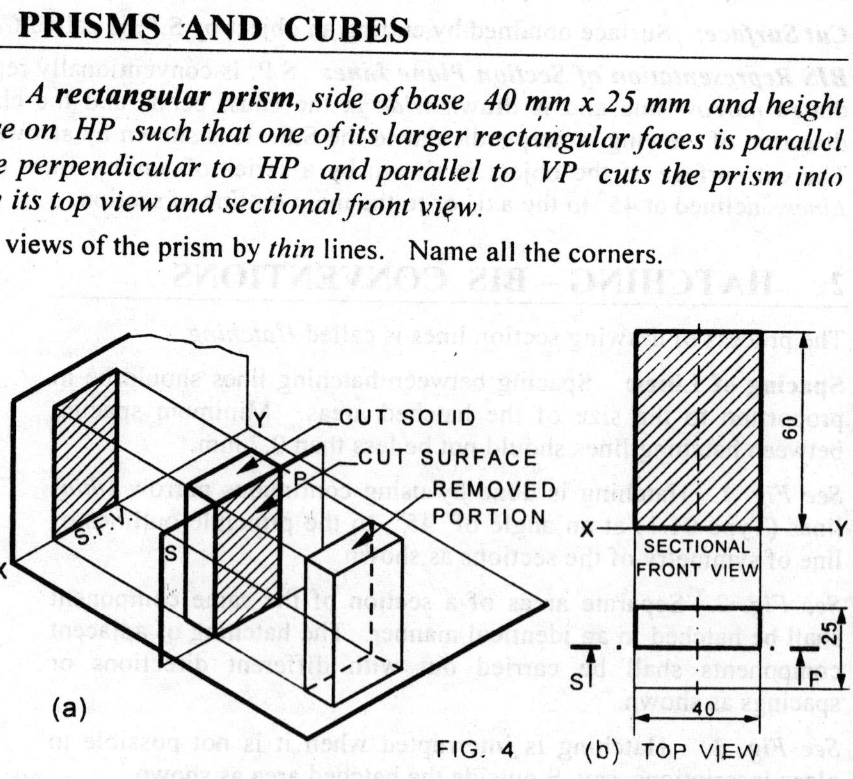

FIG. 4 (b) TOP VIEW

3. Assume that the portion of the prism between the S.P. and the observer is removed. The remaining portion is then shown by *thick* (wide) lines in the top view.

Note: *(i)* *Within the solid portion, the S.P. coincides with the cut surface which is to be shown by full wide line necessarily.*

(ii) *When S.P. is parallel to HP or VP, the Sectional Front View or Sectional Top View of the cut surface itself shows the* ***True Shape of Section.***

---

4. Then draw the sectional front view for the remaining portion of the prism by *thick* lines (H). In this view the cut surface is shown by hatching (*thin*) lines using 2H.

Problem 2: *(Fig. 5)* ***A hexagonal prism,*** *side of base 30 mm and axis 60 mm long, rests with its base on HP such that one of its rectangular faces is parallel to VP. A section plane perpendicular to HP and parallel to VP cuts the prism at a distance of 10 mm from its axis. Draw its top and sectional front views.*

1. Draw top and front views of the prism by *thin lines* and name all the corners.
2. Draw SP parallel to XY and 10 mm from the axis towards the observer in the top view.

Note: *Always major portion of the solid should be retained for projections.*

3. Name the section points 1 and 2 where the S.P. cuts the visible edges ab and cd respectively. Name the points (3) and (4) where the S.P. cuts the invisible edges $(d_1)(c_1)$ and $(b_1)(a_1)$ respectively. Show remaining portion of the prism as *thick* lines.
4. **Cut surface:** Project the above section points on the corresponding edges in the front view. From 1(4) draw a projector to cut a′ b′ at 1′ and a_1' b_1' at 4′. Similarly mark 2′ and 3′ in the front view.

   Join 1′ 2′ 3′ 4′ by *thick* lines and hatch this area to represent the front view of the cut surface.
5. Look at the remaining portion of prism in top view in the direction of arrows. 1 a (a_1) (4) and 2d (d_1) (3) are visible. Show 1′ a′ a_1' 4′ and 2′ d′ d_1' 3′ as *thick* lines.

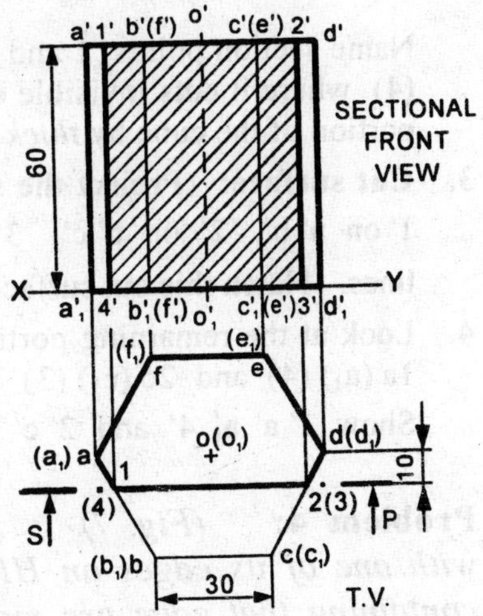

FIG. 5

Problem 3: *(Fig. 6)* ***A cube*** *of 45 mm side rests with a face on HP such that one of its vertical faces is inclined at 30° to VP. A section plane, parallel to VP cuts the cube at a distance of 15 mm from the vertical edge nearer to the observer. Draw its top and sectional front views.*

1. Draw the top and front views of the cube.
2. **Section Plane:** Draw S.P. parallel to XY and 15 mm from the vertical edge $b(b_1)$ nearer to the observer in the top view.

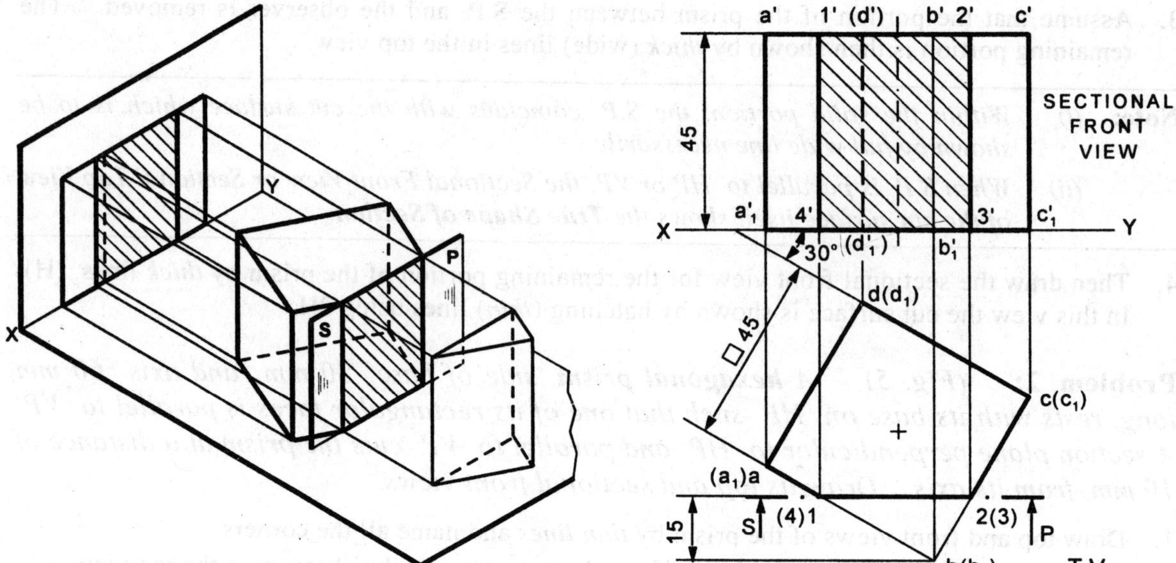

FIG. 6

Name section points 1 and 2 where S.P. cuts the visible edges ab and bc. Name (3) and (4) where it cuts invisible edges (c_1) (b_1) and (b_1) (a_1) respectively. Show the remaining portion of the cube by *thick* lines in the top view.

3. **Cut surface:** Project the section points on the corresponding edges in the front view, *i.e.*, 1′ on a′ b′, 2′ on b′ c′, 3′ on c_1' b_1' and 4′ on b_1' a_1' respectively. Join 1′ 2′ 3′ 4′ by *thick* lines. Hatch this cut surface.
4. Look at the remaining portion of the cube in the top view in the direction of arrows shown. 1a (a_1) (4) and 2c (c_1) (3) are visible.
   Show 1′ a′ a_1' 4′ and 2′ c′ c_1' 3′ as *thick* lines.

Problem 4: *(Fig. 7) A **cube** of 50 mm side rests with one of its edges on HP such that the square faces containing that edge are making equal inclinations with HP. A horizontal section plane cuts the cube at a distance of 18 mm below the horizontal edge nearer to the observer. Obtain front and sectional top views.*

1. **Section Plane:** Draw S.P. parallel to XY and 18 mm below d′ (d_1'). Mark 1′, 2′, (3′) and (4′) as shown in front view. Show the remaining portion by *thick* lines.
2. **Cut Surface:** Project the section points on the corresponding edges in the top view. Join 1 2 3 4 and hatch this area. Complete the sectional top view as shown.

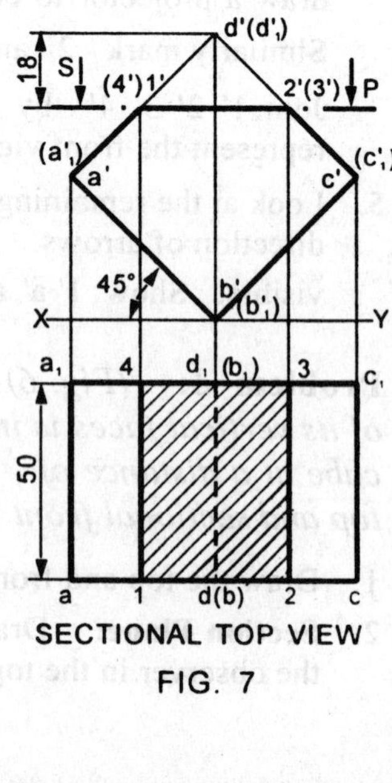

SECTIONAL TOP VIEW

FIG. 7

Problem 5: *(Fig. 8) A* ***square prism****, side of base 30 mm and axis 60 mm long, rests with its base on HP and one of its rectangular faces is inclined at 30° to VP. A section plane perpendicular to VP and inclined at 60° to HP cuts the axis of the prism at a point 20 mm from its top end. Draw sectional top view and true shape of section.* (UQ)

1. Draw S.P. at 60° to XY and passing through a point on the axis 20 mm from its top end.
2. Name the section points 1′, 2′, 3′, (4′) and (5′) in front view.
3. Project section points on the corresponding edges in top view and draw the cut surface.
4. Then complete the sectional top view. Since the S.P. is inclined to HP, cut surface obtained does not appear in true shape and is called **Apparent Shape of Section.**

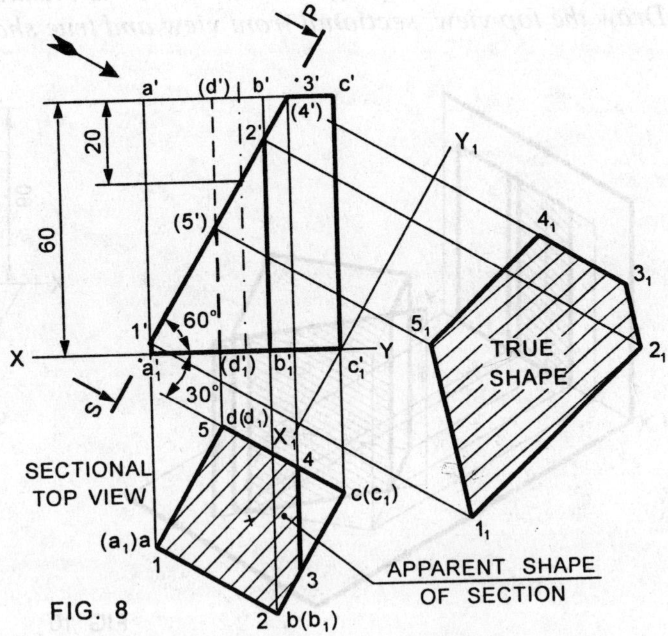

FIG. 8

True Shape of Section

5. Draw X_1Y_1 parallel to S.P. & at any convenient distance from it.
6. Draw projectors perpendicular to X_1Y_1 from 1′, 2′, 3′, (4′) and (5′).
7. Measure the distances of the points 1, 2, 3, 4 and 5 from XY and transfer the same on the corresponding projectors from X_1Y_1. Thus obtain 1_1, 2_1, 3_1, 4_1 and 5_1 and join them, which is the True Shape of Section.

Problem 6: *(Fig. 9) A* ***cube*** *of side 25 mm rests on HP on one of its faces with a vertical face inclined at 35° to VP. A plane perpendicular to HP and inclined at 50° to VP cuts the cube, 3 mm away from the axis. Draw the top view and the sectional front view. Also, find the true shape of the section.* (UQ)

See Fig. 9 for Solution.

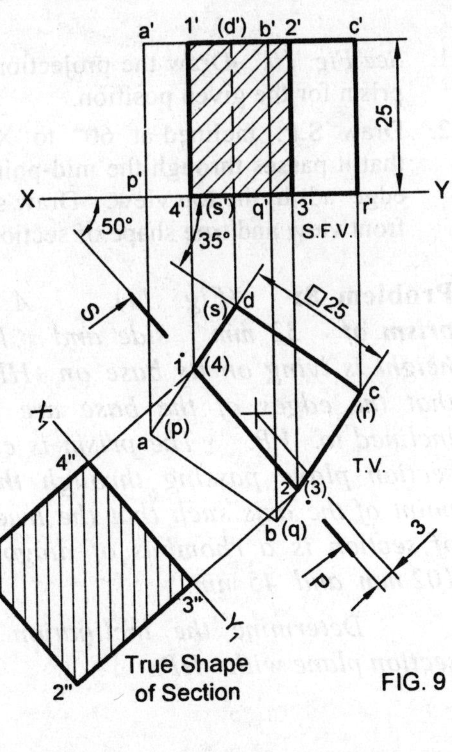

FIG. 9

Problem 7: *(Fig. 10)* *A* ***square prism****, side of base 40 mm and axis 60 mm long, rests with its base on HP such that one of its rectangular faces is inclined at 30° to VP. A section plane perpendicular to HP and inclined at 60° to VP passes through the prism such that a rectangular face which is making 60° with VP is cut into two halves. Draw the top view, sectional front view and true shape of section.* (UQ)

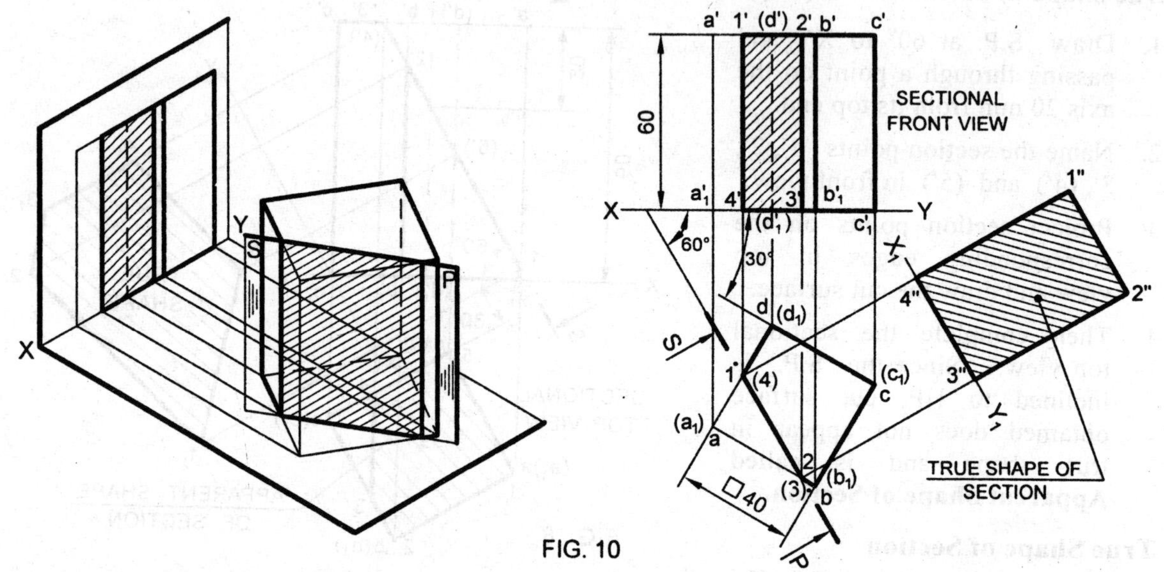

FIG. 10

1. *See Fig. 10.* Draw the projections of the prism for the given position.
2. Draw S.P. inclined at 60° to XY such that it passes through the mid-point of the edge ad in the top view. Draw sectional front view and true shape of section.

Problem 8: *(Fig. 11)* *A* ***square prism*** *of 32 mm side and 100 mm height is lying on its base on HP such that the edges of the base are equally inclined to VP. The prism is cut by a section plane passing through the mid-point of the axis such that the true shape of section is a rhombus of diagonals of 102 mm and 45 mm.*

Determine the inclination of the section plane with HP. (UQ)

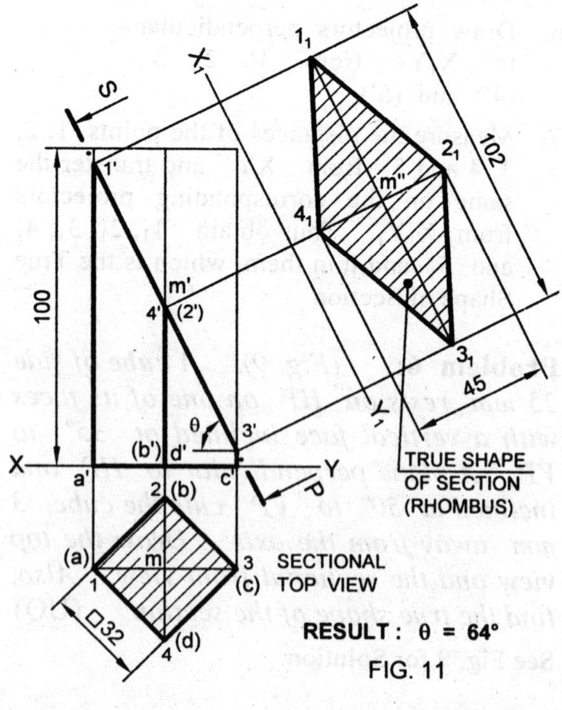

FIG. 11

Given the True Shape of Section, to draw the S.P. ***(Fig. 11)***

1. Draw the top and front views. Mark the mid-point of the axis m′ in the front view.
2. m′ as center and radius = Half of the longer diagonal (= 51 mm), draw 2 arcs cutting 2 opposite vertical sides of the prism in front view. Thus mark Section Plane position SP.
3. Project the points from the S.P. on X_1Y_1 and draw the true shape of section, *i.e.*, rhombus. Measure the inclination θ of the S.P. with HP. Result: **θ = 64°**.

Problem 9: *(Fig. 12) A cube of 70 mm long edges has its vertical faces equally inclined to the VP. It is cut by an auxiliary inclined plane in such a way that the true shape of the cut part is a regular hexagon. Determine the inclination of the cutting plane with HP. Draw front view, sectional top view and true shape of the section.* (UQ)

Note: As the true shape of the section is a hexagon, the S.P. must cut the cube at 6 points, i.e., the S.P. will cut 2 edges of the top two edges of the base and two *vertical edges*. If the true shape is a *regular* hexagon, S.P. should cut these 6 edges at *mid-points*.
The points of intersection at the 2 *vertical edges* will be farthest from each other. These points represent 2 opposite corners of the hexagon. Distance between them is equal to b d.

Location of S.P.: If the 6 edges of the cube should be cut by a S.P., it has to be placed with one of its square faces on HP and 2 of the vertical faces making equal inclination with VP.

1. First, draw the top view of the cube such that the cube is placed with one of its square faces on HP and two of the vertical faces making equal inclination with VP. Project the corresponding front view.
2. Join bd. Draw a circle of diameter bd.
3. Inscribe a hexagon 123456 in the circle as shown. Measure the distance between the lines 16 and 34 = MN.
4. In the front view, locate 2′ at the mid-point of q′ b′.
5. With 2′ as center and radius equal to ½ MN, draw arcs to cut a′b′ and q′r′ at the points 1′ and 3′ respectively.
6. Join 1′3′, which is the *required S.P.* Measure the angle **θ = 55°**.
7. Draw X_1Y_1 parallel to 1′ 3′. Redraw the hexagon 123456 as $1_12_13_14_15_16_1$ such that mn is parallel to X_1Y_1.
8. Project all the corners of the hexagon in front view. Project 1′, 2′, ..., 6′ to top view. Obtain the sectional top view.

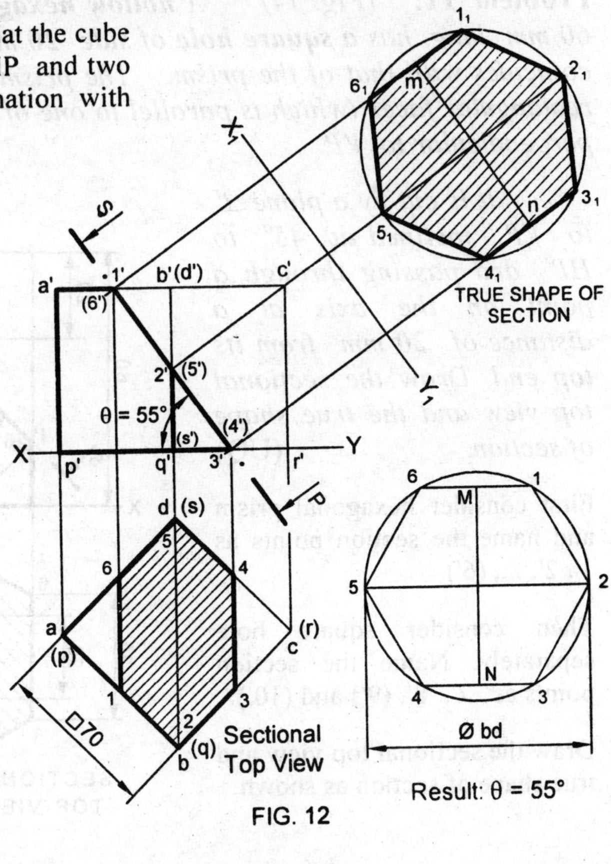

FIG. 12

Problem 10: *(Fig. 13)* ***A triangular prism** with a base side of 45 mm and height 60 mm rests on its base on HP with a rectangular face perpendicular to VP. It is cut by an Auxiliary Inclined Plane such that the true shape of the section is an isosceles triangle of maximum size.* (UQ)

Note: *For the maximum size of isosceles triangle, the S.P. must pass through a top edge of the prism and the base corner opposite to that edge.*

1. Draw the top view and front view as shown.
2. Mark S.P. through the points a′(b′) and 3′.
3. Project a′, (b′) and 3′ on the S.P. and obtain the auxiliary view a_1, b_1 and 3_1. Thus, the true shape of section is drawn as an isosceles triangle of maximum size.

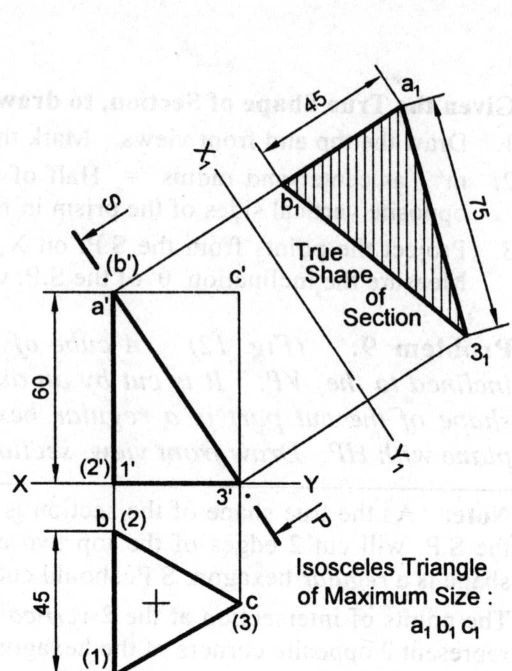

FIG. 13

Problem 11: *(Fig. 14)* *A **hollow hexagonal prism**, side of base 30 mm and axis 60 mm long, has a **square hole** of side 20 mm through it such that the axis of the hole coincides with that of the prism. The prism rests with its base on HP and one of its rectangular faces (which is parallel to one of the rectangular faces of the square hole) is perpendicular to VP.*

It is cut by a plane ⊥r to VP, inclined at 45° to HP and passing through a point on the axis at a distance of 20 mm from its top end. Draw the sectional top view and the true shape of section. (UQ)

First consider hexagonal prism and name the section points as 1′, 2′, ..., (6′).

Then consider square hole separately. Name the section points as 7′, 8′, (9′) and (10′).

Draw the sectional top view and true shape of section as shown.

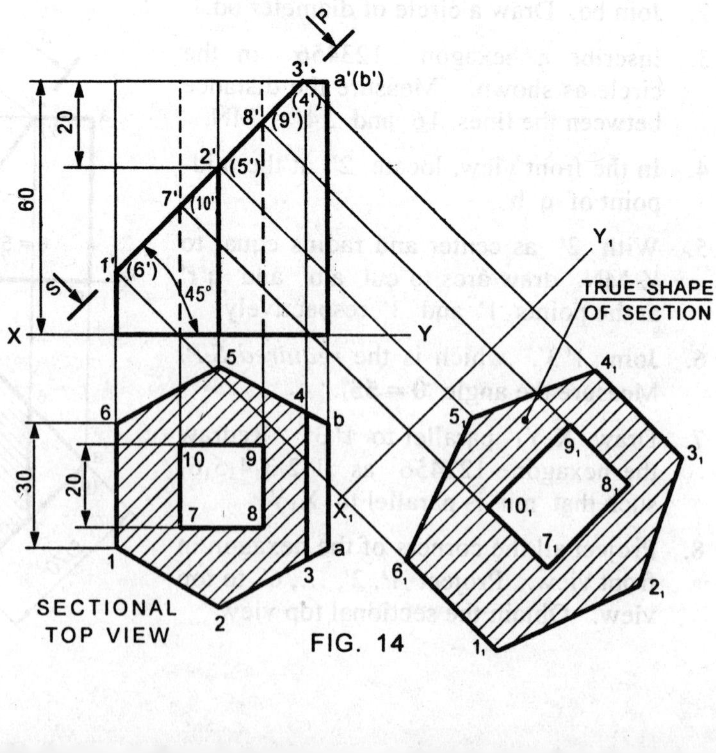

FIG. 14

Problem 12: *(Fig. 15)* ***A pentagonal prism**, side of base 25 mm and axis 60 mm long, rests with one of the edges of its base on HP. Its axis is inclined at 30° to HP and parallel to VP. It is cut by a horizontal section plane passing through the highest corner of the base. Draw the sectional top view.* (UQ)

1. For the given position of the prism, draw the top view, first and second front views as shown.
2. Draw S.P. parallel to XY through the highest corner of the base a_1'.
3. Mark 1′, 2′, 3′, (4′) and (5′).
   Here 1′ coincides with a_1'.
   2′ and 3′ are the intersections of S.P. with visible edges $b'b_1'$ and b′c′ respectively.
   (4′) and (5′) are the intersection of S.P. with invisible edges (d′)(e′) and $(e')(e_1')$ respectively.
   Show remaining portion of the prism by *thick* lines.
4. Project the section points on the corresponding edges in the II top view. Join 12345. Show the remaining portion in the sectional top view by *thick* lines.

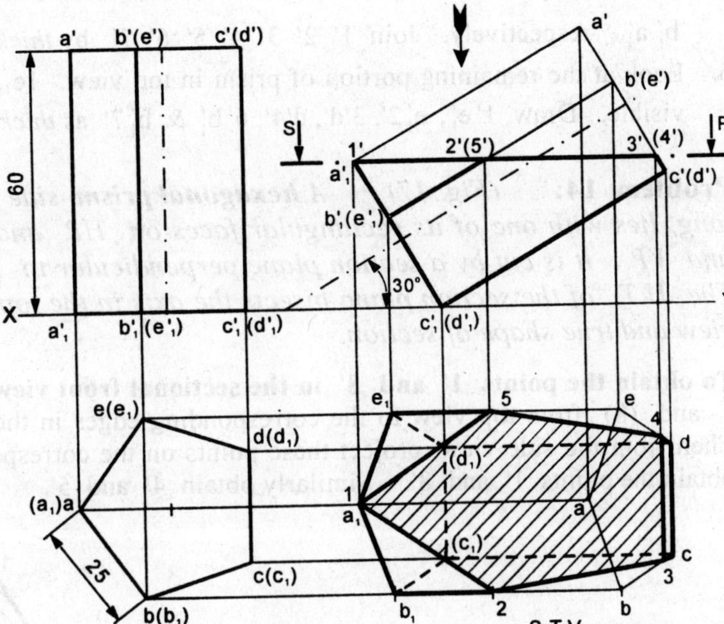

FIG. 15

Problem 13: *(Fig. 16)* ***A pentagonal prism**, side of base 30 mm and axis 60 mm long, lies with one of its rectangular faces on HP and its axis is inclined at 30° to VP. A section plane ⊥r to HP and parallel to VP cuts the prism into two halves. Obtain its top and sectional front views.* (UQ)

1. For the given position of prism, draw the front view, first and second top views.

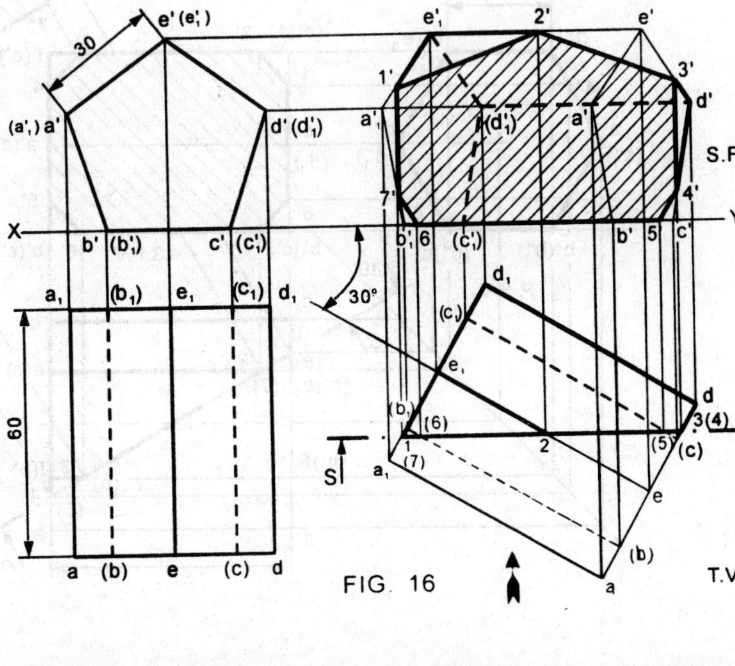

FIG. 16

2. Draw S.P. parallel to XY and passing through the mid-point of the axis in final top view.
3. Name the points 1, 2 & 3 where S.P. cuts visible edges a_1e_1, e_1e & ed respectively.
4. Similarly mark (4), (5), (6) and (7) where S.P. cuts the invisible edges d (c), (c) (c_1), (b)(b_1) and a_1(b_1) respectively. Show the remaining portion of the prism by *thick* lines.
5. **Cut surface:** Project the section points on the corresponding edges in final front view, *i.e.*, 1′ on a_1' e_1', 2′ on e_1'e′, 3′ on e′d′, 4′ on d′c′, 5′ on c′(c_1'), 6′ on b′b_1' and 7′ on b_1' a_1' respectively. Join 1′ 2′ 3′ 4′ 5′ 6′ 7′ by *thick* lines and hatch this area.
6. Look at the remaining portion of prism in top view. 1e_1, $e_1$2, 3d, d(4), (6)(b_1) & (b_1)(7) are visible. Draw 1′e_1', e_1'2′, 3′d′, d′4′, 6′b_1' & b_1'7′ as *thick* lines in Sectional F.V.

Problem 14: *(Fig. 17)* ***A hexagonal prism****, side of base 25 mm and axis 65 mm long, lies with one of its rectangular faces on HP and its axis is parallel to both HP and VP. It is cut by a section plane perpendicular to HP and inclined at 30° to VP. The H.T. of the section plane bisects the axis in the top view. Draw the sectional front view and true shape of section.* (UQ)

To obtain the points 1′ and 8′ in the sectional front view: *See Fig. 17.* Project the points 1 and (8) from top view to the corresponding edges in the side view and mark 1″ and 8″. Then from the side view, project these points on the corresponding edges in the front view and obtain the points 1′ and 8′. Similarly obtain 4′ and 5′.

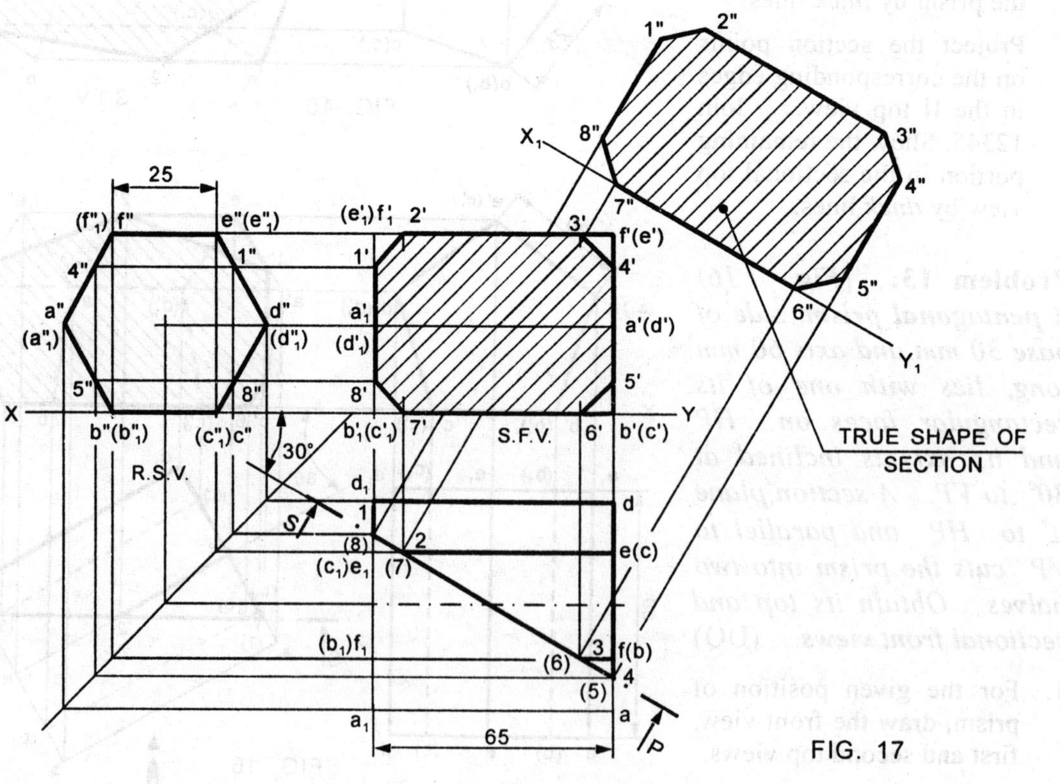

FIG. 17

Problem 15: *(Fig. 18) A* ***triangular prism*** *of 35 mm side of base and axis 56 mm long rests on its base on HP with a side of the base perpendicular to VP. It is cut by an Auxiliary Inclined Plane such that the true shape of section is a trapezium of parallel sides 28 mm and 8 mm. Draw the sectional top and front views and an auxiliary view showing the true shape of section. Determine the angle made by S.P. with HP.* (UQ)

1. Draw the top and front views of the triangular prism.
2. In the top view, mark the points a and b such that ab = 28 mm. Also, mark the points c and d such that cd = 8 mm. Complete the sectional top view.
3. Project a, b, c and d to front view and mark a′, (b′), (c′) and d′ on the corresponding edges.
4. Draw the S.P. through the points a′(b′) and (c′)d′.
5. **Measure the angle made by the S.P. with XY = θ = 73°.** Thus, complete the sectional front view.
6. Draw X_1Y_1 parallel to S.P. Project front view on X_1Y_1 and complete auxiliary view showing true shape of section.

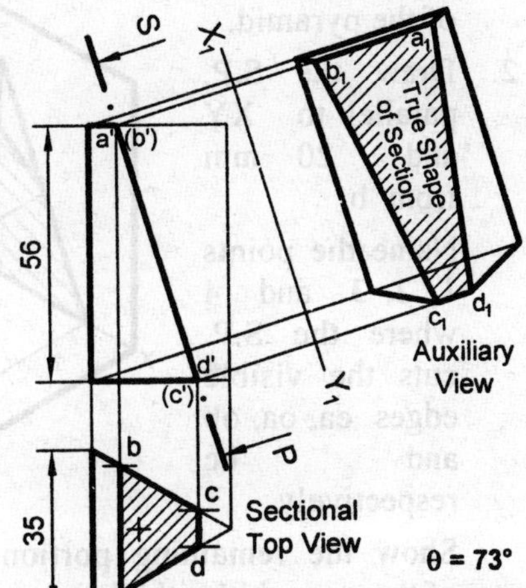

FIG. 18

Problem 16: *(Fig. 19) A* ***hexagonal prism*** *of 35 mm side and 65 mm height has a concentric hexagonal hole of 18 mm side. It rests with its base on HP such that two rectangular faces make equal inclination to VP. The prism is cut by a section plane whose vertical trace is inclined at 60° to XY. The section plane bisects the axis of the prism. Draw the sectional top and profile views. Add the true shape of section.* (UQ)

1. Draw top and front views of prism.
2. Draw the hexagonal hole in top view and project it to front view.
3. In front view, draw the S.P. at 60° to XY and bisecting the axis.
4. Draw the sectional top view. Also, draw the sectional left side view using the mitre line.
5. Draw X_1Y_1 parallel to S.P. Project the front view on X_1Y_1 and draw the true shape of section.

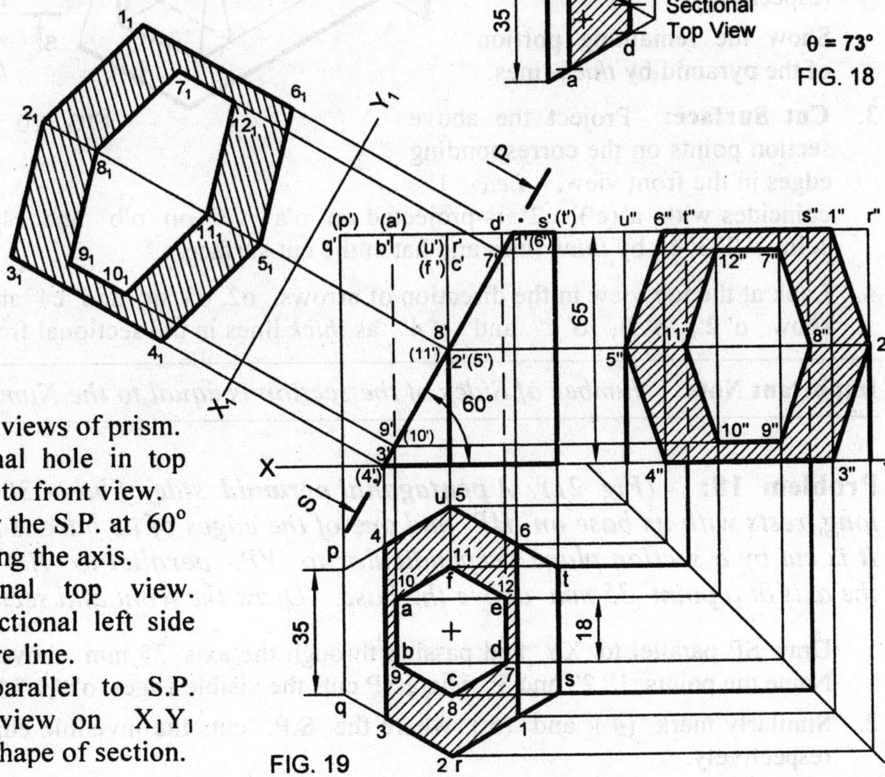

FIG. 19

4. SECTIONS OF PYRAMIDS AND TETRAHEDRONS

Problem 17: *(Fig. 20)* *A **pentagonal pyramid**, side of base 35 mm and axis 60 mm long, rests with its base on HP such that one of the edges of the base is perpendicular to VP. A section plane perpendicular to HP and parallel to VP cuts the pyramid at a distance of 20 mm from the corner of the base nearer to the observer. Draw its top and sectional front views.* (UQ)

1. Draw the top and front views of the pyramid.
2. Draw the S.P. parallel to XY and 20 mm from b.
   Name the points 1, 2, 3 and 4 where the S.P. cuts the visible edges ea, oa, ob and bc respectively.
   Show the remaining portion of the pyramid by *thick* lines.

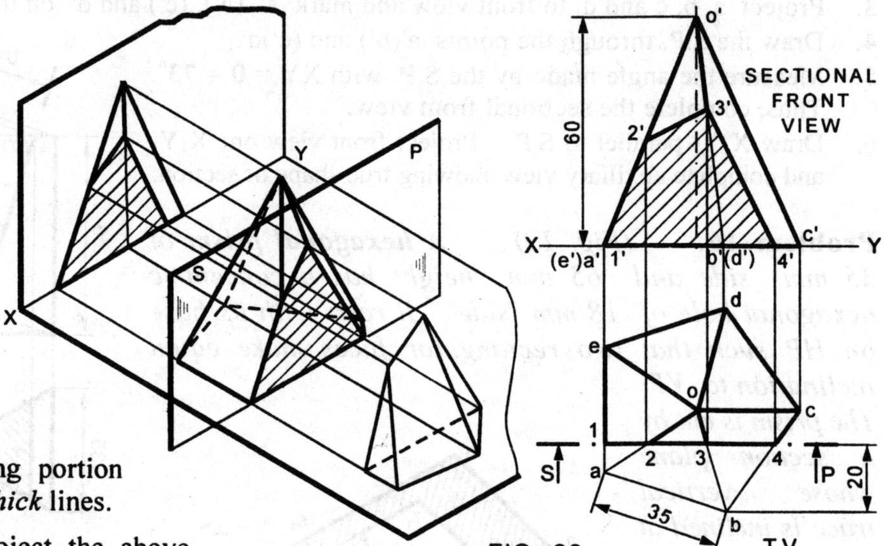

FIG. 20

3. **Cut Surface:** Project the above section points on the corresponding edges in the front view, *i.e.*, 1′ coincides with a′(e′), 2′ is projected on o′a′, 3′ on o′b′ and 4′ on b′c′ respectively. Join 1′ 2′ 3′ 4′ by *thick* lines and hatch the cut surface.
4. Look at the top view in the direction of arrows. o2, o3, oc and c4 are visible edges. Hence show o′ 2′, o′ 3′, o′ c′ and c′ 4′ as *thick* lines in the sectional front view.

Important Note: *Number of Sides of the section is equal to the Number of Edges cut.*

Problem 18: *(Fig. 21)* *A **pentagonal pyramid**, side of base 30 mm and axis 60 mm long, rests with its base on HP and one of the edges of its base is perpendicular to VP. It is cut by a section plane perpendicular to VP, parallel to HP and passing through the axis at a point 35 mm above the base. Draw the front and sectional top views.*

1. Draw SP parallel to XY and passing through the axis 35 mm above the base in front view. Name the points 1′, 2′ and 3′ where SP cuts the visible edges o′a′, o′b′ and o′c′ respectively.
2. Similarly mark (4′) and (5′) where the S.P. cuts the invisible edges o′ (d′) and o′ (e′) respectively.

3. Show the remaining portion of the pyramid a′ b′ c′ (d′) (e′) and the slant edges 1′ a′, 2′ b′, 3′ c′, (4′)(d′) and (5′)(e′) as *thick* lines in the front view.
4. **Cut surface:** Project the section points on the corresponding edges in the top view, *i.e.*, mark 1 on oa, 2 on ob and so on. Join 1 2 3 4 5 by *thick* lines and hatch this area.
5. Look at the remaining portion of the pyramid in the front view in the direction of arrows. Show abcde, 1a, 2b, 3c, 4d and 5e as *thick* lines in sectional top view.

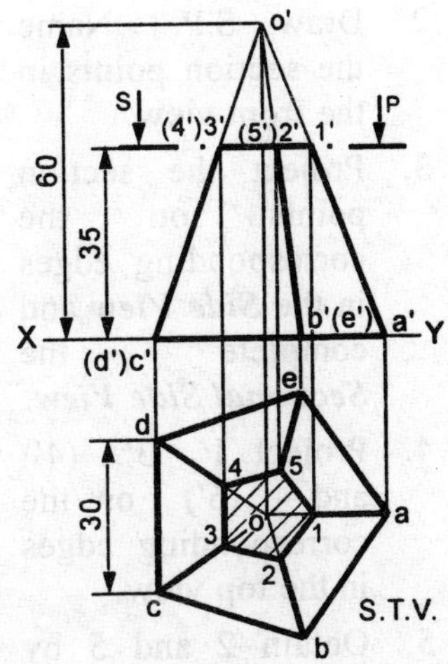

FIG. 21

Problem 19: *(Fig. 22)* *A* ***pentagonal pyramid****, side of base 30 mm and axis 60 mm long, rests with its base on HP and an edge of its base is parallel to VP. A section plane perpendicular to VP and inclined at 45° to HP passes through the axis at a point 35 mm above the base. Draw the sectional top view.* (UQ)

See Fig. 22. To obtain the point 2 in the sectional top view, draw a horizontal line from 2′ to cut o′ c′ at p′ in the front view. Project p′ and obtain p on oc.

With o as center and op as radius draw an arc to cut ob at 2.

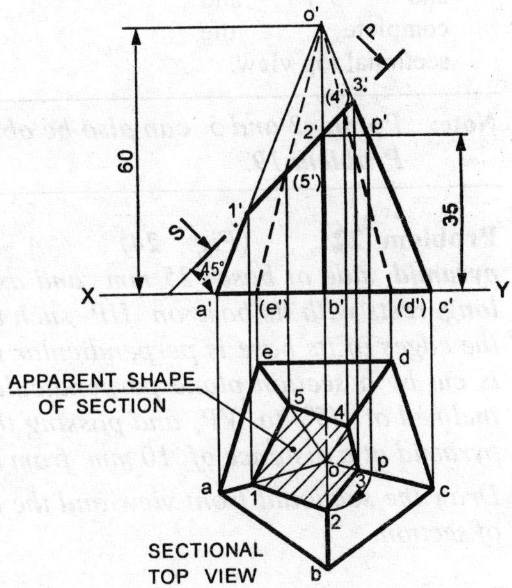

FIG. 22

Problem 20: *(Exercise)* *A* ***hexagonal pyramid****, side of base 30 mm and height 65 mm is resting on its base on HP with two edges parallel to VP. It is cut by a plane perpendicular to VP, inclined at 45° to HP and intersecting the axis at a point 25 mm above the base. Draw the sectional top view and true shape of section.* (UQ)

Hint: In this Problem, the base is also cut. Therefore the true shape of section obtained is a seven-sided irregular polygon called *Heptagon*.

Problem 21: *(Fig. 23)* *A* ***hexagonal pyramid****, side of base 25 mm and axis 50 mm long, rests with its base on HP and an edge of its base is perpendicular to VP. It is cut by a section plane perpendicular to VP, inclined at 30° to HP and passing through a point on the axis 20 mm below the apex. Draw the sectional side view and sectional top view.* (UQ)

1. Draw the top, front and left side views.

2. Draw S.P. Name the section points in the front view.
3. Project the section points on the corresponding edges in the *Side View* and complete the *Sectional Side View*.
4. Project 1′, 3′, (4′) and (6′) on the corresponding edges in the top view.
5. Obtain 2 and 5 by projecting them from the sectional side view (*i.e.*, from 2″ and 5″) and complete the sectional top view.

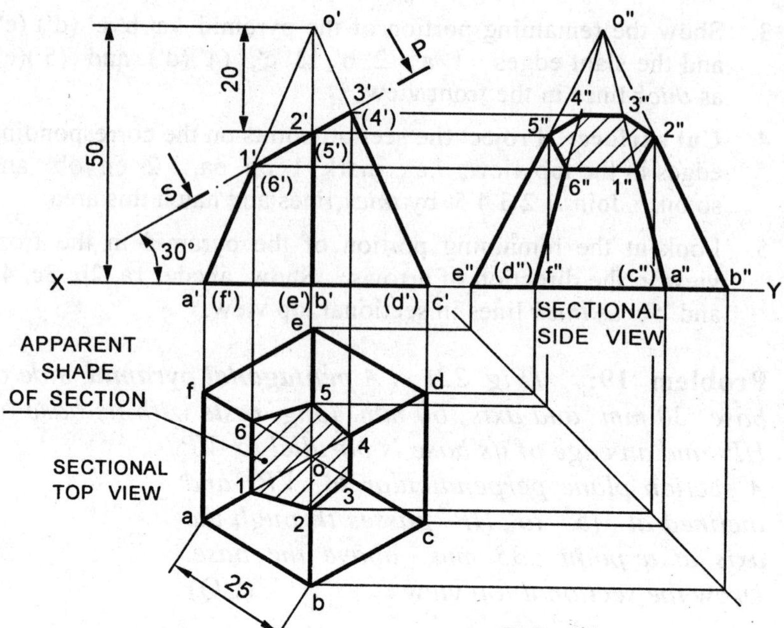

FIG. 23

Note: ***Points 2 and 5 can also be obtained directly from the Front View as explained in Problem 19.***

Problem 22: *(Fig. 24)* ***A hexagonal pyramid****, side of base 25 mm and axis 55 mm long, rests with its base on HP such that one of the edges of its base is perpendicular to VP. It is cut by a section plane perpendicular to HP, inclined at 45° to VP and passing through the pyramid at a distance of 10 mm from the axis. Draw the sectional front view and the true shape of section.* (UQ)

1. Draw the projections of the pyramid for the given position. Draw S.P. and name the section points.
2. Point 3 is on ob. To locate this point in the front view, with o as center and o3 as radius draw an arc to intersect oc at p.
3. Project p and obtain p′ on o′ c′. From p′ draw a horizontal line to intersect o′b′ at 3′.
4. Complete the sectional front view and the true shape of section as shown.

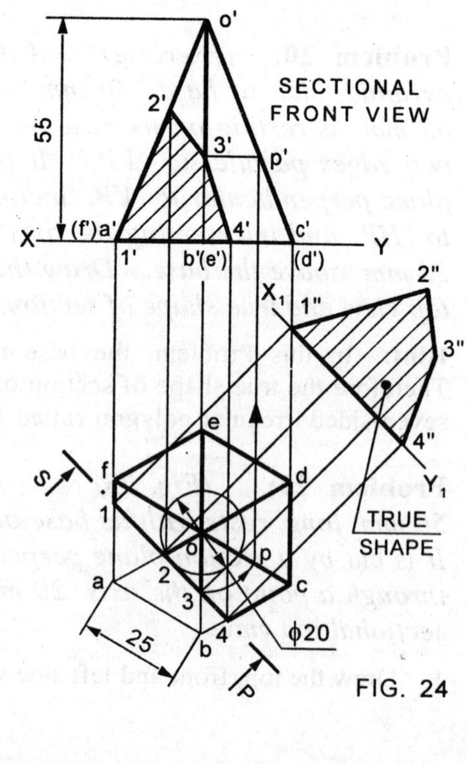

FIG. 24

Problem 23: *(Exercise)* *A **pentagonal pyramid**, side of base 30 mm and axis 60 mm long, rests with its base on HP such that one of the edges of its base is parallel to VP. It is cut by a section plane perpendicular to HP, inclined at 45° to VP and passing through the pyramid at a distance of 8 mm from its axis. Draw the sectional front view and the true shape of section.* (UQ)

Hint: To draw the S.P., first draw a circle with the top view of the axis as center and 8 mm as radius. Then draw a tangent to the circle at 45° to XY.

Problem 24: *(Fig. 25)* *A **pentagonal pyramid**, side of base 26 mm and 52 mm height, rests with its base on HP. One of the edges of its base is perpendicular to VP. A section plane perpendicular to VP and inclined at 45° to HP bisects the axis. Draw the sectional top view and project another top view on an A.I.P. parallel to the section plane showing true shape of section.* (UQ)

See Fig. 25. Draw the sectional top view. Project the remaining portion of the pyramid along with cut surface on an AIP parallel to S.P. Area $1_1 2_1 3_1 4_1 5_1$ is the true shape of section.

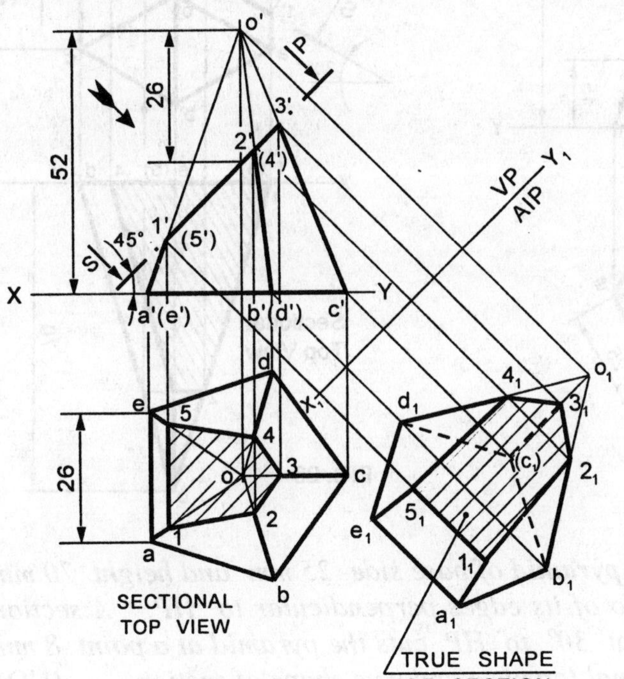

FIG. 25

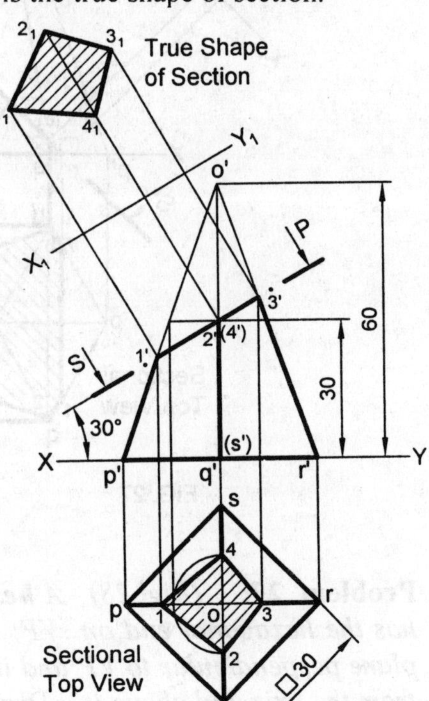

FIG. 26

Problem 25: *(Fig. 26)* *A **square pyramid** of base 30 mm and axis 60 mm long is standing on HP with its base edges equally inclined to VP. It is cut by a section plane perpendicular to VP and inclined at 30° to HP, bisecting the axis. Draw the sectional top view and the true shape of the section, if the upper portion is removed.* (UQ)

See Fig. 26 for Solution.

Problem 26: *(Fig. 27) A* ***hexagonal pyramid****, side of base 25 mm and altitude 70 mm long, rests with its base on HP with two of its base sides parallel to VP. It is cut by a section plane perpendicular to VP and inclined at 45° to HP and passing through a point 15 mm above the base and located on the axis. Draw the sectional top view and the true shape of the section.* (UQ)

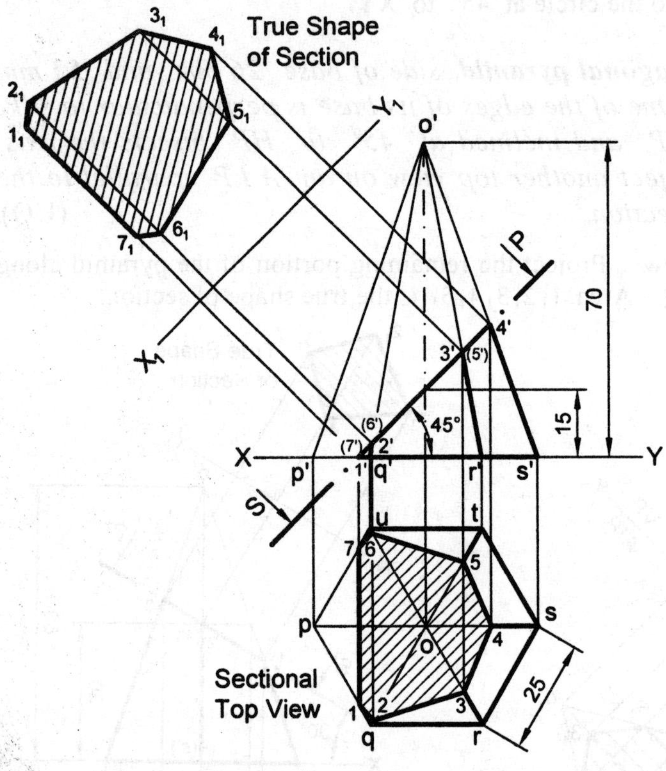

FIG. 27

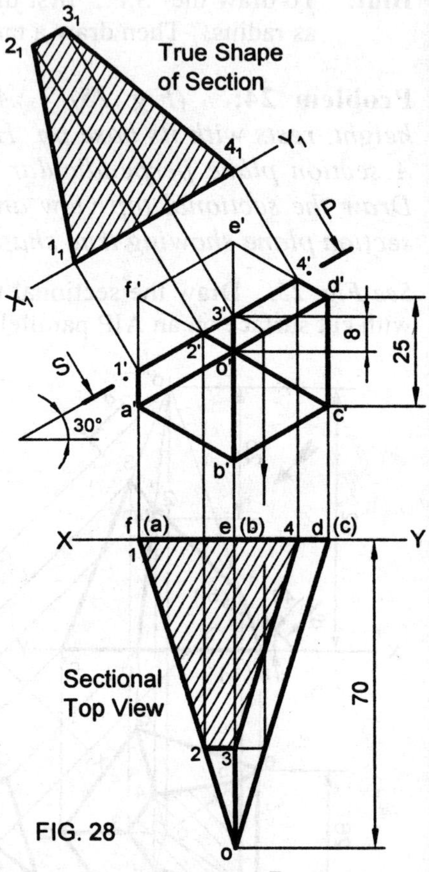

FIG. 28

Problem 27: *(Fig. 28) A* ***hexagonal pyramid*** *of base side 25 mm and height 70 mm has the hexagonal end on VP, with two of its edges perpendicular to HP. A section plane perpendicular to VP and inclined at 30° to HP, cuts the pyramid at a point 8 mm from the axis and above it. Draw sectional top view and true shape of section.* (UQ)

See Fig. 28, which is self-explanatory.

Problem 28: *(Fig. 29) A* ***pentagonal pyramid*** *side of base 25 mm and length of axis 50 mm, lies on one of its triangular faces on HP with its axis parallel to VP. A section plane perpendicular to HP and inclined to VP at 30° cuts the pyramid bisecting its axis. Draw its top view, sectional front view and true shape of section.* (UQ)

See Fig. 29, which is self-explanatory.

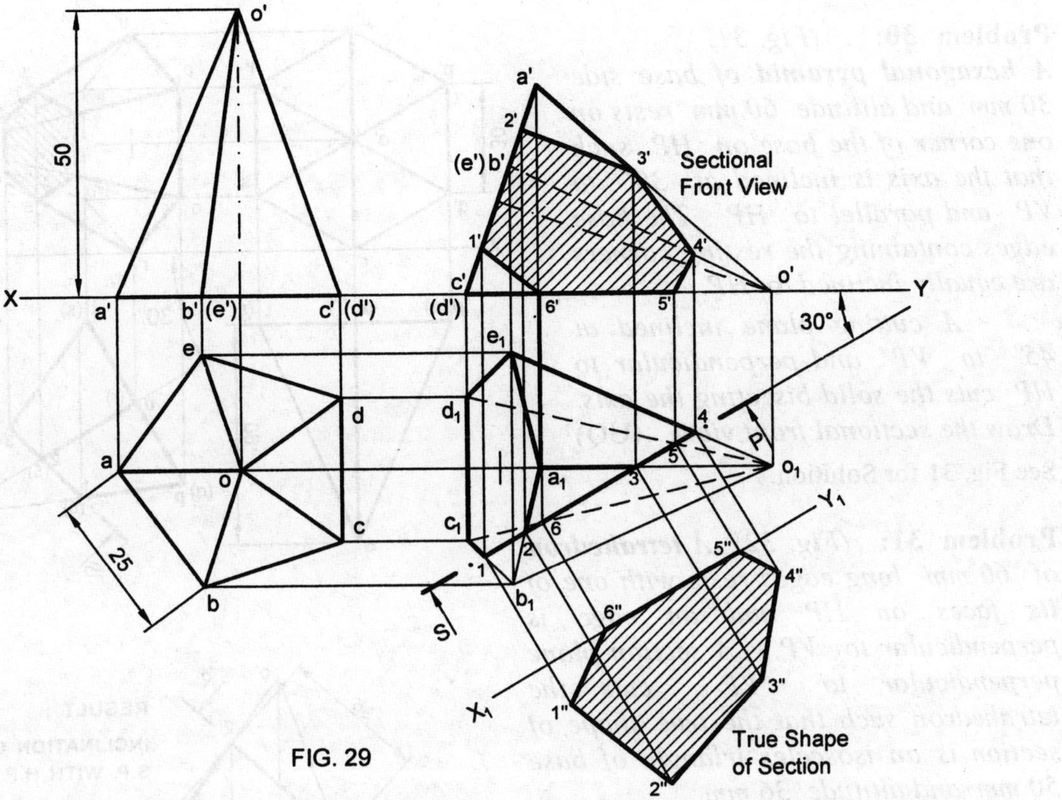

FIG. 29

Problem 29: *(Fig. 30)*
***A hexagonal pyramid** of base 30 mm and axis 70 mm long lies on HP on its base edge such that one of the slant faces is perpendicular to both the planes.*

Draw its projections, when it is cut by a section plane parallel to HP and at a distance of 15 mm from apex point.

(UQ)

See Fig. 30 for Solution.

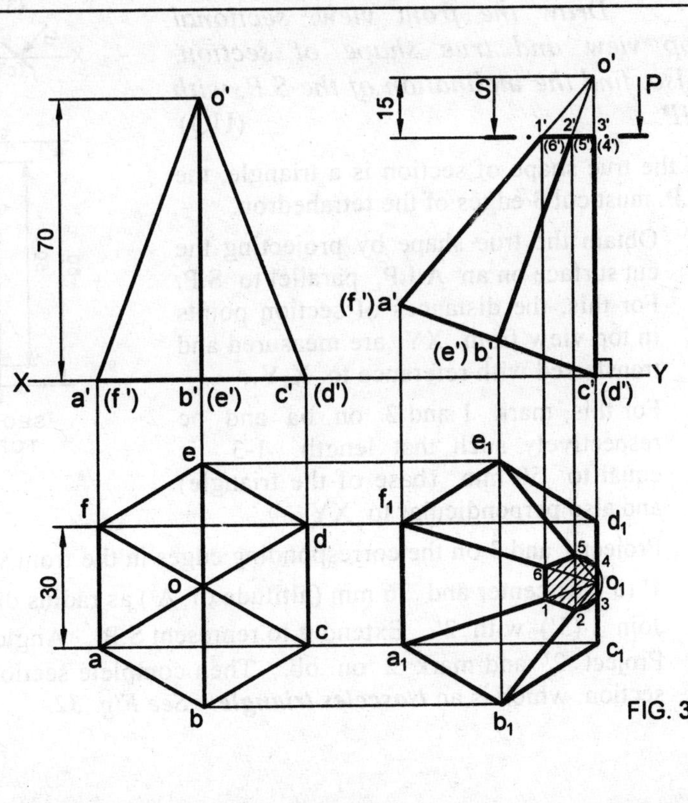

FIG. 30

Problem 30: *(Fig. 31)*
*A **hexagonal pyramid** of base side 30 mm and altitude 60 mm rests on one corner of the base on HP such that the axis is inclined at 30° to VP and parallel to HP. The base edges containing the resting corners are equally inclined to HP.*

A cutting plane inclined at 45° to VP and perpendicular to HP cuts the solid bisecting the axis. Draw the sectional front view. (UQ)

See Fig. 31 for Solution.

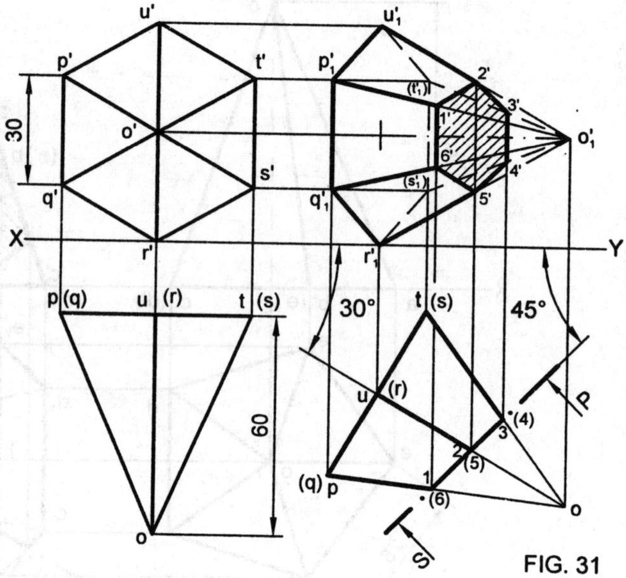

FIG. 31

Problem 31: *(Fig. 32) A **tetrahedron** of 60 mm long edges rests with one of its faces on HP and an edge is perpendicular to VP. A section plane perpendicular to VP cuts the tetrahedron such that the true shape of section is an isosceles triangle of base 50 mm and altitude 36 mm.*

Draw the front view, sectional top view and true shape of section. Also, find the inclination of the S.P. with HP. (UQ)

If the true shape of section is a triangle, the S.P. must cut 3 edges of the tetrahedron.

1. Obtain the true shape by projecting the cut surface on an A.I.P. parallel to S.P. For this, the distances of section points in top view from XY are measured and transferred with reference to X_1Y_1.
   For this, mark 1 and 3 on oa and oc respectively such that length 1-3 is equal to 50 mm (base of the triangle) and also perpendicular to XY.

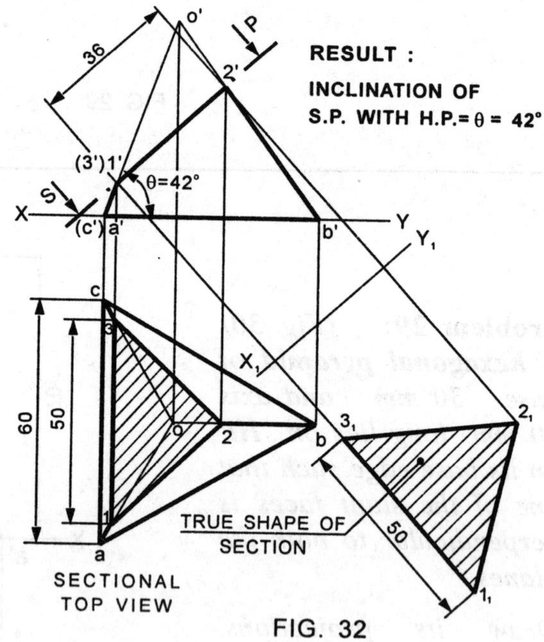

RESULT : INCLINATION OF S.P. WITH H.P.= θ = 42°

SECTIONAL TOP VIEW

TRUE SHAPE OF SECTION

FIG. 32

2. Project 1 and 3 on the corresponding edges in the front view.
3. 1′ (3′) as center and 36 mm (altitude of Δ) as radius draw an arc to cut o′ b′ at 2′.
4. Join 1′(3′) with 2′. Extend it to represent S.P. Angle **θ** between S.P. and XY is **42°**.
5. Project 2′ and mark 2 on ob. Then complete sectional top view. Draw the true shape of section, which is an ***isosceles triangle***. *See Fig. 32.*

Problem 32: *(Fig. 33) A* ***pentagonal pyramid*** *of side of base 25 mm and altitude 50 mm is resting on its base on HP with one side of the base perpendicular to VP.*

It is cut by a plane perpendicular to VP such that the true shape of section is an isosceles triangle of maximum possible base and maximum height.

Draw the front view, sectional top view and true shape of section. Measure the inclination of the cutting plane with HP. (UQ)

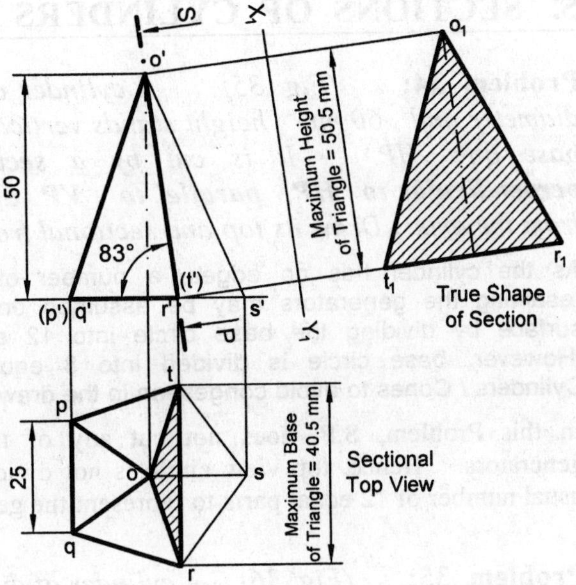

FIG. 33

Results: *See Fig. 33.*
Inclination of S.P. with HP = θ = 83°
Maximum base of isosceles triangle = 40.5 mm
Maximum height of isosceles triangle = 50.5 mm

Problem 33: *(Fig. 34) A* ***square pyramid*** *of 45 mm side of base and altitude 60 mm rests on its base on the ground. A side of base is perpendicular to VP. It is cut by two section planes. One is parallel to its extreme left face and 8 mm away from it, while the other is parallel to the extreme right face and 8 mm away from it. Both the planes intersect each other on the axis of the pyramid.*

Draw the sectional top and front views and Sectional Right Side View. (UQ)

1. Draw the top and front views of the square pyramid.
2. Draw the section planes S_1P_1 and S_2P_2 in the front view as shown. Both the planes intersect at points 2′ and 3′. Mark the other cutting points as shown.
3. Project the corresponding sectional top view. Points 2′ and 3′ are transferred on the slant edge o′r′ and then projected to the top view as shown. Draw the hatching lines in the opposite directions for the left and right faces.
4. Draw the Sectional Right Side View of the remaining portion of the square pyramid as shown.

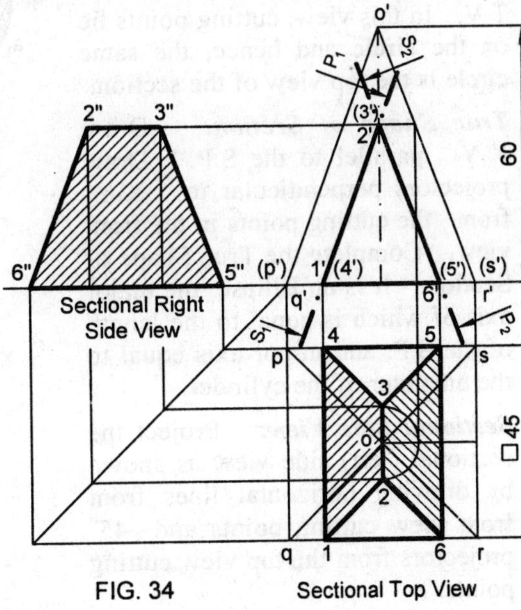

FIG. 34

5. SECTIONS OF CYLINDERS

Problem 34: *(Fig. 35) A cylinder of 50 mm diameter and 60 mm height stands vertically with its base on HP. It is cut by a* ***section plane perpendicular to HP, parallel to VP*** *and 15 mm from the axis. Draw its top and sectional front views.*

As the cylinder has no edges, a number of lines representing the generators may be assumed on its curved surface by dividing the base circle into 12 equal parts. [However, base circle is divided into 8 equal parts in Cylinders / Cones to avoid congestion in the drawings.]

In this Problem, S.P. does not cut any of the cylinder generators. Hence, top view circle is not divided into the usual number of 12 equal parts to represent the generators.

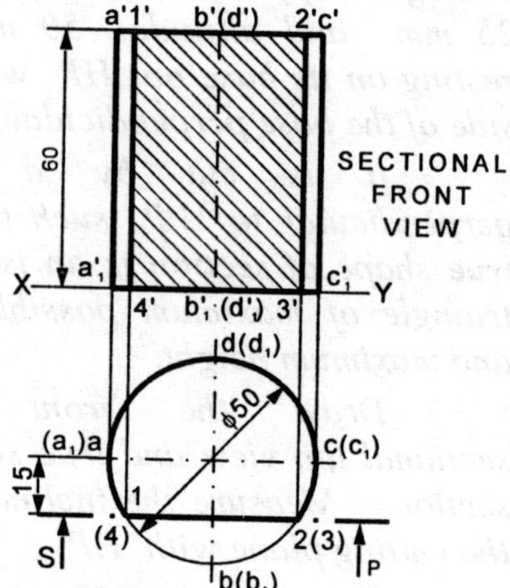

FIG. 35

Problem 35: *(Fig. 36) A cylinder of diameter 40 mm and height 60 mm is having its axis vertical. It is cut by a* ***plane perpendicular to VP*** *and* ***inclined at 30° to HP****. The plane bisects the axis of the cylinder. Draw its front view, sectional top view, sectional side view and true shape of section.* (UQ)

1. Draw the top and front views of the cylinder. In front view, draw a S.P. at 30° and bisecting the axis of the cylinder.
2. Name the cutting points in the front view. Project these cutting points to top view and obtain Sectional T.V. In this view, cutting points lie on the circle and hence, the same circle is the top view of the section.
3. ***True Shape of Section:*** Draw X_1Y_1 parallel to the S.P. Draw projectors perpendicular to X_1Y_1 from the cutting points in the front view. Complete the True Shape of Section. It is an Ellipse, the major axis of which is equal to the length of the S.P. and minor axis equal to the diameter of the cylinder.
4. ***Sectional Side View:*** Project the sectional (left) side view as shown by drawing horizontal lines from front view cutting points and 45° projectors from the top view cutting points.

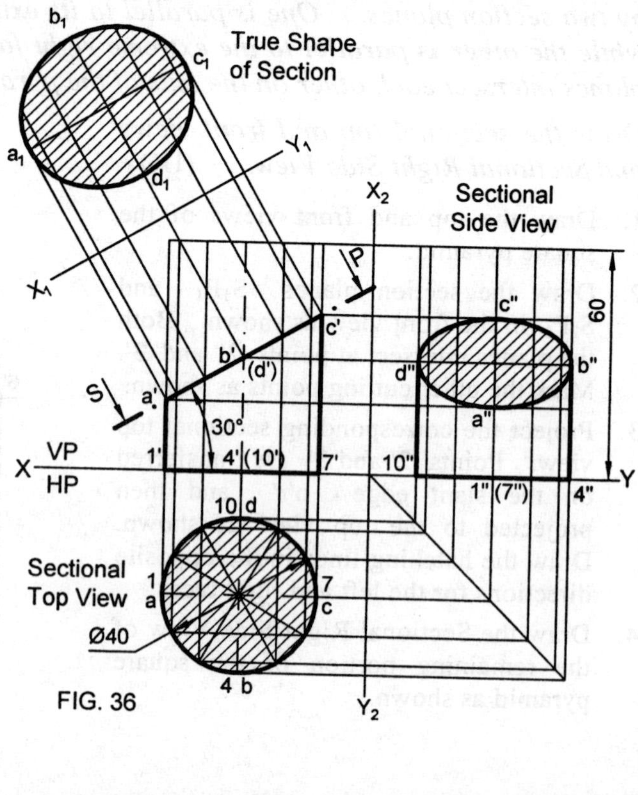

FIG. 36

Section is seen as an ellipse in side view. Major axis is equal to the diameter of the cylinder.

Note: *The sectional side view will be a circle, if the S.P. is inclined at 45° with HP.*

Problem 36: *(Fig. 37)*
A ***horizontal cylinder*** *of 50 mm diameter and 75 mm length has its axis parallel to VP. It is cut by an* ***A.I.P.*** *such that the true shape of section is an ellipse of major axis 70 mm.*

Locate the cutting plane and draw the sectional side view and true shape of section. (UQ)

Length of S.P. between two extreme generators of the cylinder should be equal to 70 mm.

With the midpoint of the axis as center and 35 mm (=70/2) as radius, draw two arcs to cut the two extreme generators as shown.

Join these two points to obtain S.P. Mark cutting points as shown in Fig. 37.

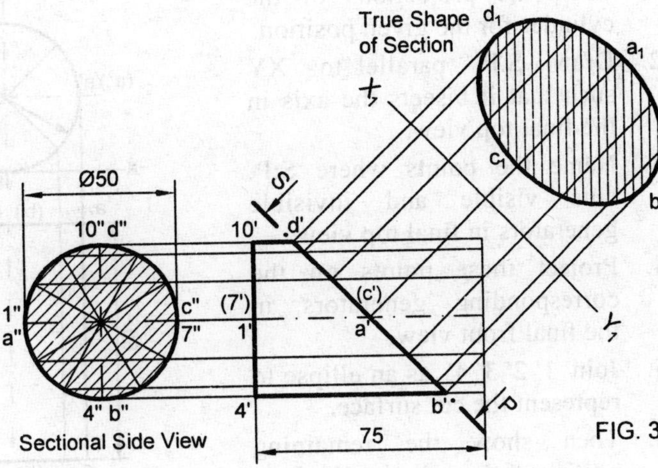

FIG. 37

Problem 37: *(Fig. 38)*
A cylinder 45 mm diameter and 55 mm height, is resting on its base on HP. It is cut by a ***plane perpendicular to VP****, the V.T. of which cuts the axis 35 mm from the base and makes* ***45° with HP****.*

Draw its front view, sectional top view and another sectional top view on an A.I.P. parallel to the section plane. (UQ)

See Fig. 38. In this Problem, the top face of the cylinder is also sectioned. Therefore, the true shape of section is a part of an ellipse as shown in the Auxiliary Top View.

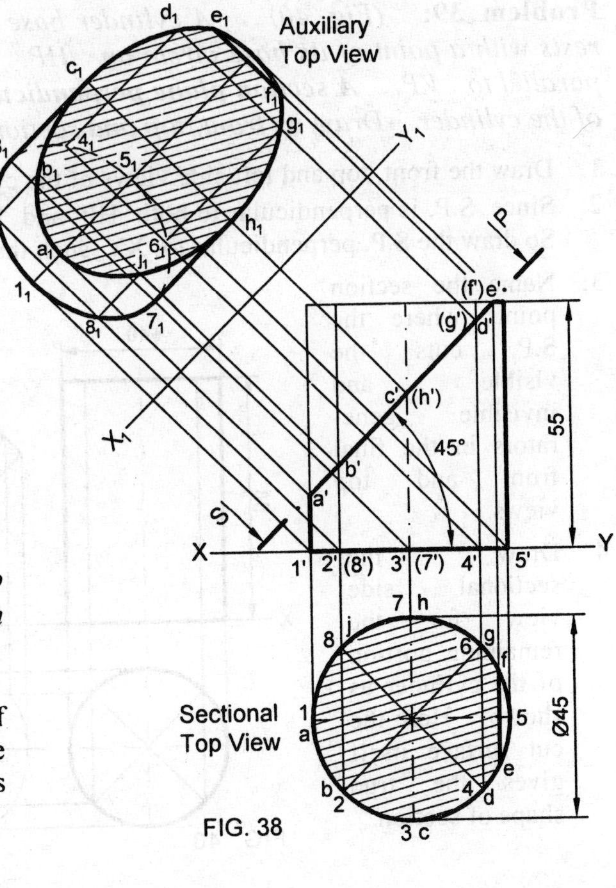

FIG. 38

Problem 38: *(Fig. 39) A cylinder of 50 mm diameter and axis 80 mm long, lies with one of its generators on HP such that its axis is inclined at 45° to VP.* ***A section plane parallel to VP*** *bisects axis of the cylinder. Draw the top and sectional front views.* (UQ)

1. Draw the projections of the cylinder for the given position.
2. Draw S.P. parallel to XY such that it bisects the axis in the final top view.
3. Name the points where S.P. cuts visible and invisible generators in final top view.
4. Project these points on the corresponding generators in the final front view.
5. Join 1′ 2′ 3′ 4′ as an ellipse to represent the cut surface.
6. Then show the remaining portion of the cylinder *thick*.

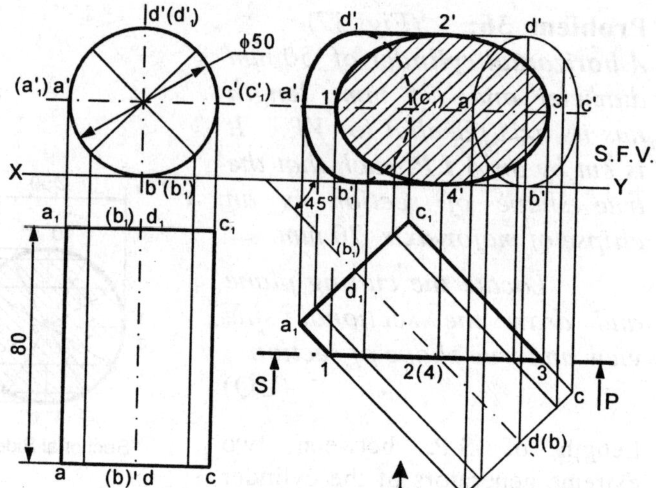

FIG. 39

Problem 39: *(Fig. 40) A cylinder base 40 mm diameter and axis 58 mm long, rests with a point of its base circle on HP. Its axis is inclined at 45° to HP and parallel to VP.* ***A section plane perpendicular to both HP and VP*** *bisects the axis of the cylinder. Draw its front, top and sectional side views.* (UQ)

1. Draw the front, top and left side views of the cylinder for the given position.
2. Since S.P. is perpendicular to both HP and VP, it appears as a straight line in both views. So draw the S.P. perpendicular to XY such that it bisects the axis in both the views.
3. Name the section points where the S.P. cuts the visible and invisible generators in the final front and top views.
4. Draw the sectional side view for the remaining portion of the cylinder as shown. Here the cut surface itself gives the true shape of section.

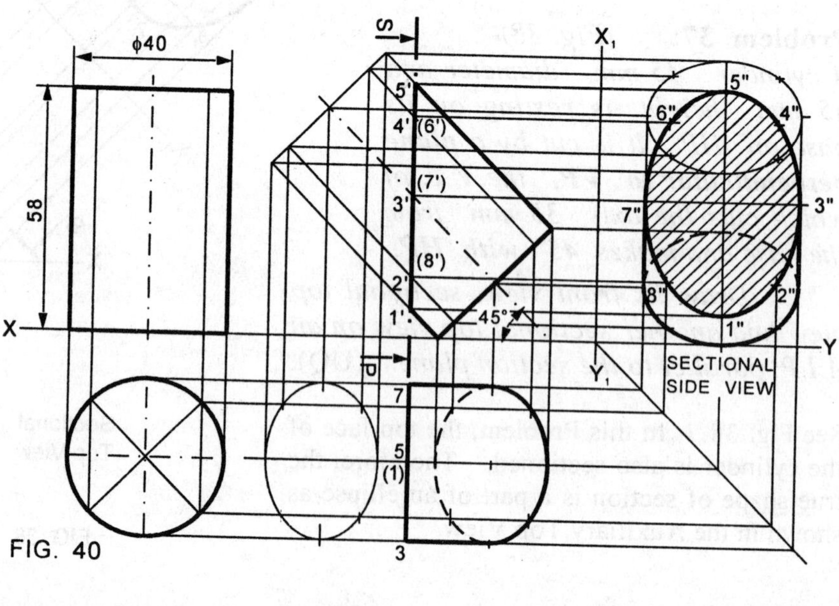

FIG. 40

Problem 40: *(Fig. 41) A cylinder of 45 mm diameter and 60 mm long rests with one of its generators on HP such that its axis is parallel to both HP and VP.* ***A section plane perpendicular to VP** and **inclined at 30° to HP*** *passes through a point on the axis 25 mm from one of its ends. Draw the sectional top view.* (UQ)

1. Draw the top and front views of the cylinder in simple position.
2. Tilt the front view such that the cylinder is lying with one of its generators on HP and obtain the second front view. Project the corresponding Second Top View.
3. Draw S.P. at 30° to XY in Second F.V., passing through c′ 25 mm from right end. S.P. cuts both the circular ends and generators of the cylinder.
4. Project all cutting points from Second F.V. to Second T.V. except a′, f′, (g′) and (m′) lying on left and right circular ends.
5. Cuttings points a′, f′, (g′) and (m′) cannot be projected directly because the two circular ends of the cylinder appear ⊥^{r} to XY in both Second top and front views. Mark these points in the top view as follows: Transfer the distances f′ 1′ and a′ p′ from the Second F.V. to First F.V. Project the cutting points so obtained in the First F.V. to the First T.V. Project these cutting points in the First T.V. to the Second T.V. Complete the Sectional Top View.

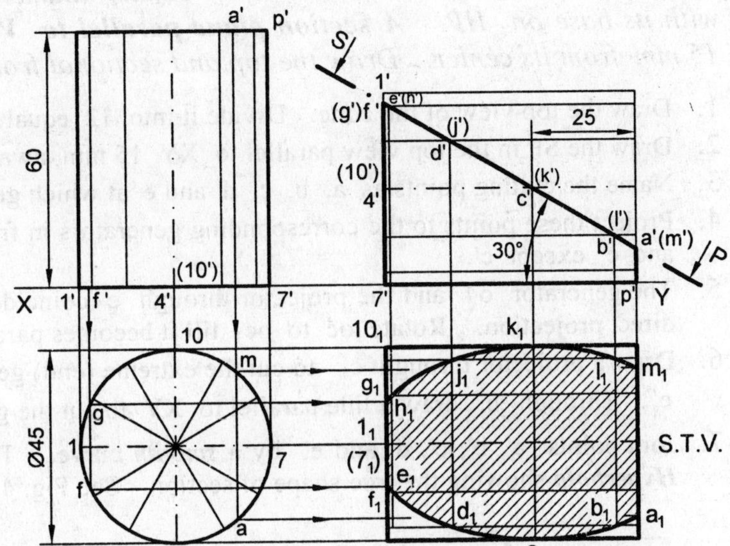

FIG. 41

Problem 41: *(Fig. 42) A cylinder of 55 mm diameter and axis 65 mm long has a coaxial pentagonal hole of 22 mm sides. The cylinder is placed on HP with its axis perpendicular to VP. A* ***vertical section plane*** *cuts the cylinder such that it meets the bases at 7 mm from the diametrically opposite ends. Draw the sectional front view and true shape of section.* (UQ)

See Fig. 42 which is self-explanatory.

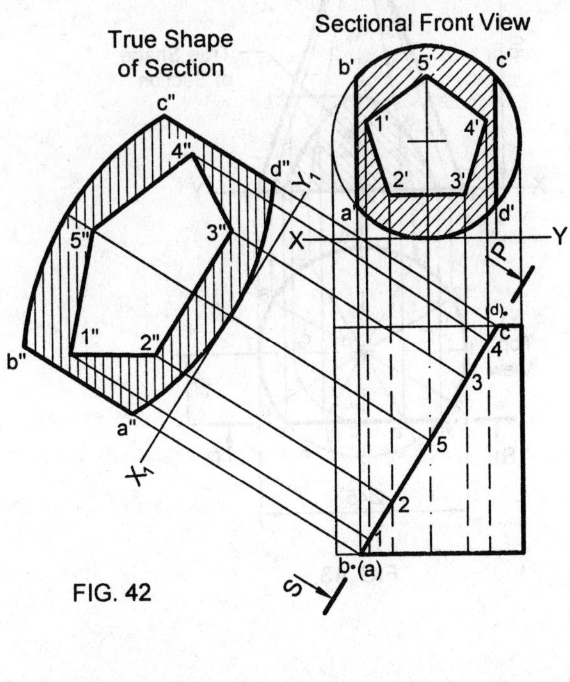

FIG. 42

6. SECTIONS OF CONES

Sectional views of cones are drawn by using any one of the *two* methods, viz., ***Circle Method*** and ***Generator Method***. Circle Method is more time-consuming and is rather more difficult when the cone is in inclined position. Generator Method is more suitable, particularly when the cone is inclined. Therefore, Generator Method is followed in the following problems.

Problem 42: *(Fig. 43) A cone of 45 mm diameter and altitude 60 mm is resting with its base on HP.* ***A section plane parallel to VP*** *cuts the cone at a distance of 15 mm from its center. Draw the top and sectional front views.*

1. Draw the top view of the cone. Divide it into 12 equal parts. Project the front view.
2. Draw the SP in the top view parallel to XY 15 mm away from the center o.
3. Name the cutting points as a, b, c, d and e at which generators are cut in the top view.
4. Project these points to the corresponding generators in front view. Mark them as a′, b′, d′ and e′ except c′.
5. The generator o4 and the projector through c coincide. Hence, c′ cannot be located by direct projection. Rotate oc to oc_1 till it becomes parallel to XY.
6. Draw a projector through c_1 to cut the extreme (end) generator in the front view at the point c_1'. Through c_1' draw a line parallel to XY to cut the generator o′4′ at c′.
7. Join points a′, b′, c′, d′ and e′ by a *smooth* curve. The curve obtained is a ***Rectangular Hyperbola*** showing the true shape of section. See Fig. 43.

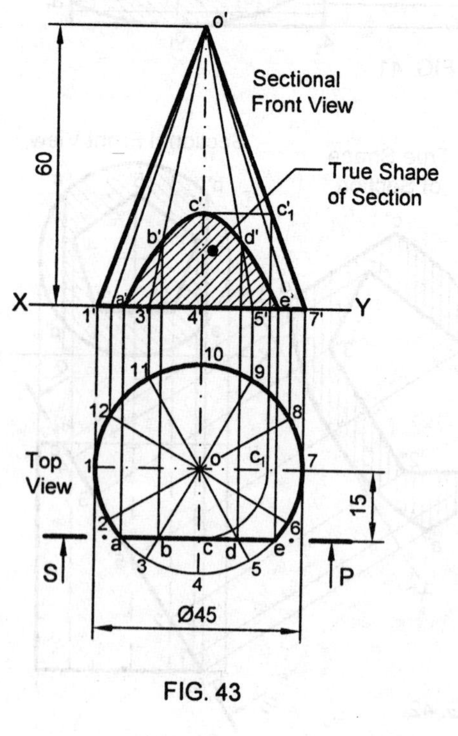

FIG. 43

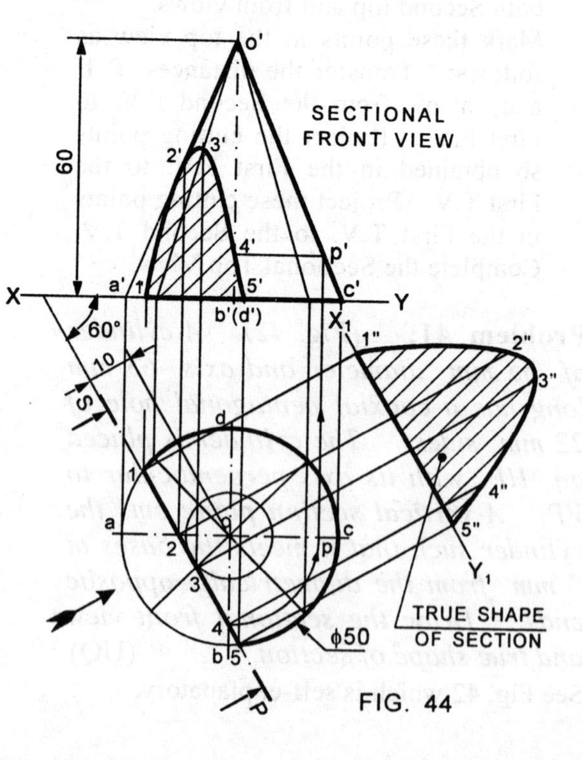

FIG. 44

Problem 43: *(Fig. 44) A* **cone**, *base 50 mm diameter and axis 60 mm long, rests with its base on HP. It is cut by a section plane perpendicular to HP, inclined at 60° to VP and at a distance of 10 mm from its axis. Draw the sectional front view and the true shape of section.* (UQ)

See Fig. 44. Section point 4 is on ob. How to locate 4′ in the sectional front view? For this, with o as center and o4 as radius draw an arc to cut oc at p. Project p and obtain p′ on o′c′. From p′ draw a horizontal to cut o′ b′ at 4′.

Problem 44: *(Fig. 45) A* **cone**, *base 50 mm diameter and axis 65 mm long, rests with its base on HP. It is cut by a section plane perpendicular to VP, inclined at 45° to HP and passing through a point on the axis 35 mm above the base. Draw the sectional top view, sectional side view and true shape of section.* (UQ)

In this Problem, S.P. is inclined to the base of the cone at an angle smaller than the angle of inclination of the generators with the base.

Note: *Points 3′ and (7′) cannot be projected directly. Therefore, the same method as in the case of pyramids should be used to find the positions of 3 and 7 in top view.*

To obtain the points 3 and 7, draw horizontal from 3′ (7′) to cut o′e′ at p′. Project p′ on oe. With o as center and op as radius draw a semi-circle to cut oc and og at 3 and 7 respectively.

True Shape of Section

Draw X_1Y_1 parallel to S.P. Draw projectors through 1′, 2′, 3′, etc.

Obtain the true shape of section (***Ellipse***).

True shape of the section is an ***Ellipse*** with major axis equal to the length of the section and minor axis equal to the width of the section at its center.

Sectional Side View

Draw the sectional side view by projecting the points on corresponding generators.

Apparent Section

The section shown in the top view or front view is called Apparent Section, since the S.P. is not parallel to VP or HP.

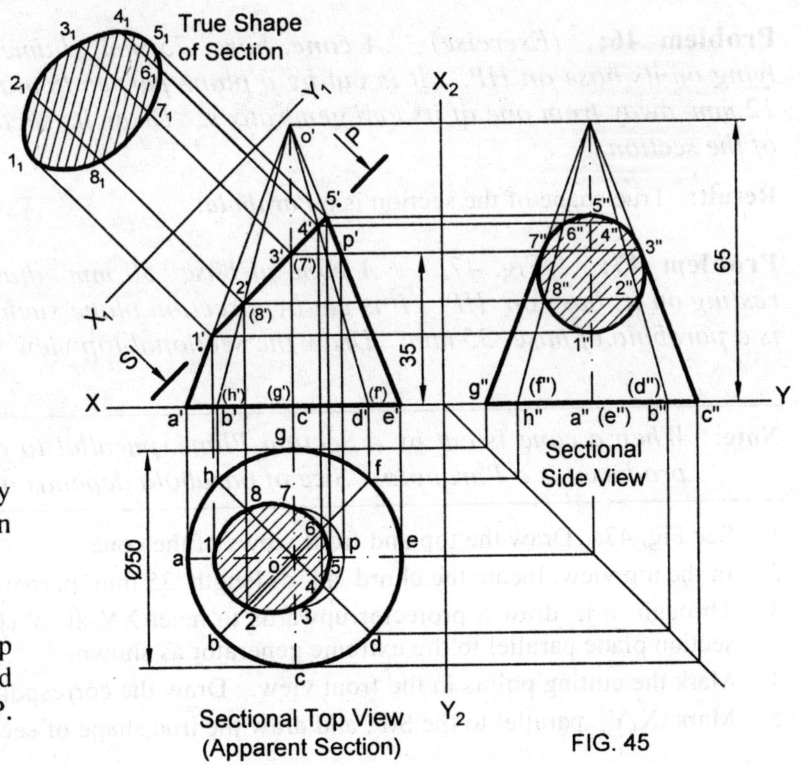

Sectional Top View (Apparent Section)

FIG. 45

Problem 45: *(Fig. 46) A* ***cone****, base 40 mm diameter and axis 60 mm long, rests with its base on HP. It is cut by a section plane perpendicular to VP, parallel to one of the end generators and passing through a point on the axis 25 mm from the apex. Draw sectional top view and true shape of the section.* (UQ)

Here, the S.P. is parallel to an end/extreme generator of the cone.

In this Problem, S.P. cuts not only the generators, but also the base of the cone at 1′(7′) in the front view. Hence project these points to the top view.

Obtain other section points in sectional top view as explained in previous Problem. Draw X_1Y_1 parallel to S.P. and obtain true shape of section (***Parabola***).

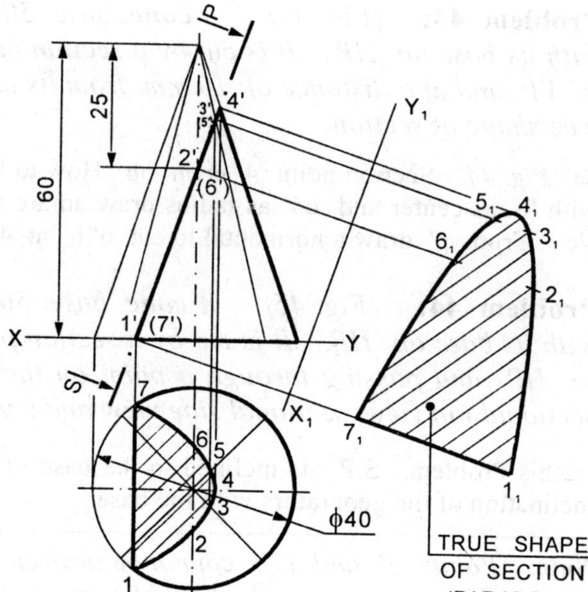

SECTIONAL TOP VIEW

FIG. 46

Note: *For a section to be a Parabola, the S.P. must be parallel to the End Generator.*

Problem 46: *(Exercise) A* ***cone****, base 75 mm diameter and axis 80 mm long is lying on its base on HP. It is cut by a plane perpendicular to VP and parallel to and 12 mm away from one of its end generators. Draw its sectional top view and true shape of the section.* (UQ)

Result: True shape of the section is a ***Parabola***.

Problem 47: *(Fig. 47) A cone of base 50 mm diameter and 65 mm height is resting on its base on HP. It is cut by a section plane such that the true shape produced is a parabola of base 35 mm. Draw the sectional top view. Also, find its true shape.* (UQ)

Note: *When a cone is cut by a Section Plane, parallel to an end generator, the section produced is a Parabola. Size of parabola depends on the position of S.P.*

---

1. See Fig. 47. Draw the top and front views of the cone.
2. In the top view, locate the chord a j of length 35 mm perpendicular to XY.
3. Through a j, draw a projector upwards to meet XY at a′ (j′). Through a′ (j′), draw the section plane parallel to the extreme generator as shown.
4. Mark the cutting points in the front view. Draw the corresponding sectional top view.
5. Mark X_1Y_1 parallel to the S.P. and draw the true shape of section as a parabola.

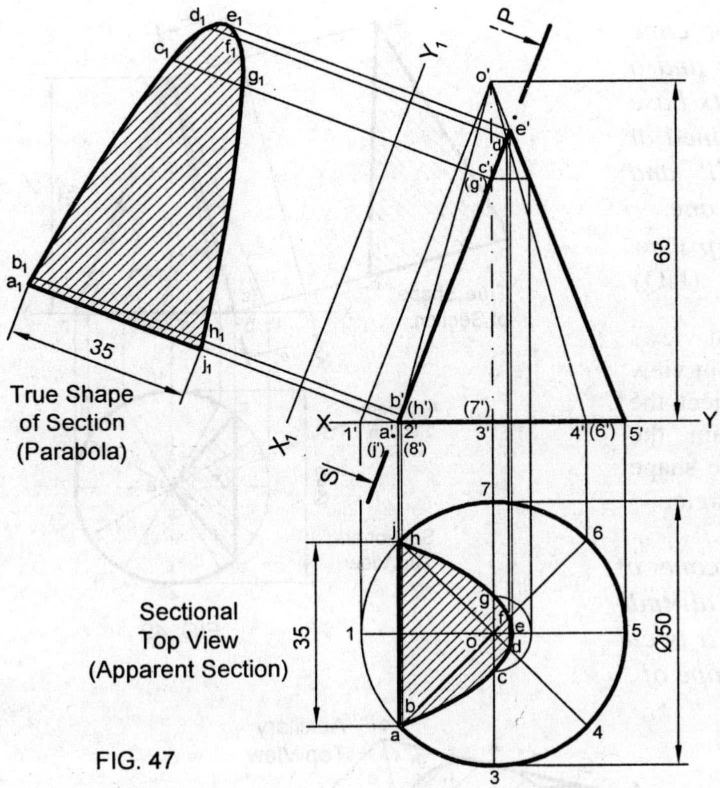

FIG. 47

Problem 48: *(Exercise)* *A **cone** of base 70 mm diameter, standing upright is cut by a section plane such that, the true shape is a parabola of maximum double ordinate 50 mm and the vertex of the parabola is 70 mm away from this ordinate.*

Draw the front view, top view and the true shape of section. (UQ)

Hint: First find the height of the cone. For this, draw the base circle and mark the double ordinate 50 mm of the parabola equally about the horizontal center line.

Problem 49: *(Fig. 48)* *A **cone** of base 50 mm diameter and axis 65 mm long, lies on HP on one of its generators with its axis parallel to VP. A horizontal section plane bisects the axis of the cone.*

Draw the front view and sectional top view. (UQ)

1. Draw S.P. parallel to XY such that it bisects the axis in the final front view.
2. Name the section points at which the base and the generators are cut by the S.P. in correct sequence as shown.
3. Project the section points to the final top view and draw the cut surface.
4. Then complete the sectional top view.

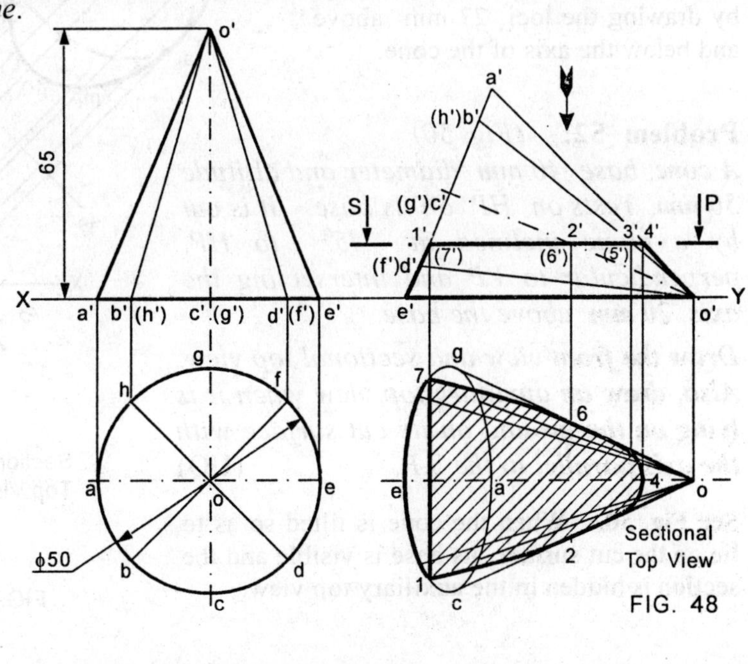

FIG. 48

Problem 50: *(Fig. 49) A cone with a 45 mm diameter of base and a 55 mm length of axis, rests on its base on HP. It is cut by a plane inclined at 75° to HP, perpendicular to VP and passing through the vertex of the cone. Draw the front view, sectional top view and true shape of section.* (UQ)

See Fig. 49. Draw the top and front views of the cone. Mark the S.P. in the front view and name the cutting points. Project the same to the top view and obtain the Sectional Top View. Draw the true shape of section, which is an *Isosceles Triangle*.

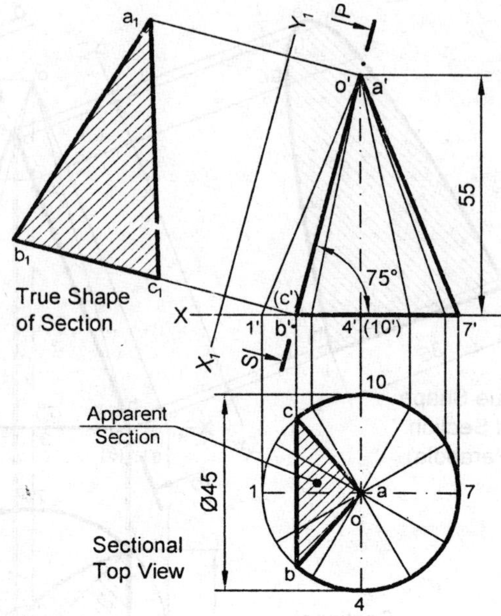

FIG. 49

Problem 51: *(Exercise) A cone of base diameter 70 mm and altitude 90 mm rests on its base on HP. It is cut by an A.I.P. such that the true shape of section is an isosceles triangle with a 54 mm base. Draw the S.P., sectional top view and true shape of section.*

Hint: In the top view, locate base of the triangle 54 mm in top view, by drawing the loci 27 mm above and below the axis of the cone.

Problem 52: *(Fig. 50)*
A cone, base 40 mm diameter and altitude 50 mm, rests on HP on its base. It is cut by a plane inclined at 45° to HP, perpendicular to VP and intersecting the axis 20 mm above the base.

Draw the front view and sectional top view. Also, draw an auxiliary top view when it is lying on the ground on its cut surface with the axis parallel to the S.P. (UQ)

See Fig. 50. When the cone is tilted so as to lie on the cut surface, its base is visible and the section is hidden in the auxiliary top view.

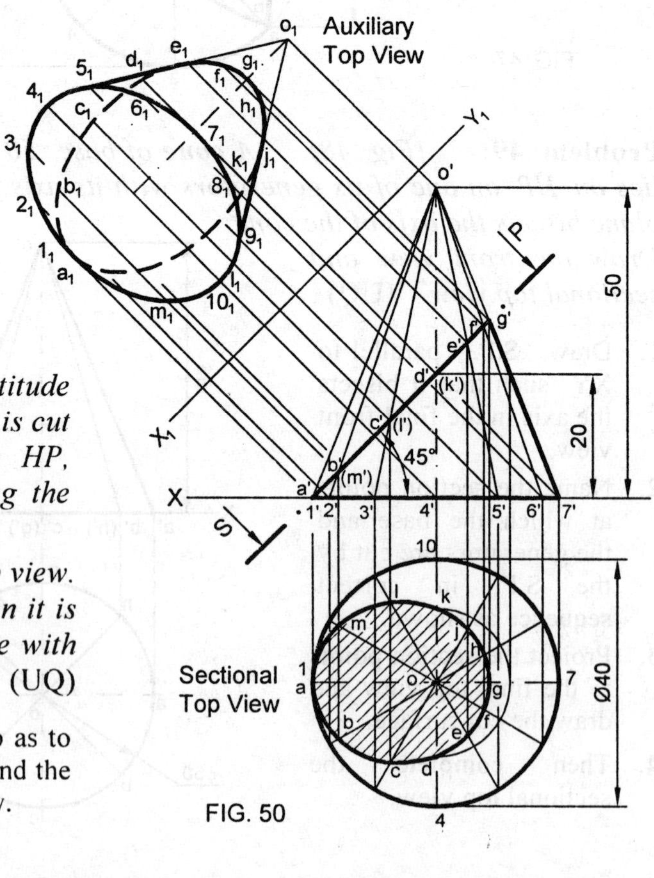

FIG. 50

22

DEVELOPMENT OF SURFACES

In industrial world, an engineer is frequently confronted with problems where the development of surfaces of an object has to be made in the fabrication of furnaces, chimneys, boilers, automobile body building, ship building, aircraft, etc. Development of surface of an object means the unrolling or unfolding of all surfaces of the object on a plane. A typical example is your tooth paste carton of the shape of a square / rectangular prism. Open its two bases and cut the carton along a longer edge. Place it on a plane surface to obtain the development of the carton.

Layout of only the sides of the object without its top or bottom is called *Lateral Surface.*

Principle of Development: Every line on the development should show the *true length* of the corresponding line on the surface which is developed. Surface generated by moving a straight line is called *Plane Surface* and that generated by a curve is called *Curved Surface*.

Solids bounded by Plane Surfaces : Prism, Pyramid, Tetrahedron
Solids bounded by Single-curved Surfaces : Cylinder, Cone
Solids bounded by Double-curved Surfaces : Sphere, Torus, Paraboloid

Sheet Metal Work: To fabricate a sheet metal product, first cut the metal sheet to the shape formed by laying all the lateral surfaces one adjacent to other, bend, fold and then join by welding or riveting. It is called *Sheet Metal Work*.

1. METHODS OF DEVELOPMENT OF LATERAL SURFACES

(i) Parallel Line Development
See Figs. 1(a) and (b). This method is used to develop *cubes*, *prisms* and *cylinders*, in which all the edges / generators of lateral surfaces are *parallel* to each other. Prism and cylinder roll out into rectangular stretch-outs by this method.

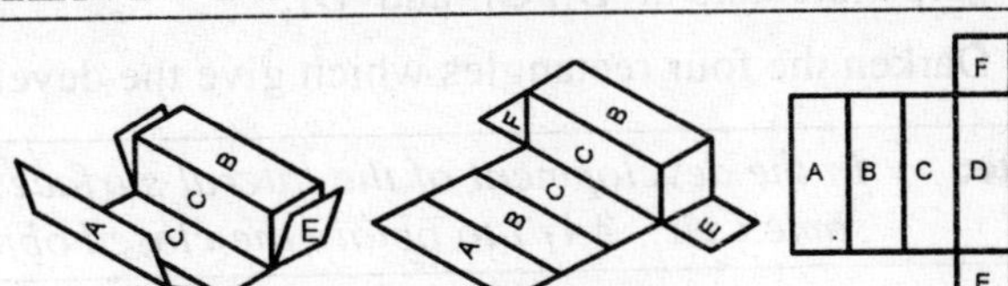

FIG. 1(a) PRISM

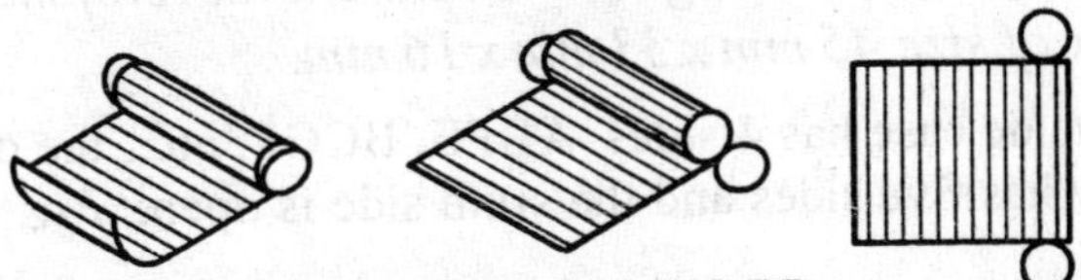

FIG. 1(b) CYLINDER

(ii) Radial Line Development
See Figs. 1(c) and (d). Radial This method is employed for *pyramids* and *cones* in which the apex is taken as center and the true length of the slant edge or generator as radius for its development.

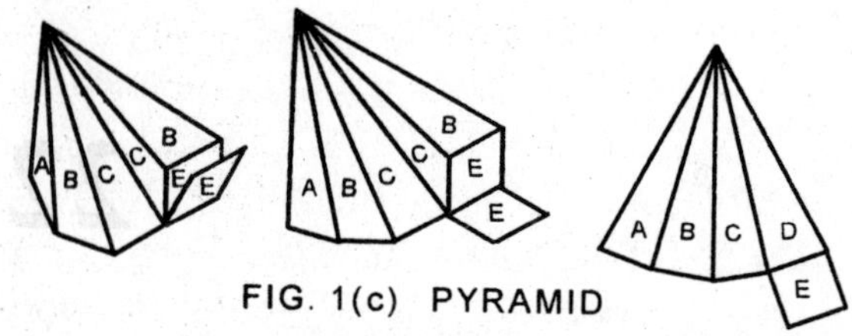

FIG. 1(c) PYRAMID

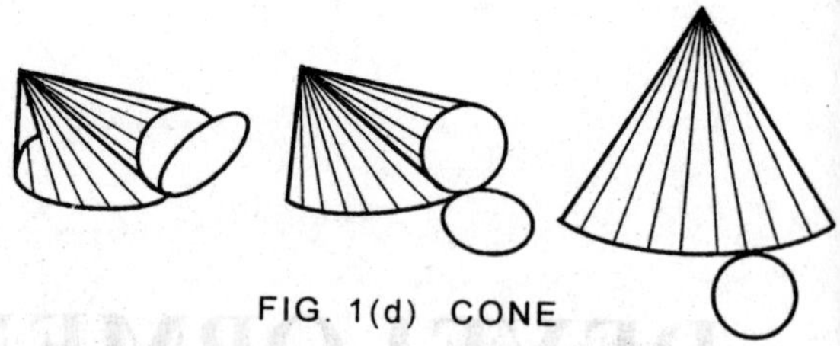

FIG. 1(d) CONE

Pyramids and cones roll out into isosceles triangles and sectors of a circle respectively.

(iii) **Approximate Development:** This method is employed for double-curved surfaces like spheres, as they are theoretically un-developable in a single plane.

(iv) **Triangulation Development:** This method is used for developing transition pieces.

2. PRISMS AND CYLINDERS – Parallel Line Development

Problem 1: *(Fig. 2) Draw the development of the lateral surfaces of a right* ***square prism*** *of edge of base 30 mm and axis 50 mm long.*

1. Draw the top and front views.
2. Development consists of four equal rectangles of size 50 mm x 30 mm in contact and in sequence. Hence draw a rectangle $A_1 A_1 AA$ such that $A_1 A_1$ = Perimeter of the base of the prism = Length of the stretch-out line = AA. AA_1 = its Height.
3. On A_1A_1, mark four equal divisions A_1B_1, B_1C_1, etc., each equal to the side of base of the prism 30 mm. Erect perpendiculars at B_1, C_1 and D_1.
4. Darken the four rectangles which give the development of the lateral surface of the prism.

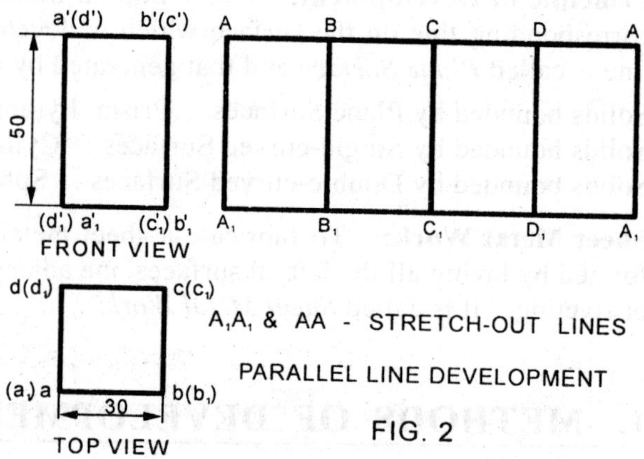

FIG. 2

Note: *In the development of the lateral surfaces, the starting and closing edges should be the same (viz., AA_1) to obtain the closed object.*

Problem 2: *(Fig. 3) Draw the development of the outside case and tray of a match-box of size 45 mm x 33 mm x 16 mm.*

Outside case has 4 sides ABFE, BCGF, etc., the development of which is shown in *Fig. (i)*. The tray has five sides and the sixth side is open. *Fig. (ii)* shows the development of the tray.

Problem 3: *(Fig. 4) Draw the development of the* ***complete surface*** *of a G.I.* ***cylindrical drum*** *with lid. Diameter is 30 cm and height is 1.6 times the diameter.*

Since the drum has bottom base and lid, the complete development is as shown.

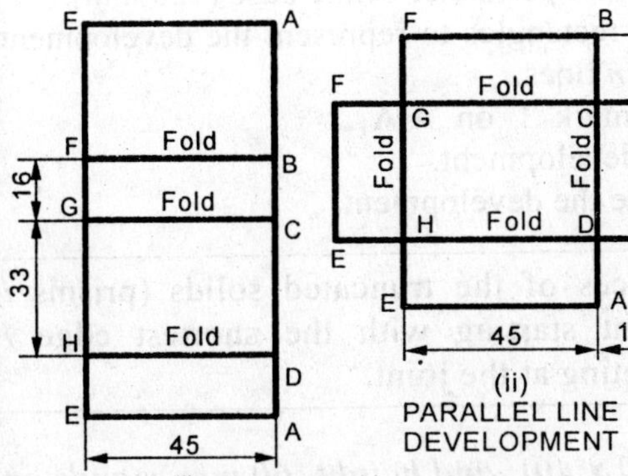

(i)

(ii) PARALLEL LINE DEVELOPMENT

FIG. 3

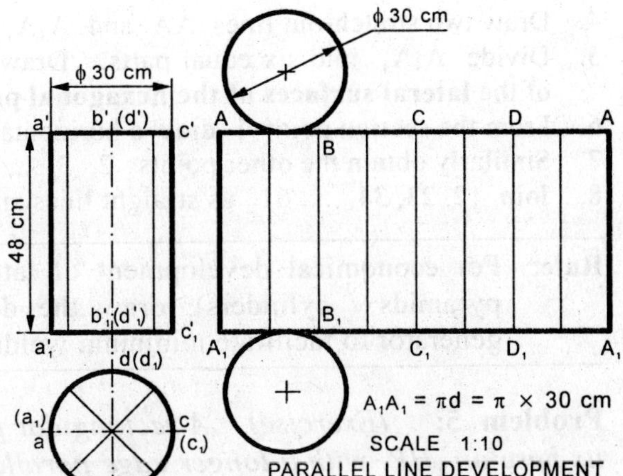

PARALLEL LINE DEVELOPMENT

FIG. 4

Note: *1. It is advisable to draw the top and bottom circles at the* ***extreme end*** *as shown, to facilitate* ***economical way of cutting*** *the sheet metal.*

2. Usually the latest surfaces of solids are developed and the ends or bases are omitted from the developments. They can easily be added whenever required.

3. TRUNCATED PRISMS AND CYLINDERS

Problem 4: *(Fig. 5) A hexagonal prism, edge of base 20 mm and axis 50 mm long, rests with its base on HP such that one of its rectangular faces is parallel to VP. It is cut by a plane perpendicular to VP, inclined at 45° to HP and passing through the right corner of the top face of the prism. (i) Draw the* ***sectional*** *top view. (ii) Develop the lateral surfaces of the* ***truncated prism****.* (UQ)

(i) Sectional Top View

1. Draw top and front views of the prism for the given position.
2. Draw the S.P. in front view. Mark section points.
3. Draw sectional top view and hatch it.

(ii) To draw the development of the Lateral Surfaces of the Truncated Hexagonal Prism

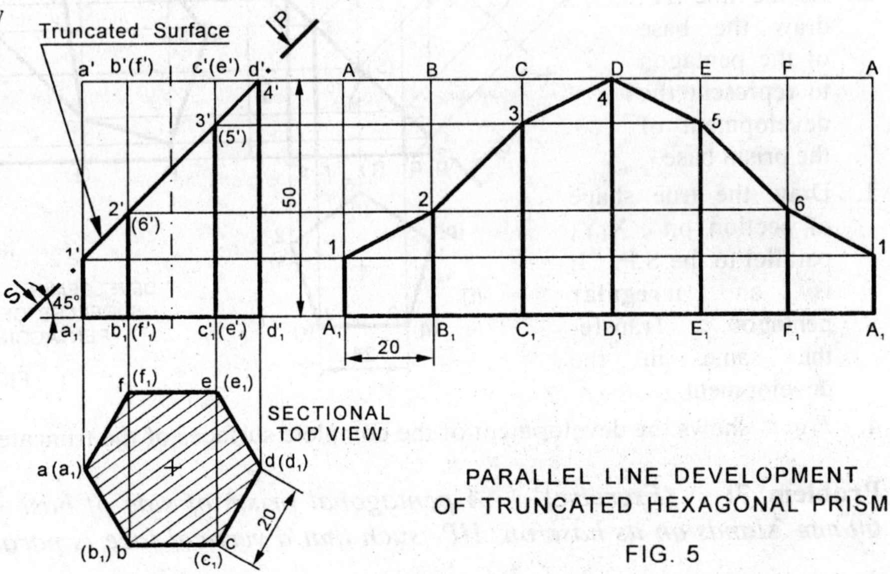

PARALLEL LINE DEVELOPMENT OF TRUNCATED HEXAGONAL PRISM

FIG. 5

4. Draw two stretch-out lines AA and A_1A_1 each equal to perimeter of the base (120 mm).
5. Divide A_1A_1 into six equal parts. Draw six equal rectangles to represent the development of the **lateral surfaces of the hexagonal prism** in *thin* lines.
6. From the section point 1 draw a horizontal line and mark 1 on $A A_1$.
7. Similarly obtain the other points 2, 3, ..., 6 in the development.
8. Join 12, 23, 34, ... 61 as straight lines and complete the development.

Rule: For economical development of lateral surfaces of the truncated solids (prisms / pyramids / cylinders), draw the development starting with the shortest edge / generator to facilitate minimum welding or riveting at the joint.

Problem 5: *(Exercise) A rectangular prism (60 x 40) and height 60 mm stands on its base on HP with a longer edge parallel to VP. It is cut by a plane normal to VP, inclined at 60° to HP and passing through left corner of the top face of the prism. Develop the lateral surfaces of the truncated prism.*

Problem 6: *(Fig. 6) A pentagonal prism, side of base 25 mm and altitude 50 mm, rests on its base on the HP such that an edge of the base is parallel to VP and nearer to the observer. It is cut by a plane inclined at 45° to HP, perpendicular to VP and passing through the center of the axis. (i) Draw the true shape of section. (ii) Draw the development of the complete surfaces of the truncated prism.* (UQ)

1. Draw the development of lateral surfaces of truncated pentagonal prism.
2. At the line TP, draw the base of the pentagon to represent the development of the prism base.
3. Draw the true shape of section on X_1Y_1 parallel to the S.P. It is an **irregular** pentagon. Transfer the same in the development.

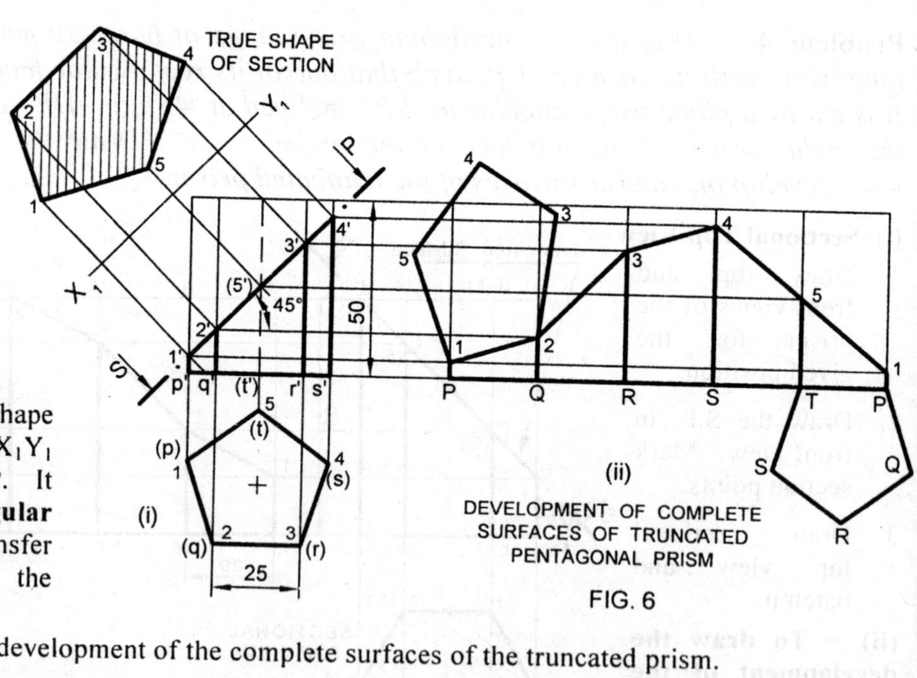

FIG. 6

4. *Fig. 6* shows the development of the complete surfaces of the truncated prism.

Problem 7: *(Exercise) A pentagonal prism of side of base 30 mm and altitude 60 mm stands on its base on HP such that a vertical face is parallel to VP and away* *from observer. It is cut by a plane perpendicular to VP, inclined at an angle of 50° to HP and passing through the axis 35 mm above the base. Draw the development of the lower portion of the prism.*

Problem 8: *(Fig. 7) A cube of side 33 mm rests on its base on HP with a vertical face inclined at 30° to VP. It is cut by a cutting plane perpendicular to VP and inclined at 50° to HP. Cutting plane bisects the axis of the cube. Develop the complete surfaces of the right portion of the cut cube.* (UQ)

1. Draw top view of the cube resting on its base with a vertical face inclined at 30° to VP.
2. Project the front view.
3. Draw S.P. at 50° to HP and bisecting axis of cube.
4. Mark section points in front view. Project these points to the top view.
5. Develop the lateral surface of the complete cube in thin lines.
6. Project horizontally from the section points of the front view to the development of the cube. Join all the intersecting points in the correct sequence.
7. Complete the development of the surfaces of the right portion of the cut cube as shown.

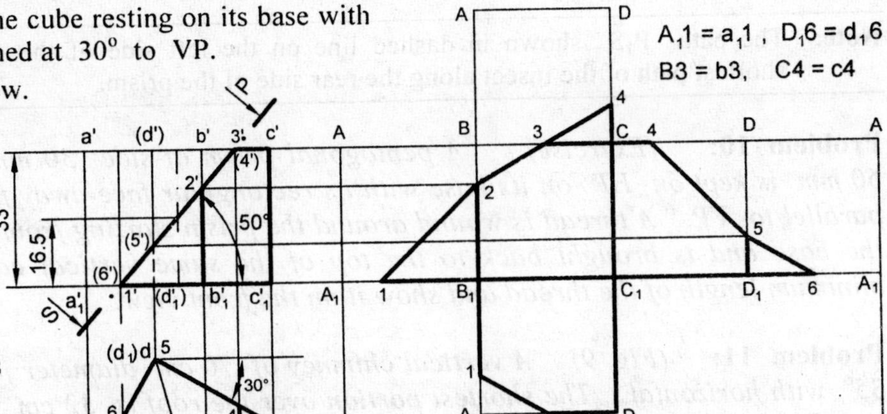

DEVELOPMENT OF COMPLETE SURFACES OF TRUNCATED CUBE

FIG. 7

Problem 9: *(Fig. 8) Draw the development of the lateral surface of a hexagonal prism of 24 mm base edge and 68 mm height. An insect moves on its surface from a corner on the base to the diametrically opposite corner of the top face by the shortest route. Trace the path of the insect in front view.* (UQ)

1. Draw the top view of the given hexagonal prism. Project the corresponding front view.

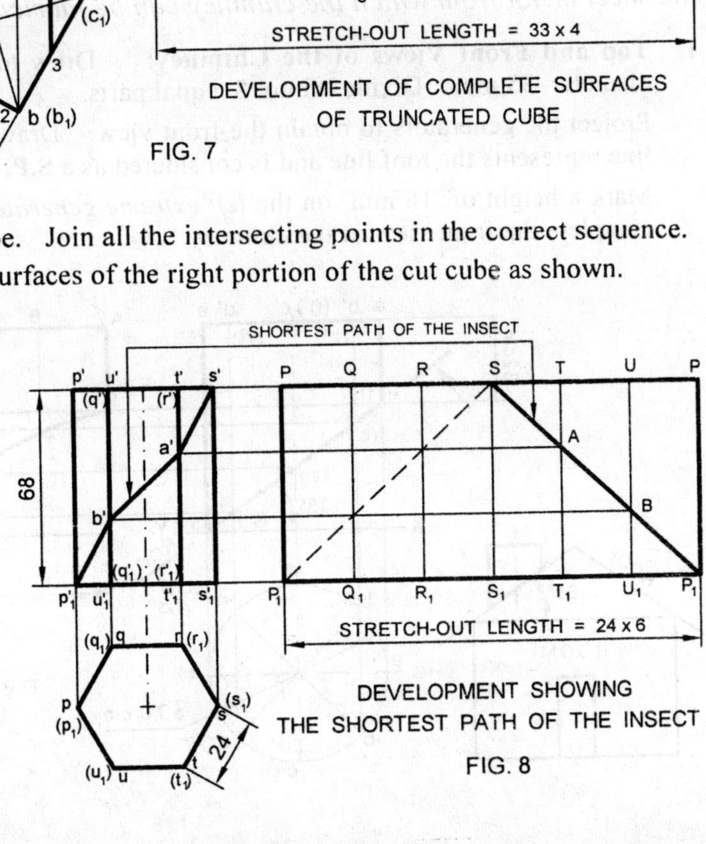

DEVELOPMENT SHOWING THE SHORTEST PATH OF THE INSECT

FIG. 8

2. Draw the stretch-out length = 24 mm x 6 = 144 mm. Divide it into 6 equal parts.
3. Join P_1S which represents the shortest path of the insect moving from one corner at the **base** of the prism to its diametrically opposite corner at the **top** of the prism.
4. Similarly join SP_1 to represent the path of the insect from the top S to the base P_1 cutting the edges TT_1, UU_1 at the points A and B respectively.
5. From A, draw a horizontal line to meet the vertical drawn through t′ at the point a′.
6. Similarly, mark the point b′ on the vertical line through u′.
7. Join p_1' b′ a′ s′ which represents the **shortest path of the insect** in the front view.

Note: The path P_1S shown in dashed line on the left side of the development shows the shortest path of the insect along the rear side of the prism.

Problem 10: *(Exercise)* *A pentagonal prism of side 30 mm and length of axis 60 mm is kept on HP on its base with its rectangular face away from the observer and parallel to VP. A thread is wound around the prism starting from the nearest corner of the base and is brought back to the top of the same vertical edge. Determine the minimum length of the thread and show it on the front view.* (UQ)

Problem 11: *(Fig. 9) A vertical chimney of 70 cm diameter joins a roof sloping at 35° with horizontal. The shortest portion over the roof is 32 cm. Obtain the shape of the sheet metal from which the chimney can be fabricated.* (UQ)

1. **Top and Front Views of the Chimney:** Draw top view of the chimney as a circle of diameter 35 mm. Divide it into 8 equal parts.

   Project the generators to obtain the front view. Draw a line inclined at 35° to XY. This line represents the roof line and is considered as a **S.P.**

   Mark a height of 16 mm on the *left extreme generator* (*shortest height of the chimney*) and complete the front view of the chimney.

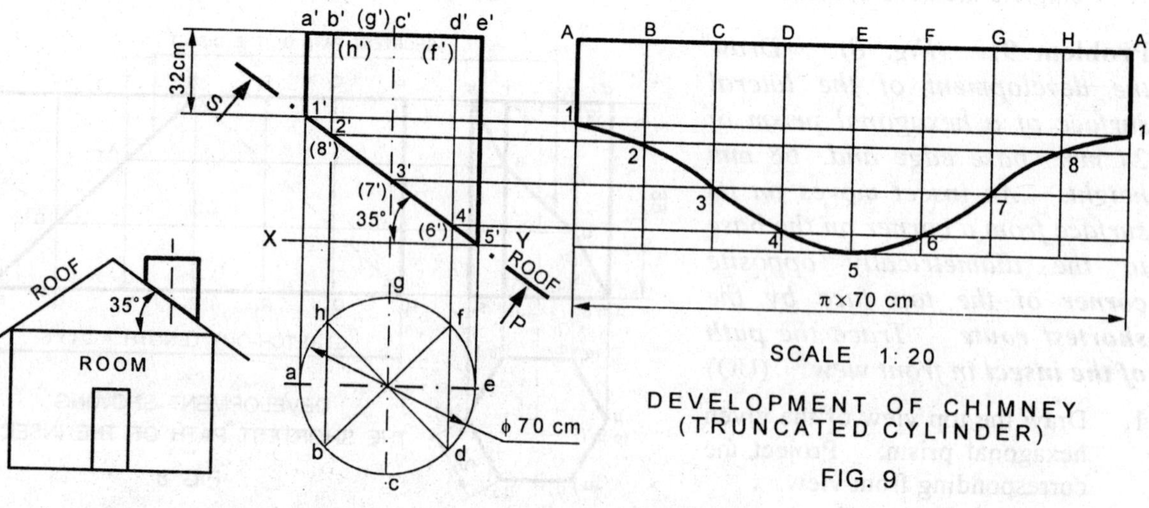

FIG. 9

2. **Shape of the Sheet Metal to Fabricate the Chimney:** Draw a stretch-out line of length AA = π x 35 mm. Divide it into 8 equal parts. Show the generators in the development. Project the section points on the corresponding generators in the development.

Note: ***1. Stretch-out Length of the Cylinder:*** *Stretch-out line AA of length π x 70 can be divided into 8 equal parts by Geometrical Construction. Instead, the chord length ab in the top view can be had on the small bow divider and 8 equal divisions can be marked on the line AA. This method results in reduction of the circumference by about 1% only, which is within limits.*

2. *The stretching-out can be started from either of the extreme ends, whatever be the solid. Both the forms of development with different shapes yield the same object after proper rolling or folding and hence are accepted.*

Problem 12: *(Fig. 10)*
Draw the development of the lateral surface of the lower portion of a cylinder of diameter 50 mm and axis 70 mm when sectioned by a plane inclined at 40° to HP and ⊥ʳ to VP and bisecting the axis. (UQ)

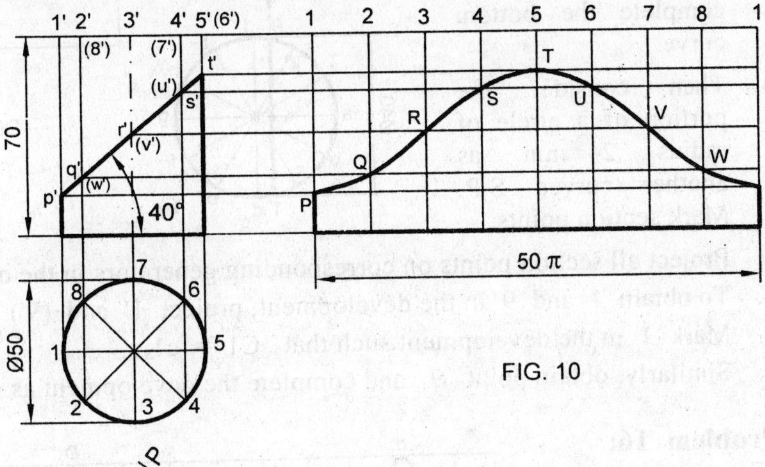

FIG. 10

Problem 13: *(Fig. 11)*
A cylinder of diameter 40 mm, height 75 mm is cut by a plane perpendicular to VP and inclined at 55° to HP meeting the axis at top face. Draw the lateral development of the solid. (UQ)

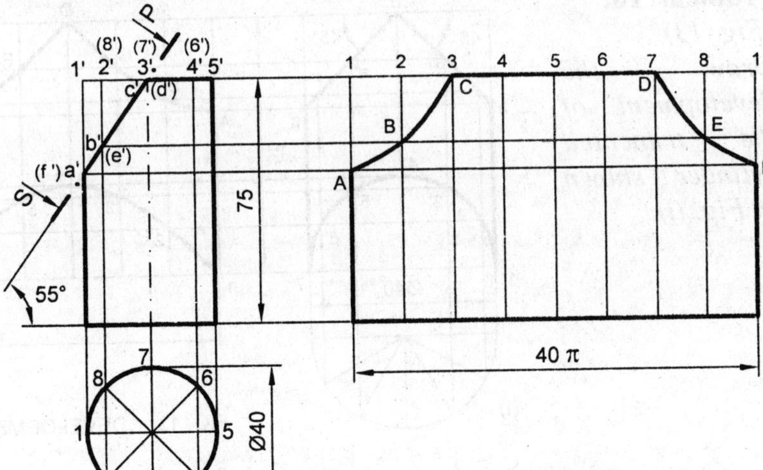

FIG. 11

Problem 14: *(Exercise)*
A cylinder of 45 mm diameter and 55 mm long rests with its base on HP.
It is cut by a plane perpendicular to VP, inclined at 60° to HP and passing through a point on the axis, 12 mm from its top. Draw the sectional top view and the development of the lateral surface of the ***truncated cylinder****.*

Problem 15: *(Fig. 12) Draw the development of the lateral surface of the Part I of the cylinder shown in Fig. (i).* (UQ)

1. Draw the development of the cylinder.
2. Consider the 30° line as a S.P. perpendicular to VP and inclined at 30° to HP.
3. Project section points on the corresponding generators in the development and complete the bottom curve.
4. Then consider the portion of a circle of radius 24 mm as another curved S.P. Mark section points.

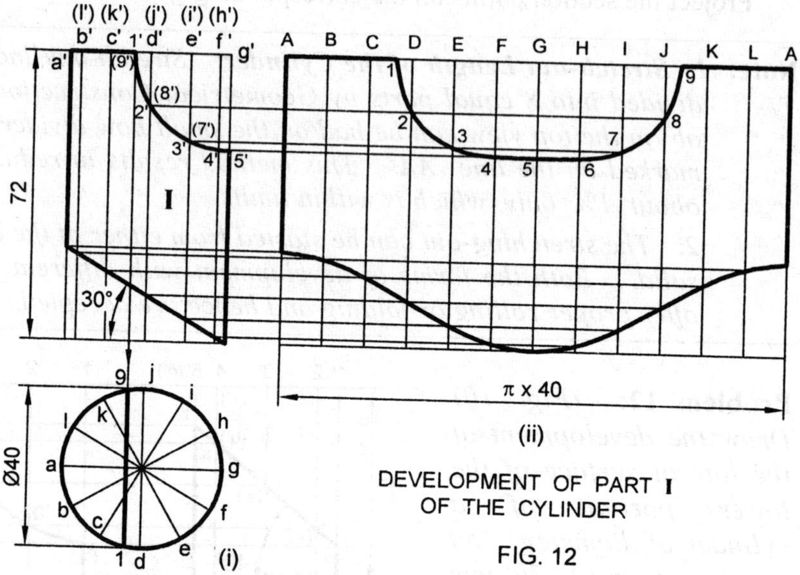

(i) (ii)

DEVELOPMENT OF PART I OF THE CYLINDER

FIG. 12

5. Project all section points on corresponding generators in the development except 1′ & (9′).
6. To obtain 1 and 9 in the development, project 1′ and (9′) to the top view.
7. Mark 1 in the development such that C1 = c1.
8. Similarly, obtain point 9 and complete the development as shown in *Fig. (ii)*.

Problem 16: *(Fig. 13) Draw the development of the truncated cylinder shown in Fig. (i).*

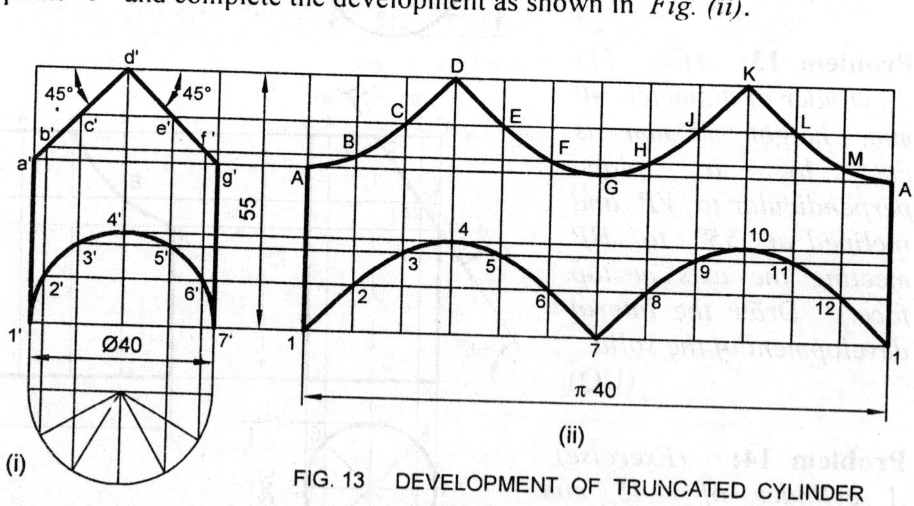

(i) (ii)

FIG. 13 DEVELOPMENT OF TRUNCATED CYLINDER

Problem 17: *(Exercise) A cylinder of 40 mm diameter and 75 mm altitude is cut by a plane perpendicular to VP, inclined at 55° to HP and meeting the axis at the top face. Another plane of 25 mm radius with right bottom corner as center cuts the cylinder. Develop the lateral surface of truncated cylinder.* (UQ)

Problem 18: *(Fig. 14)* ***Draw the development of a two-piece Cylindrical 90° Elbow shown in Fig. (i).***

1. Consider the given elbow as two truncated cylinders I and II. Draw the development of horizontal cylinder as a rectangle of size π 50 mm x 85 mm.
2. Draw the development of the truncated horizontal cylinder I in the above rectangle. Similarly, draw development of truncated vertical cylinder II.

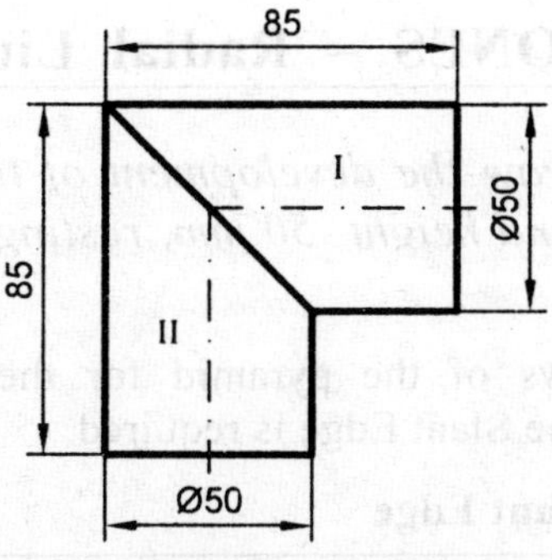

(i) TWO-PIECE CYLINDRICAL 90° ELBOW

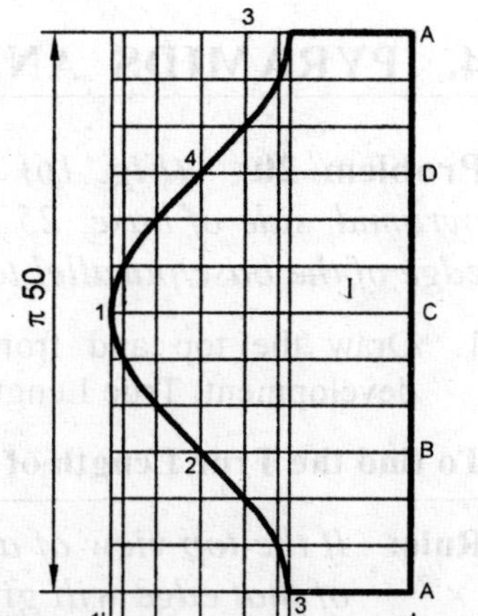

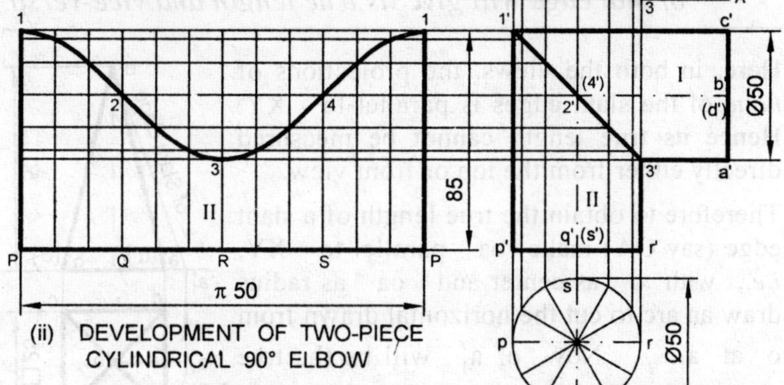

(ii) DEVELOPMENT OF TWO-PIECE CYLINDRICAL 90° ELBOW

FIG. 14

Problem 19: *(Fig. 15)* *The development of a cylinder is a rectangle of size 110 mm x 55 mm. A semi-circle of 55 mm radius is inscribed in the rectangle. Draw the projections of the cylinder showing the semi-circle on them.* (UQ)

1. Draw development of cylinder as a rectangle 110 x 55. *[Fig. (i)]*. Inscribe the given semi-circle of 55 mm radius in the above rectangle.

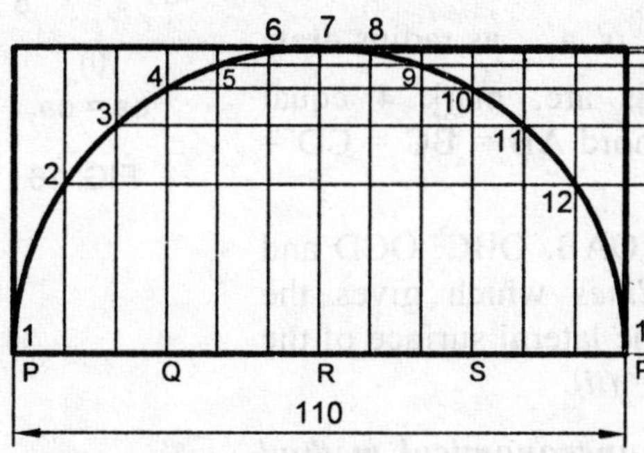

(i) DEVELOPMENT OF GIVEN CYLINDER

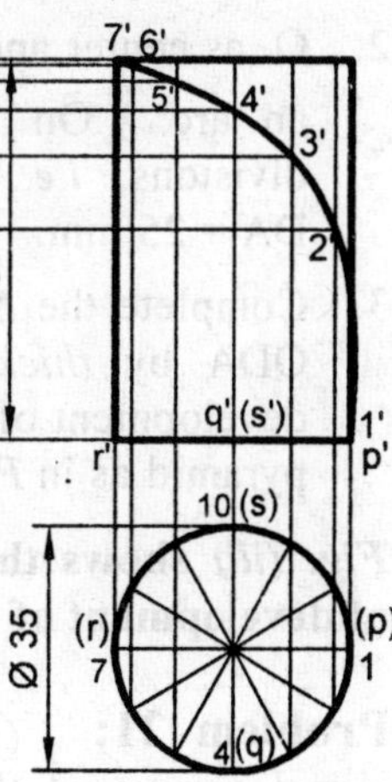

(ii) PROJECTIONS OF CYLINDER

FIG. 15

Projections of Cylinder – Reverse Procedure

2. Diameter of cylinder = 110/π = 35 mm.
3. Height of cylinder = Height of given rectangle = 55 mm.
4. Draw front view of the cylinder as a rectangle 35 x 55. *(Fig. (ii)*
5. Draw the corresponding top view as a circle of 35 mm diameter.
6. Divide the above circle into 12 equal parts. Name them and project to the front view.
7. Divide the rectangle into 12 equal parts. Name the points of intersection of the semi-circle with the above divisions.
8. Project above points to front view to meet vertical projectors drawn from top view. Draw a smooth curve to represent the semi-circle in front view.

4. PYRAMIDS AND CONES – Radial Line Development

Problem 20: *(Fig. 16) Draw the development of the lateral surfaces of a square pyramid, side of base 25 mm and height 50 mm, resting with its base on HP and an edge of the base parallel to VP.*

1. Draw the top and front views of the pyramid for the given position. To draw its development, True Length of the Slant Edge is required.

To find the True Length of the Slant Edge

Rule: *If the top view of a slant edge of a pyramid is parallel to XY, then the front view of that edge will give its true length and vice-versa.*

Here, in both the views, the projections of *none* of the slant edges is parallel to XY. Hence its true length cannot be measured directly either from the top or front view.

Therefore to obtain the true length of a slant edge (say OA) make oa parallel to XY, *i.e.*, with o as center and oa as radius draw an arc to cut the horizontal drawn from o at a_1. Now $o'a_1'$ will be the true length of the slant edge OA.

2. O as center and $o'a_1'$ as radius draw an arc. On this arc, mark 4 equal divisions, *i.e.*, chord AB = BC = CD = DA = 25 mm.
3. Complete the Δs OAB, OBC, OCD and ODA by *thick lines* which gives the development of the lateral surface of the pyramid as in *Fig. (ii)*.

***Fig. (iii)* shows the *uneconomical method* of development of the pyramid.**

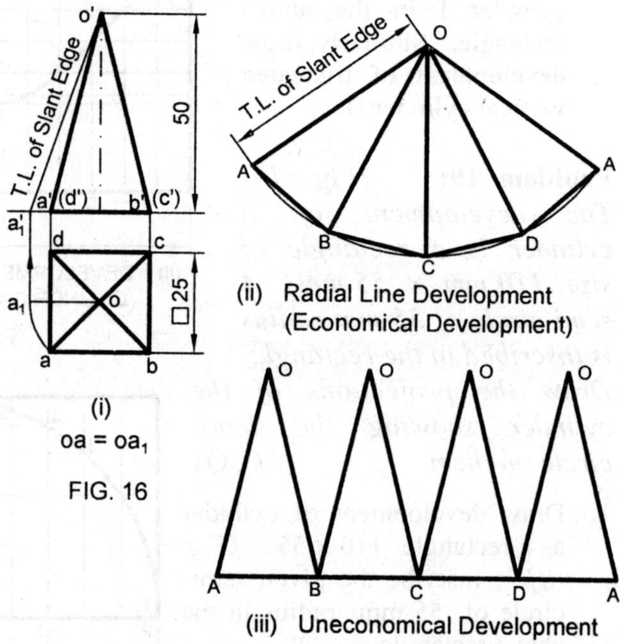

(i)

oa = oa_1

FIG. 16

(ii) Radial Line Development (Economical Development)

(iii) Uneconomical Development

Problem 21: *(Fig. 17) Draw the development of the lateral surface of a* ***cone*** *of base diameter 48 mm and altitude 55 mm.*

1. Draw the top view and front view of the cone. In top view, oa is parallel to XY. Hence, o'a' = L = True Length of the Slant Generator.

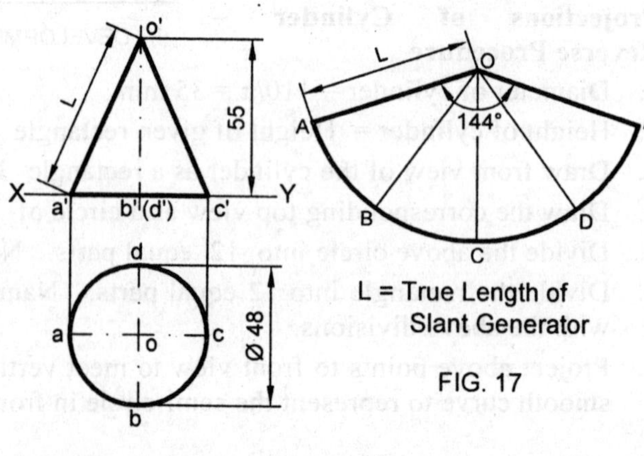

FIG. 17

2. Development of the lateral surface of the cone is a sector of a circle of radius = L. Length of the arc of the sector = AA = Circumference of the base circle = $2\pi r$. Angle subtended by the arc at the center is θ. Thus, $L\theta = 2\pi r = 360^\circ \times r$. So, $\theta = 360^\circ \times r/L$.
3. O as center and L as radius, draw an arc subtending angle θ at O.

5. FRUSTUM AND TRUNCATED PYRAMIDS AND CONES

Problem 22: *(Fig. 18) A regular hexagonal pyramid of side of base 30 mm and height 60 mm is resting vertically on its base on HP such that two of the sides of the base are perpendicular to VP. It is cut by a plane inclined at 40° to HP and perpendicular to VP. The cutting plane bisects the axis of the pyramid. Obtain the development of the lateral surface of the* ***truncated pyramid****.* (UQ)

1. Draw the top and front views of the given pyramid. Mark the cut surface as shown.
2. Draw the development of the lateral surfaces of the pyramid.
3. In the front view, mark the meeting points of the cut surface with the edges of the pyramid as p′, q′, r′, (s′), (t′) and (u′).
4. Find the true length o′ m′ of the slant edge graphically as shown.
5. Draw horizontals through p′, q′, r′, (s′), (t′) and (u′) to cut o′ m_1' at p_1', q_1', etc.
6. On the development, mark P on OA and U on OF such that OP = o′ p_1' = OU.
7. Similarly mark the points Q, R, S and T.
8. Join PQ, QR, …, UP and complete the development of the truncated pyramid.

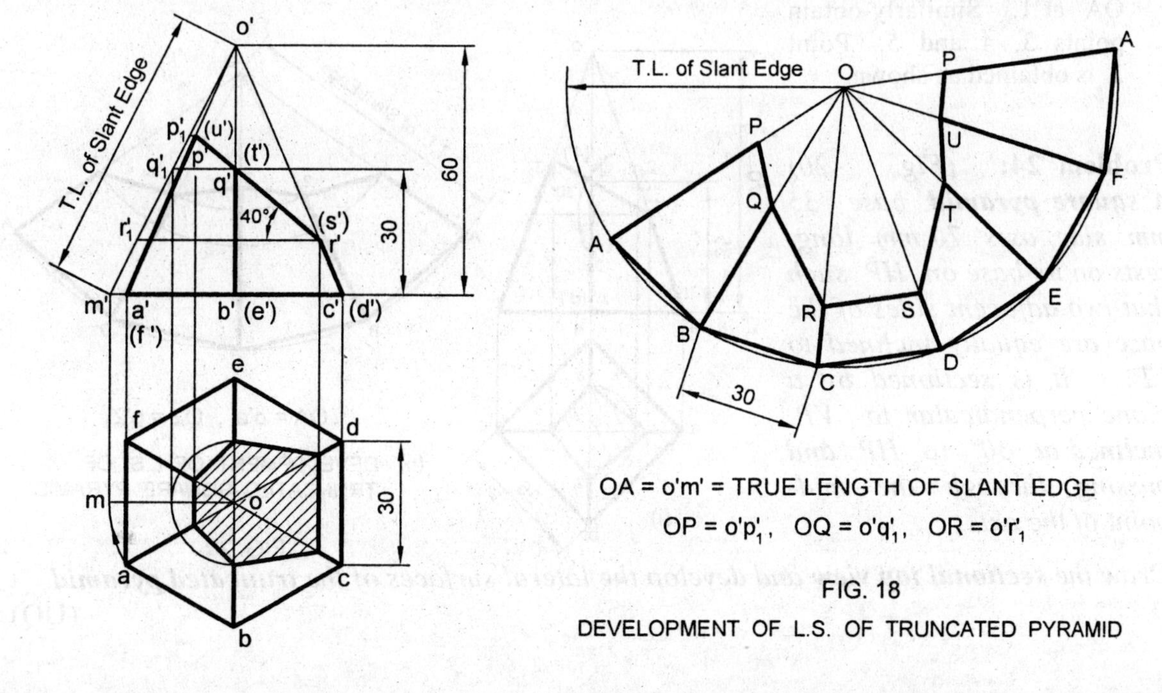

FIG. 18

DEVELOPMENT OF L.S. OF TRUNCATED PYRAMID

Problem 23: *(Fig. 19) A pentagonal pyramid, side of base 30 mm and height 52 mm, stands with its base on HP and an edge of the base is parallel to VP and nearer to it. It is cut by a plane perpendicular to VP, inclined at 40° to HP and passing through a point on the axis, 32 mm above the base. Draw the* ***sectional top view****. Develop the lateral surface of the* ***truncated pyramid****.* (UQ)

1. Draw top and front views of the pyramid. Find the true length of slant edge.
2. Draw the development of lateral surface of pyramid.
3. Draw S.P. in front view. Mark the section points. Draw sectional top view and hatch the cut surface.
4. **True Length of remaining portion of Slant Edges:** Draw horizontals through section points 1′, 2′, etc., to cut o′a′$_1$ (=True Length of Slant Edge OA) at 1′$_1$, 2′$_1$, etc.
5. O as center and o′1′$_1$ as radius draw an arc to cut OA at 1. Similarly obtain points 3, 4 and 5. Point 2 is obtained as shown.

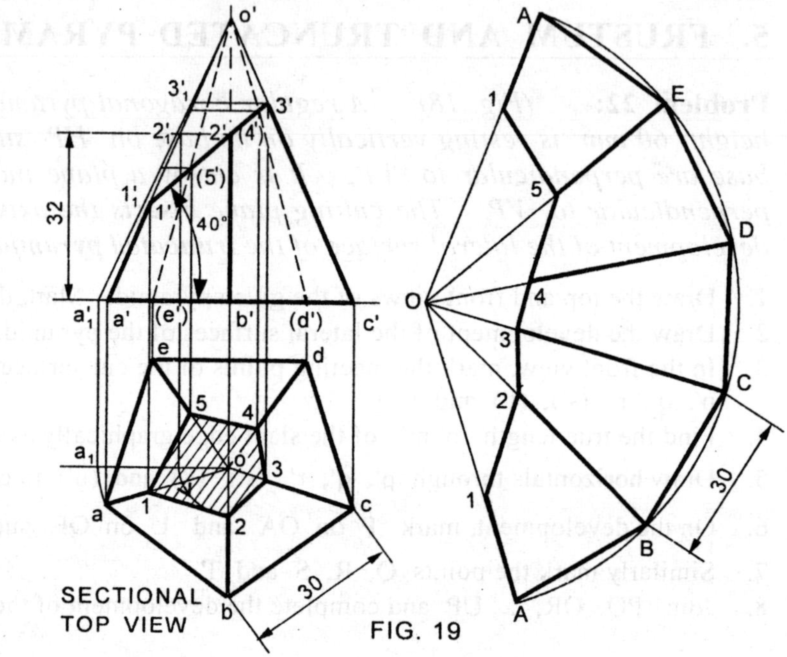

FIG. 19

Problem 24: *(Fig. 20) A* ***square pyramid****, base 35 mm side, axis 70 mm long, rests on its base on HP such that two adjacent sides of the base are equally inclined to VP. It is sectioned by a plane perpendicular to VP, inclined at 30° to HP and passing through the mid-point of the axis.*

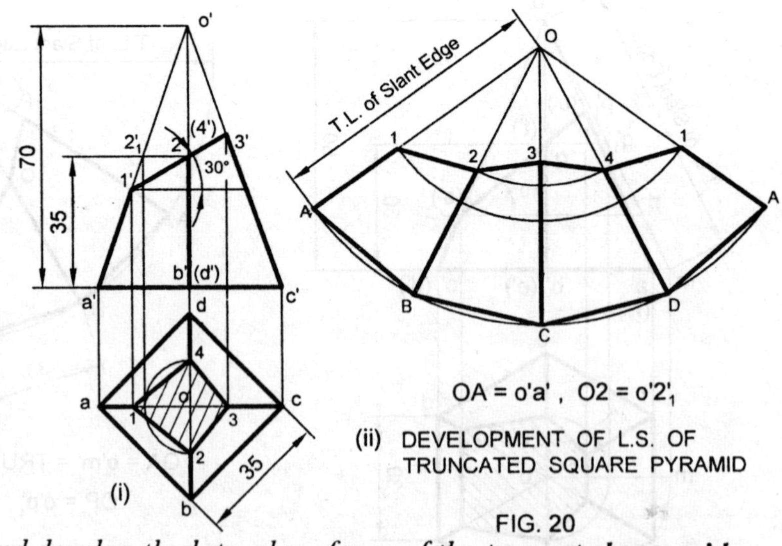

(i)

OA = o'a' , O2 = o'2'$_1$

(ii) DEVELOPMENT OF L.S. OF TRUNCATED SQUARE PYRAMID

FIG. 20

Draw the ***sectional top view*** *and develop the lateral surfaces of the* ***truncated pyramid****.*

(UQ)

Problem 25: *(Fig. 21)* *A square pyramid of base side 30 mm and altitude 65 mm is resting on HP on its base with a side of base inclined at 25° to VP. It is cut by a plane inclined at 35° to HP, perpendicular to VP and bisecting the axis.*

Draw the development of remaining lower portion of the pyramid. (UQ)

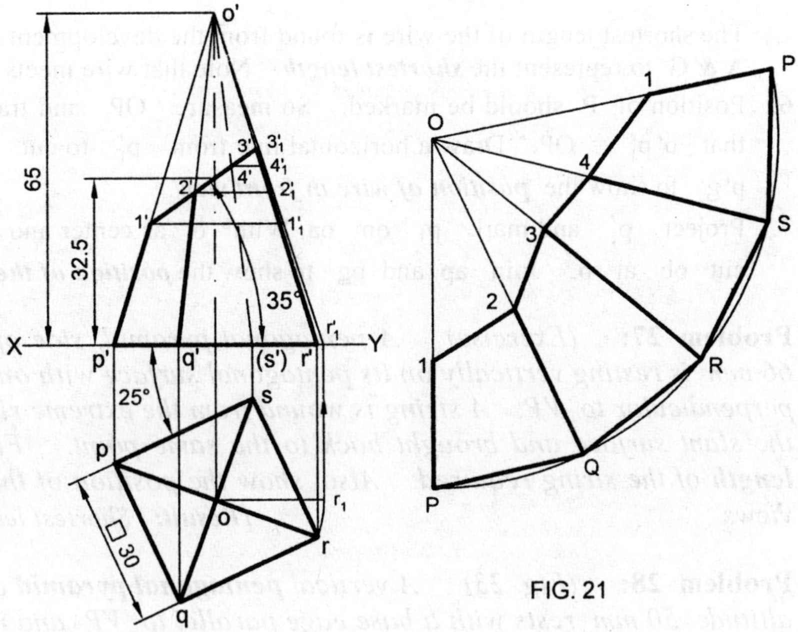

Problem 26: *(Fig. 22)* *A **monument** is in the form of frustum of a **square pyramid** of base 1.2 m side, top 0.5 m side and height 1.0 m. An electrical connection is to be made along the surface of this monument between one corner of the base and the diagonally opposite corner on the top. Find the shortest length of the wire required and show the position of the wire in the top and front views.* (UQ)

1. To obtain the true length of the slant edge directly in the front view, imagine the edges of the base making equal inclinations with VP. For this position draw the top view.
2. Project the front view of the frustum of pyramid. Extend a'e' and c'g' to fix o'.
3. Since oa is parallel to XY in the top view, o'a' will directly give True Length of that Slant Edge. Also e'a' is true length of EA (slant edge of the frustum). So O as center and o'a' as radius draw an arc and ***complete the development of the pyramid.***
4. With O as center and o'e' as radius draw an arc and obtain E, F, G and H on the respective edges and ***complete the development of the frustum.***
5. Let electrical connection be given from A (one corner of the base) to G (diagonally opposite corner on the top face).

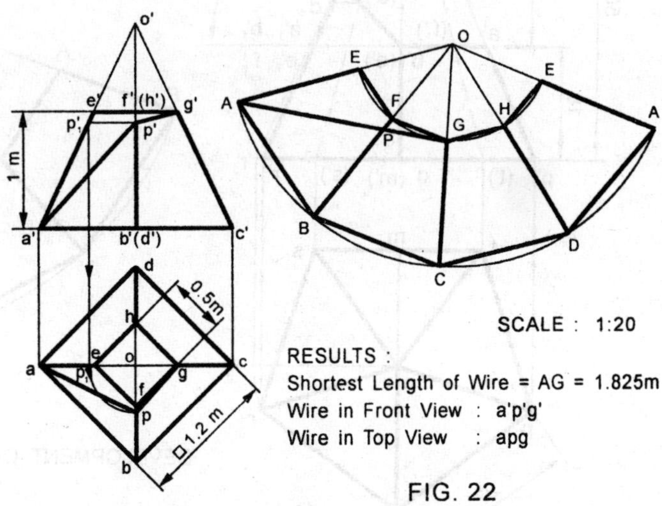

SCALE : 1:20

RESULTS :

Shortest Length of Wire = AG = 1.825m

Wire in Front View : a'p'g'

Wire in Top View : apg

FIG. 22

The shortest length of the wire is found from the development drawn in true lengths. So join A & G to represent the ***shortest length***. Note that wire meets slant edge FB at P.

6. Position of P should be marked. So measure OP and transfer the same on o'a' such that $o'p_1' = OP$. Draw a horizontal line from p_1' to cut o' b' at p'. Join a'p' and p'g' to show the ***position of wire in front view***.
7. Project p_1' and mark p_1 on oa. With o as center and op_1 as radius draw an arc to cut ob at p. Join ap and pg to show the ***position of the wire in top view***.

Problem 27: *(Exercise) A pentagonal pyramid, side of base 28 mm and altitude 66 mm is resting vertically on its pentagonal surface with one of the sides of that surface perpendicular to VP. A string is wound from the extreme right point in the base around the slant surface and brought back to the same point. Find graphically the* ***shortest length*** *of the string required. Also, show the position of the string in the top and front views.* *[**Result:** Shortest length of the string = 118 mm.]*

Problem 28: *(Fig. 23) A vertical* ***pentagonal pyramid*** *of side of base 27 mm and altitude 50 mm rests with a base edge parallel to VP and nearer to it. It is cut by two planes perpendicular to VP. One is horizontal and cuts the portion of the pyramid on the left of the axis at a height of 18 mm above the base of the pyramid. The other plane inclined at 45° to HP cuts the portion of the pyramid to the right of the axis passing through a point on it 18 mm above the base and leans upwards.*

Draw the development of the lateral surfaces of the ***truncated pyramid****.* (UQ)

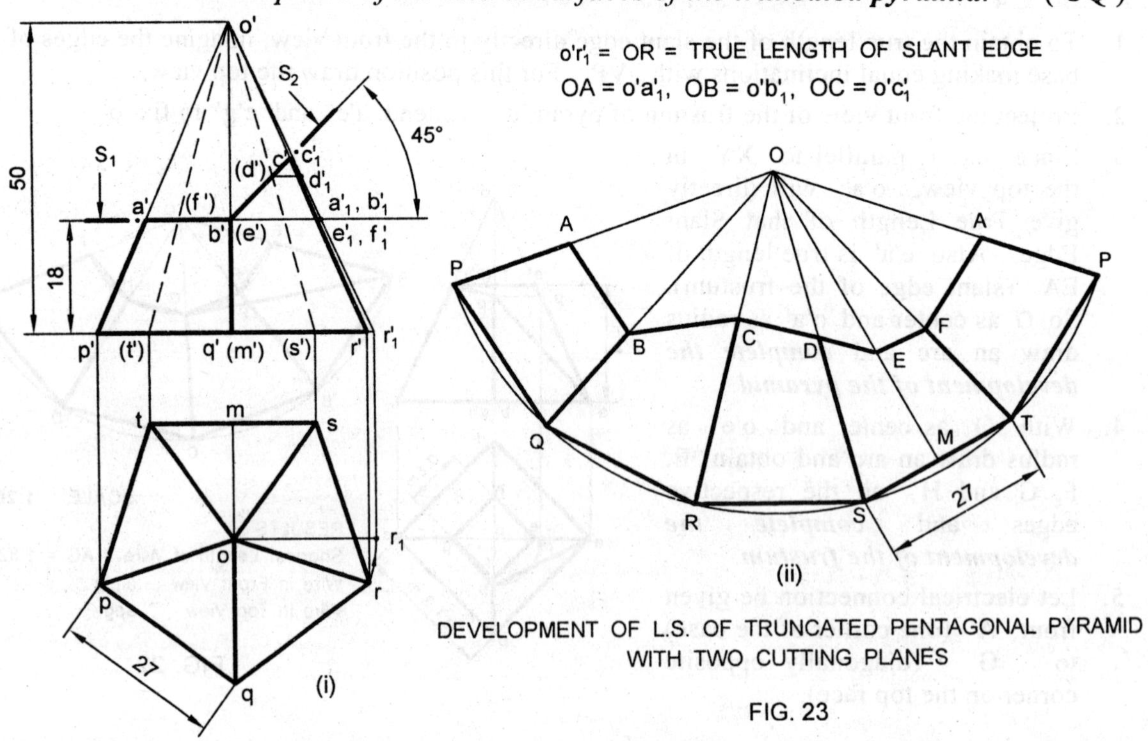

DEVELOPMENT OF L.S. OF TRUNCATED PENTAGONAL PYRAMID WITH TWO CUTTING PLANES

FIG. 23

1. *See Fig. (i).* Draw the top view of the pentagonal pyramid such that the base edge s t is parallel and nearer to XY. Project the corresponding front view of the pyramid.
2. Note that in the front view, none of the slant edges is in true length. Find the True Length of the slant edge o′ r_1' as shown.
3. Draw the development of the lateral surface of the full pyramid.
4. Mark the two given cutting planes S_1 and S_2 in the front view.
5. Mark the cutting points in the front view as shown. Mark the cutting point (e′) on an imaginary line o′ (m′) which is lying on the rear side triangular face o′ (t′) (s′).
6. Complete the development of the lateral surfaces of the truncated pyramid as in *Fig. (ii)*.

Problem 29: *(Fig. 24) A hexagonal pyramid of side of base 25 mm and altitude 65 mm rests on its base on HP such that a base edge is parallel to VP. It is sectioned by a plane normal to VP, inclined at 30° to HP and passing through a point on the axis 22 mm from the vertex leaning downwards. It is cut by another plane normal to VP, passing through the same point and perpendicular to the previous section plane. The portion containing the vertex is removed.*

(i) Draw the sectional top view of the cut solid.

(ii) Develop the lateral surfaces of the remaining solid such that the developed surface is symmetrical. (UQ)

See Fig. 24 for Solution.

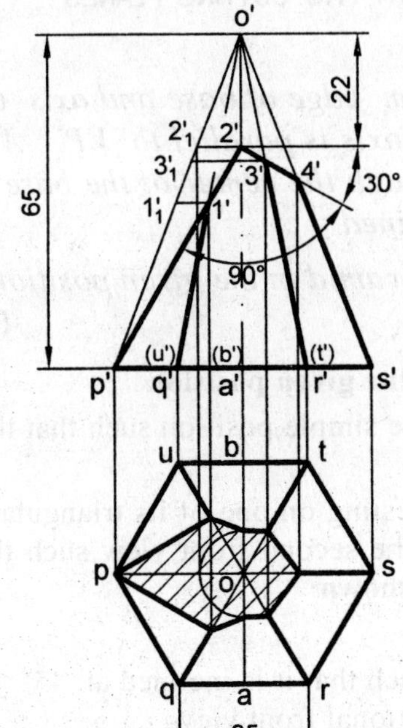

(i) SECTIONAL TOP VIEW

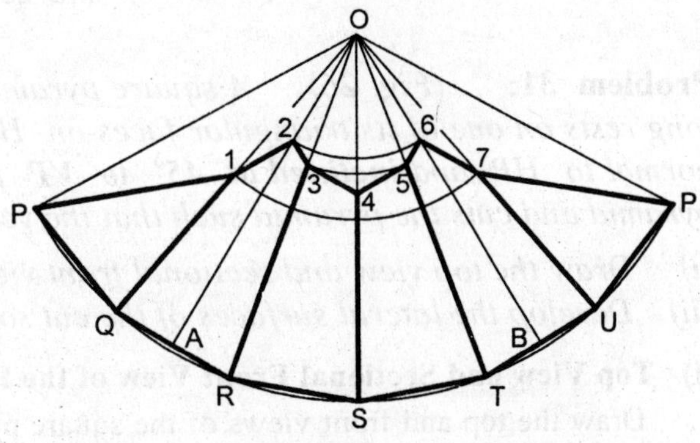

OP = o'p' = TRUE LENGTH OF SLANT EDGE

O1 = $o'1_1'$, O2 = $o'2_1'$, O3 = $o'3_1'$

(ii)

DEVELOPMENT OF L.S. OF TRUNCATED HEXAGONAL PYRAMID WITH TWO CUTTING PLANES

(DEVELOPED SURFACE IS SYMMETRICAL)

FIG. 24

Problem 30: *(Fig. 25) A hexagonal pyramid of side of base 25 mm and altitude 70 mm rests on its base on HP such that a base edge is parallel to VP. It is cut by two cutting planes, perpendicular to VP. One of the planes is inclined at 30° to HP and meeting the axis at a point 40 mm from the base. The other plane is curved of 25 mm radius with the right corner of the base as center.*

Develop the lateral surfaces of the remaining portion of the pyramid. (UQ)

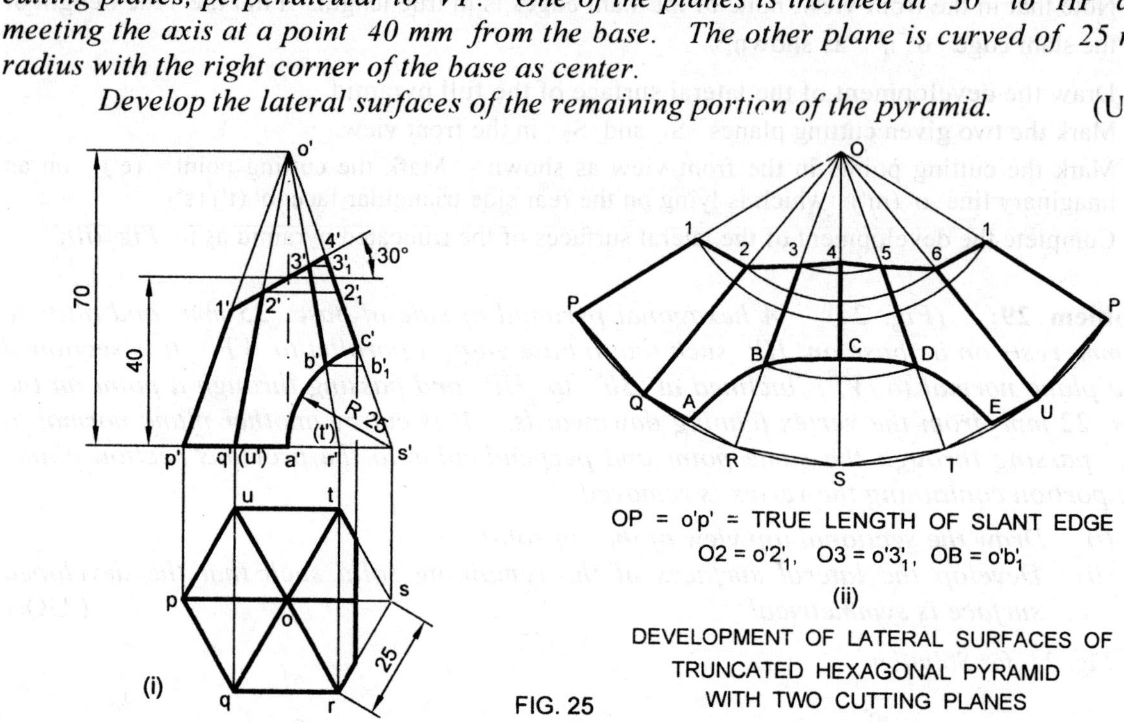

OP = o'p' = TRUE LENGTH OF SLANT EDGE

O2 = o'2'$_1$, O3 = o'3'$_1$, OB = o'b'$_1$

(ii)

DEVELOPMENT OF LATERAL SURFACES OF TRUNCATED HEXAGONAL PYRAMID WITH TWO CUTTING PLANES

FIG. 25

Problem 31: *(Fig. 26) A square pyramid of 35 mm edge of base and axis 60 mm long rests on one of its triangular faces on HP and the axis is parallel to VP. A plane normal to HP and* ***inclined at 45° to VP*** *passes through the center of the base of the pyramid and cuts the pyramid such that the vertex is retained.*

(i) Draw the top view and sectional front view of the pyramid in the given position.

(ii) Develop the lateral surfaces of the cut solid. (UQ)

(i) Top View and Sectional Front View of the Pyramid in the given position

1. Draw the top and front views of the square pyramid in the simple position such that the axis of the pyramid is vertical.
2. The axis of the pyramid is parallel to VP. Also, it is resting on one of its triangular faces on HP and hence the axis is inclined to HP. Draw the second front view such that the pyramid is lying on one of its triangular faces on HP as shown.
3. Project the corresponding second top view.
4. Draw the cutting plane as a line in the second top view such that it is inclined at 45° to XY and passing through the center of the base. Draw the sectional front view.

(ii) Development of lateral surfaces of the cut solid

5. Find the true length of the slant edge of the pyramid in the top and front views.

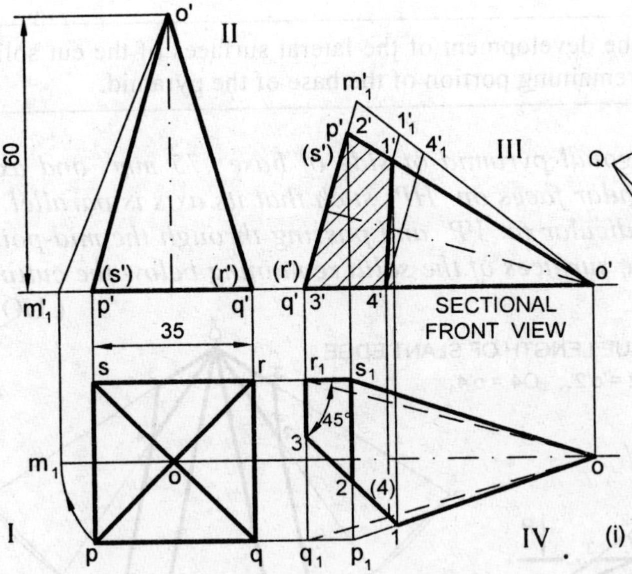

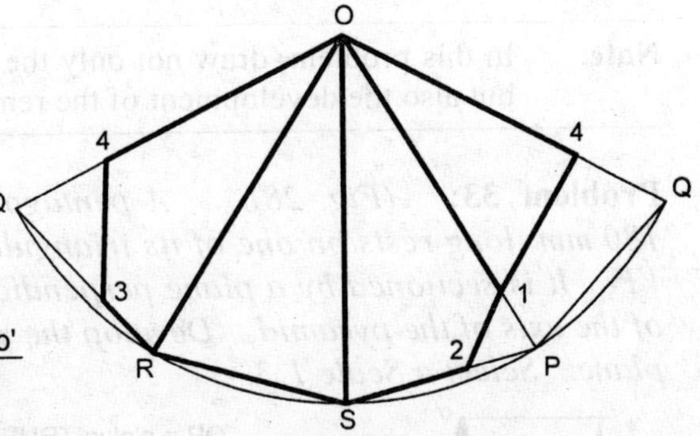

OP = o'm'$_1$ = TRUE LENGTH OF SLANT EDGE

O1 = o'1'$_1$, O4 = o'4'$_1$

(ii)

DEVELOPMENT OF LATERAL SURFACES OF CUT SQUARE PYRAMID

FIG. 26

6. Transfer the true length of the slant edge to the second front view.
7. Draw the development of the square pyramid using the true length.
8. Mark the development of the lateral surfaces of the cut solid as shown.

Problem 32: *(Fig. 27) A hexagonal pyramid, edge of base 25 mm and axis 60 mm long is lying on one of its triangular faces on HP. It is sectioned by a horizontal plane 15 mm from the highest point of the pyramid. (i) Draw the projections of the solid. (ii) Develop the surfaces of the solid remaining below the section plane.* (UQ)

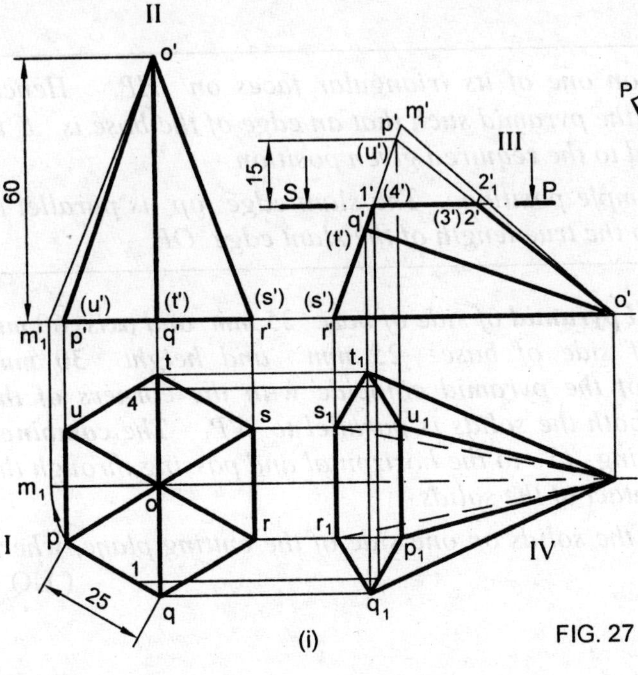

(i)

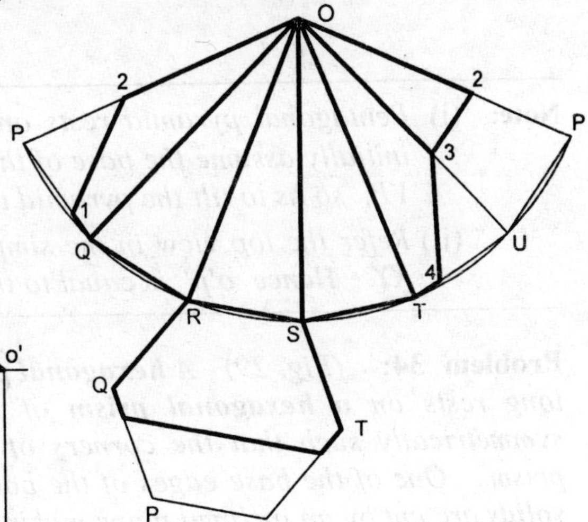

OP = o'm'$_1$ = TRUE LENGTH OF SLANT EDGE

O2 = o'2'$_1$, O3 = o'3'$_1$

FIG. 27

(ii) DEVELOPMENT OF CUT HEXAGONAL PYRAMID

Note: In this problem, draw not only the development of the lateral surfaces of the cut solid, but also the development of the remaining portion of the base of the pyramid.

Problem 33: *(Fig. 28) A pentagonal pyramid of side of base 75 mm and axis 180 mm long rests on one of its triangular faces on HP such that its axis is parallel to VP. It is sectioned by a plane perpendicular to VP and passing through the mid-point of the axis of the pyramid. Develop the surfaces of the solid remaining below the cutting plane. Select a Scale 1:3.* (UQ)

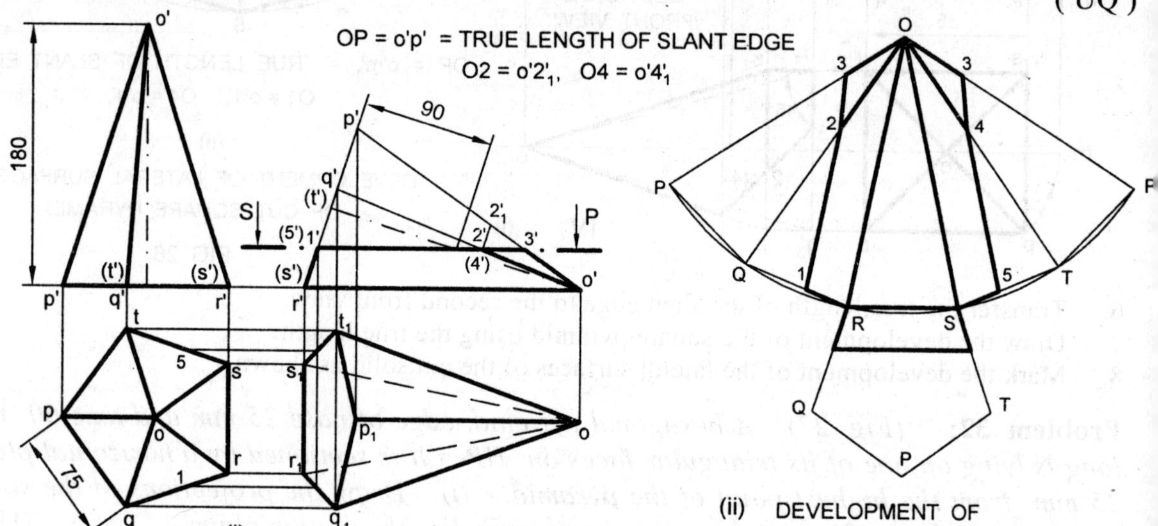

FIG. 28

Note: (i) *Pentagonal pyramid rests on one of its triangular faces on HP. Hence, initially assume the base of the pyramid such that an edge of the base is ⊥r to VP, so as to tilt the pyramid to the required given position.*

(ii) *Refer the top view in the simple position. The slant edge op is parallel to XY. Hence o'p' is equal to the true length of the slant edge OP.*

Problem 34: *(Fig. 29) A* ***hexagonal pyramid*** *of side of base 25 mm and axis 60 mm long rests on a* ***hexagonal prism*** *of side of base 25 mm and height 30 mm, symmetrically such that the corners of the pyramid coincide with the corners of the prism. One of the base edges of the both the solids is parallel to VP. The* ***combined solids*** *are cut by an inclined plane making 45° to the horizontal and passing through the center of the hexagonal faces of the contact of the solids.*

Develop the lateral surfaces of the solids on one side of the cutting plane, where the apex of the pyramid is existing. (UQ)

See Fig. 29 for Solution.

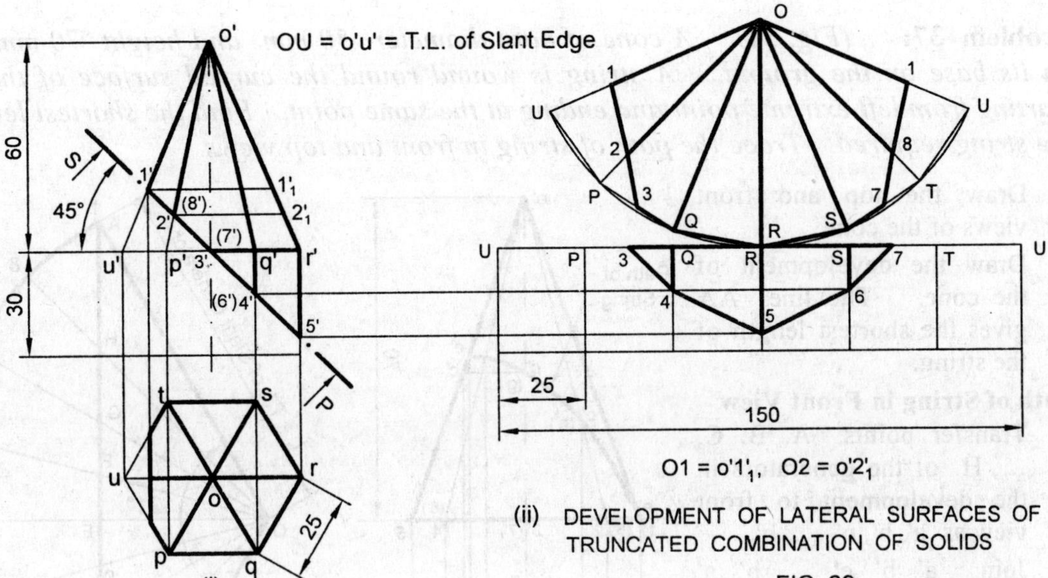

(ii) DEVELOPMENT OF LATERAL SURFACES OF TRUNCATED COMBINATION OF SOLIDS

FIG. 29

Problem 35: *(Fig. 30)* *Determine graphically the shortest length measured along the surface of a frustum of a cone, between two points A and B. Point A is on the base of the frustum which is 60 mm in diameter. Point B is a diametrically opposite point on the top surface which is 30 mm in diameter. Height of the frustum is 40 mm.* (UQ)

1. Draw the top and front views of the frustum of the cone.
2. Extend the two extreme generators to meet at o′.
3. L = True Length of Slant Generator. Find the subtended angle $\theta = (r/L) \times 360^\circ = (30/85.4) \times 360^\circ = 126.4^\circ$.
4. Draw the development of the frustum of the cone as shown.
5. Mark the points a′ and b′ in the front view as they are diametrically opposite. Mark a and b in top view.

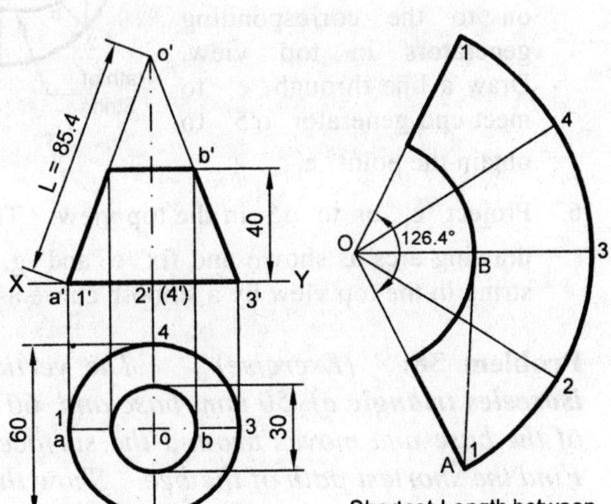

FIG. 30

Problem 36: *(Exercise)* *The inside of the **hopper** of a flour mill is lined with tin sheet. Top and bottom of the hopper are **circles** of 50 cm and 30 cm diameters respectively. Height of the hopper is 48 cm. Draw the shape of the sheet to which it is to be cut to fit-in the hopper.* (UQ)

Problem 37: *(Fig. 31) A cone of base diameter 50 mm and height 70 mm rests on its base on the ground. A string is wound round the curved surface of the cone starting from left extreme point and ending at the same point. Find the shortest length of the string required. Trace the path of string in front and top views.* (UQ)

1. Draw the top and front views of the cone.
2. Draw the development of the cone. The line AA gives the shortest length of the string.

Path of String in Front View

3. Transfer points A, B, C, ..., H of the generators in the development to front view as a′, b′, c′, ..., h′.
4. Join a′ b′ c′ ... h′ a′ representing the path of string by a smooth curve.

Path of String in Top View

5. Project the points a′, b′, d′, e′, f′, h′ (except c′ and g′) on to the corresponding generators in top view. Draw a line through c′ to meet end generator o′5′ to obtain the point c_1'.

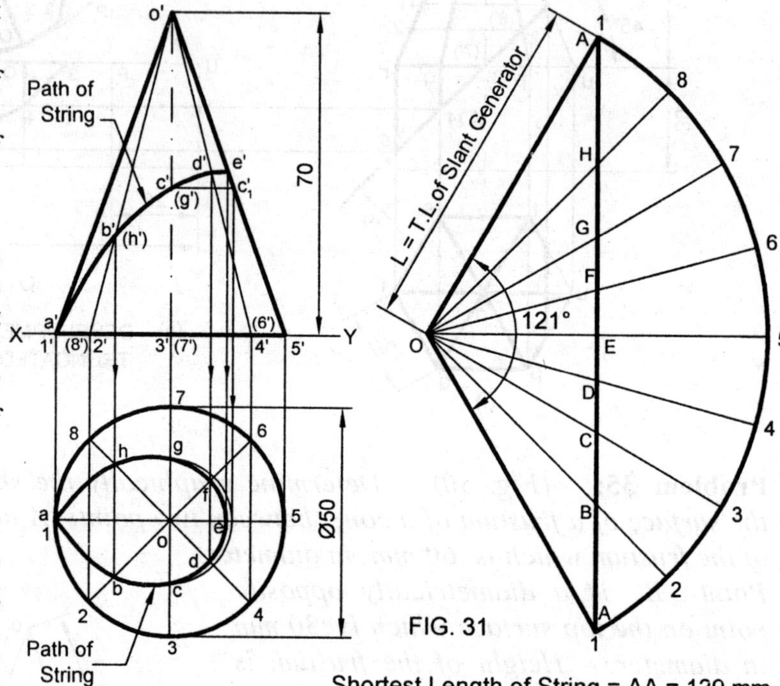

FIG. 31

Shortest Length of String = AA = 129 mm

6. Project c_1' on to o5 in the top view. Transfer the same on to the 3rd and 7th generators by drawing arcs as shown and fix c and g. Join abcdefgha representing the path of the string in the top view by a smooth curve as shown.

Problem 38: *(Exercise) The vertical section of a* **cone** *through the axis is an* ***isosceles triangle*** *of 50 mm base and 60 mm height. A bee sits on the extreme left end of the base and moves around the surface of the cone and returns to the starting point. Find the shortest path of the bee. Show the path in front and top views.* (UQ)

Results: θ = 138.5°; Shortest path of the bee = 121.5 mm.

Problem 39: *(Fig. 32) A* **cone** *of diameter of base 50 mm and altitude 65 mm is lying on its base on HP. It is sectioned by a plane normal to both HP and VP at a distance of 10 mm to the right of the axis of the cone. Draw the development of the lateral surface of the left side remaining portion.* (UQ)

Note: *Cutting plane normal to HP and VP is vertical.*

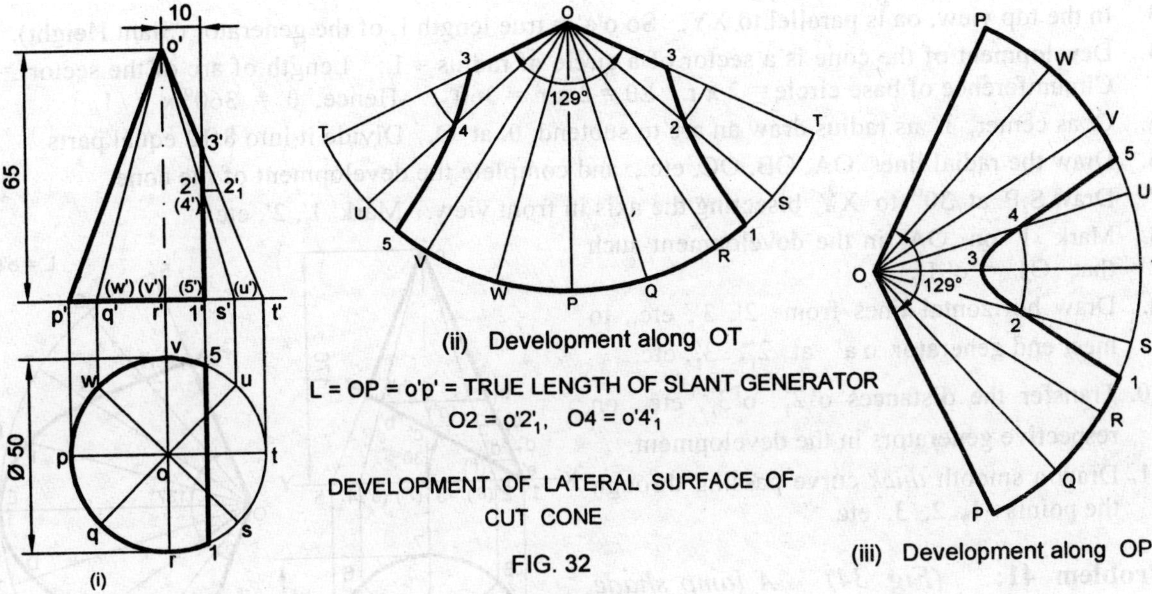

L = OP = o'p' = TRUE LENGTH OF SLANT GENERATOR

$O2 = o'2'_1$, $O4 = o'4'_1$

DEVELOPMENT OF LATERAL SURFACE OF CUT CONE

FIG. 32

1. Draw the top and front views of the cone as in Fig. (i).
2. Mark the vertical S.P. 10 mm to the right of the axis in front view. Project it to top view.
3. Draw the development of the lateral surface of cone. $\theta = (25/69.6) \times 360^\circ = 129^\circ$.
4. Transfer the cutting points from the top view to the development of the cone.
5. Complete the development of lateral surface of left side remaining portion of the cone.

Note: *The required development may be as shown in Fig. (ii) or in Fig. (iii).*

Problem 40: *(Fig. 33)*
A cone of base 50 mm diameter and height 65 mm rests with its base on HP. A section plane ⊥^{r} to VP and inclined at 30° to HP bisects the axis of the cone. Draw the development of its lateral surface. (UQ)

1. Draw the top and front views of cone.
2. Divide the base circle into 8 / 12 equal parts. Show the generators in both views.

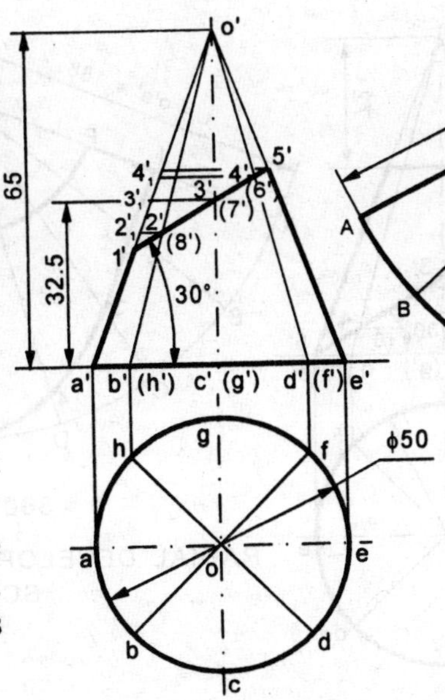

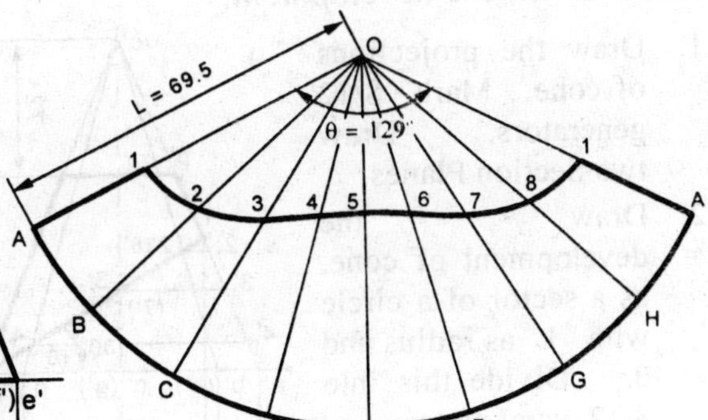

$\theta = 360^\circ \times (r / L) = 129^\circ$

L = o'a' = OA
= True Length of Slant Generator

RADIAL DEVELOPMENT OF TRUNCATED CONE

FIG 33

3. In the top view, oa is parallel to XY. So o'a' = true length L of the generator (Slant Height).
4. Development of the cone is a sector of a circle of radius = L; Length of arc of the sector = Circumference of base circle = $2\pi r$. $L\theta = 2\pi r = 360^\circ r$. Hence, $\theta = 360^\circ \times r / L$.
5. O as center, L as radius draw an arc to subtend θ at O. Divide it into 8/12 equal parts.
6. Draw the radial lines OA, OB, OC, etc., and complete the development of the cone.
7. Draw S.P. at 30° to XY bisecting the axis in front view. Mark 1′, 2′, etc.
8. Mark 1 on OA in the development such that O1 = o′1′.
9. Draw horizontal lines from 2′, 3′, etc., to meet end generator o′a′ at $2'_1$, $3'_1$, etc.
10. Transfer the distances $o'2'_1$, $o'3'_1$, etc., on respective generators in the development.
11. Draw a smooth *thick* curve passing through the points 1, 2, 3, etc.

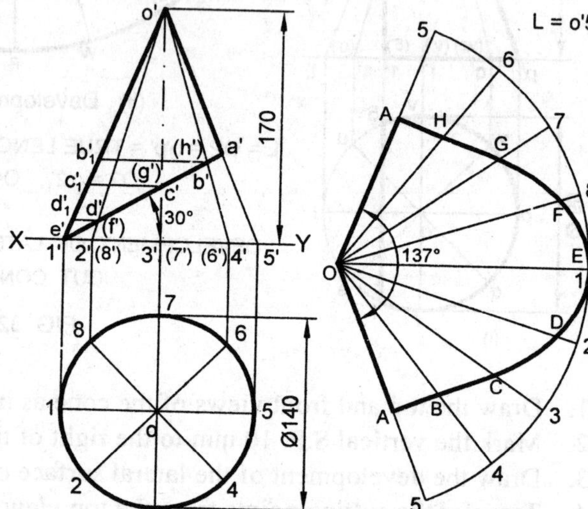

FIG. 34

Problem 41: *(Fig. 34) A lamp shade is formed by cutting a cone of base 140 mm diameter and 170 mm height by a horizontal plane inclined at 30° to HP, passing through one extremity of the base. Draw the development.* (UQ)

Problem 42: *(Fig. 35) A **lamp shade** is formed by cutting a **cone** of base 144 mm diameter and 174 mm height by a horizontal plane at a distance of 72 mm from the apex and another plane inclined at 30° to HP, passing through one extremity of the base. Draw the development.* (UQ)

1. Draw the projections of cone. Mark 8/12 generators. Draw two Section Planes.
2. Draw the development of cone, as a sector of a circle with L as radius and θ. Divide this into 8/12 equal parts.
3. Horizontal S.P. gives frustum of the cone. To develop this, with O as center and o′ p′ as radius, draw an arc cutting OA at P.

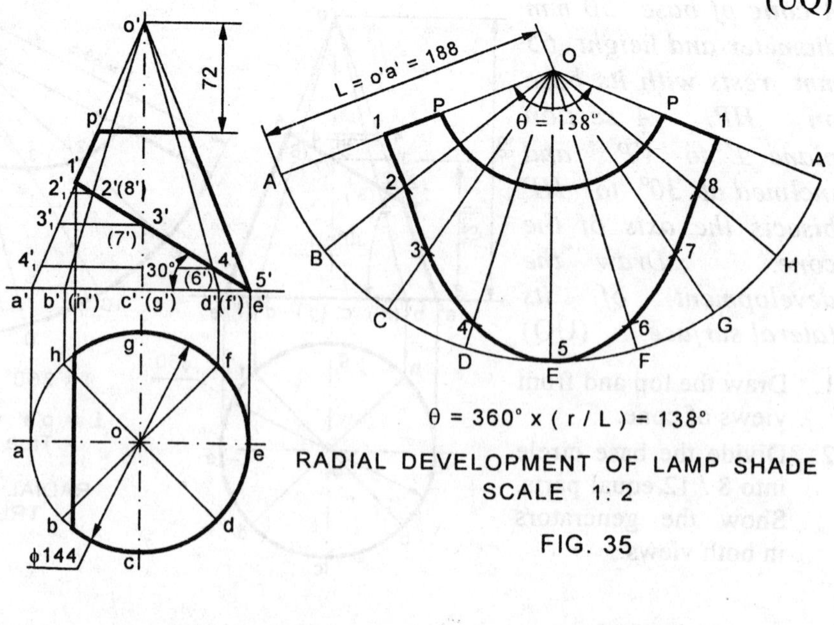

θ = 360° x (r / L) = 138°

RADIAL DEVELOPMENT OF LAMP SHADE

SCALE 1: 2

FIG. 35

4. *Inclined S.P.:* In the development mark 1 on OA such that O1 = o′ 1′.
5. Draw horizontal line from 2′ to cut o′a′ at $2'_1$. Mark 2 on OB such that O2 = o′ $2'_1$.
6. Similarly obtain 3, 4, ..., 8 in the development. Join all the points by a smooth curve.

Problem 43: *(Exercise) A cone of base diameter 60 mm and height 70 mm is resting on its base on HP. It is cut by a plane perpendicular to VP and parallel to HP at a distance of 20 mm from the vertex. It is also cut by a plane inclined at 40° to the base and meeting the axis 20 mm above the base. Draw the development.* (UQ)

Results: L = 76 mm; θ = 142°.

Problem 44: *(Fig. 36) A cone of base 50 mm diameter and 60 mm height, rests with its base on HP. It is cut by a section plane perpendicular to VP, parallel to one of the generators and passing through a point on the axis at a distance of 22 mm from the apex. Draw the sectional top view and develop the lateral surface of the remaining portion of the cone.* (UQ)

1. Draw the projections of the cone. Draw SP parallel to the extreme generator and passing through a point on the axis, 22 mm from the apex in the front view. Mark the section points. Project them to top view and complete the Sectional Top View.
2. Draw the development of the cone, showing 12 generators.
3. Mark 2, 3, 4, ..., 10 (except 1 & 11) on the corresponding generators in the development.
4. Section point 1′ is on the portion of the base b′ c′ in the front view. b′ c′ does not represent the true length of that portion of the base. Hence 1 cannot be marked in the development directly from the front view.

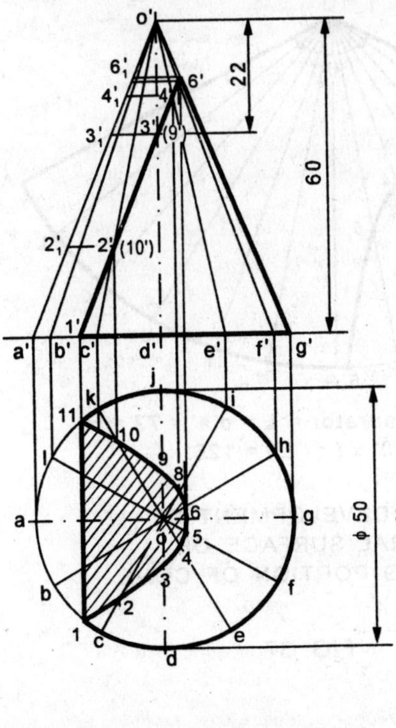

SECTIONAL TOP VIEW

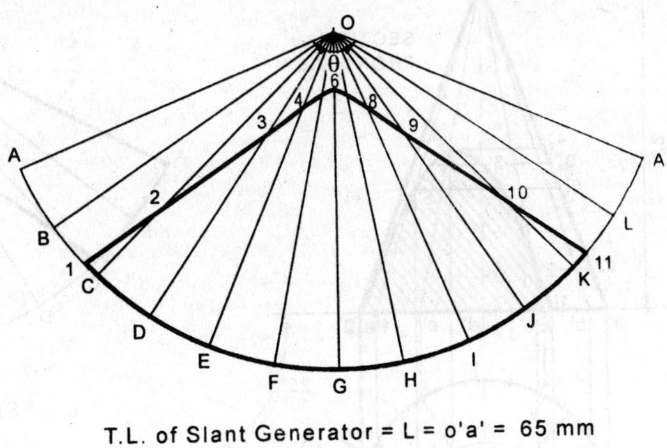

T.L. of Slant Generator = L = o'a' = 65 mm
θ = 360° x (r / L) = 138.5°

RADIAL DEVELOPMENT OF LATERAL SURFACE OF REMAINING PORTION OF CONE

FIG 36

5. For this, project 1′ to the top view and obtain 1 on bc which is in true length. So mark 1 in the development such that B1 = b1.
6. Similarly mark 11 such that K-11 = k-11. Complete the development as shown.

Problem 45: *(Exercise) A cone, base 50 mm diameter and axis 60 mm long, rests with its base on HP. A section plane perpendicular to VP and inclined at 45° to HP bisects the axis of the cone. Draw the sectional top view and the development of the lateral surface of the remaining portion of the cone.*

Problem 46: *(Fig. 37) A* **cone,** *base 54 mm diameter and height 72 mm, rests with its base on HP. A section plane perpendicular to HP and inclined at 25° to VP cuts the cone at a distance of 13.5 mm from the axis. Draw the sectional front view and develop the lateral surface of the remaining portion of the cone.* (UQ)

1. Draw the projections of the cone. Draw the SP in the top view and project the sectional front view. Draw the development of the cone.
2. In the top view, base is in true size. Section points 1 and 6 lie on the base. Hence mark 1 and 6 in the development directly from top view such that B1 = b1 and F6 = f6.
3. In the top view oc, od, oe and of are *not in true size.*
   Hence 2, 3, 4 and 5 cannot be directly marked in the development.
   Now project these points to the front view and obtain $2'_1$, $3'_1$, $4'_1$ and $5'_1$.
4. Mark 2 in the development such that O2 = o′ $2'_1$.
5. Similarly obtain other points 3, 4 and 5. Complete the development as shown.

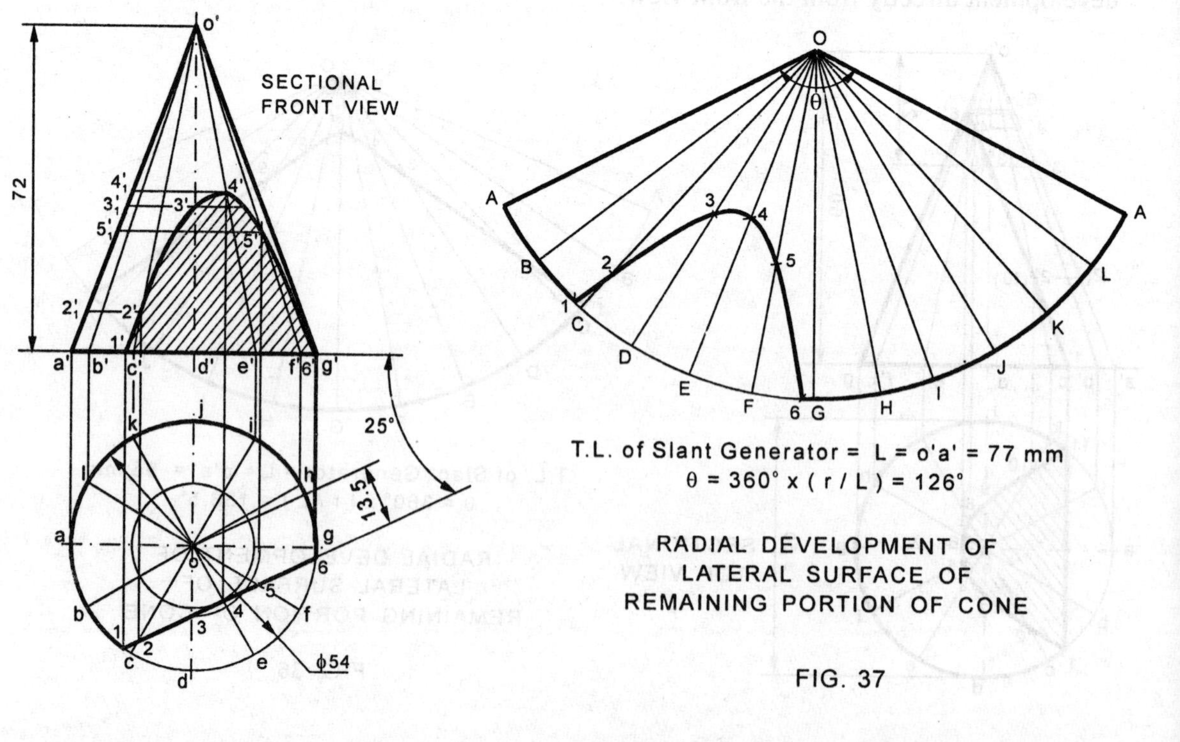

FIG. 37

Problem 47: *(Fig. 38) A cone of 50 mm diameter of base and 65 mm altitude is lying on one of its generators on HP with its axis parallel to VP. It is sectioned by a plane parallel to VP and 5 mm in front of the axis of the cone. (i) Draw the sectional front view and (ii) Draw the development of the lateral surface of the remaining portion of the cone.* (UQ)

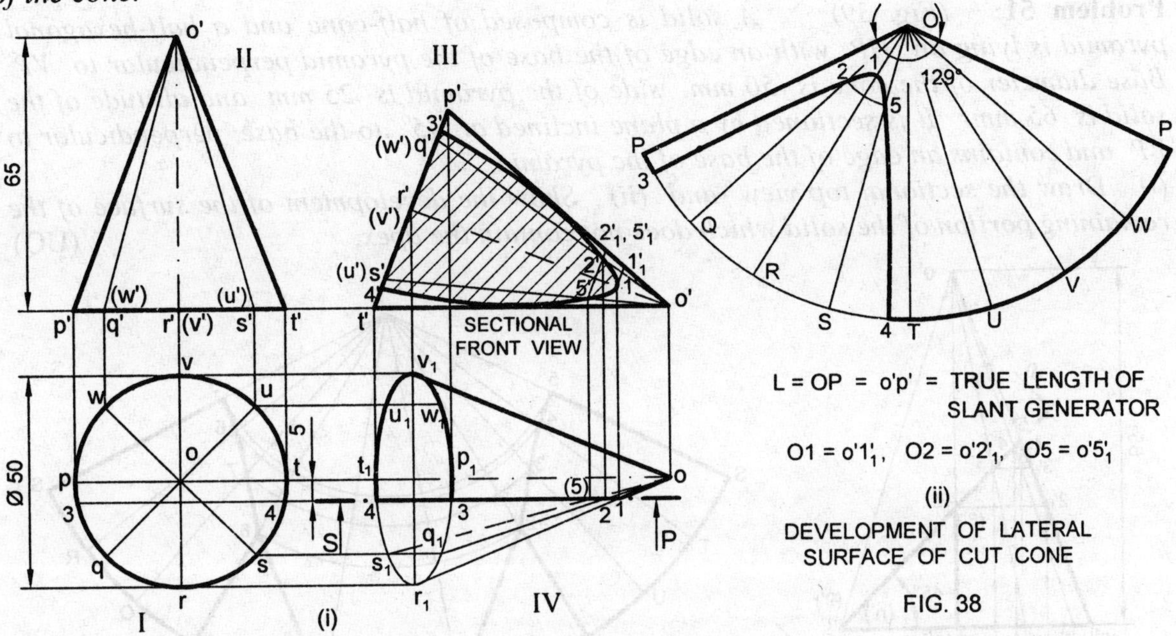

FIG. 38

1. Draw the top and front views of the cone in the simple position.
2. Draw the second front view and second top view of the cone in the tilted position.
3. S.P. is parallel to VP (given). Hence it will be perpendicular to HP. Mark the S.P. in the second top view 5 mm below the axis. Project the cutting points from the second top view to the sectional front view. Complete the sectional front view as shown.
4. Draw the development of the lateral surface of the cut cone.

Problem 48: *(Exercise) A cone of 50 mm diameter of base and 60 mm height is lying on HP on one of its generators with its axis parallel to VP. It is sectioned by a plane perpendicular to VP, parallel to and 30 mm above HP. Draw the development of the lateral surface of the cut cone.* (UQ)

Problem 49: *(Exercise) The development of the conical surface is a sector of the semi-circular plane. Radius of the sector is 90 mm. Angle subtended by the arc at the center is 120°. Construct the cone and draw its projections when the vertex is 42 mm above HP and its axis is parallel to VP.* (UQ)

Hint: Draw the top view of the cone as a circle of 30 mm radius. [Arc Length = A = $R\theta$ = $2\pi r$ and R = 90 mm, $\theta = (\pi \times 120) / 180$. So r = 30 mm.] Draw the front view of height 90 mm. Tilt it so that vertex is 42 mm above XY. Complete the projection.

Problem 50: *(Exercise) The development of a right circular cone is a semi-circle of 110 mm diameter. A square having a diagonal of length 110 mm is removed. Draw the front and top views of the cone showing the views of the portion removed.*

Results: $\theta = 180^\circ$; r = 55 mm.

Problem 51: *(Fig. 39) A solid is composed of half-cone and a half-hexagonal pyramid is lying on HP with an edge of the base of the pyramid perpendicular to VP. Base diameter of the cone is 50 mm, side of the pyramid is 25 mm and altitude of the solid is 65 mm. It is sectioned by a plane inclined at 45° to the base, perpendicular to VP and contains an edge of the base of the pyramid.*

(i) Draw the sectional top view and (ii) Show the development of the surface of the remaining portion of the solid which does not contain the apex. (UQ)

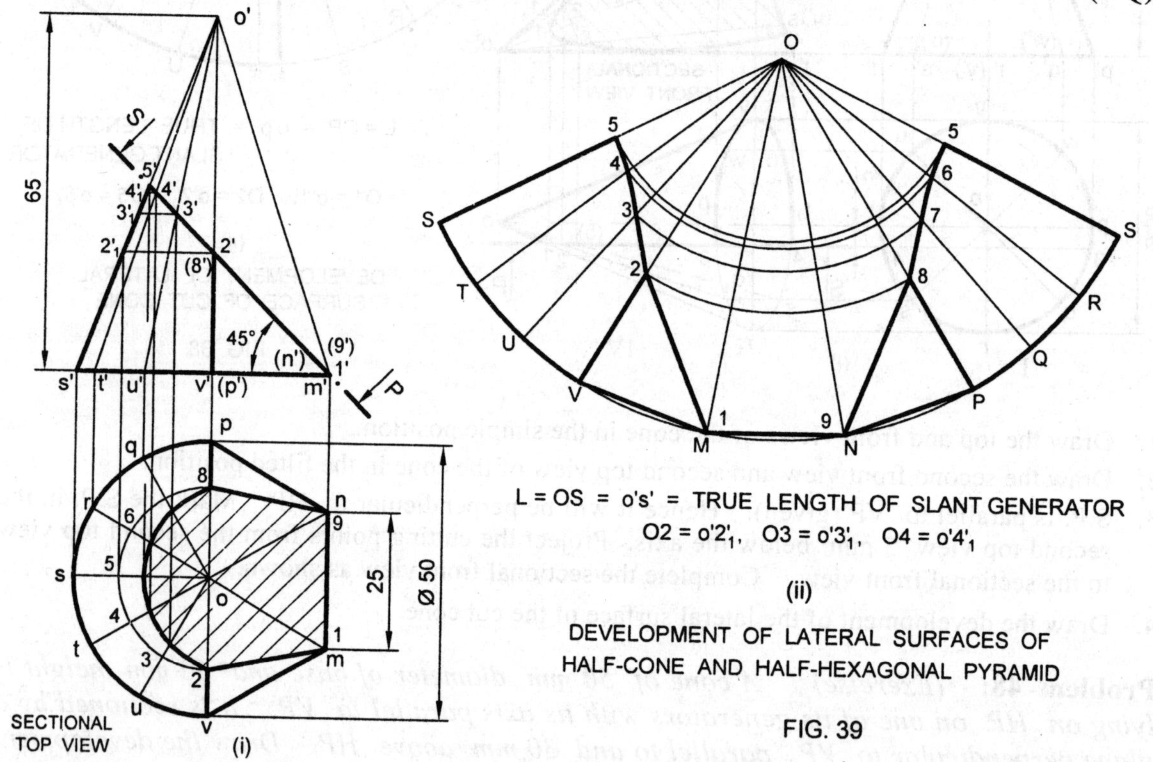

FIG. 39

Problem 52: *(Exercise) A solid is composed of half-cone and a half-hexagonal pyramid is lying on HP with an edge of the base of the pyramid perpendicular to VP. Base diameter of the cone is 50 mm, side of the pyramid is 25 mm and altitude of the solid is 65 mm. It is sectioned by a plane inclined at 60° to the base, perpendicular to VP and contains an edge of the base of the cone.*

(i) Draw the sectional top view and (ii) Show the development of the surface of the remaining portion of the solid which does not contain the apex.

23

INTERSECTION OF SURFACES

1. INTERSECTION OF SURFACES OF SOLIDS

When one solid enters or penetrates into the other of the same or different cross-section, the surfaces of both the solids come in contact with one another. Thus the outline of the penetrating solid is seen. This is known as ***Interpenetration of Solids***.

In engineering practice, when two solids are joined to make a single object, for example, a dome fitted on a boiler, their surfaces will intersect each other. The line of intersection of both the surfaces will be common.

When two solids of the same or different cross-section come in contact with each other, the surfaces of both the solids intersect one another. This sort of surface contact is termed as ***Intersection of Surfaces of Solids***.

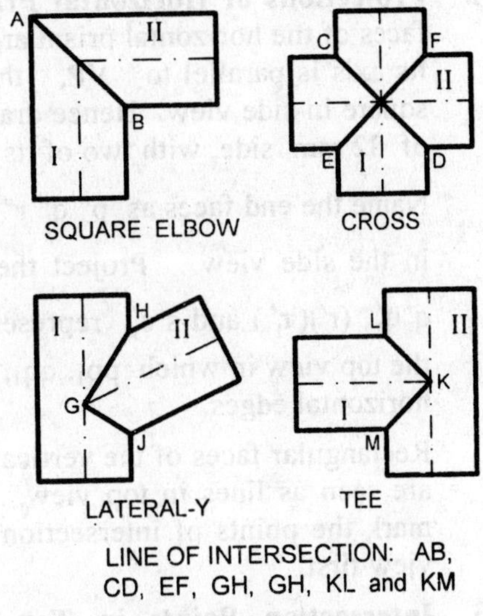

FIG. 1 PIPE FITTINGS

Engineering Applications of Intersection of Surfaces: *See Fig. 1.* Pipe fittings like elbow, T-joint, Y-joint, etc., Fitting of steam dome of a cylindrical boiler, Fabrication of chimney / hopper, Wings of airplane, Automobile body building, Sheet metal works, Ship building, etc.

Classifications of Intersection of Surfaces of Solids

1. Intersection of two plane surfaces (Prism Vs Prism)
2. Intersection between a plane surface and a curved surface (Prism Vs Cylinder)
3. Intersection of two curved surfaces (Cylinder Vs Cylinder, Cylinder Vs Cone)

Line or Curve of Intersection: Line of intersection of two surfaces of solids is a line common to both. When two solids bounded by plane surfaces such as prism and pyramid penetrate each other, we obtain ***straight lines*** as their lines of intersection. Line of intersection between a plane surface and a curved surface of two solids such as prism and cylinder is a ***curve***. The line of intersection between two curved surfaces of solids such as cylinder and cone is a ***curve***.

1.1 METHODS OF DETERMINING THE LINE / CURVE OF INTERSECTION

1. Line / Generator Method / Piercing Point Method / Edgewise View Method: Used for Prisms, Cylinders, etc.
2. Section Plane Method: Used for Pyramids, Cones, Cylinders, Spheres, etc.

2. INTERSECTION OF PRISM Vs PRISM

Problem 1: *(Fig. 2) A square prism, side of base 40 mm and height 75 mm stands with its base on HP and two of its rectangular faces are equally inclined to VP. It is completely penetrated by a horizontal square prism, side of base 32 mm and axis 75 mm long, such that the axes of the two prisms intersect each other at right angles. The two rectangular faces of the horizontal prism are equally inclined to HP and its axis is parallel to VP. Draw the projection of the prisms showing the lines of intersection.*

1. **Projections of Vertical Prism:** Draw the top view of vertical prism as a square of 40 mm side such that two of its sides are equally inclined to XY. Name the top and bottom square faces as abcd and (a_1) (b_1) (c_1) (d_1) respectively. Project its corresponding front view in which $a'\,a_1'$, $b'\,b_1'$, $c'\,c_1'$ and $(d')\,(d_1')$ represent vertical edges. Draw Left Side View in which $d''\,d_1''$, $a''\,a_1''$, $b''\,b_1''$ and $(c'')(c_1'')$ represent vertical edges.

2. **Projections of Horizontal Prism:** Since two rectangular faces of the horizontal prism are equally inclined to HP and its axis is parallel to VP, the end faces will be seen as a square in side view. Hence draw Left Side View as a square of 32 mm side, with two of its sides equally inclined to XY.

   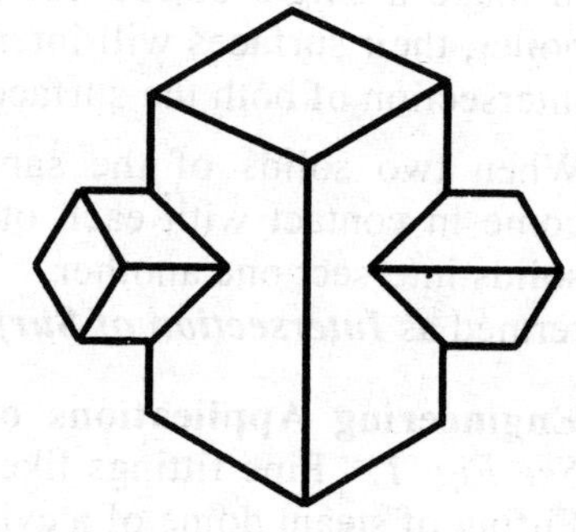

   Name the end faces as $p''\,q''\,r''\,s''$ and $(p_1'')\,(q_1'')\,(r_1'')\,(s_1'')$ in the side view. Project the front view in which $p'\,p_1'$, $q'\,q_1'$, $(r')(r_1')$ and $s'\,s_1'$ represent horizontal edges. Project the top view in which pp_1, qq_1, rr_1 and $(s)(s_1)$ represent the horizontal edges.

   Rectangular faces of the vertical prism are seen as lines in top view. Hence mark the points of intersection in top view first.

3. **Intersection Points in Top View:** In the top view the horizontal edges pp_1 pierces rectangular face $ab(b_1)(a_1)$ at 1, qq_1 intersects the edge $a(a_1)$ at 2, rr_1 pierces the rectangular face $ad(d_1)(a_1)$ at 3 and $(s)(s_1)$ intersects the edges $a(a_1)$ at (4).

   Similarly mark 5, 6, 7 and (8) in the top view.

4. **Intersection Points in Front View:** From 1 and 2 draw projectors to intersect $p'\,p_1'$ and $q'\,q_1'$ at 1′ and 2′ respectively.

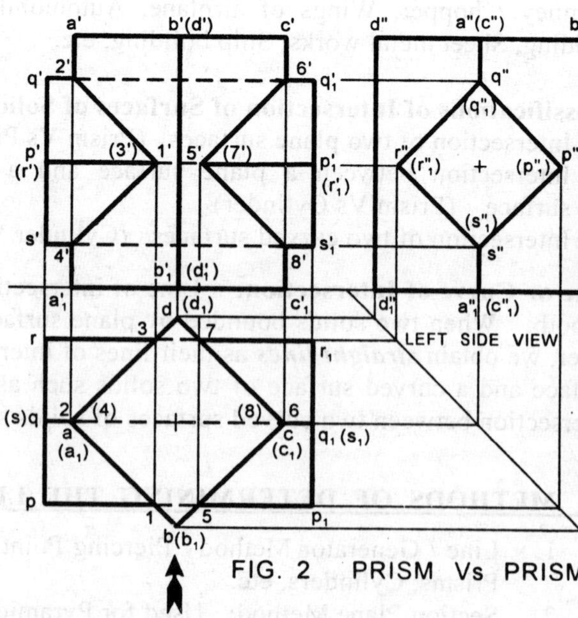

FIG. 2 PRISM Vs PRISM

Also project (3′) and 4′ on (r′)(r_1') and s′ s_1' respectively. Similarly obtain 5′, 6′, (7′) and 8′ in the front view.

5. **Visible and Invisible Lines of Intersection in Front View:** Look at the top view in the direction of arrow. Lines of intersection 1-2, 1-(4), 5-6 and 5-(8) are visible. Hence draw 1′2′, 1′4′, 5′6′ and 5′8′ as thick lines in the front view.
   Lines of intersection 2-3, 3-(4), 6-7, 7-(8) are hidden by 1-2, 1-(4), 5-6 and 5-(8) respectively in the top view. Hence show the lines 2′(3′), (3′) 4′, 6′(7′), (7′) 8′ to coincide with 1′2′, 1′4′, 5′6′ and 5′8′ in the front view.

Rules of Visibility

- ☞ If a point lies on a visible edge or faces of both the intersecting surfaces, it is visible.
- ☞ Line of intersection connecting two visible points is visible.
- ☞ Line of intersection connecting one visible and one invisible point is invisible.
- ☞ Line of intersection changes from visible to invisible or vice-versa, only on crossing the visible edge or face of one of the intersecting surfaces.

Problem 2: *(Fig. 3) A square prism, side of base 45 mm stands with its base on HP and two of its rectangular faces are equally inclined to VP. It is completely penetrated by a horizontal square prism of side of base 35 mm such that their axes are 8 mm apart. The rectangular faces of the horizontal prism are equally inclined to VP and its axis is parallel to VP. Draw the lines of intersection.* (UQ)

Projections of the Vertical and Horizontal Prisms

1. Draw three views of vertical prism for the given position.
2. Axis of the horizontal prism is 8 mm from the axis of the vertical prism. Hence draw the side view of the horizontal prism as a square of 35 mm side such that its axis is 8 mm from the axis of the vertical prism.
3. From the side view, project the front and top views of the horizontal prism.

Intersection Points in Top View

4. In top view, mark 1 where the horizontal edge pp_1 pierces the front vertical rectangular face $ab(b_1)(a_1)$.

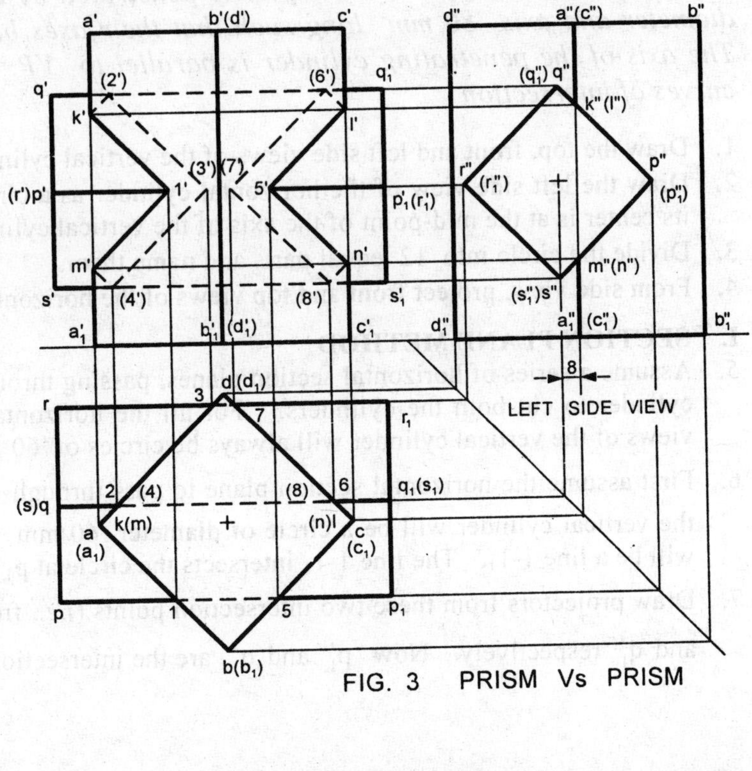

FIG. 3 PRISM Vs PRISM

5. Mark 2, 3 and (4) where the horizontal edges qq_1, rr_1 and (s) (s_1) intersect the rear vertical rectangular face ad (d_1) (a_1) respectively. Similarly mark 5,6,7 and (8) in top view.
6. In the side view, the rectangular face $p'' q'' (q_1'') (p_1'')$ intersects the vertical edges $a'' a_1''$ and $(c'')(c_1'')$ at k'' and $(1'')$ respectively. Similarly the rectangular face $s'' p'' (p_1'') (s_1'')$ intersects the vertical edges $a'' a_1''$ and $(c'')(c_1'')$ at m'' and (n'') respectively. Hence project these intersection points to the top view and obtain k (m) and 1 (n) as shown.

Intersection Points in Front View

7. From 1, 2, 3 and (4) draw projectors to intersect the corresponding edges of the horizontal prism in the front view at 1′, (2′), (3′) and (4′) respectively.
8. Draw horizontal lines from k'' and m'' from side view to intersect the projector drawn from k (m) at k′ and m′ respectively.

Visible and Invisible Lines of Intersection in Front View

9. In top view 1k and 1 (m) are visible. Hence draw 1′k′ and 1′m′ in front view.
10. Draw k′(2′), (2′)(3′), (3′)(4′) and (4′)m′ as dashed lines. Thus complete the lines of intersection on left side. Similarly draw the lines of intersection on right side in front view.

3. INTERSECTION OF CYLINDER Vs CYLINDER

Problem 3: *(Fig. 4) A cylinder of 60 mm diameter and axis 80 mm long stands with its base on HP. It is completely penetrated by a horizontal cylinder of 40 mm diameter and axis 80 mm long such that their axes bisect each other at right angles. The axis of the penetrating cylinder is parallel to VP. Draw the projections showing curves of intersection.*

(UQ)

1. Draw the top, front and left side views of the vertical cylinder.
2. Draw the left side view of the horizontal cylinder as a circle of diameter 40 mm such that its center is at the mid-point of the axis of the vertical cylinder.
3. Divide the circle into 12 equal parts and name them.
4. From side view, project front and top views of the horizontal cylinder.

I. SECTION PLANE METHOD

5. Assume a series of horizontal section planes, passing through the generators of the horizontal cylinder to cut both the cylinders. For all the horizontal section planes, the sectional top views of the vertical cylinder will always be circles of 60 mm diameter.
6. First assume the horizontal section plane to pass through $1''(1_1'')$. The sectional top view of the vertical cylinder will be a circle of diameter 60 mm and that of the horizontal cylinder will be a line $1\text{-}1_1$. The line $1\text{-}1_1$ intersects the circle at p_1 and q_1 in the top view.
7. Draw projectors from these two intersection points (*i.e.*, from p_1 and q_1) to cut $1'1_1'$ at p_1' and q_1' respectively. Now p_1' and q_1' are the intersection points in front view.

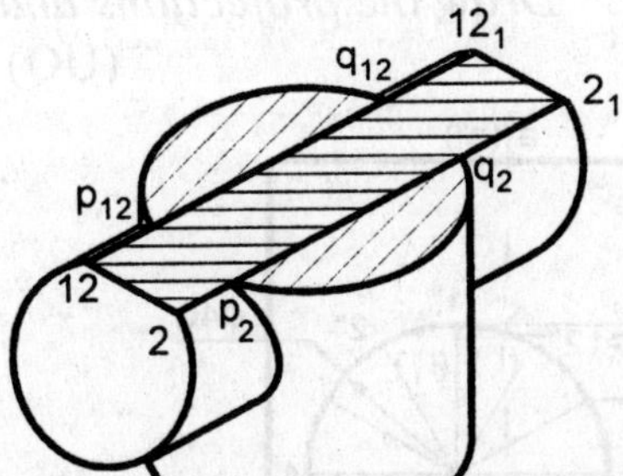

(i)

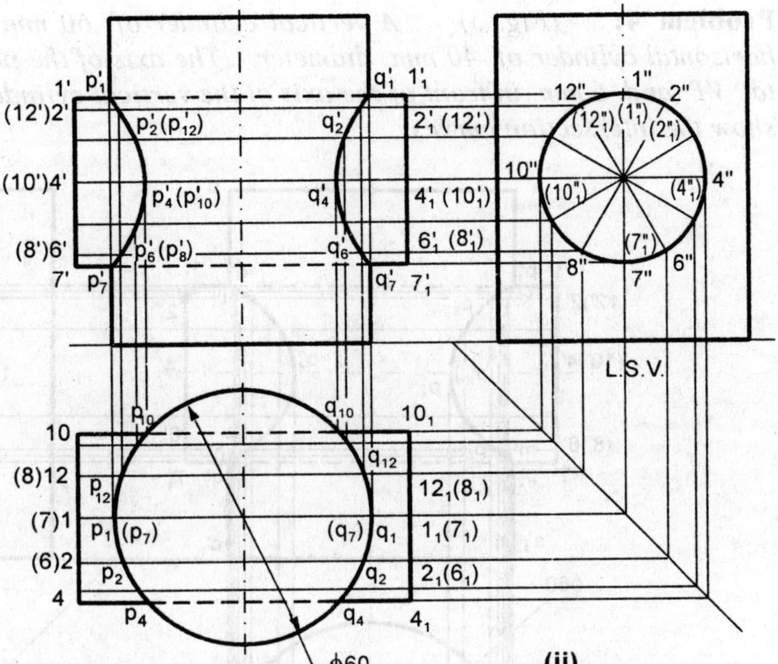

(ii)

FIG. 4
CYLINDER Vs CYLINDER

8. Assume second horizontal section plane to pass through $2''$ ($2''_1$) and $12''$ ($12''_1$).
9. Sectional top view of the vertical cylinder is again a circle of same diameter (60 mm).
10. Sectional top view of the horizontal cylinder will be a rectangle of length $2\text{-}2_1$ and breadth 2-12 (*Fig. i*). This rectangle intersects the circle at p_2, q_2, q_{12} and p_{12} in top view.
11. Project these four intersection points to the corresponding generators in the front view and obtain p'_2, (p'_{12}), (q'_{12}) and q'_2.
12. Similarly assume series of section planes passing through other generators and obtain the intersection points both in the top and front views.

Visible and Invisible Curves of Intersection in Front View

13. In top view the portion of intersection curve $p_1 p_2$ ..... (p_7) is nearer to the observer and visible. So show the curve p'_1, p'_2, ..., p'_7 as continuous thick line in front view.
14. The rear portion of the curve is hidden and coincides with the visible portion of the curve.
15. Similarly obtain the intersection curve on the right side.

II. GENERATOR METHOD

1. Mark p_1, p_2, etc., where the generators $1\text{-}1_1$, $2\text{-}2_1$, etc., cut the circle in top view.
2. Project the above points. Obtain p'_1, p'_2, etc., on the corresponding generators in front view.
3. Join these points in correct sequence. Obtain the intersection curve on the left side.
4. Similarly mark q_1, q_2, etc., in the top view.
5. Project these points to get q'_1, q'_2 in front view. Complete the right side intersection curve.

Problem 4: *(Fig. 5) A vertical cylinder of 60 mm diameter is penetrated by a horizontal cylinder of 40 mm diameter. The axis of the penetrating cylinder is parallel to VP and 6 mm in front of the axis of the vertical cylinder. Draw the projections and show the intersection curves.* (UQ)

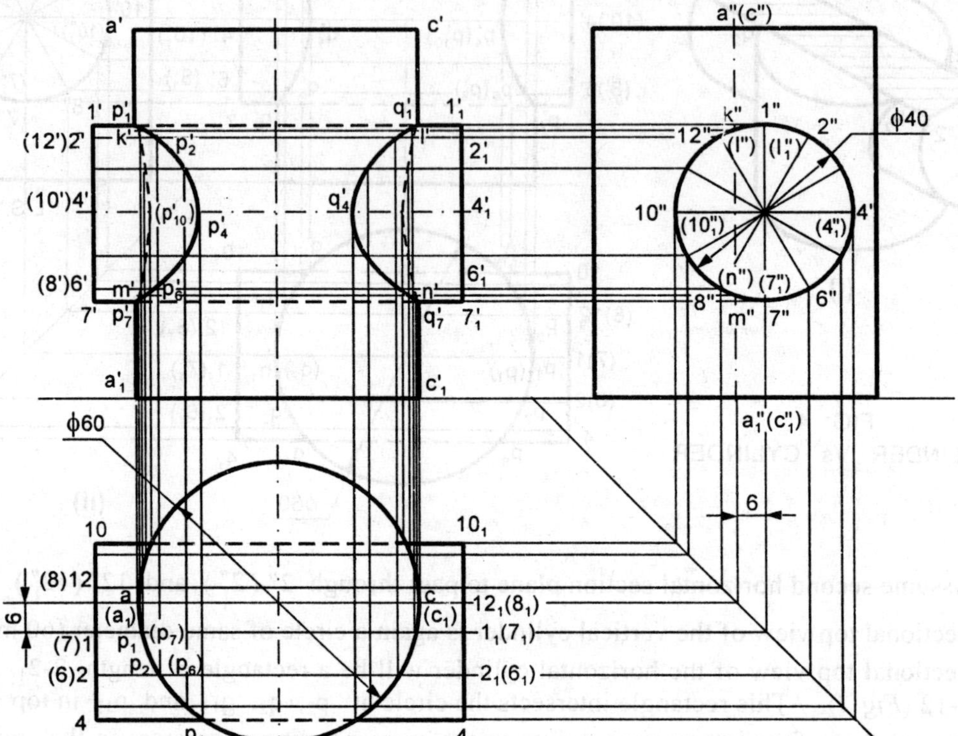

FIG. 5

1. Draw the three views of the vertical cylinder.
2. Axis of the horizontal cylinder is parallel to VP and 6 mm in front of the axis of the vertical cylinder. Hence draw side view of horizontal cylinder as a circle of 40 mm diameter such that its center is 6 mm to right side of the axis of vertical cylinder.
3. Project its front and top views and name the generators as explained in Problem 3.
4. Obtain 12 (twelve) intersection points p_1', p_2', ..., p_{12}' on the left side and another 12 intersection points q_1', q_2', ..., q_{12}' on the right side.

To obtain Critical Intersection Points: These are the intersection points at which the intersection curve changes its direction.

5. The circle in the side view intersects the generator $a'' a_1''$ at k'' and m'' and (c'') (c_1'') at (l'') and (n'') respectively. Project these points to the front view.
6. Draw the curve through p_1', p_2', ..., p_7' as thick since this portion is visible.
7. Draw the curve through p_1', k', (p_{12}'), (p_{10}'), (p_8'), m', p_7' as dashed since this portion is hidden.

Problem 5: *(Fig. 6) A vertical cylinder of 60 mm diameter is penetrated by a horizontal cylinder of 40 mm diameter. The axis of the horizontal cylinder is parallel to VP and 10 mm in front of the axis of the vertical cylinder. Draw the projections showing the curves of intersection.* (UQ)

Note: When the horizontal cylinder is tangential to the vertical cylinder at a point, the form of intersection curve will be a continuous curve crossed at the tangent point.

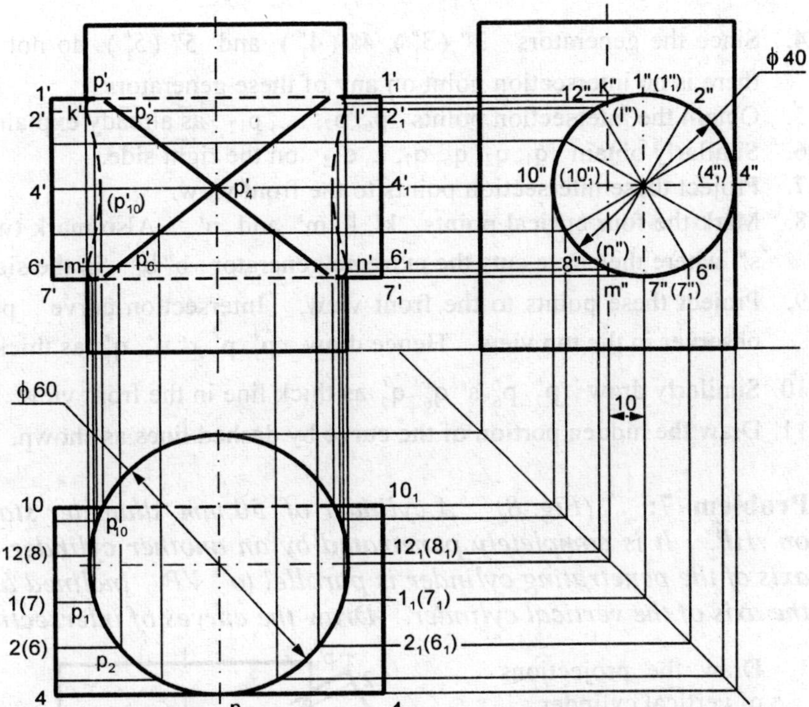

FIG. 6

Problem 6: *(Fig. 7) A vertical cylinder of 60 mm diameter is penetrated by a horizontal cylinder of 50 mm diameter. Axis of horizontal cylinder is parallel to VP and 10 mm in front of the axis of vertical cylinder. Draw the intersection curves.*

1. Draw the projections of both cylinders for the given position.
2. Name the generators of the penetrating (horizontal) cylinder in all the three views.
3. Mark p_1, p_2 where the generators $1\text{-}1_1$ and $2\text{-}2_1$ intersect the circle in the top view.

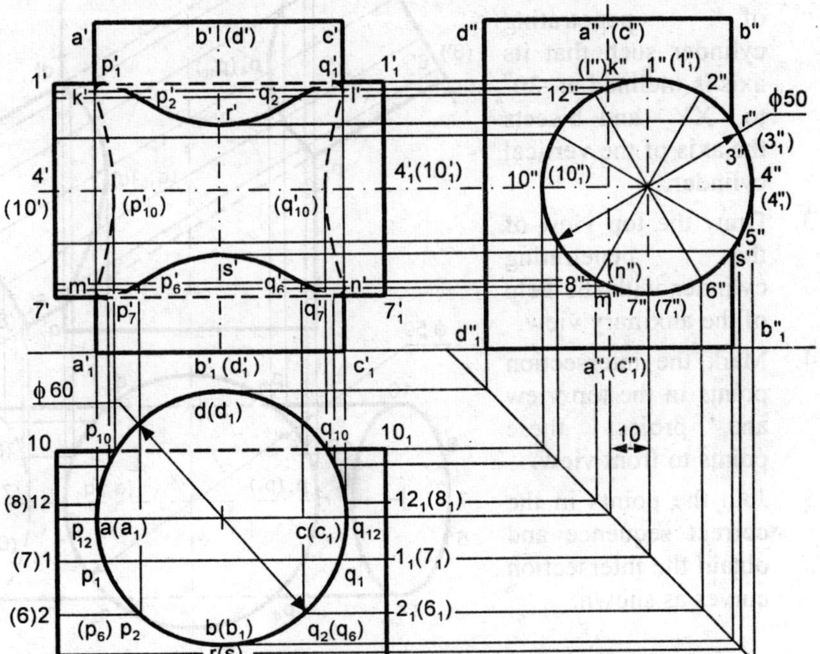

FIG. 7

4. Since the generators $3''\ (3_1'')$, $4''\ (4_1'')$ and $5''\ (5_1'')$ do not intersect the vertical cylinder there is no intersection point on any of these generators.
5. Obtain the intersection points $p_6, p_7, \ldots, p_{12}$ as already explained.
6. Similarly obtain $q_1, q_2, q_6, q_7, \ldots q_{12}$ on the right side.
7. Project these intersection points to the front view.
8. Mark the four critical points k′, l′, m′ and n′. Also mark two more critical points r″ and s″ where the circle cuts the extreme generator $b''\ b_1''$ in the side view.
9. Project these points to the front view. Intersection curve $p_1\ p_2\ r\ q_2\ q_1$ is nearer to the observer in the top view. Hence draw $p_1'\ p_2'\ r'\ q_2'\ q_1'$ as thick line in the front view.
10. Similarly draw $p_7'\ p_6'\ s'\ q_6'\ q_7'$ as thick line in the front view.
11. Draw the hidden portion of the curve by dashed lines as shown.

Problem 7: *(Fig. 8) A cylinder of 50 mm diameter stands vertically with its base on HP. It is completely penetrated by an another cylinder of 35 mm diameter. The axis of the penetrating cylinder is parallel to VP, inclined at 30° to HP and bisects the axis of the vertical cylinder. Draw the curves of intersection.* **(UQ)**

1. Draw the projections of vertical cylinder.
2. Draw the front view of penetrating cylinder such that its axis is inclined at 30° to XY and bisects the axis of the vertical cylinder.
3. Draw the top view of the penetrating cylinder with the help of the auxiliary view.
4. Mark the intersection points in the top view and project these points to front view.
5. Join the points in the correct sequence and obtain the intersection curves as shown.

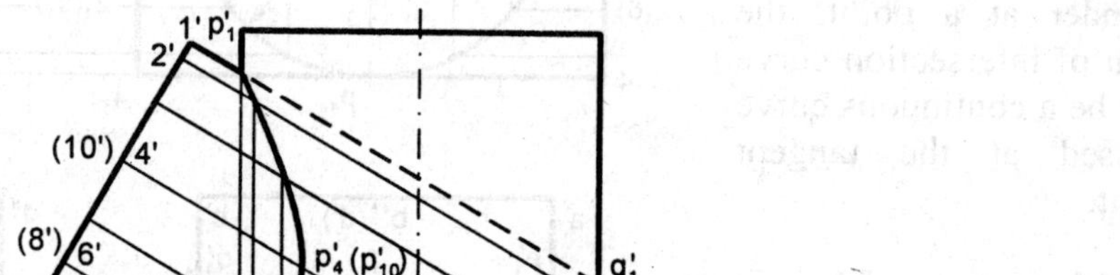

FIG. 8

Problem 8: *(Exercise) A pipe of 25 mm diameter is joined to an another vertical pipe of 50 mm diameter at an angle of 45°. Their axes are 12.5 mm apart. Draw the projections of the pipes showing curves of intersection.*

4. INTERSECTION OF PRISM Vs CYLINDER

Problem 9: *(Fig. 9) A cylinder of 60 mm diameter and axis 80 mm long is standing vertically on its base on HP. It is pierced by a square prism of 30 mm side and 100 mm length, the axis of which is parallel to both the reference planes and the faces equally inclined to HP. The axes of the solids intersect at right angles. The height of the axis of the prism above HP is 40 mm. Draw the projections of the solids showing the curves of intersection in the front view.* (UQ)

1. Draw the top view of the cylinder and square prism in the given position.
2. Project the front view.
3. Draw the corresponding left side view of the intersecting cylinder and prism. From the side view, it is clear that the cylinder envelops the square prism.
4. Apply the Horizontal Section Plane Method for solving this problem.
5. Obtain the curves of intersection in the front view as shown.

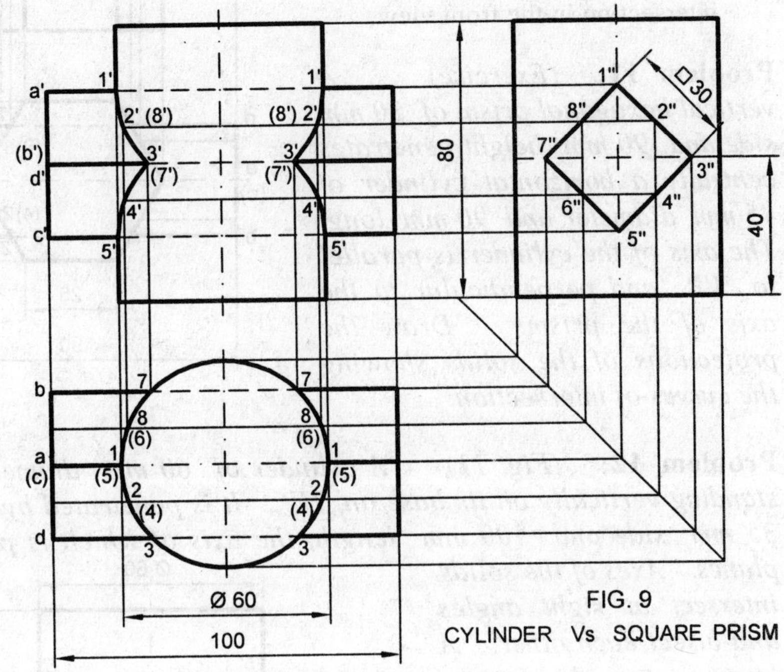

FIG. 9
CYLINDER Vs SQUARE PRISM

Problem 10: *(Fig. 10) A horizontal cylinder of 30 mm diameter and 90 mm long penetrates centrally a vertical hexagonal prism of 25 mm side and 80 mm height. The axis of the cylinder is parallel to VP and perpendicular to the axis of the prism. Draw the projections of the solids showing the curves of intersection in the front view.* (UQ)

1. Draw the top view of the given intersecting cylinder and hexagonal prism.
2. Project the front view as shown.
3. Draw the left side view of the intersecting cylinder and hexagonal prism.
4. Divide the cylinder circle in the side view into 8 or 12 equal parts.
5. Project the generators to the top and front views.

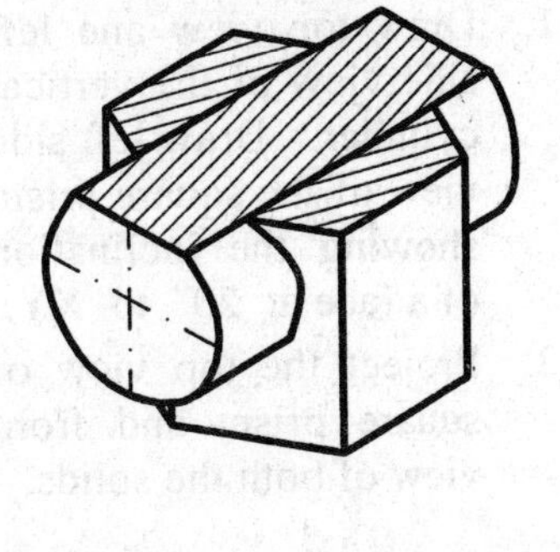

6. Obtain the intersecting points in the top view at the points at which the generators of the cylinder intersect the rectangular faces of the hexagonal prism.
7. Project the intersection points from the top view to front view.
8. Draw smooth curves of intersection in the front view.

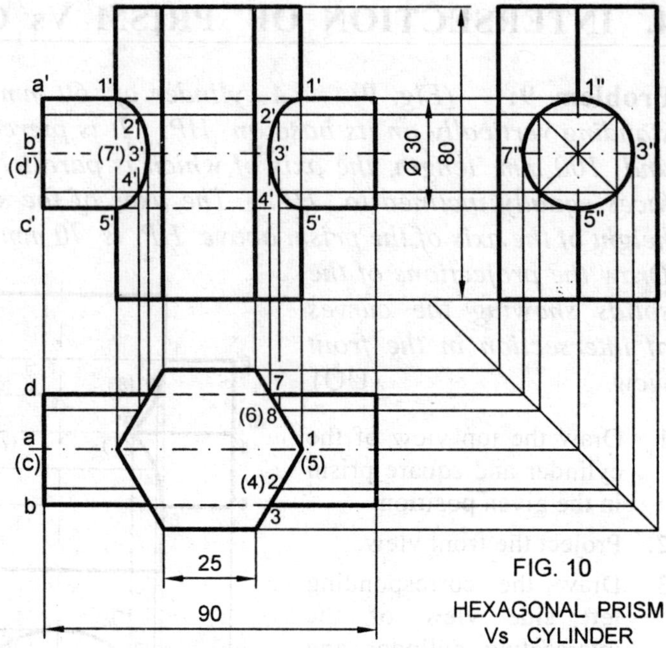

FIG. 10
HEXAGONAL PRISM Vs CYLINDER

Problem 11: *(Exercise) A vertical hexagonal prism of 30 mm side and 90 mm height penetrates centrally a horizontal cylinder of 45 mm diameter and 90 mm long. The axis of the cylinder is parallel to VP and perpendicular to the axis of the prism. Draw the projections of the solids showing the curves of intersection.*

Problem 12: *(Fig. 11) A cylinder of 60 mm diameter and axis 80 mm long is standing vertically on its base on HP. It is penetrated by a horizontal square prism of 35 mm side and 100 mm length, the axis of which is parallel to both the reference planes. Axes of the solids intersect at right angles and bisect each other. A lower rectangular face of the horizontal prism is inclined at an angle of 20° to HP. Draw the projections showing the intersection curve in the front view.* (UQ)

1. Draw top view and left side view of the vertical cylinder. Draw left side view of the square prism showing the inclination of a face at 20° to XY.
2. Project the top view of square prism and front view of both the solids.

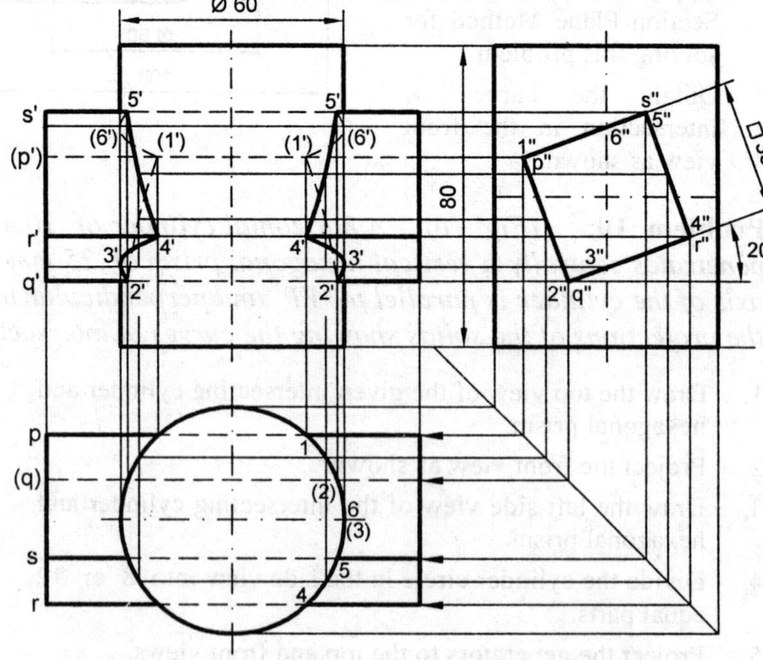

FIG. 11 CYLINDER Vs INCLINED SQUARE PRISM

3. In the top view mark 1 on the circle where the edge line p cuts it. Project it in the front as 1′ on the p′ edge line. Similarly mark other points for curve of intersection.
4. Join the above points in proper sequence. Show the hidden curve in *dashed lines*.
5. The prism is split into two separate parts. Therefore, obtain the curve on the other side of intersection of the solids in the same way.

Problem 13: *(Fig. 12) A vertical cylinder of 45 mm diameter is penetrated by horizontal square prism of side of base 30 mm. Axis of the prism is 7.5 mm away from the axis of the cylinder and is parallel to VP. Rectangular faces of the prism are equally inclined to both VP and HP. Axis of the prism is in front of the axis of the cylinder. Draw their projections showing intersection curves in front view.* (UQ)

1. Draw top view of the intersecting prism and cylinder. One longer edge of the prism remains outside the cylinder.
2. Project corresponding front view.
3. Draw left side view such that the prism remains outside the cylinder. Mark the points of intersection of the edges of the prism with the cylinder in the top and side views.
4. Project these points of intersection to the front view.
5. Also mark the key points of the extreme generators of the cylinder with the surface of the prism as shown in the top and side views.
6. Project these key points of intersection to the front view.
7. Draw smooth curves passing through all the intersection points. Mark the hidden portion by *dashed lines*.

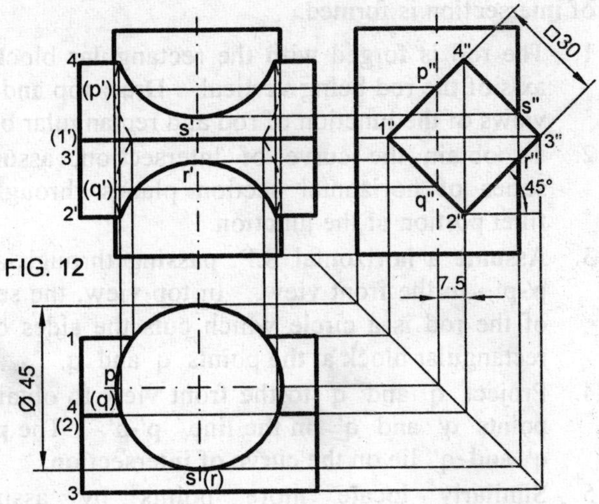

FIG. 12

Problem 14: *(Fig. 13) A square prism of side of base 55 mm and 100 mm height rests with its base on HP having its faces equally inclined to VP. It is pierced by a horizontal cylinder of diameter 40 mm and 100 mm long. The axis of the cylinder is parallel to VP and 8 mm away from the axis of the prism. Draw the projections showing the curves of intersection.* (UQ)

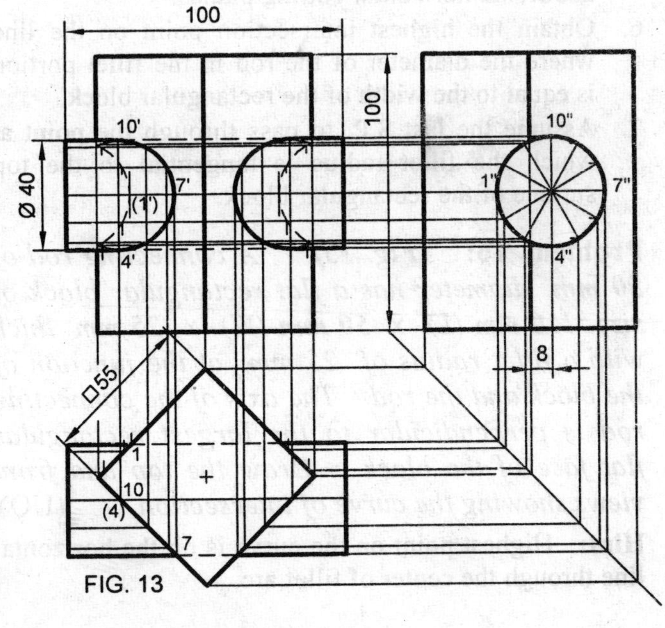

FIG. 13

Problem 15: *(Fig. 14) A connecting rod of 50 mm diameter has a flat rectangular block of size 110 mm (L) x 62 mm (W) x 25 mm thick with a fillet radius of 25 mm at the junction of the block and the rod. The axis of the connecting rod is perpendicular to the largest rectangular flat face of the block. Draw the top and front views showing the curve of intersection.* (UQ)

Connecting rod increases in diameter as it approaches the rectangular block. This forms what is called a **fillet** of 25 mm radius. Note that the fillet gradually increases the diameter of connecting rod. Width of the rectangular block is less than the sum of diameter of the rod and twice the fillet radius. Hence a curve of intersection is formed.

1. The rod is forged with the rectangular block, the axis of the rod being vertical. Draw top and front views of the junction of rod and rectangular block.
2. To obtain the curve of intersection, assume a series of horizontal section planes through the fillet portion of the junction.
3. Assume a horizontal S.P. passing through a line p′-p′ in the front view. In top view, the section of the rod is a circle which cuts the sides of the rectangular block at the points q and q.
4. Project q and q to the front view to obtain the points q′ and q′ on the line p′-p′. The points q′ and q′ lie on the curve of intersection.
5. Similarly locate more points by assuming additional horizontal cutting planes.
6. Obtain the highest intersection point on the line where the diameter of the rod in the fillet portion is equal to the width of the rectangular block.
7. Assume the last S.P. to pass through the point at which the fillet radius is tangential to the top surface of the rectangular block.

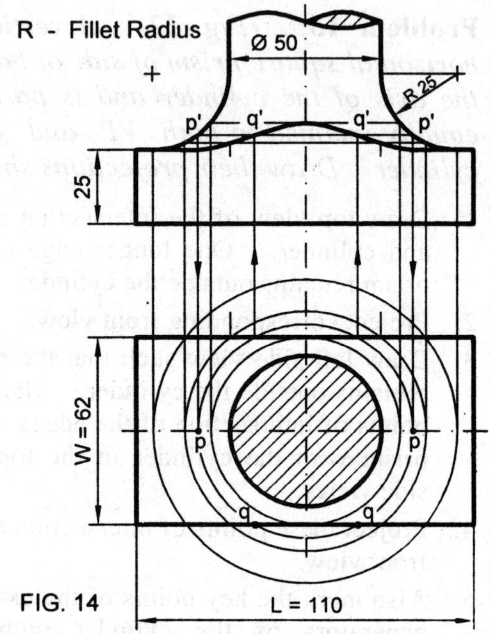

FIG. 14

Problem 16: *(Fig. 15) A connecting rod of* ***50 mm*** *diameter has a flat rectangular block of size 110 mm (L) x* ***50 mm*** *(W) x 25 mm thick with a fillet radius of 25 mm at the junction of the block and the rod. The axis of the connecting rod is perpendicular to the largest rectangular flat face of the block. Draw the top and front views showing the curve of intersection.* (UQ)

Hint: Highest point on the curve is on the horizontal line through the center of fillet arc.

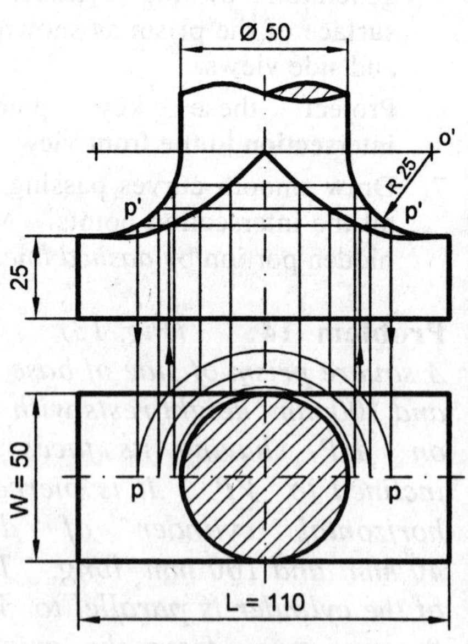

FIG. 15

5. INTERSECTION OF CYLINDER Vs CONE

Problem 17: *(Fig. 16) A cone of base 60 mm diameter and axis 70 mm long, rests with its base on HP. It is completely penetrated by a horizontal cylinder of 30 mm diameter such that both the axes intersect each other at right angles. The axis of the cylinder is parallel to VP and 20 mm above the base of the cone. Draw the projections of the solids showing the curves of intersection.* (UQ)

1. Draw the three views of the cone. On the axis of the cone in the Side View mark o″ at 20 mm above the base of the cone to represent the axis of the horizontal cylinder.
2. With o″ as center and 30 mm as diameter draw a circle to represent the cylinder.
3. Project the corresponding front and top views of the cylinder and name the generators.

CUTTING PLANE METHOD : *See Fig. (i).*

4. First assume a horizontal S.P. passing through $1''$ $(1_1'')$. The sectional top view of the cone is a circle of diameter a′a′ and that of the horizontal cylinder is a line $1\text{-}1_1$. Hence draw a circle with o as center and a′a′ as diameter in the top view.
5. Mark the intersection points p_1 and q_1 where the line $1\text{-}1_1$ intersects this circle.
6. Project p_1 and q_1 and obtain p_1' and q_1' on $1'1_1'$ in the front view.

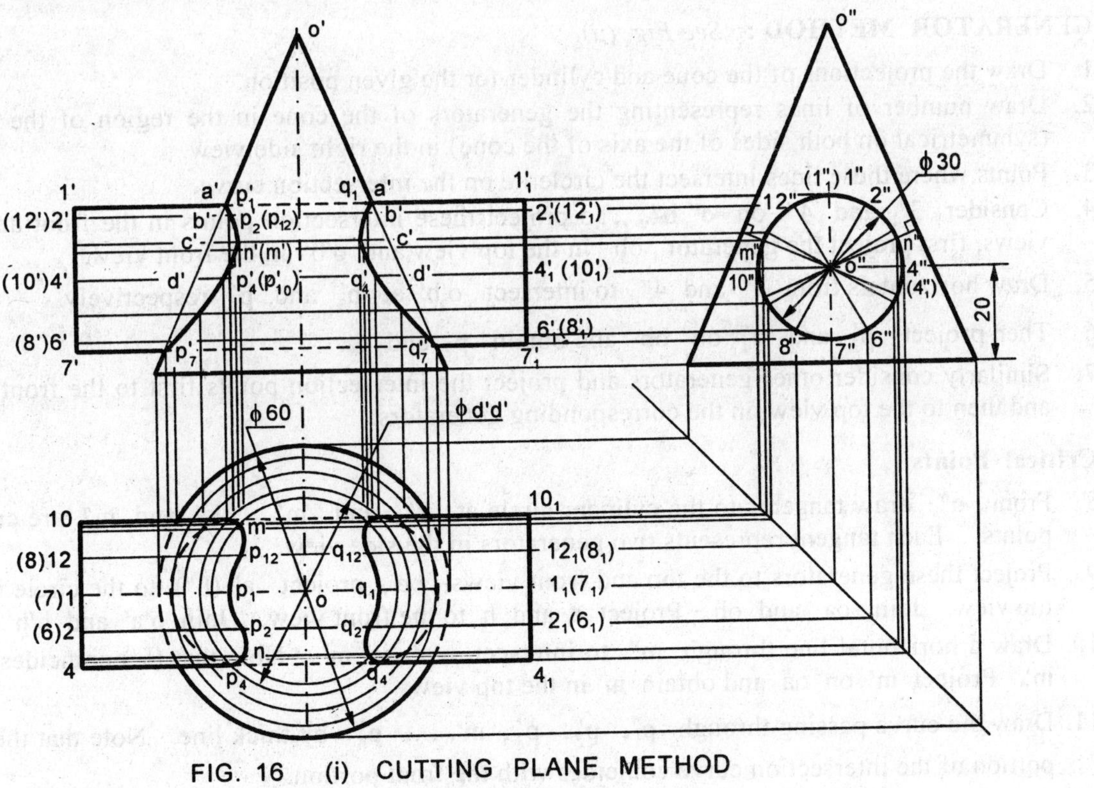

FIG. 16 (i) CUTTING PLANE METHOD

7. Assume second horizontal S.P. passing through 2″ ($2''_1$) and 12″ ($12''_1$). The sectional top view of the cone will be a circle of diameter b′b′ and that of the cylinder will be a rectangle of length $2\text{-}2_1$ and breadth 2-12. The circle and the rectangle intersect at 4 intersection points. Mark them as p_2 q_2 q_{12} and p_{12} in the top view.
8. Project these points on the corresponding generators of the cylinder in the front view and obtain p'_2, q'_2, (q'_{12}) and (p'_{12}) respectively.
9. Similarly assume series of horizontal section planes passing through 3″($3''_1$), 4″ ($4''_1$), etc.
10. Obtain the intersection points in top and front views.

Critical Points

11. From o″ draw perpendiculars to the end generators of the cone in the side view. They cut the circle at m″ and n″. Assume a horizontal S.P. passing through m″ and n″.
12. Project these critical points to the top and front views and obtain m, n and m′, n′.

To Mark Visible and Invisible Portions of the Intersection Curves

13. In the front view draw curve passing through p'_1, p'_2, p'_3, n′, p'_4, ..., p'_7 as *thick* lines.
14. The rear portion of the intersection curve coincides with the front portion. Hence it is not shown as dashed line in this view.
15. In the top view join n, p_2, p_1, p_{12} and m to represent the visible portion of the curve. The hidden portion is shown as *dashed* lines.

GENERATOR METHOD : *See Fig. (ii).*

1. Draw the projections of the cone and cylinder for the given position.
2. Draw number of lines representing the generators of the cone in the region of the circle (symmetrical on both sides of the axis of the cone) in the right side view.
3. Points where these lines intersect the circle lie on the intersection curve.
4. Consider 3″ and 4″ on o″ b″. To project these intersection points in the front and top views, first project the generator ob in the top view and o′b′ in the front view.
5. Draw horizontals from 3″ and 4″ to intersect o′b′ at p'_3 and p'_4 respectively.
6. Then project p'_3 and p'_4 on ob and obtain p_3 and p_4.
7. Similarly consider other generators and project the intersection points first to the front view and then to the top view on the corresponding generators.

Critical Points

8. From o″ draw tangents to the cylinder circle at k″ and m″. k″ and m″ are critical points. Each tangent represents two generators in the side view.
9. Project these generators to the top and front views, *i.e.*, project a″ (h″) to the circle in the top view. Join oa and oh. Project a and h to the front view. Join o′a′ and o′h′.
10. Draw a horizontal line through m″ to intersect o′a′ at m′. Note that (k′) coincides with m′. Project m′ on oa and obtain m in the top view.
11. Draw the curve passing through p'_1, p'_2, p'_3, m′, ..., p'_6 by thick line. Note that the rear portion of the intersection curve coincides with the front portion.

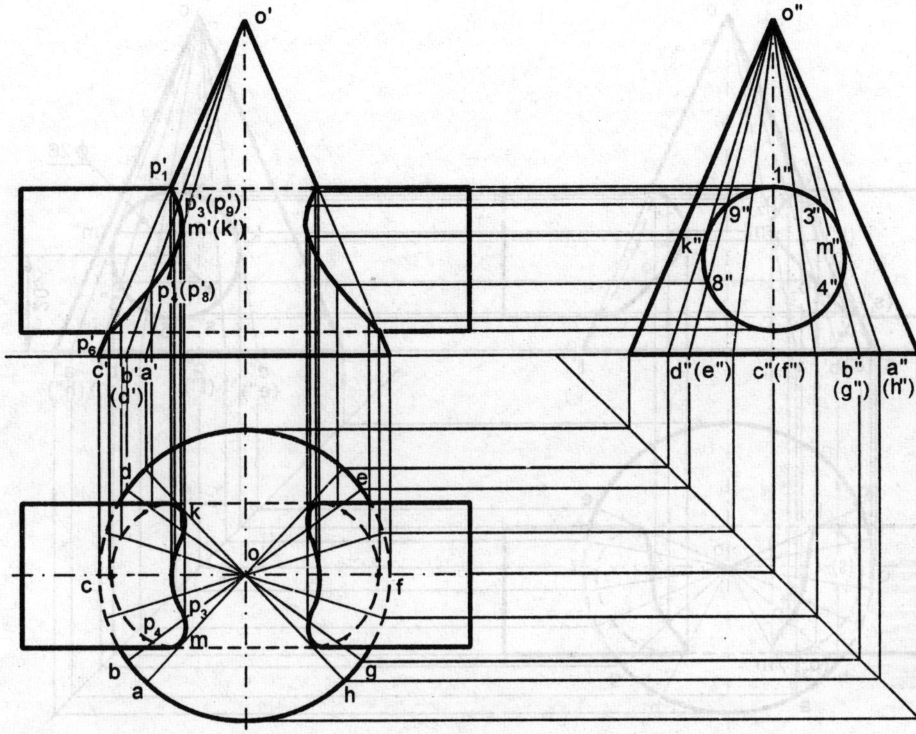

FIG. 16 (ii) GENERATOR METHOD

12. Similarly obtain the intersection curve on the right side in the front view. The hidden portion of the curve is shown dashed in the top view.

Note: *This method is also a type of Cutting Plane Method in which cutting planes pass through the apex and parallel to the axis of the cylinder.*

Problem 18: *(Fig. 17) A cone of base 60 mm diameter and axis 70 mm long, stands vertically with its base on HP. It is penetrated by a horizontal cylinder of 26 mm diameter. The axis of the cylinder is parallel to VP, 20 mm above the base and 5 mm in front of the axis of the cone. Draw the projections of solids showing the curves of intersection.* (UQ)

1. Draw the projections of the cone. Draw side view of the horizontal cylinder as a circle such that its axis is 20 mm above the base and 5 mm to right side of axis of cone.
2. Draw the top and front views of the horizontal cylinder.
3. Draw number of lines, representing the generators of the cone in the region of the circle in left side view. Points where these lines intersect the circle lie on intersection curve.
4. Project these intersection points to front and top views. For example 2″ and 3″ are on o″ b″. Project these points on o′b′ and obtain p_2' and p_3' in front view.

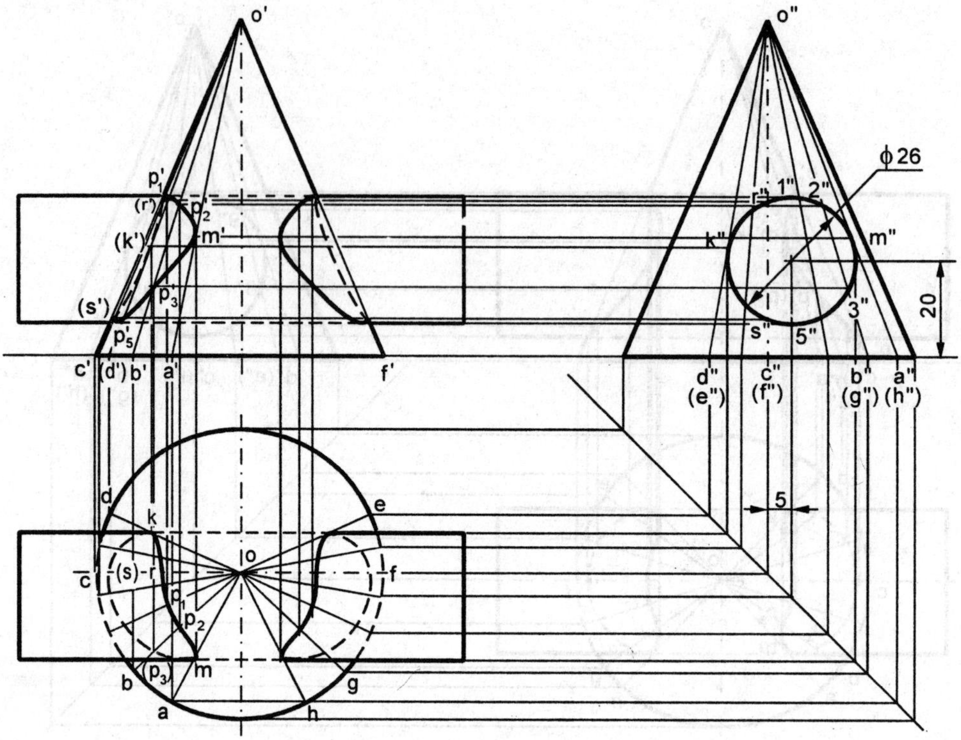

FIG. 17

5. Project these points to the top view and obtain p_2 and (p_3) on ob.
6. Draw two tangents to the circle in the side view and obtain the critical points m″ and k″ on the generators o″ a″ and o″ d″ respectively.
7. Project m″ and k″ on o′a′ and o′ (d′) and obtain m′ and (k′) respectively. Project these points to the top view and obtain m and k on oa and od respectively.
8. Two more critical points r″ and s″ are on o″c″. Project these points to front and top views.
9. Note that 1″ and 5″ are also critical points. To locate these points consider the generators passing through these points in the side view and project them to the other views. Construction is shown to obtain p_1' and p_1 only.
10. Join the intersection points in the correct sequence and obtain the curves on both sides.
11. Show the hidden portion of the curve as *dashed lines* in both views.

Problem 19: *(Fig. 18) A funnel consists of a cylindrical pipe and a conical part, both enveloping an imaginary common sphere of 40 mm diameter, with their axes intersecting at right angles. The diameter of the mouth of the funnel is 80 mm and its total height is 60 mm. The funnel is placed with its mouth on HP and two axes parallel to VP. Draw the curves of intersection.* (UQ)

See Fig. 18 for Solution.

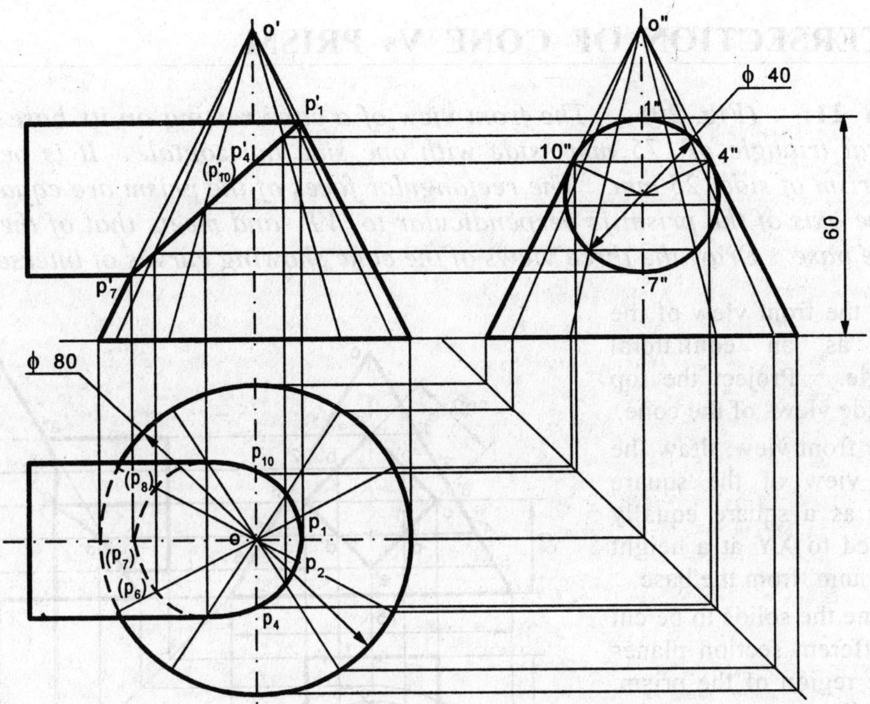

FIG. 18

Problem 20: *(Fig. 19) A vertical cylinder of 50 mm diameter and 80 mm height is penetrated by a cone of base 60 mm diameter and axis 70 mm long, the axes of the two solids bisecting each other at right angles. Draw the projections showing the curves of intersection.* (UQ)

1. Name the intersection points p_1, p_2, …, p_{12} in top view wherever the corresponding generators of the cone cut the cylinder circle.
2. Project these points and obtain p_1', p_2', …, p_{12}' on the corresponding generators in the front view. Join these points in the correct sequence.
3. Obtain the intersection curve on one side.
4. Similarly obtain the curve on the other side.

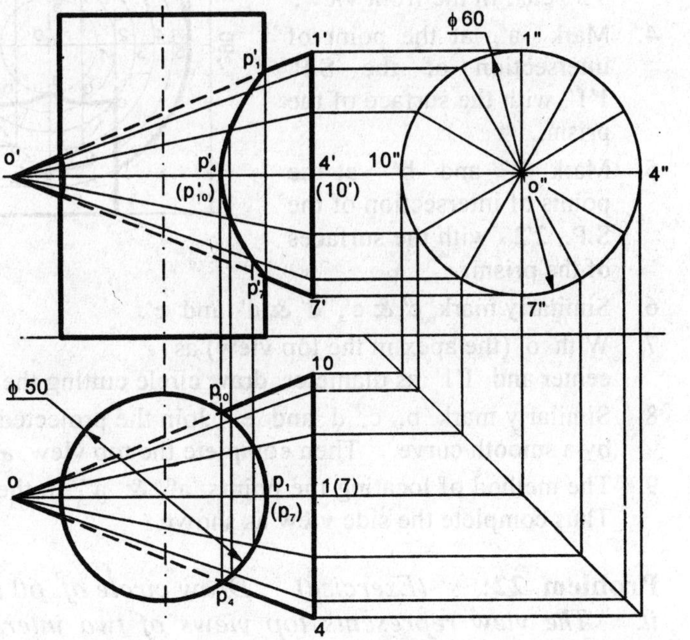

FIG. 19

6. INTERSECTION OF CONE Vs PRISM

Problem 21: *(Fig. 20) The front view of a cone resting on its base on HP is an equilateral triangle of 75 mm side with one side horizontal. It is penetrated by a square prism of side 25 mm. The rectangular faces of the prism are equally inclined to HP. The axis of the prism is perpendicular to VP and meets that of the cone 25 mm above the base. Draw the three views of the cone showing curves of intersection.* (UQ)

1. Draw the front view of the cone as an equilateral triangle. Project the top and side views of the cone.
2. In the front view, draw the front view of the square prism as a square equally inclined to XY at a height of 25 mm from the base.
3. Assume the solids to be cut by different section planes in the region of the prism. Accordingly, mark the section planes as 1′1′, 2′2′, 3′3′, etc., in the front view.
4. Mark a′ at the point of intersection of the S.P. 1′1′ with the surface of the prism.
5. Mark b′ and b′ at the points of intersection of the S.P. 2′2′ with the surfaces of the prism.

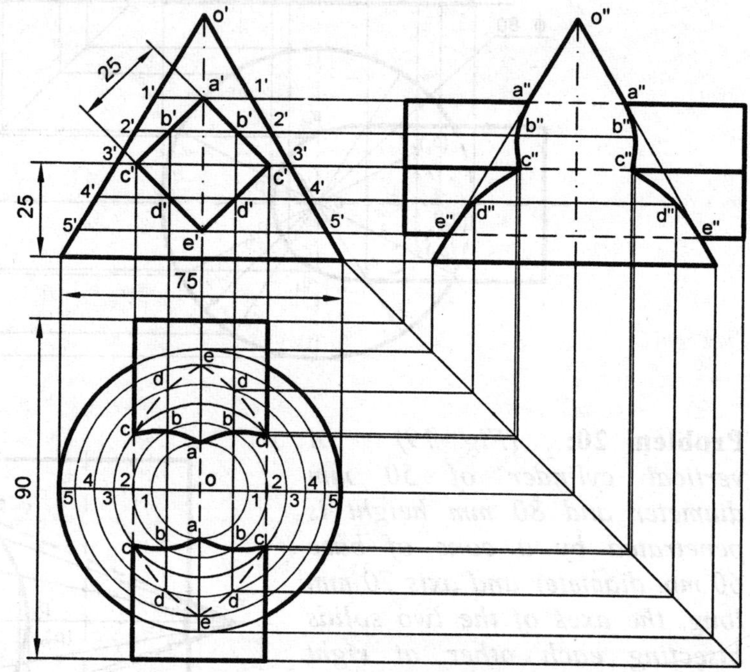

FIG. 20

6. Similarly mark c′ & c′, d′ & d′ and e′.
7. With o (the apex in the top view) as center and 1′1′ as diameter, draw circle cutting the projector through a′ at a.
8. Similarly mark b, c, d and e. Join the projected points in the top view in proper sequence by a smooth curve. Then complete the top view, assuming suitable length of the prism.
9. The method of locating the points a″ & a″ in the **side view** is clearly shown by mitre line. Thus complete the side view as shown.

Problem 22: *(Exercise) Draw circle of 60 mm diameter. Inscribe a pentagon in it. The view represents top views of two intersecting solids, namely, a cone and a pentagonal prism. Both the solids share a common axis length of 60 mm. The base of the cone is on HP. A rectangular face of the prism is parallel and nearer to VP. Draw the front view of the solids and show the curves of intersection.* (UQ)

24

PERSPECTIVE PROJECTION

1. PRINCIPLE OF PERSPECTIVE PROJECTION / VIEW

Perspective Projection is the graphic representation of an object on a single plane called *Picture Plane* (PP), as it appears to an observer. The object is placed behind the PP. The observer is stationed in front of PP. Therefore, visual rays from the eyes of the observer to the object are cut by the PP, *i.e.*, the rays pierce the picture plane and form an image on it. This image is the perspective view of the object.

The arrangement of the eye, picture plane and the object is the same as in the case of ***Third Angle Projection***. Therefore, the top view of the object is drawn above the front view.

Fig. 1 shows the perspective projection of a rail track with electric posts as seen by the observer. Picture taken by a camera is a perspective.

FIG. 1

Applications: As the photographic representation, it is extensively used by architects to show the appearance of proposed buildings, by drafters for illustrations of huge machinery or equipments and also by artists for producing advertisement drawings.

1.1 PERSPECTIVE ELEMENTS – DEFINITIONS

Ground Plane (GP): *See Fig. 2.* Ground Plane is a horizontal plane representing ground on which the object is assumed to be placed.

Picture Plane (PP): It is a vertical transparent plane on which the perspective view of the object is projected. It is placed in between the object and observer. The front view of the perspective elements and of the object (if necessary) may also be projected on the PP.

- Position of PP relative to the object determines the size of the perspective projection.
- When the object is in the PP, its perspective will be in true size.
- When the object is placed in front of PP, its perspective is of enlarged size.
- When the object is placed behind PP, the perspective is in reduced size.

Horizon Plane / Horizontal Plane (HP): It is an imaginary horizontal plane at the level of the observer's eye or S. It is above and parallel to the GP and perpendicular to the PP.

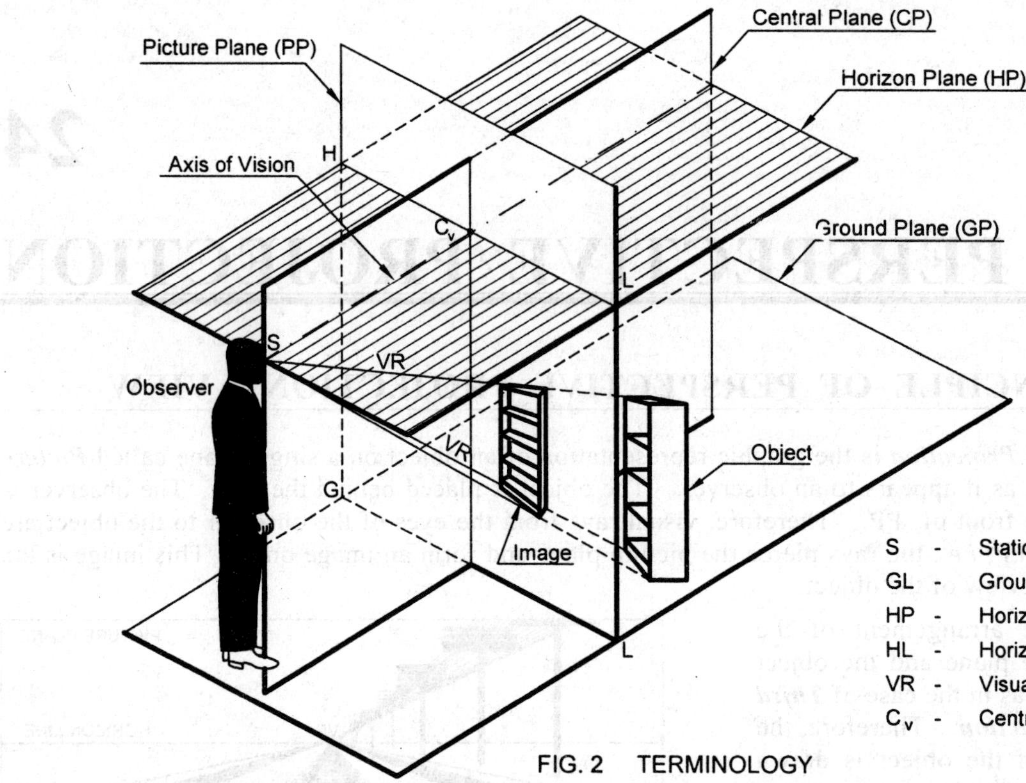

FIG. 2 TERMINOLOGY

Station point (S) ***(i.e., eye level)*:** Station point is the position of the observer's eye. Its position is fixed at any convenient distance either from the picture plane or from the object. Shape and visibility of edges of perspective projection depends upon the position of S.

- For large objects like buildings, S is taken at eye level of height 1.8 m.
- For small objects, fix S slightly above the top of the object to have a complete view of its top and side surfaces.
- Take the distance of S from PP = twice the width of the object to avoid distorted view.
- Locate S in front of the objects so that CP passes through the center of the object.

Ground Line (GL): It is the line of intersection of the GP with the PP.

Horizon Line (HL): It is the line of intersection of HP with PP. It is parallel to GL.

Axis of Vision (AV) / Perpendicular Axis (PA) / Line of Sight: It is a line drawn perpendicular to the PP from S.

Center of vision (C_v): It is the point at which the axis of vision pierces the PP. It is also the point of intersection of HL with AV.

Central Plane (CP): It is an imaginary vertical plane which passes through the S and C. It is perpendicular to both the PP and the GP. It contains the Axis of Vision (AV).

Auxiliary Ground Plane (AGP) *(Not shown)*: It is a horizontal plane parallel to GP. Top view of the object and of the perspective elements is projected on this plane.

Vanishing Point (VP): All the projections converge to a point known as Vanishing Point.

2. PERSPECTIVE VIEWS OF POINTS, LINES AND PLANES

Problem 1: *(Fig. 3) Draw the perspective projection of a* ***point*** *A situated 20 mm behind the picture plane and 15 mm above the ground plane. The station point is 30 mm in front of the picture plane, 40 mm above the ground plane and lies in a central plane which is 35 mm to the left of the given point.*

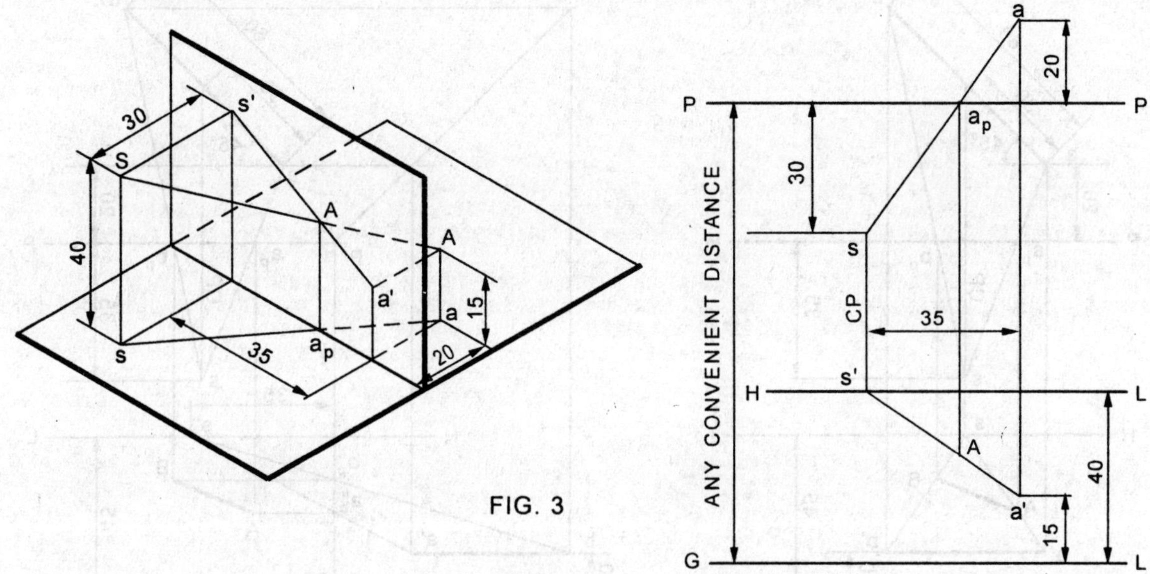

FIG. 3

VISUAL RAY METHOD

In this method, points on the perspective projection are obtained by drawing visual rays from the station point S to both top view and *either* front view *or* side view of the object.

Top view

1. Draw a horizontal line PP to represent the top view of the picture plane.
2. Point A is 20 mm behind PP. Hence mark the point a 20 mm above PP.
3. Station point S lies in a central plane CP which is 35 mm to the left of point A. So, draw a vertical line to represent the top view of CP at 35 mm to the left of a.
4. S is 30 mm in front of PP. Therefore, on CP, mark s 30 mm below PP.
5. Join a and s to represent the top view of the visual ray. It pierces the PP at a_p.

Front view

6. Draw a ground line GL at any convenient distance below PP. Mark a′, the front view of A, 15 mm above GL and in projection with a. To avoid overlap of visual rays and get a clear perspective, select GL such that HL lies below s. S is 40 mm above GP. So, draw HL 40 mm above GL.
7. Further CP also represents front view of CP. Hence mark s′ at the intersection of CP with HL. Join a′s′, the front view of the visual ray. From the piercing point a_p, erect a vertical to cut a′s′ at point A, which is the required perspective projection.

Problem 2: *(Fig. 4) Draw the perspective projection of a* **line** *AB, 60 mm long. It is parallel to and 10 mm above the ground plane and inclined at 45° to PP. The end A is 20 mm behind the picture plane. Station point is 35 mm in front of the picture plane and 45 mm above the ground plane and lies in a central plane passing through the mid-point of AB.*

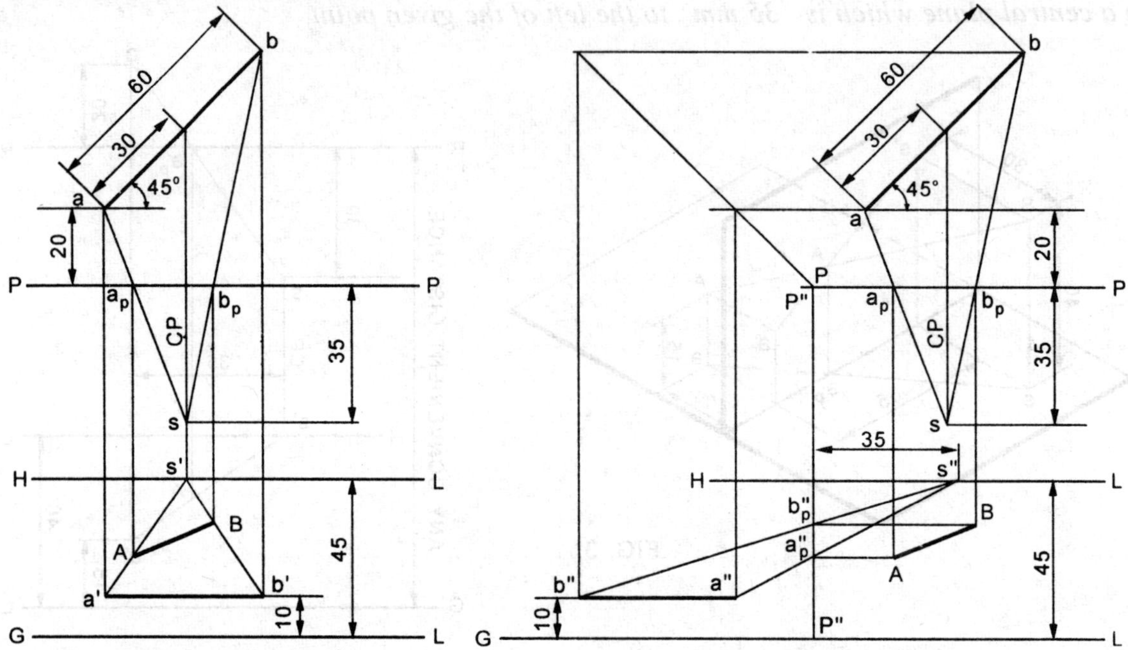

(a) VISUAL RAY METHOD

(b) ALITER (USING SIDE VIEW)

FIG. 4

1. **Top View:** *See Fig. (a).* Draw PP and mark a 20 mm above it. Draw ab = 60 mm (True length of AB) inclined at 45° to PP. From the mid-point of ab erect a vertical line to represent CP. Along the central plane mark s 35 mm below PP. Join a and b with s to represent the top view of the visual rays. Mark the piercing points a_p and b_p on as and bs respectively.
2. **Front View:** Draw GL at any convenient distance below PP. Draw a′ b′ parallel to and 10 mm above GL. Draw HL 45 mm above GL. Mark s′ at the intersection of CP and HL. Join s′ with a′ and b′. From a_p and b_p (piercing points) erect verticals to intersect s′ a′ and s′ b′ (the front view of the visual rays) at A and B respectively. AB is the required perspective projection.

ALITER (using SIDE VIEW): *[Fig. (b)]*

Draw top view as in previous method. Draw GL at any convenient distance below PP. Draw P″ P″, the side view of PP. Draw a″ b″ (Side View of AB) 10 mm above GL. Draw HL 45 mm above GL. As S is 35 mm in front of PP, mark s″ 35 mm from P″ P″ and on HL. Join s″ with a″ & b″ to cut p″ p″ at a''_p & b''_p. From a_p erect vertical to cut a horizontal line drawn from a''_p at A. Similarly obtain B. Join A & B.

Problem 3: *(Fig. 5)* *A* ***square lamina*** *of 40 mm side lies on the ground plane. One of its corners is touching the PP and an edge is inclined at 60° to PP. The station point is 40 mm in front of PP, 60 mm above GP and lies in a central plane which is at a distance of 35 mm to the right of the corner touching the PP. Draw the perspective projection of the lamina.*

VISUAL RAY METHOD : [*Fig. 5(a)*]

Top View

1. Draw the top view of the lamina as a square such that the corner b is touching PP and the edge bc is inclined at 60° to PP.
2. Draw CP, 35 mm from b on right side. Along CP mark s 40 mm below PP.
3. Join s with all the four corners of the square lamina in the top view. Obtain the corresponding piercing points on PP.

Front View

4. Draw GL and obtain the front view of the lamina on it. Draw HL 60 mm above GL and obtain s′ on it. Join s′ with all the corners of the lamina in the front view.

Perspective Projection

5. Since the corner b touches the PP, its perspective will be in its true position.
6. Since the lamina lies on the ground plane, b′ is on GL and is also the perspective projection of B.
7. From a_p draw vertical to intersect a′ s′ at A.
8. Similarly obtain B, C and D. Join ABCD and complete the perspective projection.

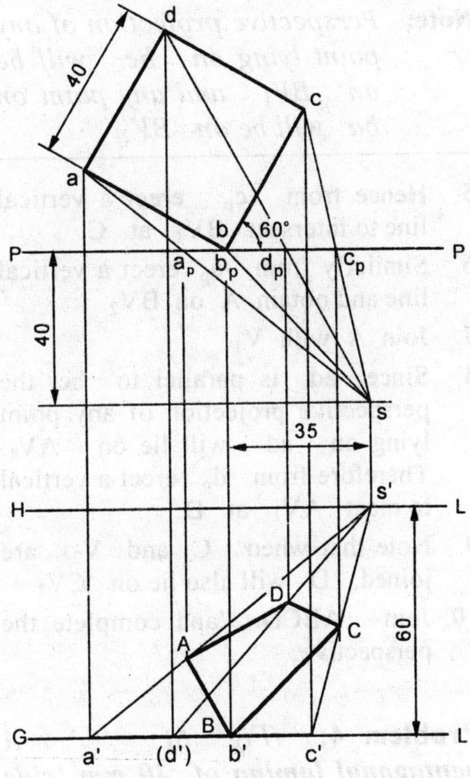

FIG. 5(a) VISUAL RAY METHOD

VANISHING POINT METHOD : [*Fig. 5(b)*]

Vanishing Point is an imaginary point infinite distance away from station point. The point at which the visual ray from the eye to that infinitely distant vanishing point pierces the picture plane is termed as *Vanishing Point*. When the observer views an object, all its parallel edges converge to one / two / three points depending on the locations of the object and the observer.

1. Draw the top view as explained in *Steps 1 to 3* in the above method. Draw GL and HL.

Vanishing Points

2. From s draw a line parallel to bc to cut PP at v_1. Erect vertical from v_1 to cut HL at vanishing point V_1.
3. Similarly from s draw a line parallel to ba to cut PP at v_2. Erect vertical from v_2 and obtain the other vanishing point V_2 on HL.

Perspective Projection

4. Since b touches PP, draw a vertical line from b and obtain B on GL. Join B with V_1 and V_2.

Note: *Perspective projection of any point lying on bc will be on BV_1 and any point on ba will be on BV_2.*

---

5. Hence from c_p erect a vertical line to intersect BV_1 at C.
6. Similarly from a_p erect a vertical line and obtain A on BV_2.
7. Join A with V_1.
8. Since ad is parallel to bc, the perspective projection of any point lying on ad will lie on AV_1. Therefore from d_p erect a vertical to meet AV_1 at D.
9. Note that when C and V_2 are joined, D will also lie on CV_2.
10. Join ABCD and complete the perspective.

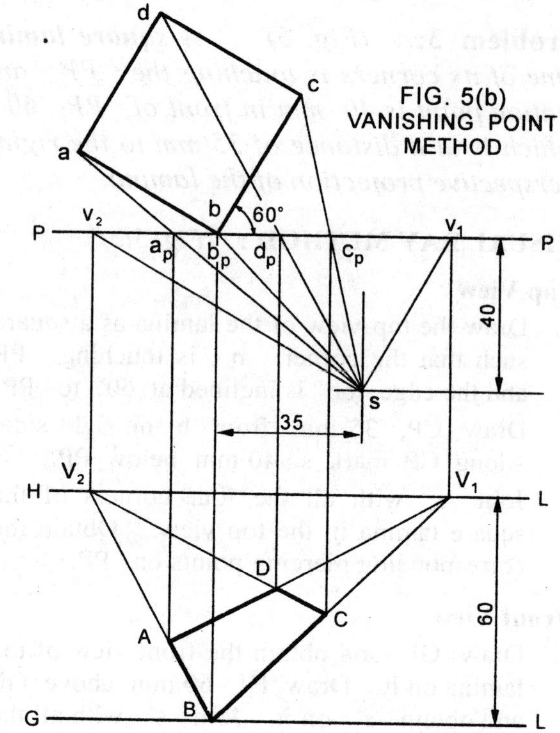

FIG. 5(b) VANISHING POINT METHOD

Problem 4: *(Fig. 6) A **pentagonal lamina** of 40 mm side lies on the ground. The corner which is nearest to PP is 15 mm behind it and an edge containing that corner is making 45° with PP.*

The station point is 40 mm in front of PP, 50 mm above GP and lies in a central plane which is at a distance of 70 mm to the left of the corner nearest to the PP.

Draw the perspective projection of the lamina.

Hint: See Fig. 6. Perspective projection is drawn by Visual Ray Method using the top and front views.

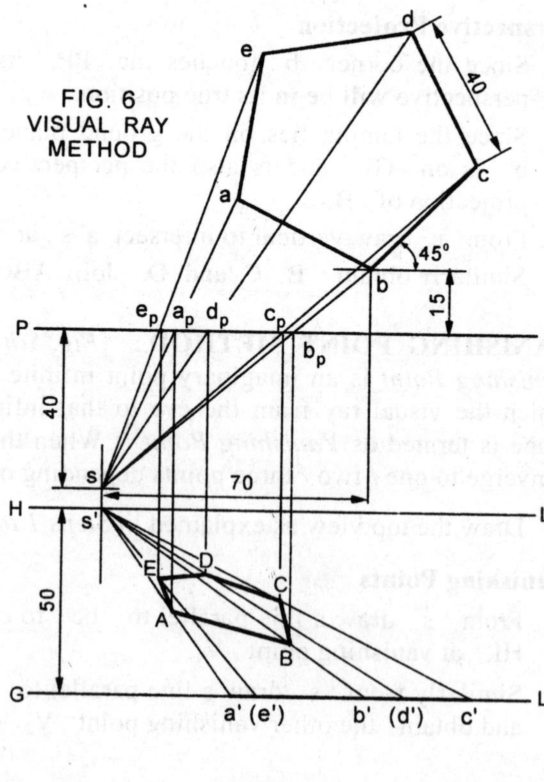

FIG. 6 VISUAL RAY METHOD

Problem 5: *(Fig. 7) A **circular lamina** of 45 mm diameter lies on the ground plane and touches the picture plane. The center plane passes through the center of the circle. Station point is 70 mm in front of picture plane and 60 mm above the ground plane. Draw the perspective projection.*

Visual Ray Method

1. Draw the top view of the lamina as a circle touching PP. Draw the vertical and horizontal diameters of the circle.
2. Enclose the circle in a square and join the diagonals.
3. Draw the front view of the lamina. Then draw the perspective projection of the square.
4. Join the diagonals to intersect at O.
5. Through O, draw a vertical and a horizontal line.
6. Obtain the points 1, 5 and 3, 7 on the respective sides of the square in the perspective.
7. To obtain 2 and 4 in the perspective, mark 2′ and 4′.
8. Join 2′(4′) with s′ which intersects the diagonals BD and AC at 2 and 4 respectively.
9. Similarly obtain 6 and 8 in the perspective as shown.
10. Join 1, 2, ..., 8 by a smooth ellipse to represent the perspective projection of the lamina.

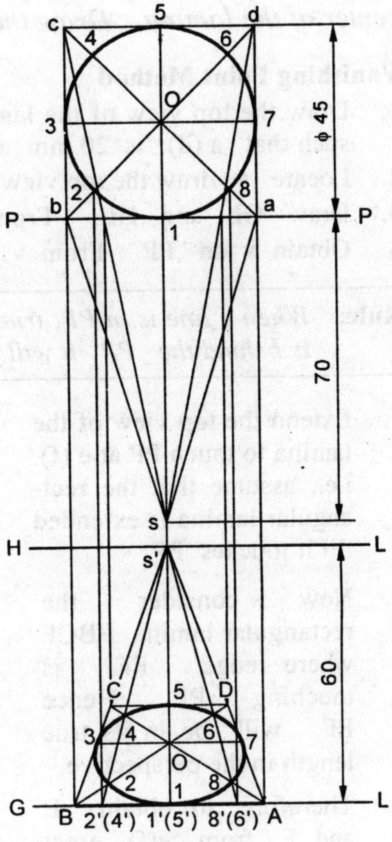

FIG. 7

Note: *1. In the perspective, the center of the ellipse does not coincide with the center of the perspective of the square.*

2. A circle will appear as a circle in its perspective projection, when it is parallel to the PP. In other positions, its perspective will be an ellipse (except when it lies in the CP).

3. Circular features of an object inclined to PP will appear as ellipse in perspective. To obtain the points for the ellipse, enclose the circle in a square. The midpoints of the sides of the square are the four points for the ellipse. Points of intersection of diagonals with the circle are the other four points. Draw lines through these points parallel to the sides of the square. Then draw the perspective of the square along with the parallel lines, locating eight points for drawing the ellipse.

Problem 6: *(Fig. 8)* ***A rectangular lamina*** *of sides 50 mm x 25 mm stands vertically with one of its longer edges on the ground and inclined at 45° to PP. The vertical edge nearest to PP is 20 mm behind it. The station point is 30 mm in front of the PP, 45 mm above the ground and lies in a central plane which passes through the center of the lamina. Draw the perspective projection of the lamina.*

Vanishing Point Method

1. Draw the top view of the lamina as a straight line of length 50 mm inclined at 45° to PP such that a (d) is 20 mm above PP.
2. Locate s, draw the top view of the visual rays and obtain the piercing points.
3. Draw GL and HL. From s draw a line parallel to ab.
4. Obtain v on PP. From v erect a vertical and obtain the vanishing point V on HL.

Rule: *When a line is in PP, true length will be seen in its perspective projection. When a line is behind the PP it will be foreshortened in its perspective.*

---

5. Extend the top view of the lamina to touch PP at e (f). i.e., assume that the rectangular lamina is extended till it touches PP.
6. Now consider the rectangular lamina EBCF where edge EF is touching PP. Hence EF will be in its true length in the perspective.

   Therefore, to obtain E and F from e(f) erect a vertical line to intersect GL. This line is called **measuring line** or **line of heights**.

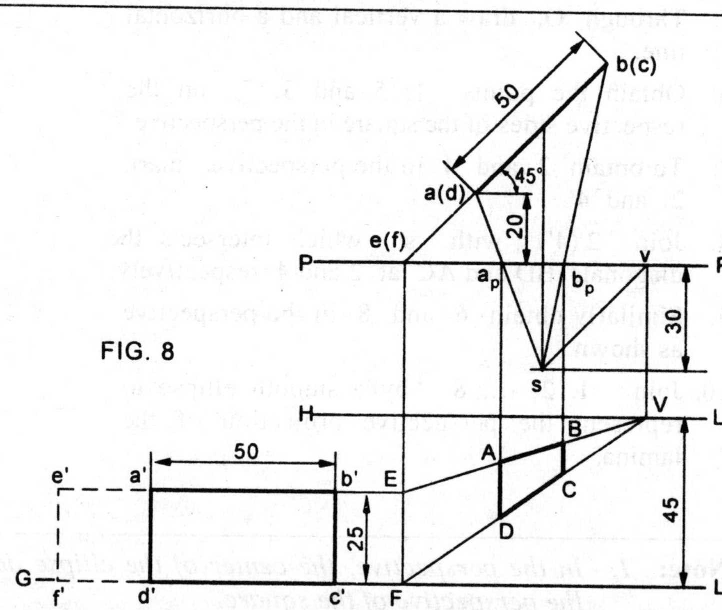

FIG. 8

   On this line the heights of all the points in the lamina can be marked. Hence project e' and f' horizontally to intersect the measuring line at E and F. Now EF is the perspective projection and is in its true length = 25 mm.
7. Join F and V. FC is on the ground. Hence, the perspective projection of any point on the edge FC will lie on FV. Hence obtain D and C on FV by drawing vertical lines from the corresponding piercing points.
8. Edge EB is at 25 mm above the ground. Perspective of E has been already marked. Join E and V. Now perspective of any point on edge EB can be obtained on EV. Therefore, obtain A and B on EV by erecting vertical lines from the corresponding piercing points.

9. Join ABCD by *thick* lines and complete the perspective projection of the given lamina. Note that in the perspective, BC is smaller than AD since the edge BC is farther away from PP than AD.

Problem 7: *(Fig. 9) A* ***hexagonal lamina*** *of 25 mm side stands vertically on the ground plane and inclined at 50° to PP. The corner nearest to PP is 20 mm behind it. The station point is 45 mm in front of PP, 50 mm above the ground plane and lies in a central plane which passes through the center of the lamina. Draw the perspective projection.*

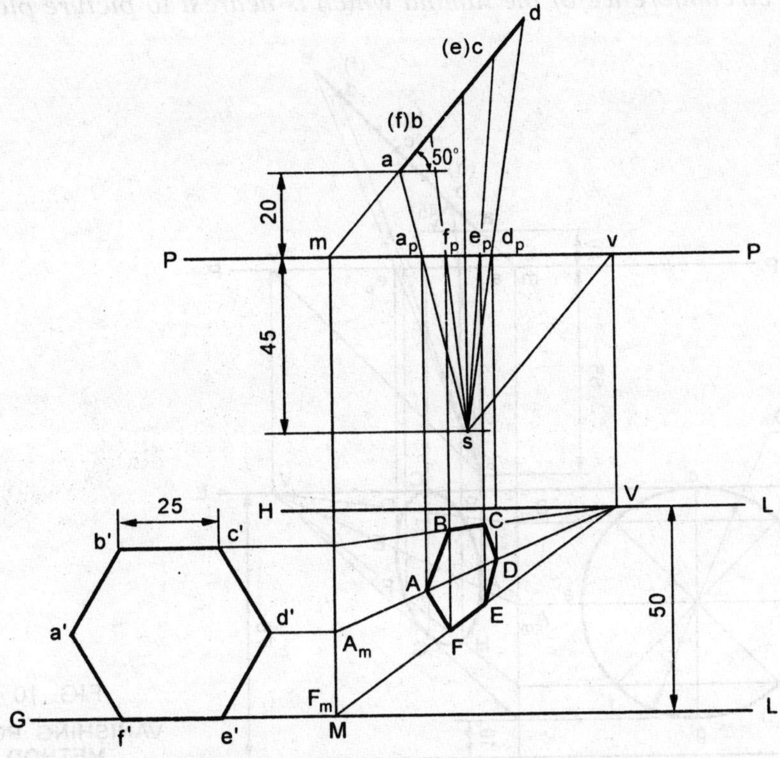

FIG. 9

VANISHING POINT METHOD

1. Draw PP and draw a line inclined at 50° to it.
2. Draw a regular hexagon of side 25 mm such that one of its sides is on GL and name all the corners. Now complete the top view of the lamina with the help of this hexagon.
3. Locate s, draw the visual rays and obtain the piercing points.
4. Draw GL and obtain the **vanishing point V** on HL.
5. Extend the top view of the lamina to meet PP at m. From m draw a vertical line to meet GL at M (mM is the measuring line).
6. The edge FE is on the ground. Project f' and e' horizontally to meet the measuring line at F_m. Join F_m with V.
7. From the piercing points f_p and e_p erect verticals to intersect F_mV at F and E respectively.
8. Through a' and d' draw a horizontal line and obtain A_m on mM. Join A_m with V.

9. Since the corners A and D of the lamina are at the same height from the ground, obtain the perspective of A and D on A_m V.
10. Similarly obtain B and C and complete the perspective projection.

Problem 8: *(Fig. 10) Draw the perspective projection of a* ***circular lamina*** *of 60 mm diameter having its surface vertical and inclined at 45° to the picture plane. The center of the lamina is 40 mm above the ground plane. Station point is 70 mm above ground plane and 55 mm in front of the picture plane. Central plane passes through a point on the circumference of the lamina which is nearest to picture plane and 10 mm behind it.*

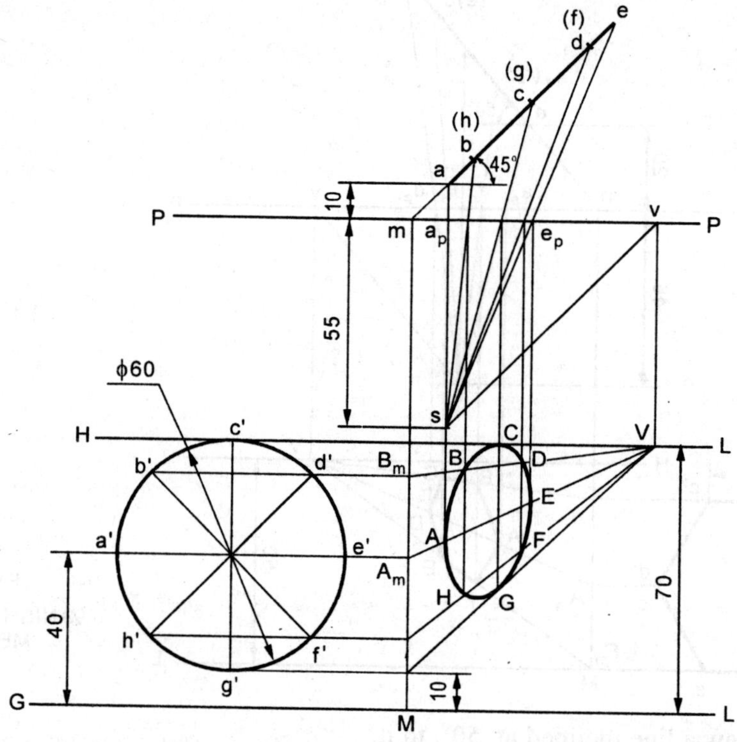

FIG. 10
VANISHING POINT METHOD

1. Draw the top view of the lamina.
2. Draw a circle of 60 mm diameter on the left hand side as shown. Divide it into 8 equal parts and name the division points as a′, b′, etc.
3. Mark these points on the top view also.
4. Mark s and obtain the piercing points on PP.
5. Obtain the **vanishing point V** and draw the measuring line mM.
6. Project the division points horizontally and obtain A_m, B_m, etc., on the measuring line.
7. Join them with V. From the piercing points, erect verticals and obtain A, B, etc.
8. Join these points by an ellipse to represent the perspective projection of circular lamina.

3. PERSPECTIVE PROJECTION OF SOLIDS

Problem 9: *(Fig. 11)* *A **square prism**, side of base 40 mm and height 60 mm rests with its base on the ground such that one of its rectangular faces is parallel to and 10 mm behind the picture plane. The station point is 30 mm in front of PP, 80 mm above the ground plane and lies in a central plane 45 mm to the right of the center of the prism. Draw the perspective projection of the square prism.* (UQ)

VISUAL RAY METHOD

Top View

1. *See Fig. 11.* Draw the top view of the prism as a square of side 40 mm such that ab is parallel to and 10 mm above PP.
2. Locate s. Draw top view of the visual rays. Mark piercing points.

Front View

3. Draw front view of the prism for given position. Locate s' and draw front view of the visual rays.
4. From piercing points erect vertical lines to cut the corresponding visual rays in the front view. Thus obtain all corners in perspective projection.

To mark the visible and invisible edges in the perspective

5. Draw boundary lines (*thick lines*).
6. The faces $ab(b_1)(a_1)$ and $bc(c_1)(b_1)$ are nearer to s and visible. Hence draw BB_1, BA and BC as *thick lines.*
7. Edge $d(d_1)$ is farther away from s. So draw DD_1, D_1A_1 and D_1C_1 as *dashed* lines.

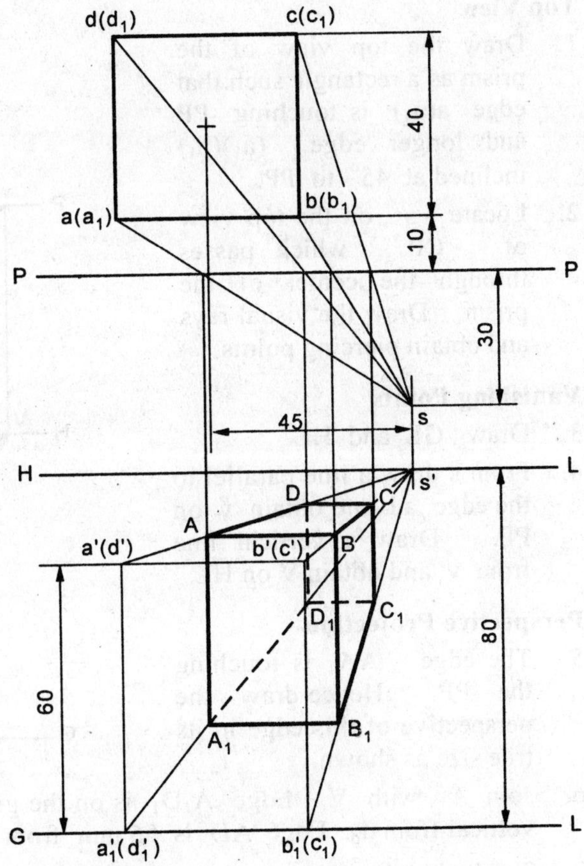

FIG. 11 VISUAL RAY METHOD

Problem 10: *(Exercise)* *A **square prism** of 30 cm side and 50 cm length is lying on the ground plane on one of its rectangular faces, in such a way that one of its square faces is parallel to and 10 cm behind the picture plane. The station point is located 60 cm in front of the picture plane and 40 cm above the ground plane. The central plane is 50 cm away from the axis of the prism towards the left. Draw the perspective projection of the prism.*

[In case of large objects, perspective view may partly overlap front view causing confusion.]

Problem 11: *(Fig. 12)* ***A rectangular prism**, sides of base 50 mm x 30 mm and height 55 mm, rests with its base on the ground plane. A vertical edge is in the picture plane and one of the longer edges of its base is inclined at 45° to PP and behind it. The station point is 50 mm in front of PP, 75 mm above the ground plane and lies in a central plane which passes through the center of the prism. Draw the perspective projection of the solid.* (UQ)

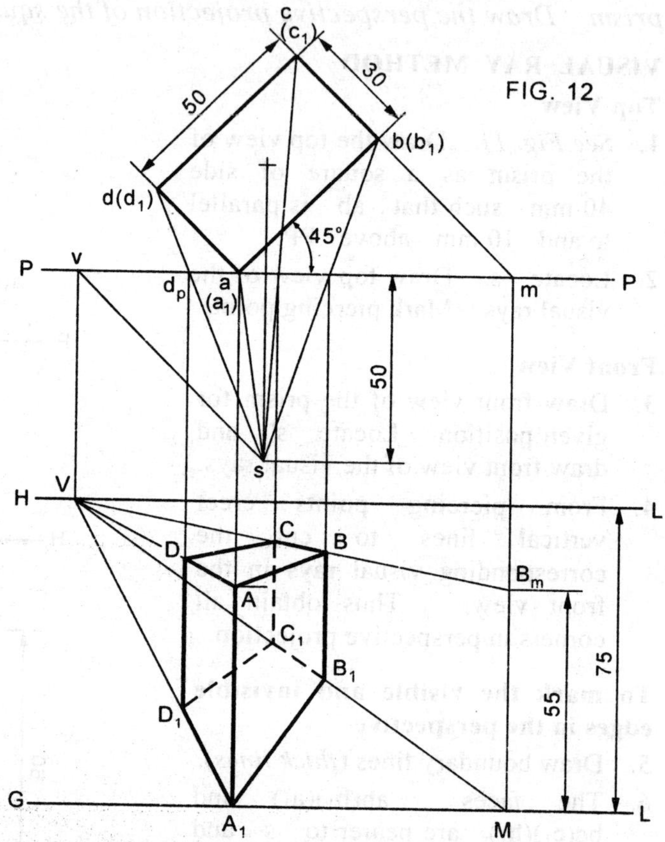

FIG. 12

VANISHING POINT METHOD

Top View

1. Draw the top view of the prism as a rectangle such that edge $a(a_1)$ is touching PP and longer edge $(a_1)(b_1)$ inclined at 45° to PP.
2. Locate s on the top view of CP which passes through the center of the prism. Draw the visual rays and obtain piercing points.

Vanishing Points

3. Draw GL and HL.
4. From s draw a line parallel to the edge ad and obtain v on PP. Draw a vertical line from v and obtain V on HL.

Perspective Projection

5. The edge AA_1 is touching the PP. Hence draw the perspective of this edge in its true size as shown.
6. Join A_1 with V. Edge A_1D_1 is on the ground. Hence obtain D_1 on A_1V by erecting vertical from d_p. Edge AD is 55 mm from ground. Hence obtain D on AV.
7. To obtain other corners of the prism in the perspective, extend cb to meet PP at m.
8. Draw the measuring line. Join M and V. Obtain the perspective of B_1 and C_1 on MV.
9. Mark B_m on the measuring line such that MB_m = 55 mm = height of the prism.
10. Join B_m with V. Obtain the perspective of B and C on B_mV.

To mark the visible and invisible edges

11. Draw boundary lines thick. Look at the top view and visual rays. The edge $a(a_1)$ is nearer to the station point. Hence draw AA_1, AD and AB as thick lines.
12. The edge $c(c_1)$ is farther away from s. Hence draw C_1D_1, C_1B_1 and C_1C as dashed lines.

Problem 12: *(Exercise)* *A* ***cube*** *of edge 30 mm rests with one of its faces on the ground plane such that a vertical edge touches the picture plane. The vertical faces of the cube are equally inclined to the PP and behind it. A station point is 40 mm in front of the PP, 50 mm above the ground plane and lies in a central plane 15 mm to the right of the axis of the cube. Draw the perspective projection of the cube.*

Note: ***Distance Points*** *– Vanishing points for all horizontal lines inclined at 45° to PP are known as Distance Points because of their definite positions. They are equidistant from center of vision. The distance of each from the center of vision is equal to the distance of S from PP.*

Problem 13: *(Fig. 13)* *A* ***square pyramid*** *of base edge 40 mm and altitude 50 mm, rests with its base on the ground plane such that all the edges of the base are equally inclined to the PP. One of the corners of the base is touching the PP. The station point is 60 mm in front of the PP, 80 mm above the ground plane and lies in a central plane which passes through the axis of the pyramid. Draw the perspective projection.* (UQ)

VANISHING POINT METHOD

1. Draw the top view of the square pyramid and the visual rays. From s draw a line parallel to ab. Obtain the vanishing point V on HL.
2. Corner A is touching the PP and on the ground. Hence erect a vertical line from a and mark the perspective of A on GL. Join A with V.
3. As edge AB is on the ground, obtain the perspective of B on AV.
4. To obtain the perspective of D and C extend cd to meet PP at m.
5. Draw the measuring line mM.
6. Since the edge DC is on the ground, join M with V.
7. Obtain the perspective of D and C on this line MV.
8. To mark the perspective of the apex O, draw a line parallel to ab and passing through o to meet PP at h.
9. Draw another measuring line hH.
10. On this line mark the height of the apex as O_hH = 50 mm.
11. Join O_h with V. Obtain the perspective of O on this line.
12. Then complete perspective as shown.

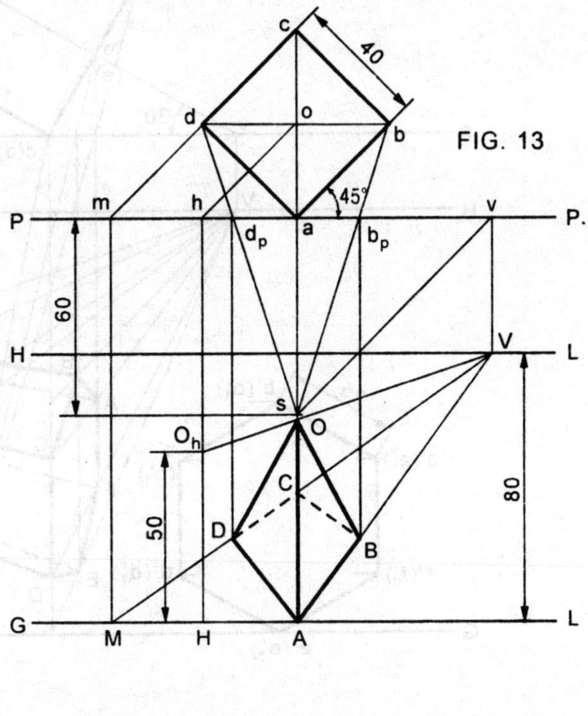

FIG. 13

Problem 14: *(Exercise) A rectangular pyramid, sides of base 55 mm x 20 mm and height 60 mm rests with its base on the ground plane such that one of the longer edges of the base is parallel to and 20 mm behind the picture plane. The station point is 40 mm in front of the picture plane, 60 mm above the ground plane and lies in a central plane which passes through the axis of the pyramid. Draw the perspective projection.*

(UQ)

Hint: This problem can be easily solved by Visual Ray Method using the top and side views of the visual rays.

Problem 15: *(Fig. 14) Draw the perspective projection of a* **hexagonal prism** *lying on the ground plane on one of its longer edges such that one of its rectangular faces is perpendicular to the ground plane.*

The axis is inclined at 30° to the picture plane and an edge of the base is touching the picture plane. The station point is 110 mm in front of the PP, 95 mm above the ground plane and lies in a central plane which bisects the axis. For the prism, side of base is 25 mm and height 75 mm. (UQ)

VANISHING POINT METHOD

FIG. 14

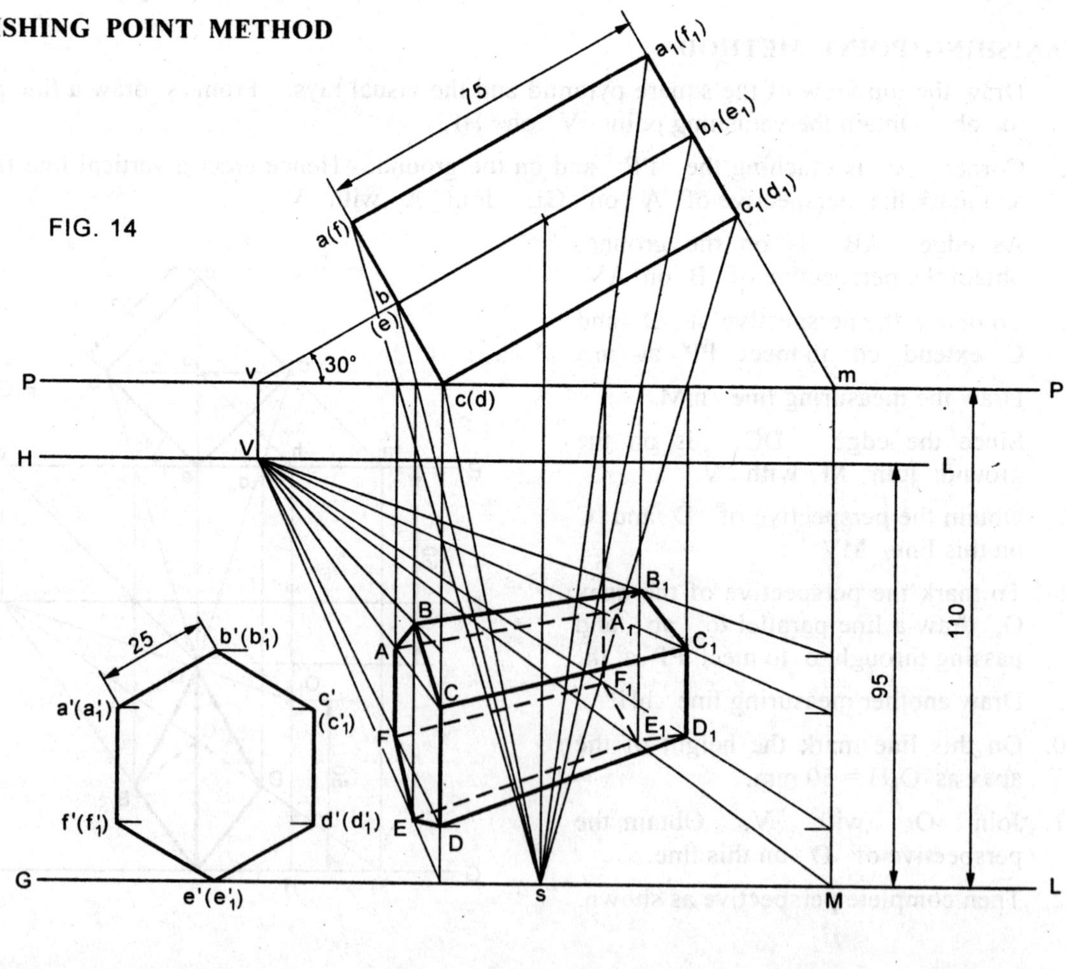

1. Draw the top view of the hexagonal prism, such that edge c(d) is touching the PP and axis inclined at 30° to PP, with the help of the hexagon drawn on left-hand side.
2. Locate s on CP which bisects the axis. Join s with all corners of prism in top view.
3. Draw GL and HL.
4. Consider two parallel lines ca and c_1a_1. Each line contains the top view of six corners. Hence from s draw a line parallel to ca. Obtain V on HL.
5. From c(d) draw a vertical line (measuring line). Mark the heights of all the six corners of the base on it. Since edge CD is touching PP, it will be in true size in its perspective.
6. From all the points on this measuring line, draw vanishing lines to converge at V. Draw vertical lines from the piercing points.
7. Obtain the perspective of E, F, A and B on the corresponding vanishing lines.
8. Obtain the perspective of 6 corners of other end face. Complete the perspective projection.

Problem 16: *(Exercise)* *A* ***pentagonal prism****, side of base 25 mm and axis 60 mm long, lies with one of its rectangular faces on the ground plane such that a pentagonal face is touching the PP. The station point is 20 mm in front of the picture plane, 55 mm above the ground plane and lies in a central plane which is at 80 mm to the right of the center of the prism. Draw the perspective projection.* (UQ)

Problem 17: *(Fig. 15)*
A ***pentagonal prism****, side of base 30 mm and height 50 mm rests with its base on the ground plane such that one of its rectangular faces is inclined at 45° to PP and the vertical edge nearer to PP is 10 mm behind it.*

The station point is 45 mm in front of the picture plane, 90 mm above the ground plane and lies in a central plane which is 15 mm to the left of the vertical edge nearer to PP. Draw the perspective projection.

(UQ)

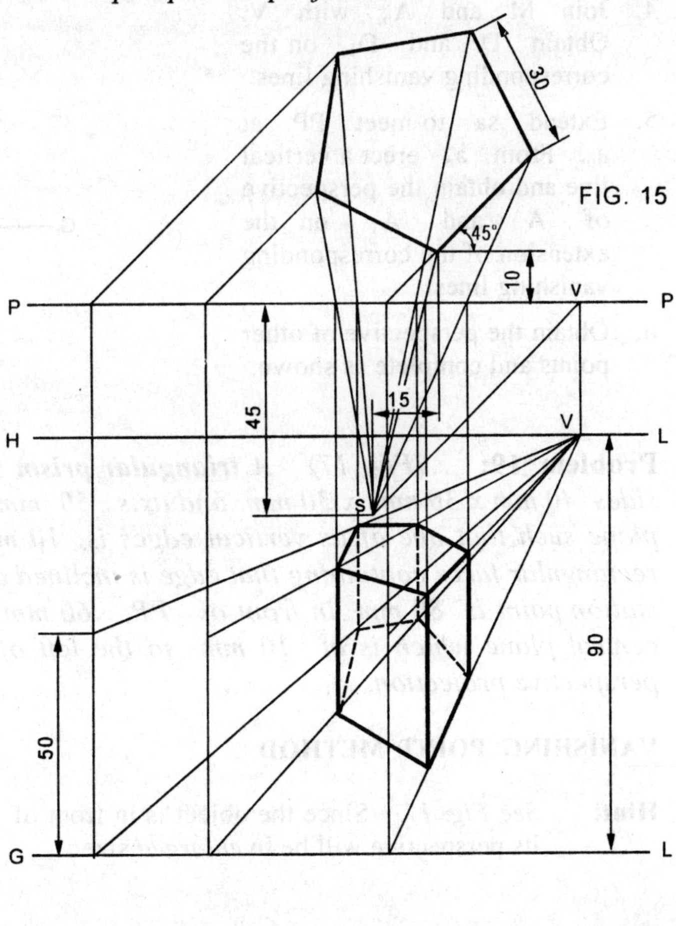

FIG. 15

Problem 18: *(Fig. 16) A **cube** of 35 mm long edges rests with a square face on the ground plane such that one of the vertical edges of the cube is 8 mm in front of the picture plane and a vertical face containing that edge is inclined at 30° to PP. The station point is 40 mm in front of the PP, 70 mm above the ground plane and lies in a central plane which is 45 mm to the left of the center of the cube. Draw the perspective projection.* (UQ)

1. Draw the top view of the cube for the given position. Locate s and join it with all the corners of the cube in top view.
2. Draw GL and HL. Then obtain V on HL.
3. Note that the picture plane intersects ad at m. Draw the measuring line mM and mark the height of the cube.
4. Join M and A_m with V. Obtain D and D_1 on the corresponding vanishing lines.
5. Extend sa to meet PP at a_p. From a_p erect a vertical line and obtain the perspective of A and A_1 on the extension of the corresponding vanishing lines.
6. Obtain the perspective of other points and complete as shown.

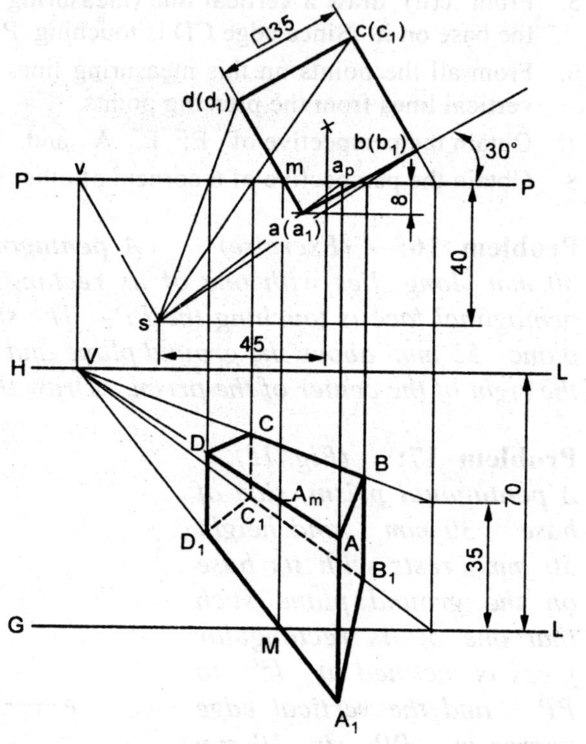

FIG. 16

Problem 19: *(Fig. 17) A **triangular prism** whose base is an isosceles triangle of sides 40 mm x 30 mm x 30 mm and axis 50 mm long rests with its base on the ground plane such that one of its vertical edges is 10 mm in front of the PP. One of the rectangular faces containing that edge is inclined at 45° to PP and in front of it. The station point is 80 mm in front of PP, 60 mm above the ground plane and lies in a central plane which is at 10 mm to the left of the center of the prism. Draw the perspective projection.* (UQ)

VANISHING POINT METHOD

Hint: *See Fig. 17.* Since the object is in front of PP, (between the PP and the observer) its perspective will be in *enlarged* size.

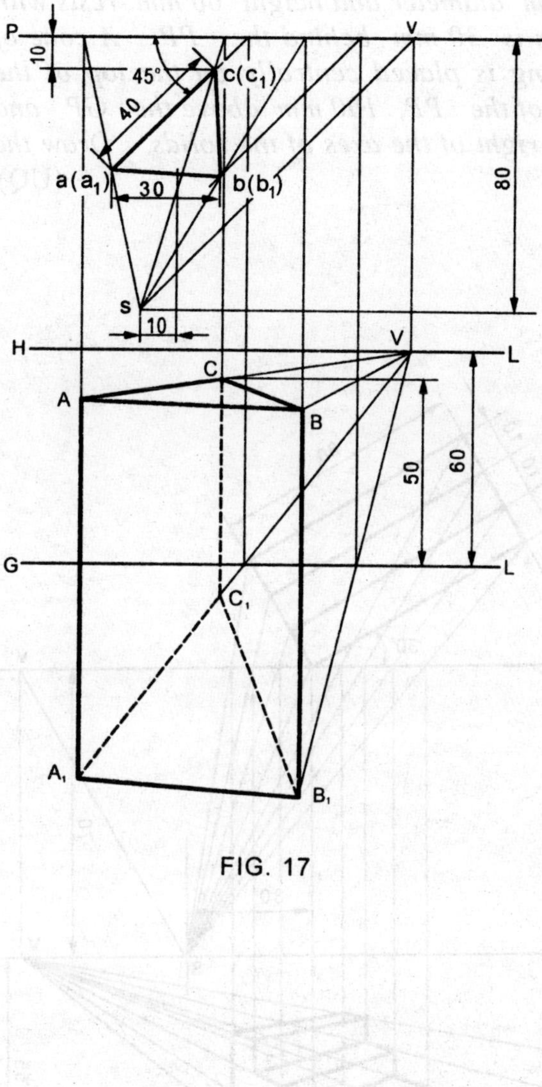

FIG. 17

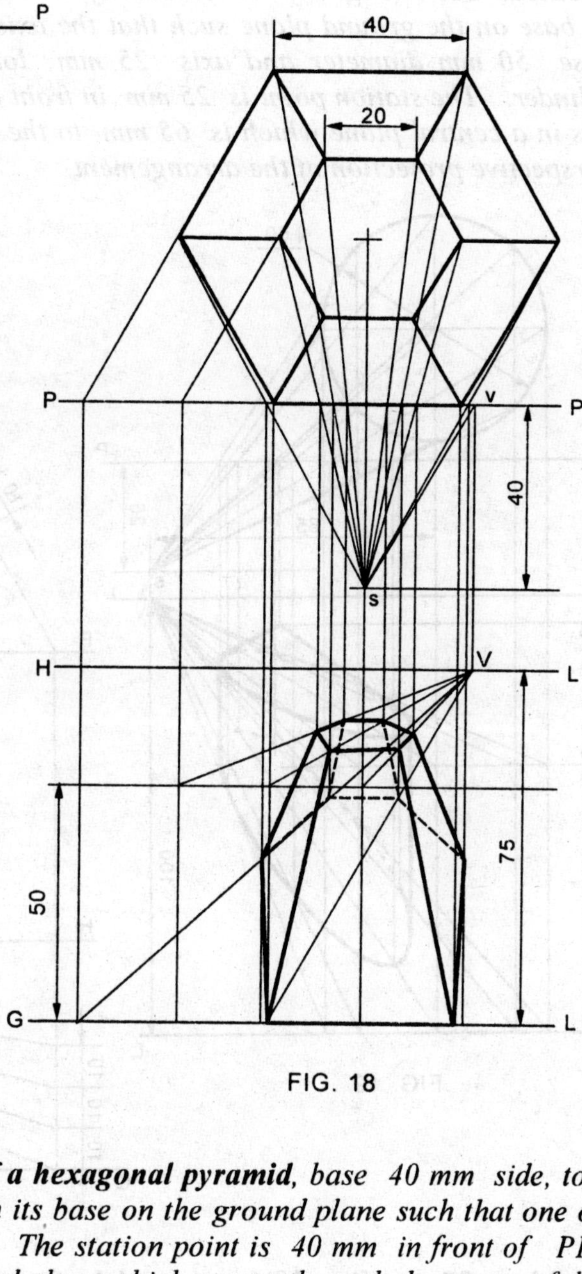

FIG. 18

Problem 20: *(Fig. 18) A **frustum of a hexagonal pyramid**, base 40 mm side, top 20 mm side and height 50 mm rests with its base on the ground plane such that one of the edges of its base is touching the PP. The station point is 40 mm in front of PP, 75 mm above the GP and lies in a central plane which passes through the center of the frustum. Draw the perspective projection.* (UQ)

Hint: *See Fig. 18.* Follow the principle of symmetry to minimize the congestion of projector lines and to have clarity of the figure; that is, locate a corner by drawing the projector from the piercing point on the Picture Plane to cut the corresponding converging line. Mark the other corresponding point by symmetry.

Problem 21: *(Fig. 19) A* ***cylinder*** *of 50 mm diameter and height 60 mm rests with its base on the ground plane such that the axis is 30 mm behind the PP. A cone of base 50 mm diameter and axis 25 mm long is placed centrally on the top of the cylinder. The station point is 25 mm in front of the PP, 100 mm above the GP and lies in a central plane which is 65 mm to the right of the axes of the solids. Draw the perspective projection of the arrangement.* (UQ)

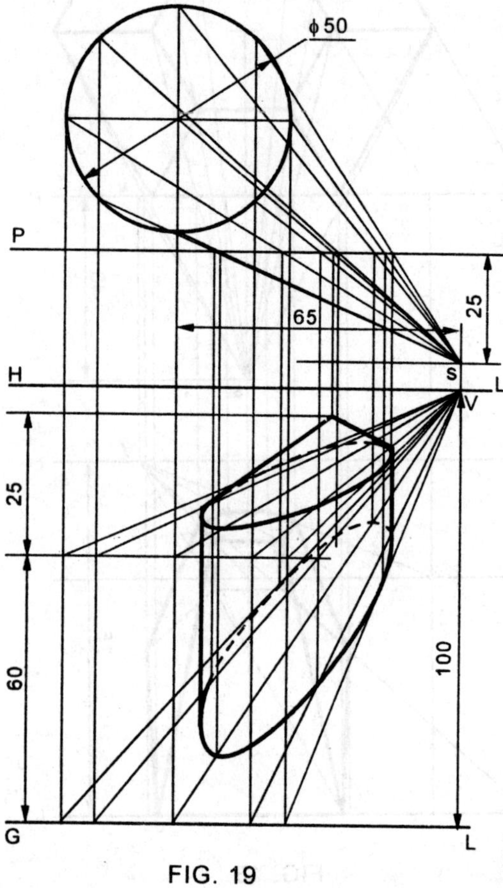

FIG. 19

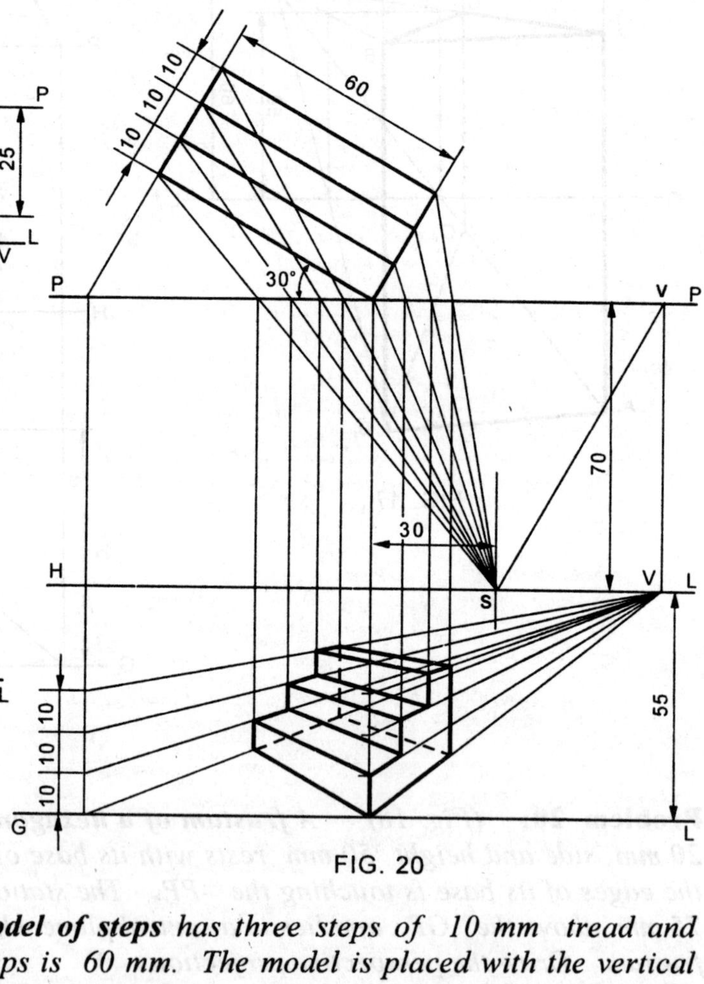

FIG. 20

Problem 22: *(Fig. 20) A* ***model of steps*** *has three steps of 10 mm tread and 10 mm rise. The length of the steps is 60 mm. The model is placed with the vertical edge of the first step touching the PP and its longer edge inclined at 30° to PP. The station point is 70 mm in front of PP, 55 mm above the ground plane and lies in a central plane which is at 30 mm to the right of the vertical edge touching the PP. Draw the perspective projection.* (UQ)

Problem 23: *(Fig. 21) Draw the perspective projection of a* ***shed*** *whose top and side views are given.*

The roof of the shed is inclined at 30° to the ground plane. The horizontal long edges of the shed are inclined at 45° to PP. One of the vertical edges of the wall touches the picture plane and 2.6 m to the left of the central plane. The station point is 9 m in front of PP and 6.5 m above the ground plane. The height and width of the door are 2.2 m and 1.4 m respectively. (UQ)

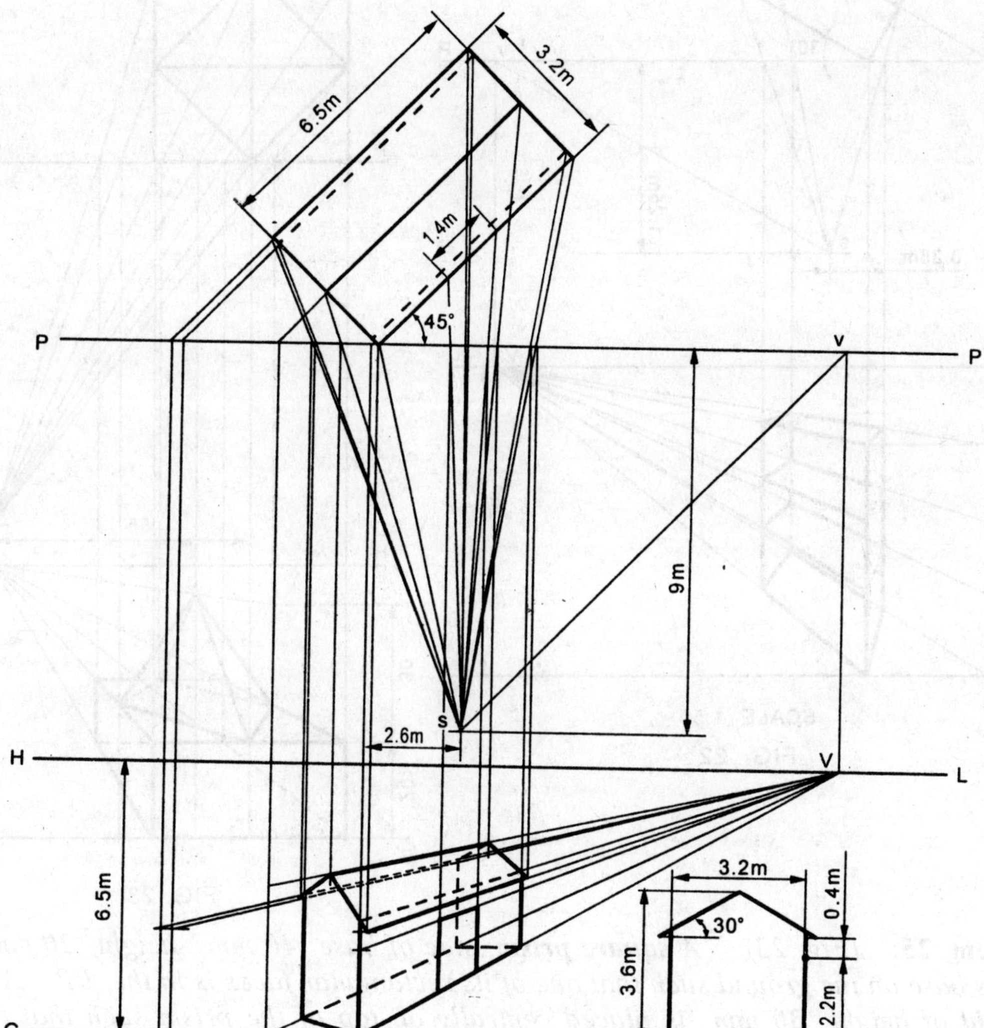

SCALE 1:20

FIG. 21

Problem 24: *(Fig. 22) A* ***steel rack*** *of size 1 m x 0.4 m x 1.5 m has 3 shelves equally spaced. It stands on the ground plane such that one of its vertical edges touches the PP and a longer horizontal edge is inclined at 30° to PP. The station point is 1.25 m in front of PP, 2m above the ground plane and lies in a central plane which is at 0.25 m to the right of the edge, touching the picture plane. Draw the perspective projection of the rack when its front portion is open.* (UQ)

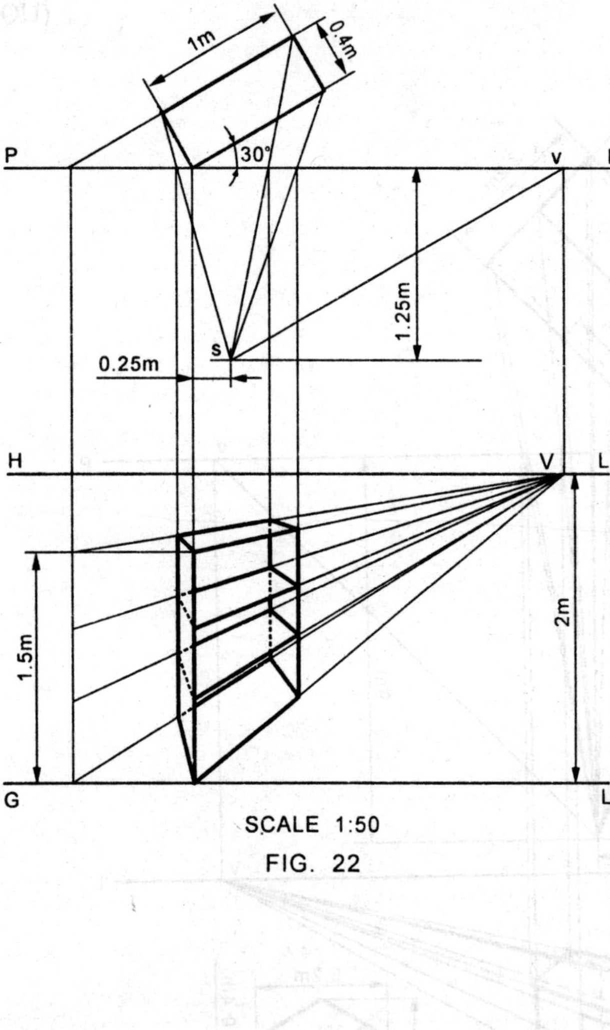

SCALE 1:50

FIG. 22

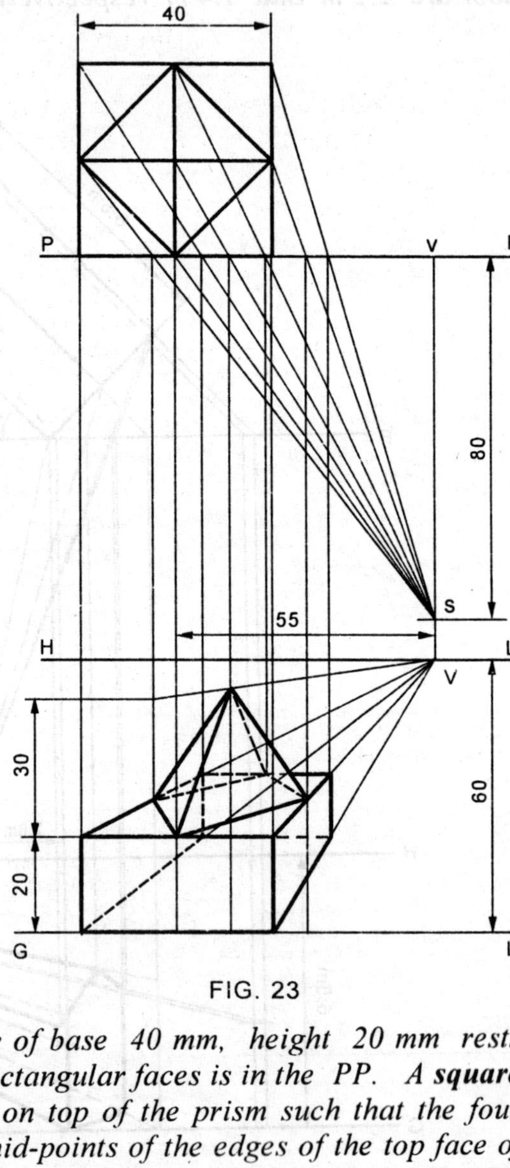

FIG. 23

Problem 25: *(Fig. 23) A* ***square prism****, side of base 40 mm, height 20 mm rests with its base on the ground such that one of its rectangular faces is in the PP. A* ***square pyramid*** *of height 30 mm is placed centrally on top of the prism such that the four corners of the base of the pyramid touches the mid-points of the edges of the top face of the prism. Station point is 80 mm in front of the picture plane, 60 mm above the GP and lies in a central plane which is 55 mm to the right of the axes of the solids. Draw the perspective projection of the* ***combination of solids.*** (UQ)